Fundamental constants

Quantity	Symbol	Value	Power of ten	Units
Speed of light	c	2.997 925 58*	10^{8}	$m\ s^{-1}$
Elementary charge	e	1.602 176	10^{-19}	C
Boltzmann's constant	k	1.380 65	10^{-23}	$J\ K^{-1}$
Planck constant	h	6.626 08	10^{-34}	J s
	$\hbar = h/2\pi$	1.054 57	10^{-34}	J s
Avogadro's constant	N_A	6.022 14	10^{23}	mol^{-1}
Atomic mass constant	m_u	1.660 54	10^{-27}	kg
Mass				
electron	m_e	9.109 38	10^{-31}	kg
proton	m_p	1.672 62	10^{-27}	kg
neutron	m_n	1.674 93	10^{-27}	kg
Vacuum permittivity	$\varepsilon_0 = 1/c^2\mu_0$	8.854 19	10^{-12}	$J^{-1}\ C^2\ m^{-1}$
	$4\pi\varepsilon_0$	1.112 65	10^{-10}	$J^{-1}\ C^2\ m^{-1}$
Vacuum permeability	μ_0	4π	10^{-7}	$J\ s^2\ C^{-2}\ m^{-1}$ ($= T^2\ J^{-1}\ m^3$)
Magneton				
Bohr	$\mu_B = e\hbar/2m_e$	9.274 01	10^{-24}	$J\ T^{-1}$
nuclear	$\mu_N = e\hbar/2m_p$	5.050 78	10^{-27}	$J\ T^{-1}$
g value of the electron	g_e	2.002 32		
Bohr radius	$a_0 = 4\pi\varepsilon_0\hbar^2/m_e e^2$	5.291 77	10^{-11}	m
Rydberg constant	$R = m_e e^4/8h^3 c\varepsilon_0^2$	1.097 37	10^{5}	cm^{-1}
Standard acceleration of free fall	g	9.806 65*		$m\ s^{-2}$

**Exact value*

Elements

Of Physical Chemistry

Peter Atkins
University of Oxford

Julio de Paula
Lewis & Clark College

Fifth edition

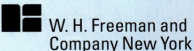

W. H. Freeman and
Company New York

Library of Congress Control Number: 2008934074

Elements of Physical Chemistry, Fifth Edition

ISBN-13: 978–1–4292–1813–9
ISBN-10: 1–4292–1813–4

Published in Great Britain by Oxford University Press

This edition has been authorized by Oxford University Press for sale in the United States and Canada only and not for export therefrom.

First printing

W. H. Freeman and Company
41 Madison Avenue
New York, New York 10010
www.whfreeman.com

About the book

We pay particular attention to the needs of the student, and provide many pedagogical features to make the learning process more enjoyable and effective. This section reviews these features. Paramount among them, though, is something that pervades the entire text: we try throughout to *interpret* the mathematical expressions, for mathematics is a language, and it is crucially important to be able to recognize what it is seeking to convey. We pay particular attention to the level at which we introduce information, the possibility of progressively deepening one's understanding, and providing background information to support the development in the text. We are also very alert to the demands associated with problem solving, and provide a variety of helpful procedures.

Organizing the information

Checklist of key ideas

We summarize the principal concepts introduced in each chapter as a checklist at the end of the chapter. We suggest checking off the box that precedes each entry when you feel confident about the topic.

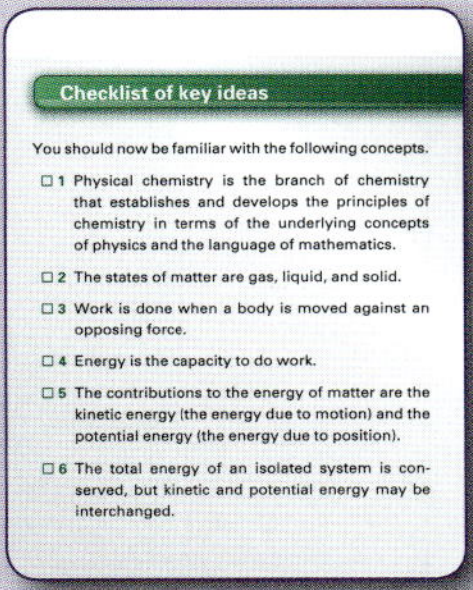

Checklist of key ideas

You should now be familiar with the following concepts.

- ☐ 1 Physical chemistry is the branch of chemistry that establishes and develops the principles of chemistry in terms of the underlying concepts of physics and the language of mathematics.
- ☐ 2 The states of matter are gas, liquid, and solid.
- ☐ 3 Work is done when a body is moved against an opposing force.
- ☐ 4 Energy is the capacity to do work.
- ☐ 5 The contributions to the energy of matter are the kinetic energy (the energy due to motion) and the potential energy (the energy due to position).
- ☐ 6 The total energy of an isolated system is conserved, but kinetic and potential energy may be interchanged.

Molecular Interpretation icons

Although thermo-dynamics is a self-contained subject, it is greatly enriched when its concepts are explained in terms of atoms and molecules.

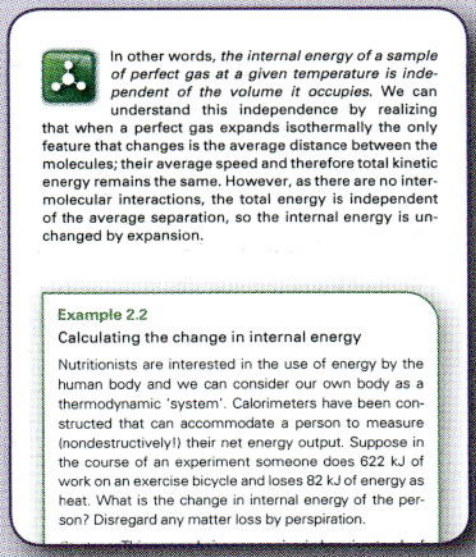

In other words, *the internal energy of a sample of perfect gas at a given temperature is independent of the volume it occupies.* We can understand this independence by realizing that when a perfect gas expands isothermally the only feature that changes is the average distance between the molecules; their average speed and therefore total kinetic energy remains the same. However, as there are no intermolecular interactions, the total energy is independent of the average separation, so the internal energy is unchanged by expansion.

Example 2.2

Calculating the change in internal energy

Nutritionists are interested in the use of energy by the human body and we can consider our own body as a thermodynamic 'system'. Calorimeters have been constructed that can accommodate a person to measure (nondestructively!) their net energy output. Suppose in the course of an experiment someone does 622 kJ of work on an exercise bicycle and loses 82 kJ of energy as heat. What is the change in internal energy of the person? Disregard any matter loss by perspiration.

This icon indicates where we are introducing a molecular interpretation.

Table of key equations

We summarize the most important equations introduced in each chapter as a checklist that follows the chapter's *Table of key ideas*. When appropriate, we describe the physical conditions under which an equation applies.

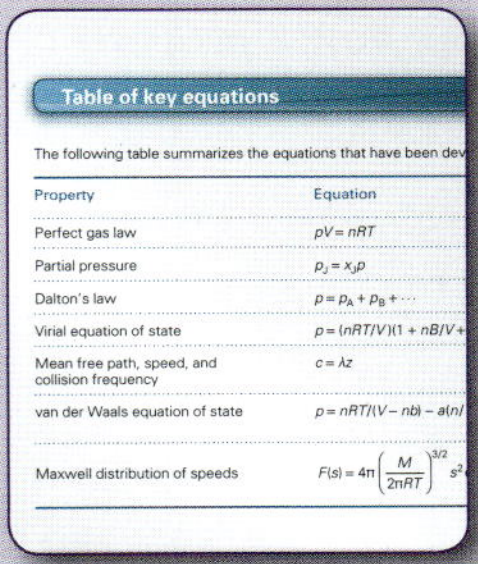

Table of key equations

The following table summarizes the equations that have been dev

Property	Equation
Perfect gas law	$pV = nRT$
Partial pressure	$p_J = x_J p$
Dalton's law	$p = p_A + p_B + \cdots$
Virial equation of state	$p = (nRT/V)(1 + nB/V +$
Mean free path, speed, and collision frequency	$c = \lambda z$
van der Waals equation of state	$p = nRT/(V - nb) - a(n/$
Maxwell distribution of speeds	$F(s) = 4\pi\left(\frac{M}{2\pi RT}\right)^{3/2} s^2$

Notes on good practice

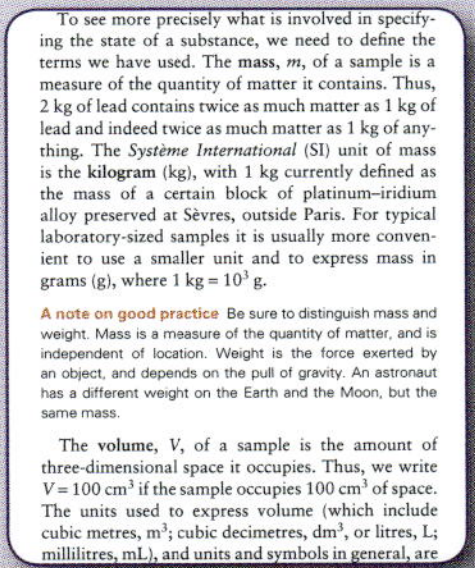

To see more precisely what is involved in specifying the state of a substance, we need to define the terms we have used. The **mass**, *m*, of a sample is a measure of the quantity of matter it contains. Thus, 2 kg of lead contains twice as much matter as 1 kg of lead and indeed twice as much matter as 1 kg of anything. The *Système International* (SI) unit of mass is the **kilogram** (kg), with 1 kg currently defined as the mass of a certain block of platinum–iridium alloy preserved at Sèvres, outside Paris. For typical laboratory-sized samples it is usually more convenient to use a smaller unit and to express mass in grams (g), where 1 kg = 10^3 g.

A note on good practice Be sure to distinguish mass and weight. Mass is a measure of the quantity of matter, and is independent of location. Weight is the force exerted by an object, and depends on the pull of gravity. An astronaut has a different weight on the Earth and the Moon, but the same mass.

The **volume**, *V*, of a sample is the amount of three-dimensional space it occupies. Thus, we write $V = 100\ cm^3$ if the sample occupies $100\ cm^3$ of space. The units used to express volume (which include cubic metres, m^3; cubic decimetres, dm^3, or litres, L; millilitres, mL), and units and symbols in general, are

Science is a precise activity and its language should be used accurately. We use this feature to help encourage the use of the language and procedures of science in conformity to international practice (as specified by IUPAC, the International Union of Pure and Applied Chemistry) and to help avoid common mistakes.

Boxes

Where appropriate, we separate the principles from their applications: the principles are constant; the applications come and go as the subject progresses. The *Boxes*, about one in each chapter, show how the principles developed in the chapter are currently being applied in a variety of modern contexts, especially biology and materials science.

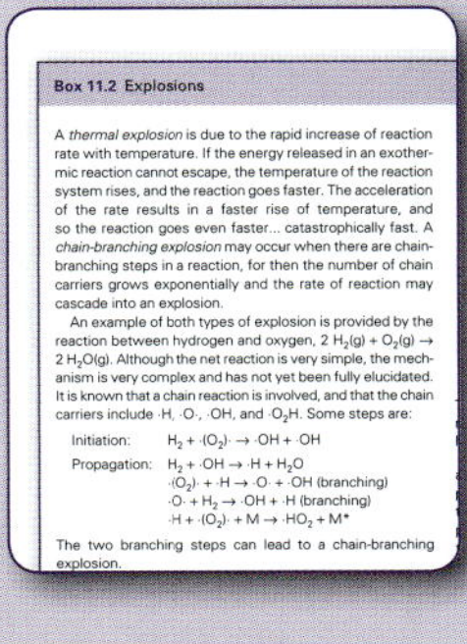

Box 11.2 Explosions

A *thermal explosion* is due to the rapid increase of reaction rate with temperature. If the energy released in an exothermic reaction cannot escape, the temperature of the reaction system rises, and the reaction goes faster. The acceleration of the rate results in a faster rise of temperature, and so the reaction goes even faster... catastrophically fast. A *chain-branching explosion* may occur when there are chain-branching steps in a reaction, for then the number of chain carriers grows exponentially and the rate of reaction may cascade into an explosion.

An example of both types of explosion is provided by the reaction between hydrogen and oxygen, $2\ H_2(g) + O_2(g) \rightarrow 2\ H_2O(g)$. Although the net reaction is very simple, the mechanism is very complex and has not yet been fully elucidated. It is known that a chain reaction is involved, and that the chain carriers include $\cdot H$, $\cdot O\cdot$, $\cdot OH$, and $\cdot O_2H$. Some steps are:

Initiation: $H_2 + \cdot(O_2)\cdot \rightarrow \cdot OH + \cdot OH$

Propagation: $H_2 + \cdot OH \rightarrow \cdot H + H_2O$

$\cdot(O_2)\cdot + \cdot H \rightarrow \cdot O\cdot + \cdot OH$ (branching)

$\cdot O\cdot + H_2 \rightarrow \cdot OH + \cdot H$ (branching)

$\cdot H + \cdot(O_2)\cdot + M \rightarrow \cdot HO_2 + M^*$

The two branching steps can lead to a chain-branching explosion.

Derivations

On first reading it might be sufficient simply to appreciate the 'bottom line' rather than work through detailed development of a mathematical expression. However, mathematical development is an intrinsic part of physical chemistry, and to achieve full

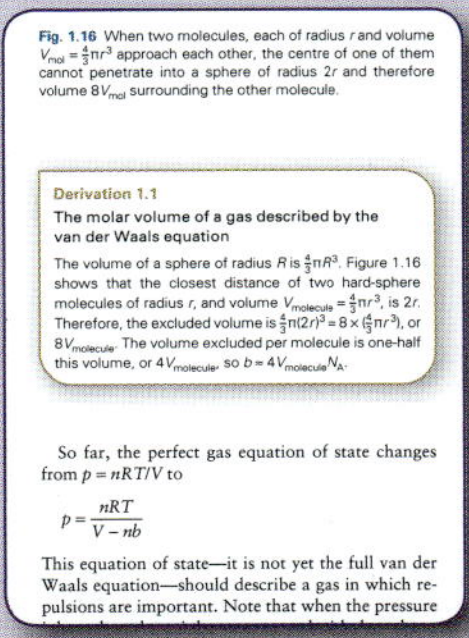

Fig. 1.16 When two molecules, each of radius *r* and volume $V_{mol} = \frac{4}{3}\pi r^3$ approach each other, the centre of one of them cannot penetrate into a sphere of radius $2r$ and therefore volume $8V_{mol}$ surrounding the other molecule.

Derivation 1.1

The molar volume of a gas described by the van der Waals equation

The volume of a sphere of radius R is $\frac{4}{3}\pi R^3$. Figure 1.16 shows that the closest distance of two hard-sphere molecules of radius r, and volume $V_{molecule} = \frac{4}{3}\pi r^3$, is $2r$. Therefore, the excluded volume is $\frac{4}{3}\pi(2r)^3 = 8 \times (\frac{4}{3}\pi r^3)$, or $8V_{molecule}$. The volume excluded per molecule is one-half this volume, or $4V_{molecule}$, so $b = 4V_{molecule}N_A$.

So far, the perfect gas equation of state changes from $p = nRT/V$ to

$$p = \frac{nRT}{V - nb}$$

This equation of state—it is not yet the full van der Waals equation—should describe a gas in which repulsions are important. Note that when the pressure

understanding it is important to see how a particular expression is obtained. The *Derivations* let you adjust the level of detail that you require to your current needs, and make it easier to review material. All the calculus in the book is confined within these *Derivations*.

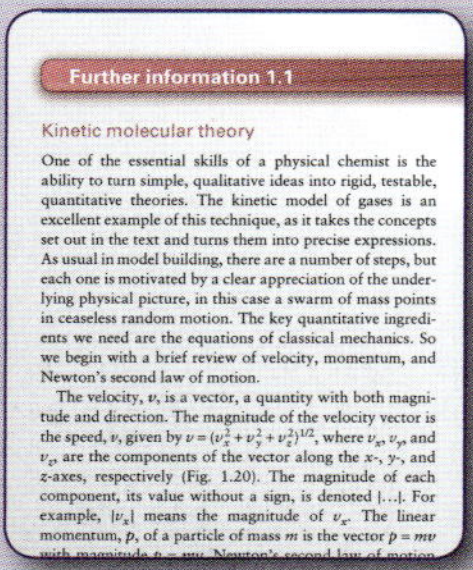
Further information 1.1

Kinetic molecular theory

One of the essential skills of a physical chemist is the ability to turn simple, qualitative ideas into rigid, testable, quantitative theories. The kinetic model of gases is an excellent example of this technique, as it takes the concepts set out in the text and turns them into precise expressions. As usual in model building, there are a number of steps, but each one is motivated by a clear appreciation of the underlying physical picture, in this case a swarm of mass points in ceaseless random motion. The key quantitative ingredients we need are the equations of classical mechanics. So we begin with a brief review of velocity, momentum, and Newton's second law of motion.

The velocity, *v*, is a vector, a quantity with both magnitude and direction. The magnitude of the velocity vector is the speed, *v*, given by $v = (v_x^2 + v_y^2 + v_z^2)^{1/2}$, where v_x, v_y, and v_z are the components of the vector along the *x*-, *y*-, and *z*-axes, respectively (Fig. 1.20). The magnitude of each component, its value without a sign, is denoted |...|. For example, $|v_x|$ means the magnitude of v_x. The linear momentum, *p*, of a particle of mass *m* is the vector *p* = *mv*

Further information

In some cases, we have judged that a derivation is too long, too detailed, or too different in level for it to be included in the text. In these cases, the derivations are found less obtrusively at the end of the chapter.

Mathematics support

Bubbles

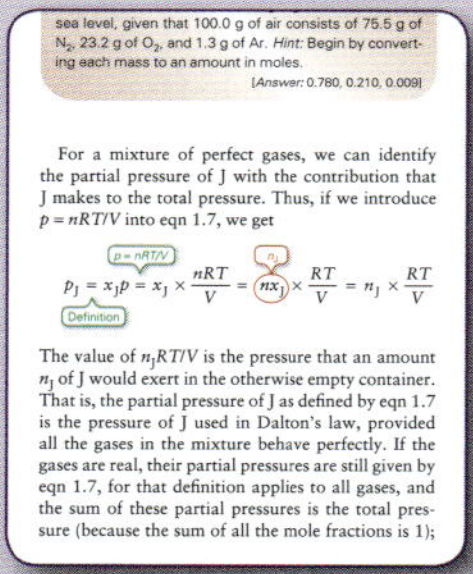
sea level, given that 100.0 g of air consists of 75.5 g of N_2, 23.2 g of O_2, and 1.3 g of Ar. *Hint*: Begin by converting each mass to an amount in moles.
[Answer: 0.780, 0.210, 0.009]

For a mixture of perfect gases, we can identify the partial pressure of J with the contribution that J makes to the total pressure. Thus, if we introduce $p = nRT/V$ into eqn 1.7, we get

$$p_J = x_J p = x_J \times \frac{nRT}{V} = (n x_J) \times \frac{RT}{V} = n_J \times \frac{RT}{V}$$

The value of n_JRT/V is the pressure that an amount n_J of J would exert in the otherwise empty container. That is, the partial pressure of J as defined by eqn 1.7 is the pressure of J used in Dalton's law, provided all the gases in the mixture behave perfectly. If the gases are real, their partial pressures are still given by eqn 1.7, for that definition applies to all gases, and the sum of these partial pressures is the total pressure (because the sum of all the mole fractions is 1);

You often need to know how to develop a mathematical expression, but how do you go from one line to the next? A green 'bubble' is a little reminder about the substitution used, the approximation made, the terms that have been assumed constant, and so on. A red 'bubble' is a reminder of the significance of an individual term in an expression.

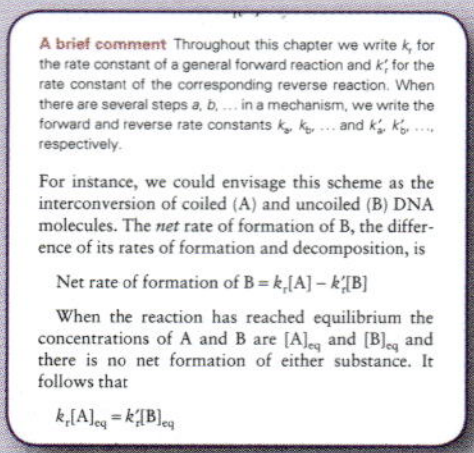
A brief comment Throughout this chapter we write k_r for the rate constant of a general forward reaction and k_r' for the rate constant of the corresponding reverse reaction. When there are several steps *a*, *b*, ... in a mechanism, we write the forward and reverse rate constants k_a, k_b, ... and k_a', k_b', ..., respectively.

For instance, we could envisage this scheme as the interconversion of coiled (A) and uncoiled (B) DNA molecules. The *net* rate of formation of B, the difference of its rates of formation and decomposition, is

Net rate of formation of B = $k_r[A] - k_r'[B]$

When the reaction has reached equilibrium the concentrations of A and B are $[A]_{eq}$ and $[B]_{eq}$ and there is no net formation of either substance. It follows that

$k_r[A]_{eq} = k_r'[B]_{eq}$

A brief comment

A topic often needs to draw on a mathematical procedure or a concept of physics; *A brief comment* is a quick reminder of the procedure or concept.

Visualizing the information

Artwork

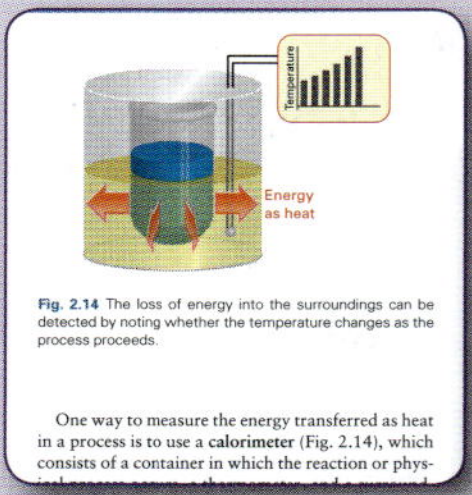

Fig. 2.14 The loss of energy into the surroundings can be detected by noting whether the temperature changes as the process proceeds.

One way to measure the energy transferred as heat in a process is to use a **calorimeter** (Fig. 2.14), which consists of a container in which the reaction or phys-

In many instances, a concept is easier to understand if it is presented in visual, as well as written, form. Every piece of artwork in this new edition has been carefully rendered in full colour, to help you master the concepts presented.

Living Graphs

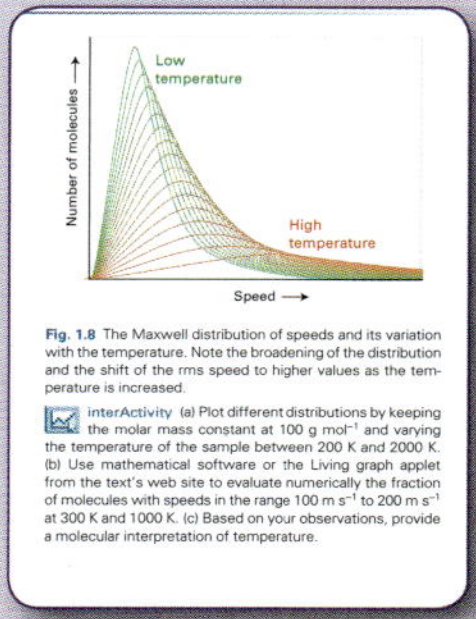

Fig. 1.8 The Maxwell distribution of speeds and its variation with the temperature. Note the broadening of the distribution and the shift of the rms speed to higher values as the temperature is increased.

interActivity (a) Plot different distributions by keeping the molar mass constant at 100 g mol^{-1} and varying the temperature of the sample between 200 K and 2000 K. (b) Use mathematical software or the Living graph applet from the text's web site to evaluate numerically the fraction of molecules with speeds in the range 100 m s^{-1} to 200 m s^{-1} at 300 K and 1000 K. (c) Based on your observations, provide a molecular interpretation of temperature.

In some cases, the trends or properties presented in a graph are difficult to interpret when the graph is viewed as a static figure. In such cases, a dynamic *Living graph* is available in the eBook version of the text. A *Living graph* can be used to explore how a property changes as a variety of parameters are changed.

The figures in the book with associated *Living graphs* are flagged with icons in the figure legends as shown here.

Animations

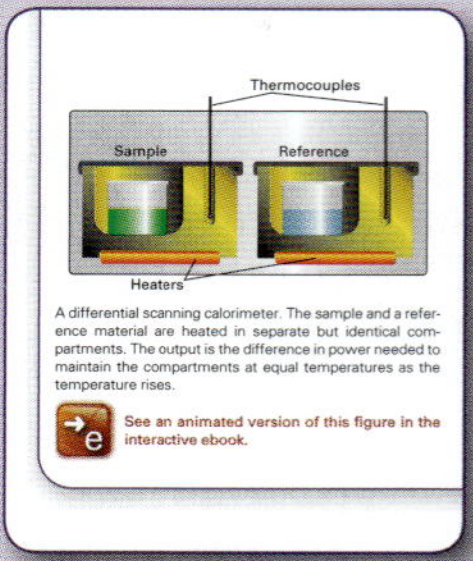

A differential scanning calorimeter. The sample and a reference material are heated in separate but identical compartments. The output is the difference in power needed to maintain the compartments at equal temperatures as the temperature rises.

See an animated version of this figure in the interactive ebook.

In some cases, it is difficult to communicate a dynamic process in a static figure. In such instances, animated versions of selected artwork are available in the eBook version of the text. Where animated versions of figures are available, these are flagged in the text as shown below.

Problem solving

A brief illustration

A brief illustration is a short example of how to use an equation that has just been introduced in the text. In particular, we show how to use data and how to manipulate units correctly.

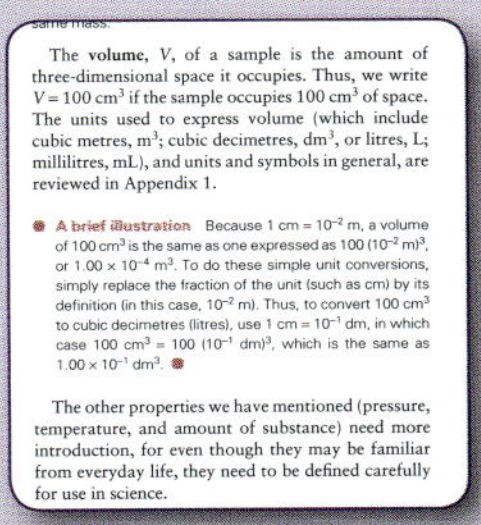

same mass.

The **volume**, V, of a sample is the amount of three-dimensional space it occupies. Thus, we write $V = 100\ cm^3$ if the sample occupies $100\ cm^3$ of space. The units used to express volume (which include cubic metres, m^3; cubic decimetres, dm^3, or litres, L; millilitres, mL), and units and symbols in general, are reviewed in Appendix 1.

A brief illustration Because $1\ cm = 10^{-2}\ m$, a volume of $100\ cm^3$ is the same as one expressed as $100\ (10^{-2}\ m)^3$, or $1.00 \times 10^{-4}\ m^3$. To do these simple unit conversions, simply replace the fraction of the unit (such as cm) by its definition (in this case, 10^{-2} m). Thus, to convert $100\ cm^3$ to cubic decimetres (litres), use $1\ cm = 10^{-1}\ dm$, in which case $100\ cm^3 = 100\ (10^{-1}\ dm)^3$, which is the same as $1.00 \times 10^{-1}\ dm^3$.

The other properties we have mentioned (pressure, temperature, and amount of substance) need more introduction, for even though they may be familiar from everyday life, they need to be defined carefully for use in science.

Worked examples

Each *Worked example* has a *Strategy* section to suggest how to set up the problem (another way might seem more natural: setting up problems is a highly personal business) and use or find the necessary data. Then there is the worked-out *Answer*, where we emphasize the importance of using units correctly.

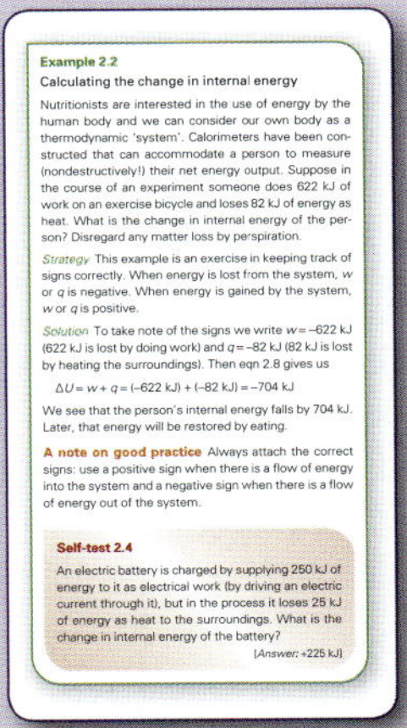

Example 2.2

Calculating the change in internal energy

Nutritionists are interested in the use of energy by the human body and we can consider our own body as a thermodynamic 'system'. Calorimeters have been constructed that can accommodate a person to measure (nondestructively!) their net energy output. Suppose in the course of an experiment someone does 622 kJ of work on an exercise bicycle and loses 82 kJ of energy as heat. What is the change in internal energy of the person? Disregard any matter loss by perspiration.

Strategy This example is an exercise in keeping track of signs correctly. When energy is lost from the system, w or q is negative. When energy is gained by the system, w or q is positive.

Solution To take note of the signs we write $w = -622$ kJ (622 kJ is lost by doing work) and $q = -82$ kJ (82 kJ is lost by heating the surroundings). Then eqn 2.8 gives us

$$\Delta U = w + q = (-622\ kJ) + (-82\ kJ) = -704\ kJ$$

We see that the person's internal energy falls by 704 kJ. Later, that energy will be restored by eating.

A note on good practice Always attach the correct signs: use a positive sign when there is a flow of energy into the system and a negative sign when there is a flow of energy out of the system.

Self-test 2.4

An electric battery is charged by supplying 250 kJ of energy to it as electrical work (by driving an electric current through it), but in the process it loses 25 kJ of energy as heat to the surroundings. What is the change in internal energy of the battery?

[*Answer*: +225 kJ]

Self-tests

Each *Worked example* has a *Self-test* with the answer provided as a check that the procedure has been mastered. There are also a number of free-standing *Self-tests* that are located where we thought it a good idea to provide a question to check your understanding. Think of *Self-tests* as in-chapter Exercises designed to help you monitor your progress.

Discussion questions

The end-of-chapter material starts with a short set of questions that are intended to encourage reflection on the material and to view it in a broader context than is obtained by solving numerical problems.

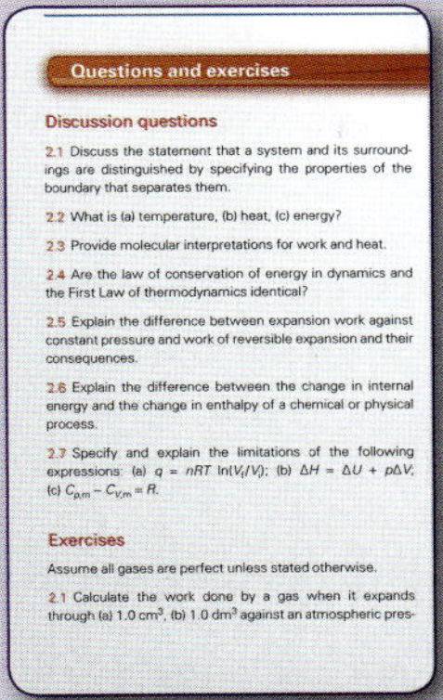

Questions and exercises

Discussion questions

2.1 Discuss the statement that a system and its surroundings are distinguished by specifying the properties of the boundary that separates them.

2.2 What is (a) temperature, (b) heat, (c) energy?

2.3 Provide molecular interpretations for work and heat.

2.4 Are the law of conservation of energy in dynamics and the First Law of thermodynamics identical?

2.5 Explain the difference between expansion work against constant pressure and work of reversible expansion and their consequences.

2.6 Explain the difference between the change in internal energy and the change in enthalpy of a chemical or physical process.

2.7 Specify and explain the limitations of the following expressions: (a) $q = nRT \ln(V_f/V_i)$; (b) $\Delta H = \Delta U + p\Delta V$; (c) $C_{p,m} - C_{V,m} = R$.

Exercises

Assume all gases are perfect unless stated otherwise.

2.1 Calculate the work done by a gas when it expands through (a) $1.0\ cm^3$, (b) $1.0\ dm^3$ against an atmospheric pres-

Exercises

The core of testing understanding is the collection of end-of-chapter *Exercises*. At the end of the *Exercises* you will find a small collection of *Projects* that bring together a lot of the foregoing material, may call for the use of calculus, and are typically based on material introduced in the *Boxes*.

The Book Companion Site

The Book Companion Site provides teaching and learning resources to augment the printed book. It is free of charge, complements the textbook, and offers additional materials which can be downloaded. The resources it provides are fully customizable and can be incorporated into a virtual learning environment.

The Book Companion Site can be accessed by visiting
http://www.whfreeman.com/elements5e

For students

Answers to exercises

The final answers to most end-of-chapter exercises are available for you to check your work.

Web links

Links to a range of useful and relevant physical chemistry web sites.

For lecturers

Artwork

A lecturer may wish to use the illustrations from this text in a lecture. Almost all the illustrations are available in PowerPoint® format and can be used for lectures without charge (but not for commercial purposes without specific permission).

Tables of data

All the tables of data that appear in the chapter text are available and may be used under the same conditions as the illustrations.

On-line quizzing

New for this edition, on line quizzing available on the book companion site offers multiple-choice questions for use within a virtual learning environment, with feedback referred back to relevant sections of the book. This feature is a valuable tool for either formative or summative assessment.

Elements of Physical Chemistry eBook

The *eBook*, which is a complete version of the textbook itself, provides a rich learning experience by taking full advantage of the electronic medium integrating all student media resources and adds features unique to the *eBook*. The *eBook* also offers lecturers unparalleled flexibility and customization options. Access to the *eBook* is either provided in the form of an access code packaged with the text or it can be purchased at **http://ebooks.bfwpub.com/elements5e**. Key features of the *eBook* include:

- Living Graphs

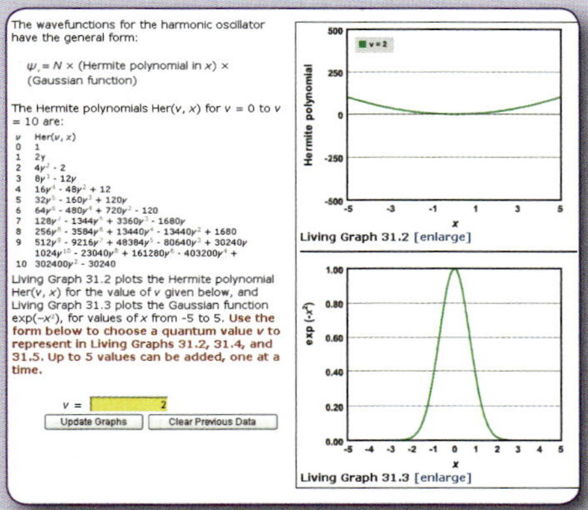

- Dynamic figures: animated versions of figures from the book

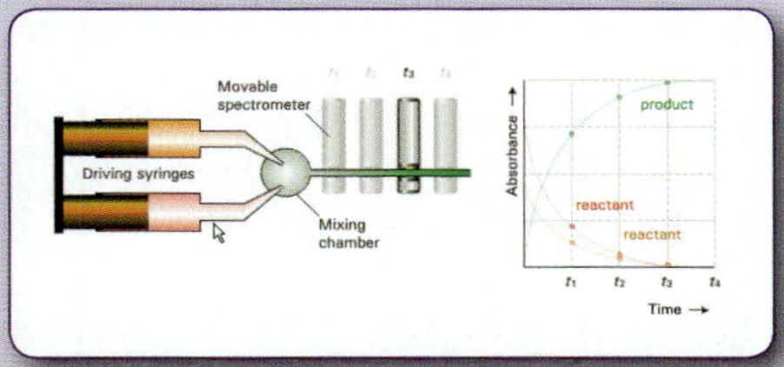

- Interactive equations: extra annotations, extra interim steps, and explanatory comments

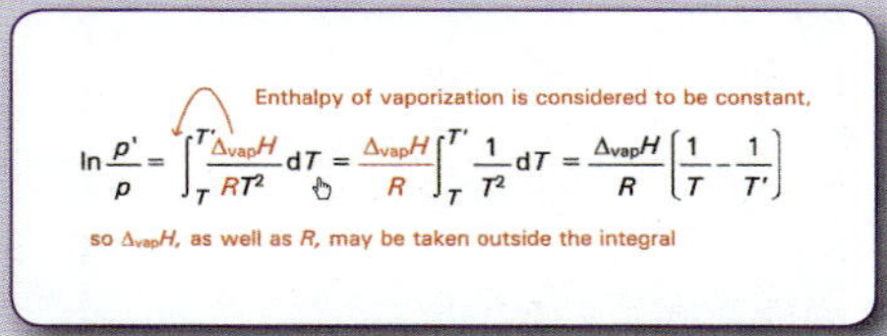

- Hidden answers to self tests and the questions from the end of the chapter
- Full text search, highlighting, and bookmarks
- Quick navigation from key terms to glossary definitions, and from maths and physics comments to fuller explanations

Tailor the book to your own needs:

- Users are able to add, share, and print their own notes
- Registered adopters may add sections to customise the text to match their course

Other resources

Explorations in Physical Chemistry by Valerie Walters, Julio de Paula, and Peter Atkins.

Explorations in Physical Chemistry consists of interactive Mathcad® worksheets and interactive Excel® workbooks, complete with thought-stimulating exercises. They motivate students to simulate physical, chemical, and biochemical phenomena with their personal computers. Harnessing the computational power of Mathcad® by Mathsoft, Inc. and Excel® by Microsoft Corporation, students can manipulate over 75 graphics, alter simulation parameters, and solve equations to gain deeper insight into physical chemistry. *Explorations in Physical Chemistry* can be purchased at **http://ebooks.bfwpub.com/explorations.php**.

Solutions manual

Charles Trapp and Marshall Cady have produced a solutions manual to accompany the book, which features full worked solutions to all end-of-chapter discussion questions and exercises, and is available free-of-charge to registered adopters of the text. (ISBN 1-4292-2400-2).

Preface

When a book enters its fifth edition you might expect a certain maturity and a settling down into a comfortable middle if not old age. We hope you will identify the former but not the latter. We learn enormously from each new edition and like to refresh the exposition and introduce new ideas at every opportunity. We hope that you will see maturity certainly but also a new vibrancy in this edition.

The structure of the book remains much the same as in the fourth edition, but with a small reorganization of chapters, such as the reversal of the order of the groups of chapters on Materials. We have also brought together under various umbrella titles the related chapters to give a greater sense of cohesion. Thus there is a Chemical Equilibrium family, a Chemical Kinetics family, a Quantum Chemistry family, a Materials family, and a Spectroscopy family. Throughout the text we have had in mind one principal objective: to ensure that the coverage is appropriate to a single compact physical chemistry course. As a result, we have eliminated some material but (with our eyes alert to the dangers of expanding the text unduly) have strengthened the discussion of a wide range of topics.

One aspect of the vibrancy of presentation that we have sought to achieve is that the entire art programme has been redrawn in full colour. As a result, we hope that not only will you enjoy using the book more than earlier editions but find the illustrations much more informative. We have paid more attention to the presentation of mathematics in this edition. We introduced 'bubbles' in the fourth edition: they contain remarks about the steps being taken to develop an equation. We have taken this popular feature much further in this edition, and have added many more bubbles. The green bubbles indicate how to proceed across an equals sign; the red bubbles indicate the meaning of terms in an expression. In this edition we have introduced another new feature that should help you with your studies: each chapter now has a *Checklist of key equations* following the *Checklist of key ideas*, which now summarizes only the concepts.

A source of confusion in the fourth edition was the use of the term *Illustration*: some thought it meant a diagram; others a short example. We have renamed all the short examples *A brief illustration*, so that confusion should now be avoided. These brief illustrations have been joined by *A brief comment* and we have retained and expanded the popular *Notes on good practice*. A good proportion of the end-of-chapter *Exercises* have been modified or replaced; we have added *Projects*, rather involved exercises that often call for the use of calculus. The new features are summarized in the following *About the book* section.

As always in the preparation of a new edition we have relied heavily on advice from users throughout the world, our numerous translators into other languages, and colleagues who have given their time in the reviewing process. We are greatly indebted to them, and have learned a lot from them. They are identified and thanked in the *Acknowledgements* section.

PWA
JdeP

About the authors

Peter Atkins is a fellow of Lincoln College in the University of Oxford and the author of more than sixty books for students and a general audience. His texts are market leaders around the globe. A frequent lecturer in the United States and throughout the world, he has held visiting professorships in France, Israel, Japan, China, and New Zealand. He was the founding chairman of the Committee on Chemistry Education of the International Union of Pure and Applied Chemistry and was a member of IUPAC's Physical and Biophysical Chemistry Division.

Julio de Paula is Professor of Chemistry and Dean of the College of Arts & Sciences at Lewis & Clark College. A native of Brazil, Professor de Paula received a B.A. degree in chemistry from Rutgers, The State University of New Jersey, and a Ph.D. in biophysical chemistry from Yale University. His research activities encompass the areas of molecular spectroscopy, biophysical chemistry, and nanoscience. He has taught courses in general chemistry, physical chemistry, biophysical chemistry, instrumental analysis, and writing.

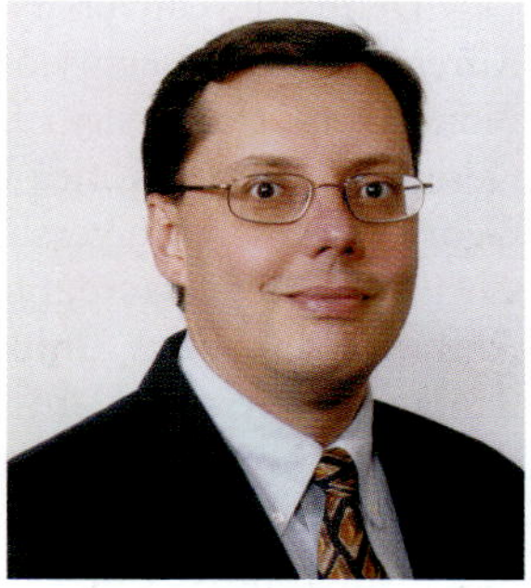

Acknowledgements

The authors have received a great deal of help during the preparation and production of this text and wish to thank all their colleagues who have made such thought-provoking and useful suggestions. In particular, we wish to record publicly our thanks to:

I think formal names s/b used, not familiars
David Andrews, University of East Anglia
Richard Ansell, University of Leeds
Nicholas Brewer, University of Dundee
Melanie Britton, University of Birmingham
Gerrit ten Brinke, University of Groningen
Guy Denuault, University of Southampton
Karen Edler, University of Bath
Fiona Gray, University of St. Andrews
Gerhard Grobner, Umeå University
Georg Haehner, University of St. Andrews
Christopher Hardacre, Queens University Belfast
Anthony Harriman, University of Newcastle
Benjamin Horrocks, University of Newcastle
Robert Jackson, University of Keele
Phillip John, Heriot-Watt University
Peter Karadakov, University of York
Peter Knowles, University of Cardiff
Adam Lee, University of York
Dónal Leech, National University of Ireland, Galway
Göran Lindblom, Umeå University
Lesley Lloyd, University of Birmingham
Michael Lyons, Trinity College Dublin
Alexander Lyubartsev, Stockholm University
Arnold Maliniak, Stockholm University
David McGarvey, University of Keele
Anthony Meijer, University of Sheffield
Marcelo de Miranda, University of Leeds
Damien Murphy, University of Cardiff
Gavin Reid, University of Leeds
Stephen Roser, University of Bath
Karl Ryder, University of Leicester
Sven Schroeder, University of Manchester
David Steytler, University of East Anglia
Michael Stockenhuber, University of Newcastle, New South Wales
Svein Stolen, University of Oslo
Jeremy Titman, University of Nottingham
Palle Waage Jensen, University of Southern Denmark
Jay Wadhawan, University of Hull
Darren Walsh, University of Nottingham
Kjell Waltersson, Mälardalen University
Richard Wells, University of Aberdeen

David Smith of the University of Bristol, has played a central role in the reviewing process, and we would like to thank him for his detailed and insightful remarks, all of which have helped to shape the book. He has also developed many of the interactive components of the eBook, in the process adding a valuable educational dimension to this new resource.

Last, but by no means least, we wish to acknowledge the whole-hearted and unstinting support of our two commissioning editors, Jonathan Crowe of Oxford University Press and Jessica Fiorillo of W.H. Freeman & Co., and our development editor, Leonie Sloman, who—in other projects as well as this—have helped the authors to realize their vision and have done so in such an agreeable and professional a manner.

Brief contents

Detailed contents

Introduction

Chemistry is the science of matter and the changes it can undergo. The branch of the subject called **physical chemistry** is concerned with the physical principles that underlie chemistry. Physical chemistry seeks to account for the properties of matter in terms of fundamental concepts such as atoms, electrons, and energy. It provides the basic framework for all other branches of chemistry—for inorganic chemistry, organic chemistry, biochemistry, geochemistry, and chemical engineering. It also provides the basis of modern methods of analysis, the determination of structure, and the elucidation of the manner in which chemical reactions occur. To do all this, it draws on two of the great foundations of modern physical science, thermodynamics and quantum mechanics.

This text introduces the central concepts of these two subjects and shows how they are used in chemistry. This chapter reviews material fundamental to the whole of physical chemistry, much of which will be familiar from introductory courses. We begin by thinking about matter in bulk. The broadest classification of matter is into one of three **states of matter**, or forms of bulk matter, namely gas, liquid, and solid. Later we shall see how this classification can be refined, but these three broad classes are a good starting point.

0.1 The states of matter

We distinguish the three states of matter by noting the behaviour of a substance enclosed in a container:

A **gas** is a fluid form of matter that fills the container it occupies.

A **liquid** is a fluid form of matter that possesses a well-defined surface and (in a gravitational field) fills the lower part of the container it occupies.

A **solid** retains its shape regardless of the shape of the container it occupies.

One of the roles of physical chemistry is to establish the link between the properties of bulk matter and the behaviour of the particles—atoms, ions, or molecules—of which it is composed. A physical chemist formulates a **model**, a simplified description, of each physical state and then shows how the state's properties can be understood in terms of this model. The existence of different states of matter is a first illustration of this procedure, as the properties of the three states suggest that they are composed of particles with different degrees of freedom of movement. Indeed, as we work through this text, we shall gradually establish and elaborate the following models:

A gas is composed of widely separated particles in continuous rapid, disordered motion. A particle travels several (often many) diameters before colliding with another particle. For most of the time the particles are so far apart that they interact with each other only very weakly.

A liquid consists of particles that are in contact but are able to move past each other in a restricted manner. The particles are in a continuous state of motion, but travel only a fraction of a diameter before bumping into a neighbour. The overriding image is one of movement, but with molecules jostling one another.

A solid consists of particles that are in contact and only rarely able to move past one another. Although the particles oscillate at an average location, they are essentially trapped in their initial positions, and typically lie in ordered arrays.

The essential difference between the three states of matter is the freedom of the particles to move past one another. If the average separation of the particles is large, there is hardly any restriction on their motion and the substance is a gas. If the particles interact so strongly with one another that they are locked together rigidly, then the substance is a solid. If the particles have an intermediate mobility between these extremes, then the substance is a liquid. We can understand the melting of a solid and the vaporization of a liquid in terms of the progressive increase in the liberty of the particles as a sample is heated and the particles become able to move more freely.

0.2 Physical state

The term 'state' has many different meanings in chemistry, and it is important to keep them all in mind. We have already met one meaning in the expression 'the states of matter' and specifically 'the gaseous state'. Now we meet a second: by **physical state** (or just 'state') we shall mean a specific condition of a sample of matter that is described in terms of its physical form (gas, liquid, or solid) and the volume, pressure, temperature, and amount of substance present. (The precise meanings of these terms are described below.) So, 1 kg of hydrogen gas in a container of volume 10 dm^3 at a specified pressure and temperature is in a particular state. The same mass of gas in a container of volume 5 dm^3 is in a different state. Two samples of a given substance are in the same state if they are the same state of matter (that is, are both present as gas, liquid, or solid) *and* if they have the same mass, volume, pressure, and temperature.

To see more precisely what is involved in specifying the state of a substance, we need to define the terms we have used. The **mass**, m, of a sample is a measure of the quantity of matter it contains. Thus, 2 kg of lead contains twice as much matter as 1 kg of lead and indeed twice as much matter as 1 kg of anything. The *Système International* (SI) unit of mass is the **kilogram** (kg), with 1 kg currently defined as the mass of a certain block of platinum–iridium alloy preserved at Sèvres, outside Paris. For typical laboratory-sized samples it is usually more convenient to use a smaller unit and to express mass in grams (g), where 1 kg = 10^3 g.

A note on good practice Be sure to distinguish mass and weight. Mass is a measure of the quantity of matter, and is independent of location. Weight is the force exerted by an object, and depends on the pull of gravity. An astronaut has a different weight on the Earth and the Moon, but the same mass.

The **volume**, V, of a sample is the amount of three-dimensional space it occupies. Thus, we write $V = 100\ cm^3$ if the sample occupies 100 cm^3 of space. The units used to express volume (which include cubic metres, m^3; cubic decimetres, dm^3, or litres, L; millilitres, mL), and units and symbols in general, are reviewed in Appendix 1.

A brief illustration Because 1 cm = 10^{-2} m, a volume of 100 cm^3 is the same as one expressed as 100 $(10^{-2}\ m)^3$, or $1.00 \times 10^{-4}\ m^3$. To do these simple unit conversions, simply replace the fraction of the unit (such as cm) by its definition (in this case, 10^{-2} m). Thus, to convert 100 cm^3 to cubic decimetres (litres), use 1 cm = 10^{-1} dm, in which case 100 cm^3 = 100 $(10^{-1}\ dm)^3$, which is the same as $1.00 \times 10^{-1}\ dm^3$.

The other properties we have mentioned (pressure, temperature, and amount of substance) need more introduction, for even though they may be familiar from everyday life, they need to be defined carefully for use in science.

0.3 Force

One of the most basic concepts of physical science is that of **force**, F. In classical mechanics, the mechanics originally formulated by Isaac Newton at the end of the seventeenth century, a body of mass m travels in a straight line at constant speed until a force acts on it. Then it undergoes an acceleration a, a rate of change of velocity, given by Newton's second law of motion:

$$\text{Force} = \text{mass} \times \text{acceleration} \qquad F = ma$$

Force is actually a 'vector' quantity, a quantity with direction as well as magnitude, so it could be represented by an arrow pointing in the direction in which the force is applied. The acceleration is also a vector, and Newton's law captures the sense that if a force is applied in the direction of increasing x (in one dimension), then the acceleration is in that direction too. In most instances in this text we need consider only the magnitude explicitly, but we shall need to keep in mind the often unstated direction in which it is applied.

A brief illustration The acceleration of a freely falling body at the surface of the Earth is close to 9.81 m s^{-2}, so the magnitude of the gravitational force acting on a mass of 1.0 kg is

$$F = (1.0 \text{ kg}) \times (9.81 \text{ m s}^{-2}) = 9.8 \text{ kg m s}^{-2}$$

and directed towards the centre of mass of the Earth. The derived unit of force is the newton, N:

$$1 \text{ N} = 1 \text{ kg m s}^{-2}$$

Therefore, we can report that $F = 9.8$ N. It might be helpful to note that a force of 1 N is approximately the gravitational force exerted on a small apple (of mass 100 g).

A note on good practice A unit raised to a negative power (such as the s^{-2} in m s^{-2}) is the same as writing it after a slash (as in m/s^2). In this sense, units behave like numbers (where 10^{-2} is the same as $1/10^2$). Negative powers are unambiguous: thus, a combination such as kg m^{-1} s^{-2} is much easier to interpret than when it is written kg/m/s^2.

When an object is moved through a distance s against an opposing force, we say that **work** is done. The magnitude of the work is the product of the distance moved and the magnitude of the opposing force:

$$\text{Work} = \text{force} \times \text{distance}$$

This expression applies when the force is constant; if it varies along the path, then we use it for each segment of the path and then add together the resulting values.

A brief illustration To raise a body of mass 1.0 kg on the surface of the Earth through a vertical distance (against the direction of the force) of 1.0 m requires us to do the following amount of work:

$$\text{Work} = (9.8 \text{ N}) \times (1.0 \text{ m}) = 9.8 \text{ N m}$$

As we see more formally in the next section, the unit 1 N m (or, in terms of base units, 1 kg m^2 s^{-2}) is called 1 joule (1 J). So, 9.8 J is needed to raise a mass of 1.0 kg through 1.0 m on the surface of the Earth.

The same expression applies to **electrical work**, the work associated with the motion of electrical charge, with the force on a charge Q (in coulombs, C) equal to $Q\mathcal{E}$, where $\mathcal{E}$ is the strength of the electric field (in volts per metre, V m^{-1}). However, it is normally converted by using relations encountered in electrostatics to an expression in terms of the charge and the 'potential difference' $\Delta\phi$ (delta phi, in volts, V) between the initial and final locations:

$$\text{Work} = \text{charge} \times \text{potential difference, or Work} = Q\Delta\phi$$

We shall need this expression—and develop it further —when we discuss electrochemistry in Chapter 9.

0.4 Energy

A property that will occur in just about every chapter of the following text is the **energy**, E. Everyone uses the term 'energy' in everyday language, but in science it has a precise meaning, a meaning that we shall draw on throughout the text. Energy is the capacity to do work. A fully wound spring can do more work than a half-wound spring (that is, it can raise a weight through a greater height, or move a greater weight through a given height. A hot object, when attached to some kind of heat engine (a device for converting heat into work) can do more work than the same object when it is cool, and therefore a hot object has a higher energy than the same cool object.

The SI unit of energy is the joule (J), named after the nineteenth-century scientist James Joule, who helped to establish the concept of energy (see Chapter 2). It is defined as

$$1 \text{ J} = 1 \text{ kg m}^2 \text{ s}^{-2}$$

A joule is quite a small unit, and in chemistry we often deal with energies of the order of kilojoules (1 kJ = 10^3 J).

There are two contributions to the total energy of a particle. The **kinetic energy**, E_k, is the energy of a body due to its motion. For a body of mass m moving at a speed v,

$$E_k = \tfrac{1}{2}mv^2 \qquad (0.1)$$

That is, a heavy object moving at the same speed as a light object has a higher kinetic energy, and doubling the speed of any object increases its kinetic energy by a factor of 4. A ball of mass 1 kg travelling at 1 m s^{-1} has a kinetic energy of 0.5 J.

The **potential energy**, E_p, of a body is the energy it possesses due to its position. The precise dependence on position depends on the type of force acting on the body. For a body of mass m on the surface of the Earth, the potential energy depends on its height, h, above the surface as

$$E_p = mgh \qquad (0.2)$$

where g is a constant known as the **acceleration of free fall**, which is close to 9.81 m s^{-2} at sea level. Thus, doubling the height, doubles the potential energy. This expression is based on the convention of taking the potential energy to be zero at sea level. A ball of mass 1.0 kg at 1.0 m above the surface of the Earth has a potential energy of 9.8 J. Another type of potential energy is the **Coulombic potential energy** of one electric charge Q_1 (typically in coulombs, C) at a distance r from another electric charge Q_2:

$$E_p = \frac{Q_1 Q_2}{4\pi\varepsilon_0 r} \qquad (0.3)$$

The quantity ε_0 (epsilon zero), the **vacuum permittivity**, is a fundamental constant with the value 8.854×10^{-12} J^{-1} C^2 m^{-1}. As we shall see as the text develops, most contributions to the potential energy that we need consider in chemistry are due to this Coulombic interaction.

The **total energy**, E, of a body is the sum of its kinetic and potential energies:

$$E = E_k + E_p \qquad (0.4)$$

Provided no external forces are acting on the body, its total energy is constant. This remark is elevated to a central statement of classical physics known as the **law of the conservation of energy**. Potential and kinetic energy may be freely interchanged: for instance, a falling ball loses potential energy but gains kinetic energy as it accelerates), but their total remains constant provided the body is isolated from external influences.

0.5 Pressure

Pressure, p, is force, F, divided by the area, A, on which the force is exerted:

$$\text{Pressure} = \frac{\text{force}}{\text{area}}, \qquad p = \frac{F}{A} \qquad (0.5)$$

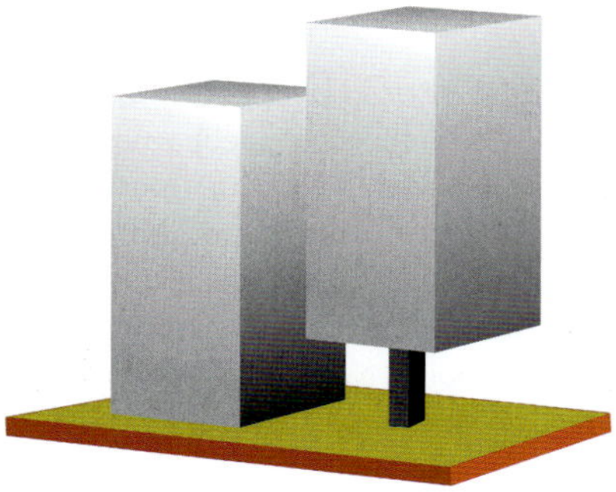

Fig. 0.1 These two blocks of matter have the same mass. They exert the same force on the surface on which they are standing, but the block on the right exerts a stronger pressure because it exerts the same force over a smaller area than the block on the left.

When you stand on ice, you generate a pressure on the ice as a result of the gravitational force acting on your mass and pulling you towards the centre of the Earth. However, the pressure is low because the downward force of your body is spread over the area equal to that of the soles of your shoes. When you stand on skates, the area of the blades in contact with the ice is much smaller, so although your downward *force* is the same, the *pressure* you exert is much greater (Fig. 0.1).

Pressure can arise in ways other than from the gravitational pull of the Earth on an object. For example, the impact of gas molecules on a surface gives rise to a force and hence to a pressure. If an object is immersed in the gas, it experiences a pressure over its entire surface because molecules collide with it from all directions. In this way, the atmosphere exerts a pressure on all the objects in it. We are incessantly battered by molecules of gas in the atmosphere, and experience this battering as the atmospheric pressure. The pressure is greatest at sea level because the density of air, and hence the number of colliding molecules, is greatest there. The atmospheric pressure is very considerable: it is the same as would be exerted by loading 1 kg of lead (or any other material) on to a surface of area 1 cm^2. We go through our lives under this heavy burden pressing on every square centimetre of our bodies. Some deep-sea creatures are built to withstand even greater pressures: at 1000 m below sea level the pressure is 100 times greater than at the surface. Creatures and submarines that operate at these depths must withstand the equivalent of 100 kg of lead loaded on to each square centimetre of their surfaces. The pressure of the air in our lungs helps us withstand the relatively low but still substantial pressures that we experience close to sea level.

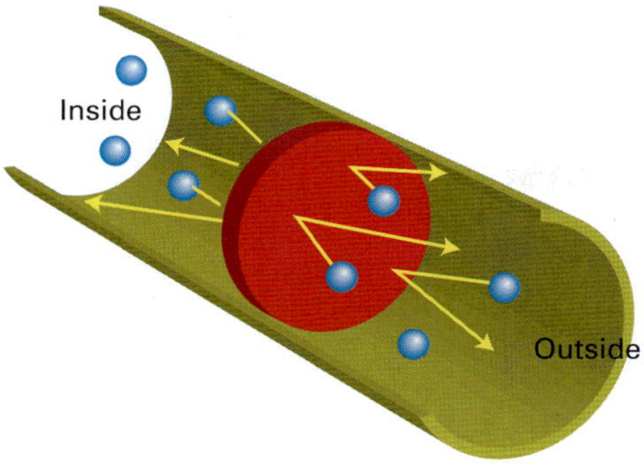

Fig. 0.2 A system is in mechanical equilibrium with its surroundings if it is separated from them by a movable wall and the external pressure is equal to the pressure of the gas in the system.

When a gas is confined to a cylinder fitted with a movable piston, the position of the piston adjusts until the pressure of the gas inside the cylinder is equal to that exerted by the atmosphere. When the pressures on either side of the piston are the same, we say that the two regions on either side are in **mechanical equilibrium.** The pressure of the confined gas arises from the impact of the particles: they batter the inside surface of the piston and counter the battering of the molecules in the atmosphere that is pressing on the outside surface of the piston (Fig. 0.2). Provided the piston is weightless (that is, provided we can neglect any gravitational pull on it), the gas is in mechanical equilibrium with the atmosphere whatever the orientation of the piston and cylinder, because the external battering is the same in all directions.

The SI unit of pressure is called the **pascal** (Pa):

$$1\ \text{Pa} = 1\ \text{N m}^{-2} = 1\ \text{kg m}^{-1}\ \text{s}^{-2}$$

The pressure of the atmosphere at sea level is about 10^5 Pa (100 kPa). This fact lets us imagine the magnitude of 1 Pa, for we have just seen that 1 kg of lead resting on 1 cm^2 on the surface of the Earth exerts about the same pressure as the atmosphere; so $1/10^5$ of that mass, or 10 mg (1 mg = 10^{-3} g), will exert about 1 Pa, we see that the pascal is rather a small unit of pressure. Table 0.1 lists the other units commonly used to report pressure. One of the most important in modern physical chemistry is the **bar**, where 1 bar = 10^5 Pa exactly; the bar is not an SI unit, but it is an accepted and widely used abbreviation for 10^5 Pa. The atmospheric pressure that we normally experience is close to 1 bar; meteorological information on weather maps is commonly reported in millibars (1 mbar = 10^{-3} bar = 10^2 Pa). **Standard pressure**, which is used to report the values of pressure-sensitive properties systematically in a standard way (as we explain in later chapters), is denoted $p^{\ominus}$ and defined as exactly 1 bar.

Table 0.1
*Pressure units and conversion factors**

pascal, Pa	1 Pa = 1 N m^{-2}
bar	1 bar = 10^5 Pa
atmosphere, atm	1 atm = **101.325** kPa = **1.013 25** bar
torr, Torr†	**760** Torr = 1 atm
	1 Torr = 133.32 Pa

* Values in bold are exact.
† The name of the unit is torr, its symbol is Torr.

Example 0.1
Converting between units

A scientist was exploring the effect of atmospheric pressure on the rate of growth of a lichen, and measured a pressure p of 1.115 bar. What is the pressure in atmospheres?

Strategy Write the relation between the 'old units' (the units to be replaced) and the 'new units' (the units required) in the form

1 old unit = x new units

then replace the 'old unit' everywhere it occurs by 'x new units', and multiply out the numerical expression.

Solution From Table 0.1 we have 1.013 25 bar = 1 atm, with atm the 'new unit' and bar the 'old unit'. As a first step we write

$$\underbrace{1\ \text{bar}}_{1\ \text{old unit}} = \underbrace{\frac{1}{1.013\ 25}\ \text{atm}}_{x\ \text{new units}}$$

Then we replace bar wherever it appears by (1/1.013 25) atm:

$$p = 1.115\ \text{bar} = 1.115 \times \underbrace{\frac{1}{1.013\ 25}\ \text{atm}}_{1\ \text{bar}} = 1.100\ \text{atm}$$

A note on good practice The number of significant figures in the answer (four in this instance) is the same as the number of significant figures in the data; the relation between old and new numbers in this case is exact.

Self-test 0.1

The pressure in the eye of a hurricane was recorded as 723 Torr. What is the pressure in kilopascals?

[*Answer:* 96.4 kPa]

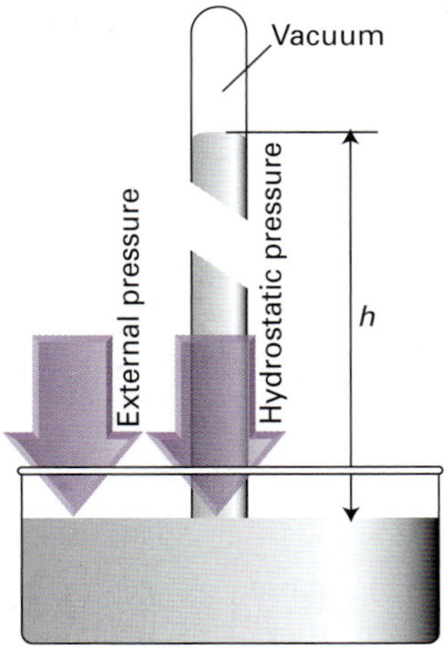

Fig. 0.3 The operation of a mercury barometer. The space above the mercury in the vertical tube is a vacuum, so no pressure is exerted on the top of the mercury column; however, the atmosphere exerts a pressure on the mercury in the reservoir, and pushes the column up the tube until the pressure exerted by the mercury column is equal to that exerted by the atmosphere. The height *h* reached by the column is proportional to the external pressure, so the height can be used as a measure of this pressure.

Atmospheric pressure (a property that varies with altitude and the weather) is measured with a **barometer.** A mercury barometer consists of an inverted tube of mercury that is sealed at its upper end and stands with its lower end in a bath of mercury. The mercury falls until the pressure it exerts at its base is equal to the atmospheric pressure (Fig. 0.3). As shown in the following Derivation, we can determine the atmospheric pressure p by measuring the height h of the mercury column by using the relation

$$p = \rho gh \qquad (0.6)$$

where ρ (rho) is the mass density (commonly just 'density'), the mass of a sample divided by the volume it occupies:

$$\rho = \frac{m}{V} \qquad (0.7)$$

With the mass measured in kilograms and the volume in cubic metres, density is reported in kilograms per cubic metre (kg m^{-3}); however, it is equally acceptable and often more convenient to report mass density in grams per cubic centimetre (g cm^{-3}). The relation between these units is

$$1\ \text{g cm}^{-3} = 10^3\ \text{kg m}^{-3}$$

Thus, the density of mercury may be reported as either 13.6 g cm^{-3} or as 1.36×10^4 kg m^{-3}.

A brief illustration The pressure at the foot of a column of mercury of height 760 mm (0.760 m) and density 13.6 g cm^{-3} (1.36×10^4 kg m^{-3}) is

$$p = (1.36 \times 10^4\ \text{kg m}^{-3}) \times (9.81\ \text{m s}^{-2}) \times (0.760\ \text{m})$$
$$= 1.01 \times 10^5\ \text{kg m}^{-1}\ \text{s}^{-2} = 1.01 \times 10^5\ \text{Pa}$$

For the last equality, we have used 1 kg m^{-1} s^{-2} = 1 Pa. This pressure corresponds to 101 kPa or 1.01 bar (equivalent, with three significant figures, to 1.00 atm).

A note on good practice Write units at *every* stage of a calculation and do not simply attach them to a final numerical value. Also, it is often sensible to express all numerical quantities in terms of base units when carrying out a calculation.

Derivation 0.1

Hydrostatic pressure

The strategy of the calculation is to relate the mass of the column to its height, to calculate the downward force exerted by that mass, and then to divide the force by the area over which it is exerted.

Consider Fig. 0.4. The volume V of a cylinder of liquid of height h and cross-sectional area A is the product of the area and height:

$$V = hA$$

The mass m of this cylinder of liquid is the volume multiplied by the density ρ of the liquid:

$$m = \rho \times V = \rho \times hA$$

The downward force exerted by this mass is mg, where g is the acceleration of free fall. Therefore, the force exerted by the column (its 'weight') is

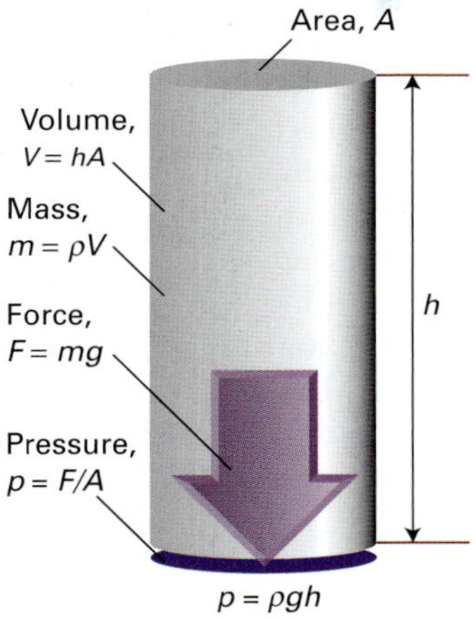

Fig. 0.4 The calculation of the hydrostatic pressure exerted by a column of height h and cross-sectional area A.

$$F = mg = \rho hA \times g$$

This force acts over the area A at the foot of the column, so according to eqn 0.5, the pressure at the base is

$$p = \frac{F}{A} = \frac{\rho hAg}{A} = \rho hg$$

which is eqn 0.6.

0.6 Temperature

In everyday terms, the temperature is an indication of how 'hot' or 'cold' a body is. In science, **temperature**, T, is the property of an object that determines in which direction energy will flow when it is in contact with another object. Energy flows from higher temperature to lower temperature. When the two bodies have the same temperature, there is no net flow of energy between them. In that case we say that the bodies are in **thermal equilibrium** (Fig. 0.5).

A note on good practice Never confuse temperature with heat. Everyday language comes close to confusing them, by equating 'high temperature' with 'hot', but they are entirely different concepts. Heat—as we shall see in detail in Chapter 2—is a mode of transfer of energy; temperature is a property that determines the direction of flow of energy as heat.

Temperature in science is measured on either the Celsius scale or the Kelvin scale. On the **Celsius scale,**

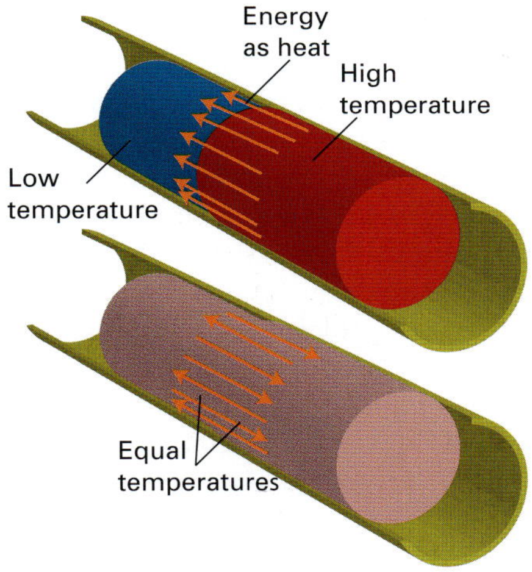

Fig. 0.5 The temperatures of two objects act as a signpost showing the direction in which energy will flow as heat through a thermally conducting wall: (a) heat always flows from high temperature to low temperature. (b) When the two objects have the same temperature, although there is still energy transfer in both directions, there is no net flow of energy.

in which the temperature is expressed in degrees Celsius (°C), the freezing point of water at 1 atm corresponds to 0°C and the boiling point at 1 atm corresponds to 100°C. This scale is in widespread everyday use. Temperatures on the Celsius scale are denoted by the Greek letter θ (theta) throughout this text. However, it turns out to be much more convenient in many scientific applications to adopt the **Kelvin scale** and to express the temperature in kelvin (K; note that the degree sign is not used for this unit). *Whenever we use T to denote a temperature, we mean a temperature on the Kelvin scale.* The Celsius and Kelvin scales are related by

$$T \text{ (in kelvin)} = \theta \text{ (in degrees Celsius)} + 273.15$$

That is, to obtain the temperature in kelvins, add 273.15 to the temperature in degrees Celsius. Thus, water at 1 atm freezes at 273 K and boils at 373 K; a warm day (25°C) corresponds to 298 K.

A more sophisticated way of expressing the relation between T and θ, and one that we shall use in other contexts, is to regard the value of T as the product of a number (such as 298) and a unit (K), so that T/K (that is, the temperature divided by K) is a pure number. For example, if $T = 298$ K, then $T/\text{K} = 298$. Likewise, $\theta/°\text{C}$ is also a pure number. For example, if $\theta = 25°\text{C}$, then $\theta/°\text{C} = 25$. With this convention, we can write the relation between the two scales as

$$T/\text{K} = \theta/°\text{C} + 273.15 \qquad (0.8)$$

This expression is a relation between pure numbers. Equation 0.8, in the form $\theta/°\text{C} = T/\text{K} - 273.15$, also *defines* the Celsius scale in terms of the more fundamental Kelvin scale.

A note on good practice All physical quantities have the form

physical quantity = numerical value × unit

as in $T = 298 \times (1\text{ K})$, abbreviated to 298 K, and $m = 65 \times (1\text{ kg})$, abbreviated to 65 kg. Units are treated like algebraic quantities and so may be multiplied and divided. Thus, the same information could be reported as $T/\text{K} = 298$ and $m/\text{kg} = 65$. It might seem unfamiliar to manipulate units in this way, but it is perfectly legitimate and widely used. By international convention, all physical quantities are represented by sloping symbols; all units are roman (upright).

Self-test 0.2

Use eqn 0.8 to express body temperature, 37°C, in kelvins.

[*Answer:* 310 K]

0.7 Amount of substance

Mass is a measure of the quantity of matter in a sample regardless of its chemical identity. Thus, 1 kg of lead is the same quantity of matter as 1 kg of butter. In chemistry, where we focus on the behaviour of atoms, it is usually more useful to know the numbers of each specific kind of atom, molecule, or ion in a sample rather than the quantity of matter (the mass) itself. However, because even 10 g of water consists of about 10^{23} H_2O molecules, it is clearly appropriate to define a new unit that can be used to express such large numbers simply. As will be familiar from introductory chemistry, chemists have introduced the **mole** (the abbreviation for this unit is mol; the name is derived, ironically, from the Latin word meaning 'massive heap') which is defined as follows:

> 1 mol of specified particles is equal to the number of atoms in exactly 12 g of carbon-12 (^{12}C).

This number is determined experimentally by dividing 12 g by the mass of one atom of carbon-12. Because the mass of one carbon-12 atom is measured by using a mass spectrometer as $1.992\,65 \times 10^{-23}$ g, the number of atoms in exactly 12 g of carbon-12 is

$$\text{Number of atoms} = \frac{\text{total mass of sample}}{\text{mass of one atom}} = \frac{12\text{ g}}{1.992\,65 \times 10^{-23}\text{ g}} = 6.022 \times 10^{23}$$

This number is the number of particles in 1 mol of any substance. For example, a sample of hydrogen gas that contains 6.022×10^{23} hydrogen molecules consists of 1.000 mol H_2, and a sample of water that contains 1.2×10^{24} ($= 2.0 \times 6.022 \times 10^{23}$) water molecules consists of 2.0 mol H_2O.

A note on good practice Always specify the identity of the particles when using the unit mole, for that avoids any ambiguity. If, improperly, we report that a sample consisted of 1 mol of hydrogen, it would not be clear whether it consisted of 6×10^{23} hydrogen atoms (1 mol H) or 6×10^{23} hydrogen molecules (1 mol H_2).

The mole is the unit used when reporting the value of the physical property called the **amount of substance**, n, in a sample. Thus, we can write $n = 1$ mol H_2 or $n_{H_2} = 1$ mol, and say that the amount of hydrogen molecules in a sample is 1 mol. The term 'amount of substance', however, has been slow to find wide acceptance among chemists and in casual conversation they commonly refer to 'the number of moles' in a sample. The term **chemical amount**, however, is becoming more widely used as a convenient synonym for amount of substance, and we shall often use it in this book.

There are various useful concepts that stem from the introduction of the chemical amount and its unit the mole. One is **Avogadro's constant**, N_A, the number of particles (of any kind) per mole of substance:

$$N_A = 6.022 \times 10^{23}\text{ mol}^{-1}$$

Avogadro's constant makes it very simple to convert from the number of particles N (a pure number) in a sample to the chemical amount n (in moles) it contains:

Number of particles = chemical amount × number of particles per mole

That is,

$$N = n \times N_A \tag{0.9}$$

A brief illustration From eqn 0.9 in the form $n = N/N_A$, a sample of copper containing 8.8×10^{22} Cu atoms corresponds to

$$n_{Cu} = \frac{N}{N_A} = \frac{8.8 \times 10^{22}}{6.022\ 10^{23}\text{ mol}^{-1}} = 0.15\text{ mol}$$

Notice how much easier it is to report the amount of Cu atoms present rather than their actual number.

A note on good practice As remarked above, always ensure that the use of the unit mole refers unambiguously to the entities intended. This may be done in a variety of ways: here we have labelled the amount n with the entities (Cu atoms), as in n_{Cu}.

The second very important concept that should be familiar from introductory courses is the **molar mass**, M, the mass per mole of substance: that is, the mass of a sample of the substance divided by the chemical amount of atoms, molecules, or formula units it contains. When we refer to the molar mass of an element we always mean the mass per mole of its atoms. When we refer to the molar mass of a compound, we always mean the molar mass of its molecules or, in the case of solid compounds in general, the mass per mole of its formula units (such as NaCl for sodium chloride and Cu_2Au for a specific alloy of copper and gold). The molar mass of a typical sample of carbon, the mass per mole of carbon atoms (with carbon-12 and carbon-13 atoms in their typical abundances), is 12.01 g mol^{-1}. The molar mass of water is the mass per mole of H_2O molecules, with the isotopic abundances of hydrogen and oxygen those of typical samples of the elements, and is 18.02 g mol^{-1}.

The terms *atomic weight* (AW) or *relative atomic mass* (RAM) and *molecular weight* (MW) or *relative molar mass* (RMM) are still commonly used to signify the numerical value of the molar mass of an element or compound, respectively. More precisely (but equivalently), the RAM of an element or the RMM of a compound is its average atomic or molecular mass relative to the mass of an atom of carbon-12 set equal to 12. The atomic weight (or RAM) of a natural sample of carbon is 12.01 and the molecular weight (or RMM) of water is 18.02.

The molar mass of an element is determined by mass spectrometric measurement of the mass of its atoms and then multiplication of the mass of one atom by Avogadro's constant (the number of atoms per mole). Care has to be taken to allow for the isotopic composition of an element, so we must use a suitably weighted mean of the masses of the isotopes present. The values obtained in this way are printed on the periodic table inside the back cover. The molar mass of a compound of known composition is calculated by taking a sum of the molar masses of its constituent atoms. The molar mass of a compound of unknown composition is determined experimentally by using mass spectrometry in a similar way to the determination of atomic masses.

Molar mass is used to convert from the mass m of a sample (which we can measure) to the amount of substance n (which, in chemistry, we often need to know):

Mass of sample = chemical amount × molar mass

That is,

$$m = n \times M \tag{0.10}$$

A brief illustration To find the amount of C atoms present in 21.5 g of carbon, given the molar mass of carbon is 12.01 g mol^{-1}, from eqn 0.10 in the form $n = m/M$ we write (taking care to specify the species)

$$n_C = \frac{m}{M_C} = \frac{21.5\text{ g}}{12.01\text{ g mol}^{-1}} = 1.79\text{ mol}$$

That is, the sample contains 1.79 mol C.

Self-test 0.3

What amount of H_2O molecules is present in 10.0 g of water?

[*Answer:* 0.555 mol H_2O]

0.8 Extensive and intensive properties

A distinction is made in chemistry between extensive properties and intensive properties. An **extensive property** is a property that depends on the amount of substance in the sample. An **intensive property** is a property that is independent of the amount of substance in the sample. Two examples of extensive properties are mass and volume. Examples of intensive properties are temperature and pressure.

Some intensive properties are ratios of two extensive properties. Consider the mass density of a substance, the ratio of two extensive properties—the mass and the volume (eqn 0.7). The mass density of a substance is independent of the size of the sample because doubling the volume also doubles the mass, so the ratio of mass to volume remains the same. The mass density is therefore an intensive property. A **molar quantity** is the value of a property of a sample divided by the amount of substance in a sample (the 'molar concentration', described below, is an exception to this usage). Thus, the molar mass, M, of an element is the mass of a sample of the element divided by the amount of atoms in the sample: $M = m/n$. In general, molar quantities are denoted X_m, where X is the property of interest. Thus, the molar volume of a substance is denoted V_m and calculated from $V_m = V/n$. Molar quantities are intensive properties.

A note on good practice Distinguish a molar quantity, such as the molar volume, with units of cubic metres per mole (m^3 mol^{-1}), from the quantity *for* 1 mole, such as the volume occupied by 1 mole of the substance, with units cubic metres (m^3).

0.9 Measures of concentration

There are three measures of concentration commonly used to describe the composition of mixtures. One, the *molar concentration*, is used when we need to know the amount of **solute** (the dissolved substance) in a sample of known volume of solution. The other two, the *molality* and the *mole fraction*, are used when we need to know the relative numbers of solute and solvent molecules in a sample.

The **molar concentration**, [J] or c_J, of a solute J in a solution (more formally, the 'amount of substance concentration') is the chemical amount of J divided by the volume of the solution:

$$[J] = \frac{n_J}{V} \tag{0.11a}$$

n_J: Amount of J (mol)
V: Volume of solution (dm^3)

A note on good practice Be alert to the fact that the V in this expression is the volume of *solution*, not the volume of solvent used to make up the solution. That is, to prepare a solution of known molar concentration, a known amount of solute is dissolved in some solvent (usually water), and then more solvent is added to reach the desired total volume.

Molar concentration, still commonly called 'molarity', is typically reported in moles per cubic decimetre ($mol\ dm^{-3}$; more informally, as moles per litre, $mol\ L^{-1}$). The unit $1\ mol\ dm^{-3}$ is commonly denoted 1 M (and read 'molar'). Once we know the molar concentration of a solute, we can calculate the amount of that substance in a given volume, V, of solution by writing the last equation in the form

$$n_J = [J]V \qquad (0.11b)$$

Self-test 0.4

Suppose that 0.282 g of glycine, NH_2CH_2COOH, is dissolved in enough water to make 250 cm^3 of solution. What is the molar concentration of the solution?

[*Answer:* 0.0150 M $NH_2CH_2COOH(aq)$]

The **molality**, b_J, of a solute J in a solution is the amount of substance divided by the mass of solvent used to prepare the solution:

$$b_J = \frac{n_J}{m_{solvent}} \qquad (0.12)$$

n_J: Amount of J (mol); $m_{solvent}$: Mass of solute (kg)

Molality is typically reported in moles of solute per kilogram of solvent ($mol\ kg^{-1}$). This unit is sometimes (but unofficially and potentially confusingly) denoted m, with $1\ m = 1\ mol\ kg^{-1}$. An important distinction between molar concentration and molality is that whereas the former is defined in terms of the volume of the *solution*, the molality is defined in terms of the mass of *solvent* used to prepare the solution. A distinction to remember is that molar concentration varies with temperature as the solution expands and contracts but the molality does not. For dilute solutions in water, the numerical values of the molality and molar concentration differ very little because 1 dm^3 of solution is mostly water and has a mass close to 1 kg; for concentrated aqueous solutions and for all nonaqueous solutions with densities different from $1\ g\ cm^{-3}$, the two values are very different.

As we have indicated, we use molality when we need to emphasize the relative amounts of solute and solvent molecules. To see why this is so, we note that the mass of solvent is proportional to the amount of solvent molecules present, so from eqn 0.12 we see that the molality is proportional to the ratio of the amounts of solute and solvent molecules. For example, any $1.0\ mol\ kg^{-1}$ aqueous nonelectrolyte solution contains 1.0 mol solute particles per 55.5 mol H_2O molecules, so in each case there is 1 solute molecule per 55.5 solvent molecules.

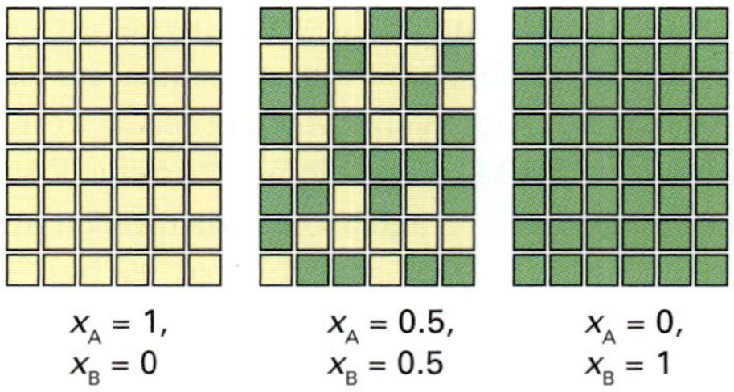

Fig. 0.6 The mole fraction is an indication of the fraction of molecules in a sample that are of the specified identity. Note that in a binary (two-component) mixture, $x_A + x_B = 1$.

Closely related to the molality of a solute is the **mole fraction**, x_J:

$$x_J = \frac{n_J}{n} \qquad (0.13)$$

n_J: Amount of J (mol); n: Total amount of molecules (mol)

$x_J = 0$ corresponds to the absence of J molecules and $x_J = 1$ corresponds to pure J (Fig. 0.6). Note that mole fractions are pure numbers without units ('dimensionless numbers').

Example 0.2

Relating mole fraction and molality

What is the mole fraction of glucose molecules, $C_6H_{12}O_6$, in 0.140 m $C_6H_{12}O_6(aq)$?

Strategy We consider a sample that contains (exactly) 1 kg of solvent, and hence an amount $n_G = b_G \times (1\ kg)$ of glucose molecules. The amount of water molecules in exactly 1 kg of water is $n_W = (1\ kg)/M_W$, where M_W is the molar mass of water. We refer to *exactly* 1 kg of water to avoid problems with significant figures. Once these two amounts are available, we can calculate the mole fraction by using eqn 0.13 with $n = n_G + n_W$.

Solution It follows from the discussion in the Strategy that the amount of glucose molecules in exactly 1 kg of water is

$$n_G = (0.140\ mol\ kg^{-1}) \times (1\ kg) = 0.140\ mol$$

The amount of water molecules in exactly 1 kg (10^3 g) of water is

$$n_{water} = \frac{10^3\ g}{18.02\ g\ mol^{-1}} = \frac{10^3}{18.02}\ mol$$

The total amount of molecules present is

$$n = 0.140\ mol + \frac{10^3}{18.02}\ mol$$

The mole fraction of glucose molecules is therefore

$$x_G = \frac{0.140\ \text{mol}}{0.140 + (10^3/18.02)\ \text{mol}} = 2.52 \times 10^{-3}$$

Self-test 0.5

Calculate the mole fraction of sucrose molecules, $C_{12}H_{22}O_{11}$, in 1.22 m $C_{12}H_{22}O_{11}$(aq).

[*Answer:* 2.15×10^{-2}]

0.10 Reaction stoichiometry

It should be familiar from elementary chemistry that a chemical reaction is **balanced**, in the sense that the same numbers of atoms of each element appear on both sides of the arrow, as in

$$CH_4(g) + 2\ O_2(g) \rightarrow CO_2(g) + 2\ H_2O(l)$$

The numbers multiplying each chemical formula are called **stoichiometric coefficients** (from the Greek words for 'element' and 'measure'). The stoichiometric coefficients in this equation are 1, 2, 1, and 2, respectively, for CH_4, O_2, CO_2, and H_2O. They indicate that for each CH_4 molecule that is consumed, two O_2 molecules are consumed, one CO_2 molecule is formed and two H_2O molecules are formed. It is often convenient to multiply these numbers by 6.022×10^{23} (the number of entities in 1 mol), and to interpret the equation as indicating that when 1 mol CH_4 is consumed, 2 mol O_2 are also consumed and 1 mol CO_2 and 2 mol H_2O are produced. That is, the stoichiometric coefficients indicate the amount of each substance (in moles) that are consumed and produced when the reaction goes to completion.

The interpretation of stoichiometric coefficients as amount in moles provides a simple route to the calculation of yields of chemical reactions (provided the reaction proceeds as written and goes to completion).

A brief illustration To calculate the mass of carbon dioxide produced when 22.0 g of methane burns in a plentiful supply of air, we note that the molar mass of CH_4 is 16.0 g mol^{-1} and therefore that the amount of CH_4 consumed is (22.0 g)/(16.0 g mol^{-1}) = 1.38 mol. Because 1 mol CO_2 is produced when 1 mol CH_4 is consumed, when 1.38 mol CH_4 is consumed, 1.38 mol CO_2 is produced. The molar mass of CO_2 is 44.0 g mol^{-1}, so the mass of CO_2 produced is 1.38 mol × 44.0 g mol^{-1} = 60.7 g. Because 2 mol H_2O is produced when 1 mol CH_4 burns, 2 × 1.38 mol H_2O is produced in the same reaction. The molar mass of H_2O is 18.02 g mol^{-1}, so the mass of water produced is (2 × 1.38 mol) × (18.02 g mol^{-1}) = 49.7 g.

Checklist of key ideas

You should now be familiar with the following concepts.

☐ **1** Physical chemistry is the branch of chemistry that establishes and develops the principles of chemistry in terms of the underlying concepts of physics and the language of mathematics.

☐ **2** The states of matter are gas, liquid, and solid.

☐ **3** Work is done when a body is moved against an opposing force.

☐ **4** Energy is the capacity to do work.

☐ **5** The contributions to the energy of matter are the kinetic energy (the energy due to motion) and the potential energy (the energy due to position).

☐ **6** The total energy of an isolated system is conserved, but kinetic and potential energy may be interchanged.

☐ **7** Two systems in contact through movable walls are in mechanical equilibrium when their pressures are equal.

☐ **8** Two systems in contact through thermally conducting walls are in thermal equilibrium when their temperatures are equal.

☐ **9** Chemical amounts, *n*, are expressed in moles of specified entities.

☐ **10** An extensive property is a property that depends on the amount of substance in the sample. An intensive property is a property that is independent of the amount of substance in the sample.

Table of key equations

The following table summarizes the equations that have been developed in this chapter.

Description	Equation	Comment
Newton's law of motion	$F = ma$	
Work	Work = force × distance	Constant force
Electrical work	Work = $q\Delta\phi$	
Kinetic energy	$E_k = \frac{1}{2}m\boldsymbol{v}^2$	
Potential energy	$E_p = mgh$	Close to surface of the Earth
Coulomb potential energy	$E_p = Q_1Q_2/4\pi\varepsilon_0 r$	Charges separated by a vacuum
Pressure	$p = F/A$	
Hydrostatic pressure	$p = \rho gh$	Vertical tube
Mass density	$\rho = m/V$	
Temperature conversion	$T/\text{K} = \theta/°\text{C} + 273.15$	273.15 is exact
Number and amount	$N = nN_A$	
Molar mass	$M = m/n$	
Molar concentration	$[\text{J}] = n/V$	V is the volume of solution
Molality	$b_J = n/m$	m is the mass of solvent
Mole fraction	$x_J = n_J/n$	

Questions and exercises

Discussion questions

0.1 Explain the differences between gases, liquids, and solids.

0.2 Define the terms: force, work, energy, kinetic energy, and potential energy.

0.3 Distinguish between mechanical and thermal equilibrium. In what sense are these equilibria dynamic?

0.4 Identify whether the following properties are extensive or intensive: (a) volume, (b) mass density, (c) temperature, (d) molar volume, (e) amount of substance.

0.5 Identify and define the various uses of the term 'state' in chemistry.

Exercises

0.1 What is the gravitational force that you are currently experiencing?

0.2 Calculate the percentage change in your weight as you move from the North Pole, where $g = 9.832$ m s^{-2}, to the Equator, where $g = 9.789$ m s^{-2}.

0.3 Calculate the work that a person of mass 65 kg must do to climb between two floors of a building separated by 4.0 m.

0.4 What is the kinetic energy of a tennis ball of mass 58 g served at 35 m s^{-1}?

0.5 A car of mass 1.5 t (1 t = 10^3 kg) travelling at 50 km h^{-1} must be brought to a stop. How much kinetic energy must be dissipated?

0.6 Consider a region of the atmosphere of volume 25 dm^3, which at 20°C contains about 1.0 mol of molecules. Take the average molar mass of the molecules as 29 g mol^{-1} and their average speed as about 400 m s^{-1}. Estimate the energy stored as molecular kinetic energy in this volume of air.

0.7 What is the difference in potential energy of a mercury atom between the top and bottom of a column of mercury in a barometer when the pressure is 1.0 atm?

0.8 Calculate the minimum energy that a bird of mass 25 g must expend in order to reach a height of 50 m.

0.9 The unit 1 electronvolt (1 eV) is defined as the energy acquired by an electron as it moves through a potential difference of 1 V. Express 1 eV in (a) joules, (b) kilojoules per mole.

0.10 Calculate the work done by (a) one electron, (b) 1 mol e^- as they move between the electrodes of a commercial cell rated at 1.5 V.

0.11 You need to assess the fuel needed to send the robot explorer *Spirit*, which has a mass of 185 kg, to Mars. (a) What was the energy needed to raise the vehicle itself from the surface of the Earth to a distant point where the Earth's gravitation field was effectively zero? The mean radius of the Earth is 6371 km and its average mass density is 5.517 g cm^{-3}. *Hint*: Use the full expression for gravitational potential energy in Exercise 0.35.

0.12 Express (a) 108 kPa in torr, (b) 0.975 bar in atmospheres, (c) 22.5 kPa in atmospheres, (d) 770 Torr in pascals.

0.13 Calculate the pressure in the Mindañao trench, near the Philippines, the deepest region of the oceans. Take the depth there as 11.5 km and for the average mass density of sea water use 1.10 g cm^{-3}.

0.14 The atmospheric pressure on the surface of Mars, where $g = 3.7$ m s^{-2}, is only 0.0060 atm. To what extent is that low pressure due to the low gravitational attraction and not to the thinness of the atmosphere? What pressure would the same atmosphere exert on Earth, where $g = 9.81$ m s^{-2}?

0.15 What pressure difference must be generated across the length of a 15 cm vertical drinking straw in order to drink a water-like liquid of mass density 1.0 g cm^{-3} (a) on Earth, (b) on Mars. For data, see Example 0.14.

0.16 The unit '1 millimetre of mercury' (1 mmHg) has been replaced by the unit 1 torr (1 Torr): 1 mmHg is defined as the pressure at the base of a column of mercury exactly 1 mm high when its density is 13.5951 g cm^{-3} and the acceleration of free fall is 9.806 65 m s^{-2}. What is the relation between the two units?

0.17 Suppose that the pressure unit '1 millimetre of water' (1 mmH_2O) is defined as the pressure at the base of a column of water of mass density 1000 kg m^{-3} in a standard gravitational field. Express 1 mmH_2O in (a) pascals, (b) torr.

0.18 Given that the Celsius and Fahrenheit temperature scales are related by $\theta_{\text{Celsius}}/°\text{C} = \frac{5}{9}(\theta_{\text{Fahrenheit}}/°\text{F} - 32)$, what is the temperature of absolute zero ($T = 0$) on the Fahrenheit scale?

0.19 In his original formulation, Anders Celsius identified 0 with the boiling point of water and 100 with its freezing point. Find a relation (expressed like eqn 0.8) between this original scale (denote it $\theta'/°\text{C}'$) and (a) the current Celsius scale ($\theta/°\text{C}$), (b) the Fahrenheit scale.

0.20 Imagine that Pluto is inhabited and that its scientists use a temperature scale in which the freezing point of liquid nitrogen is 0°P (degrees Plutonium) and its boiling point is 100°P. The inhabitants of Earth report these temperatures as −209.9°C and −195.8°C, respectively. What is the relation between temperatures on (a) the Plutonium and Kelvin scales, (b) the Plutonium and Fahrenheit scales?

0.21 The *Rankine scale* is used in some engineering applications. On it, the absolute zero of temperature is set at zero but the size of the Rankine degree (°R) is the same as that of the Fahrenheit degree (°F). What is the boiling point of water on the Rankine scale?

0.22 Calculate the amount of $C_6H_{12}O_6$ molecules in 10.0 g of glucose.

0.23 The density of octane (which we take to model gasoline) is 0.703 g cm^{-3}; what amount (in moles) of octane molecules do you get when you buy 1.00 dm^3 (1.00 litre) of gasoline?

0.24 The molar mass of the oxygen-storage protein myoglobin is 16.1 kg mol^{-1}. How many myoglobin molecules are present in 1.0 g of the compound?

0.25 The mass of a red blood cell is about 33 pg (where 1 pg = 10^{-12} g), and it contains typically 3×10^8 haemoglobin molecules. Each haemoglobin molecule is a tetramer of myoglobin (see preceding exercise). What fraction of the mass of the cell is due to haemoglobin?

0.26 Express the mass density of a compound, which is defined as $\rho = m/V$, in terms of its molar mass and its molar volume.

0.27 A sugar (sucrose, $C_{12}H_{22}O_{11}$) cube has a mass of 5.0 g. What is the molar concentration of sucrose when one sugar cube is dissolved in a cup of coffee of volume 200 cm^3?

0.28 What mass of sodium chloride should be dissolved in enough water to make 300 cm^3 of 1.00 M NaCl(aq)?

0.29 Use the following data to calculate (a) the molar concentration of B in (i) water, (ii) benzene, (b) the molality of B in (i) water, (ii) benzene.

Mass of B used to make up 100 cm^3 of solution: 2.11 g
Molar mass of B: 234.01 g mol^{-1}
Density of solution in water: 1.01 g cm^{-3}
Density of solution in benzene: 0.881 g cm^{-3}

0.30 Calculate the mole fractions of the molecules of a mixture that contains 56 g of benzene and 120 g of methylbenzene (toluene).

0.31 A simple model of dry air at sea level is that it consists of 75.53 per cent (by mass) of nitrogen, 23.14 per cent of oxygen, and 1.33 per cent of other substances (principally argon and carbon dioxide). Calculate the mole fractions of the three principal substances. Treat 'other substances' as argon.

0.32 Treat air (see the preceding exercise) as a solution of oxygen in nitrogen. What is the molality of oxygen in air?

0.33 Calculate the mass of carbon dioxide produced by the combustion of 1.00 dm^3 of gasoline treated as octane of mass density 0.703 g cm^{-3}.

0.34 What mass of carbon monoxide is needed to reduce 1.0 t of iron(III) oxide to the metal?

Projects

0.35 The gravitational potential energy of a body of mass m at a distance r from the centre of the Earth is $-Gmm_E/r$, where m_E is the mass of the Earth and G is the gravitational constant (see inside front cover). Consider the difference in potential energy of the body when it is moved from the surface of the Earth (radius r_E) to a height h above the surface, with $h \ll r_E$, and find an expression for the acceleration of free fall, g, in terms of the mass and radius of the Earth. *Hint*: Use the approximation $(1 + h/r_E)^{-1} = 1 - h/r_E + \cdots$. See Appendix 2 for more information on series expansions.

0.37 Use the same approach as in the preceding exercise to find an approximate expression for moving an electric charge Q_1 through a distance h from a point r_0 from another charge Q_2, with $h \ll r_0$.

Chapter 1

The properties of gases

Although gases are simple, both to describe and in terms of their internal structure, they are of immense importance. We spend our whole lives surrounded by gas in the form of air and the local variation in its properties is what we call the 'weather'. To understand the atmospheres of this and other planets we need to understand gases. As we breathe, we pump gas in and out of our lungs, where it changes composition and temperature. Many industrial processes involve gases, and both the outcome of the reaction and the design of the reaction vessels depend on a knowledge of their properties.

Equations of state

We can specify the state of any sample of substance by giving the values of the following properties (all of which are defined in the Introduction):

V, the volume of the sample
p, the pressure of the sample
T, the temperature of the sample
n, the amount of substance in the sample

However, an astonishing experimental fact is that *these four quantities are not independent of one another*. For instance, we cannot arbitrarily choose to have a sample of 0.555 mol H_2O in a volume of 100 cm^3 at 100 kPa and 500 K: it is found *experimentally* that that state simply does not exist. If we select the amount, the volume, and the temperature, then we find that we have to accept a particular pressure (in this case, close to 23 MPa). The same is true of all substances, but the pressure in general will be different for each one. This experimental generalization is summarized by saying the substance obeys an **equation of state**, an equation of the form

$$p = f(n,V,T) \tag{1.1}$$

This expression tells us that the pressure is some function of amount, volume, and temperature and that if we know those three variables, then the pressure can have only one value.

The equations of state of most substances are not known, so in general we cannot write down an explicit expression for the pressure in terms of the other variables. However, certain equations of state are known. In particular, the equation of state of a low-pressure gas is known, and proves to be very simple and very useful. This equation is used to describe the behaviour of gases taking part in reactions, the behaviour of the atmosphere, as a starting point for problems in chemical engineering, and even in the description of the structures of stars.

1.1 The perfect gas equation of state

The equation of state of a low-pressure gas was among the first results to be established in physical chemistry. The original experiments were carried out by Robert Boyle in the seventeenth century and there was a resurgence in interest later in the century when people began to fly in balloons. This technological progress demanded more knowledge about the response of gases to changes of pressure and temperature and, like technological advances in other fields today, that interest stimulated a lot of experiments.

The experiments of Boyle and his successors led to the formulation of the following **perfect gas equation of state:**

$$pV = nRT \tag{1.2}$$

In this equation (which has the form of eqn 1.1 when we rearrange it into $p = nRT/V$), the **gas constant,** R, is an experimentally determined quantity that turns out to have the same value for all gases. It may be determined by evaluating $R = pV/nRT$ as the pressure is allowed to approach zero or by measuring the speed of sound (which depends on R). Values of R in different units are given in Table 1.1.

The perfect gas equation of state—more briefly, the 'perfect gas law'—is so-called because it is an idealization of the equations of state that gases actually obey. Specifically, it is found that all gases obey the equation ever more closely as the pressure is reduced towards zero. That is, eqn 1.2 is an example of a **limiting law,** a law that becomes increasingly valid as the pressure is reduced and is obeyed exactly in the limit of zero pressure.

Table 1.1

The gas constant in various units

$R =$	8.314 47	$J\ K^{-1}\ mol^{-1}$
	8.314 47	$dm^3\ kPa\ K^{-1}\ mol^{-1}$
	$8.205\ 74 \times 10^{-2}$	$dm^3\ atm\ K^{-1}\ mol^{-1}$
	62.364	$dm^3\ Torr\ K^{-1}\ mol^{-1}$
	1.987 21	$cal\ K^{-1}\ mol^{-1}$

$1\ dm^3 = 10^{-3}\ m^3$

The text's website contains links to online databases of properties of gases.

A hypothetical substance that obeys eqn 1.2 at *all* pressures is called a **perfect gas.** From what has just been said, an actual gas, which is termed a **real gas,** behaves more and more like a perfect gas as its pressure is reduced towards zero. In practice, normal atmospheric pressure at sea level ($p \approx 100$ kPa) is already low enough for most real gases to behave almost perfectly, and unless stated otherwise we shall always assume in this text that the gases we encounter behave like a perfect gas. The reason why a real gas behaves differently from a perfect gas can be traced to the attractions and repulsions that exist between actual molecules and that are absent in a perfect gas (Chapter 15).

A note on good practice A perfect gas is widely called an 'ideal gas' and the perfect gas equation of state is commonly called 'the ideal gas equation' We use 'perfect gas' to imply the absence of molecular interactions; we use 'ideal' in Chapter 6 to denote mixtures in which all the molecular interactions are the same but not necessarily zero.

The perfect gas law summarizes three sets of experimental observations. One is **Boyle's law:**

> At constant temperature, the pressure of a fixed amount of gas is inversely proportional to its volume.

Mathematically:

$$\textit{Boyle's law: at constant temperature, } p \propto \frac{1}{V}$$

We can easily verify that eqn 1.2 is consistent with Boyle's law: by treating n and T as constants, the perfect gas law becomes pV = constant, and hence $p \propto 1/V$. Boyle's law implies that if we compress (reduce the volume of) a fixed amount of gas at constant temperature into half its original volume, then its pressure will double. Figure 1.1 shows the graph obtained by plotting experimental values of p against V for a fixed amount of gas at different temperatures

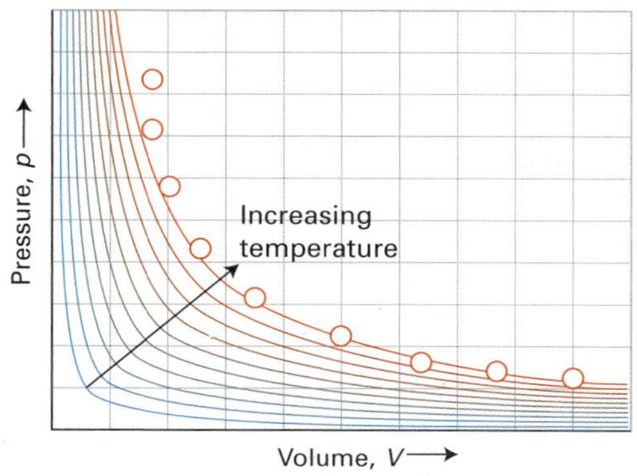

Fig. 1.1 The volume of a gas decreases as the pressure on it is increased. For a sample that obeys Boyle's law and that is kept at constant temperature, the graph showing the dependence is a hyperbola, as shown here. Each curve corresponds to a single temperature, and hence is an isotherm. The isotherms are *hyperbolas*, graphs of xy = constant, or y = constant/x (see Appendix 2).

interActivity Explore how the pressure of 1.5 mol CO_2(g) varies with volume as it is compressed at (a) 273 K, (b) 373 K from 30 dm^3 to 15 dm^3.
Hint: To solve this and other interActivities, use either mathematical software or the Living graphs from the text's web site.

and the curves predicted by Boyle's law. Each curve is called an **isotherm** because it depicts the variation of a property (in this case, the pressure) at a single constant temperature. It is hard, from this graph, to judge how well Boyle's law is obeyed. However, when we plot p against $1/V$, we get straight lines, just as we would expect from Boyle's law (Fig. 1.2).

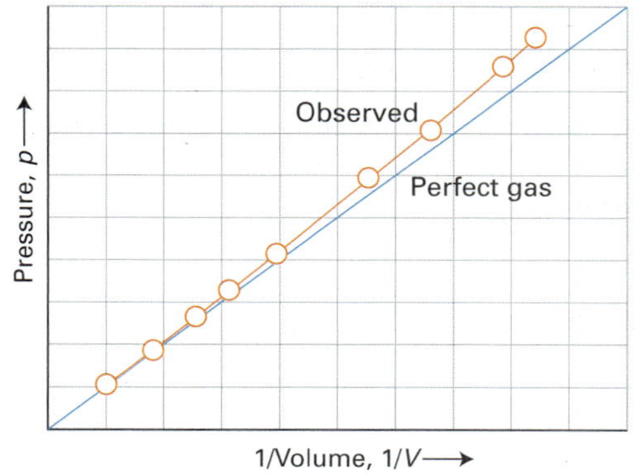

Fig. 1.2 A good test of Boyle's law is to plot the pressure against $1/V$ (at constant temperature), when a straight line should be obtained. This diagram shows that the observed pressures (the blue line) approach a straight line as the volume is increased and the pressure reduced. A perfect gas would follow the straight line at all pressures; real gases obey Boyle's law in the limit of low pressures.

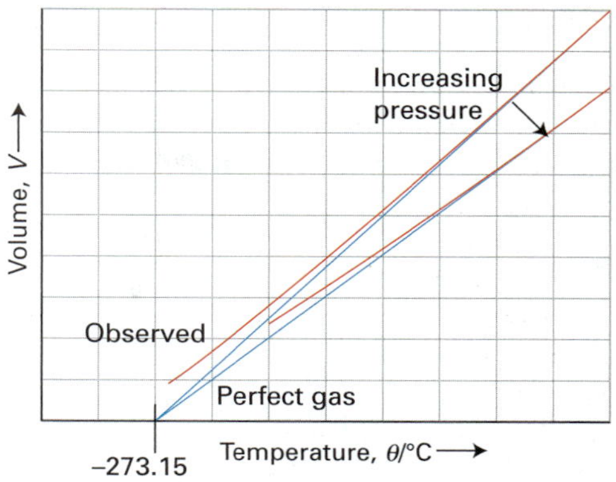

Fig. 1.3 This diagram illustrates the content and implications of Charles's law, which asserts that the volume occupied by a gas (at constant pressure) varies linearly with the temperature. When plotted against Celsius temperatures (as here), all gases give straight lines that extrapolate to $V = 0$ at −273.15°C. This extrapolation suggests that −273.15°C is the lowest attainable temperature.

A note on good practice It is generally the case that a proposed relation is easier to verify if the experimental data are plotted in a form that should give a straight line. That is, the expression being plotted should have the form $y = mx + b$, where m and b are the slope and y-intercept of the line, respectively. For more information, see Appendix 2.

The second experimental observation summarized by eqn 1.2 is **Charles's law:**

> At constant pressure, the volume of a fixed amount of gas varies linearly with the temperature.

Mathematically:

Charles's law: at constant pressure, $V = A + B\theta$

where θ (theta) is the temperature on the Celsius scale and A and B are constants that depend on the amount of gas and the pressure. Figure 1.3 shows typical plots of volume against temperature for a series of samples of gases at different pressures and confirms that (at low pressures, and for temperatures that are not too low) the volume varies linearly with the Celsius temperature. We also see that all the volumes extrapolate to zero as θ approaches the same very low temperature (−273.15°C, in fact), regardless of the identity of the gas. Because a volume cannot be negative, this common temperature must represent the **absolute zero** of temperature, a temperature below which it is impossible to cool an object. Indeed, the 'thermodynamic' scale ascribes the value $T = 0$ to this absolute zero of temperature. In terms of the thermodynamic temperature, therefore, Charles's law takes the simpler form

Charles's law: at constant pressure, $V \propto T$

It follows that doubling the temperature (such as from 300 K to 600 K, corresponding to an increase from 27°C to 327°C) doubles the volume, provided the pressure remains the same. Now we can see that eqn 1.2 is consistent with Charles's law. First, we rearrange it into $V = nRT/p$, and then note that when the amount n and the pressure p are both constant, we can write $V \propto T$, as required.

The third feature of gases summarized by eqn 1.2 is **Avogadro's principle:**

At a given temperature and pressure, equal volumes of gas contain the same numbers of molecules.

That is, 1.00 dm^3 of oxygen at 100 kPa and 300 K contains the same number of molecules as 1.00 dm^3 of carbon dioxide, or any other gas, at the same temperature and pressure. The principle implies that if we double the number of molecules, but keep the temperature and pressure constant, then the volume of the sample will double. We can therefore write:

Avogadro's principle: at constant temperature and pressure, $V \propto n$

This result follows easily from eqn 1.2 if we treat p and T as constants. Avogadro's suggestion is a principle rather than a law (a direct summary of experience), because it is based on a model of a gas, in this case as a collection of molecules. Even though there is no longer any doubt that molecules exist, this relation remains a principle rather than a law.

The **molar volume,** V_m, of any substance (not just a gas) is the volume it occupies per mole of molecules. It is calculated by dividing the volume of the sample by the amount of molecules it contains:

$$V_m = \frac{V}{n} \qquad (1.3)$$

(V: Volume of sample; n: Amount of substance in sample)

With volume in cubic decimetres and amount in moles, the units of molar volume are cubic decimetres per mole ($dm^3\ mol^{-1}$). Avogadro's principle implies that the molar volume of a gas should be the same for all gases at the same temperature and pressure. The data in Table 1.2 show that this conclusion is approximately true for most gases under normal conditions (normal atmospheric pressure of about 100 kPa and room temperature).

Table 1.2

The molar volumes of gases at standard ambient temperature and pressure (SATP: 298.15 K and 1 bar)

Gas	$V_m/(dm^3\ mol^{-1})$
Perfect gas	24.7896*
Ammonia	24.8
Argon	24.4
Carbon dioxide	24.6
Nitrogen	24.8
Oxygen	24.8
Hydrogen	24.8
Helium	24.8

* At STP (0°C, 1 atm), $V_m = 24.4140\ dm^3\ mol^{-1}$.

1.2 Using the perfect gas law

Here we review three elementary applications of the perfect gas equation of state. The first is the prediction of the pressure of a gas given its temperature, its chemical amount, and the volume it occupies. The second is the prediction of the change in pressure arising from changes in the conditions. The third is the calculation of the molar volume of a perfect gas under any conditions. Calculations like these underlie more advanced considerations, including the way that meteorologists understand the changes in the atmosphere that we call the weather (Box 1.1).

Example 1.1

Predicting the pressure of a sample of gas

A chemist is investigating the conversion of atmospheric nitrogen to usable form by the bacteria that inhabit the root systems of certain legumes, and needs to know the pressure in kilopascals exerted by 1.25 g of nitrogen gas in a flask of volume 250 cm^3 at 20°C.

Strategy For this calculation we need to arrange eqn 1.2 ($pV = nRT$) into a form that gives the unknown (the pressure, p) in terms of the information supplied:

$$p = \frac{nRT}{V}$$

To use this expression, we need to know the amount of molecules (in moles) in the sample, which we can obtain from the mass and the molar mass (by using $n = m/M$) and to convert the temperature to the Kelvin scale (by adding 273.15 to the Celsius temperature). Select the value of R from Table 1.1 using the units that match the data and the information required (pressure in kilopascals and volume in litres).

Solution The amount of N_2 molecules (of molar mass 28.02 g mol^{-1}) present is

$$n_{N_2} = \frac{m}{M_{N_2}} = \frac{1.25\text{ g}}{28.02\text{ g mol}^{-1}} = \frac{1.25}{28.02}\text{ mol}$$

The temperature of the sample is

$T/\text{K} = 20 + 273.15$

Therefore, from $p = nRT/V$,

n | R | $T = 293$ K

$$p = \frac{(1.25/28.02)\text{ mol} \times (8.314\,47\text{ J K}^{-1}\text{ mol}^{-1}) \times (20 + 273.15)\text{ K}}{2.50 \times 10\text{ m}^3}$$

$V = 250$ cm^{-4}

$$= \frac{(1.25/28.02) \times 8.314\,47 \times 293}{2.50 \times 10^{-4}} \frac{\text{J}}{\text{m}^3}$$

$\frac{\text{Pa m}^3}{\text{m}^3}$

$= 4.35 \times 10^5$ Pa $= 435$ kPa

We have used the relation 1 J = 1 Pa m^3 and 1 kPa = 10^3 Pa. Note how the units cancel like ordinary numbers.

A note on good practice It is best to postpone the actual numerical calculation to the last possible stage, and carry it out in a single step. This procedure avoids rounding errors.

Self-test 1.1

Calculate the pressure exerted by 1.22 g of carbon dioxide confined to a flask of volume 500 cm^3 at 37°C.

[*Answer:* 143 kPa]

In some cases, we are given the pressure under one set of conditions and are asked to predict the pressure of the same sample under a different set of conditions. We use the perfect gas law as follows. Suppose the initial pressure is p_1, the initial temperature is T_1, and the initial volume is V_1. Then by dividing both sides of eqn 1.2 by the temperature we can write

$$\frac{p_1V_1}{T_1} = nR$$

Suppose now that the conditions are changed to T_2 and V_2, and the pressure changes to p_2 as a result. Then under the new conditions eqn 1.2 tells us that

$$\frac{p_2V_2}{T_2} = nR$$

The nR on the right of these two equations is the same in each case, because R is a constant and the amount of gas molecules has not changed. It follows that we can combine the two equations into a single equation:

$$\frac{p_1V_1}{T_1} = \frac{p_2V_2}{T_2} \qquad (1.4)$$

This expression is known as the **combined gas equation.** We can rearrange it to calculate any one unknown (such as p_2, for instance) in terms of the other variables.

Self-test 1.2

What is the final volume of a sample of gas that has been heated from 25°C to 1000°C and its pressure increased from 10.0 kPa to 150.0 kPa, given that its initial volume was 15 cm^3?

[*Answer:* 4.3 cm^3]

Finally, we see how to use the perfect gas law to calculate the molar volume of a perfect gas at any temperature and pressure. Equation 1.3 expresses the molar volume in terms of the volume of a sample; eqn 1.2 in the form $V = nRT/p$ expresses the volume in terms of the pressure. When we combine the two, we get

$V = nRT/p$

$$V_m = \frac{V}{n} = \frac{nRT/p}{n} = \frac{RT}{p} \qquad (1.5)$$

Definition

This expression lets us calculate the molar volume of any gas (provided it is behaving perfectly) from its pressure and its temperature. It also shows that, for a given temperature and pressure, provided they are behaving perfectly, all gases have the same molar volume.

Chemists have found it convenient to report much of their data at a particular set of 'standard' conditions. By **standard ambient temperature and pressure** (SATP) they mean a temperature of 25°C (more precisely, 298.15 K) and a pressure of exactly 1 bar (100 kPa). The **standard pressure** is denoted $p^{\ominus}$, so $p^{\ominus} = 1$ bar exactly. The molar volume of a perfect gas at SATP is 24.79 dm^3 mol^{-1}, as can be verified by substituting the values of the temperature and pressure into eqn 1.5. This value implies that at SATP, 1 mol of perfect gas molecules occupies about 25 dm^3 (a cube of about 30 cm on a side). An earlier set of standard conditions, which is still encountered, is **standard temperature and pressure** (STP), namely 0°C and 1 atm. The molar volume of a perfect gas at STP is 22.41 dm^3 mol^{-1}.

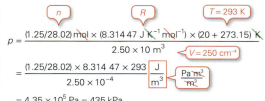

Box 1.1 The gas laws and the weather

The biggest sample of gas readily accessible to us is the atmosphere, a mixture of gases with the composition summarized in the table. The composition is maintained moderately constant by diffusion and convection (winds, particularly the local turbulence called *eddies*) but the pressure and temperature vary with altitude and with the local conditions, particularly in the troposphere (the 'sphere of change'), the layer extending up to about 11 km.

One of the most variable constituents of air is water vapour, and the humidity it causes. The presence of water vapour results in a *lower* density of air at a given temperature and pressure, as we may conclude from Avogadro's principle. The numbers of molecules in 1 m^3 of moist air and dry air are the same (at the same temperature and pressure), but the mass of an H_2O molecule is less than that of all the other major constituents of air (the molar mass of H_2O is 18 g mol^{-1}, the average molar mass of air molecules is 29 g mol^{-1}), so the density of the moist sample is less than that of the dry sample.

The pressure and temperature vary with altitude. In the troposphere the average temperature is 15°C at sea level, falling to −57°C at the bottom of the tropopause at 11 km. This variation is much less pronounced when expressed on the Kelvin scale, ranging from 288 K to 216 K, an average of 268 K. If we suppose that the temperature has its average value all the way up to the edge of the troposphere, then the pressure varies with altitude, *h*, according to the *barometric formula*:

$$p = p_0 e^{-h/H}$$

where p_0 is the pressure at sea level and H is a constant approximately equal to 8 km. More specifically, $H = RT/Mg$,

The composition of the Earth's atmosphere

Substance	Percentage By volume	Percentage By mass
Nitrogen, N_2	78.08	75.53
Oxygen, O_2	20.95	23.14
Argon, Ar	0.93	1.28
Carbon dioxide, CO_2	0.031	0.047
Hydrogen, H_2	5.0×10^{-3}	2.0×10^{-4}
Neon, Ne	1.8×10^{-3}	1.3×10^{-3}
Helium, He	5.2×10^{-4}	7.2×10^{-5}
Methane, CH_4	2.0×10^{-4}	1.1×10^{-4}
Krypton, Kr	1.1×10^{-4}	3.2×10^{-4}
Nitric oxide, NO	5.0×10^{-5}	1.7×10^{-6}
Xenon, Xe	8.7×10^{-6}	3.9×10^{-5}
Ozone, O_3 Summer:	7.0×10^{-6}	1.2×10^{-5}
Winter:	2.0×10^{-6}	3.3×10^{-6}

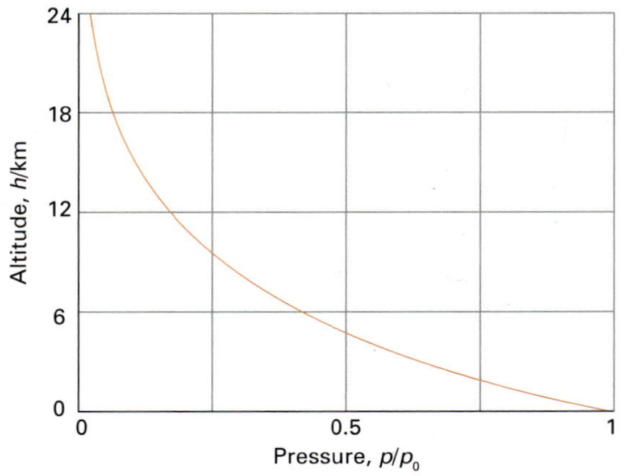

The variation of atmospheric pressure with altitude as predicted by the barometric formula.

interActivity How would the graph shown in the illustration change if the temperature variation with altitude were taken into account? Construct a graph allowing for a linear decrease in temperature with altitude.

where M is the average molar mass of air and T is the temperature. The barometric formula fits the observed pressure distribution quite well even for regions well above the troposphere (see the illustration). It implies that the pressure of the air and its density fall to half their sea-level value at $h = H \ln 2$, or 6 km.

Local variations of pressure, temperature, and composition in the troposphere are manifest as 'weather'. A small region of air is termed a *parcel*. First, we note that a parcel of warm air is less dense than the same parcel of cool air. As a parcel rises, it expands without transfer of heat from its surroundings and it cools. Cool air can absorb lower concentrations of water vapour than warm air, so the moisture forms clouds. Cloudy skies can therefore be associated with rising air and clear skies are often associated with descending air.

The motion of air in the upper altitudes may lead to an accumulation in some regions and a loss of molecules from other regions. The former result in the formation of regions of high pressure ('highs' or anticyclones) and the latter result in regions of low pressure ('lows', depressions, or cyclones). These regions are shown as H and L on the accompanying weather map. The lines of constant pressure—differing by 4 mbar (400 Pa, about 3 Torr)—marked on it are called *isobars*. The elongated regions of high and low pressure are known, respectively, as *ridges* and *troughs*.

In meteorology, large-scale vertical movement is called *convection*. Horizontal pressure differentials result in the flow of air that we call *wind*. Because the Earth is rotating from west to east, winds are deflected towards the right in

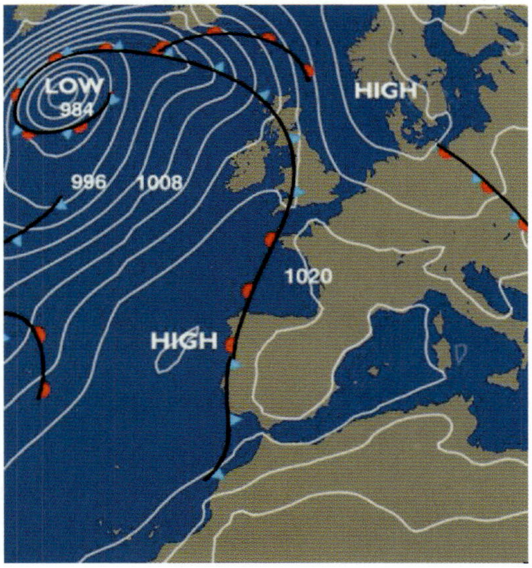

A typical weather map; this one for Western Europe on 3 May 2008.

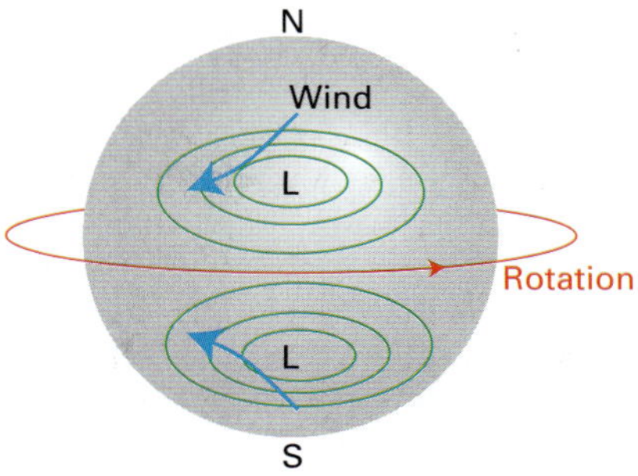

The horizontal flow of air relative to an area of low pressure in the northern and southern hemispheres.

the northern hemisphere and towards the left in the southern hemisphere. Winds travel nearly parallel to the isobars, with low pressure to their left in the northern hemisphere and to the right in the southern hemisphere. At the surface, where wind speeds are lower, the winds tend to travel perpendicular to the isobars from high to low pressure. This differential motion results in a spiral outward flow of air clockwise in the northern hemisphere around a high and an inward counterclockwise flow around a low.

The air lost from regions of high pressure is restored as an influx of air converges into the region and descends. As we have seen, descending air is associated with clear skies. It also becomes warmer by compression as it descends, so regions of high pressure are associated with high surface temperatures. In winter, the cold surface air may prevent the complete fall of air, and result in a temperature *inversion*, with a layer of warm air over a layer of cold air. Geographical conditions may also trap cool air, as in Los Angeles, and the photochemical pollutants we know as *smog* may be trapped under the warm layer. A less dramatic manifestation of an inversion layer is the presence of hazy skies, particularly in industrial areas. Hazy skies also form over vegetation that generate aerosols of terpenes or other plant transpiration products. These hazes give rise to the various 'Blue Mountains' of the world, such as the Great Dividing Range in New South Wales, the range in Jamaica, and the range stretching from central Oregon into south-eastern Washington, which are dense with eucalyptus, tree ferns, and pine and fir, respectively. The Blue Ridge section of the Appalachians is another example.

1.3 Mixtures of gases: partial pressures

We are often concerned with mixtures of gases, such as when we are considering the properties of the atmosphere in meteorology, the composition of exhaled air in medicine, or the mixtures of hydrogen and nitrogen used in the industrial synthesis of ammonia. We need to be able to assess the contribution that each component of a gaseous mixture makes to the total pressure.

In the early nineteenth century, John Dalton carried out a series of experiments that led him to formulate what has become known as **Dalton's law**:

> The pressure exerted by a mixture of perfect gases is the sum of the pressures that each gas would exert if it were alone in the container at the same temperature:
>
> $$p = p_A + p_B + \cdots \tag{1.6}$$

In this expression, p_J is the pressure that the gas J would exert if it were alone in the container at the same temperature. Dalton's law is strictly valid only for mixtures of perfect gases (or for real gases at such low pressures that they are behaving perfectly), but it can be treated as valid under most conditions we encounter.

A brief illustration Suppose we were interested in the composition of inhaled and exhaled air, and we knew that a certain mass of carbon dioxide exerts a pressure of

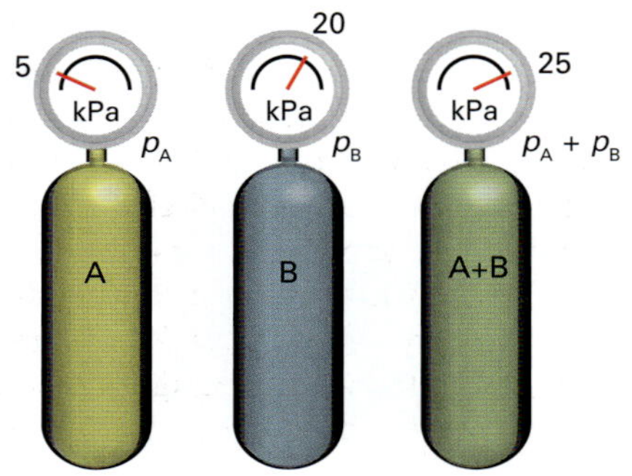

Fig. 1.4 The partial pressure p_A of a perfect gas A is the pressure that the gas would exert if it occupied a container alone; similarly, the partial pressure p_B of a perfect gas B is the pressure that the gas would exert if it occupied the same container alone. The total pressure p when both perfect gases simultaneously occupy the container is the sum of their partial pressures.

5 kPa when present alone in a container, and that a certain mass of oxygen exerts 20 kPa when present alone in the same container at the same temperature. Then, when both gases are present in the container, the carbon dioxide in the mixture contributes 5 kPa to the total pressure and oxygen contributes 20 kPa; according to Dalton's law, the total pressure of the mixture is the sum of these two pressures, or 25 kPa (Fig. 1.4). ■

For any type of gas (real or perfect) in a mixture, the **partial pressure**, p_J, of the gas J is defined as

$$p_J = x_J p \qquad (1.7)$$

where x_J is the **mole fraction** of the gas J in the mixture. The mole fraction of J is the amount of J molecules expressed as a fraction of the total amount of molecules in the mixture. In a mixture that consists of n_A A molecules, n_B B molecules, and so on (where the n_J are amounts in moles), the mole fraction of J (where J = A, B, ...) is

$$x_J = \frac{n_J}{n} \qquad (1.8a)$$

(n_J: amount of J molecules in the mixture; n: total amount of molecules in the mixture)

where $n = n_A + n_B + \cdots$. Mole fractions are unitless because the unit mole in numerator and denominator cancels. For a **binary mixture**, one that consists of two species, this general expression becomes

$$x_A = \frac{n_A}{n_A + n_B} \qquad x_B = \frac{n_B}{n_A + n_B} \qquad x_A + x_B = 1 \qquad (1.8b)$$

When only A is present, $x_A = 1$ and $x_B = 0$. When only B is present, $x_B = 1$ and $x_A = 0$. When both are present in the same amounts, $x_A = \frac{1}{2}$ and $x_B = \frac{1}{2}$.

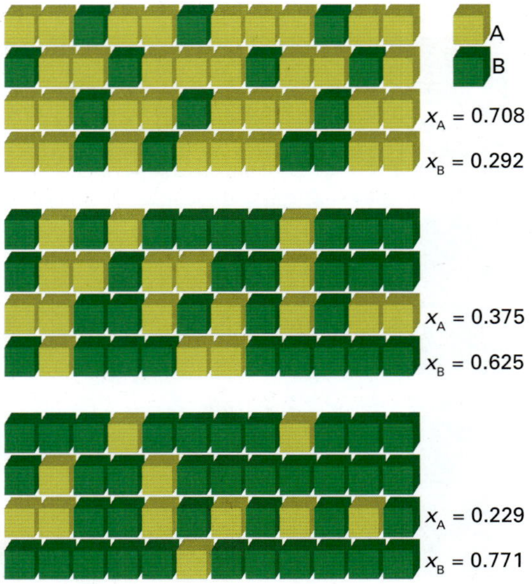

Fig. 1.5 A representation of the meaning of mole fraction. In each case, a small square represents one molecule of A (yellow squares) or B (green squares). There are 48 squares in each sample.

Self-test 1.3

Calculate the mole fractions of N_2, O_2, and Ar in dry air at sea level, given that 100.0 g of air consists of 75.5 g of N_2, 23.2 g of O_2, and 1.3 g of Ar. *Hint:* Begin by converting each mass to an amount in moles.

[*Answer:* 0.780, 0.210, 0.009]

For a mixture of perfect gases, we can identify the partial pressure of J with the contribution that J makes to the total pressure. Thus, if we introduce $p = nRT/V$ into eqn 1.7, we get

$$p_J = x_J p = x_J \times \frac{nRT}{V} = (nx_J) \times \frac{RT}{V} = n_J \times \frac{RT}{V}$$

(Definition; $p = nRT/V$; $nx_J = n_J$)

The value of $n_J RT/V$ is the pressure that an amount n_J of J would exert in the otherwise empty container. That is, the partial pressure of J as defined by eqn 1.7 is the pressure of J used in Dalton's law, provided all the gases in the mixture behave perfectly. If the gases are real, their partial pressures are still given by eqn 1.7, for that definition applies to all gases, and the sum of these partial pressures is the total pressure (because the sum of all the mole fractions is 1);

however, each partial pressure is no longer the pressure that the gas would exert when alone in the container.

A brief illustration From Self-test 1.3, we have $x_{N_2} = 0.780$, $x_{O_2} = 0.210$, and $x_{Ar} = 0.009$ for dry air at sea level. It then follows from eqn 1.7 that when the total atmospheric pressure is 100 kPa, the partial pressure of nitrogen is

$$p_{N_2} = x_{N_2}p = 0.780 \times (100\ \text{kPa}) = 78.0\ \text{kPa}$$

Similarly, for the other two components we find p_{O_2} = 21.0 kPa and p_{Ar} = 0.9 kPa. Provided the gases are perfect, these partial pressures are the pressures that each gas would exert if it were separated from the mixture and put in the same container on its own.

Self-test 1.4

The partial pressure of molecular oxygen in air plays an important role in the aeration of water, to enable aquatic life to thrive, and in the absorption of oxygen by blood in our lungs (see Box 6.1). Calculate the partial pressures of a sample of gas consisting of 2.50 g of oxygen and 6.43 g of carbon dioxide with a total pressure of 88 kPa.

[*Answer:* 31 kPa, 57 kPa]

The kinetic model of gases

We remarked in the Introduction that a gas may be pictured as a collection of particles in ceaseless, random motion (Fig. 1.6). Now we develop this model of the gaseous state of matter to see how it accounts for the perfect gas law. One of the most important functions of physical chemistry is to convert qualitative notions into quantitative statements that can be tested experimentally by making measurements and comparing the results with predictions. Indeed, an important component of science as a whole is its technique of proposing a qualitative model and then expressing that model mathematically. The 'kinetic model' (or the 'kinetic molecular theory', KMT) of gases is an excellent example of this procedure: the model is very simple, and the quantitative prediction (the perfect gas law) is experimentally verifiable.

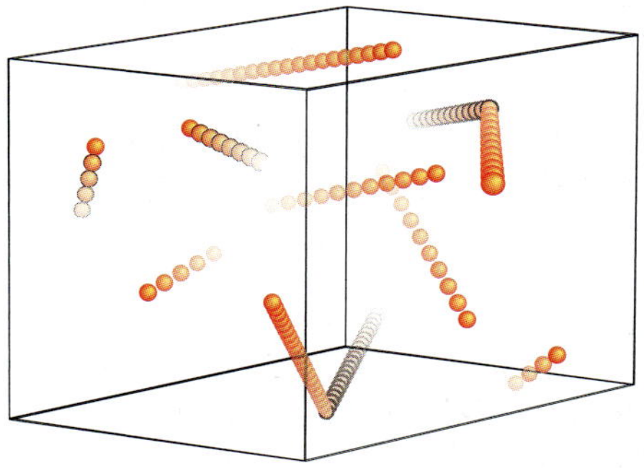

Fig. 1.6 The model used for discussing the molecular basis of the physical properties of a perfect gas. The point-like molecules move randomly with a wide range of speeds and in random directions, both of which change when they collide with the walls or with other molecules.

The **kinetic model of gases** is based on three assumptions:

1. A gas consists of molecules in ceaseless random motion.
2. The size of the molecules is negligible in the sense that their diameters are much smaller than the average distance travelled between collisions.
3. The molecules do not interact, except during collisions.

The assumption that the molecules do not interact unless they are in contact implies that the potential energy of the molecules (their energy due to their position) is independent of their separation and may be set equal to zero. The total energy of a sample of gas is therefore the sum of the kinetic energies (the energy due to motion) of all the molecules present in it. It follows that the faster the molecules travel (and hence the greater their kinetic energy), the greater the total energy of the gas.

1.4 The pressure of a gas according to the kinetic model

The kinetic model accounts for the steady pressure exerted by a gas in terms of the collisions the molecules make with the walls of the container. Each impact gives rise to a brief force on the wall, but as billions of collisions take place every second, the walls experience a virtually constant force, and hence the gas exerts a steady pressure. On the basis of this model, the pressure exerted by a gas of molar mass M in a volume V is

$$p = \frac{nMc^2}{3V} \qquad (1.9)$$

(See Further information 1.1 for a derivation of this equation.) Here c is the **root-mean-square speed** (rms speed) of the molecules. This quantity is defined as the square root of the mean value of the squares of the speeds, v, of the molecules. That is, for a sample consisting of N molecules with speeds $v_1, v_2, \ldots, v_N$, we square each speed, add the squares together, divide by the total number of molecules (to get the mean, denoted by $\langle \ldots \rangle$), and finally take the square root of the result:

$$c = \langle v^2 \rangle^{1/2} = \left(\frac{v_1^2 + v_2^2 + \cdots + v_N^2}{N} \right)^{1/2} \quad (1.10)$$

The rms speed might at first sight seem to be a rather peculiar measure of the mean speeds of the molecules, but its significance becomes clear when we make use of the fact that the kinetic energy of a molecule of mass m travelling at a speed v is $E_k = \frac{1}{2}mv^2$, which implies that the mean kinetic energy, $\langle E_k \rangle$, is the average of this quantity, or $\frac{1}{2}mc^2$. It follows from the relation $\frac{1}{2}mc^2 = \langle E_k \rangle$ that

$$c = \left(\frac{2\langle E_k \rangle}{m} \right)^{1/2} \quad (1.11)$$

Therefore, wherever c appears, we can think of it as a measure of the mean kinetic energy of the molecules of the gas. The rms speed is quite close in value to another and more readily visualized measure of molecular speed, the **mean speed**, $\bar{c}$, of the molecules:

$$\bar{c} = \frac{v_1 + v_2 + \cdots + v_N}{N} \quad (1.12)$$

For samples consisting of large numbers of molecules, the mean speed is slightly smaller than the rms speed. The precise relation is

$$\bar{c} = \left(\frac{8}{3\pi} \right)^{1/2} c \approx 0.921c \quad (1.13)$$

For elementary purposes, and for qualitative arguments, we do not need to distinguish between the two measures of average speed, but for precise work the distinction is important.

Self-test 1.5

Cars pass a point travelling at 45.00 (5), 47.00 (7), 50.00 (9), 53.00 (4), 57.00 (1) km h^{-1}, where the number of cars is given in parentheses. Calculate (a) the rms speed and (b) the mean speed of the cars. (*Hint*: Use the definitions directly; the relation in eqn 1.13 is unreliable for such small samples.)

[*Answer:* (a) 49.06 km h^{-1}, (b) 48.96 km h^{-1}]

Equation 1.9 already resembles the perfect gas equation of state, for we can rearrange it into

$$pV = \tfrac{1}{3}nMc^2 \quad (1.14)$$

and compare it to $pV = nRT$. This conclusion is a major success of the kinetic model, for the model implies an experimentally verified result.

1.5 The average speed of gas molecules

We now suppose that the expression for pV derived from the kinetic model is indeed the equation of state of a perfect gas. That being so, we can equate the expression on the right of eqn 1.14 to nRT,

$$pV = nRT$$
$$pV = \tfrac{1}{3}nMc^2$$

which gives

$$\tfrac{1}{3}nMc^2 = nRT$$

The ns now cancel, to give

$$\tfrac{1}{3}Mc^2 = RT$$

The great usefulness of this expression is that we can rearrange it into a formula for the rms speed of the gas molecules at any temperature:

$$c = \left(\frac{3RT}{M} \right)^{1/2} \quad (1.15)$$

Substitution of the molar mass of O_2 (32.0 g mol^{-1}) and a temperature corresponding to 25°C (that is, 298 K) gives an rms speed for these molecules of 482 m s^{-1}. The same calculation for nitrogen molecules gives 515 m s^{-1}. Both these values are not far off the speed of sound in air (346 m s^{-1} at 25°C). That similarity is reasonable, because sound is a wave of pressure variation transmitted by the movement of molecules, so the speed of propagation of a wave should be approximately the same as the speed at which molecules can adjust their locations.

The important conclusion to draw from eqn 1.15 is that

> The rms speed of molecules in a gas is proportional to the square root of the temperature.

Because the mean speed is proportional to the rms speed, the same is true of the mean speed too (because the two quantities are proportional to each other). Therefore, doubling the thermodynamic temperature (that is, doubling the temperature on the Kelvin scale) increases the mean and the rms speed of molecules by a factor of $2^{1/2} = 1.414\ldots$.

A brief illustration Cooling a sample of air from 25°C (298 K) to 0°C (273 K) reduces the original rms speed of the molecules by a factor of

$$\left(\frac{273\ \text{K}}{298\ \text{K}} \right)^{1/2} = \left(\frac{273}{298} \right)^{1/2} = 0.957$$

So, on a cold day, the average speed of air molecules (which is changed by the same factor) is about 4 per cent less than on a warm day.

1.6 The Maxwell distribution of speeds

So far, we have dealt only with the *average* speed of molecules in a gas. Not all molecules, however, travel at the same speed: some move more slowly than the average (until they collide, and get accelerated to a high speed, like the impact of a bat on a ball), and others may briefly move at much higher speeds than the average, but be brought to a sudden stop when they collide. There is a ceaseless redistribution of speeds among molecules as they undergo collisions. Each molecule collides once every nanosecond (1 ns = 10^{-9} s) or so in a gas under normal conditions.

The mathematical expression that tells us the fraction of molecules that have a particular speed at any instant is called the **distribution of molecular speeds.** Thus, the distribution might tell us that at 20°C 19 out of 1000 O_2 molecules have a speed in the range between 300 and 310 m s^{-1}, that 21 out of 1000 have a speed in the range 400 to 410 m s^{-1}, and so on. The precise form of the distribution was worked out by James Clerk Maxwell towards the end of the nineteenth century, and his expression is known as the **Maxwell distribution of speeds.** According to Maxwell, the fraction f of molecules that have a speed in a narrow range between s and $s + \Delta s$ (for example, between 300 m s^{-1} and 310 m s^{-1}, corresponding to s = 300 m s^{-1} and Δs = 10 m s^{-1}) is

$$f = F(s)\Delta s \quad \text{with} \quad F(s) = 4\pi\left(\frac{M}{2\pi RT}\right)^{3/2} s^2\, e^{-Ms^2/2RT} \qquad (1.16)$$

This formula was used to calculate the numbers quoted above.

Although eqn 1.16 looks complicated, its features can be picked out quite readily. One of the skills to develop in physical chemistry is the ability to interpret the message carried by equations. Equations convey information, and it is far more important to be able to read that information than simply to remember the equation. Let's read the information in eqn 1.16 piece by piece.

Before we begin, and in preparation for their occurrence throughout the text, it will be useful to know the shape of exponential functions. Here, we deal with two types, e^{-ax} and e^{-ax^2}.

- An **exponential function,** a function of the form e^{-ax}, starts off at 1 when $x = 0$ and decays toward zero, which it reaches as x approaches infinity (Fig. 1.7). This function approaches zero more rapidly as a increases.

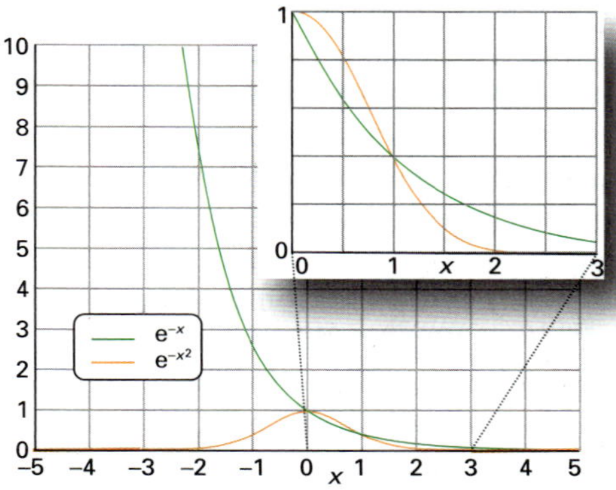

Fig. 1.7 The exponential function, e^{-x}, and the bell-shaped Gaussian function, e^{-x^2}. Note that both are equal to 1 at $x = 0$ but the exponential function rises to infinity as $x \to -\infty$. The enlargement on the right shows the behaviour for $x > 0$ in more detail.

- A **Gaussian function,** a function of the form e^{-ax^2}, also starts off at 1 when $x = 0$ and decays to zero as x increases, however, its decay is initially slower but then plunges down to zero more rapidly than an exponential function (Fig. 1.7).

The illustration also shows the behaviour of the two functions for negative values of x. The exponential function e^{-ax} rises rapidly to infinity, but the Gaussian function falls back to zero and traces out a bell-shaped curve.

Now let's consider the content of eqn 1.16.

- Because f is proportional to the range of speeds Δs, we see that the fraction in the range Δs increases in proportion to the width of the range. If at a given speed we double the range of interest (but still ensure that it is narrow), then the fraction of molecules in that range doubles too.
- Equation 1.16 includes a decaying exponential function, the term $F(s) = 4\pi(M/2\pi RT)^{3/2} s^2 \boxed{e^{-Ms^2/2RT}}$. Its presence implies that the fraction of molecules with very high speeds will be very small because e^{-x^2} becomes very small when x^2 is large.
- The factor $M/2RT$ multiplying s^2 in the exponent, $F(s) = 4\pi(M/2\pi RT)^{3/2} s^2\, e^{-\boxed{Ms^2/2RT}}$, is large when the molar mass, M, is large, so the exponential factor goes most rapidly towards zero when M is large. That tells us that heavy molecules are unlikely to be found with very high speeds.
- The opposite is true when the temperature, T, is high: then the factor $M/2RT$ in the exponent is small, so the exponential factor falls towards zero

is valid only in the absence of intermolecular forces.) We can therefore expect the reduction in pressure to be proportional to the *square* of the molar concentration, one factor of n/V reflecting the reduction in frequency of collisions and the other factor the reduction in the strength of their impulse. If the constant of proportionality is written a, we can write

$$\text{Reduction in pressure} = a \times \left(\frac{n}{V}\right)^2$$

It follows that the equation of state allowing for both repulsions and attractions is

$$p = \frac{nRT}{V - nb} - a\left(\frac{n}{V}\right)^2 \qquad (1.23a)$$

This expression is the **van der Waals equation of state.** To show the resemblance of this equation to the perfect gas equation $pV = nRT$, eqn 1.23a is sometimes rearranged by bringing the term proportional to a to the left and multiplying throughout by $V - nb$:

$$\left(p + \frac{an^2}{V^2}\right)(V - nb) = nRT \qquad (1.23b)$$

We have built the van der Waals equation by using physical arguments about the volumes of molecules and the effects of forces between them. It can be derived in other ways, but the present method has the advantage of showing how to derive the form of an equation out of general ideas. The derivation also has the advantage of keeping imprecise the significance of the **van der Waals parameters**, the constants a and b: they are much better regarded as empirical parameters than as precisely defined molecular properties. The van der Waals parameters depend on the gas, but are taken as independent of temperature (Table 1.5). It follows from the way we have constructed the equation that a (the parameter representing the role of attractions) can be expected to be large when the molecules attract each other strongly, whereas b (the parameter representing the role of repulsions) can be expected to be large when the molecules are large.

We can judge the reliability of the van der Waals equation by comparing the isotherms it predicts, which are shown in Fig. 1.17, with the experimental isotherms already shown in Fig. 1.13. Apart from the waves below the critical temperature they do resemble experimental isotherms quite well. The waves, which are called **van der Waals' loops**, are unrealistic because they suggest that under some conditions compression results in a decrease of pressure. The loops are therefore trimmed away and replaced by horizontal lines (Fig. 1.18). The van der Waals parameters in Table 1.5 were found by fitting the calculated curves to experimental isotherms.

Two important features of the van der Waals equation should be noted. First, perfect-gas isotherms are obtained from the van der Waals equation at high temperatures and low pressures. To confirm this remark, we need to note that when the temperature is high, RT may be so large that the first term on the right in eqn 1.23a greatly exceeds the second, so the latter may be ignored. Furthermore, at low pressures, the molar volume is so large that $V - nb$ can be replaced by V. Hence, under these conditions (of high temperature and low pressure), eqn 1.23a reduces to $p = nRT/V$, the perfect gas equation. Second, and as shown in Derivation 1.2, the critical constants are related to the van der Waals coefficients as follows:

Table 1.5

van der Waals parameters of gases

Substance	a/(atm dm^6 mol^{-2})	b/(10^{-2} dm^3 mol^{-1})
Air	1.4	0.039
Ammonia, NH_3	4.225	3.71
Argon, Ar	1.337	3.20
Carbon dioxide, CO_2	3.610	4.29
Ethane, C_2H_6	5.507	6.51
Ethene, C_2H_4	4.552	5.82
Helium, He	0.0341	2.38
Hydrogen, H_2	0.2420	2.65
Nitrogen, N_2	1.352	3.87
Oxygen, O_2	1.364	3.19
Xenon, Xe	4.137	5.16

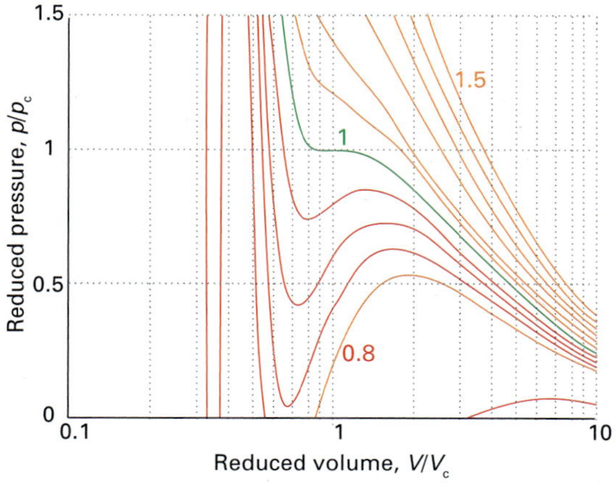

Fig. 1.17 Isotherms calculated by using the van der Waals equation of state. The axes are labelled with the 'reduced pressure', p/p_c, and 'reduced volume', V/V_c, where $p_c = a/27b^2$ and $V_c = 3b$. The individual isotherms are labelled with the 'reduced temperature', T/T_c, where $T_c = 8a/27Rb$. The isotherm labelled 1 is the critical isotherm (the isotherm at the critical temperature).

interActivity (a) Show that the van der Waals equation may be written as $p = RT/(V_m - b) - a/V_m^2$. (b) Use your result to show that the van der Waals equation may also be written as $V_m^3 - (b + RT/p)V_m^2 + (a/p)V_m - ab/p = 0$. (c) Calculate the molar volume of carbon dioxide gas at 500 K and 150 kPa by using mathematical software to find the physically acceptable roots of the equation from part (b). (c) Calculate the percentage difference between the value you calculated in part (b) and the value predicted by the perfect gas equation.

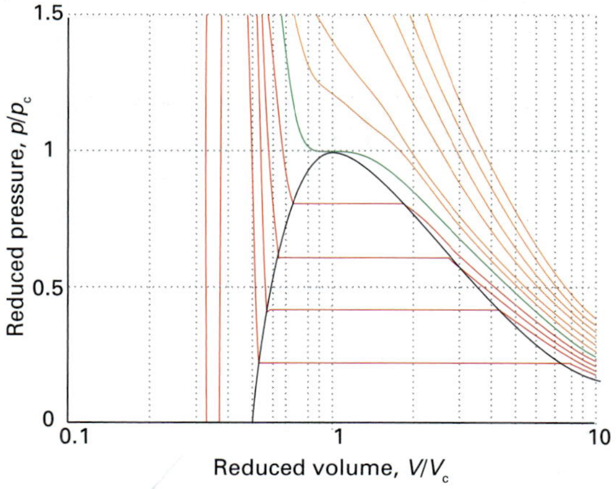

Fig. 1.18 The unphysical van der Waals loops are eliminated by drawing straight lines that divide the loops into areas of equal size. With this procedure, the isotherms strongly resemble the observed isotherms.

$$V_c = 3b \qquad T_c = \frac{8a}{27Rb} \qquad p_c = \frac{a}{27b^2} \tag{1.24}$$

The first of these relations shows that the critical volume is about three times the volume occupied by the molecules themselves.

Derivation 1.2

Relating the critical constants to the van der Waals parameters

We see from Fig. 1.17 that, for $T < T_c$, the calculated isotherms oscillate, and each one passes through a minimum followed by a maximum. These extrema converge as $T \rightarrow T_c$ and coincide at $T = T_c$; at the critical point the curve has a flat inflexion (**1**). From the properties of curves, we know that an inflexion of this type occurs when both the first and second derivatives are zero. Hence, we can find the critical temperature by calculating these derivatives and setting them equal to zero. First, we use $V_m = V/n$ to write eqn 1.23a as

$$p = \frac{RT}{V_m - b} - \frac{a}{V_m^2}$$

The first and second derivatives of p with respect to V_m are, respectively:

$$\frac{dp}{dV_m} = -\frac{RT}{(V_m - b)^2} + \frac{2a}{V_m^3}$$

$$\frac{d^2p}{dV_m^2} = \frac{2RT}{(V_m - b)^3} - \frac{6a}{V_m^4}$$

At the critical point $T = T_c$, $V_m = V_c$, and both derivatives are equal to zero:

$$-\frac{RT_c}{(V_c - b)^2} + \frac{2a}{V_c^3} = 0$$

$$\frac{2RT_c}{(V_c - b)^3} - \frac{6a}{V_c^4} = 0$$

Solving this pair of equations gives (as you should verify) the expressions for V_c and T_c in eqn 1.24. When they are inserted in the van der Waals equation itself, we find the expression for p_c given there too.

1.14 The liquefaction of gases

A gas may be liquefied by cooling it below its boiling point at the pressure of the experiment. For example, chlorine at 1 atm can be liquefied by cooling it to below −34°C in a bath cooled with dry ice (solid carbon dioxide). For gases with very low boiling points (such as oxygen and nitrogen, at −183°C and −186°C, respectively), such a simple technique is not practicable unless an even colder bath is available.

One alternative and widely used commercial technique makes use of the forces that act between molecules. We saw earlier that the rms speed of molecules in a gas is proportional to the square root of the temperature (eqn 1.15). It follows that reducing the rms speed of the molecules is equivalent to cooling the gas. If the speed of the molecules can be reduced to the point that neighbours can capture each other by their intermolecular attractions, then the cooled gas will condense to a liquid.

To slow the gas molecules, we make use of an effect similar to that seen when a ball is thrown into the air: as it rises it slows in response to the gravitational attraction of the Earth and its kinetic energy is converted into potential energy. Molecules attract each other, as we have seen (the attraction is not gravitational, but the effect is the same), and if we can cause them to move apart from each other, like a ball rising from a planet, then they should slow. It is very easy to move molecules apart from each other: we simply allow the gas to expand, which increases the average separation of the molecules. To cool a gas, therefore, we allow it to expand without allowing any heat to enter from outside. As it does so, the molecules move apart to fill the available volume, struggling as they do so against the attraction of their neighbours. Because some kinetic energy must be converted into potential energy to reach greater separations, the molecules travel more slowly as their separation increases. Therefore, because the average speed of the molecules has been reduced, the gas is now cooler than before the expansion. This process of cooling a real gas by expansion through a narrow opening called a 'throttle' is called the **Joule–Thomson effect.** The effect was first observed and analysed by James Joule (whose name is commemorated in the unit of energy) and William Thomson (who later became Lord Kelvin). The procedure works only for real gases in which the attractive interactions are dominant, because the molecules have to climb apart against the attractive force in order for them to travel more slowly. For molecules under conditions when repulsions are dominant (corresponding to $Z > 1$), the Joule–Thomson effect results in the gas becoming warmer.

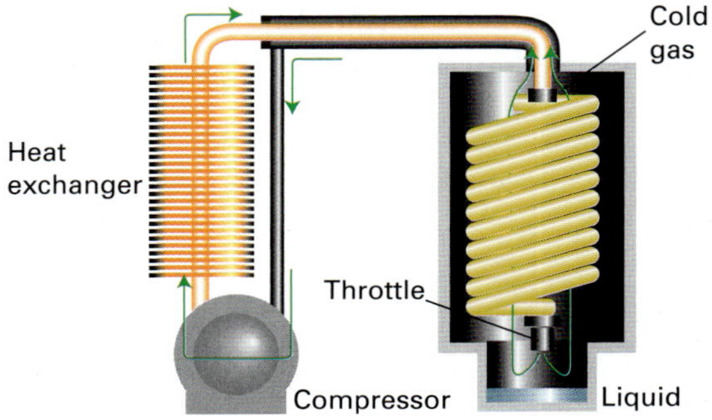

Fig. 1.19 The principle of the Linde refrigerator. The gas is recirculated and cools the gas that is about to undergo expansion through the throttle. The expanding gas cools still further. Eventually, liquefied gas drips from the throttle.

In practice, the gas is allowed to expand several times by recirculating it through a device called a *Linde refrigerator* (Fig. 1.19). On each successive expansion the gas becomes cooler, and as it flows past the incoming gas, the latter is cooled further. After several successive expansions, the gas becomes so cold that it condenses to a liquid.

Checklist of key ideas

You should now be familiar with the following concepts.

☐ **1** An equation of state is an equation relating pressure, volume, temperature, and amount of a substance.

☐ **2** The perfect-gas equation of state is based on Boyle's law ($p \propto 1/V$), Charles' law ($V \propto T$), and Avogadro's principle ($V \propto n$).

☐ **3** Dalton's law states that the total pressure of a mixture of perfect gases is the sum of the pressures that each gas would exert if it were alone in the container at the same temperature.

☐ **4** The partial pressure of any gas is defined as $p_J = x_J p$, where x_J is its mole fraction in a mixture and p is the total pressure.

☐ **5** The kinetic model of gases expresses the properties of a perfect gas in terms of a collection of mass points in ceaseless random motion.

☐ **6** The mean speed and root-mean-square speed of molecules is proportional to the square root of the (absolute) temperature and inversely proportional to the square root of the molar mass.

☐ **7** The properties of the Maxwell distribution of speeds are summarized in Figs. 1.8 and 1.9.

☐ **8** Diffusion is the spreading of one substance through another; effusion is the escape of a gas through a small hole.

☐ **9** Graham's law states that the rate of effusion is inversely proportional to the square root of the molar mass.

☐ **10** The Joule–Thomson effect is the cooling of gas that occurs when it expands through a throttle without the influx of heat.

Table of key equations

The following table summarizes the equations that have been developed in this chapter.

Property	Equation	Comment
Perfect gas law	$pV = nRT$	A limiting law for real gases as $p \to 0$
Partial pressure	$p_J = x_J p$	Definition
Dalton's law	$p = p_A + p_B + \cdots$	
Virial equation of state	$p = (nRT/V)(1 + nB/V + n^2C/V^2 + \cdots)$	
Mean free path, speed, and collision frequency	$c = \lambda z$	Perfect gas (kinetic model)
van der Waals equation of state	$p = nRT/(V - nb) - a(n/V)^2$	a: attractive effects b: repulsive effects
Maxwell distribution of speeds	$F(s) = 4\pi\left(\frac{M}{2\pi RT}\right)^{3/2} s^2 e^{-Ms^2/2RT}$	Perfect gas (kinetic model)

Further information 1.1

Kinetic molecular theory

One of the essential skills of a physical chemist is the ability to turn simple, qualitative ideas into rigid, testable, quantitative theories. The kinetic model of gases is an excellent example of this technique, as it takes the concepts set out in the text and turns them into precise expressions. As usual in model building, there are a number of steps, but each one is motivated by a clear appreciation of the underlying physical picture, in this case a swarm of mass points in ceaseless random motion. The key quantitative ingredients we need are the equations of classical mechanics. So we begin with a brief review of velocity, momentum, and Newton's second law of motion.

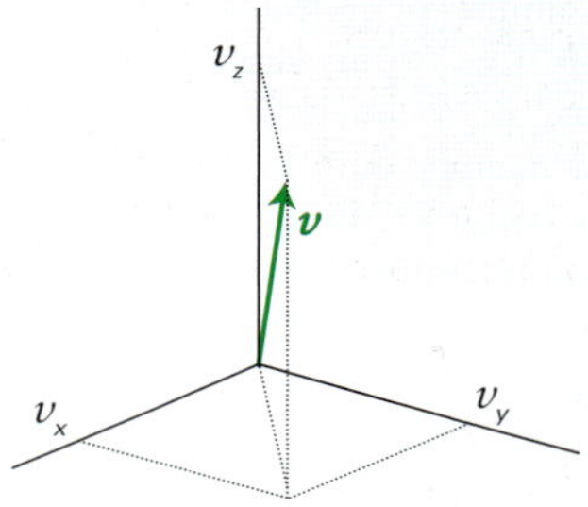

Fig. 1.20 A vector $\boldsymbol{v}$ and its three components on a set of perpendicular axes.

The velocity, $\boldsymbol{v}$, is a vector, a quantity with both magnitude and direction. The magnitude of the velocity vector is the speed, v, given by $v = (v_x^2 + v_y^2 + v_z^2)^{1/2}$, where v_x, v_y, and v_z, are the components of the vector along the x-, y-, and z-axes, respectively (Fig. 1.20). The magnitude of each component, its value without a sign, is denoted $|\ldots|$. For example, $|v_x|$ means the magnitude of v_x. The linear momentum, $\boldsymbol{p}$, of a particle of mass m is the vector $\boldsymbol{p} = m\boldsymbol{v}$ with magnitude $p = mv$. Newton's second law of motion states that the force acting on a particle is equal to the rate of change of the momentum, the change of momentum divided by the interval during which that change occurs.

Now we begin the derivation of eqn 1.9 by considering the arrangement in Fig. 1.21. When a particle of mass m that is travelling with a component of velocity v_x parallel to the x-axis ($v_x > 0$ corresponding to motion to the right and $v_x < 0$ to motion to the left) collides with the wall on the right and is reflected, its linear momentum changes from $+m|v_x|$ before the collision to $-m|v_x|$ after the collision (when it is travelling in the opposite direction at the same speed). The x-component of the momentum therefore changes by $2m|v_x|$ on each collision (the y- and z-components are unchanged). Many molecules collide with the wall in an interval Δt, and the total change of momentum is the product of the change in momentum of each molecule multiplied by the number of molecules that reach the wall during the interval.

Next, we need to calculate that number. Because a molecule with velocity component v_x can travel a distance $|v_x|\Delta t$ along the x-axis in an interval Δt, all the molecules within a distance $|v_x|\Delta t$ of the wall will strike it if they are travelling towards it. It follows that if the wall has area A, then all the particles in a volume $A \times |v_x|\Delta t$ will reach the

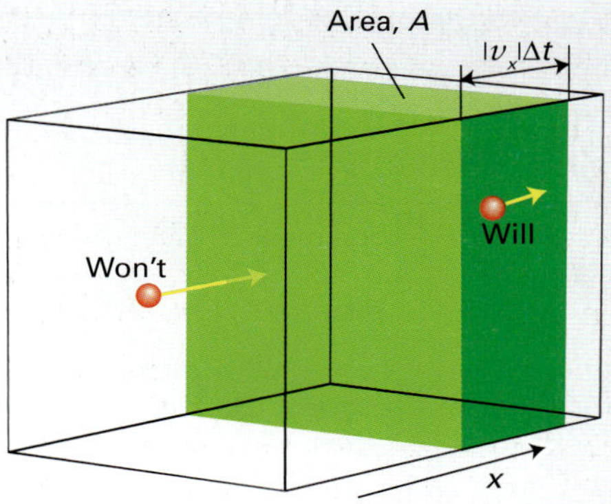

Fig. 1.21 The model used for calculating the pressure of a perfect gas according to the kinetic molecular theory. Here, for clarity, we show only the *x*-component of the velocity (the other two components are not changed when the molecule collides with the wall). All molecules within the shaded area will reach the wall in an interval Δt provided they are moving towards it.

wall (if they are travelling towards it). The number density, the number of particles divided by the total volume, is nN_A/V (where n is the total amount in moles of molecules in the container of volume V and N_A is Avogadro's constant), so the number of molecules in the volume $A|v_x|\Delta t$ is $(nN_A/V) \times A|v_x|\Delta t$. At any instant, half the particles are moving to the right and half are moving to the left. Therefore, the average number of collisions with the wall during the interval Δt is $\frac{1}{2}nN_A A|v_x|\Delta t/V$.

The total momentum change in the interval Δt is the product of the number we have just calculated and the change $2m|v_x|$:

Number of impacts in the interval Δt — Momentum change on each impact

$$\text{Momentum change} = \frac{nN_A A|v_x|\Delta t}{2V} \times 2m|v_x| = \frac{nmAN_A v_x^2 \Delta t}{V} = \frac{nMAv_x^2 \Delta t}{V} \qquad (M = mN_A)$$

Next, to find the force, we calculate the rate of change of momentum:

$$\text{Force} = \frac{\text{Change of momentum}}{\text{Time interval}} = \frac{nMAv_x^2 \Delta t/V}{\Delta t} = \frac{nMAv_x^2}{V}$$

It follows that the pressure, the force divided by the area, is

$$\text{Pressure} = \frac{nMAv_x^2/V}{A} = \frac{nMv_x^2}{V}$$

Not all the molecules travel with the same velocity, so the detected pressure, p, is the average (denoted $\langle\ldots\rangle$) of the quantity just calculated:

$$p = \frac{nM\langle v_x^2\rangle}{V}$$

To write an expression of the pressure in terms of the root mean square speed, c, we begin by writing the speed of a single molecule, v, as $v^2 = v_x^2 + v_y^2 + v_z^2$. Because the root-mean-square speed, c, is defined as $c = \langle v^2\rangle^{1/2}$ (eqn 1.10), it follows that

$$c^2 = \langle v^2\rangle = \langle v_x^2\rangle + \langle v_y^2\rangle + \langle v_z^2\rangle$$

However, because the molecules are moving randomly, all three averages are the same. It follows that $c^2 = 3\langle v_x^2\rangle$. Equation 1.9 now follows by substituting $\langle v_x^2\rangle = \frac{1}{3}c^2$ into $p = nM\langle v_x^2\rangle/V$.

Questions and exercises

Discussion questions

1.1 Explain how the experiments of Boyle, Charles, and Avogadro led to the formulation of the perfect gas equation of state.

1.2 Explain the term 'partial pressure' and why Dalton's law is a limiting law.

1.3 Use the kinetic model of gases to explain why light gases, such as H_2 and He, are rare in the Earth's atmosphere but heavier gases, such as O_2, CO_2, and N_2 are abundant.

1.4 Provide a molecular interpretation for the variation of the rates of diffusion and effusion of gases with temperature.

1.5 Explain how the compression factor varies with pressure and temperature and describe how it reveals information about intermolecular interactions in real gases.

1.6 What is the significance of the critical constants?

1.7 Describe the formulation of the van der Waals equation of state.

1.8 Describe the Joule–Thomson effect and its application to the liquefaction of gases.

Exercises

Treat all gases as perfect unless instructed otherwise.

1.1 What pressure is exerted by a sample of nitrogen gas of mass 3.055 g in a container of volume 3.00 dm^3 at 32°C?

1.2 A sample of neon of mass 425 mg occupies 6.00 dm^3 at 77 K. What pressure does it exert?

1.3 Much to everyone's surprise, nitrogen monoxide (NO) has been found to act as a neurotransmitter. To prepare to study its effect, a sample was collected in a container of volume 300.0 cm^3. At 14.5°C its pressure is found to be 34.5 kPa. What amount (in moles) of NO has been collected?

1.4 A domestic water-carbonating kit uses steel cylinders of carbon dioxide of volume 250 cm^3. They weigh 1.04 kg when full and 0.74 kg when empty. What is the pressure of gas in the cylinder at 20°C?

1.5 The effect of high pressure on organisms, including humans, is studied to gain information about deep-sea diving and anaesthesia. A sample of air occupies 1.00 dm^3 at 25°C and 1.00 atm. What pressure is needed to compress it to 100 cm^3 at this temperature?

1.6 You are warned not to dispose of pressurized cans by throwing them on to a fire. The gas in an aerosol container exerts a pressure of 125 kPa at 18°C. The container is thrown on a fire, and its temperature rises to 700°C. What is the pressure at this temperature?

1.7 Until we find an economical way of extracting oxygen from sea-water or lunar rocks, we have to carry it with us to inhospitable places, and do so in compressed form in tanks. A sample of oxygen at 101 kPa is compressed at constant temperature from 7.20 dm^3 to 4.21 dm^3. Calculate the final pressure of the gas.

1.8 To what temperature must a sample of helium gas be cooled from 22.2°C to reduce its volume from 1.00 dm^3 to 100 cm^3?

1.9 Hot-air balloons gain their lift from the lowering of density of air that occurs when the air in the envelope is heated. To what temperature should you heat a sample of air, initially at 315 K, to increase its volume by 25 per cent?

1.10 At sea level, where the pressure was 104 kPa and the temperature 21.1°C, a certain mass of air occupied 2.0 m^3. To what volume will the region expand when it has risen to an altitude where the pressure and temperature are (a) 52 kPa, –5.0°C, (b) 880 Pa, –52.0°C?

1.11 A diving bell has an air space of 3.0 m^3 when on the deck of a boat. What is the volume of the air space when the bell has been lowered to a depth of 50 m? Take the mean density of sea water to be 1.025 g cm^{-3} and assume that the temperature is the same as on the surface.

1.12 Balloons were used to obtain much of the early information about the atmosphere and continue to be used today to obtain weather information. In 1782, Jacques Charles used a hydrogen-filled balloon to fly from Paris 25 km into the French countryside. What is the mass density of hydrogen relative to air at the same temperature and pressure? What mass of payload can be lifted by 10 kg of hydrogen, neglecting the mass of the balloon?

1.13 Atmospheric pollution is a problem that has received much attention. Not all pollution, however, is from industrial sources. Volcanic eruptions can be a significant source of air pollution. The Kilauea volcano in Hawaii emits 200–300 t of SO_2 per day. If this gas is emitted at 800°C and 1.0 atm, what volume of gas is emitted?

1.14 A meteorological balloon had a radius of 1.5 m when released at sea level at 20°C and expanded to a radius of 3.5 m when it had risen to its maximum altitude where the temperature was –25°C. What is the pressure inside the balloon at that altitude?

1.15 A gas mixture being used to simulate the atmosphere of another planet consists of 320 mg of methane, 175 mg of argon, and 225 mg of nitrogen. The partial pressure of nitrogen at 300 K is 15.2 kPa. Calculate (a) the volume and (b) the total pressure of the mixture.

1.16 The vapour pressure of water at blood temperature is 47 Torr. What is the partial pressure of dry air in our lungs when the total pressure is 760 Torr?

1.17 A determination of the density of a gas or vapour can provide a quick estimate of its molar mass even though for practical work mass spectrometry is far more precise. The density of a gaseous compound was found to be 1.23 g dm^{-3} at 330 K and 25.5 kPa. What is the molar mass of the compound?

1.18 In an experiment to measure the molar mass of a gas, 250 cm^3 of the gas was confined in a glass vessel. The pressure was 152 Torr at 298 K and the mass of the gas was 33.5 mg. What is the molar mass of the gas?

1.19 A vessel of volume 22.4 dm^3 contains 2.0 mol H_2 and 1.0 mol N_2 at 273.15 K. Calculate (a) their partial pressures and (b) the total pressure.

1.20 Calculate the mean speed of (a) He atoms, (b) CH_4 molecules at (i) 79 K, (ii) 315 K, (iii) 1500 K.

1.21 A 1.0 dm^3 glass bulb contains 1.0×10^{23} H_2 molecules. If the pressure exerted by the gas is 100 kPa, what is (a) the temperature of the gas, (b) the root-mean-square speeds of the molecules. (c) Would the temperature be different if they were O_2 molecules?

1.22 At what pressure does the mean free path of argon at 25°C become comparable to the diameter of a spherical vessel of volume 1.0 dm^3 that contains it? Take $\sigma = 0.36$ nm^2.

1.23 At what pressure does the mean free path of argon at 25°C become comparable to 10 times the diameters of the atoms themselves? Take $\sigma = 0.36$ nm^2.

1.24 When we are studying the photochemical processes that can occur in the upper atmosphere, we need to know how often atoms and molecules collide. At an altitude of 20 km the temperature is 217 K and the pressure is 0.050 atm. What is the mean free path of N_2 molecules? Take $\sigma = 0.43$ nm^2.

1.25 How many collisions does a single Ar atom make in 1.0 s when the temperature is 25°C and the pressure is (a) 10 bar, (b) 100 kPa, (c) 1.0 Pa?

1.26 Calculate the total number of collisions per second in 1.0 dm^3 of argon under the same conditions as in Exercise 1.25.

1.27 How many collisions per second does an N_2 molecule make at an altitude of 20 km? (See Exercise 1.24 for data.)

1.28 The spread of pollutants through the atmosphere is governed partly by the effects of winds but also by the natural tendency of molecules to diffuse. The latter depends on how far a molecule can travel before colliding with another molecule. Calculate the mean free path of diatomic molecules in air using $\sigma = 0.43$ nm^2 at 25°C and (a) 10 bar, (b) 103 kPa, (c) 1.0 Pa.

1.29 How does the mean free path in a sample of a gas vary with temperature in a constant-volume container?

1.30 Calculate the pressure exerted by 1.0 mol C_2H_6 behaving as (a) a perfect gas, (b) a van der Waals gas when it is confined under the following conditions: (i) at 273.15 K in 22.414 dm^3, (ii) at 1000 K in 100 cm^3. Use the data in Table 1.5.

1.31 How reliable is the perfect gas law in comparison with the van der Waals equation? Calculate the difference in pressure of 10.00 g of carbon dioxide confined to a container of volume 100 cm^3 at 25.0°C between treating it as a perfect gas and a van der Waals gas.

1.32 Express the van der Waals equation of state as a virial expansion in powers of $1/V_m$ and obtain expressions for B and C in terms of the parameters a and b. *Hint.* The expansion you will need is $(1-x)^{-1} = 1 + x + x^2 + \cdots$. Series expansions are discussed in Appendix 2.

1.33 A certain gas obeys the van der Waals equation with $a = 0.50$ m^6 Pa mol^{-2}. Its volume is found to be 5.00×10^{-4} m^3 mol^{-1} at 273 K and 3.0 MPa. From this information calculate the van der Waals constant b. What is the compression factor for this gas at the prevailing temperature and pressure?

1.34 Measurements on argon gave $B = -21.7$ cm^3 mol^{-1} and $C = 1200$ cm^6 mol^{-2} for the virial coefficients at 273 K. What are the values of a and b in the corresponding van der Waals equation of state?

1.35 Show that there is a temperature at which the second virial coefficient, B, is zero for a van der Waals gas, and calculate its value for carbon dioxide. *Hint*: Use the expression for B derived in Exercise 1.32.

1.36 The critical constants of ethane are $p_c = 48.20$ atm, $V_c = 148$ cm^3 mol^{-1}, and $T_c = 305.4$ K. Calculate the van der Waals parameters of the gas and estimate the radius of the molecules.

Projects

The symbol ‡ indicates that calculus is required.

1.37‡ In the following exercises you will explore the Maxwell distribution of speeds in more detail.

(a) Confirm that the mean speed of molecules of molar mass M at a temperature T is equal to $(8RT/\pi M)^{1/2}$. *Hint*: You will need an integral of the form $\int_0^\infty x^3 e^{-ax^2}\,dx = n!/2a^2$.

(b) Confirm that the root-mean-square speed of molecules of molar mass M at a temperature T is equal to $(3RT/M)^{1/2}$ and hence confirm eqn 1.13. *Hint*: You will need an integral of the form $\int_0^\infty x^4 e^{-ax^2}\,dx = (3/8a^2)(\pi/a)^{1/2}$.

(c) Find an expression for the most probable speed of molecules of molar mass M at a temperature T. *Hint:* Look for a maximum in the Maxwell distribution (the maximum occurs as $dF/ds = 0$).

(d) Estimate the fraction of N_2 molecules at 500 K that have speeds in the range 290 to 300 m s^{-1}.

1.38‡ Here we explore the van der Waals equation of state. Using the language of calculus, the critical point of a van der Waals gas occurs where the isotherm has a flat inflexion, which is where $dp/dV_m = 0$ (zero slope) and $d^2p/dV_m^2 = 0$ (zero curvature).

(a) Evaluate these two expressions using eqn 1.23b, and find expressions for the critical constants in terms of the van der Waals parameters.

(b) Show that the value of the compression factor at the critical point is $\frac{3}{8}$.

1.39 The kinetic model of gases is valid when the size of the particles is negligible compared with their mean free path. It may seem absurd, therefore, to expect the kinetic theory and, as a consequence, the perfect gas law, to be applicable to the dense matter of stellar interiors. In the Sun, for instance, the density is 150 times that of liquid water at its centre and comparable to that of water about half-way to its surface. However, we have to realize that the state of matter is that of a *plasma,* in which the electrons have been stripped from the atoms of hydrogen and helium that make up the bulk of the matter of stars. As a result, the particles making up the plasma have diameters comparable to those of nuclei, or about 10 fm. Therefore, a mean free path of only 0.1 pm satisfies the criterion for the validity of the kinetic model and the perfect gas law. We can therefore use $pV = nRT$ as the equation of state for the stellar interior.

(a) Calculate the pressure half-way to the centre of the Sun, assuming that the interior consists of ionized hydrogen atoms, the temperature is 3.6 MK, and the mass density is 1.20 g cm^{-3} (slightly higher than the density of water).

(b) Combine the result from part (a) with the expression for the pressure from kinetic model to show that the pressure of the plasma is related to its *kinetic energy density* $\rho_k = E_k/V$, the kinetic energy of the molecules in a region divided by the volume of the region, by

$$p = \tfrac{2}{3}\rho_k$$

(c) What is the kinetic energy density half-way to the centre of the Sun? Compare your result with the (translational) kinetic energy density of the Earth's atmosphere on a warm day (25°C): 1.5×10^5 J m^{-3} (corresponding to 0.15 J cm^{-3}).

(d) A star eventually depletes some of the hydrogen in its core, which contracts and results in higher temperatures. The increased temperature results in an increase in the rates of nuclear reaction, some of which result in the formation of heavier nuclei, such as carbon. The outer part of the star expands and cools to produce a red giant. Assume that half-way to the centre a red giant has a temperature of 3500 K, is composed primarily of fully ionized carbon atoms and electrons, and has a mass density of 1200 kg m^{-3}. What is the pressure at this point?

(e) If the red giant in part (d) consisted of neutral carbon atoms, what would be the pressure at the same point under the same conditions?

Chapter 2

Thermodynamics: the first law

The branch of physical chemistry known as **thermodynamics** is concerned with the study of the transformations of energy and, in particular, the transformation of heat into work and vice versa. That concern might seem remote from chemistry; indeed, thermodynamics was originally formulated by physicists and engineers interested in the efficiency of steam engines. However, thermodynamics has proved to be of immense importance in chemistry. Not only does it deal with the energy output of chemical reactions but it also helps to answer questions that lie right at the subject's heart, such as why reactions reach equilibrium, their composition at equilibrium, and how reactions in electrochemical (and biological) cells can be used to generate electricity.

Classical thermodynamics, the thermodynamics developed during the nineteenth century, stands aloof from any models of the internal constitution of matter: we could develop and use thermodynamics without ever mentioning atoms and molecules. However, the subject is greatly enriched by acknowledging that atoms and molecules do exist and interpreting thermodynamic properties and relations in terms of them. Wherever it is appropriate, we shall cross back and forth between thermodynamics, which provides useful relations between observable properties of bulk matter, and the properties of atoms and molecules, which are ultimately responsible for these bulk properties. When we call on a molecular interpretation, we shall signal it with the icon shown here. The theory of the connection between atomic and bulk thermodynamic properties is called **statistical thermodynamics** and is treated in Chapter 22.

Chemical thermodynamics is a tree with many branches. **Thermochemistry** is the branch that deals with the heat output of chemical reactions. As we elaborate the content of thermodynamics, we shall see that we can also discuss the output of energy in the form of work. This connection leads us into the fields of **electrochemistry**, the interaction between electricity and chemistry, and **bioenergetics**, the deployment of energy in living organisms. The whole of equilibrium chemistry—the formulation of equilibrium constants, and the very special case of the equilibrium composition of solutions of acids and bases—is an aspect of thermodynamics.

The conservation of energy

Almost every argument and explanation in chemistry boils down to a consideration of some aspect of a single property: the *energy*. Energy determines what molecules may form, what reactions may occur, how fast they may occur, and—with a refinement in our conception of energy that we explore in Chapter 4—in which direction a reaction has a tendency to occur.

As we saw in the Introduction:

Energy is the capacity to do work.

Work is done to achieve motion against an opposing force.

These definitions imply that a raised weight of a given mass has more energy than one of the same mass resting on the ground because the former has a greater capacity to do work: it can do work as it falls to the level of the lower weight. The definition also implies that a gas at high temperature has more energy than the same gas at a low temperature: the hot gas has a higher pressure and can do more work in driving out a piston.

People struggled for centuries to create energy from nothing, for they believed that if they could create energy, then they could produce work (and wealth) endlessly. However, without exception, despite strenuous efforts, many of which degenerated into deceit, they failed. As a result of their failed efforts, we have come to recognize that energy can be neither created nor destroyed but merely converted from one form into another or moved from place to place. This 'law of the conservation of energy' is of great importance in chemistry. Most chemical reactions release energy or absorb it as they occur; so according to the law of the conservation of energy, we can be confident that all such changes must result only in the conversion of energy from one form into another or its transfer from place to place, not its creation or annihilation. The detailed study of that conversion and transfer is the domain of thermodynamics.

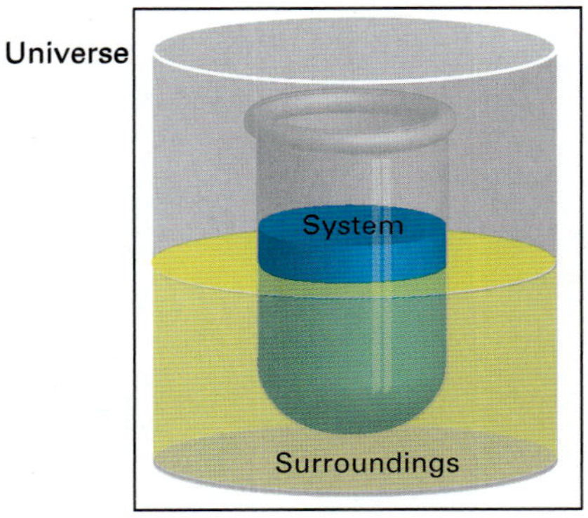

Fig. 2.1 The sample is the system of interest; the rest of the world is its surroundings. The surroundings are where observations are made on the system. They can often be modelled, as here, by a large water bath. The universe consists of the system and surroundings.

2.1 Systems and surroundings

In thermodynamics, a **system** is the part of the world in which we have a special interest. The **surroundings** are where we make our observations (Fig. 2.1). The surroundings, which can be modelled as a large water bath, remain at constant temperature regardless of how much energy flows into or out of them. They are so huge that they also have either constant volume or constant pressure regardless of any changes that take place to the system. Thus, even though the system might expand, the surroundings remain effectively the same size.

We need to distinguish three types of systems (Fig. 2.2):

An **open system** can exchange both energy and matter with its surroundings.

A **closed system** can exchange energy but not matter with its surroundings.

An **isolated system** can exchange neither matter nor energy with its surroundings.

An example of an open system is a flask that is not stoppered and to which various substances can be added. A biological cell is an open system because nutrients and waste can pass through the cell wall. You and I are open systems: we ingest, respire, perspire, and excrete. An example of a closed system is a stoppered flask: energy can be exchanged with the contents of the flask because the walls may be able to conduct heat. An example of an isolated system is a sealed flask that is thermally, mechanically, and electrically insulated from its surroundings.

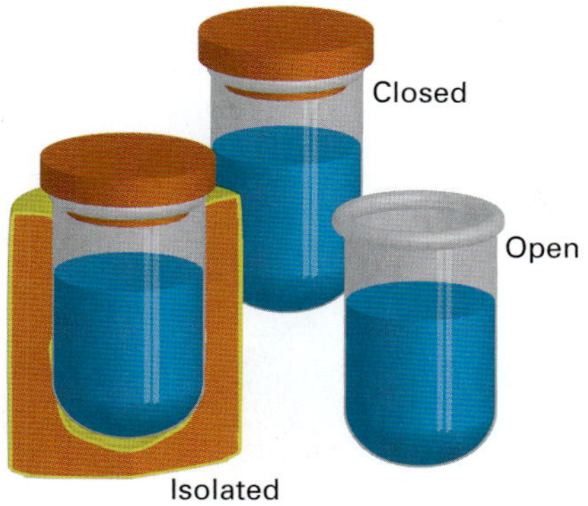

Fig. 2.2 A system is *open* if it can exchange energy and matter with its surroundings, *closed* if it can exchange energy but not matter, and *isolated* if it can exchange neither energy nor matter.

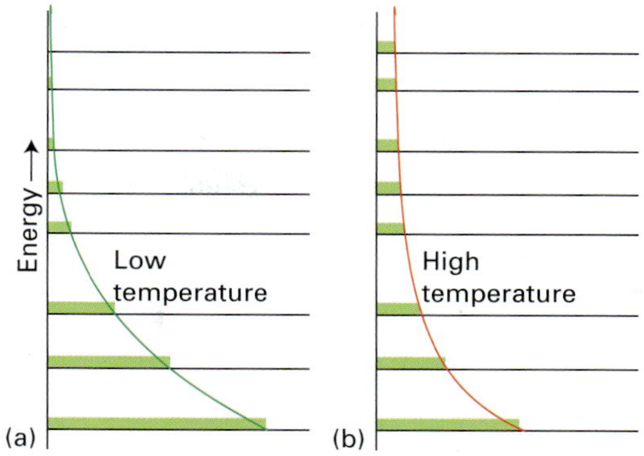

Fig. 2.3 The temperature is a parameter that indicates the extent to which the exponentially decaying Boltzmann distribution reaches up into the higher energy levels of a system. (a) When the temperature is low, only the lower energy states are occupied (as indicated by the green rectangles). (b) At higher temperatures, more higher states are occupied. In each case, the populations decay exponentially with increasing temperature, with the total population of all levels a constant.

2.2 Work and heat

Energy can be exchanged between a closed system and its surroundings by doing work or by the process called 'heating'. A system does work when it causes motion against an opposing force. We can identify when a system does work by noting whether the process can be used to change the height of a weight somewhere in the surroundings. **Heating** is the process of transferring energy as a result of a temperature difference between the system and its surroundings. To avoid a lot of awkward circumlocution, it is common to say that 'energy is transferred as work' when the system does work, and 'energy is transferred as heat' when the system heats or is heated by its surroundings. However, we should always remember that 'work' and 'heat' are *modes of transfer* of energy, not *forms* of energy.

Although in everyday language the terms 'temperature' and 'heat' are sometimes not distinguished, they are entirely different entities:

Heat, q, is energy in transit as a result of a temperature difference.

Temperature, T, is an intensive property that is used to define the state of a system and determines the direction in which energy flows as heat.

Later in the text we shall discuss the molecular interpretation of temperature. At this stage, all we need to appreciate is that *the temperature is the single parameter that tells us the relative populations of the available energy levels in a system.* We need to recall from introductory chemistry courses that molecules can possess only certain energies (the energy is 'quantized'). At any given temperature, the numbers of molecules that occupy the available energy levels depend on the temperature: at low temperatures, most molecules are in the lowest energy states; as the temperature is raised, more molecules occupy states of higher energy, so the population spreads into these upper states (Fig. 2.3). Temperature is the parameter that summarizes this spread of populations.

Walls that permit heating as a mode of transfer of energy are called **diathermic** (Fig. 2.4). A metal container is diathermic. Walls that do not permit heating even though there is a difference in temperature are

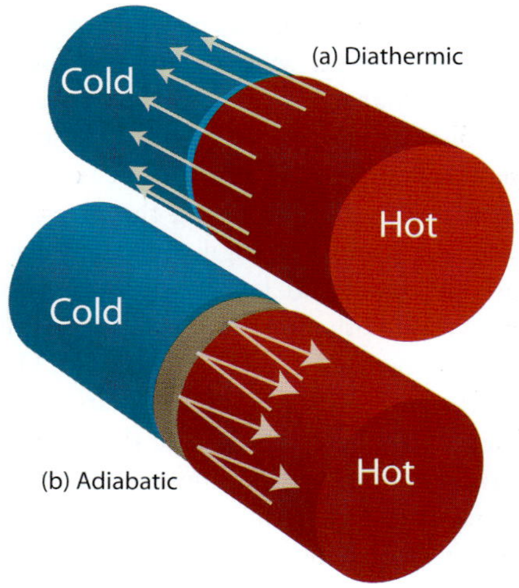

Fig. 2.4 (a) A diathermic wall permits the passage of energy as heat; (b) an adiabatic wall does not, even if there is a temperature difference across the wall.

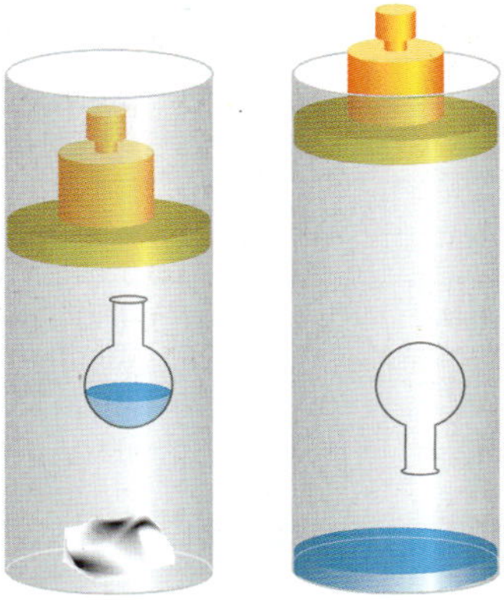

Fig. 2.5 When hydrochloric acid reacts with zinc, the hydrogen gas produced must push back the surrounding atmosphere (represented by the weight resting on the piston), and hence must do work on its surroundings. This is an example of energy leaving a system as work.

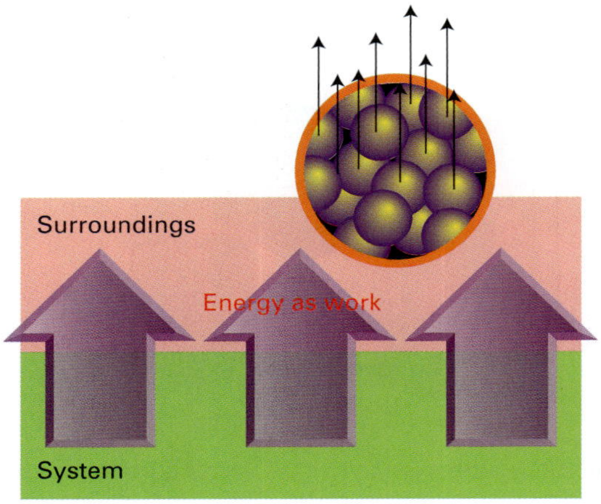

Fig. 2.6 Work is transfer of energy that causes or utilizes uniform motion of atoms in the surroundings. For example, when a weight is raised, all the atoms of the weight (shown magnified) move in unison in the same direction.

called **adiabatic** (from the Greek words for 'not passing through'). The double walls of a vacuum flask are adiabatic to a good approximation.

As an example of the different ways of transferring energy, consider a chemical reaction that produces gases, such as the reaction of an acid with zinc:

$$Zn(s) + 2\,HCl(aq) \rightarrow ZnCl_2(aq) + H_2(g)$$

Suppose first that the reaction takes place inside a cylinder fitted with a piston, then the gas produced drives out the piston and raises a weight in the surroundings (Fig. 2.5). In this case, energy has migrated to the surroundings as a result of the system doing work because a weight has been raised in the surroundings: that weight can now do more work, so it possesses more energy. Some energy also migrates into the surroundings as heat. We can detect that transfer of energy by immersing the reaction vessel in an ice bath and noting how much ice melts. Alternatively, we could let the same reaction take place in a vessel with a piston locked in position. No work is done, because no weight is raised. However, because it is found that more ice melts than in the first experiment, we can conclude that more energy has migrated to the surroundings as heat.

A process in a system that releases energy as heat is called **exothermic.** A process in a system that absorbs energy as heat is called **endothermic.** An example of an exothermic reaction is any combustion of an organic compound. Endothermic reactions are much less common. The endothermic dissolution of ammonium nitrate in water is the basis of the instant cold-packs that are included in some first-aid kits. They consist of a plastic envelope containing water dyed blue (for psychological reasons) and a small tube of ammonium nitrate, which is broken when the pack is to be used.

The clue to the molecular nature of work comes from thinking about the motion of a weight in terms of its component atoms. When a weight is raised, all its atoms move in the same direction. This observation suggests that *work is the mode of transfer of energy that achieves or utilizes uniform motion in the surroundings* (Fig. 2.6). Whenever we think of work, we can always think of it in terms of uniform motion of some kind. Electrical work, for instance, corresponds to electrons being pushed in the same direction through a circuit. Mechanical work corresponds to atoms being pushed in the same direction against an opposing force.

Now consider the molecular nature of heat. When energy is transferred as heat to the surroundings, the atoms and molecules oscillate more vigorously around their positions or move more rapidly from place to place. The key point is that the motion stimulated by the arrival of energy from the system as heat is disorderly, not uniform as in the case of doing work. This observation suggests that *heat is the mode of transfer of energy that achieves or utilizes disorderly motion in the surroundings* (Fig. 2.7). A fuel burning, for example, generates disorderly molecular motion in its vicinity.

An interesting historical point is that the molecular difference between work and heat correlates with the chronological order of their application. The release of

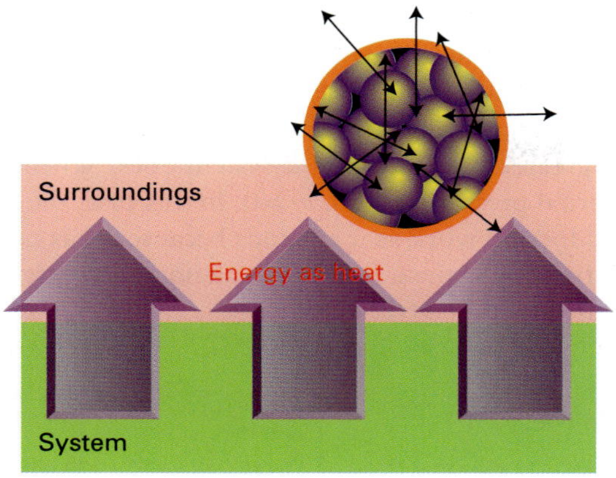

Fig. 2.7 Heat is the transfer of energy that causes or utilizes chaotic motion in the surroundings. When energy leaves the system (the green region), it generates chaotic motion in the surroundings (shown magnified).

energy when a fire burns is a relatively unsophisticated procedure because the energy emerges in a disordered fashion from the burning fuel. It was developed—stumbled upon—early in the history of civilization. The generation of work by a burning fuel, in contrast, relies on a carefully controlled transfer of energy so that vast numbers of molecules move in unison. Apart from Nature's achievement of work through the evolution of muscles, the large-scale transfer of energy by doing work was achieved thousands of years later than the transfer of energy by heating, as it had to await the development of the steam engine.

2.3 The measurement of work

We saw in Section 0.3 that the force opposing the raising of a mass m vertically on the surface of the Earth is mg, where g is the acceleration of free fall (9.81 m s^{-2}). Therefore, the work done to raise the mass through a height h is

$$\text{Work} = mgh \tag{2.1}$$

It follows that we have a simple way of measuring the work done by or on a system: we measure the height through which a weight is raised or lowered in the surroundings and then use eqn 2.1.

A brief illustration To raise a book like this one (of mass about 1.0 kg) from the floor to the table 75 cm above requires

$$\text{Work} = (1.0\ \text{kg}) \times (9.81\ \text{m s}^{-2}) \times (0.75\ \text{m})$$
$$= 7.4\ \text{kg m}^2\ \text{s}^{-2} = 7.4\ \text{J}$$

(We saw in Section 0.4 that 1 J = 1 kg m^2 s^{-2}.)

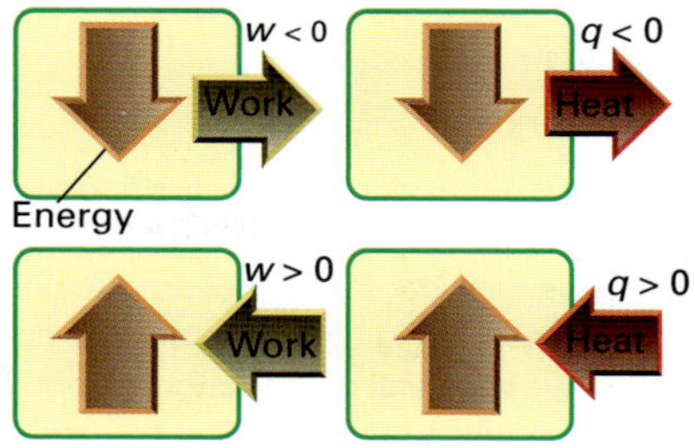

Fig. 2.8 The sign convention in thermodynamics: w and q are positive if energy enters the system (as work and heat, respectively), but negative if energy leaves the system.

When a system does work, such as by raising a weight in the surroundings or forcing an electric current through a circuit, the energy transferred, w, is reported as a negative quantity. For instance, if a system raises a weight in the surroundings and in the process does 100 J of work (that is, 100 J of energy leaves the system by doing work), then we write $w = -100$ J. When work is done on the system—for example, when we wind a spring inside a clockwork mechanism—w is reported as a positive quantity. We write $w = +100$ J to signify that 100 J of work has been done on the system (that is, 100 J of energy had been transferred to the system by doing work). The sign convention is easy to follow if we think of changes to the energy of the system: its energy decreases (w is negative) if energy leaves it and its energy increases (w is positive) if energy enters it (Fig. 2.8).

We use the same convention for energy transferred as heat, q. We write $q = -100$ J if 100 J of energy leaves the system as heat, so reducing the energy of the system, and $q = +100$ J if 100 J of energy enters the system as heat.

Because many chemical reactions produce gas, one very important type of work in chemistry is **expansion work**, the work done when a system expands against an opposing pressure. The action of acid on zinc illustrated in Fig. 2.5 is an example of a reaction in which expansion work is done in the process of making room for the gaseous product, hydrogen in this case. We show in Derivation 2.1 that when a system expands through a volume ΔV against a constant external pressure p_{ex} the work done is

$$w = -p_{ex}\,\Delta V \tag{2.2}$$

Derivation 2.1

Expansion work

To calculate the work done when a system expands from an initial volume V_i to a final volume V_f, a change $\Delta V = V_f - V_i$, we consider a piston of area A moving out through a distance h (Fig. 2.9). There need not be an actual piston: we can think of the piston as representing the boundary between the expanding gas and the surrounding atmosphere. However, there may be an actual piston, such as when the expansion takes place inside an internal combustion engine.

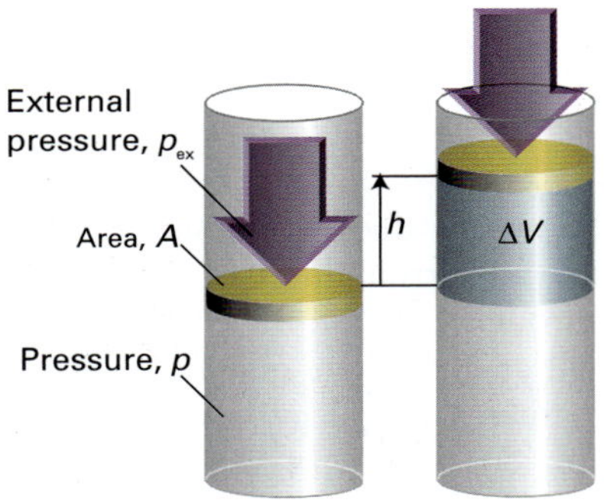

Fig. 2.9 When a piston of area A moves out through a distance h, it sweeps out a volume $\Delta V = Ah$. The external pressure p_{ex} opposes the expansion with a force $p_{ex}A$.

The force opposing the expansion is the constant external pressure p_{ex} multiplied by the area of the piston (because force is pressure times area, Section 0.5), so $F = p_{ex}A$. The work done is therefore

$$\text{Work done} = \underbrace{h}_{\text{Distance moved}} \times \overbrace{(p_{ex}A)}^{\text{Opposing force}} = p_{ex} \times \underbrace{hA}_{\text{Volume change}}$$

$$= p_{ex} \times \Delta V$$

The last equality follows from the fact that hA is the volume of the cylinder swept out by the piston as the gas expands, so we can write $hA = \Delta V$. That is, for expansion work,

$$\text{Work done} = p_{ex}\Delta V$$

Now consider the sign. A system does work and thereby loses energy (that is, w is negative) when it expands (when ΔV is positive). Therefore, we need a negative sign in the equation to ensure that w is negative when ΔV is positive), so we obtain eqn 2.2.

A note on good practice Keep track of signs by considering whether the stored energy has decreased when the system does work (w is then negative) or has increased when work has been done on the system (w is then positive).

According to eqn 2.2, the *external* pressure determines how much work a system does when it expands through a given volume: the greater the external pressure, the greater the opposing force and the greater the work that a system does. When the external pressure is zero, $w = 0$. In this case, the system does no work as it expands because it has nothing to push against. Expansion against zero external pressure is called **free expansion**.

Self-test 2.1

Calculate the work done by a system in which a reaction results in the formation of 1.0 mol CO_2(g) at 25°C and 100 kPa. *Hint.* The increase in volume will be 25 dm^3 under these conditions if the gas is treated as perfect; use the relations 1 $dm^3 = 10^{-3}$ m^3 and 1 Pa m^3 = 1 J.

[*Answer:* 2.5 kJ]

Equation 2.2 shows us how to get the *least* expansion work from a system: we just reduce the external pressure to zero. But how can we achieve the *greatest* work for a given change in volume? According to eqn 2.2, the system does maximum work when the external pressure has its maximum value. The force opposing the expansion is then the greatest and the system must exert most effort to push the piston out. However, that external pressure cannot be greater than the pressure, p, of the gas inside the system, for otherwise the external pressure would compress the gas instead of allowing it to expand. Therefore, *maximum work is obtained when the external pressure is only infinitesimally less than the pressure of the gas in the system.* In effect, the two pressures must be adjusted to be the same at all stages of the expansion. In the Introduction we called this balance of pressures a state of mechanical equilibrium. Therefore, we can conclude that *a system that remains in mechanical equilibrium with its surroundings at all stages of the expansion does maximum expansion work.*

There is another way of expressing this condition. Because the external pressure is infinitesimally less than the pressure of the gas at some stage of the expansion, the piston moves out. However, suppose we increase the external pressure so that it became infinitesimally greater than the pressure of the gas; now the piston moves in. That is, *when a system is in a state of mechanical equilibrium, an infinitesimal change in the pressure results in opposite directions of motion.* A process that can be reversed by an *infinitesimal* change in a variable—in this case, the pressure—is said to be **reversible**. In everyday life 'reversible' means a process that can be reversed; in thermodynamics it has a stronger meaning—it means that a process can be reversed by an *infinitesimal* modification in some variable (such as the pressure).

We can summarize this discussion by the following remarks:

- A system does maximum expansion work when the external pressure is equal to that of the system at every stage of the expansion ($p_{ex} = p$).
- A system does maximum expansion work when it is in mechanical equilibrium with its surroundings at every stage of the expansion.
- Maximum expansion work is achieved in a reversible change.

All three statements are equivalent, but they reflect different degrees of sophistication in the way the point is expressed.

We cannot write down the expression for maximum expansion work simply by replacing p_{ex} in eqn 2.2 by p (the pressure of the gas in the cylinder) because, as the piston moves out, the pressure inside the system falls. To make sure the entire process occurs reversibly, we have to adjust the external pressure to match the changing internal pressure. Suppose that we conduct the expansion isothermally (that is, at constant temperature) by immersing the system in a water bath held at a specified temperature. As we show in Derivation 2.2, the work of isothermal, reversible expansion of a perfect gas from an initial volume V_i to a final volume V_f at a temperature T is

$$w = -nRT \ln \frac{V_f}{V_i} \qquad (2.3)$$

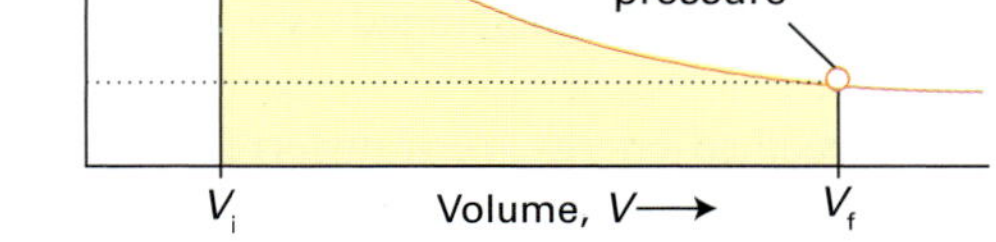

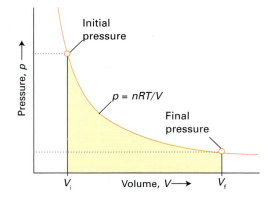

Fig. 2.10 The work of reversible isothermal expansion of a gas is equal to the area beneath the corresponding isotherm evaluated between the initial and final volumes (the tinted area). The isotherm shown here is that of a perfect gas, but the same relation holds for any gas.

where n is the amount of gas molecules in the system. As explained in Derivation 2.2, this result is equal to the area beneath a graph of $p = nRT/V$ between the limits V_i and V_f (Fig. 2.10).

Derivation 2.2

Reversible, isothermal expansion work

Because (to ensure reversibility) the external pressure must be adjusted in the course of the expansion, we have to think of the process as taking place in a series of small steps during each one of which the external pressure is constant. We calculate the work done in each step for the prevailing external pressure, and then add all these values together. To ensure that the overall result is accurate, we have to make the steps as small as possible—infinitesimal, in fact—so that the pressure is truly constant during each one. In other words, we have to use calculus, in which case the sum over an infinite number of infinitesimal steps becomes an integral.

When the system expands through an infinitesimal volume dV, the infinitesimal work, dw, done is the infinitesimal version of eqn 2.2:

$$dw = -p_{ex}dV$$

A brief comment For a review of calculus, see Appendix 2. As indicated there, the replacement of Δ by d always indicates an infinitesimal change: dV is positive for an infinitesimal increase in volume and negative for an infinitesimal decrease.

At each stage, we ensure that the external pressure is the same as the current pressure, p, of the gas (Fig. 2.11). So, we set $p_{ex} = p$ and obtain

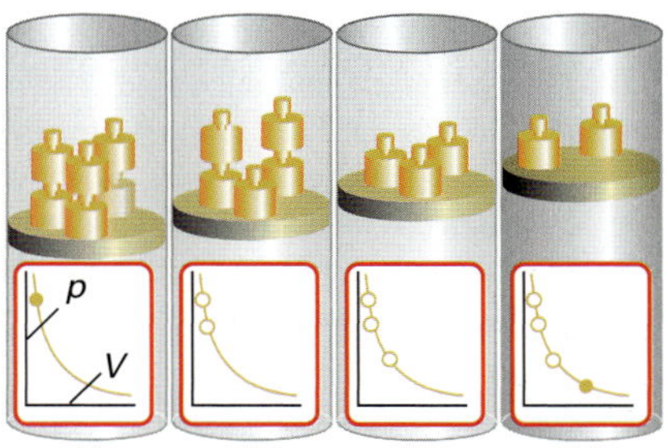

Fig. 2.11 For a gas to expand reversibly, the external pressure must be adjusted to match the internal pressure at each stage of the expansion. This matching is represented in this illustration by gradually unloading weights from the piston as the piston is raised and the internal pressure falls. The procedure results in the extraction of the maximum possible work of expansion.

$$dw = -p dV$$

The total work when the system expands from V_i to V_f is the sum (integral) of all the infinitesimal changes between the limits V_i and V_f, which we write

For a reversible expansion, $w = -\int_{V_i}^{V_f} p dV$

To evaluate the integral, we need to know how p, the pressure of the gas in the system, changes as it expands. For this step, we suppose that the gas is perfect, in which case we can use the perfect gas $pV = nRT$ in the form $p = nRT/V$. At this stage we have

For the reversible expansion of a perfect gas,

$$w = -\int_{V_i}^{V_f} \frac{nRT}{V} dV$$

In general, the temperature might change as the gas expands, so in general T depends on V, and T changes as V changes. For isothermal expansion, however, the temperature is held constant and we can take n, R, and T outside the integral and write

For the isothermal, reversible expansion of a perfect gas, $w = -nRT \int_{V_i}^{V_f} \frac{dV}{V}$

A standard result of calculus is

$$\int \frac{dx}{x} = \ln x + \text{constant}$$

where $\ln x$ is the natural logarithm of x. It follows that

$$\int_{V_i}^{V_f} \frac{dV}{V} = (\ln V_f + \text{constant}) - (\ln V_i + \text{constant}) = \ln \frac{V_f}{V_i}$$

($\ln x - \ln y = \ln(x/y)$)

When we insert this result into the preceding one, we obtain eqn 2.3. The interpretation of eqn 2.3 as an area follows from the fact, as explained in Appendix 2, that a definite integral is equal to the area beneath a graph of the function lying between the two limits of the integral.

A note of good practice Introduce (and keep note of) the restrictions (in this case, in succession: reversible process, perfect gas, isothermal) only as they prove necessary, as you might be able to use an intermediate formula without needing to restrict it in the way that was necessary to achieve the final expression.

Equation 2.3 will turn up in various disguises throughout this text. Once again, it is important to be able to interpret it rather than just remember it:

- In an expansion $V_f > V_i$, so $V_f/V_i > 1$ and the logarithm is positive ($\ln x$ is positive if $x > 1$). Therefore, in an expansion, w is negative. That is what we should expect: energy *leaves* the system as the system does expansion work.
- For a given change in volume, we get more work the higher the temperature of the confined gas (Fig. 2.12). That is also what we should expect: at high temperatures, the pressure of the gas is high, so we have to use a high external pressure, and therefore a stronger opposing force, to match the internal pressure at each stage.

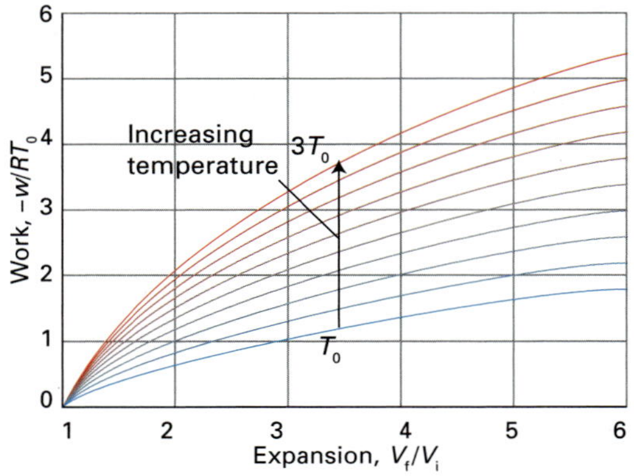

Fig. 2.12 The work of reversible, isothermal expansion of a perfect gas. Note that for a given change of volume and fixed amount of gas, the work is greater the higher the temperature.

Self-test 2.2

Calculate the work done when 1.0 mol Ar(g) confined in a cylinder of volume 1.0 dm^3 at 25°C expands isothermally and reversibly to 2.0 dm^3.

[*Answer:* $w = -1.7$ kJ]

2.4 The measurement of heat

When a substance is heated, its temperature typically rises. We say 'typically' because the temperature does not always rise. The temperature of boiling water, for instance, remains unchanged as it is heated (see Chapter 5).

For a specified energy, q, transferred by heating, the size of the resulting temperature change, ΔT, depends on the 'heat capacity' of the substance. The **heat capacity**, C, is defined as:

$$C = \frac{q}{\Delta T} \qquad (2.4a)$$

(q: Energy supplied as heat; ΔT: Change in temperature)

It follows that we have a simple way of measuring the heat absorbed or released by a system: we measure a temperature change and then use the appropriate value of the heat capacity of the system and eqn 2.4a rearranged into

$$q = C\Delta T \tag{2.4b}$$

A brief illustration If the heat capacity of a beaker of water is 0.50 kJ K^{-1}, and we observe a temperature rise of 4.0 K, then we can infer that the heat transferred to the water is

$$q = (0.50\ \mathrm{kJ\ K^{-1}}) \times (4.0\ \mathrm{K}) = 2.0\ \mathrm{kJ}$$

Heat capacities occur frequently in the following sections and chapters, and we need to be aware of their properties and how their values are reported. First, we note that the heat capacity is an extensive property (a property that depends on the amount of substance in the sample, Section 0.8): 2 kg of iron has twice the heat capacity of 1 kg of iron, so twice as much heat is required to raise its temperature by a given amount. It is more convenient to report the heat capacity of a substance as an intensive property (a property that is independent of the amount of substance in the sample). We therefore use either the **specific heat capacity**, C_s, the heat capacity divided by the mass of the sample ($C_s = C/m$, in joules per kelvin per gram, J K^{-1} g^{-1}) or the **molar heat capacity**, C_m, the heat capacity divided by the amount of substance ($C_m = C/n$, in joules per kelvin per mole, J K^{-1} mol^{-1}). In common usage, the specific heat capacity is often called the *specific heat*. To obtain the heat capacity of a sample of known mass or that contains a known amount of substance, we use these definitions in the form $C = mC_s$ or $C = nC_m$, respectively.

For reasons that will be explained shortly, the heat capacity of a substance depends on whether the sample is maintained at constant volume (like a gas in a sealed, rigid vessel) as it is heated, or whether the sample is maintained at constant pressure (like water in an open container), and free to change its volume. The latter is a more common arrangement, and the values given in Table 2.1 are for the **heat capacity at constant pressure**, C_p. The **heat capacity at constant volume** is denoted C_V. The respective molar values are denoted $C_{p,m}$ and $C_{V,m}$.

A brief illustration The molar heat capacity of water at constant pressure, $C_{p,m}$, is 75 J K^{-1} mol^{-1}. Suppose a 1.0 kW kettle (where 1 W = 1 J s^{-1}) is turned on for 100 s. The energy supplied in that interval (the product of the power and the time) is

$$q = (1.0\ \mathrm{kW}) \times (100\ \mathrm{s}) = (1.0 \times 10^3\ \mathrm{J\ s^{-1}}) \times (100\ \mathrm{s}) = 1.0 \times 10^5\ \mathrm{J}$$

It follows that the increase in temperature of 1.0 kg of water (55.5 mol H_2O, from $n = m/M$) in the kettle is approximately

$$\Delta T = \frac{q}{C_p} = \frac{q}{nC_{p,\mathrm{m}}} = \frac{1.0 \times 10^5\ \mathrm{J}}{(55.5\ \mathrm{mol}) \times (75\ \mathrm{J\ K^{-1}\ mol^{-1}})} = +24\ \mathrm{K}$$

The reason why different substances have different molar heat capacities can be traced to differences in the separations of their energy levels. We shall understand those differences when we get to quantum theory (in Chapter 12) but, as remarked earlier, from introductory chemistry courses we know that molecules can exist with only certain energies. If those energy levels are close together, the incoming energy can be accommodated with little change in their populations. The change in temperature (which determines the populations) is therefore also small and the heat capacity is correspondingly large. In other words, closely spaced energy levels correlate with a high heat capacity (Fig. 2.13). The translational energy levels of molecules in a gas are very close together, and all monatomic gases have similar molar heat capacities. The separation of the vibrational energies of atoms bound together in solids depend on the stiffness of the bonds between them and on the masses of the atoms, and solids show a wide range of molar heat capacities. We shall return to this topic in Chapter 22 after we learn more about energy levels and their populations.

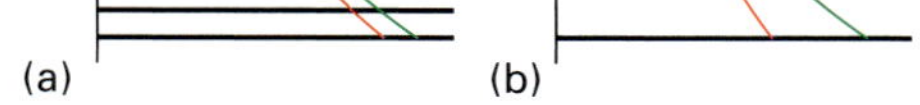

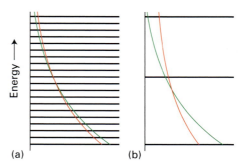

Fig. 2.13 The heat capacity depends on the availability of levels. (a) When the levels are close together, a given amount of energy arriving as heat can be accommodated with little adjustment of the populations and hence the temperature that occurs in the Boltzmann distribution. This system has a high heat capacity. (b) When the levels are widely separated, the same incoming energy has to be accommodated by making use of higher energy levels, with a consequent greater change in the 'reach' of the Boltzmann distribution, and therefore a greater change in temperature. This system therefore has a low heat capacity. In each case the green line is the distribution at low temperature and the red line that at higher temperature.

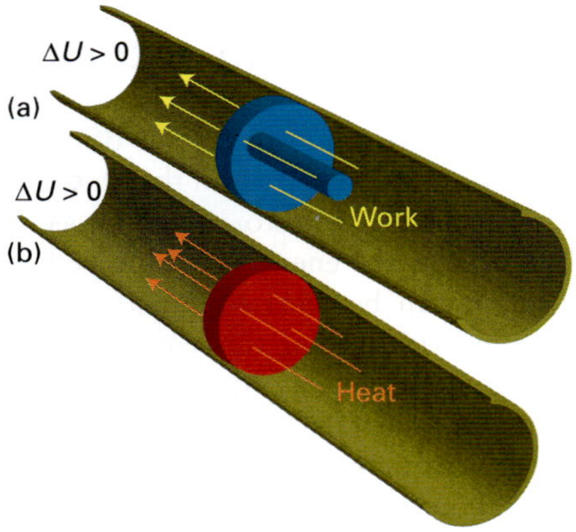

Fig. 2.15 Provided there are no other transfers of energy, when work is done on a system, its internal energy rises ($\Delta U > 0$). Likewise, provided there are no other transfers of energy, the internal energy also rises when energy is transferred into the system as heat.

of work': q and w are the quantities of energy transferred as heat and work, respectively, that result in the change ΔU.

A brief illustration When a system releases 10 kJ of energy into the surroundings by doing work (that is, when $w = -10$ kJ), the internal energy of the system decreases by 10 kJ, and we write $\Delta U = -10$ kJ. The minus sign signifies the reduction in internal energy that has occurred. If the system loses 20 kJ of energy by heating its surroundings (so $q = -20$ kJ), we write $\Delta U = -20$ kJ. If the system loses 10 kJ as work *and* 20 kJ as heat, as in an inefficient internal combustion engine, the internal energy falls by a total of 30 kJ, and we write $\Delta U = -30$ kJ. On the other hand, if we do 10 kJ of work on the system ($w = +10$ kJ), for instance, by winding a spring it contains or pushing in a piston to compress a gas (Fig. 2.15), then the internal energy of the system increases by 10 kJ, and we write $\Delta U = +10$ kJ. Likewise, if we supply 20 kJ of energy by heating the system ($q = +20$ kJ), then the internal energy increases by 20 kJ, and we write $\Delta U = +20$ kJ.

Another note on good practice Notice that ΔU always carries a sign explicitly, even if it is positive: we never write $\Delta U = 20$ kJ, for instance, but always +20 kJ.

We have seen that a feature of a perfect gas is that for any *isothermal* expansion, the total energy of the sample remains the same, and therefore, because $\Delta U = 0$, that $q = -w$. That is, any energy lost as work is restored by an influx of energy as heat. We can express this property in terms of the internal energy, for it implies that the internal energy remains constant when a perfect gas expands isothermally: from eqn 2.8 we can write

Isothermal expansion of a perfect gas: $\Delta U = 0$ (2.9)

In other words, *the internal energy of a sample of perfect gas at a given temperature is independent of the volume it occupies.* We can understand this independence by realizing that when a perfect gas expands isothermally the only feature that changes is the average distance between the molecules; their average speed and therefore total kinetic energy remains the same. However, as there are no intermolecular interactions, the total energy is independent of the average separation, so the internal energy is unchanged by expansion.

Example 2.2

Calculating the change in internal energy

Nutritionists are interested in the use of energy by the human body and we can consider our own body as a thermodynamic 'system'. Calorimeters have been constructed that can accommodate a person to measure (nondestructively!) their net energy output. Suppose in the course of an experiment someone does 622 kJ of work on an exercise bicycle and loses 82 kJ of energy as heat. What is the change in internal energy of the person? Disregard any matter loss by perspiration.

Strategy This example is an exercise in keeping track of signs correctly. When energy is lost from the system, w or q is negative. When energy is gained by the system, w or q is positive.

Solution To take note of the signs we write $w = -622$ kJ (622 kJ is lost by doing work) and $q = -82$ kJ (82 kJ is lost by heating the surroundings). Then eqn 2.8 gives us

$$\Delta U = w + q = (-622\text{ kJ}) + (-82\text{ kJ}) = -704\text{ kJ}$$

We see that the person's internal energy falls by 704 kJ. Later, that energy will be restored by eating.

A note on good practice Always attach the correct signs: use a positive sign when there is a flow of energy into the system and a negative sign when there is a flow of energy out of the system.

Self-test 2.4

An electric battery is charged by supplying 250 kJ of energy to it as electrical work (by driving an electric current through it), but in the process it loses 25 kJ of energy as heat to the surroundings. What is the change in internal energy of the battery?

[*Answer:* +225 kJ]

2.7 Internal energy as a state function

An important characteristic of the internal energy is that it is a **state function**, a physical property that depends only on the present state of the system and is independent of the path by which that state was reached. If we were to change the temperature of the system, then change the pressure, then adjust the temperature and pressure back to their original values, the internal energy would return to its original value too. A state function is very much like altitude: each point on the surface of the Earth can be specified by quoting its latitude and longitude, and (on land areas, at least) there is a unique property, the altitude, that has a fixed value at that point. In thermodynamics, the role of latitude and longitude is played by the pressure and temperature (and any other variables needed to specify the state of the system), and the internal energy plays the role of the altitude, with a single, fixed value for each state of the system.

The fact that U is a state function implies that *a change, ΔU, in the internal energy between two states of a system is independent of the path between them* (Fig. 2.16). Once again, the altitude is a helpful analogy. If we climb a mountain between two fixed points, we make the same change in altitude regardless of the path we take between the two points. Likewise, if we compress a sample of gas until it reaches a certain pressure and then cool it to a certain temperature, the change in internal energy has a particular value. If, on the other hand, we change the temperature and then the pressure, but ensure that the two final values are the same as in the first experiment, then the overall change in internal energy is exactly the same as before. This path independence

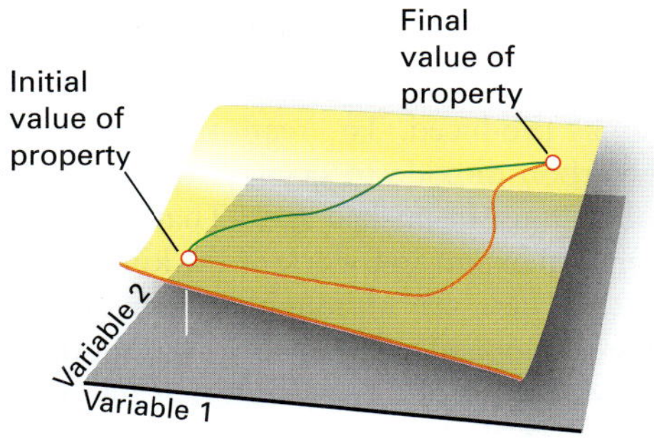

Fig. 2.16 The curved sheet shows how a property (for example, the altitude) changes as two variables (for example, latitude and longitude) are changed. The altitude is a state property, because it depends only on the current state of the system. The change in the value of a state property is independent of the path between the two states. For example, the difference in altitude between the initial and final states shown in the diagram is the same whatever path (as depicted by the dark and light lines) is used to travel between them.

of the value of ΔU is of the greatest importance in chemistry, as we shall soon see.

Suppose we now consider an isolated system. Because an isolated system can neither do work nor heat its surroundings (or acquire energy by either process), it follows that its internal energy cannot change. That is,

> *The internal energy of an isolated system is constant.*

This statement is the **First Law of thermodynamics**. It is closely related to the law of conservation of energy, but allows for transfers of energy as heat as well as by doing work. Unlike thermodynamics, mechanics does not deal with the concept of heat.

The experimental evidence for the First Law is the impossibility of making a 'perpetual motion machine', a device for producing work without consuming fuel. As we have already remarked, try as people might, they have never succeeded. No device has ever been made that creates internal energy to replace the energy drawn off as work. We cannot extract energy as work, leave the system isolated for some time, and hope that when we return the internal energy will have become restored to its original value.

The definition of ΔU in terms of w and q points to a very simple method for measuring the change in internal energy of a system when a reaction takes place. We have seen already that the work done by a system when it pushes against a fixed external pressure is proportional to the change in volume. Therefore, if we carry out a reaction in a container of constant volume, the system can do no expansion work and provided it can do no other kind of work (so-called 'nonexpansion work', such as electrical work) we can set $w = 0$. Then eqn 2.8 simplifies to

At constant volume, no nonexpansion work:

$$\Delta U = q \qquad (2.10a)$$

This relation is commonly written

$$\Delta U = q_V \qquad (2.10b)$$

The subscript V signifies that the volume of the system is constant. An example of a chemical system that can be approximated as a constant-volume container is an individual biological cell.

To measure a change in internal energy, we should use a calorimeter that has a fixed volume and monitor the energy released as heat ($q < 0$) or supplied ($q > 0$). A **bomb calorimeter** is an example of a constant-volume calorimeter: it consists of a sturdy, sealed, constant-volume vessel in which the reaction

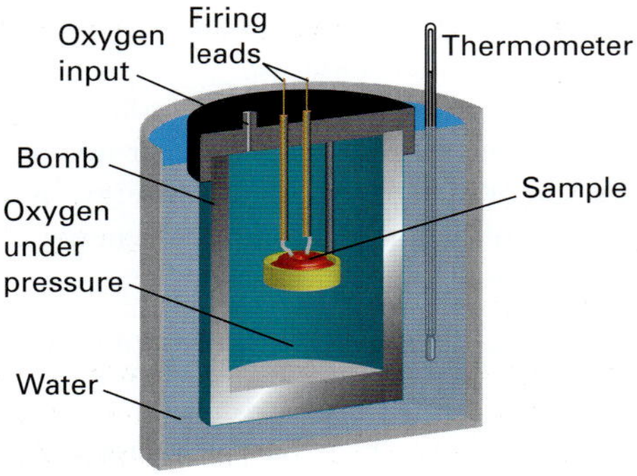

Fig. 2.17 A constant-volume bomb calorimeter. The 'bomb' is the central, sturdy vessel, which is strong enough to withstand moderately high pressures. The calorimeter is the entire assembly shown here. To ensure that no heat escapes into the surroundings, the calorimeter may be immersed in a water bath with a temperature that is continuously adjusted to that of the calorimeter at each stage of the combustion.

takes place, and a surrounding water bath (Fig. 2.17). To ensure that no heat escapes unnoticed from the calorimeter, it is immersed in a water bath with a temperature adjusted to match the rising temperature of the calorimeter. The fact that the temperature of the bath is the same as that of the calorimeter ensures that no heat flows from one to the other. That is, the arrangement is adiabatic.

We can use eqn 2.10 to obtain more insight into the heat capacity of a substance. The definition of heat capacity is given in eqn 2.4 ($C = q/\Delta T$). At constant volume, q may be replaced by the change in internal energy of the substance, so

$$C_V = \frac{\Delta U}{\Delta T} \text{ at constant volume.} \tag{2.11}$$

The expression on the right is the slope of the graph of internal energy plotted against temperature, with the volume of the system held constant, so C_V tells us how the internal energy of a constant-volume system varies with temperature. If, as is generally the case, the graph of internal energy against temperature is not a straight line, we interpret C_V as the slope of the tangent to the curve at the temperature of interest (Fig. 2.18).

A brief comment A more precise definition of heat capacity is constructed as follows. First, we consider an infinitesimal change in temperature, dT, and the accompanying infinitesimal change in internal energy, dU, and replace eqn 2.11 by

$$C_V = \frac{dU}{dT}$$

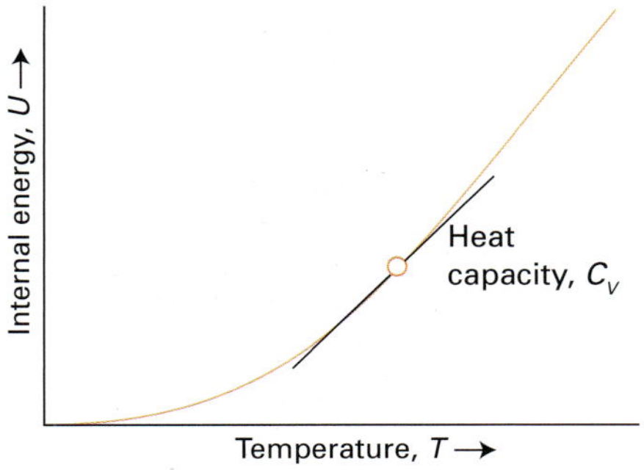

Fig. 2.18 The constant-volume heat capacity is the slope of a curve showing how the internal energy varies with temperature. The slope, and therefore the heat capacity, may be different at different temperatures.

To signify that the system is constrained to have constant volume, we attach a subscript V to the derivative, and write

$$C_V = \left(\frac{\partial U}{\partial T}\right)_V$$

The change from d to ∂ ('curly d') is used in mathematics to signal that a constraint has been applied. This is the form that you will commonly see used to define the heat capacity at constant volume.

2.8 The enthalpy

Much of chemistry, and most of biology, takes place in vessels that are open to the atmosphere and subjected to constant pressure, not constrained to constant volume in a rigid, sealed container. In general, when a change takes place in a system open to the atmosphere, the volume of the system changes. For example, the thermal decomposition of 1.0 mol $CaCO_3(s)$ at 1 bar results in an increase in volume of nearly 90 dm^3 at 800°C on account of the carbon dioxide gas produced. To create this large volume for the carbon dioxide to occupy, the surrounding atmosphere must be pushed back. That is, the system must perform expansion work of the kind treated in Section 2.3. Therefore, although a certain quantity of heat may be supplied to bring about the endothermic decomposition, the increase in internal energy of the system is not equal to the energy supplied as heat because some energy has been used to do work of expansion (Fig. 2.19). In other words, because the volume has increased, some of the heat supplied to the system has leaked back into the surroundings as work.

Another example is the oxidation of a fat, such as tristearin, to carbon dioxide in the body. The overall reaction is

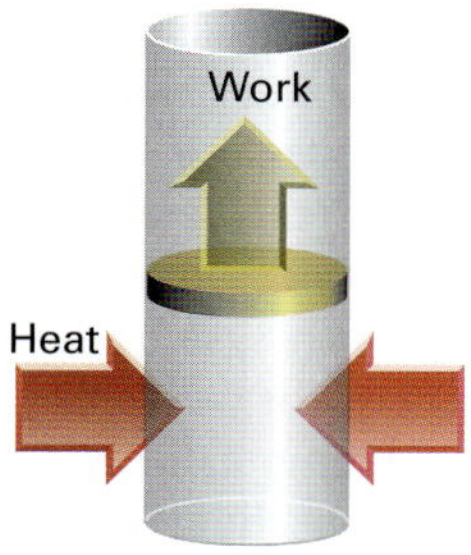

Fig. 2.19 The change in internal energy of a system that is free to expand or contract is not equal to the energy supplied as heat because some energy may escape back into the surroundings as work. However, the change in enthalpy of the system under these conditions is equal to the energy supplied as heat.

$$2\,C_{57}H_{110}O_6(s) + 163\,O_2(g) \rightarrow 114\,CO_2(g) + 110\,H_2O(l)$$

In this exothermic reaction there is a net *decrease* in volume equivalent to the elimination of (163 − 114) mol = 49 mol of gas molecules for every 2 mol of tristearin molecules that react. The decrease in volume at 25°C is about 600 cm^3 for the consumption of 1 g of fat. Because the volume of the system decreases, the atmosphere does work *on* the system as the reaction proceeds. That is, energy is transferred *to* the system as it contracts. In effect, a weight has been lowered in the surroundings, so the surroundings can do less work after the reaction has occurred. Some of their energy has been transferred into the system. For this reaction, the decrease in the internal energy of the system is less than the energy released as heat because some energy has been restored by doing work.

We can avoid the complication of having to take into account the work of expansion by introducing a new property that will be at the centre of our attention throughout the rest of the chapter and will recur throughout the book. The **enthalpy**, H, of a system is defined as

$$H = U + pV \qquad (2.12)$$

That is, the enthalpy differs from the internal energy by the addition of the product of the pressure, p, and the volume, V, of the system. This expression applies to *any* system or individual substance: don't be mislead by the 'pV' term into thinking that eqn 2.12 applies only to a perfect gas.

Enthalpy is an extensive property. The **molar enthalpy**, $H_m = H/n$, of a substance, an intensive property, differs from the molar internal energy by an amount proportional to the molar volume, V_m, of the substance:

$$H_m = U_m + pV_m \qquad (2.13a)$$

This relation is valid for all substances. For a perfect gas we can go on to write $pV_m = RT$, and obtain

$$\textit{For a perfect gas: } H_m = U_m + RT \qquad (2.13b)$$

At 25°C, $RT = 2.5$ kJ mol^{-1}, so the molar enthalpy of a perfect gas is greater than its molar internal energy by 2.5 kJ mol^{-1}. Because the molar volume of a solid or liquid is typically about 1000 times less than that of a gas, we can also conclude that the molar enthalpy of a solid or liquid is only about 2.5 J mol^{-1} (note: joules, not kilojoules) more than its molar internal energy, so the numerical difference is negligible.

A change in enthalpy (the only quantity we can measure in practice) arises from a change in the internal energy and a change in the product pV:

$$\Delta H = \Delta U + \Delta(pV) \qquad (2.14a)$$

where $\Delta(pV) = p_f V_f - p_i V_i$. If the change takes place at constant pressure p, the second term on the right simplifies to

$$\Delta(pV) = pV_f - pV_i = p(V_f - V_i) = p\Delta V$$

and we can write

$$\textit{At constant pressure: } \Delta H = \Delta U + p\Delta V \qquad (2.14b)$$

We shall often make use of this important relation for processes occurring at constant pressure, such as chemical reactions taking place in containers open to the atmosphere.

Although the enthalpy and internal energy of a sample may have similar numerical values, the introduction of the enthalpy has very important consequences in thermodynamics. First, notice that because H is defined in terms of state functions (U, p, and V), *the enthalpy is a state function*. The implication is that the change in enthalpy, ΔH, when a system changes from one state to another is independent of the path between the two states. Secondly, we show in Derivation 2.3 that the change in enthalpy of a system can be identified with the heat transferred to it at constant pressure:

At constant pressure, no nonexpansion work:
$$\Delta H = q \qquad (2.15a)$$

This relation is commonly written

$$\Delta H = q_p \qquad (2.15b)$$

The subscript p signifies that the pressure is held constant.

Derivation 2.3

Heat transfers at constant pressure

Consider a system open to the atmosphere, so that its pressure p is constant and equal to the external pressure p_{ex}. From eqn 2.14b we can write

$$\Delta H = \Delta U + p\Delta V = \Delta U + p_{ex}\Delta V$$

However, we know that the change in internal energy is given by eqn 2.8 ($\Delta U = w + q$) with $w = -p_{ex}\Delta V$ (provided the system does no other kind of work). When we substitute that expression into this one we obtain

$$\Delta H = (-p_{ex}\Delta V + q) + p_{ex}\Delta V = q$$

which is eqn 2.15.

The result expressed by eqn 2.15, that *at constant pressure, with no nonexpansion work, we can identify the energy transferred by heating with a change in enthalpy of the system*, is enormously powerful. It relates a quantity we can measure (the energy transferred as heat at constant pressure) to the change in a state function (the enthalpy). Dealing with state functions greatly extends the power of thermodynamic arguments, because we don't have to worry about how we get from one state to another: all that matters are the initial and final states.

A brief illustration Equation 2.15 implies that if 10 kJ of energy is supplied as heat to the system that is free to change its volume at constant pressure, then the enthalpy of the system increases by 10 kJ, regardless of how much energy enters or leaves by doing work, and we write $\Delta H = +10$ kJ. On the other hand, if the reaction is exothermic and releases 10 kJ of energy as heat when it occurs, then $\Delta H = -10$ kJ regardless of how much work is done. For the particular case of the combustion of tristearin mentioned at the beginning of the section, in which 90 kJ of energy is released as heat, we would write $\Delta H = -90$ kJ.

An endothermic reaction ($q > 0$) taking place at constant pressure results in an increase in enthalpy ($\Delta H > 0$) because energy enters the system as heat. On the other hand, an exothermic process ($q < 0$) taking place at constant pressure corresponds to a decrease in enthalpy ($\Delta H < 0$) because energy leaves the system as heat. All combustion reactions, including the controlled combustions that contribute to respiration, are exothermic and are accompanied by a decrease in enthalpy. These relations are consistent with the name 'enthalpy', which is derived from the Greek words meaning 'heat inside': the 'heat inside' the system is increased if the process is endothermic and absorbs energy as heat from the surroundings; it is decreased if the process is exothermic and releases energy as heat into the surroundings. However, never forget that heat does not actually 'exist' inside: only energy exists in a system; heat is a means of recovering that energy or increasing it.

Differential scanning calorimetry, discussed in Box 2.1, is a common technique for the measurement of the enthalpy change that accompanies a physical or chemical change occurring at constant pressure.

2.9 The temperature variation of the enthalpy

We have seen that the internal energy of a system rises as the temperature is increased. The same is true of the enthalpy, which also rises when the temperature is increased (Fig. 2.20). For example, the enthalpy of 100 g of water is greater at 80°C than at 20°C. We can measure the change by monitoring the energy that we must supply as heat to raise the temperature through 60°C when the sample is open to the atmosphere (or subjected to some other constant pressure); it is found that $\Delta H \approx +25$ kJ in this instance.

Just as we saw that the constant-volume heat capacity tells us about the temperature dependence of the internal energy at constant volume, so the constant-pressure heat capacity tells us how the enthalpy of a system changes as its temperature is raised at constant pressure. To derive the relation, we combine the definition of heat capacity in eqn 2.4 ($C = q/\Delta T$) with eqn 2.15 and obtain

$$C_p = \frac{\Delta H}{\Delta T} \text{ at constant pressure} \qquad (2.16)$$

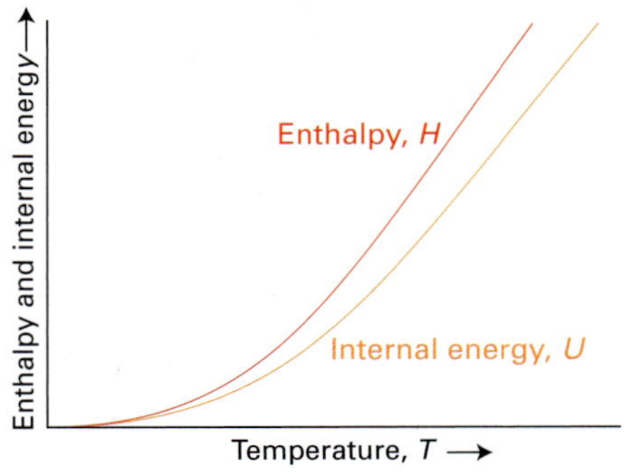

Fig. 2.20 The enthalpy of a system increases as its temperature is raised. Note that the enthalpy is always greater than the internal energy of the system, and that the difference increases with temperature.

Box 2.1 Differential scanning calorimetry

Biological samples might be very small, sometimes as small as 0.5 mg, and a highly sensitive technique is needed to determine the heat output when a sample undergoes a physical or chemical change. To achieve this sensitivity a *differential scanning calorimeter* (DSC, see the first illustration) may be used. A DSC is used in biochemistry laboratories to assess the stability of proteins, nucleic acids, and membranes. Large molecules, such as biological polymers, attain complex three-dimensional structures due to intra- and intermolecular interactions, such as hydrogen bonding and hydrophobic interactions. The disruption of these interactions is an endothermic process that can be studied with a DSC.

A DSC consists of two small compartments that are heated electrically at a constant rate (the 'scanning' part of the technique). The temperature T at time t during a linear scan is $T = T_0 + \alpha t$, where T_0 is the initial temperature and α is the temperature scan rate (in kelvin per second, K s^{-1}). Note that the rate of increase of temperature is $dT/dt = \alpha$. A computer controls the electrical power output to each compartment in order to maintain the same temperature in the two compartments throughout the analysis. The term 'differential' refers to the fact that the behaviour of the sample is compared to that of a reference material that does not undergo a physical or chemical change during the analysis. The temperature of the sample would change relative to that of the reference material if a chemical or physical process involving heat transfer occurs in the sample during the scan. To maintain the same temperature in both compartments, excess heat is transferred to the sample during the process.

If an endothermic process occurs in the sample at a particular temperature, we have to supply additional 'excess' heat, dq_{ex}, to the sample to achieve the same temperature as the reference. We can express this excess heat in terms of an 'excess' heat capacity at each stage of the scan, C_{ex}, by writing $dq_{ex} = C_{ex}dT$. It follows that, because $dT = \alpha dt$, then

$$C_{ex} = \frac{dq_{ex}}{dT} = \frac{dq_{ex}}{\alpha dt} = \frac{P_{ex}}{\alpha}$$

where $P_{ex} = dq_{ex}/dt$ is the excess electrical power (the rate of supply of energy, in watts, where 1 W = 1 J s^{-1}) necessary to equalize the temperature of the sample and reference compartments at each stage of the scan. A *thermogram* is a plot of C_{ex} (which is obtained from the value of P_{ex}/α and the power monitored throughout the scan) against T. The second illustration is a thermogram that indicates that the protein ubiquitin retains its native structure up to about 45°C. At higher temperatures, the protein undergoes an endothermic conformational change that results in the loss of its three-dimensional structure.

To find the total heat transferred, we need to integrate both sides of $dq_{ex} = C_{ex}dT$ from an initial temperature T_1 to a final temperature T_2:

$$q_{ex} = \int_{T_1}^{T_2} C_{ex}\,dT$$

The integral is the area under the thermogram between T_1 and T_2. Because the apparatus is at constant pressure, we can identify q_{ex} with the change in enthalpy accompanying the process.

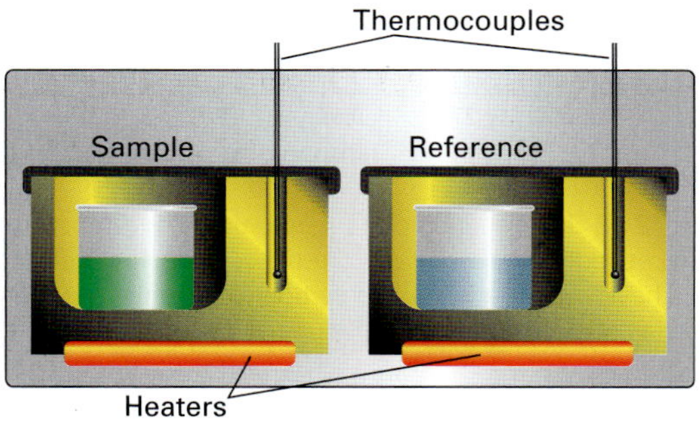

A differential scanning calorimeter. The sample and a reference material are heated in separate but identical compartments. The output is the difference in power needed to maintain the compartments at equal temperatures as the temperature rises.

See an animated version of this figure in the interactive ebook.

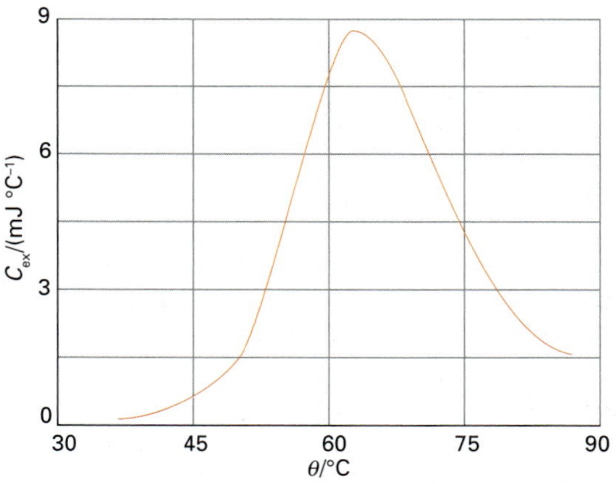

A thermogram for the protein ubiquitin. The protein retains its native structure up to about 45°C and then undergoes an endothermic conformational change. (Adapted from B. Chowdhry and S. LeHarne, *J. Chem. Educ.* **74**, 236 (1997).)

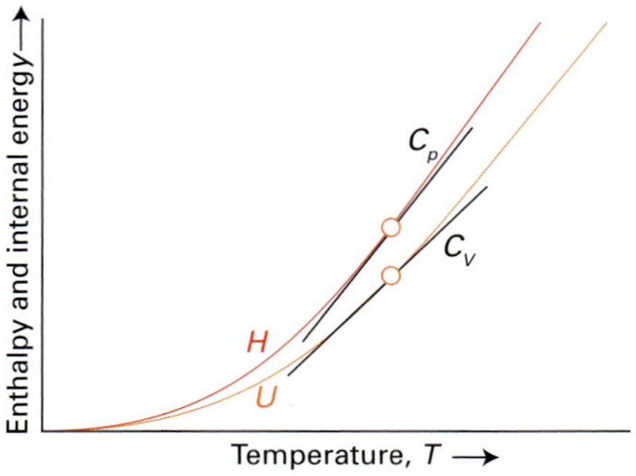

Fig. 2.21 The heat capacity at constant pressure is the slope of the curve showing how the enthalpy varies with temperature; the heat capacity at constant volume is the corresponding slope of the internal energy curve. Note that the heat capacity varies with temperature (in general), and that C_p is greater than C_V.

A brief comment The more precise definition of the constant-pressure heat capacity, by analogy with the discussion in Comment 2.3, is

$$C_p = \left(\frac{\partial H}{\partial T}\right)_p$$

The constant-pressure heat capacity is the slope of a plot of enthalpy against temperature of a system kept at constant pressure (Fig. 2.21). In general, the constant-pressure heat capacity depends on the temperature, and some values at 298 K are given in Table 2.1 (many more will be found in the Data section). Values at other temperatures not too different from room temperature are commonly estimated from the expression

$$C_{p,\mathrm{m}} = a + bT + \frac{c}{T^2} \qquad (2.17a)$$

with values of the constants a, b, and c obtained by fitting this expression to experimental data. Some values of the constants are given in Table 2.2 and a typical temperature variation is shown in Fig. 2.22. At very low temperatures, nonmetallic solids are found to have heat capacities that are proportional to T^3:

$$C_{p,\mathrm{m}} = aT^3 \qquad (2.17b)$$

where a is another constant (not the same as in eqn 2.17a). The reason for this behaviour was unclear until quantum mechanics provided an explanation (see Chapter 12).

A brief illustration Provided the heat capacity is constant over the range of temperatures of interest, we can write eqn 2.16 as $\Delta H = C_p\Delta T$. This relation means that when the temperature of 100 g of liquid water (5.55 mol H_2O) is raised from 20°C to 80°C (so $\Delta T = +60$ K) at constant pressure, the enthalpy of the sample changes by

$$\Delta H = C_p\Delta T = nC_{p,\mathrm{m}}\Delta T$$
$$= (5.55\ \mathrm{mol}) \times (75.29\ \mathrm{J\ K^{-1}\ mol^{-1}}) \times (60\ \mathrm{K})$$
$$= +25\ \mathrm{kJ}$$

The calculation in the illustration is only approximate, because the heat capacity depends on the

Table 2.1
Heat capacities of common materials

Substance	Heat capacity Specific, $C_{p,\mathrm{s}}$ /(J K^{-1} g^{-1})	Molar, $C_{p,\mathrm{m}}$ /(J K^{-1} mol^{-1})*
Air	1.01	29
Benzene, C_6H_6(l)	1.74	136.1
Brass (Cu/Zn)	0.37	
Copper, Cu(s)	0.38	24.44
Ethanol, C_2H_5OH(l)	2.42	111.46
Glass (Pyrex)	0.78	
Granite	0.80	
Marble	0.84	
Polyethylene	2.3	
Stainless steel	0.51	
Water, H_2O(s)	2.03	37
H_2O(l)	4.18	75.29
H_2O(g)	2.01	33.58

* Molar heat capacities are given only for air and well-defined pure substances; see also the Data section. The text's website contains links to online databases of heat capacities.

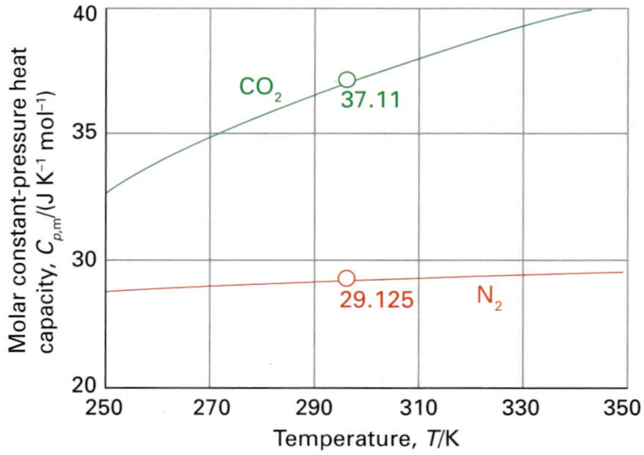

Fig. 2.22 The variation of heat capacity with temperature as expressed by the empirical formula in eqn 2.17a. For this plot we have used the values for carbon dioxide and nitrogen: the circles show the measured values at 298 K.

Table 2.2

*Temperature dependence of heat capacities**

Substance	a/ (J K^{-1} mol^{-1})	b/ (J K^{-2} mol^{-1})	c/ (J K mol^{-1})
C(s, graphite)	16.86	4.77×10^{-3}	-8.54×10^{5}
CO_2(g)	44.22	8.79×10^{-3}	-8.62×10^{5}
H_2O(l)	75.29	0	0
N_2(g)	28.58	3.77×10^{-3}	-5.0×10^{4}
Cu(s)	22.64	6.28×10^{-3}	0
NaCl(s)	45.94	16.32×10^{-3}	0

* The constants are for use in the expression $C_{p,m} = a + bT + c/T^2$.

temperature, and we have used an average value for the temperature range of interest. In Exercise 2.30 you are invited to show that if we allow for the variation in heat capacity with temperature by using eqn 2.17a, then

$$\Delta H = \left(a + bT_{\text{a. mean}} + \frac{c}{T^2_{\text{g. mean}}} \right) \Delta T \qquad (2.18)$$

where $T_{\text{a. mean}}$ is the 'arithmetic mean' of the initial and final temperatures, $T_{\text{a. mean}} = \frac{1}{2}(T_i + T_f)$, and $T_{\text{g. mean}}$ is their 'geometrical mean', $T_{\text{g. mean}} = (T_i T_f)^{1/2}$.

Because we know that the difference between the enthalpy and internal energy of a perfect gas depends in a very simple way on the temperature (eqn 2.13), we can suspect that there is a simple relation between the heat capacities at constant volume and constant pressure. We show in Derivation 2.4 that, in fact,

$$\textit{For a perfect gas: } C_{p,m} - C_{V,m} = R \qquad (2.19)$$

Derivation 2.4

The relation between heat capacities

The molar internal energy and enthalpy of a perfect gas are related by eqn 2.13b ($H_m = U_m + RT$), which we can write as $H_m - U_m = RT$. When the temperature increases by ΔT, the molar enthalpy increases by ΔH_m and the molar internal energy increases by ΔU_m, so

$$\Delta H_m - \Delta U_m = R\Delta T$$

Now divide both sides by ΔT, which gives

$$\frac{\Delta H_m}{\Delta T} - \frac{\Delta U_m}{\Delta T} = R$$

We recognize the first term on the left as the molar constant-pressure heat capacity, $C_{p,m}$ and the second term as the molar constant-volume heat capacity, $C_{V,m}$. Therefore, this relation can be written as in eqn 2.19.

Equation 2.19 shows that the molar heat capacity of a perfect gas is greater at constant pressure than at constant volume. This difference is what we should expect. At constant volume, all the energy supplied as heat to the system remains inside and the temperature rises accordingly. At constant pressure, though, some of the energy supplied as heat escapes back into the surroundings when the system expands and does work. As less energy remains in the system, the temperature does not rise as much, which corresponds to a greater heat capacity. The difference is significant for gases (for oxygen, $C_{V,m} = 20.8$ J K^{-1} mol^{-1} and $C_{p,m} = 29.1$ J K^{-1} mol^{-1}), which undergo large changes of volume when heated, but is negligible for most solids and liquids under normal conditions, for they expand much less and therefore lose less energy to the surroundings as work.

Checklist of key ideas

You should now be familiar with the following concepts.

☐ **1** A system is classified as open, closed, or isolated.

☐ **2** The surroundings remain at constant temperature and either constant volume or constant pressure when processes occur in the system.

☐ **3** An exothermic process releases energy as heat, q, an endothermic process absorbs energy as heat.

☐ **4** Maximum expansion work is achieved in a reversible change.

☐ **5** The First Law of thermodynamics states that the internal energy of an isolated system is constant.

☐ **6** Energy transferred as heat at constant volume is equal to the change in internal energy of the system; heat transferred at constant pressure is equal to the change in enthalpy.

Table of key equations

The following table summarizes the equations developed in this chapter.

Description	Equation	Comment
Work of expansion	$w=-p_{ex}\Delta V$	Against constant external pressure
	$w=-nRT\ln(V_f/V_i)$	Reversible, isothermal expansion of a perfect gas
Change in internal energy	$\Delta U=w+q$	Definition
	$\Delta U=q_V$	Constant volume, no nonexpansion work
Enthalpy	$H=U+pV$	Definition
	$\Delta H=q_p$	Constant pressure, no nonexpansion work
Heat capacity	$C=q/\Delta T$, $C_m=C/n$	Definition
	$C_V=\Delta U/\Delta T$ more formally: $C_V=(\partial U/\partial T)_V$	At constant volume
	$C_p=\Delta H/\Delta T$ more formally: $C_p=(\partial H/\partial T)_p$	At constant pressure
	$C_{p,m}-C_{V,m}=R$	For perfect gases

Questions and exercises

Discussion questions

2.1 Discuss the statement that a system and its surroundings are distinguished by specifying the properties of the boundary that separates them.

2.2 What is (a) temperature, (b) heat, (c) energy?

2.3 Provide molecular interpretations for work and heat.

2.4 Are the law of conservation of energy in dynamics and the First Law of thermodynamics identical?

2.5 Explain the difference between expansion work against constant pressure and work of reversible expansion and their consequences.

2.6 Explain the difference between the change in internal energy and the change in enthalpy of a chemical or physical process.

2.7 Specify and explain the limitations of the following expressions: (a) $q = nRT\ln(V_f/V_i)$; (b) $\Delta H = \Delta U + p\Delta V$; (c) $C_{p,m} - C_{V,m} = R$.

Exercises

Assume all gases are perfect unless stated otherwise.

2.1 Calculate the work done by a gas when it expands through (a) 1.0 cm^3, (b) 1.0 dm^3 against an atmospheric pressure of 100 kPa. What work must be done to compress the gas back to its original state in each case?

2.2 Calculate the work done by 2.0 mol of a gas when it expands reversibly and isothermally from 1.0 dm^3 to 3.0 dm^3 at 300 K.

2.3 A sample of methane of mass 4.50 g occupies 12.7 dm^3 at 310 K. (a) Calculate the work done when the gas expands isothermally against a constant external pressure of 30.0 kPa until its volume has increased by 3.3 dm^3. (b) Calculate the work that would be done if the same expansion occurred isothermally and reversibly.

2.4 In the isothermal reversible compression of 52.0 mmol of a perfect gas at 260 K, the volume of the gas is reduced from 300 cm^3 to 100 cm^3. Calculate w for this process.

2.5 A sample of blood plasma occupies 0.550 dm^3 at 0°C and 1.03 bar, and is compressed isothermally by 0.57 per cent by being subjected to a constant external pressure of 95.2 bar. Calculate w.

2.6 A strip of magnesium metal of mass 12.5 g is dropped into a beaker of dilute hydrochloric acid. Given that the magnesium is the limiting reactant, calculate the work done by the system as a result of the reaction. The atmospheric pressure is 1.00 atm and the temperature is 20.2°C.

2.7 Calculate the work of expansion accompanying the complete combustion of 10.0 g of sucrose ($C_{12}H_{22}O_{11}$) to carbon dioxide and (a) liquid water, (b) water vapour at 20°C when the external pressure is 1.20 atm.

2.8 We are all familiar with the general principles of operation of an internal combustion reaction: the combustion of fuel drives out the piston. It is possible to imagine engines that use reactions other than combustions, and we need to assess the work they can do. A chemical reaction takes place in a container of cross-sectional area 100 cm^2; the container has a piston at one end. As a result of the reaction, the piston is pushed out through 10.0 cm against a constant external pressure of 100 kPa. Calculate the work done by the system.

2.9 What is the heat capacity of a sample of liquid that rose in temperature by 5.23°C when supplied with 124 J of energy as heat?

2.10 A cube of iron was heated to 70°C and transferred to a beaker containing 100 g of water at 20°C. The final temperature of the water and the iron was 23°C. What is (a) the heat capacity, (b) the specific heat capacity, and (c) the molar heat capacity of the iron cube? Ignore heat losses from the assembly.

2.11 The high heat capacity of water is ecologically benign because it stabilizes the temperatures of lakes and oceans: a large quantity of energy must be lost or gained before there is a significant change in temperature. Conversely, it means that a lot of heat must be supplied to achieve a large rise in temperature. The molar heat capacity of water is 75.3 J K^{-1} mol^{-1}. What energy is needed to heat 250 g of water (a cup of coffee, for instance) through 40°C?

2.12 A current of 1.55 A from a 110 V source was passed through a heater for 8.5 min. The heater was immersed in a water bath. What quantity of energy was transferred to the water as heat?

2.13 When 229 J of energy is supplied as heat to 3.00 mol Ar(g), the temperature of the sample increases by 2.55 K. Calculate the molar heat capacities at constant volume and constant pressure of the gas.

2.14 The heat capacity of air is much smaller than that of water, and relatively modest amounts of heat are needed to change its temperature. This is one of the reasons why desert regions, though very hot during the day, are bitterly cold at night. The heat capacity of air at room temperature and pressure is approximately 21 J K^{-1} mol^{-1}. How much energy is required to raise the temperature of a room of dimensions 5.5 m × 6.5 m × 3.0 m by 10°C? If losses are neglected, how long will it take a heater rated at 1.5 kW to achieve that increase given that 1 W = 1 J s^{-1}?

2.15 The transfer of energy from one region of the atmosphere to another is of great importance in meteorology for it affects the weather. Calculate the heat needed to be supplied to a parcel of air containing 1.00 mol air molecules to maintain its temperature at 300 K when it expands reversibly and isothermally from 22 dm^3 to 30.0 dm^3 as it ascends.

2.16 The temperature of a block of iron ($C_{V,m}$ = 25.1 J K^{-1} mol^{-1}) of mass 1.4 kg fell by 65°C as it cooled to room temperature. What is its change in internal energy?

2.17 In an experiment to determine the calorific value of a food, a sample of the food was burned in an oxygen atmosphere and the temperature rose by 2.89°C. When a current of 1.27 A from a 12.5 V source flowed through the same calorimeter for 157 s, the temperature rose by 3.88°C. What energy is released as heat by the combustion?

2.18 A laboratory animal exercised on a treadmill that, through pulleys, raised a 250 g mass through 1.85 m. At the same time, the animal lost 10.0 J of energy as heat. Disregarding all other losses, and regarding the animal as a closed system, what is its change in internal energy?

2.19 In preparation for a study of the metabolism of an organism, a small, sealed calorimeter was prepared. In the initial phase of the experiment, a current of 22.22 mA from a 11.8 V source was passed for 162 s through a heater inside the calorimeter. What is the change in internal energy of the calorimeter?

2.20 In a computer model of the atmosphere, 20 kJ of energy was transferred as heat to a parcel of air of initial volume 1.0 m^3. What is the change in enthalpy of the parcel of air?

2.21 The internal energy of a perfect gas does not change when the gas undergoes isothermal expansion. What is the change in enthalpy?

2.22 Carbon dioxide, although only a minor component of the atmosphere, plays an important role in determining the weather and the composition and temperature of the atmosphere. Calculate the difference between the molar enthalpy and the molar internal energy of carbon dioxide regarded as a real gas at 298.15 K. For this calculation treat carbon dioxide as a van der Waals gas and use the data in Table 1.5.

2.23 A sample of a serum of mass 25 g is cooled from 290 K to 275 K at constant pressure by the extraction of 1.2 kJ of energy as heat. Calculate q and ΔH and estimate the heat capacity of the sample.

2.24 When 3.0 mol O_2(g) is heated at a constant pressure of 3.25 atm, its temperature increases from 260 K to 285 K. Given that the molar heat capacity of O_2 at constant pressure is 29.4 J K^{-1} mol^{-1}, calculate q, ΔH, and ΔU.

2.25 The molar heat capacity at constant pressure of carbon dioxide is 29.14 J K^{-1} mol^{-1}. What is the value of its molar heat capacity at constant volume?

2.26 Use the information in Exercise 2.25 to calculate the change in (a) molar enthalpy, (b) molar internal energy when carbon dioxide is heated from 15°C (the temperature when air is inhaled) to 37°C (blood temperature, the temperature in our lungs).

Projects

The sign ‡ indicates that calculus is required.

2.27‡ Here we explore the van der Waals equation of state in more detail. (a) Repeat Derivation 2.2 for a gas that obeys the equation of state $p = nRT/(V - nb)$, which is appropriate when molecular repulsions are important. Does the gas do more or less work than a perfect gas for the same change of volume? (b) Now repeat the preceding exercise for a gas that obeys the equation of state $p = nRT/V - n^2a/V^2$, which is appropriate when molecular attractions are important. Does the gas do more or less work than a perfect gas for the same change of volume?

2.28‡ Derivation 2.2 showed how to calculate the work of reversible, isothermal expansion of a perfect gas. Suppose that the expansion is reversible but not isothermal and that the temperature decreases as the expansion proceeds. (a) Find an expression for the work when $T = T_i - c(V - V_i)$, with c a positive constant. (b) Is the work greater or smaller than for isothermal expansion?

2.29‡ We now explore the effect of the temperature dependence of the heat capacity on the internal energy. (a) The heat capacity of a nonmetallic solid at very low temperatures (close to $T = 0$) typically varies as aT^3, where a is a constant. How does its internal energy vary? (b) Suppose that the molar internal energy of a substance over a limited temperature range can be expressed as a polynomial in T as $U_m(T) = a + bT + cT^2$. Find an expression for the constant-volume molar heat capacity at a temperature T.

2.30‡ Now we explore the implications of the temperature dependence of the heat capacity for the enthalpy. (a) The heat capacity of a substance is often reported in the form $C_{p,m} = a + bT + c/T^2$. Use this expression to make a more accurate estimate of the change in molar enthalpy of carbon dioxide when it is heated from 15°C to 37°C (as in the preceding exercise), given $a = 44.22\ \mathrm{J\ K^{-1}\ mol^{-1}}$, $b = 8.79 \times 10^{-3}\ \mathrm{J\ K^{-2}\ mol^{-1}}$, and $c = -8.62 \times 10^5\ \mathrm{J\ K\ mol^{-1}}$. *Hint.* You will need to integrate $dH = C_p dT$. (b) Use the expression from part (a) to determine how the molar enthalpy of the substance changes over that limited temperature range. Plot the molar enthalpy as a function of temperature.

2.31‡ The exact expression for the relation between the heat capacities at constant volume and constant pressure is $C_p - C_V = \alpha^2 TV/\kappa$, where α is the expansion coefficient, $\alpha = (dV/dT)/V$ at constant pressure and κ (kappa) is the isothermal compressibility, $\kappa = -(dV/dp)/V$. Confirm that this general expression reduces to that in eqn 2.19 for a perfect gas.

2.32 This exercise explores differential scanning calorimetry in more detail (a) In many experimental thermograms, such as that shown in Box 2.1, the baseline below T_1 is at a different level from that above T_2. Explain this observation. (b) You have at your disposal a sample of pure polymer P and a sample of P that has just been synthesized in a large chemical reactor and that might contain impurities. Describe how you would use differential scanning calorimetry to determine the mole percentage composition of P in the allegedly impure sample.

Chapter 3

Thermodynamics: applications of the First Law

This chapter is an extended illustration of the role of enthalpy in chemistry. There are three properties of enthalpy to keep in mind. One is that a change in enthalpy can be identified with the heat supplied at constant pressure ($\Delta H = q_p$). Secondly, enthalpy is a state function, so we can calculate the change in its value ($\Delta H = H_f - H_i$) between two specified initial and final states by selecting the most convenient path between them. Thirdly, the slope of a plot of enthalpy against temperature is the constant-pressure heat capacity of the system ($C_p = \Delta H/\Delta T$). All the material in this chapter is based on these three properties.

Physical change

First, we consider physical change, such as when one form of a substance changes into another form of the same substance, as when ice melts to water. We shall also include changes of a particularly simple kind, such as the ionization of an atom or the breaking of a bond in a molecule.

The numerical value of a thermodynamic property depends on the conditions, such as the states of the substances involved, the pressure, and the temperature. Chemists have therefore found it convenient to report their data for a set of standard conditions at the temperature of their choice:

> The **standard state** of a substance is the pure substance at exactly 1 bar.

(Recall that 1 bar = 10^5 Pa.) We denote the standard state value by the superscript $^{\ominus}$ on the symbol for the property, as in $H_m^{\ominus}$ for the standard molar enthalpy of a substance and $p^{\ominus}$ for the standard pressure of 1 bar. For example, the standard state of hydrogen gas is the pure gas at 1 bar and the standard state of solid calcium carbonate is the pure solid at 1 bar, with either the calcite or aragonite form specified. The physical state needs to be specified because we can speak of the standard states of the solid, liquid, and vapour forms of water, for instance, which are the pure solid, the pure liquid, and the pure vapour, respectively, at 1 bar in each case.

In older texts you might come across a standard state defined for 1 atm (101.325 kPa) in place of 1 bar. That is the old convention. In most cases, data for 1 atm differ only a little from data for 1 bar. You might also come across standard states defined as referring to 298.15 K. That is incorrect: temperature is not a part of the definition of a standard state, and standard states may refer to any temperature (but it should be specified). Thus, it is possible to speak of the standard state of water vapour at 100 K, 273.15 K, or any other temperature. It is conventional, though, for data to be reported at the so-called 'conventional temperature' of 298.15 K (25.00°C), and from now on, unless specified otherwise, all data will be for that temperature. For simplicity, we shall often refer to 298.15 K as '25°C'. Finally, a standard state need not be a stable state and need not be realizable in practice. Thus, the standard state of water vapour at 25°C is the vapour at 1 bar, but water vapour at that temperature and pressure would immediately condense to liquid water.

3.1 The enthalpy of phase transition

A **phase** is a specific state of matter that is uniform throughout in composition and physical state. The liquid and vapour states of water are two of its phases. The term 'phase' is more specific than 'state of matter' because a substance may exist in more than one solid form, each one of which is a solid phase. Thus, the element sulfur may exist as a solid. However, as a solid it may be found as rhombic sulfur or as monoclinic sulfur; these two solid phases differ in the manner in which the crown-like S_8 molecules stack together. No substance has more than one gaseous phase, so 'gas phase' and 'gaseous state' are effectively synonyms. The only substance that exists in more than one liquid phase is helium. Most substances exist in a variety of solid phases. Carbon, for instance, exists as graphite, diamond, and a variety of forms based on fullerene structures; calcium carbonate exists as calcite and aragonite; there are at least twelve forms of ice; one more was discovered in 2006.

The conversion of one phase of a substance to another phase is called a **phase transition**. Thus, vaporization (liquid → gas) is a phase transition, as is a transition between solid phases (such as rhombic sulfur → monoclinic sulfur). Most phase transitions are accompanied by a change of enthalpy, for the rearrangement of atoms or molecules usually requires or releases energy. 'Evaporation' is virtually synonymous with vaporization, but commonly denotes vaporization to dryness.

The vaporization of a liquid, such as the conversion of liquid water to water vapour when a pool of water evaporates at 20°C or a kettle boils at 100°C is an endothermic process ($\Delta H > 0$), because heat must be supplied to bring about the change. At a molecular level, molecules are being driven apart from the grip they exert on one another, and this process requires energy. One of the body's strategies for maintaining its temperature at about 37°C is to use the endothermic character of the vaporization of water, because the evaporation of perspiration requires heat and withdraws it from the skin.

The energy that must be supplied as heat at constant pressure per mole of molecules that are vaporized under standard conditions (that is, pure liquid at 1 bar changing to pure vapour at 1 bar) is called the **standard enthalpy of vaporization** of the liquid, and is denoted $\Delta_{vap}H^\ominus$ (Table 3.1). For example, 44 kJ of heat is required to vaporize 1 mol $H_2O(l)$ at 1 bar and 25°C, so $\Delta_{vap}H^\ominus = 44\ kJ\ mol^{-1}$. All enthalpies of vaporization are positive, so the sign is not normally given. Alternatively, we can report the same information by writing the **thermochemical equation**

$$H_2O(l) \rightarrow H_2O(g) \qquad \Delta H^\ominus = +44\ kJ$$

A thermochemical equation shows the standard enthalpy change (including the sign) that accompanies the conversion of an amount of reactant equal to its stoichiometric coefficient in the accompanying chemical equation (in this case, 1 mol H_2O). If the stoichiometric coefficients in the chemical equation are multiplied through by 2, then the thermochemical equation would be written

$$2\ H_2O(l) \rightarrow 2\ H_2O(g) \qquad \Delta H^\ominus = +88\ kJ$$

A note on good practice The attachment of the subscript vap to the Δ is the modern convention; however, the older convention in which the subscript is attached to the *H*, as in ΔH_{vap}, is still widely used.

This equation signifies that 88 kJ of heat is required to vaporize 2 mol $H_2O(l)$ at 1 bar and (recalling our convention) at 298.15 K. Unless otherwise stated, all data in this text are for 298.15 K.

Table 3.1

*Standard enthalpies of transition at the transition temperature**

Substance	Freezing point, T_f/K	$\Delta_{fus}H^{\ominus}$/(kJ mol^{-1})	Boiling point, T_b/K	$\Delta_{vap}H^{\ominus}$/(kJ mol^{-1})
Ammonia, NH_3	195.4	5.65	239.7	23.4
Argon, Ar	83.8	1.2	87.3	6.5
Benzene, C_6H_6	278.6	10.59	353.2	30.8
Ethanol, C_2H_5OH	158.7	4.60	351.5	43.5
Helium, He	3.5	0.02	4.22	0.08
Mercury, Hg	234.3	2.292	629.7	59.30
Methane, CH_4	90.7	0.94	111.7	8.2
Methanol, CH_3OH	175.2	3.16	337.2	35.3
Propanone, CH_3COCH_3	177.8	5.72	329.4	29.1
Water, H_2O	273.15	6.01	373.2	40.7

* For values at 298.15 K, use the information in the Data section. The text's website also contains links to online databases of thermochemical data.

Example 3.1

Determining the enthalpy of vaporization of a liquid

Ethanol, C_2H_5OH, is brought to the boil at 1 atm. When an electric current of 0.682 A from a 12.0 V supply is passed for 500 s through a heating coil immersed in the boiling liquid, it is found that the temperature remains constant but 4.33 g of ethanol is vaporized. What is the enthalpy of vaporization of ethanol at its boiling point at 1 atm?

Strategy Because the heat is supplied at constant pressure, we can identify the heat supplied, q, with the change in enthalpy of the ethanol when it vaporizes. We need to calculate the heat supplied and the amount of ethanol molecules vaporized. Then the enthalpy of vaporization is the heat supplied divided by the amount. The heat supplied is given by eqn 2.5 ($q = I\mathcal{V}t$; recall that 1 A V s = 1 J). The amount of ethanol molecules is determined by dividing the mass of ethanol vaporized by its molar mass ($n = m/M$).

Solution The energy supplied by heating is

$$q = I\mathcal{V}t = (0.682\ \text{A}) \times (12.0\ \text{V}) \times (500\ \text{s})$$
$$= 0.682 \times 12.0 \times 500\ \text{J}$$

This value is the change in enthalpy of the sample. The amount of ethanol molecules (of molar mass 46.07 g mol^{-1}) vaporized is

$$n = \frac{m}{M} = \frac{4.33\ \text{g}}{46.07\ \text{g mol}^{-1}} = \frac{4.33}{46.07}\ \text{mol}$$

The molar enthalpy change is therefore

$$\Delta_{vap}H = \frac{0.682 \times 12.0 \times 500\ \text{J}}{(4.33/46.07)\ \text{mol}} = 4.35 \times 10^4\ \text{J mol}^{-1}$$

corresponding to 43.5 kJ mol^{-1}.

Because the pressure is 1 atm, not 1 bar, the enthalpy of vaporization calculated here is not the standard value. However, 1 atm differs only slightly from 1 bar, so we can expect the standard enthalpy of vaporization to be very close to the value found here. Note also that the value calculated here is for the boiling point of ethanol, which is 78°C (351 K): we convey this information by writing $\Delta_{vap}H^{\ominus}(351\ \text{K}) = 43.5$ kJ mol^{-1}.

A note on good practice Molar quantities are expressed as a quantity per mole (as in kilojoules per mole, kJ mol^{-1}). Distinguish them from the magnitude of a property *for* 1 mol of substance, which is expressed as the quantity itself (as in kilojoules, kJ). All enthalpies of transition, denoted $\Delta_{trs}H$, are molar quantities.

Self-test 3.1

In a similar experiment, it was found that 1.36 g of boiling benzene, C_6H_6, is vaporized when a current of 0.835 A from a 12.0 V source is passed for 53.5 s. What is the enthalpy of vaporization of benzene at its boiling point?

[*Answer:* 30.8 kJ mol^{-1}]

There are some striking differences in standard enthalpies of vaporization: although the value for water is 44 kJ mol^{-1}, that for methane, CH_4, at its boiling point is only 8 kJ mol^{-1}. Even allowing for the fact that vaporization is taking place at different temperatures, the difference between the enthalpies of vaporization signifies that water molecules are held together in the bulk liquid much more tightly than methane molecules are in liquid methane. We shall see in Chapter 15 that the interaction responsible for the low volatility of water is the hydrogen bond.

The high enthalpy of vaporization of water has profound ecological consequences, for it is partly responsible for the survival of the oceans and the generally low humidity of the atmosphere. If only a small amount of heat had to be supplied to vaporize the oceans, the atmosphere would be much more heavily saturated with water vapour than is in fact the case.

Another common phase transition is **fusion**, or melting, as when ice melts to water or iron becomes molten. The change in molar enthalpy that accompanies fusion under standard conditions (pure solid at 1 bar changing to pure liquid at 1 bar) is called the **standard enthalpy of fusion**, $\Delta_{fus}H^{\ominus}$. Its value for water at 0°C is 6.01 kJ mol^{-1} (all enthalpies of fusion are positive, and the sign need not be given), which signifies that 6.01 kJ of energy is needed to melt 1 mol $H_2O(s)$ at 0°C and 1 bar. Notice that the enthalpy of fusion of water is much less than its enthalpy of vaporization. In vaporization the molecules become completely separated from each other, whereas in melting the molecules are merely loosened without separating completely (Fig. 3.1).

The reverse of vaporization is **condensation** and the reverse of fusion (melting) is **freezing**. The molar enthalpy changes are, respectively, the negative of the enthalpies of vaporization and fusion, because the heat that is supplied to vaporize or melt the substance is released when it condenses or freezes. It is always the case that *the enthalpy change of a reverse transition is the negative of the enthalpy change of the forward transition* (under the same conditions of temperature and pressure):

$$H_2O(s) \rightarrow H_2O(l) \qquad \Delta H^{\ominus} = +6.01\ \text{kJ}$$
$$H_2O(l) \rightarrow H_2O(s) \qquad \Delta H^{\ominus} = -6.01\ \text{kJ}$$

and in general

$$\Delta_{forward}H^{\ominus} = -\Delta_{reverse}H^{\ominus} \tag{3.1}$$

This relation follows directly from the fact that H is a state property, for H must return to the same value if a forward change is followed by the reverse of that change (Fig. 3.2). The high standard enthalpy of vaporization of water (44 kJ mol^{-1}), signifying a strongly endothermic process, implies that the condensation of water (−44 kJ mol^{-1}) is a strongly exothermic process. That exothermicity is the origin of the ability of steam to scald severely, because the energy is passed on to the skin.

The direct conversion of a solid to a vapour is called **sublimation**. The reverse process is called **vapour deposition**. Sublimation can be observed on a cold, frosty morning, when frost vanishes as vapour without first melting. The frost itself forms by vapour deposition from cold, damp air. The vaporization of solid carbon dioxide ('dry ice') is another example of sublimation. The standard molar enthalpy change accompanying sublimation is called the **standard enthalpy of sublimation**, $\Delta_{sub}H^{\ominus}$. Because enthalpy is a state property, the same change in enthalpy must be obtained both in the *direct* conversion of solid to vapour and in the *indirect* conversion, in which the solid first melts to the liquid and then that liquid vaporizes (Fig. 3.3):

$$\Delta_{sub}H^{\ominus} = \Delta_{fus}H^{\ominus} + \Delta_{vap}H^{\ominus} \tag{3.2}$$

The two enthalpies that are added together must be for the same temperature, so to get the enthalpy of sublimation of water at 0°C we must add together

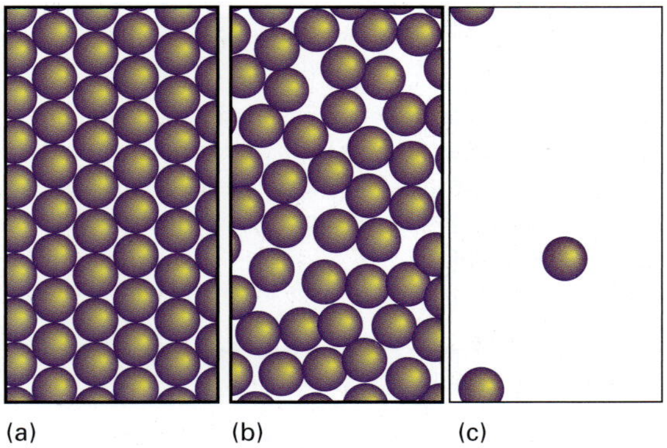

Fig. 3.1 When a solid (a) melts to a liquid (b), the molecules separate from one another only slightly, the intermolecular interactions are reduced only slightly, and there is only a small change in enthalpy. When a liquid vaporizes (c), the molecules are separated by a considerable distance, the intermolecular forces are reduced almost to zero, and the change in enthalpy is much greater. The text's website contains links to animations illustrating this point.

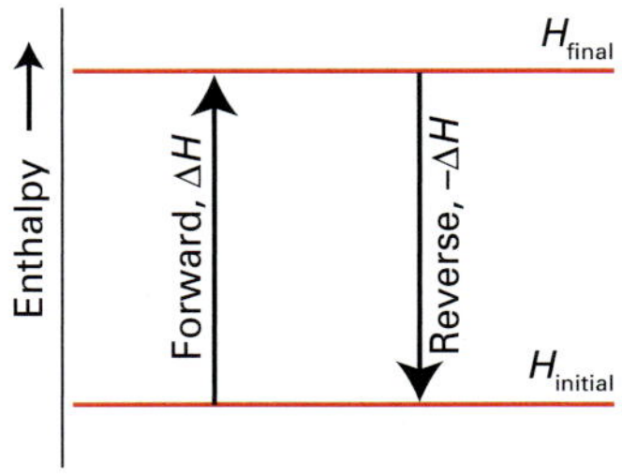

Fig. 3.2 An implication of the First Law is that the enthalpy change accompanying a reverse process is the negative of the enthalpy change for the forward process.

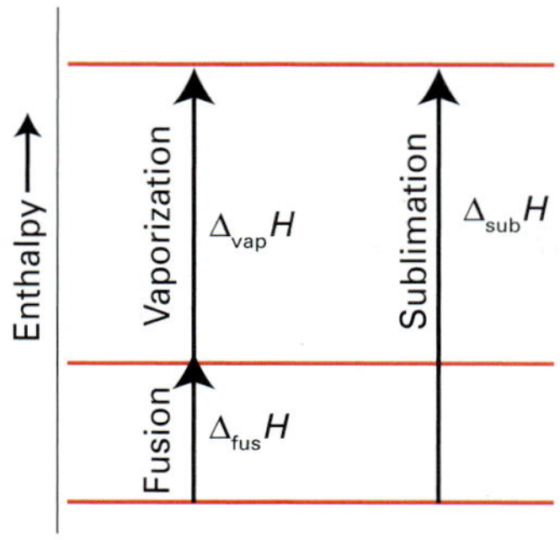

Fig. 3.3 The enthalpy of sublimation at a given temperature is the sum of the enthalpies of fusion and vaporization at that temperature. Another implication of the First Law is that the enthalpy change of an overall process is the sum of the enthalpy changes for the possibly hypothetical steps into which it may be divided.

the enthalpies of fusion and vaporization for this temperature. Adding together enthalpies of transition for different temperatures gives a meaningless result.

Self-test 3.2

Calculate the standard enthalpy of sublimation of ice at 0°C from its standard enthalpy of fusion at 0°C (6.01 kJ mol^{-1}) and the standard enthalpy of vaporization of water at 0°C (45.07 kJ mol^{-1}).

[*Answer:* 51.08 kJ mol^{-1}]

The result expressed by eqn 3.2 is an example of a more general statement that will prove useful time and again during our study of thermochemistry:

The enthalpy change of an overall process is the sum of the enthalpy changes for the steps (observed or hypothetical) into which it may be divided.

3.2 Atomic and molecular change

A group of enthalpy changes we employ quite often in the following pages are those accompanying changes to individual atoms and molecules. Among the most important is the **standard enthalpy of ionization**, $\Delta_{ion}H^{\ominus}$, the standard molar enthalpy change accompanying the removal of an electron from a gas-phase atom (or ion). For example, because

$$H(g) \rightarrow H^+(g) + e^-(g) \qquad \Delta H^{\ominus} = +1312\ kJ$$

the standard enthalpy of ionization of hydrogen atoms is reported as 1312 kJ mol^{-1}. This value signifies that 1312 kJ of energy must be supplied as heat to ionize 1 mol H(g) at 1 bar (and 298.15 K). Table 3.2 gives values of the ionization enthalpies for a number of elements; note that all enthalpies of ionization are positive.

A brief comment The enthalpy of ionization is closely related to—but is not identical to—the ionization energy. Briefly, the ionization energy corresponds to the standard ionization enthalpy at $T = 0$. The ionization enthalpy at a different temperature can be calculated from the ionization energy by using the concepts developed in Section 3.7.

We often need to consider a succession of ionizations, such as the conversion of magnesium atoms to Mg^+ ions, the ionization of these Mg^+ ions to Mg^{2+} ions, and so on. The successive molar enthalpy changes are called, respectively, the **first ionization enthalpy**, the **second ionization enthalpy**, and so on. For magnesium, these enthalpies refer to the processes

$$Mg(g) \rightarrow Mg^+(g) + e^-(g) \qquad \Delta H^{\ominus} = +738\ kJ$$
$$Mg^+(g) \rightarrow Mg^{2+}(g) + e^-(g) \qquad \Delta H^{\ominus} = +1451\ kJ$$

Note that the second ionization enthalpy is larger than the first: more energy is needed to separate an electron from a positively charged ion than from the neutral atom. Note also that enthalpies of ionization refer to the ionization of the gas phase atom or ion, not to the ionization of an atom or ion in a solid. To determine the latter, we need to combine two or more enthalpy changes.

Example 3.2

Combining enthalpy changes

The standard enthalpy of sublimation of magnesium at 25°C is 148 kJ mol^{-1}. How much energy as heat (at constant temperature and pressure) must be supplied to 1.00 g of solid magnesium metal to produce a gas composed of Mg^{2+} ions and electrons?

Strategy The enthalpy change for the overall process is a sum of the steps, sublimation followed by the two stages of ionization, into which it can be divided. Then, the heat required for the specified process is the product of the overall molar enthalpy change and the amount of atoms; the latter is calculated from the given mass and the molar mass of the substance.

Solution The overall process is

$$Mg(s) \rightarrow Mg^{2+}(g) + 2\ e^-(g)$$

The thermochemical equation for this process is the sum of the following thermochemical equations:

		$\Delta H^{\ominus}$/kJ
Sublimation:	$Mg(s) \rightarrow Mg(g)$	+148
First ionization:	$Mg(g) \rightarrow Mg^+(g) + e^-(g)$	+738
Second ionization:	$Mg^+(g) \rightarrow Mg^{2+}(g) + e^-(g)$	+1451
Overall (sum):	$Mg(s) \rightarrow Mg^{2+}(g) + 2\ e^-(g)$	+2337

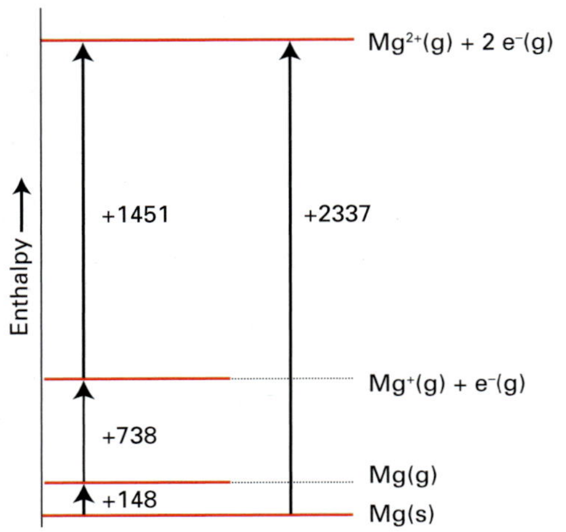

Fig. 3.4 The contributions to the enthalpy change treated in Example 3.2.

These processes are illustrated diagrammatically in Fig. 3.4. It follows that the overall enthalpy change per mole of Mg is 2337 kJ mol^{-1}. Because the molar mass of magnesium is 24.31 g mol^{-1}, 1.0 g of magnesium corresponds to

$$n_{Mg} = \frac{m_{Mg}}{M_{Mg}} = \frac{1.00\ \text{g}}{24.31\ \text{g mol}^{-1}} = \frac{1.00}{24.31}\ \text{mol}$$

Therefore, the energy that must be supplied as heat (at constant pressure) to ionize 1.00 g of magnesium metal is

$q_p = \Delta H$ and $\Delta H = n\Delta H_m$; ΔH_m

$$q_p = \left(\frac{1.00}{24.31}\ \text{mol}\right) \times (2337\ \text{kJ mol}^{-1}) = 96.1\ \text{kJ}$$

n_{Mg}

This quantity of energy is approximately the same as that needed to vaporize about 43 g of boiling water.

Self-test 3.3

The enthalpy of sublimation of aluminium is 326 kJ mol^{-1}. Use this information and the ionization enthalpies in Table 3.2 to calculate the energy that must be supplied as heat (at constant pressure) to convert 1.00 g of solid aluminium metal to a gas of Al^{3+} ions and electrons at 25°C.

[*Answer:* +202 kJ]

The reverse of ionization is **electron gain**, and the corresponding molar enthalpy change under standard conditions is called the **standard electron gain enthalpy, $\Delta_{eg}H^{\ominus}$**. This quantity is closely related to the electron affinity, as we shall see in Section 13.16. For example, because experiments show that

$$Cl(g) + e^-(g) \rightarrow Cl^-(g) \qquad \Delta H^{\ominus} = -349\ \text{kJ}$$

it follows that the electron gain enthalpy of Cl atoms is −349 kJ mol^{-1}. Notice that electron gain by Cl is an exothermic process, so heat is released when a Cl atom captures an electron and forms an ion. It can be seen from Table 3.3, which lists a number of electron gain enthalpies, that some electron gains are exothermic and others are endothermic, so we need to include their sign. For example, electron gain by an O^- ion is strongly endothermic because it takes energy to push an electron on to an already negatively charged species:

$$O^-(g) + e^-(g) \rightarrow O^{2-}(g) \qquad \Delta H^{\ominus} = +844\ \text{kJ}$$

The final atomic and molecular process to consider at this stage is the **dissociation**, or breaking, of a chemical bond, as in the process

$$HCl(g) \rightarrow H(g) + Cl(g) \qquad \Delta H^{\ominus} = +431\ \text{kJ}$$

The corresponding standard molar enthalpy change is called the **bond enthalpy**, so we would report the H—Cl bond enthalpy as 431 kJ mol^{-1}. All bond enthalpies are positive.

Some bond enthalpies are given in Table 3.4. Note that the nitrogen–nitrogen bond in molecular nitrogen, N_2, is very strong, at 945 kJ mol^{-1}, which helps to account for the chemical inertness of nitrogen and its ability to dilute the oxygen in the atmosphere without reacting with it. In contrast, the fluorine–fluorine bond in molecular fluorine, F_2, is relatively weak, at 155 kJ mol^{-1}; the weakness of this bond contributes to the high reactivity of elemental fluorine. However, bond enthalpies alone do not account for reactivity because, although the bond in molecular iodine is even weaker, I_2 is less reactive than F_2, and the bond in CO is stronger than the bond in N_2, but CO forms many carbonyl compounds, such as $Ni(CO)_4$. The types and strengths of the bonds that the elements can make to other elements are additional factors.

A complication when dealing with bond enthalpies is that their values depend on the molecule in which the two linked atoms occur. For instance, the total standard enthalpy change for the atomization (the complete dissociation into atoms) of water

$$H_2O(g) \rightarrow 2\,H(g) + O(g) \qquad \Delta H^{\ominus} = +927\ \text{kJ}$$

is not twice the O—H bond enthalpy in H_2O even though two O—H bonds are dissociated. There are in fact two different dissociation steps. In the first step, an O—H bond is broken in an H_2O molecule:

$$H_2O(g) \rightarrow HO(g) + H(g) \qquad \Delta H^{\ominus} = +499\ \text{kJ}$$

Table 3.2

First and second (and some higher) standard enthalpies of ionization, $\Delta_{ion}H/(kJ\ mol^{-1})$

1	2	13	15	15	16	17	18
H 1312							**He** 2370 5250
Li 519 7300	**Be** 900 1760	**B** 799 2420 14 800	**C** 1090 2350 3660 25 000	**N** 1400 2860	**O** 1310 3390	**F** 1680 3370	**Ne** 2080 3950
Na 494 4560	**Mg** 738 1451 7740	**Al** 577 1820 2740 11 600	**Si** 786	**P** 1060	**S** 1000	**Cl** 1260	**Ar** 1520
K 418 3070	**Ca** 590 1150 4940	**Ga** 577	**Ge** 762	**As** 966	**Se** 941	**Br** 1140	**Kr** 1350
Rb 402 2650	**Sr** 548 1060 4120	**In** 556	**Sn** 707	**Sb** 833	**Te** 870	**I** 1010	**Xe** 1170
Cs 376 2420 3300	**Ba** 502 966 3390	**Tl** 812	**Pb** 920	**Bi** 1040	**Po** 812	**At** 920	**Rn** 1040

Strictly, these are the values of $\Delta_{ion}U(0)$. For more precise work, use $\Delta_{ion}H(T) = \Delta_{ion}U(0) + \frac{5}{2}RT$, with $\frac{5}{2}RT = 6.20\ J\ mol^{-1}$ at 298 K.

Table 3.3

*Electron gain enthalpies of the main-group elements, $\Delta_{eg}H/(kJ\ mol^{-1})$**

1	2	13	14	15	16	17	18
H −73							**He** +21
Li −60	**Be** +18	**B** −27	**C** −122	**N** +7	**O** −141 +844	**F** −328	**Ne** +29
Na −53	**Mg** +232	**Al** −43	**Si** −134	**P** −44	**S** −200 +532	**Cl** −349	**Ar** +35
K −48	**Ca** +186	**Ga** −29	**Ge** −116	**As** −78	**Se** −195	**Br** −325	**Kr** +39
Rb −47	**Sr** +146	**In** −29	**Sn** −116	**Sb** −103	**Te** −190	**I** −295	**Xe** +41
Cs −46	**Ba** +46	**Tl** −19	**Pb** −35	**Bi** −91	**Po** −183	**At** −270	**Ra**

* Where two values are given, the first refers to the formation of the ion X^- from the neutral atom X; the second, to the formation of X^{2-} from X^-.

Table 3.4

Selected bond enthalpies, ΔH(A–B)/(kJ mol^{-1})

Diatomic molecules							
H—H	436	O=O	497	F—F	155	H—F	565
		N≡N	945	Cl—Cl	242	H—Cl	431
		O—H	428	Br—Br	193	H—Br	366
		C≡O	1074	I—I	151	H—I	299
Polyatomic molecules							
H—CH_3	435	H—NH_2	460	H—OH	492		
H—C_6H_5	469	O_2N—NO_2	54	HO—OH	213		
H_3C—CH_3	368	O=CO	531	HO—CH_3	377		
H_2C=CH_2	720			Cl—CH_3	352		
HC≡CH	962			Br—CH_3	293		
				I—CH_3	237		

Table 3.5

Mean bond enthalpies, ΔH_B/(kJ mol^{-1})

	H	C	N	O	F	Cl	Br	I	S	P	Si
H	436										
C	412	348(1) 612(2) 838(3) 518 (a)									
N	388	305(1) 613 (2) 890 (3)	163 (1) 409 (2) 945 (3)								
O	463	360 (1) 743 (2)	157	146(1) 497(2)							
F	565	484	270	185	155						
Cl	431	338	200	203	254	242					
Br	366	276				219	193				
I	299	238				210	178	151			
S	338	259			496	250	212		264		
P	322									200	
Si	318		374	466							226

Values are for single bonds except where otherwise stated (in parentheses).
(a) Denotes aromatic.

In the second step, the O—H bond is broken in an OH radical:

$$\mathrm{HO(g)} \rightarrow \mathrm{H(g)} + \mathrm{O(g)} \qquad \Delta H^{\ominus} = +428\ \mathrm{kJ}$$

The sum of the two steps is the atomization of the molecule. As can be seen from this example, the O—H bonds in H_2O and HO have similar but not identical bond enthalpies.

Although accurate calculations must use bond enthalpies for the molecule in question and its successive fragments, when such data are not available there is no choice but to make estimates by using **mean bond enthalpies**, ΔH_B, which are the averages of bond enthalpies over a related series of compounds (Table 3.5). For example, the mean HO bond enthalpy, $\Delta H_B(\mathrm{H—O}) = 463\ \mathrm{kJ\ mol^{-1}}$, is the mean of the O—H bond enthalpies in H_2O and several other similar compounds, including methanol, CH_3OH.

Example 3.3

Using mean bond enthalpies

Estimate the standard enthalpy change for the reaction

$$C(s, graphite) + 2\,H_2(g) + \tfrac{1}{2}\,O_2(g) \rightarrow CH_3OH(l)$$

in which liquid methanol is formed from its elements at 25°C. Use information from the Data section and bond enthalpy data from Tables 3.4 and 3.5.

Strategy In calculations of this kind, the procedure is to break the overall process down into a sequence of steps such that their sum is the chemical equation required. Always ensure, when using bond enthalpies, that all the species are in the gas phase. That may mean including the appropriate enthalpies of vaporization or sublimation. One approach is to atomize all the reactants and then to build the products from the atoms so produced. When explicit bond enthalpies are available (that is, data are given in the tables available), use them; otherwise, use mean bond enthalpies to obtain estimates. It is often helpful to display the enthalpy changes diagrammatically.

Solution The following steps are required (Fig. 3.5):

		$\Delta H^{\ominus}$/ kJ
Atomization of graphite:	C(s, graphite) → C(g)	+716.68
Dissociation of 2 mol H_2(g):	$2\,H_2(g) \rightarrow 4\,H(g)$	+871.88
Dissociation of $\tfrac{1}{2}\,O_2$(g):	$\tfrac{1}{2}\,O_2(g) \rightarrow O(g)$	+249.17
Overall, so far:	$C(s) + 2\,H_2(g) + \tfrac{1}{2}\,O_2(g) \rightarrow C(g) + 4\,H(g) + O(g)$	+1837.73

These values are accurate. In the second step, three CH bonds, one CO bond, and one OH bond are formed, and we estimate their enthalpies from mean values. The standard enthalpy change for bond formation (the reverse of dissociation) is the negative of the mean bond enthalpy (obtained from Table 3.5):

	$\Delta H^{\ominus}$/kJ
Formation of 3 C—H bonds:	−1236
Formation of 1 C—O bond:	−360
Formation of 1 O—H bond:	−463
Overall, in this step:	−2059

$$C(g) + 4\,H(g) + O(g) \rightarrow CH_3OH(g)$$

These values are estimates. The final stage of the reaction is the condensation of methanol vapour:

$$CH_3OH(g) \rightarrow CH_3OH(l) \qquad \Delta H^{\ominus} = -38.00\text{ kJ}$$

The sum of the enthalpy changes is

$$\Delta H^{\ominus} = (+1837.73\text{ kJ}) + (-2059\text{ kJ}) + (-38.00\text{ kJ}) = -259\text{ kJ}$$

The experimental value is −239.00 kJ.

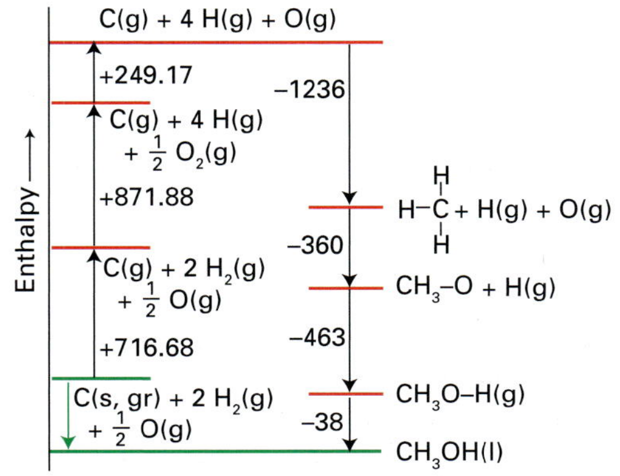

Fig. 3.5 The enthalpy changes used to estimate the enthalpy change accompanying the formation of liquid methanol from its elements. The bond enthalpies are mean values, so the final value is only approximate.

Self-test 3.4

Estimate the enthalpy change for the combustion of liquid ethanol to carbon dioxide and liquid water under standard conditions by using the bond enthalpies, mean bond enthalpies, and the appropriate standard enthalpies of vaporization.

[*Answer:* −1348 kJ; the experimental value is −1368 kJ]

Chemical change

In the remainder of this chapter we concentrate on enthalpy changes accompanying chemical reactions, such as the hydrogenation of ethene:

$$CH_2{=}CH_2(g) + H_2(g) \rightarrow CH_3CH_3(g) \qquad \Delta H^{\ominus} = -137\text{ kJ}$$

The value of $\Delta H^{\ominus}$ given here signifies that the enthalpy of the system decreases by 137 kJ (and, if the reaction takes place at constant pressure, that 137 kJ of energy is released by heating the surroundings) when 1 mol $CH_2{=}CH_2$(g) at 1 bar combines with 1 mol H_2(g) at 1 bar to give 1 mol CH_3CH_3(g) at 1 bar, all at 25°C.

3.3 Enthalpies of combustion

One commonly encountered reaction is **combustion**, the complete reaction of a compound, most commonly an organic compound, with oxygen, as in the combustion of methane in a natural gas flame:

$$CH_4(g) + 2\,O_2(g) \rightarrow CO_2(g) + 2\,H_2O(l) \qquad \Delta H^{\ominus} = -890\text{ kJ}$$

Table 3.6
Standard enthalpies of combustion

Substance	$\Delta_c H^{\ominus}/(\text{kJ mol}^{-1})$
Benzene, C_6H_6(l)	−3268
Carbon monoxide, CO(g)	−394
Carbon, C(s,graphite)	−394
Ethanol, C_2H_5OH(l)	−1368
Ethyne, C_2H_2(g)	−1300
Glucose, $C_6H_{12}O_6$(s)	−2808
Hydrogen, H_2(g)	−286
iso-Octane, *C_8H_{18}(l)	−5461
Methane, CH_4(g)	−890
Methanol, CH_3OH(l)	−726
Methylbenzene, $C_6H_5CH_3$(l)	−3910
Octane, C_8H_{18}(l)	−5471
Propane, C_3H_8(g)	−2220
Sucrose, $C_{12}H_{22}O_{11}$(s)	−5645
Urea, $CO(NH_2)_2$(s)	−632

* 2,2,4-Trimethylpentane.

By convention, combustion of an organic compound results in the formation of carbon dioxide gas, liquid water, and—if the compound contains nitrogen—nitrogen gas. The **standard enthalpy of combustion**, $\Delta_c H^{\ominus}$, is the change in standard enthalpy per mole of combustible substance. In this example, we would write $\Delta_c H^{\ominus}(CH_4, g) = -890 \text{ kJ mol}^{-1}$. Some typical values are given in Table 3.6. Note that $\Delta_c H^{\ominus}$ is a molar quantity, and is obtained from the value of $\Delta H^{\ominus}$ by dividing by the amount of organic reactant consumed (in this case, by 1 mol CH_4). We see in Box 3.1 that the enthalpy of combustion is a useful measure of the efficiency of fuels.

Enthalpies of combustion are commonly measured by using a bomb calorimeter, a device in which energy is transferred as heat at constant volume. According to the discussion in Section 2.7 and the relation $\Delta U = q_V$, the energy transferred as heat at constant volume is equal to the change in internal energy, ΔU, not ΔH. To convert from ΔU, to ΔH we need to note that the molar enthalpy of a substance is related to its molar internal energy by $H_m = U_m + pV_m$ (eqn 2.13a). For condensed phases, pV_m is so small it may be ignored. For example, the molar volume of liquid water is 18 $\text{cm}^3 \text{ mol}^{-1} = 18 \times 10^{-6} \text{ m}^3 \text{ mol}^{-1}$, and at 1.0 bar

$$\begin{aligned} pV_m &= (1.0 \times 10^5 \text{ Pa}) \times (18 \times 10^{-6} \text{ m}^3 \text{ mol}^{-1}) \\ &= 1.8 \text{ Pa m}^3 \text{ mol}^{-1} = 1.8 \text{ J mol}^{-1} \\ &= 1.8 \times 10^{-3} \text{ kJ mol}^{-1} \end{aligned}$$

However, the molar volume of a gas, and therefore the value of pV_m, is about 1000 times greater and cannot be ignored. For gases treated as perfect, pV_m may be replaced by RT. Therefore, if in the chemical equation the difference (products − reactants) in the stoichiometric coefficients of *gas phase* species is $\Delta\nu_{gas}$, we can write

$$\Delta_c H = \Delta_c U + \Delta\nu_{gas} RT \qquad (3.3)$$

Note that $\Delta\nu_{gas}$ (where ν is nu) is a dimensionless quantity.

A brief illustration The energy released as heat when glycine is burned in a bomb calorimeter is 969.6 kJ mol^{-1} at 298.15 K, so $\Delta_c U = -969.6 \text{ kJ mol}^{-1}$. From the chemical equation

$$NH_2CH_2COOH(s) + \tfrac{9}{4} O_2(g) \rightarrow 2\,CO_2(g) + \tfrac{5}{2} H_2O(l) + \tfrac{1}{2} N_2(g)$$

we find that $\Delta\nu_{gas} = (2 + \tfrac{1}{2}) - \tfrac{9}{4} = \tfrac{1}{4}$. Therefore, assuming that the difference between $\Delta_c U$ and $\Delta_c H$ is due to contributions from reactants and products in the gas phase, we use eqn 3.3 to obtain

$$\begin{aligned} \Delta_c H &= \Delta_c U + \tfrac{1}{4} RT \\ &= -969.6 \text{ kJ mol}^{-1} + \tfrac{1}{4} \times (8.3145 \times 10^{-3} \text{ kJ K}^{-1} \text{ mol}^{-1}) \\ &\quad \times (298.15 \text{ K}) \\ &= -969.6 \text{ kJ mol}^{-1} + 0.62 \text{ kJ mol}^{-1} \\ &= -969.0 \text{ kJ mol}^{-1} \end{aligned}$$

3.4 The combination of reaction enthalpies

The **reaction enthalpy** (or the 'enthalpy of reaction'), $\Delta_r H$, is the change in enthalpy that accompanies a chemical reaction: the enthalpy of combustion is just a special case. The reaction enthalpy is the difference between the molar enthalpies of the reactants and the products, with each term weighted by the stoichiometric coefficient, ν (nu), in the chemical equation:

$$\Delta_r H = \sum \nu H_m(\text{products}) - \sum \nu H_m(\text{reactants}) \qquad (3.4a)$$

The **standard reaction enthalpy** (the 'standard enthalpy of reaction'), $\Delta_r H^{\ominus}$, is the value of the reaction enthalpy when all the reactants and products are in their standard states:

$$\Delta_r H^{\ominus} = \sum \nu H_m^{\ominus}(\text{products}) - \sum \nu H_m^{\ominus}(\text{reactants}) \qquad (3.4b)$$

Because the $H_m^{\ominus}$ are molar quantities and the stoichiometric coefficients are pure numbers, the units of $\Delta_r H^{\ominus}$ are kilojoules per mole. The standard reaction enthalpy is the change in enthalpy of the system when the reactants in their standard states (pure, 1 bar) are

Box 3.1 Fuels, food and energy reserves

We shall see in Chapter 4 that the best assessment of the ability of a compound to act as a fuel to drive many of the processes occurring in the body makes use of the 'Gibbs energy'. However, a useful guide to the resources provided by a fuel, and the only one that matters when its heat output is being considered, is the enthalpy, particularly the enthalpy of combustion. The thermochemical properties of fuels and foods are commonly discussed in terms of their *specific enthalpy*, the enthalpy of combustion divided by the mass of material (typically in kilojoules per gram), or the *enthalpy density*, the magnitude of the enthalpy of combustion divided by the volume of material (typically in kilojoules per cubic decimetre). Thus, if the standard enthalpy of combustion is $\Delta_c H^{\ominus}$ and the molar mass of the compound is M, then the specific enthalpy is $\Delta_c H^{\ominus}/M$. Similarly, the enthalpy density is $\Delta_c H^{\ominus}/V_m$, where V_m is the molar volume of the material.

The table lists the specific enthalpies and enthalpy densities of several fuels. The most suitable fuels are those with high specific enthalpies, as the advantage of a high molar enthalpy of combustion may be eliminated if a large mass of fuel is to be transported. We see that H_2 gas compares very well with more traditional fuels such as methane (natural gas), iso-octane (gasoline), and methanol. Furthermore, the combustion of H_2 gas does not generate CO_2 gas, a pollutant implicated in the mechanism of global warming. As a result, H_2 gas has been proposed as an efficient, clean alternative to fossil fuels, such as natural gas and petroleum. However, we also see that H_2 gas has a very low enthalpy density, which arises from the fact that hydrogen is a very light gas. So, the advantage of a high specific enthalpy is undermined by the large volume of fuel to be transported and stored. Strategies are being developed to solve the storage problem. For example, the small H_2 molecules can travel through holes in the crystalline lattice of a sample of metal, such as titanium, where they bind as metal hydrides. In this way it is possible to increase the effective density of hydrogen atoms to a value that is higher than that of liquid H_2. Then the fuel can be released on demand by heating the metal.

We now assess the factors that optimize the heat output of carbon-based fuels, with an eye toward understanding such biological fuels as carbohydrates, fats, and proteins. Let's consider the combustion of 1 mol $CH_4(g)$, the main constituent of natural gas. The reaction involves changes in the oxidation numbers of carbon from −4 to +4, an oxidation, and of oxygen from 0 to −2, a reduction (see Appendix 4 for a review of oxidation numbers and oxidation–reduction reactions). From the thermochemical equation, we see that 890 kJ of energy is released as heat per mole of carbon that is oxidized. Now consider the oxidation of 1 mol $CH_3OH(g)$:

$$CH_3OH(g) + \tfrac{3}{2}O_2(g) \rightarrow CO_2(g) + 2\,H_2O(l) \quad \Delta H^{\ominus} = -764\ \text{kJ}$$

This reaction is also exothermic, but now only 764 kJ of energy is released as heat per mole of carbon that undergoes oxidation. Much of the observed change in energy output between the reactions can be explained by noting that the carbon in methanol has an oxidation number of −2, and not −4 as in methane. That is, the replacement of a C—H bond by a C—O bond renders the carbon in methanol more oxidized than the carbon in methane, so it is reasonable to expect that less energy is released to complete the oxidation of carbon to CO_2 in methanol. In general, we find that the presence of partially oxidized carbon atoms (that is, carbon atoms bonded to oxygen atoms) in a material makes it a less suitable fuel than a similar material containing more highly reduced carbon atoms.

Another factor that determines the heat output of combustion reactions is the number of carbon atoms in

Box 3.1

Thermochemical properties of some fuels

Fuel	Combustion equation	$\Delta_c H^{\ominus}$/ (kJ mol^{-1})	Specific enthalpy/ (kJ g^{-1})	Enthalpy density*/ (kJ dm^{-3})
Hydrogen	$2\,H_2(g) + O_2(g) \rightarrow 2\,H_2O(l)$	−286	142	13
Methane	$CH_4(g) + 2\,O_2(g) \rightarrow CO_2(g) + 2\,H_2O(l)$	−890	55	40
iso-Octane	$2\,C_8H_{18}(l) + 25\,O_2(g) \rightarrow 16\,CO_2(g) + 18\,H_2O(l)$	−5471	48	3.8×10^4
Methanol	$2\,CH_3OH(l) + 3\,O_2(g) \rightarrow 2\,CO_2(g) + 4\,H_2O(l)$	−726	23	1.8×10^4

* At atmospheric pressures and room temperature.
† 2,2,4-Trimethylpentane.

hydrocarbon compounds. For example, from the value of the standard enthalpy of combustion for methane we know that for each mole of CH_4 supplied to a furnace, 890 kJ of heat can be released, whereas for each mole of iso-octane molecules (C_8H_{18}, 2,2,4-trimethylpentane, a typical component of gasoline) supplied to an internal combustion engine, 5461 kJ of heat is released (see the table). The much larger value for iso-octane is a consequence of each molecule having eight C atoms to contribute to the formation of carbon dioxide, whereas methane has only one.

Now we turn our attention to biological fuels, the foods we ingest to meet the energy requirements of daily life. A typical 18–20 year old man requires a daily input of about 12 MJ (1 MJ = 10^6 J); a woman of the same age needs about 9 MJ. If the entire consumption were in the form of glucose, which has a specific enthalpy of 16 kJ g^{-1}, meeting energy needs would require the consumption of 750 g of glucose by a man and 560 g by a woman. In fact, the complex carbohydrates (polymers of carbohydrate units, such as starch) more commonly found in our diets have slightly higher specific enthalpies (17 kJ g^{-1}) than glucose itself, so a carbohydrate diet is slightly less daunting than a pure glucose diet, as well as being more appropriate in the form of fibre, the indigestible cellulose that helps move digestion products through the intestine.

The specific enthalpy of fats, which are long-chain esters like tristearin (beef fat), is much greater than that of carbohydrates, at around 38 kJ g^{-1}, slightly less than the value for the hydrocarbon oils used as fuel (48 kJ g^{-1}). The reason for this lies in the fact that many of the carbon atoms in carbohydrates are bonded to oxygen atoms and are already partially oxidized, whereas most of the carbon atoms in fats are bonded to hydrogen and other carbon atoms and hence have lower oxidation numbers. As we saw above, the presence of partially oxidized carbons lowers the heat output of a fuel.

Fats are commonly used as an energy store, to be used only when the more readily accessible carbohydrates have fallen into short supply. In Arctic species, the stored fat also acts as a layer of insulation; in desert species (such as the camel), the fat is also a source of water, one of its oxidation products.

Proteins are also used as a source of energy, but their components, the amino acids, are often too valuable to squander in this way, and are used to construct other proteins instead. When proteins are oxidized (to urea, $CO(NH_2)_2$), the equivalent enthalpy density is comparable to that of carbohydrates.

We have already remarked that not all the energy released by the oxidation of foods is converted to work. The heat that is also released needs to be discarded in order to maintain body temperature within its typical range of 35.6–37.8°C. A variety of mechanisms contribute to this aspect of homeostasis, the ability of an organism to counteract environmental changes with physiological responses. The general uniformity of temperature throughout the body is maintained largely by the flow of blood. When heat needs to be dissipated rapidly, warm blood is allowed to flow through the capillaries of the skin, so producing flushing. Radiation is one means of discarding heat; another is evaporation and the energy demands of the enthalpy of vaporization of water. Evaporation removes about 2.4 kJ per gram of water perspired. When vigorous exercise promotes sweating (through the influence of heat selectors on the hypothalamus), 1–2 dm^3 of perspired water can be produced per hour, corresponding to a heat loss of 2.4–5.0 MJ h^{-1}.

completely converted into products in their standard states (pure, 1 bar), with the change expressed in kilojoules per mole of reaction as written. Thus, if for the reaction $2\,H_2(g) + O_2(g) \rightarrow 2\,H_2O(l)$ we report that $\Delta_r H^\ominus = -572$ kJ mol^{-1}, then the 'per mole' means that the reaction releases 572 kJ of heat per mole of O_2 consumed or per 2 mol H_2O formed (and therefore 286 kJ per mole of H_2O formed).

It is often the case that a reaction enthalpy is needed but is not available in tables of data. Now the fact that enthalpy is a state function comes in handy, because it implies that we can construct the required reaction enthalpy from the reaction enthalpies of known reactions. We have already seen a primitive example when we calculated the enthalpy of sublimation from the sum of the enthalpies of fusion and vaporization. The only difference is that we now apply the technique to a sequence of chemical reactions. The procedure is summarized by **Hess's law**:

> The standard enthalpy of a reaction is the sum of the standard enthalpies of the reactions into which the overall reaction may be divided.

Although the procedure is given the status of a law, it hardly deserves the title because it is nothing more than a consequence of enthalpy being a state function, which implies that an overall enthalpy change can be expressed as a sum of enthalpy changes for each step in an indirect path. The individual steps need not be actual reactions that can be carried out in the laboratory—they may be entirely hypothetical reactions, the only requirement being that their equations should balance. Each step must correspond to the same temperature.

Example 3.4

Using Hess's law

Given the thermochemical equations

$C_3H_6(g) + H_2(g) \rightarrow C_3H_8(g) \qquad \Delta H^{\ominus} = -124\ kJ$

$C_3H_8(g) + 5\ O_2(g) \rightarrow 3\ CO_2(g) + 4\ H_2O(l) \qquad \Delta H^{\ominus} = -2220\ kJ$

where C_3H_6 is propene and C_3H_8 is propane, calculate the standard enthalpy of combustion of propene.

Strategy We need to add or subtract the thermochemical equations, together with any others that are needed (from the Data section), so as to reproduce the thermochemical equation for the reaction required. In calculations of this type, it is often necessary to use the synthesis of water to balance the hydrogen or oxygen atoms in the overall equation. Once again, it may be helpful to express the changes diagrammatically.

Solution The overall reaction is

$$C_3H_6(g) + \tfrac{9}{2}\ O_2(g) \rightarrow 3\ CO_2(g) + 3\ H_2O(l) \qquad \Delta H^{\ominus}$$

We can recreate this thermochemical equation from the following sum (Fig. 3.6):

	$\Delta H^{\ominus}$/kJ
$C_3H_6(g) + H_2(g) \rightarrow C_3H_8(g)$	−124
$C_3H_8(g) + 5\ O_2(g) \rightarrow 3\ CO_2(g) + 4\ H_2O(l)$	−2220
$H_2O(l) \rightarrow H_2(g) + \tfrac{1}{2}\ O_2(g)$	+286
Overall: $C_3H_6(g) + \tfrac{9}{2}\ O_2(g) \rightarrow 3\ CO_2(g) + 3\ H_2O(l)$	−2058

It follows that the standard enthalpy of combustion of propene is −2058 kJ mol^{-1}.

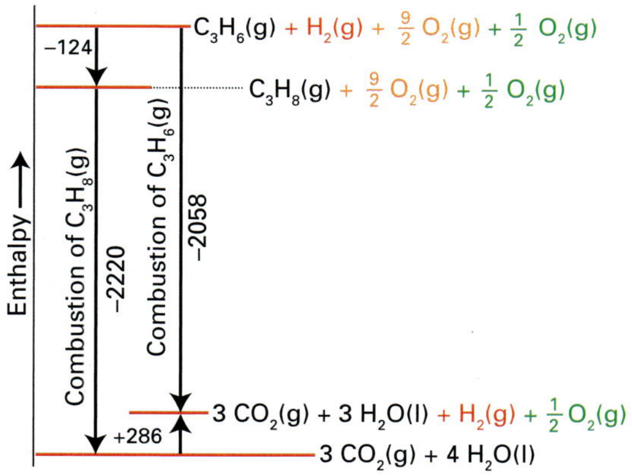

Fig. 3.6 The enthalpy changes used in Example 3.4 to illustrate Hess's law.

Self-test 3.5

Calculate the standard enthalpy of the reaction $C_6H_6(l) + 3\ H_2(g) \rightarrow C_6H_{12}(l)$ from the standard enthalpies of combustion of benzene (Table 3.6) and cyclohexane (−3930 kJ mol^{-1}).

[*Answer:* −196 kJ]

3.5 Standard enthalpies of formation

The problem with eqn 3.4 is that we have no way of knowing the absolute enthalpies of the substances. To avoid this problem, we can imagine the reaction as taking place by an indirect route, in which the reactants are first broken down into the elements and then the products are formed from the elements (Fig. 3.7). Specifically, the **standard enthalpy of formation**, $\Delta_f H^{\ominus}$, of a substance is the standard enthalpy (per mole of the substance) for its formation from its elements in their reference states. The **reference state** of an element is its most stable form under the prevailing conditions (Table 3.7). Don't confuse 'reference state' with 'standard state': the reference state of carbon at 25°C is graphite; the standard state of carbon is any specified phase of the element at 1 bar. For example, the standard enthalpy of formation

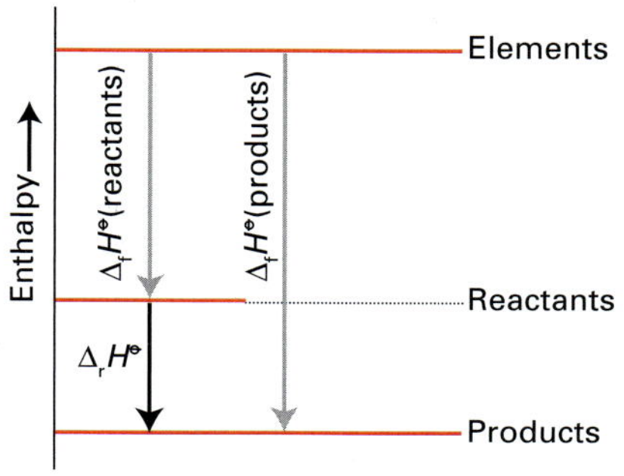

Fig. 3.7 An enthalpy of reaction may be expressed as the difference between the enthalpies of formation of the products and the reactants.

Table 3.7
Reference states of some elements

Element	Reference state
Arsenic	grey arsenic
Bromine	liquid
Carbon	graphite
Hydrogen	gas
Iodine	solid
Mercury	liquid
Nitrogen	gas
Oxygen	gas
Phosphorus	white phosphorus
Sulfur	rhombic sulfur
Tin	white tin, α-tin

of liquid water (at 25°C, as always in this text) is obtained from the thermochemical equation

$$H_2(g) + \tfrac{1}{2}O_2(g) \rightarrow H_2O(l) \qquad \Delta H^\ominus = -286 \text{ kJ}$$

and is $\Delta_f H^\ominus(H_2O, l) = -286 \text{ kJ mol}^{-1}$. Note that enthalpies of formation are molar quantities, so to go from $\Delta H^\ominus$ in a thermochemical equation to $\Delta_f H^\ominus$ for that substance, divide by the amount of substance formed (in this instance, by 1 mol H_2O).

With the introduction of standard enthalpies of formation, we can write

$$\Delta_r H^\ominus = \Sigma \nu \Delta_f H^\ominus(\text{products}) - \Sigma \nu \Delta_f H^\ominus(\text{reactants}) \qquad (3.5)$$

The first term on the right is the enthalpy of formation of all the products from their elements; the second term on the right is the enthalpy of formation of all the reactants from their elements. The fact that the enthalpy is a state function means that a reaction enthalpy calculated in this way is identical to the value that would be calculated from eqn 3.4 if absolute enthalpies were available.

The values of some standard enthalpies of formation at 25°C are given in Table 3.8, and a longer list is given in the Data section. The standard enthalpies of formation of elements in their reference states are zero by definition (because their formation is the null reaction: element → element). Note, however, that the standard enthalpy of formation of an element in a state other than its reference state is not zero:

$$C(s, \text{graphite}) \rightarrow C(s, \text{diamond}) \qquad \Delta H^\ominus = +1.895 \text{ kJ}$$

Therefore, although $\Delta_f H^\ominus(C, \text{graphite}) = 0$, $\Delta_f H^\ominus(C, \text{diamond}) = +1.895 \text{ kJ mol}^{-1}$.

Example 3.5

Using standard enthalpies of formation

Calculate the standard enthalpy of combustion of liquid benzene from the standard enthalpies of formation of the reactants and products.

Strategy We write the chemical equation, identify the stoichiometric numbers of the reactants and products, and then use eqn 3.5. Note that the expression has the form 'products − reactants'. Numerical values of standard enthalpies of formation are given in the Data section. The standard enthalpy of combustion is the enthalpy change per mole of substance, so we need to interpret the enthalpy change accordingly.

Solution The chemical equation is

$$C_6H_6(l) + \tfrac{15}{2}O_2(g) \rightarrow 6\,CO_2(g) + 3\,H_2O(l)$$

It follows that

$$\begin{aligned}\Delta_r H^\ominus &= \{6\Delta_f H^\ominus(CO_2,g) + 3\Delta_f H^\ominus(H_2O,l)\} \\ &\quad - \{\Delta_f H^\ominus(C_6H_6, l) + \tfrac{15}{2}\Delta_f H^\ominus(O_2, g)\} \\ &= \{6 \times (-393.51 \text{ kJ mol}^{-1}) + 3 \times (-285.83 \text{ kJ mol}^{-1})\} \\ &\quad - \{(49.0 \text{ kJ mol}^{-1}) + 0\} \\ &= -3268 \text{ kJ mol}^{-1}\end{aligned}$$

Inspection of the chemical equation shows that, in this instance, the 'per mole' is per mole of C_6H_6, which is exactly what we need for an enthalpy of combustion. It follows that the standard enthalpy of combustion of liquid benzene is $-3268 \text{ kJ mol}^{-1}$.

A note on good practice The standard enthalpy of formation of an element in its reference state (oxygen gas in this example) is written 0 not 0 kJ mol^{-1}, because it is zero whatever units we happen to be using.

Self-test 3.6

Use standard enthalpies of formation to calculate the enthalpy of combustion of propane gas to carbon dioxide and water vapour.

[*Answer:* $-2044 \text{ kJ mol}^{-1}$]

The reference states of the elements define a thermochemical 'sea level', and enthalpies of formation can be regarded as thermochemical 'altitudes' above or below sea level (Fig. 3.8). Compounds that have negative standard enthalpies of formation (such as water) are classified as **exothermic compounds**, for they lie at a lower enthalpy than their component elements (they lie below thermochemical sea level). Compounds that have positive standard enthalpies of formation (such as carbon disulfide) are classified as **endothermic compounds**, and possess a higher enthalpy than their component elements (they lie above sea level).

3.6 Enthalpies of formation and molecular modelling

It is difficult to estimate standard enthalpies of formation of different conformations of molecules. For example, we obtain the same enthalpy of formation for the equatorial (**1**) and axial (**2**) conformations of methylcyclohexane if we proceed as in Example 3.3. However, it has been observed experimentally that

Table 3.8

*Standard enthalpies of formation at 298.15 K**

Substance	$\Delta_f H°$/(kJ mol^{-1})
Inorganic compounds	
Ammonia, NH_3(g)	–46.11
Ammonium nitrate, NH_4NO_3(s)	–365.56
Carbon monoxide, CO(g)	–110.53
Carbon disulfide, CS_2(l)	+89.70
Carbon dioxide, CO_2(g)	–393.51
Dinitrogen tetroxide, N_2O_4(g)	+9.16
Dinitrogen monoxide, N_2O(g)	+82.05
Hydrogen chloride, HCl(g)	–92.31
Hydrogen fluoride, HF(g)	–271.1
Hydrogen sulfide, H_2S(g)	–20.63
Nitric acid, HNO_3(l)	–174.10
Nitrogen dioxide, NO_2(g)	+33.18
Nitrogen monoxide, NO(g)	+90.25
Sodium chloride, NaCl(s)	–411.15
Sulfur dioxide, SO_2(g)	–296.83
Sulfur trioxide, SO_3(g)	–395.72
Sulfuric acid, H_2SO_4(l)	–813.99
Water, H_2O(l)	–285.83
H_2O(g)	–241.82
Organic compounds	
Benzene, C_6H_6(l)	+49.0
Ethane, C_2H_6(g)	–84.68
Ethanol, C_2H_5OH(l)	–277.69
Ethene, C_2H_4(g)	+52.26
Ethyne, C_2H_2(g)	+226.73
Glucose, $C_6H_{12}O_6$(s)	–1268
Methane, CH_4(g)	–74.81
Methanol, CH_3OH(l)	–238.86
Sucrose, $C_{12}H_{22}O_{11}$(s)	–2222.86

* A longer list is given in the Data section at the end of the book. The text's website also contains links to additional data.

molecules in these two conformations have different standard enthalpies of formation as a result of the greater steric repulsion when the methyl group is in an axial position than when it is equatorial.

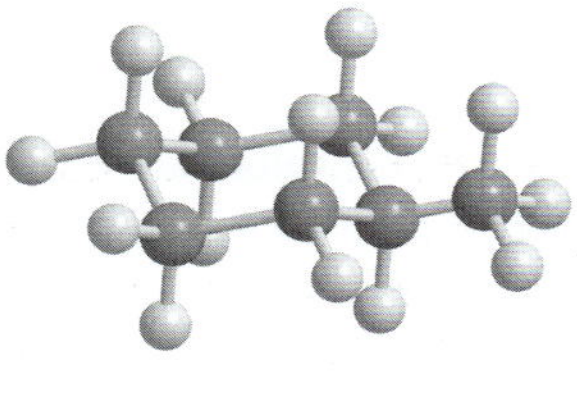

1 Equatorial

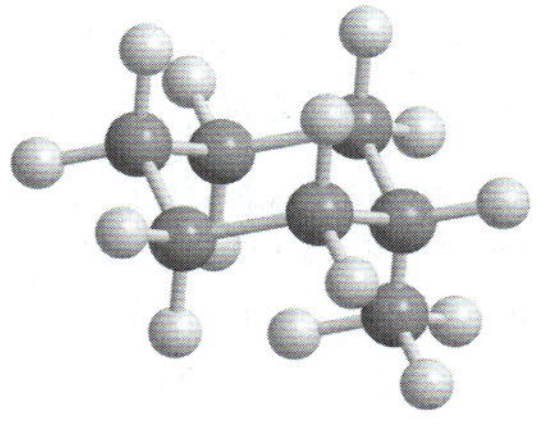

2 Axial

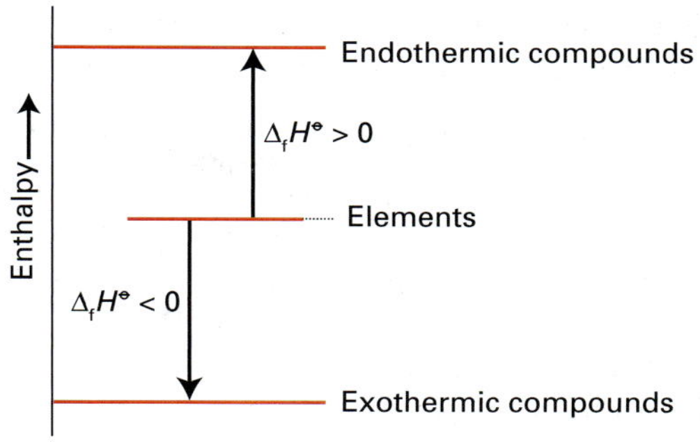

Fig. 3.8 The enthalpy of formation acts as a kind of thermochemical 'altitude' of a compound with respect to the 'sea level' defined by the elements from which it is made. Endothermic compounds have positive enthalpies of formation; exothermic compounds have negative energies of formation.

Computer-aided molecular modelling using commercially available software is now widely used to estimate standard enthalpies of formation of molecules with complex three-dimensional structures, and can distinguish between different conformations of the same molecule. In the case of methylcyclohexane, for instance, the calculated difference in enthalpy of formation ranges from 5.9 to 7.9 kJ mol^{-1}, which compares favourably with the experimental value of 7.5 kJ mol^{-1}. However, good agreement between calculated and experimental values is relatively rare. Computational methods almost always predict correctly which conformation of a molecule is most stable but do not always predict the correct numerical values of the difference in enthalpies of formation.

A calculation performed in the absence of solvent molecules estimates the properties of the molecule of interest in the gas phase. Computational methods are available that allow for the inclusion of several solvent molecules around a solute molecule, thereby taking into account the effect of molecular interactions with the solvent on the enthalpy of formation of the solute. Again, the numerical results are only estimates and the primary purpose of the calculation is to predict whether interactions with the solvent increase or decrease the enthalpy of formation. As an example, consider the amino acid glycine, which can exist in a neutral or zwitterionic form, H_2NHCH_2COOH and $^+H_3NCH_2CO_2^-$, respectively, in which in the latter the amino group is protonated and the carboxyl group is deprotonated. Molecular modelling shows that in the gas phase the neutral form has a lower enthalpy of formation than the zwitterionic form. However, in water the opposite is true on account of the strong interactions between the polar solvent and the charges on the zwitter ion.

3.7 The variation of reaction enthalpy with temperature

It often happens that we have data at one temperature but need it at another temperature. For example, we might want to know the enthalpy of a particular reaction at body temperature, 37°C, but may have data available for 25°C. Another type of question that could arise might be whether the oxidation of glucose is more exothermic when it takes place inside an Arctic fish that inhabits water at 0°C than when it takes place at mammalian body temperatures. Similarly, we may need to predict whether the synthesis of ammonia is more exothermic at a typical industrial temperature of 450°C than at 25°C. In precise work, every attempt would be made to measure the reaction enthalpy at the temperature of interest, but it is useful to have a 'back-of-the-envelope' way of estimating the direction of change and even a moderately reliable numerical value.

Figure 3.9 illustrates the technique we use. As we have seen, the enthalpy of a substance increases with temperature; therefore, the total enthalpy of the reactants and the total enthalpy of the products increase as shown in the illustration. Provided the two total enthalpy increases are different, the standard reaction enthalpy (their difference at a given temperature) will change as the temperature is changed. The change in the enthalpy of a substance depends on the slope of the graph and therefore on the constant-pressure heat capacities of the substances. We can therefore expect the temperature dependence of the reaction enthalpy to be related to the difference in heat capacities of the products and the reactants.

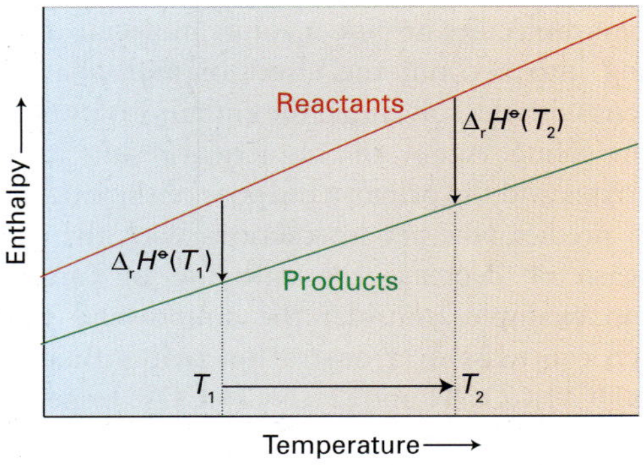

Fig. 3.9 The enthalpy of a substance increases with temperature. Therefore, if the total enthalpy of the reactants increases by a different amount from that of the products, the reaction enthalpy will change with temperature. The change in reaction enthalpy depends on the relative slopes of the two lines and hence on the heat capacities of the substances.

As a simple example, consider the reaction

$$2\,H_2(g) + O_2(g) \rightarrow 2\,H_2O(l)$$

where the standard enthalpy of reaction is known at one temperature (for example, at 25°C from the tables in this book). According to eqn 3.5, we can write

$$\Delta_r H^\ominus(T) = 2H_m^\ominus(H_2O, l) - \{2H_m^\ominus(H_2, g) + H_m^\ominus(O_2, g)\}$$

for the reaction at a temperature T. If the reaction takes place at a higher temperature T', the molar enthalpy of each substance is increased because it stores more energy and the standard reaction enthalpy becomes

$$\Delta_r H^\ominus(T') = 2H_m^{\ominus\prime}(H_2O, l) - \{2H_m^{\ominus\prime}(H_2, g) + H_m^{\ominus\prime}(O_2, g)\}$$

where the primes signify the values at the new temperature. Equation 2.16 ($C_p = \Delta H/\Delta T$) implies that the increase in molar enthalpy of a substance when the temperature is changed from T to T' is $C_{p,m}^\ominus \times (T' - T)$, where $C_{p,m}^\ominus$ is the standard molar constant-pressure heat capacity of the substance, the molar heat capacity measured at 1 bar. For example, the molar enthalpy of water changes to

$$H_m^{\ominus\prime}(H_2O, l) = H_m^\ominus(H_2O, l) + C_{p,m}^\ominus(H_2O, l) \times (T' - T)$$

if $C_{p,m}^\ominus(H_2O, l)$ is constant over the temperature range. When we substitute terms like this into the expression above, we find

$$\Delta_r H^\ominus(T') = \Delta_r H^\ominus(T) + \Delta_r C_p^\ominus \times (T' - T) \qquad (3.6)$$

where

$$\Delta_r C_p^\ominus = 2C_{p,m}^\ominus(H_2O, l) - \{2C_{p,m}^\ominus(H_2, g) + C_{p,m}^\ominus(O_2, g)\}$$

Note that this combination has the same pattern as the reaction enthalpy and the stoichiometric numbers occur in the same way. In general, $\Delta_r C_p^\ominus$ is the difference between the weighted sums of the standard molar heat capacities of the products and the reactants:

$$\Delta_r C_p^\ominus = \sum \nu C_{p,m}^\ominus(\text{products}) - \sum \nu C_{p,m}^\ominus(\text{reactants}) \qquad (3.7)$$

Equation 3.6 is **Kirchhoff's law.** We see that, just as we anticipated, the standard reaction enthalpy at one temperature can be calculated from the standard reaction enthalpy at another temperature provided we know the standard molar constant-pressure heat capacities of all the substances. These values are given in the Data section. The derivation of Kirchhoff's law supposes that the heat capacities are constant over the range of temperature of interest, so the law is best restricted to small temperature differences (of no more than 100 K or so).

Example 3.6

Using Kirchhoff's law

The standard enthalpy of formation of gaseous water at 25°C is –241.82 kJ mol^{-1}. Estimate its value at 100°C.

Strategy First, write the chemical equation and identify the stoichiometric numbers. Then calculate the value of $\Delta_r C_p^{\ominus}$ from the data in the Data section by using eqn 3.7 and use the result in eqn 3.6.

Solution The chemical equation is

$$H_2(g) + \tfrac{1}{2}O_2(g) \rightarrow H_2O(g)$$

and the molar constant-pressure heat capacities of $H_2O(g)$, $H_2(g)$, and $O_2(g)$ are 33.58 J K^{-1} mol^{-1}, 28.84 J K^{-1} mol^{-1}, and 29.37 J K^{-1} mol^{-1}, respectively. It follows that

$$\begin{aligned}\Delta_r C_p^{\ominus} &= C_{p,m}^{\ominus}(H_2O, g) - \{C_{p,m}^{\ominus}(H_2, g) + \tfrac{1}{2}C_{p,m}^{\ominus}(O_2, g)\}\\ &= (33.58\ \mathrm{J\,K^{-1}\,mol^{-1}}) - \{(28.84\ \mathrm{J\,K^{-1}\,mol^{-1}})\\ &\quad + \tfrac{1}{2}\times(29.37\ \mathrm{J\,K^{-1}\,mol^{-1}})\}\\ &= -9.95\ \mathrm{J\,K^{-1}\,mol^{-1}} = -9.95\times10^{-3}\ \mathrm{kJ\,K^{-1}\,mol^{-1}}\end{aligned}$$

Then, because $T' - T = +75$ K, from eqn 3.6 we find

$$\begin{aligned}\Delta_r H^{\ominus\prime} &= (-241.82\ \mathrm{kJ\,mol^{-1}}) +\\ &\quad (-9.95\times10^{-3}\ \mathrm{kJ\,K^{-1}\,mol^{-1}})\times(75\ \mathrm{K})\\ &= (-241.82\ \mathrm{kJ\,mol^{-1}}) - (0.75\ \mathrm{kJ\,mol^{-1}})\\ &= -242.57\ \mathrm{kJ\,mol^{-1}}\end{aligned}$$

The experimental value is –242.58 kJ mol^{-1}.

Self-test 3.7

Estimate the standard enthalpy of formation of $NH_3(g)$ at 400 K from the data in the Data section.

[*Answer:* –48.4 kJ mol^{-1}]

The calculation in Example 3.6 shows that the standard reaction enthalpy at 100°C is only slightly different from that at 25°C. The reason is that the change in reaction enthalpy is proportional to the *difference* between the molar heat capacities of the products and the reactants, which is usually not very large. It is generally the case that, provided the temperature range is not too wide, enthalpies of reactions vary only slightly with temperature. A reasonable first approximation is that standard reaction enthalpies are independent of temperature. When the temperature range is too wide for it to be safe to assume that heat capacities are constant, the empirical temperature dependence of each heat capacity given in eqn 3.6 may be used: the resulting expression is best developed by using mathematical software.

Checklist of key ideas

You should now be familiar with the following concepts.

☐ **1** The standard state of a substance is the pure substance at 1 bar.

☐ **2** The standard enthalpy of transition, $\Delta_{trs}H^{\ominus}$, is the change in molar enthalpy when a substance in one phase changes into another phase, both phases being in their standard states.

☐ **3** The standard enthalpy of the reverse of a process is the negative of the standard enthalpy of the forward process, $\Delta_{reverse}H^{\ominus} = -\Delta_{forward}H^{\ominus}$.

☐ **4** The standard enthalpy of a process is the sum of the standard enthalpies of the individual processes into which it may be regarded as divided, as in $\Delta_{sub}H^{\ominus} = \Delta_{fus}H^{\ominus} + \Delta_{vap}H^{\ominus}$.

☐ **5** Hess's law states that the standard enthalpy of a reaction is the sum of the standard enthalpies of the reactions into which the overall reaction may be divided.

☐ **6** The standard enthalpy of formation of a compound, $\Delta_f H^{\ominus}$, is the standard reaction enthalpy for the formation of the compound from its elements in their reference states.

☐ **7** At constant pressure, exothermic compounds are those for which $\Delta_f H^{\ominus} < 0$; endothermic compounds are those for which $\Delta_f H^{\ominus} > 0$.

Table of key equations

The following table summarizes the equations developed in this chapter.

Description	Equation	Comment
Standard reaction enthalpy from enthalpies of formation	$\Delta_r H^{\ominus} = \sum \nu \Delta_f H^{\ominus}(\text{products}) - \sum \nu \Delta_f H^{\ominus}(\text{reactants})$	
Variation of the standard reaction enthalpy with temperature (Kirchhoff's law)	$\Delta_r H^{\ominus}(T') = \Delta_r H^{\ominus}(T) + \Delta_r C_p^{\ominus} \times (T' - T)$ with $\Delta_r C_p^{\ominus} = \sum \nu C_{p,m}^{\ominus}(\text{products}) - \sum \nu C_{p,m}^{\ominus}(\text{reactants})$	$\Delta_r C_p^{\ominus}$ constant in the temperature range of interest

Questions and exercises

Discussion questions

3.1 Define the terms (a) standard enthalpy of vaporization, (b) standard enthalpy of fusion; (c) standard enthalpy of sublimation; (d) standard enthalpy of ionization; (e) standard electron-gain enthalpy; (f) mean bond enthalpy, and identify an application for each one.

3.2 Define the terms (a) standard reaction enthalpy; (b) standard enthalpy of combustion; (c) standard enthalpy of formation, and identify an application for each one.

3.3 A primitive air-conditioning unit for use in places where electrical power is not available can be made by hanging up strips of linen soaked in water. Explain why this strategy is effective.

3.4 Describe at least two calculational methods by which standard reaction enthalpies may be predicted. Discuss the advantages and disadvantages of each method.

3.5 Why is it important to distinguish between the standard state and the reference state of an element?

3.6 Discuss the limitations of the expressions: (a) $\Delta_r H = \Delta_r U + \Delta \nu_{gas} RT$; (b) $\Delta_r H^{\ominus}(T') = \Delta_r H^{\ominus}(T) + \Delta_r C_p^{\ominus} \times (T' - T)$.

3.7 In the earlier literature, and still not uncommonly, you will find the expressions 'heat of combustion' and 'latent heat of vaporization'. Why are the thermodynamic expressions 'enthalpy of combustion' and 'enthalpy of vaporization' superior?

Exercises

Assume all gases are perfect unless stated otherwise. All thermochemical data are for 298.15 K.

3.1 Estimate the difference between the standard enthalpy of formation of CO_2(g) as currently defined (at 1 bar) and its value using the former definition (at 1 atm).

3.2 Liquid mixtures of sodium and potassium are used in some nuclear reactors as coolants that can survive the intense radiation inside reactor cores. Calculate the energy required as heat to melt 250 kg of sodium metal at 371 K.

3.3 Calculate the energy that must be transferred as heat to evaporate 1.00 kg of water at (a) 25°C, (b) 100°C.

3.4 Isopropanol (2-propanol) is commonly used as 'rubbing alcohol' to relieve sprain injuries in sport: its action is due to the cooling effect that accompanies its rapid evaporation when applied to the skin. In an experiment to determine its enthalpy of vaporization, a sample was brought to the boil. It was found that when an electric current of 0.812 A from a 11.5 V supply was passed for 303 s, then 4.27 g of the alcohol was vaporized. What is the (molar) enthalpy of vaporization of isopropanol at its boiling point?

3.5 Refrigerators make use of the heat absorption required to vaporize a volatile liquid. A fluorocarbon liquid being investigated to replace a chlorofluorocarbon has $\Delta_{vap} H^{\ominus} = +32.0\ \text{kJ mol}^{-1}$. Calculate q, w, ΔH, and ΔU when 2.50 mol is vaporized at 250 K and 750 Torr.

3.6 Use the information in Tables 2.1 and 2.2 to calculate the energy that must be transferred as heat to melt 100 g of ice at 0°C, increase the sample temperature to 100°C, and then vaporize it at that temperature. Sketch a graph of temperature against time on the assumption that the sample is heated at a constant rate.

3.7 The enthalpy of sublimation of calcium at 25°C is 178.2 kJ mol^{-1}. How much energy (at constant temperature and pressure) must be supplied as heat to 5.0 g of solid calcium to produce a plasma (a gas of charged particles) composed of Ca^{2+} ions and electrons?

3.8 Estimate the difference between the standard enthalpy of ionization of Ca(g) to Ca^{2+}(g) and the accompanying change in internal energy at 25°C.

3.9 Estimate the difference between the standard electron-gain enthalpy of Br(g) and the corresponding change in internal energy at 25°C.

3.10 How much energy (at constant temperature and pressure) must be supplied as heat to 10.0 g of chlorine gas (as Cl_2) to produce a plasma (a gas of charged particles, in this case ions) composed of Cl^- and Cl^+ ions? The enthalpy of ionization of Cl(g) is +1257.5 kJ mol^{-1} and its electron-gain enthalpy is −354.8 kJ mol^{-1}.

3.11 Use the data in Exercise 3.10 to identify (a) the standard enthalpy of ionization of Cl^-(g) and (b) the accompanying change in molar internal energy.

3.12 The enthalpy changes accompanying the dissociation of successive bonds in NH_3(g) are 460, 390, and 314 kJ mol^{-1}, respectively. (a) What is the mean enthalpy of an N—H bond? (b) Do you expect the mean bond internal energy to be larger or smaller than the mean bond enthalpy?

3.13 Use bond enthalpies and mean bond enthalpies to estimate the (a) the enthalpy of the glycolysis reaction adopted by anaerobic bacteria as a source of energy, $C_6H_{12}O_6$(aq) → 2 $CH_3CH(OH)COOH$(aq), lactic acid, which is produced via the formation of pyruvic acid, $CH_3COCOOH$, and the action of lactate dehydrogenase and (b) the enthalpy of combustion of glucose. Ignore the contributions of enthalpies of fusion and vaporization.

3.14 The efficient design of chemical plants depends on the designer's ability to assess and use the heat output in one process to supply another process. The standard enthalpy of reaction for N_2(g) + 3 H_2(g) → 2 NH_3(g) is −92.22 kJ mol^{-1}. What is the change in enthalpy when (a) 1.00 t of N_2(g) is consumed, (b) 1.00 t of NH_3(g) is formed?

3.15 Ethane is flamed off in abundance from oil wells, because it is unreactive and difficult to use commercially. But would it make a good fuel? The standard enthalpy of reaction for 2 C_2H_6(g) + 7 O_2(g) → 4 CO_2(g) + 6 H_2O(l) is −3120 kJ mol^{-1}. (a) What is the standard enthalpy of combustion of ethane? (b) What is the specific enthalpy of combustion of ethane? (c) Is ethane a more or less efficient fuel than methane?

3.16 Standard enthalpies of formation are widely available, but we might need a standard enthalpy of combustion instead. The standard enthalpy of formation of ethylbenzene is −12.5 kJ mol^{-1}. Calculate its standard enthalpy of combustion.

3.17 Combustion reactions are relatively easy to carry out and study, and their data can be combined to give enthalpies of other types of reaction. As an illustration, calculate the standard enthalpy of hydrogenation of cyclohexene to cyclohexane given that the standard enthalpies of combustion of the two compounds are −3752 kJ mol^{-1} (cyclohexene) and −3953 kJ mol^{-1} (cyclohexane).

3.18 Estimate the standard internal energy of formation of liquid methyl acetate (methyl ethanoate, CH_3COOCH_3) at 298 K from its standard enthalpy of formation, which is −442 kJ mol^{-1}.

3.19 The standard enthalpy of combustion of anthracene is −7163 kJ mol^{-1}. Calculate its standard enthalpy of formation.

3.20 When 320 mg of naphthalene, $C_{10}H_8$(s), was burned in a bomb calorimeter, the temperature rose by 3.05 K. Calculate the heat capacity of the calorimeter. By how much will the temperature rise when 100 mg of phenol, C_6H_5OH(s), is burned in the calorimeter under the same conditions?

3.21 The energy resources of glucose are of major concern for the assessment of metabolic processes. When 0.3212 g of glucose was burned in a bomb calorimeter of heat capacity 641 J K^{-1} the temperature rose by 7.793 K. Calculate (a) the standard molar enthalpy of combustion, (b) the standard internal energy of combustion, and (c) the standard enthalpy of formation of glucose.

3.22 The complete combustion of fumaric acid in a bomb calorimeter released 1333 kJ per mole of HOOCCH=CHCOOH(s) at 298 K. Calculate (a) the internal energy of combustion, (b) the enthalpy of combustion, (c) the enthalpy of formation of lactic acid.

3.23 The mean bond enthalpies of the C—C, C—H, C=O, and O—H bonds are 348, 412, 743, and 463 kJ mol^{-1}, respectively. The combustion of a fuel such as octane is exothermic because relatively weak bonds break to form relatively strong bonds. Use this information to justify why glucose has a lower specific enthalpy than the lipid decanoic acid ($C_{10}H_{20}O_2$) even though these compounds have similar molar masses.

3.24 Calculate the standard enthalpy of solution of AgI(s) in water from the standard enthalpies of formation of the solid and the aqueous ions.

3.25 The standard enthalpy of decomposition of the yellow complex NH_3SO_2 into NH_3 and SO_2 is +40 kJ mol^{-1}. Calculate the standard enthalpy of formation of NH_3SO_2.

3.26 Given that the enthalpy of combustion of graphite is −393.5 kJ mol^{-1} and that of diamond is −395.41 kJ mol^{-1}, calculate the standard enthalpy of the C(s, graphite) → C(s, diamond) transition.

3.27 The pressures deep within the Earth are much greater than those on the surface, and to make use of thermochemical data in geochemical assessments we need to take the differences into account. Use the information in Exercise 3.26 together with the densities of graphite (2.250 g cm^{-3}) and diamond (3.510 g cm^{-3}) to calculate the internal energy of the transition when the sample is under a pressure of 150 kbar.

3.28 A typical human produces about 10 MJ of energy transferred as heat each day through metabolic activity. The main mechanism of heat loss is through the evaporation of water. (a) If a human body were an isolated system of mass 65 kg with the heat capacity of water, what temperature rise would the body experience? (b) Human bodies are actually open systems. What mass of water should be evaporated each day to maintain constant temperature?

3.29 Camping gas is typically propane. The standard enthalpy of combustion of propane gas is −2220 kJ mol^{-1} and the standard enthalpy of vaporization of the liquid is +15 kJ mol^{-1}.

Calculate (a) the standard enthalpy and (b) the standard internal energy of combustion of the liquid.

3.30 Classify as endothermic or exothermic: (a) a combustion reaction for which $\Delta_r H^\ominus = -2020$ kJ mol^{-1}; (b) a dissolution for which $\Delta H^\ominus = +4.0$ kJ mol^{-1}; (c) vaporization; (d) fusion; and (e) sublimation.

3.31 Standard enthalpies of formation are of great usefulness, as they can be used to calculate the standard enthalpies of a very wide range of reactions of interest in chemistry, biology, geology, and industry. Use information in the Data section to calculate the standard enthalpies of the following reactions:

(a) $2\ NO_2(g) \rightarrow N_2O_4(g)$

(b) $NO_2(g) \rightarrow \frac{1}{2}\ N_2O_4(g)$

(c) $3\ NO_2(g) + H_2O(l) \rightarrow 2\ HNO_3(aq) + NO(g)$

(d) Cyclopropane(g) → propene(g)

(e) $HCl(aq) + NaOH(aq) \rightarrow NaCl(aq) + H_2O(l)$

3.32 Calculate the standard enthalpy of formation of N_2O_5 from the following data:

$2\ NO(g) + O_2(g) \rightarrow 2\ NO_2(g)$ $\quad \Delta_r H^\ominus = -114.1$ kJ mol^{-1}

$4\ NO_2(g) + O_2(g) \rightarrow 2\ N_2O_5(g)$ $\quad \Delta_r H^\ominus = -110.2$ kJ mol^{-1}

$N_2(g) + O_2(g) \rightarrow 2\ NO(g)$ $\quad \Delta_r H^\ominus = +180.5$ kJ mol^{-1}

3.33 Heat capacity data can be used to estimate the reaction enthalpy at one temperature from its value at another. Use the information in the Data section to predict the standard reaction enthalpy of $2\ NO_2(g) \rightarrow N_2O_4(g)$ at 100°C from its value at 25°C.

3.34 Estimate the enthalpy of vaporization of water at 100°C from its value at 25°C (44.01 kJ mol^{-1}) given the constant-pressure heat capacities of 75.29 J K^{-1} mol^{-1} and 33.58 J K^{-1} mol^{-1} for liquid and gas, respectively.

3.35 It is often useful to be able to anticipate, without doing a detailed calculation, whether an increase in temperature will result in a raising or a lowering of a reaction enthalpy. The constant-pressure molar heat capacity of a gas of linear molecules is approximately $\frac{7}{2}R$, whereas that of a gas of non-linear molecules is approximately $4R$. Decide whether the standard enthalpies of the following reactions will increase or decrease with increasing temperature:

(a) $2\ H_2(g) + O_2(g) \rightarrow 2\ H_2O(g)$

(b) $N_2(g) + 3\ H_2(g) \rightarrow 2\ NH_3(g)$

(c) $CH_4(g) + 2\ O_2(g) \rightarrow CO_2(g) + 2\ H_2O(g)$

3.36 The molar heat capacity of liquid water is approximately $9R$. Decide whether the standard enthalpy of the reactions (a) and (c) in Exercise 3.35 will increase or decrease with a rise in temperature if the water is produced as a liquid.

Projects

The symbol ‡ signifies that calculus is required.

3.37‡ Here we explore Kirchhoff's law (eqn 3.6) in greater detail. (a) Derive a version of Kirchhoff's law for the temperature dependence of the internal energy of reaction. (b) The formulation of Kirchhoff's law given in eqn 3.6 is valid when the difference in heat capacities is independent of temperature over the temperature range of interest. Suppose instead that $\Delta_r C_p^\ominus = a + bT + c/T^2$. Derive a more accurate form of Kirchhoff's law in terms of the parameters *a*, *b*, and *c*. *Hint*: The change in the reaction enthalpy for an infinitesimal change in temperature is $\Delta_r C_p^\ominus dT$. Integrate this expression between the two temperatures of interest.

3.38 Here we explore the thermodynamics of carbohydrates as biological fuels. It is useful to know that glucose and fructose are simple sugars with the molecular formula $C_6H_{12}O_6$. Sucrose (table sugar) is a complex sugar with molecular formula $C_{12}H_{22}O_{11}$ that consists of a glucose unit covalently bound to a fructose unit (a water molecule is eliminated as a result of the reaction between glucose and fructose to form sucrose). There are no dietary recommendations for consumption of carbohydrates. Some nutritionists recommend diets that are largely devoid of carbohydrates, with most of the energy needs being met by fats. However, the most common diets are those in which at least 65 per cent of our food calories come from carbohydrates. (a) A $\frac{3}{4}$-cup serving of pasta contains 40 g of carbohydrates. What percentage of the daily calorie requirement for a person on a 2200 Calorie diet (1 Cal = 1 kcal) does this serving represent? (b) The mass of a typical glucose tablet is 2.5 g. Calculate the energy released as heat when a glucose tablet is burned in air. (c) To what height could you climb on the energy a glucose tablet provides assuming 25% of the energy is available for work? (d) Is the standard enthalpy of combustion of glucose likely to be higher or lower at blood temperature than at 25°C? (e) Calculate the energy released as heat when a typical table sugar cube of mass 1.5 g is burned in air. (f) To what height could you climb on the energy a table sugar cube provides assuming 25 per cent of the energy is available for work?

Chapter 4

Thermodynamics: the Second Law

Some things happen; some things don't. A gas expands to fill the vessel it occupies; a gas that already fills a vessel does not suddenly contract into a smaller volume. A hot object cools to the temperature of its surroundings; a cool object does not suddenly become hotter than its surroundings. Hydrogen and oxygen combine explosively (once their ability to do so has been liberated by a spark) and form water; water left standing in oceans and lakes does not gradually decompose into hydrogen and oxygen. These everyday observations suggest that changes can be divided into two classes. A **spontaneous change** is a change that has a tendency to occur without work having to be done to bring it about. A spontaneous change has a natural tendency to occur. A **nonspontaneous change** is a change that can be brought about only by doing work. A nonspontaneous change has no natural tendency to occur. Nonspontaneous changes can be *made* to occur by doing work: gas can be compressed into a smaller volume by pushing in a piston, the temperature of a cool object can be raised by forcing an electric current through a heater attached to it, and water can be decomposed by the passage of an electric current. However, in each case we need to act in some way on the system to bring about the nonspontaneous change. There must be some feature of the world that accounts for the distinction between the two types of change.

Throughout the chapter we shall use the terms 'spontaneous' and 'nonspontaneous' in their thermodynamic sense. That is, we use them to signify that a change does or does not have a natural *tendency* to occur. In thermodynamics the term spontaneous has nothing to do with speed. Some spontaneous changes are very fast, such as the precipitation reaction that occurs when solutions of sodium chloride and silver nitrate are mixed. However, some spontaneous changes are so slow that there may be no observable change even after millions of years. For example, although the decomposition of benzene into carbon and hydrogen is spontaneous, it does not occur at a measurable rate under normal conditions, and benzene is a common laboratory commodity with a shelf life of (in principle) millions of years. Thermodynamics deals with the tendency to change; it is silent on the rate at which that tendency is realized.

Entropy

A few moments' thought is all that is needed to identify the reason why some changes are spontaneous and others are not. That reason is *not* the tendency of the system to move towards lower energy. This point is easily established by identifying an example of a spontaneous change in which there is no change in energy. The isothermal expansion of a perfect gas into a vacuum is spontaneous, but the total energy of the gas does not change because the molecules continue to travel at the same average speed and so keep their same total kinetic energy. Even in a process in which the energy of a system does decrease (as in the spontaneous cooling of a block of hot metal), the First Law requires the total energy of the system and the surroundings to be constant. Therefore, in this case the energy of another part of the world must increase if the energy decreases in the part that interests us. For instance, a hot block of metal in contact with a cool block cools and loses energy; however, the second block becomes warmer, and increases in energy. It is equally valid to say that the second block moves spontaneously to higher energy as it is to say that the first block has a tendency to go to lower energy!

4.1 The direction of spontaneous change

We shall now show that *the apparent driving force of spontaneous change is the tendency of energy and matter to become disordered*. For example, the molecules of a gas may all be in one region of a container initially, but their ceaseless disorderly motion ensures that they spread rapidly throughout the entire volume of the container (Fig. 4.1). Because their motion is so disorderly, there is a negligibly small probability that all the molecules will find their way back simultaneously into the region of the container they occupied initially. In this instance, the natural direction of change corresponds to the dispersal of matter.

A similar explanation accounts for spontaneous cooling, but now we need to consider the dispersal of energy rather than that of matter. In a block of hot metal, the atoms are oscillating vigorously and the hotter the block the more vigorous their motion. The cooler surroundings also consist of oscillating atoms, but their motion is less vigorous. The vigorously oscillating atoms of the hot block jostle their neighbours in the surroundings, and the energy of the atoms in the block is handed on to the atoms in the surroundings (Fig. 4.2). The process continues until the vigour with which the atoms in the system are oscillating has fallen to that of the surroundings. The opposite flow of energy is very unlikely. It is highly improbable that there will be a net flow of energy into the system as a result of jostling from less vigorously oscillating molecules in the surroundings. In this case, the natural direction of change corresponds to the dispersal of energy.

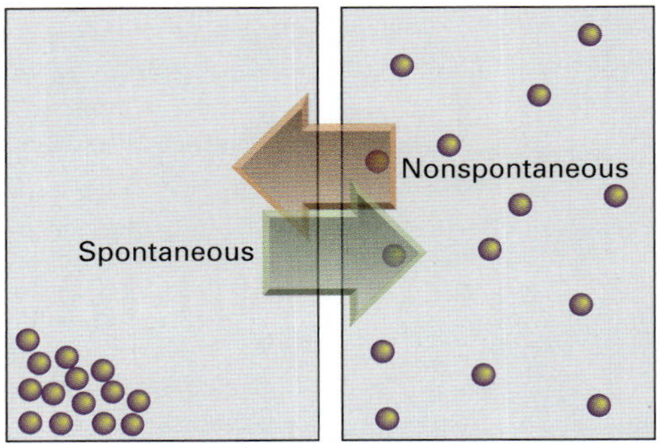

Fig. 4.1 One fundamental type of spontaneous process is the chaotic dispersal of matter. This tendency accounts for the spontaneous tendency of a gas to spread into and fill the container it occupies. It is extremely unlikely that all the particles will collect into one small region of the container. (In practice, the number of particles is of the order of 10^{23}.)

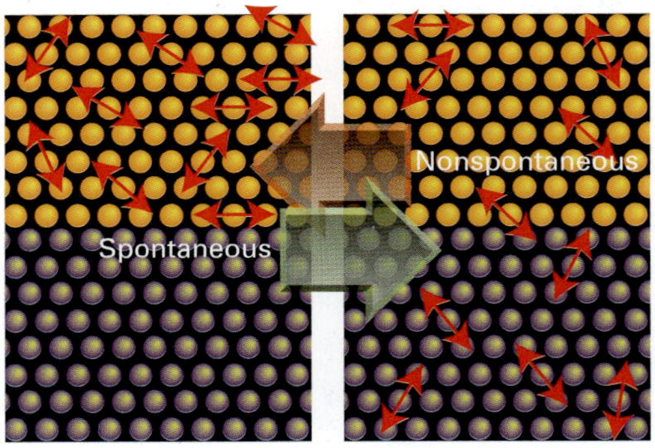

Fig. 4.2 Another fundamental type of spontaneous process is the chaotic dispersal of energy (represented by the small arrows). In these diagrams, the orange spheres represent the system and the purple spheres represent the surroundings. The double-headed arrows represent the thermal motion of the atoms.

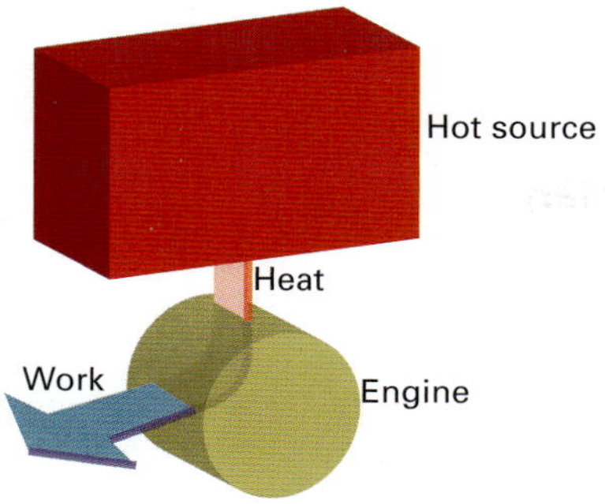

Fig. 4.3 The Second Law denies the possibility of the process illustrated here, in which heat is changed completely into work, there being no other change. The process is not in conflict with the First Law, because the energy is conserved.

The tendency toward dispersal of energy also explains the fact that, despite numerous attempts, it has proved impossible to construct an engine like that shown in Fig 4.3, in which heat, perhaps from the combustion of a fuel, is drawn from a hot reservoir and completely converted into work, such as the work of moving an automobile. All actual heat engines have both a hot region, the 'source', and a cold region, the 'sink', and it has been found that some energy must be discarded into the cold sink as heat and not used to do work. In molecular terms, only some of the energy stored in the atoms and molecules of the hot source can be used to do work and transferred to the surroundings in an orderly way. For the engine to do work, some energy must be transferred to the cold sink as heat, to stimulate disorderly motion of its atoms and molecules.

In summary, we have identified two basic types of spontaneous physical process:

1. Matter tends to become disordered.
2. Energy tends to become disordered.

We must now see how these two primitive types of spontaneous physical change result in some chemical reactions being spontaneous and others not. It may seem very puzzling that collapse into disorder can account for the formation of such ordered systems as proteins and biological cells. Nevertheless, in due course we shall see that organized structures can emerge as energy and matter disperse. We shall see, in fact, that collapse into disorder accounts for change in all its forms.

4.2 Entropy and the Second Law

The measure of disorder used in thermodynamics is called the **entropy**, S. Initially, we can take entropy to be a synonym for the extent of disorder, but shortly we shall see that it can be defined precisely and quantitatively, measured, and then applied to chemical reactions. At this point, all we need to know is that when matter and energy become disordered, the entropy increases. That being so, we can combine the two remarks above into a single statement known as the **Second Law of thermodynamics:**

The entropy of an isolated system tends to increase.

The 'isolated system' may consist of a system in which we have a special interest (a beaker containing reagents) and that system's surroundings: the two components jointly form a little 'universe' in the thermodynamic sense.

To make progress and turn the Second Law into a quantitatively useful statement, we need to define entropy precisely. We shall use the following definition of a *change* in entropy for a system maintained at constant temperature:

$$\Delta S = \frac{q_{\text{rev}}}{T} \tag{4.1}$$

That is, the change in entropy of a substance is equal to the energy transferred as heat to it *reversibly* divided by the temperature at which the transfer takes place. This definition can be justified thermodynamically, but we shall confine ourselves to showing that it is plausible and then show how to use it to obtain numerical values for a range of processes.

There are three points we need to understand about the definition in eqn 4.1: the significance of the term 'reversible', why heat (not work) appears in the numerator, and why temperature appears in the denominator.

Why reversible? We met the concept of reversibility in Section 2.3, where we saw that it refers to the ability of an infinitesimal change in a variable to change the direction of a process. Mechanical reversibility refers to the equality of pressure acting on either side of a movable wall. Thermal reversibility, the type involved in eqn 4.1, refers to the equality of temperature on either side of a thermally conducting wall. Reversible transfer of heat is smooth, careful, restrained transfer between two bodies at the same temperature. By making the transfer reversible we ensure that there are no hot spots generated in the object that later disperse spontaneously and hence add to the entropy.

Why heat and not work in the numerator? Now consider why heat and not work appears in eqn 4.1. Recall from Section 2.2 that to transfer energy as heat we make use of the disorderly motion of molecules, whereas to transfer energy as work we make use of orderly motion. It should be plausible that the change in entropy—the change in the degree of disorder—is proportional to the energy transfer that takes place by making use of disorderly motion rather than orderly motion.

Why temperature in the denominator? The presence of the temperature in the denominator in eqn 4.1 takes into account the disorder that is already present. If a given quantity of energy is transferred as heat to a hot object (one in which the atoms have a lot of disorderly thermal motion), then the additional disorder generated is less significant than if the same quantity of energy is transferred as heat to a cold object in which the atoms have less thermal motion. The difference is like sneezing in a busy street (an environment analogous to a high temperature) and sneezing in a quiet library (an environment analogous to a low temperature).

A brief illustration The transfer of 100 kJ of energy as heat to a large mass of water at 0°C (273 K) results in a change in entropy of

$$\Delta S = \frac{q_{\rm rev}}{T} = \frac{100 \times 10^3\ \rm J}{273\ \rm K} = +366\ \rm J\ K^{-1}$$

We use a large mass of water to ensure that the temperature of the sample does not change as heat is transferred. The same transfer at 100°C (373 K) results in

$$\Delta S = \frac{100 \times 10^3\ \rm J}{373\ \rm K} = +268\ \rm J\ K^{-1}$$

The increase in entropy is greater at the lower temperature.

A note on good practice The units of entropy are joules per kelvin ($\rm J\ K^{-1}$). Entropy is an extensive property. When we deal with molar entropy, an intensive property, the units will be joules per kelvin per mole ($\rm J\ K^{-1}\ mol^{-1}$).

The entropy (it can be proved) is a state function, a property with a value that depends only on the present state of the system. The entropy is a measure of the current state of disorder of the system, and how that disorder was achieved is not relevant to its current value. A sample of liquid water of mass 100 g at 60°C and 98 kPa has exactly the same degree of molecular disorder—the same entropy—regardless of what has happened to it in the past. The implication of entropy being a state function is that a change in its value when a system undergoes a change of state is independent of how the change of state is brought about. One practical application of entropy is to the discussion of the efficiencies of heat engines, refrigerators, and heat pumps (Box 4.1).

Box 4.1 Heat engines, refrigerators, and heat pumps

One practical application of entropy is to the discussion of the efficiencies of heat engines, refrigerators, and heat pumps. As remarked in the text, to achieve spontaneity—an engine is less than useless if it has to be driven—some energy must be discarded as heat into the cold sink. It is quite easy to calculate the minimum energy that must be discarded in this way by thinking about the flow of energy and the changes in entropy of the hot source and cold sink. To simplify the discussion, we shall express it in terms of the magnitudes of the heat and work transactions, which we write as $|q|$ and $|w|$, respectively (so, if $q = -100$ J, $|q| = 100$ J). Maximum work—and therefore maximum efficiency—is achieved if all energy transactions take place reversibly, so we assume that to be the case in the following.

Suppose that the hot source is at a temperature $T_{\rm hot}$. Then when energy $|q|$ is released from it reversibly as heat, its entropy changes by $-|q|/T_{\rm hot}$. Suppose that we allow an energy $|q'|$ to flow reversibly as heat into the cold sink at a temperature $T_{\rm cold}$. Then the entropy of that sink changes by $+|q'|/T_{\rm cold}$ (see the first illustration). The total change in entropy is therefore

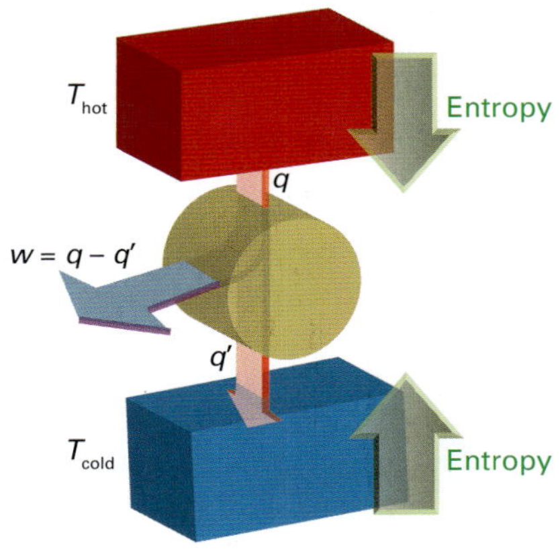

The flow of energy in a heat engine. For the process to be spontaneous, the decrease in entropy of the hot source must be offset by the increase in entropy of the cold sink. However, because the latter is at a lower temperature, not all the energy removed from the hot source need be deposited in it, leaving the difference available as work.

$$\Delta S_{total} = -\frac{|q|}{T_{hot}} + \frac{|q'|}{T_{cold}}$$

The engine will not operate spontaneously if this change in entropy is negative, and just becomes spontaneous as ΔS_{total} becomes positive. This change of sign occurs when $\Delta S_{total} = 0$, which is achieved when

$$|q'| = \frac{T_{cold}}{T_{hot}} \times |q|$$

If we have to discard an energy $|q'|$ into the cold sink, the maximum energy that can be extracted as work is $|q| - |q'|$. It follows that the *efficiency*, η (eta), of the engine, the ratio of the work produced to the heat absorbed, is

$$\eta = \frac{\text{work produced}}{\text{heat absorbed}} = \frac{|q| - |q'|}{|q|} = 1 - \frac{|q'|}{|q|}$$

$$= 1 - \frac{T_{cold}}{T_{hot}}$$

This remarkable result tells us that the efficiency of a perfect heat engine (one working reversibly and without mechanical defects such as friction) depends only on the temperatures of the hot source and cold sink. It shows that maximum efficiency (closest to $\eta = 1$) is achieved by using a sink that is as cold as possible and a source that is as hot as possible. For example, the maximum efficiency of an electrical power station using steam at 200°C (473 K) and discharging at 20°C (293 K) is

$$\eta = 1 - \frac{293\ \text{K}}{473\ \text{K}} = 1 - \frac{293}{473} = 0.381$$

or 38.1 per cent.

A refrigerator can be analysed similarly (see the second illustration). The entropy change when an energy $|q|$ is withdrawn reversibly as heat from the cold interior at a temperature T_{cold} is $-|q|/T_{cold}$. The entropy change when an energy $|q'|$ is deposited reversibly as heat in the outside world at a temperature T_{hot} is $+|q'|/T_{hot}$. The total change in entropy would be negative if $|q'| = |q|$, and the refrigerator would not work. However, if we increase the flow of energy into the warm exterior by doing work on the refrigerator, then the entropy change of the warm exterior can be increased to the point at which it overcomes the decrease in entropy of the cold interior, and the refrigerator operates. The calculation of the maximum efficiency of this process is left as an exercise (see Project 4.35a).

A heat pump is simply a refrigerator, but in which we are more interested in the supply of heat to the exterior than the cooling achieved in the interior. You are invited to show (see Project 4.35b), that the efficiency of a perfect heat pump, as measured by the heat produced divided by the work done, also depends on the ratio of the two temperatures.

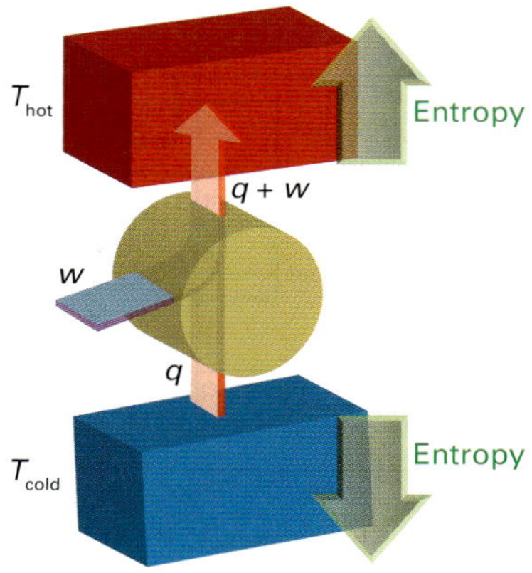

The flow of energy as heat from a cold source to a hot sink becomes feasible if work is provided to add to the energy stream. Then the increase in entropy of the hot sink can be made to cancel the entropy decrease of the hot source.

4.3 The entropy change accompanying expansion

We can often rely on intuition to judge whether the entropy increases or decreases when a substance undergoes a physical change. For instance, the entropy of a sample of gas increases as it expands because the molecules get to move in a greater volume and so have a greater degree of disorder. However, the advantage of eqn 4.1 is that it lets us express the increase *quantitatively* and make numerical calculations. For instance, as shown in Derivation 4.1, we can use the definition to calculate the change in entropy when a perfect gas expands isothermally from a volume V_i to a volume V_f, and obtain

$$\Delta S = nR \ln \frac{V_f}{V_i} \qquad (4.2)$$

We have already stressed the importance of reading equations for their physical content. In this case:

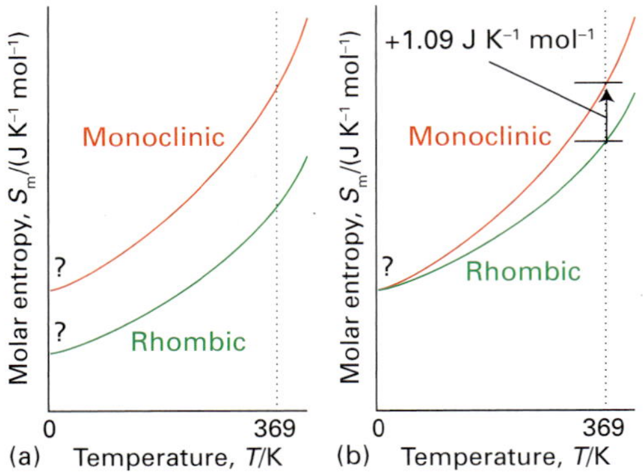

Fig. 4.8 (a) The molar entropies of monoclinic and rhombic sulfur vary with temperature as shown here. At this stage we do not know their values at $T = 0$. (b) When we slide the two curves together by matching their separation to the measured entropy of transition at the transition temperature, we find that the entropies of the two forms are the same at $T = 0$.

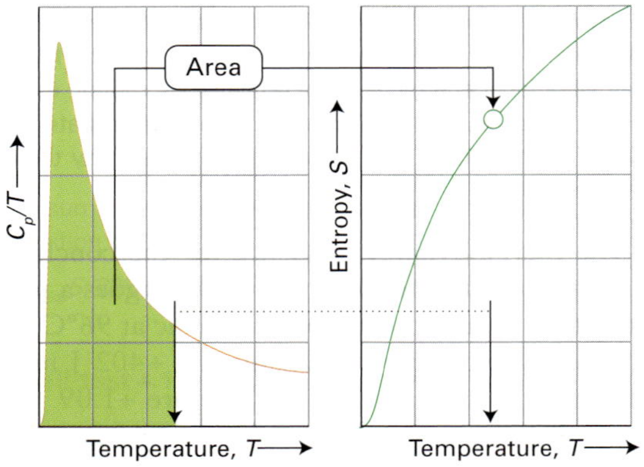

Fig. 4.9 The absolute entropy (or Third-Law entropy) of a substance is calculated by extending the measurement of heat capacities down to $T = 0$ (or as close to that value as possible), and then determining the area of the graph of C_p/T against T up to the temperature of interest. The area is equal to the absolute entropy at the temperature T.

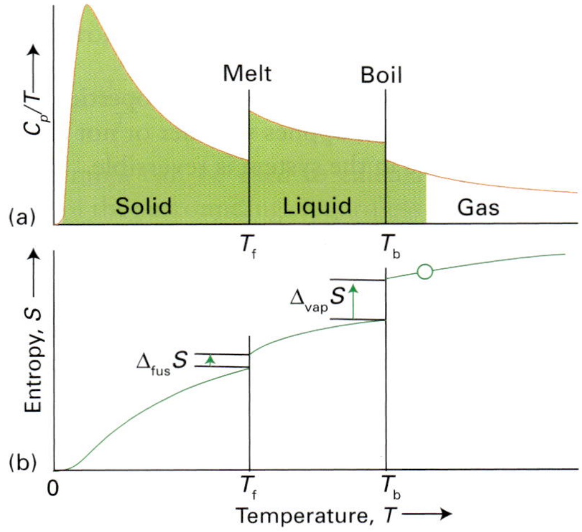

Fig. 4.10 The determination of entropy from heat capacity data. (a) Variation of C_p/T with the temperature of the sample. (b) The entropy, which is equal to the area beneath the upper curve up to the temperature of interest plus the entropy of each phase transition between $T = 0$ and the temperature of interest.

The **Third-Law entropy** at any temperature, $S(T)$, is equal to the area under the graph of C/T between $T = 0$ and the temperature T (Fig. 4.9). If there are any phase transitions (for example, melting) in the temperature range of interest, then the entropy of each transition at the transition temperature is calculated like that in eqn 4.6 and its contribution added to the contributions from each of the phases, as shown in Fig. 4.10. The Third-Law entropy, which is commonly called simply 'the entropy', of a substance depends on the pressure; we therefore select a standard pressure (1 bar) and report the **standard molar entropy**, $S_m^{\ominus}$, the molar entropy of a substance in its standard state at the temperature of interest. Some values at 298.15 K (the conventional temperature for reporting data) are given in Table 4.2.

It is worth spending a moment to look at the values in Table 4.2 to see that they are consistent with our understanding of entropy. All standard molar entropies are positive, because raising the temperature of a sample above $T = 0$ invariably increases its entropy above the value $S(0) = 0$. Another feature is that the standard molar entropy of diamond (2.4 J K^{-1} mol^{-1}) is lower than that of graphite (5.7 J K^{-1} mol^{-1}). This difference is consistent with the atoms being linked less rigidly in graphite than in diamond and their thermal disorder being correspondingly greater. The standard molar entropies of ice, water, and water vapour at 25°C are, respectively, 45, 70, and 189 J K^{-1} mol^{-1}, and the increase in values corresponds to the increasing disorder on going from a solid to a liquid and then to a gas.

Heat capacities can be measured only with great difficulty at very low temperatures, particularly close to $T = 0$. However, as remarked in Section 2.9, it has been found that many nonmetallic substances have a heat capacity that obeys the **Debye T^3-law**:

$$\text{At temperatures close to } T = 0,\ C_{p,m} = aT^3 \qquad (4.11a)$$

where a is an empirical constant that depends on the substance and is found by fitting eqn 4.11a to a series of measurements of the heat capacity close to $T = 0$.

Table 4.2

*Standard molar entropies of some substances at 298.15 K**

Substance	$S_m^\ominus$/J K^{-1} mol^{-1}
Gases	
Ammonia, NH_3	192.5
Carbon dioxide, CO_2	213.7
Helium, He	126.2
Hydrogen, H_2	130.7
Neon, Ne	146.3
Nitrogen, N_2	191.6
Oxygen, O_2	205.1
Water vapour, H_2O	188.8
Liquids	
Benzene, C_6H_6	173.3
Ethanol, CH_3CH_2OH	160.7
Water, H_2O	69.9
Solids	
Calcium oxide, CaO	39.8
Calcium carbonate, $CaCO_3$	92.9
Copper, Cu	33.2
Diamond, C	2.4
Graphite, C	5.7
Lead, Pb	64.8
Magnesium carbonate, $MgCO_3$	65.7
Magnesium oxide, MgO	26.9
Sodium chloride, NaCl	72.1
Sucrose, $C_{12}H_{22}O_{11}$	360.2
Tin, Sn (white)	51.6
Sn (grey)	44.1

* See the Data section for more values.

With a determined, it is easy to deduce, as we show in Derivation 4.3, the molar entropy at low temperatures:

$$\text{At temperatures close to } T=0,\ S_m(T)=\tfrac{1}{3}C_{p,m}(T) \qquad (4.11b)$$

That is, the molar entropy at the low temperature T is equal to one-third of the constant-pressure heat capacity at that temperature.

A brief comment The Debye T^3-law strictly applies to C_V, but C_p and C_V converge as $T \to 0$, so we can use it for estimating C_p too without significant error at low temperatures.

Derivation 4.3

Entropies close to $T=0$

Once again, we use the general expression, eqn 4.4, for the entropy change accompanying a change of temperature deduced in Derivation 4.2, with ΔS interpreted as $S(T_f) - S(T_i)$, taking molar values, and supposing that the heating takes place at constant pressure:

ΔS_m for the change in temperature from T_f to T_i

$$S_m(T_f) - S_m(T_i) = \int_{T_i}^{T_f} \frac{C_{p,m}}{T}\,dT$$

Eqn 4.4

If we set $T_i = 0$ and T_f some general temperature T, we transform this expression into

Set $T_f = T$

$$S_m(T) - S_m(0) \quad \int_0^T \frac{C_{p,m}}{T}\,dT$$

Set $T_i = 0$

According to the Third Law, $S(0) = 0$, and according to the Debye T^3-law, $C_{p,m} = aT^3$, so

$C_{p,m}$

$$S_m(T) = \int_0^T \frac{aT^3}{T}\,dT = a\int_0^T T^2 dT$$

Set $S_m(0) = 0$

At this point we can use the standard integral

$$\int x^2 dx = \tfrac{1}{3}x^3 + \text{constant}$$

to write

$$\int_0^T T^2 dT = (\tfrac{1}{3}T^3 + \text{constant})\Big|_0^T$$

$$= (\tfrac{1}{3}T^3 + \text{constant}) - \text{constant}$$

$$= \tfrac{1}{3}T^3$$

We can conclude that

$$S_m(T) = \tfrac{1}{3}aT^3 = \tfrac{1}{3}C_{p,m}(T)$$

as in eqn 4.11b.

4.8 The statistical entropy

We have referred frequently to 'molecular disorder' and have interpreted the thermodynamic quantity of entropy in terms of this so far ill-defined concept. The concept of disorder, however, can be expressed precisely and used to calculate entropies. The procedures required will be described in Chapter 22, for they draw on information that we have not yet encountered. However, it is possible to understand the basis of the approach that we use there, and see how it illuminates what we have achieved so far.

The fundamental equation that we need was originally proposed by Ludwig Boltzmann towards the end of the nineteenth century (and is carved as his epitaph on his tombstone):

$$S = k \ln W \tag{4.12}$$

The constant k is **Boltzmann's constant**, a fundamental constant with the value 1.381×10^{-23} J K^{-1}. This value is chosen so that the values of the entropy calculated from Boltzmann's formula coincide with those calculated from heat capacity data. However, it turns out that the gas constant R is equal to $N_A k$, where N_A is Avogadro's constant, and the fact that the gas constant occurs in many contexts, even those not involving gases, is due to it being Boltzmann's constant in disguise. The quantity W is the number of ways that the molecules of the sample can be arranged yet correspond to the same total energy and formally is called the 'weight' of a 'configuration' of the sample.

A brief illustration Suppose we had a tiny system of four molecules A, B, C, and D that could occupy three equally spaced levels of energies 0, ε, and 2ε, and we know that the total energy is 4ε. The 19 arrangements shown in Fig. 4.11 are possible, so $W = 19$.

First, we can readily verify that Boltzmann's formula agrees with the Third-Law value $S(0) = 0$. When $T = 0$, all the molecules must be in the lowest possible energy level. Because there is just a single arrangement of the molecules, $W = 1$, and as $\ln 1 = 0$, eqn 4.12 gives $S = 0$ too.

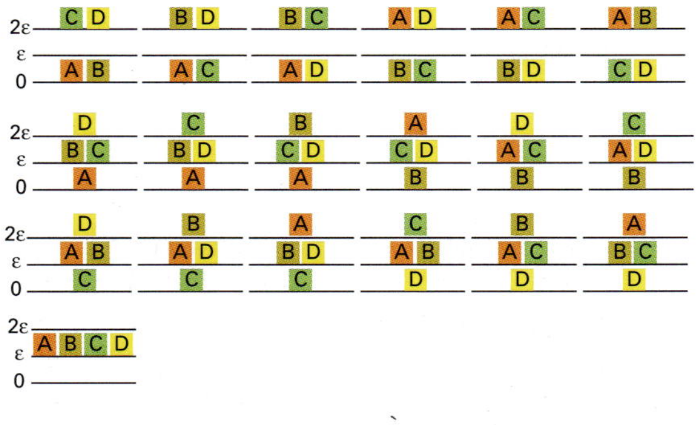

Fig. 4.11 The 19 arrangements of four molecules (represented by the blocks) in a system with three energy levels and a total energy of 4ε.

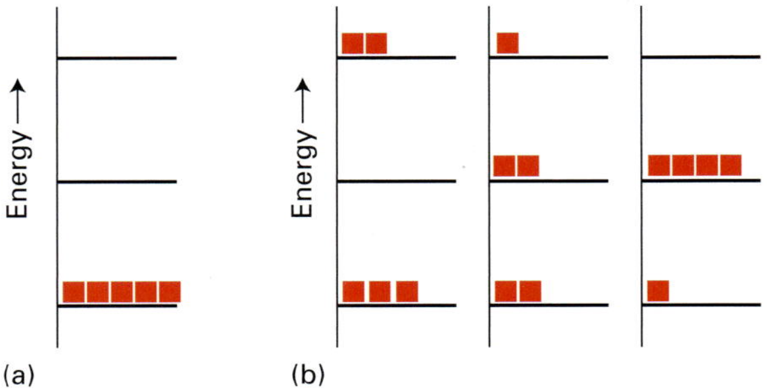

Fig. 4.12 The arrangements of molecules over the available energy levels determines the value of the statistical entropy. (a) At $T = 0$, there is only one arrangement possible: all the molecules must be in the lowest energy state. (b) When $T > 0$ several arrangements may correspond to the same total energy. In this simple case, $W = 3$.

Next, we can see that Boltzmann's formula is consistent with the entropy of a substance increasing with temperature. When $T > 0$, the molecules of a sample can occupy energy levels above the lowest one, and now many different arrangements of the molecules will correspond to the same total energy (Fig. 4.12). That is, when $T > 0$, $W > 1$ and according to eqn 4.12 the entropy rises above zero (because $\ln W > 0$ when $W > 1$).

Boltzmann's expression is also consistent with the entropy of a gas increasing as the volume it occupies is increased. When we have encountered quantum theory (in Chapter 12), we shall see that the energy levels of particles confined to a box-like region become closer together as the box expands (Fig. 4.13). If we take this model to represent a gas, then we can appreciate that as the container (the box) expands, the levels occupied by the molecules get closer together, and there are more ways of arranging the molecules for a given total energy. That is, as the container expands, W increases, and therefore S increases too. It is no coincidence that the thermodynamic expression for ΔS (eqn 4.2) is proportional to a logarithm: the logarithm in Boltzmann's formula turns out to lead to the same logarithmic expression (see Chapter 22).

4.9 Residual entropies

Boltzmann's formula provides an explanation of a rather startling conclusion: the entropy of some

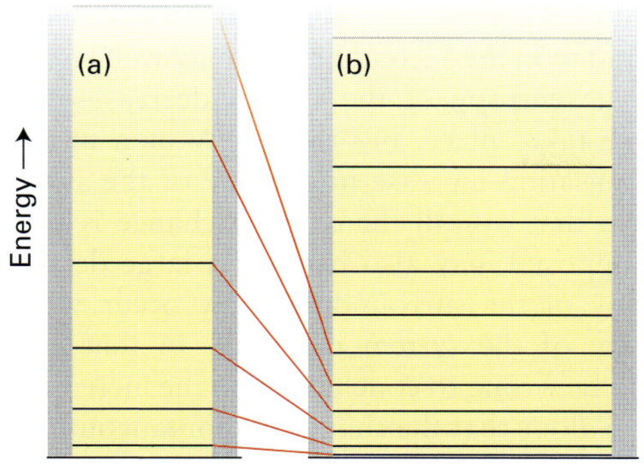

Fig. 4.13 As a box expands, the energy levels of the particles inside it come closer together. At a given temperature, the number of arrangements corresponding to the same total energy is greater when the energy levels are closely spaced than when they are far apart.

substances is greater than zero at $T = 0$, apparently contrary to the Third Law. When the entropy of carbon monoxide gas is measured thermodynamically (from heat capacity and boiling point data), it is found that $S_m^{\ominus}$ (298 K) = 192 J K^{-1} mol^{-1}. However, when Boltzmann's formula is used and the relevant molecular data included, the standard molar entropy is calculated as 198 J K^{-1} mol^{-1}. One explanation might be that the thermodynamic calculation failed to take into account a phase transition in solid carbon monoxide, which could have contributed the missing 6 J K^{-1} mol^{-1}. An alternative explanation is that the CO molecules are disordered in the solid, even at $T = 0$, and that there is a contribution to the entropy at $T = 0$ from positional disorder that is frozen in. This contribution is called the **residual entropy** of a solid.

We can estimate the value of the residual entropy by using Boltzmann's formula and supposing that at $T = 0$ each CO molecule can lie in either of two orientation (Fig. 4.14). Then the total number of ways of arranging N molecules is $(2 \times 2 \times 2 \ldots .)_{N \text{ times}} = 2^N$. Then

$$S = k \ln 2^N = Nk \ln 2$$

(We used $\ln x^a = a \ln x$.) The molar residual entropy is obtained by replacing N by Avogadro's constant:

$$S_m = N_A k \ln 2 = R \ln 2$$

This expression evaluates to 5.8 J K^{-1} mol^{-1}, in good agreement with the value needed to bring the thermodynamic value into line with the statistical value, for instead of taking $S_m(0) = 0$ in the thermodynamic calculation, we should take $S_m(0) = 5.8$ J K^{-1} mol^{-1}.

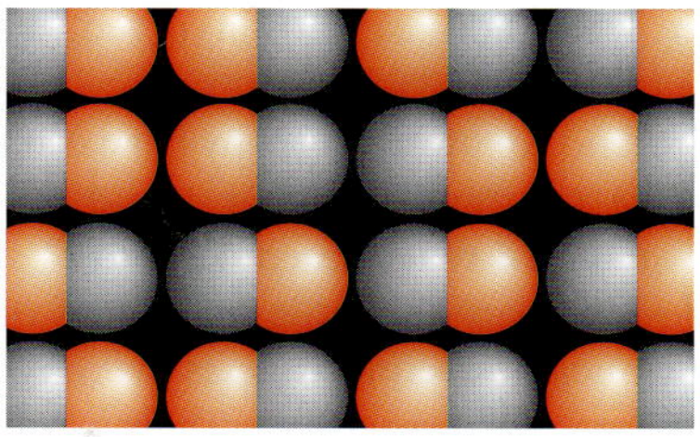

Fig. 4.14 The positional disorder of a substance that accounts for the residual entropy of molecules that can adopt either of two orientations at $T = 0$ (in this case, CO). If there are N molecules in the sample, there are 2^N possible arrangements with the same energy.

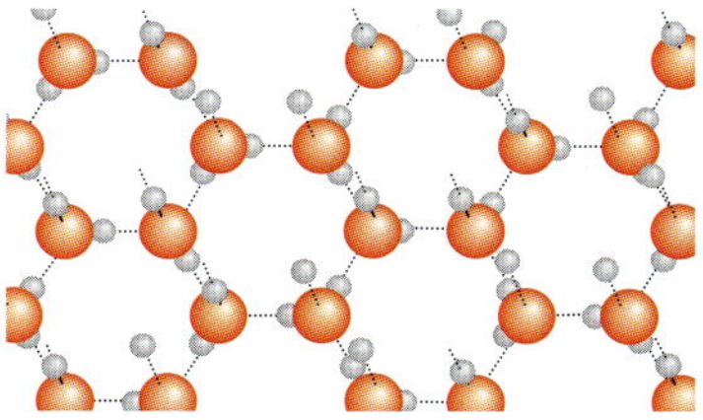

Fig. 4.15 The origin of the residual entropy of ice is the randomness in the location of the hydrogen atom in the O—H···O bonds between neigbouring molecules. Note that each molecule has two short O—H bonds and two long O···H hydrogen bonds. This schematic diagram shows one possible arrangement.

A brief illustration Ice has a residual entropy of 3.4 J K^{-1} mol^{-1}. This value can be traced to the positional disorder of the location of the H atoms that lie between neighbouring molecules. Thus, although each H_2O molecule has two short O—H covalent bonds and two long O···H—O bonds, there is a randomness in which bonds are long and which are short (Fig. 4.15). When the statistics of the disorder are analysed for a sample that contains N molecules, it turns out that $W = (\frac{3}{2})^N$. It follows that the residual entropy is expected to be $S = k \ln (\frac{3}{2})^N = Nk \ln \frac{3}{2}$, and therefore the molar residual entropy is $S_m = R \ln \frac{3}{2}$, which evaluates to 3.4 J K^{-1} mol^{-1}, in agreement with the experimental value.

4.10 The standard reaction entropy

Now we move into the arena of chemistry, where reactants are transformed into products. When there is a net formation of a gas in a reaction, as in a combustion, we can usually anticipate that the entropy increases. When there is a net consumption of gas, as in photosynthesis, it is usually safe to predict that the entropy decreases. However, for a quantitative value of the change in entropy, and to predict the sign of the change when no gases are involved, we need to make an explicit calculation.

The difference in molar entropy between the products and the reactants in their standard states is called the **standard reaction entropy**, $\Delta_r S^\ominus$. It can be expressed in terms of the molar entropies of the substances in much the same way as we have already used for the standard reaction enthalpy:

$$\Delta_r S^\ominus = \sum \nu S_m^\ominus(\text{products}) - \sum \nu S_m^\ominus(\text{reactants}) \quad (4.13)$$

where the ν are the stoichiometric coefficients in the chemical equation.

A brief illustration For the reaction $2\,H_2(g) + O_2(g) \rightarrow 2\,H_2O(l)$ we expect a negative entropy of reaction as gases are consumed. To find the explicit value we use the values in the Data section to write

$$\Delta_r S^\ominus = 2S_m^\ominus(H_2O, l) - \{2S_m^\ominus(H_2, g) + S_m^\ominus(O_2, g)\}$$
$$= 2(70\ \text{J K}^{-1}\ \text{mol}^{-1}) - \{2(131\ \text{J K}^{-1}\ \text{mol}^{-1}) + (205\ \text{J K}^{-1}\ \text{mol}^{-1})\}$$
$$= -327\ \text{J K}^{-1}\ \text{mol}^{-1}$$

A note on good practice Do not make the mistake of setting the standard molar entropies of elements equal to zero: they have nonzero values (provided $T > 0$), as we have already discussed.

Self-test 4.8

(a) Calculate the standard reaction entropy for $N_2(g) + 3\,H_2(g) \rightarrow 2\,NH_3(g)$ at 25°C. (b) What is the change in entropy when 2 mol H_2 reacts?

[*Answer:* (a) (Using values from Table 4.2) $-198.7\ \text{J K}^{-1}\ \text{mol}^{-1}$; (b) $-132.5\ \text{J K}^{-1}$]

4.11 The spontaneity of chemical reactions

The result of the calculation in the illustration should be rather surprising at first sight. We know that the reaction between hydrogen and oxygen is spontaneous and, once initiated, that it proceeds with explosive violence. Nevertheless, the entropy change that accompanies it is negative: the reaction results in less disorder, yet it is spontaneous!

The resolution of this apparent paradox underscores a feature of entropy that recurs throughout chemistry: *it is essential to consider the entropy of both the system and its surroundings when deciding whether a process is spontaneous or not.* The reduction in entropy by 327 J K^{-1} mol^{-1} relates only to the system, the reaction mixture. To apply the Second Law correctly, we need to calculate the *total* entropy, the sum of the changes in the system and the surroundings that jointly compose the 'isolated system' referred to in the Second Law. It may well be the case that the entropy of the system decreases when a change takes place, but there may be a more than compensating increase in entropy of the surroundings so that overall the entropy change is positive. The opposite may also be true: a large decrease in entropy of the surroundings may occur when the entropy of the system increases. In that case we would be wrong to conclude from the increase of the system alone that the change is spontaneous.

Whenever considering the implications of entropy, we must always consider the total change of the system and its surroundings.

To calculate the entropy change in the surroundings when a reaction takes place at constant pressure, we use eqn 4.10, interpreting the ΔH in that expression as the reaction enthalpy. $\Delta_r H$. For example, for the water formation reaction in the preceding illustration, with $\Delta_r H^\ominus = -572$ kJ mol^{-1}, the change in entropy of the surroundings (which are maintained at 25°C, the same temperature as the reaction mixture) is

Use $-572\ \text{kJ mol}^{-1} = -572 \times 10^3\ \text{J mol}^{-1}$

$$\Delta_r S_{sur} = -\frac{\Delta_r H}{T} = -\frac{(-572 \times 10^3\ \text{J mol}^{-1})}{298\ \text{K}}$$
$$= +1.92 \times 10^3\ \text{J K}^{-1}\ \text{mol}^{-1}$$

From $(-)(-) = (+)$

Now we can see that the total entropy change is positive:

$$\Delta_r S_{total} = (-327\ \text{J K}^{-1}\ \text{mol}^{-1}) + (1.92 \times 10^3\ \text{J K}^{-1}\ \text{mol}^{-1})$$
$$= +1.59 \times 10^3\ \text{J K}^{-1}\ \text{mol}^{-1}$$

This calculation confirms that the reaction is spontaneous. In this case, the spontaneity is a result of the considerable disorder that the reaction generates in the surroundings: water is dragged into existence, even though $H_2O(l)$ has a lower entropy than the gaseous reactants, by the tendency of energy to disperse into the surroundings.

The Gibbs energy

One of the problems with entropy calculations is already apparent: we have to work out two entropy changes, the change in the system and the change in the surroundings, and then consider the sign of their sum. The great American theoretician J. W. Gibbs

(1839–1903), who laid the foundations of chemical thermodynamics towards the end of the nineteenth century, discovered how to combine the two calculations into one. The combination of the two procedures in fact turns out to be of much greater relevance than just saving a little labour, and throughout this text we shall see consequences of the procedure he developed.

4.12 Focusing on the system

The total entropy change that accompanies a process is $\Delta S_{\text{total}} = \Delta S + \Delta S_{\text{sur}}$, where ΔS is the entropy change for the system; for a spontaneous change, $\Delta S_{\text{total}} > 0$. If the process occurs at constant pressure and temperature, we can use eqn 4.10 to express the change in entropy of the surroundings in terms of the enthalpy change of the system, ΔH. When the resulting expression is inserted into this one, we obtain

At constant temperature and pressure:

Entropy change of system + Entropy change of surroundings

$$\Delta S_{\text{total}} = \Delta S \left[-\frac{\Delta H}{T} \right] \qquad (4.14)$$

Use eqn 4.10

The great advantage of this formula is that it expresses the total entropy change of the system and its surroundings in terms of properties of the system alone. The only restriction is to changes at constant pressure and temperature.

Now we take a very important step. First, we introduce the **Gibbs energy**, G, which is defined as

$$G = H - TS \qquad (4.15)$$

The Gibbs energy is commonly referred to as the 'free energy' and the 'Gibbs free energy'. Because H, T, and S are state functions, G is a state function too. A change in Gibbs energy, ΔG, at constant temperature arises from changes in enthalpy and entropy, and is

At constant temperature: $\Delta G = \Delta H - T\Delta S \qquad (4.16)$

By comparing eqns 4.14 and 4.16 we obtain

At constant temperature and pressure: $\Delta G = -T\Delta S_{\text{total}} \qquad (4.17)$

We see that at constant temperature and pressure, the change in Gibbs energy of a system is proportional to the overall change in entropy of the system plus its surroundings.

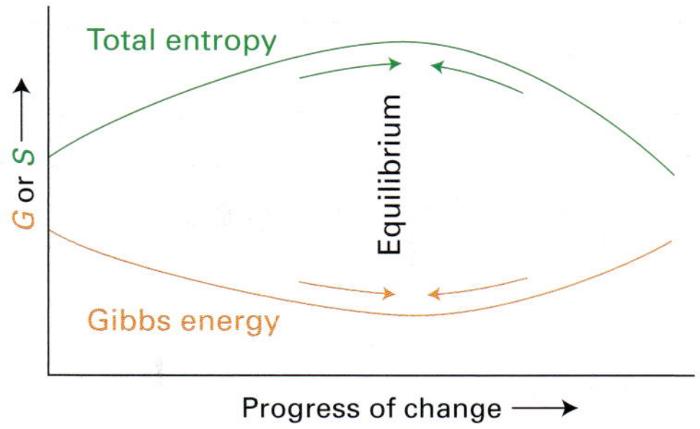

Fig. 4.16 The criterion of spontaneous change is the increase in total entropy of the system and its surroundings. Provided we accept the limitation of working at constant pressure and temperature, we can focus entirely on properties of the system, and express the criterion as a tendency to move to lower Gibbs energy.

4.13 Properties of the Gibbs energy

The difference in sign between ΔG and ΔS_{total} implies that the condition for a process being spontaneous changes from $\Delta S_{\text{total}} > 0$ in terms of the total entropy (which is universally true) to $\Delta G < 0$ in terms of the Gibbs energy (for processes occurring at constant temperature and pressure). That is, *in a spontaneous change at constant temperature and pressure, the Gibbs energy decreases* (Fig. 4.16).

It may seem more natural to think of a system as falling to a lower value of some property. However, it must never be forgotten that to say that a system tends to fall to lower Gibbs energy is only a modified way of saying that a system and its surroundings jointly tend towards a greater total entropy. The *only* criterion of spontaneous change is the total entropy of the system and its surroundings; the Gibbs energy merely contrives a way of expressing that total change in terms of the properties of the system alone, and is valid only for processes that occur at constant temperature and pressure. Every chemical reaction that is spontaneous under conditions of constant temperature and pressure, including those that drive the processes of growth, learning, and reproduction, are reactions that change in the direction of lower Gibbs energy, or—another way of expressing the same thing—result in the overall entropy of the system and its surroundings becoming greater.

A second feature of the Gibbs energy is that *the value of ΔG for a process gives the maximum nonexpansion work that can be extracted from the process at constant temperature and pressure*. By **nonexpansion work**, w', we mean any work other than that arising from the expansion of the system. It may include electrical work, if the process takes place inside an electrochemical or biological cell, or other kinds of mechanical work, such as the winding of a spring or the contraction of a muscle (we saw examples in Section 2.2). To demonstrate this property, we need to combine the First and Second Laws, and as shown in Derivation 4.4 we find

$$\text{At constant temperature and pressure: } \Delta G = w'_{\max} \tag{4.18}$$

A brief illustration Experiments show that for the formation of 1 mol $H_2O(l)$ at 25°C and 1 bar, $\Delta H = -286$ kJ and $\Delta G = -237$ kJ. It follows that up to 237 kJ of nonexpansion work can be extracted from the reaction between hydrogen and oxygen to produce 1 mol $H_2O(l)$ at 25°C. If the reaction takes place in a fuel cell—a device for using a chemical reaction to produce an electric current, like those used on the space shuttle—then up to 237 kJ of electrical energy can be generated for each mole of H_2O produced. This energy is enough to keep a 60 W light bulb shining for about 1.1 h. If no attempt is made to extract any energy as work, then 286 kJ (in general, ΔH) of energy will be produced as heat. If some of the energy released is used to do work, then up to 237 kJ (in general, ΔG) of nonexpansion work can be obtained.

Derivation 4.4

Maximum nonexpansion work

We need to consider infinitesimal changes, because dealing with reversible processes is then much easier. Our aim is to derive the relation between the infinitesimal change in Gibbs energy, dG, accompanying a process and the maximum amount of nonexpansion work that the process can do, dw'. We start with the infinitesimal form of eqn 4.16,

$$\text{At constant temperature: } dG = dH - TdS$$

where, as usual, d denotes an infinitesimal difference. A good rule in the manipulation of thermodynamic expressions is to feed in definitions of the terms that appear. We do this twice. First, we use the expression for the change in enthalpy at constant pressure (eqn 2.14b; $dH = dU + pdV$), and obtain

At constant temperature and pressure:

$$dG = dH - TdS = dU + pdV - TdS$$

Use $dH = dU + pdV$

Then we replace dU in terms of infinitesimal contributions from work and heat ($dU = dw + dq$):

$$dG = dU + pdV - TdS = dw + dq + pdV - TdS$$

Use $dU = dw + dq$

The work done on the system consists of expansion work, $-p_{ex}dV$, and nonexpansion work, dw'. Therefore,

$$dG = dw + dq + pdV - TdS$$

Use $dw = -p_{ex}dV + dw'$

$$= -p_{ex}dV + dw' + dq + pdV - TdS$$

This derivation is valid for any process taking place at constant temperature and pressure.

Now we specialize to a reversible change. For expansion work to be reversible, we need to match p and p_{ex}, in which case the first and fourth terms on the right cancel. Moreover, because the heat transfer is also reversible, we can replace dq by TdS, in which case the third and fifth terms also cancel:

$$dG = -\cancel{pdV} + dw' + \cancel{dq} + \cancel{pdV} - \cancel{TdS}$$

We are left with

$$\text{At constant temperature and pressure, for a reversible process: } dG = dw'_{rev}$$

Maximum work is done during a reversible change (Section 2.3), so another way of writing this expression is

$$\text{At constant temperature and pressure: } dG = dw'_{\max}$$

Because this relation holds for each infinitesimal step between the specified initial and final states, it applies to the overall change too. Therefore, we obtain eqn 4.18.

Example 4.3

Estimating a change in Gibbs energy

Suppose a certain small bird has a mass of 30 g. What is the minimum mass of glucose that it must consume to fly to a branch 10 m above the ground? The change in Gibbs energy that accompanies the oxidation of 1.0 mol $C_6H_{12}O_6$(s) to carbon dioxide and water vapour at 25°C is −2828 kJ.

Strategy First, we need to calculate the work needed to raise a mass *m* through a height *h* on the surface of the Earth. As we saw in eqn 2.1, this work is equal to *mgh*, where *g* is the acceleration of free fall. This work, which is nonexpansion work, can be identified with ΔG. We need to determine the amount of substance that corresponds to the required change in Gibbs energy, and then convert that amount to a mass by using the molar mass of glucose.

Solution The nonexpansion work to be done is

$$w' = mgh = (30 \times 10^{-3}\ \text{kg}) \times (9.81\ \text{m s}^{-2}) \times (10\ \text{m})$$
$$= 3.0 \times 9.81 \times 1.0 \times 10^{-1}\ \text{J}$$

(because 1 kg m^2 s^{-2} = 1 J). The amount, *n*, of glucose molecules required for oxidation to give a change in Gibbs energy of this value given that 1 mol provides 2828 kJ is

$$n = \frac{3.0\times9.81\times1.0\times10^{-1}\ \text{J}}{2.828\times10^{6}\ \text{J mol}^{-1}} = \frac{3.0\times9.81\times1.0\times10^{-7}}{2.828}\ \text{mol}$$

Therefore, because the molar mass, *M*, of glucose is 180 g mol^{-1}, the mass, *m*, of glucose that must be oxidized is

$$m = nM = \left(\frac{3.0\times9.81\times1.0\times10^{-7}}{2.828}\ \text{mol}\right)\times(180\ \text{g mol}^{-1})$$
$$= 1.9\times10^{-4}\ \text{g}$$

That is, the bird must consume at least 0.19 mg of glucose for the mechanical effort (and more if it thinks about it).

Self-test 4.9

A hard-working human brain, perhaps one that is grappling with physical chemistry, operates at about 25 W (1 W = 1 J s^{-1}). What mass of glucose must be consumed to sustain that power output for an hour?

[*Answer:* 5.7 g]

The great importance of the Gibbs energy in chemistry is becoming apparent. At this stage, we see that it is a measure of the nonexpansion work resources of chemical reactions: if we know ΔG, then we know the maximum nonexpansion work that we can obtain by harnessing the reaction in some way. In some cases, the nonexpansion work is extracted as electrical energy. This is the case when the reaction takes place in an electrochemical cell, of which a fuel cell is a special case, as we see in Chapter 9. In other cases, the reaction may be used to build other molecules. This is the case in biological cells, where the Gibbs energy available from the hydrolysis of ATP (adenosine triphosphate) to ADP is used to build proteins from amino acids, to power muscular contraction, and to drive the neuronal circuits in our brains.

Some insight into the physical significance of *G* itself comes from its definition as *H* − *TS*. The enthalpy is a measure of the energy that can be obtained from the system as heat. The term *TS* is a measure of the quantity of energy stored in the *random* motion of the molecules making up the sample. Work, as we have seen, is energy transferred in an orderly way, so we cannot expect to obtain work from the energy stored randomly. The difference between the total stored energy and the energy stored randomly, *H* − *TS*, is available for doing work, and we recognize that difference as the Gibbs energy. In other words, the Gibbs energy is the energy stored in the orderly motion and arrangement of the molecules in the system.

4.19 Octane is typical of the components of gasoline. Estimate (a) the entropy of vaporization, (b) the enthalpy of vaporization of octane, which boils at 126°C at 1 atm.

4.20 Suppose that the weight of a configuration of N molecules in a gas of volume V is proportional to V^N. Use Boltzmann's formula to deduce the change in entropy when the gas expands isothermally.

4.21 An $FClO_3$ molecule can adopt four orientations in the solid with negligible difference in energy. What is its residual molar entropy?

4.22 Without performing a calculation, estimate whether the standard entropies of the following reactions are positive or negative:

(a) Ala—Ser—Thr—Lys—Gly—Arg—Ser $\xrightarrow{\text{trypsin}}$ Ala—Ser—Thr—Lys + Gly—Arg

(b) $N_2(g) + 3\,H_2(g) \rightarrow 2\,NH_3(g)$

(c) $ATP^{4-}(aq) + 2\,H_2O(l) \rightarrow ADP^{3-}(aq) + HPO_4^{2-}(aq) + H_3O^+(aq)$

4.23 Calculate the standard reaction entropy at 298 K of

(a) $2\,CH_3CHO(g) + O_2(g) \rightarrow 2\,CH_3COOH(l)$

(b) $2\,AgCl(s) + Br_2(l) \rightarrow 2\,AgBr(s) + Cl_2(g)$

(c) $Hg(l) + Cl_2(g) \rightarrow HgCl_2(s)$

(d) $Zn(s) + Cu^{2+}(aq) \rightarrow Zn^{2+}(aq) + Cu(s)$

(e) $C_{12}H_{22}O_{11}(s) + 12\,O_2(g) \rightarrow 12\,CO_2(g) + 11\,H_2O(l)$

4.24 Suppose that when you exercise, you consume 100 g of glucose and that all the energy released as heat remains in your body at 37°C. What is the change in entropy of your body?

4.25 Calculate the standard reaction entropy and the change in entropy of the surroundings (at 298 K) of the reaction $N_2(g) + 3\,H_2(g) \rightarrow 2\,NH_3(g)$.

4.26 The constant-pressure molar heat capacities of linear gaseous molecules are approximately $\frac{7}{2}R$ and those of non-linear gaseous molecules are approximately $4R$. Estimate the change in standard reaction entropy of the following two reactions when the temperature is increased by 10 K at constant pressure:

(a) $2\,H_2(g) + O_2(g) \rightarrow 2\,H_2O(g)$

(b) $CH_4(g) + 2\,O_2(g) \rightarrow CO_2(g) + 2\,H_2O(g)$

4.27 Use the information you deduced in Exercise 4.26 to calculate the standard Gibbs energy of reaction of $N_2(g) + 3\,H_2(g) \rightarrow 2\,NH_3(g)$.

4.28 In a particular biological reaction taking place in the body at 37°C, the change in enthalpy was −135 kJ mol^{-1} and the change in entropy was −136 J K^{-1} mol^{-1}. (a) Calculate the change in Gibbs energy. (b) Is the reaction spontaneous? (c) Calculate the total change in entropy of the system and the surroundings.

4.29 The change in Gibbs energy that accompanies the oxidation of $C_6H_{12}O_6(s)$ to carbon dioxide and water vapour at 25°C is −2828 kJ mol^{-1}. How much glucose does a person of mass 65 kg need to consume to climb through 10 m?

4.30 Fuel cells are being developed that make use of organic fuels; in due course they might be used to power tiny intraveinous machines for carrying out repairs on diseased tissue. What is the maximum nonexpansion work that can be obtained from the metabolism of 1.0 mg of sucrose to carbon dioxide and water?

4.31 The formation of glutamine from glutamate and ammonium ions requires 14.2 kJ mol^{-1} of energy input. It is driven by the hydrolysis of ATP to ADP mediated by the enzyme glutamine synthetase. (a) Given that the change in Gibbs energy for the hydrolysis of ATP corresponds to $\Delta_r G = -31$ kJ mol^{-1} under the conditions prevailing in a typical cell, can the hydrolysis drive the formation of glutamine? (b) How many moles of ATP must be hydrolysed to form 1 mol of glutamine?

4.32 The hydrolysis of acetyl phosphate has $\Delta_r G = -42$ kJ mol^{-1} under typical biological conditions. If acetyl phosphate were to be synthesized by coupling to the hydrolysis of ATP, what is the minimum number of ATP molecules that would need to be involved?

4.33 Suppose that the radius of a typical cell is 10 μm and that inside it 10^6 ATP molecules are hydrolysed each second. What is the power density of the cell in watts per cubic metre (1 W = 1 J s^{-1}). A computer battery delivers about 15 W and has a volume of 100 cm^3. Which has the greater power density, the cell or the battery? (For data, see Exercise 4.32.)

Projects

The symbol ‡ indicates that calculus is required.

4.34‡ Equation 4.3 is based on the assumption that the heat capacity is independent of temperature. Suppose, instead, that the heat capacity depends on temperature as $C = a + bT + c/T^2$ (as was explored in Section 2.9). Find an expression for the change of entropy accompanying heating from T_i to T_f. *Hint:* See Derivation 4.2.

4.35 Here we explore the thermodynamics of refrigerators and heat pumps. (a) Show that the best *coefficient of cooling performance*, c_{cool}, the ratio of the energy extracted as heat at T_{cold} to the energy supplied as work in a perfect refrigerator, is $c_{cool} = T_{cold}/(T_{hot} - T_{cold})$. What is the maximum rate of extraction of energy as heat in a domestic refrigerator rated at 200 W operating at 5.0°C in a room at 22°C? (b) Show that the best *coefficient of heating performance*, c_{warm}, the ratio of the energy produced as heat at T_{hot} to the energy supplied as work in a perfect heat pump, is $c_{warm} = T_{hot}/(T_{hot} - T_{cold})$. What is the maximum power rating of a heat pump that consumes power at 2.5 kW operating at 18.0°C and warming a room at 22°C?

Chapter 5

Physical equilibria: pure substances

Boiling, freezing, and the conversion of graphite to diamond are all examples of **phase transitions**, or changes of phase without change of chemical composition. Many phase changes are common everyday phenomena and their description is an important part of physical chemistry. They occur whenever a solid changes into a liquid, as in the melting of ice, or a liquid changes into a vapour, as in the vaporization of water in our lungs. They also occur when one solid phase changes into another, as in the conversion of graphite into diamond under high pressure, or the conversion of one phase of iron into another as it is heated in the process of steelmaking. The tendency of a substance to form a liquid crystal, a distinct phase with properties intermediate between those of solid and liquid, guides the design of displays for electronic devices. Phase changes are important geologically too; for example, calcium carbonate is typically deposited as aragonite, but then gradually changes into another crystal form, calcite.

The thermodynamics of transition

The Gibbs energy, $G = H - TS$, of a substance will be at centre stage in all that follows. We need to know how its value depends on the pressure and temperature. As we work out these dependencies, we shall acquire deep insight into the thermodynamic properties of matter and the transitions it can undergo.

5.1 The condition of stability

First, we need to establish the importance of the *molar* Gibbs energy, $G_m = G/n$, in the discussion of phase transitions of a pure substance. The molar

Gibbs energy, an intensive property, depends on the phase of the substance. For instance, the molar Gibbs energy of liquid water is in general different from that of water vapour at the same temperature and pressure. When an amount n of the substance changes from phase 1 (for instance, liquid), with molar Gibbs energy $G_m(1)$ to phase 2 (for instance, vapour) with molar Gibbs energy $G_m(2)$, the change in Gibbs energy is

$$\Delta G = nG_m(2) - nG_m(1) = n\{G_m(2) - G_m(1)\}$$

We know that a spontaneous change at constant temperature and pressure is accompanied by a negative value of ΔG. This expression shows, therefore, that a change from phase 1 to phase 2 is spontaneous if the molar Gibbs energy of phase 2 is lower than that of phase 1. In other words, *a substance has a spontaneous tendency to change into the phase with the lowest molar Gibbs energy*.

If at a certain temperature and pressure the solid phase of a substance has a lower molar Gibbs energy than its liquid phase, then the solid phase is thermodynamically more stable and the liquid will (or at least has a tendency to) freeze. If the opposite is true, the liquid phase is thermodynamically more stable and the solid will melt. For example, at 1 atm, ice has a lower molar Gibbs energy than liquid water when the temperature is below 0°C, and under these conditions water converts spontaneously into ice.

Self-test 5.1

The Gibbs energy of transition from metallic white tin (α-Sn) to nonmetallic grey tin (β-Sn) is +0.13 kJ mol^{-1} at 298 K. Which is the reference state (Section 3.5) of tin at this temperature?

[*Answer:* white tin]

5.2 The variation of Gibbs energy with pressure

To discuss how phase transitions depend on the pressure, we need to know how the molar Gibbs energy varies with pressure. We show in Derivation 5.1 that when the temperature is held constant and the pressure is changed by a small amount Δp, the molar Gibbs energy of a substance changes by

$$\Delta G_m = V_m \Delta p \qquad (5.1)$$

where V_m is the molar volume of the substance. This expression is valid when the molar volume is constant in the pressure range of interest.

Derivation 5.1

The variation of G with pressure

We start with the definition of Gibbs energy, $G = H - TS$, and change the temperature, volume, and pressure by an infinitesimal amount. As a result, H changes to $H + dH$, T changes to $T + dT$, S changes to $S + dS$, and G changes to $G + dG$. After the change

G changes to | H changes to | T changes to | S changes to

$$G + dG = H + dH - (T + dT)(S + dS)$$
$$= H + dH - TS - TdS - SdT - dTdS$$

($H - TS = G$; $dTdS \to 0$)

The G on the left cancels the $H - TS$ on the right, the doubly infinitesimal $dTdS$ can be neglected, and we are left with

$$dG = dH - TdS - SdT$$

To make progress, we need to know how the enthalpy changes. From its definition $H = U + pV$, in a similar way (letting U change to $U + dU$, and so on, and neglecting the doubly infinitesimal term $dpdV$) we can write

$$dH = dU + pdV + Vdp$$

On substituting this expression into the previous one, we obtain

($dH = dU + pdV + Vdp$)

$$dG = dH - TdS - SdT = dU + pdV + Vdp - TdS - SdT$$

At this point we need to know how the internal energy changes, and write

$$dU = dq + dw$$

If initially we consider only reversible changes, we can replace dq by TdS (because $dS = dq_{rev}/T$) and dw by $-pdV$ (because $dw = -p_{ex}dV$ and $p_{ex} = p$ for a reversible change), and obtain

$$dU = TdS - pdV$$

Now we substitute this expression into the expression for dH and that expression into the expression for dG and obtain

($dU = TdS - pdV$)

$$dG = dU + pdV + Vdp - TdS - SdT$$
$$= TdS - pdV + pdV + Vdp - TdS - SdT$$

We are left with the important result that

$$dG = Vdp - SdT \qquad (5.2)$$

Now here is a subtle but important point. To derive this result we have supposed that the changes in conditions have been made reversibly. However, G is a state function, and so the change in its value is independent of path. Therefore, eqn 5.2 is valid for any change, not just a reversible change.

At this point we decide to keep the temperature constant, and set $dT = 0$ in eqn 5.2; this leaves

$$dG = Vdp$$

and, for molar quantities, $dG_m = V_m dp$. This expression is exact, but applies only to an infinitesimal change in the pressure. For an observable change, we replace dG_m and dp by ΔG_m and Δp, respectively, and obtain eqn 5.1, provided the molar volume is constant over the range of interest.

A note of good practice When confronted with a proof in thermodynamics, go back to fundamental definitions (as we did three times in succession in this derivation: first of G, then of H, and finally of U).

Equation 5.1 tells us that, because all molar volumes are positive, *the molar Gibbs energy increases* ($\Delta G_m > 0$) *when the pressure increases* ($\Delta p > 0$). We also see that, for a given change in pressure, the resulting change in molar Gibbs energy is greatest for substances with large molar volumes. Therefore, because the molar volume of a gas is much larger than that of a condensed phase (a liquid or a solid), the dependence of G_m on p is much greater for a gas than for a condensed phase. For most substances (water is an important exception), the molar volume of the liquid phase is greater than that of the solid phase. Therefore, for most substances, the slope of a graph of G_m against p is greater for a liquid than for a solid. These characteristics are illustrated in Fig. 5.1.

As we see from Fig. 5.1, when we increase the pressure on a substance, the molar Gibbs energy of the gas phase rises above that of the liquid, then the molar Gibbs energy of the liquid rises above that of the solid. Because the system has a tendency to convert into the state of lowest molar Gibbs energy, the graphs show that at low pressures the gas phase is the most stable, then at higher pressures the liquid phase becomes the most stable, followed by the solid phase. In other words, under pressure the substance condenses to a liquid, and then further pressure can result in the formation of a solid.

We can use eqn 5.1 to predict the actual shape of graphs like those in Fig. 5.1. For a solid or liquid, the molar volume is almost independent of pressure, so eqn 5.1 is an excellent approximation to the change in molar Gibbs energy and with $\Delta G_m = G_m(p_f) - G_m(p_i)$ and $\Delta p = p_f - p_i$ we can write

$$G_m(p_f) = G_m(p_i) + V_m(p_f - p_i) \tag{5.3a}$$

This equation shows that the molar Gibbs energy of a solid or liquid increases linearly with pressure. However, because the molar volume of a condensed phase is so small, the dependence is very weak, and for the typical ranges of pressure normally of interest to us we can ignore the pressure dependence of G. The molar Gibbs energy of a gas, however, does depend on the pressure, and because the molar volume of a gas is large, the dependence is significant. We show in Derivation 5.2 that

$$G_m(p_f) = G_m(p_i) + RT\ln\frac{p_f}{p_i} \tag{5.3b}$$

This equation shows that the molar Gibbs energy increases logarithmically (as $\ln p$) with the pressure (Fig. 5.2). The flattening of the curve at high pressures reflects the fact that as V_m gets smaller, G_m becomes less responsive to pressure.

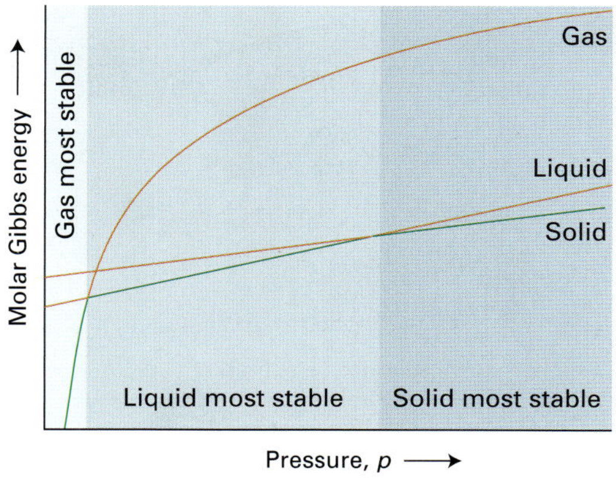

Fig. 5.1 The variation of molar Gibbs energy with pressure. The region where the molar Gibbs energy of a particular phase is least is shown by a dark line and the corresponding region of stability of each phase is indicated by the shading.

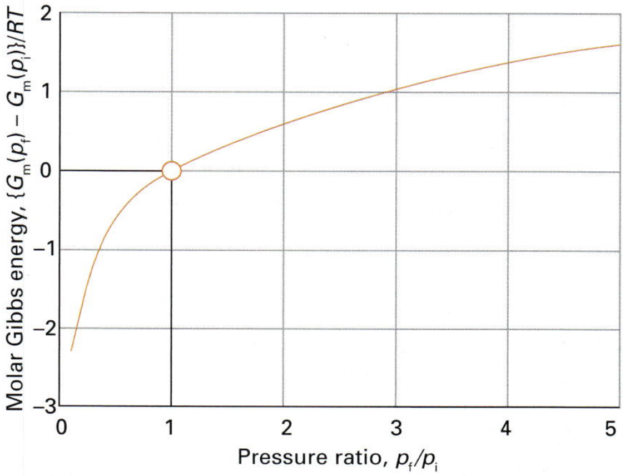

Fig. 5.2 The variation of the molar Gibbs energy of a perfect gas with pressure.

Derivation 5.2

The pressure variation of the Gibbs energy of a perfect gas

We start with the exact expression for the effect of an infinitesimal change in pressure obtained in Derivation 5.1, that $dG_m = V_m dp$. For a change in pressure from p_i to p_f, we need to add together (integrate) all these infinitesimal changes, and write

$$\Delta G_m = \int_{p_i}^{p_f} V_m dp$$

To evaluate the integral, we must know how the molar volume depends on the pressure. For a perfect gas $V_m = RT/p$. Then

Perfect gas, $V_m = RT/p$

$$\Delta G_m = \int_{p_i}^{p_f} V_m\, dp = \int_{p_i}^{p_f} \frac{RT}{p}\, dp = RT\int_{p_i}^{p_f} \frac{dp}{p}$$

Constant temperature

$$= RT \ln \frac{p_f}{p_i}$$

In the last line we have used the standard integral

$$\int \frac{dx}{x} = \ln x + \text{constant}$$

Finally, with $\Delta G_m = G_m(p_f) - G_m(p_i)$, we get eqn 5.3b.

5.3 The variation of Gibbs energy with temperature

Now we consider how the molar Gibbs energy varies with temperature. For small changes in temperature, we show in Derivation 5.3 that the change in molar Gibbs energy at constant pressure is

$$\Delta G_m = -S_m \Delta T \qquad (5.4)$$

where $\Delta G_m = G_m(T_f) - G_m(T_i)$ and $\Delta T = T_f - T_i$. This expression is valid provided the entropy of the substance is unchanged over the range of temperatures of interest.

Derivation 5.3

The variation of the Gibbs energy with temperature

The starting point for this short derivation is eqn 5.2, the expression obtained in Derivation 5.1 for the change in molar Gibbs energy when both the pressure and the temperature are changed by infinitesimal amounts. If we hold the pressure constant, $dp = 0$, and eqn 5.2 becomes (for molar quantities)

$$dG_m = -S_m dT$$

This expression is exact. If we suppose that the molar entropy is unchanged in the range of temperatures of interest, we can replace the infinitesimal changes by observable changes, and so obtain eqn 5.4.

Equation 5.4 tells us that, because molar entropy is positive, *an increase in temperature* ($\Delta T > 0$) *results in a decrease in* G_m ($\Delta G_m < 0$). We see that for a given change of temperature, the change in molar Gibbs energy is proportional to the molar entropy. For a given substance, there is more spatial disorder in the gas phase than in a condensed phase, so the molar entropy of the gas phase is greater than that for a condensed phase. It follows that the molar Gibbs energy falls more steeply with temperature for a gas than for a condensed phase. The molar entropy of the liquid phase of a substance is greater than that of its solid phase, so the slope is least steep for a solid. Figure 5.3 summarizes these characteristics.

Figure 5.3 also reveals the thermodynamic reason why substances melt and vaporize as the temperature is raised. At low temperatures, the solid phase has the lowest molar Gibbs energy and is therefore the most stable. However, as the temperature is raised, the molar Gibbs energy of the liquid phase falls below that of the solid phase, and the substance melts. At even higher temperatures, the molar Gibbs energy of the gas phase plunges down below that of the liquid phase, and the gas becomes the most stable phase. In other words, above a certain temperature, the liquid vaporizes to a gas.

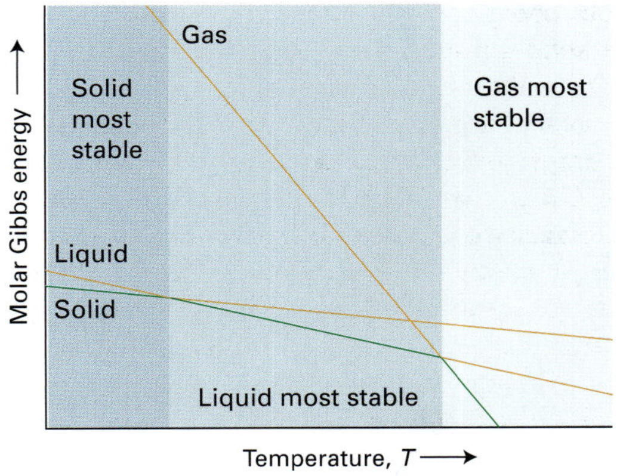

Fig. 5.3 The variation of molar Gibbs energy with temperature. All molar Gibbs energies decrease with increasing temperature. The regions of temperature over which the solid, liquid, and gaseous forms of a substance have the lowest molar Gibbs energy are indicated by the shading.

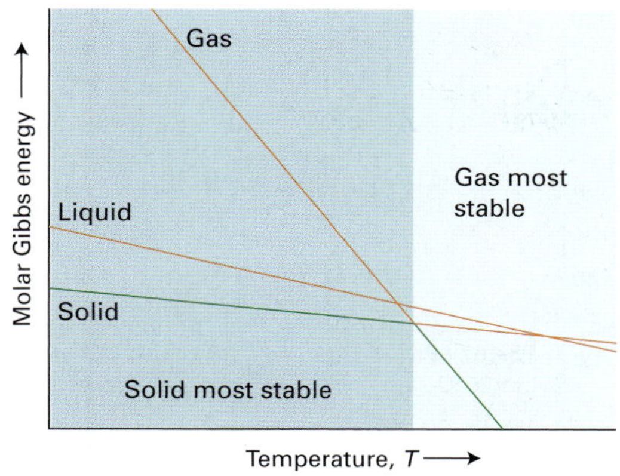

Fig. 5.4 If the line for the Gibbs energy of the liquid phase does not cut through the line for the solid phase (at a given pressure) before the line for the gas phase cuts through the line for the solid, the liquid is not stable at any temperature at that pressure. Such a substance sublimes.

We can also start to understand why some substances, such as carbon dioxide, sublime to a vapour without first forming a liquid. There is no fundamental requirement for the three lines to lie exactly in the positions we have drawn them in Fig. 5.3: the liquid line, for instance, could lie where we have drawn it in Fig. 5.4. Now we see that at no temperature (at the given pressure) does the liquid phase have the lowest molar Gibbs energy. Such a substance converts spontaneously directly from the solid to the vapour. That is, the substance sublimes.

The **transition temperature**, T_{trs}, between two phases, such as between liquid and solid or between ordered and disordered states of a protein, is the temperature, at a given pressure, at which the molar Gibbs energies of the two phases are equal. Above the solid–liquid transition temperature the liquid phase is thermodynamically more stable; below it, the solid phase is more stable. For example, at 1 atm, the transition temperature for ice and liquid water is 0°C and that for grey and white tin is 13°C. At the transition temperature itself, the molar Gibbs energies of the two phases are identical and there is no tendency for either phase to change into the other. At this temperature, therefore, the two phases are in equilibrium. At 1 atm, ice and liquid water are in equilibrium at 0°C and the two allotropes of tin are in equilibrium at 13°C.

As always when using thermodynamic arguments, it is important to keep in mind the distinction between the spontaneity of a phase transition and its rate. *Spontaneity is a tendency, not necessarily an actuality.* A phase transition predicted to be spontaneous may occur so slowly as to be unimportant in practice. For instance, at normal temperatures and pressures the molar Gibbs energy of graphite is 3 kJ mol^{-1} lower than that of diamond, so there is a thermodynamic tendency for diamond to convert into graphite. However, for this transition to take place, the carbon atoms of diamond must change their locations, and because the bonds between the atoms are so strong and large numbers of bonds must change simultaneously, this process is unmeasurably slow except at high temperatures. In gases and liquids the mobilities of the molecules normally allow phase transitions to occur rapidly, but in solids thermodynamic instability may be frozen in and a thermodynamically unstable phase may persist for thousands of years.

Phase diagrams

The **phase diagram** of a substance is a map showing the conditions of temperature and pressure at which its various phases are thermodynamically most stable (Fig. 5.5). For example, at point A in the illustration, the vapour phase of the substance is thermodynamically the most stable, but at C the liquid phase is the most stable.

The boundaries between regions in a phase diagram, which are called **phase boundaries**, show the values of p and T at which the two neighbouring phases are in equilibrium. For example, if the system

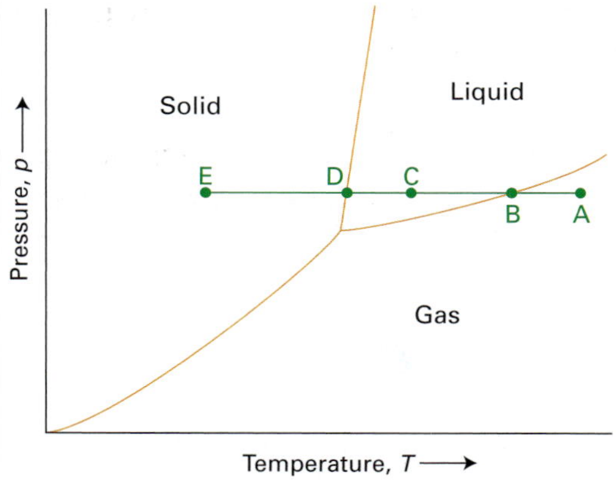

Fig. 5.5 A typical phase diagram, showing the regions of pressure and temperature at which each phase is the most stable. The phase boundaries (three are shown here) show the values of pressure and temperature at which the two phases separated by the line are in equilibrium. The significance of the letters A, B, C, D, and E (also referred to in Fig. 5.8) is explained in the text.

is arranged to have a pressure and temperature represented by point B, then the liquid and its vapour are in equilibrium (like liquid water and water vapour at 1 atm and 100°C). If the temperature is reduced at constant pressure, the system moves to point C where the liquid is stable (like water at 1 atm and at temperatures between 0°C and 100°C). If the temperature is reduced still further to D, then the solid and the liquid phases are in equilibrium (like ice and water at 1 atm and 0°C). A further reduction in temperature takes the system into the region where the solid is the stable phase.

5.4 Phase boundaries

The pressure of the vapour in equilibrium with its condensed phase is called the **vapour pressure** of the substance. Vapour pressure increases with temperature because, as the temperature is raised, more molecules have sufficient energy to leave their neighbours in the liquid.

The liquid–vapour boundary in a phase diagram is a plot of the vapour pressure against temperature. To determine the boundary, we can introduce a liquid into the near vacuum at the top of a mercury barometer and measure by how much the column is depressed (Fig. 5.6). To ensure that the pressure exerted by the vapour is truly the vapour pressure, we have to add enough liquid for some to remain after the vapour forms, for only then are the liquid and vapour phases in equilibrium. We can change the temperature and determine another point on the curve, and so on (Fig. 5.7).

Now suppose we have a liquid in a cylinder fitted with a piston. If we apply a pressure greater than the

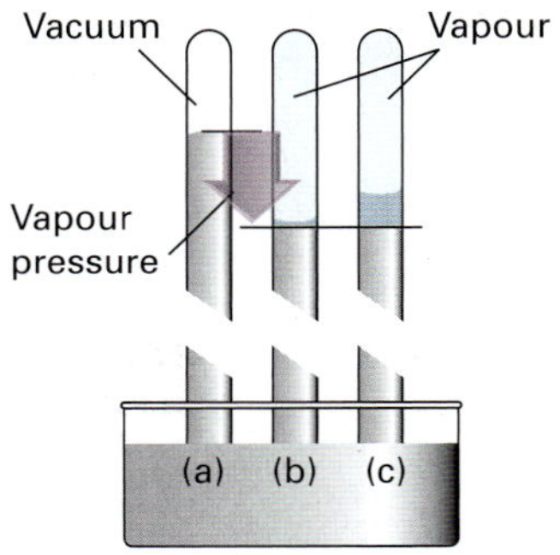

Fig. 5.6 When a small volume of water is introduced into the vacuum above the mercury in a barometer (a), the mercury is depressed (b) by an amount that is proportional to the vapour pressure of the liquid. (c) The same pressure is observed however much liquid is present (provided some is present).

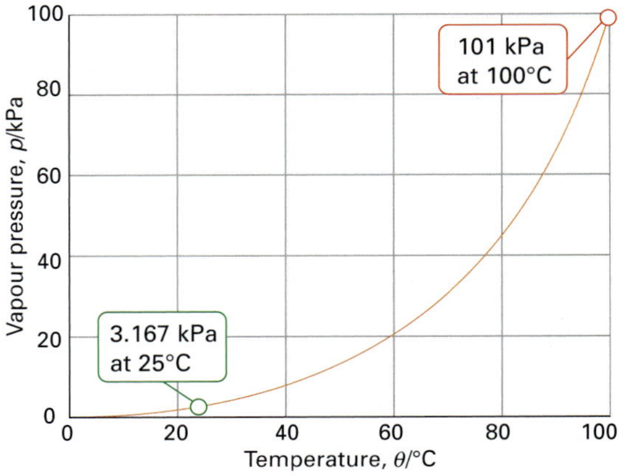

Fig. 5.7 The experimental variation of the vapour pressure of water with temperature.

vapour pressure of the liquid, the vapour is eliminated, the piston rests on the surface of the liquid, and the system moves to one of the points in the 'liquid' region of the phase diagram. Only a single phase is present. If instead we reduce the pressure on the system to a value below the vapour pressure, the system moves to one of the points in the 'vapour' region of the diagram. Reducing the pressure will involve pulling out the piston a long way, so that all the liquid evaporates; while any liquid is present, the pressure in the system remains constant at the vapour pressure of the liquid.

Self-test 5.2

What phase changes would be observed when a pressure of 7.0 kPa is applied to a sample of water in equilibrium with its vapour at 25°C, when its vapour pressure is 3.2 kPa?

[*Answer:* The sample condenses entirely to liquid.]

The same approach can be used to plot the solid–vapour boundary, which is a graph of the vapour pressure of the solid against temperature. The **sublimation vapour pressure** of a solid, the pressure of the vapour in equilibrium with a solid at a particular temperature, is usually much lower than that of a liquid.

A more sophisticated procedure is needed to determine the locations of solid–solid phase boundaries like that between calcite and aragonite, for instance, because the transition between two solid phases is more difficult to detect. One approach is to use **thermal analysis**, which takes advantage of the heat released during a transition. In a typical thermal

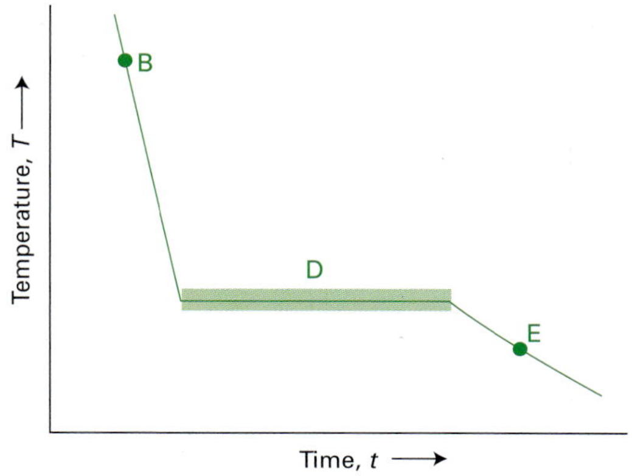

Fig. 5.8 The cooling curve for the B–E section of the horizontal line in Fig. 5.5. The halt at D corresponds to the pause in cooling while the liquid freezes and releases its enthalpy of transition. The halt lets us locate T_f even if the transition cannot be observed visually.

analysis experiment, a sample is allowed to cool and its temperature is monitored. When the transition occurs, energy is released as heat and the cooling stops until the transition is complete (Fig. 5.8). The transition temperature is obvious from the shape of the graph and is used to mark a point on the phase diagram. The pressure can then be changed, and the corresponding transition temperature determined.

Any point lying on a phase boundary represents a pressure and temperature at which there is a 'dynamic equilibrium' between the two adjacent phases. A state of **dynamic equilibrium** is one in which a reverse process is taking place at the same rate as the forward process. Although there may be a great deal of activity at a molecular level, there is no net change in the bulk properties or appearance of the sample. For example, any point on the liquid–vapour boundary represents a state of dynamic equilibrium in which vaporization and condensation continue at matching rates. Molecules are leaving the surface of the liquid at a certain rate, and molecules already in the gas phase are returning to the liquid at the same rate; as a result, there in no net change in the number of molecules in the vapour and hence no net change in its pressure. Similarly, a point on the solid–liquid curve represents conditions of pressure and temperature at which molecules are ceaselessly breaking away from the surface of the solid and contributing to the liquid. However, they are doing so at a rate that exactly matches that at which molecules already in the liquid are settling on to the surface of the solid and contributing to the solid phase.

5.5 The location of phase boundaries

Thermodynamics provides us with a way of predicting the location of the phase boundaries. Suppose two phases are in equilibrium at a given pressure and temperature. Then, if we change the pressure, we must adjust the temperature to a different value to ensure that the two phases remain in equilibrium. In other words, there must be a relation between the change in pressure, Δp, that we exert and the change in temperature, ΔT, we must make to ensure that the two phases remain in equilibrium. We show in Derivation 5.4 that the relation between the change in temperature and the change in pressure needed to maintain equilibrium is given by the **Clapeyron equation:**

$$\Delta p = \frac{\Delta_{trs}H}{T\Delta_{trs}V} \times \Delta T \qquad (5.5a)$$

where $\Delta_{trs}H$ is the enthalpy of transition and $\Delta_{trs}V$ is the volume of transition (the change in molar volume when the transition occurs). This form of the Clapeyron equation is valid for small changes in pressure and temperature, because only then can $\Delta_{trs}H$ and $\Delta_{trs}V$ be taken as constant across the range.

Derivation 5.4

The Clapeyron equation

This derivation is also based on eqn 5.2, the relation obtained in Derivation 5.1, that for infinitesimal changes in pressure and temperature, the molar Gibbs energy changes by $dG_m = V_m dp - S_m dT$.

Consider two phases 1 (for instance, a liquid) and 2 (a vapour). At a certain pressure and temperature the two phases are in equilibrium and $G_m(1) = G_m(2)$, where $G_m(1)$ is the molar Gibbs energy of phase 1 and $G_m(2)$ that of phase 2 (Fig. 5.9). Now change the pressure by an infinitesimal amount dp and the temperature by dT. The molar Gibbs energies of each phase change as follows:

Phase 1: $dG_m(1) = V_m(1)dp - S_m(1)dT$

Phase 2: $dG_m(2) = V_m(2)dp - S_m(2)dT$

where $V_m(1)$ and $S_m(1)$ are the molar volume and molar entropy of phase 1 and $V_m(2)$ and $S_m(2)$ are those of phase 2. The two phases were in equilibrium before the change, so the two molar Gibbs energies were equal. The two phases are still in equilibrium after the pressure and temperature are changed, so their two molar Gibbs energies are still equal. Therefore, the two *changes* in molar Gibbs energy must be equal, $dG_m(1) = dG_m(2)$, and we can write

$$\underbrace{V_m(1)dp - S_m(1)dT}_{dG_m(1)} = \underbrace{V_m(2)dp - S_m(2)dT}_{dG_m(2)} \qquad [dG_m(1) = dG_m(2)]$$

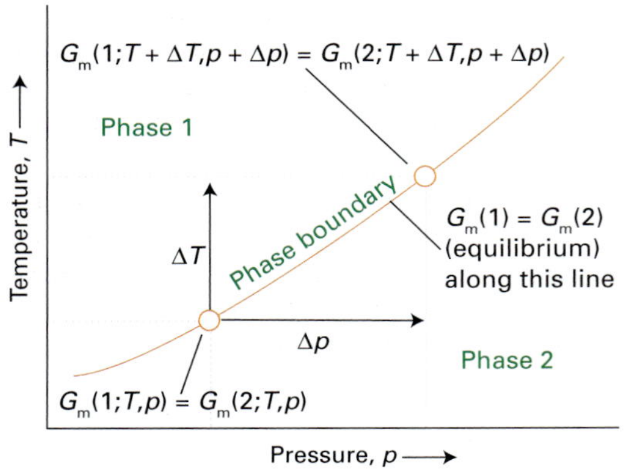

Fig. 5.9 At equilibrium, two phases have the same molar Gibbs energy. When the temperature is changed, for the two phases to remain in equilibrium, the pressure must be changed so that the Gibbs energies of the two phases remain equal.

This equation can be rearranged to

$$\{V_m(2) - V_m(1)\}dp = \{S_m(2) - S_m(1)\}dT$$

The entropy of transition, $\Delta_{trs}S$, is the difference between the two molar entropies, and the volume of transition, $\Delta_{trs}V$, is the difference between the molar volumes of the two phases:

$$\Delta_{trs}V = V_m(2) - V_m(1) \qquad \Delta_{trs}S = S_m(2) - S_m(1)$$

We can therefore write

$$\Delta_{trs}V dp = \Delta_{trs}S dT$$

or

$$dp = \frac{\Delta_{trs}S}{\Delta_{trs}V} dT$$

We saw in Chapter 4 that the transition entropy is related to the enthalpy of transition by $\Delta_{trs}S = \Delta_{trs}H/T_{trs}$, so we may also write

$$dp = \frac{\Delta_{trs}H}{T\Delta_{trs}V} dT \qquad (5.5b)$$

We have dropped the 'trs' subscript from the temperature because all the points on the phase boundary—the only points we are considering—are transition temperatures. This expression is exact. For variations in pressure and temperature small enough for $\Delta_{trs}H$ and $\Delta_{trs}V$ to be treated as constant, the infinitesimal changes dp and dT can be replaced by observable changes, and we obtain eqn 5.5a.

The Clapeyron equation tells us the slope (the value of $\Delta p/\Delta T$) of any phase boundary in terms of the enthalpy and volume of transition. For the solid–liquid phase boundary, the enthalpy of transition is the enthalpy of fusion, which is positive because melting is always endothermic. For most substances, the molar volume increases slightly on melting, so $\Delta_{trs}V$ is positive but small. It follows that the slope of the phase boundary is large and positive (up from left to right), and therefore that a large increase in pressure brings about only a small increase in melting temperature. Water, though, is quite different, for although its melting is endothermic, its molar volume decreases on melting (liquid water is denser than ice at 0°C, which is why ice floats on water), so $\Delta_{trs}V$ is small but negative. Consequently, the slope of the ice–water phase boundary is steep but negative (down from left to right). Now a large increase in pressure brings about a small lowering of the melting temperature of ice.

We cannot use eqn 5.5 to discuss the liquid–vapour phase boundary, except over very small ranges of temperature and pressure, because we cannot assume that the volume of the vapour, and therefore the volume of transition, is independent of pressure. However, if we suppose that the vapour behaves as a perfect gas, then it turns out (see Derivation 5.5) that the relation between a change in pressure and a change in temperature is given by the **Clausius–Clapeyron equation:**

$$\Delta(\ln p) = \frac{\Delta_{vap}H}{RT^2} \times \Delta T \qquad (5.6)$$

The Clausius–Clapeyron equation is an approximate equation for the slope of a plot of the logarithm of the vapor pressure against temperature ($\Delta \ln p/\Delta T$, Fig. 5.10). Moreover, it follows from eqn 5.6, as we also show in Derivation 5.5, that the vapour pressure

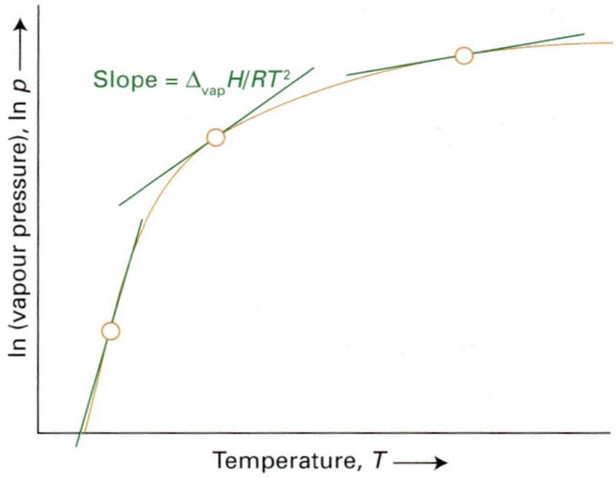

Fig. 5.10 The Clausius–Clapeyron equation gives the slope of a plot of the logarithm of the vapour pressure of a substance against the temperature. That slope at a given temperature is proportional to the enthalpy of vaporization of the substance.

p' at a temperature T' is related to the vapour pressure p at a temperature T by

$$\ln p' = \ln p + \frac{\Delta_{\text{vap}}H}{R}\left(\frac{1}{T} - \frac{1}{T'}\right) \qquad (5.7)$$

Derivation 5.5

The Clausius–Clapeyron equation

For the liquid–vapour boundary the 'trs' label in the exact form of the Clapeyron equation, eqn 5.5b in Derivation 5.4, becomes 'vap' and that equation can be written

$$\frac{dp}{dT} = \frac{\Delta_{\text{vap}}H}{T\Delta_{\text{vap}}V}$$

Because the molar volume of a gas is much larger than the molar volume of a liquid, the volume of vaporization, $\Delta_{\text{vap}}V = V_m(\text{g}) - V_m(\text{l})$, is approximately equal to the molar volume of the gas itself. Therefore, to a good approximation,

$\Delta_{\text{vap}}V = V_m(\text{g}) - V_m(\text{l})$ $\quad$ $V_m(\text{g}) >> V_m(\text{l})$

$$\frac{dp}{dT} = \frac{\Delta_{\text{vap}}H}{T\Delta_{\text{vap}}V} = \frac{\Delta_{\text{vap}}H}{T\{V_m(\text{g}) - V_m(\text{l})\}} \approx \frac{\Delta_{\text{vap}}H}{TV_m(\text{g})}$$

To make further progress, we can treat the vapour as a perfect gas and write its molar volume as $V_m = RT/p$. Then

$$\frac{dp}{dT} = \frac{\Delta_{\text{vap}}H}{TV_m(\text{g})} = \frac{\Delta_{\text{vap}}H}{T(RT/p)} = \frac{p\Delta_{\text{vap}}H}{RT^2}$$

$V_m(\text{g}) = RT/p$

and therefore

$$\frac{1}{p}\frac{dp}{dT} = \frac{\Delta_{\text{vap}}H}{RT^2}$$

A standard result of calculus is $d\ln x/dx = 1/x$, and therefore (by multiplying both sides by dx), $dx/x = d\ln x$. It follows that we may write the last equation as the Clausius–Clapeyron equation:

$$\frac{d\ln p}{dT} = \frac{\Delta_{\text{vap}}H}{RT^2}$$

Provided the range of temperature and pressure is small, the infinitesimal changes $d\ln p$ and dT can be replaced by measurable changes, and we obtain eqn 5.6.

To obtain the explicit expression for the vapour pressure at any temperature (eqn 5.7) we rearrange the equation we have just derived into

$$d\ln p = \frac{\Delta_{\text{vap}}H}{RT^2}dT$$

and integrate both sides. If the vapour pressure is p at a temperature T and p' at a temperature T', this integration takes the form

$$\int_{\ln p}^{\ln p'} d\ln p = \int_T^{T'} \frac{\Delta_{\text{vap}}H}{RT^2}dT$$

A note on good practice When setting up an integration, make sure the limits match on each side of the expression. Here, the lower limits are $\ln p$ on the left and T on the right, and the upper limits are $\ln p'$ and T', respectively.

The integral on the left evaluates to $\ln p' - \ln p$, which simplifies to $\ln(p'/p)$. To evaluate the integral on the right, we suppose that the enthalpy of vaporization is constant over the temperature range of interest, so together with R it can be taken outside the integral:

$$\ln\frac{p'}{p} = \int_T^{T'} \frac{\Delta_{\text{vap}}H}{RT^2}dT = \frac{\Delta_{\text{vap}}H}{R}\int_T^{T'} \frac{1}{T^2}dT = \frac{\Delta_{\text{vap}}H}{R}\left(\frac{1}{T} - \frac{1}{T'}\right)$$

which is eqn 5.7. To obtain this result we have used the standard integral

$$\int \frac{dx}{x^2} = -\frac{1}{x} + \text{constant}$$

Another note on good practice Keep a note of any approximations made in a derivation. In this lengthy pair of derivations we have made three: (1) the molar volume of a gas is much greater than that of a liquid, (2) the vapour behaves as a perfect gas, (3) the enthalpy of vaporization is independent of temperature in the range of interest. Approximations limit the ways in which an expression may be used to solve problems.

Equation 5.6 shows that as the temperature of a liquid is raised ($\Delta T > 0$) its vapour pressure increases (an increase in the logarithm of p, $\Delta \ln p > 0$, implies that p increases). Equation 5.7 lets us calculate the vapour pressure at one temperature provided we know it at another temperature. The equation tells us that, for a given change in temperature, the larger the enthalpy of vaporization, the greater the change in vapour pressure. The vapour pressure of water, for instance, responds more sharply to a change in temperature than that of benzene does. Note too that we can write eqn 5.7 as

$$\ln p = \ln p' + \frac{\Delta_{\text{vap}}H}{RT'} - \frac{\Delta_{\text{vap}}H}{RT}$$

Then, because $\ln x = \ln 10 \times \log x$,

$$\log p = \underbrace{\log p' + \frac{\Delta_{\text{vap}}H}{RT'\ln 10}}_{A} - \underbrace{\frac{\Delta_{\text{vap}}H}{RT\ln 10}}_{B}$$

This expression has the form

$$\log p = A - \frac{B}{T} \qquad (5.8)$$

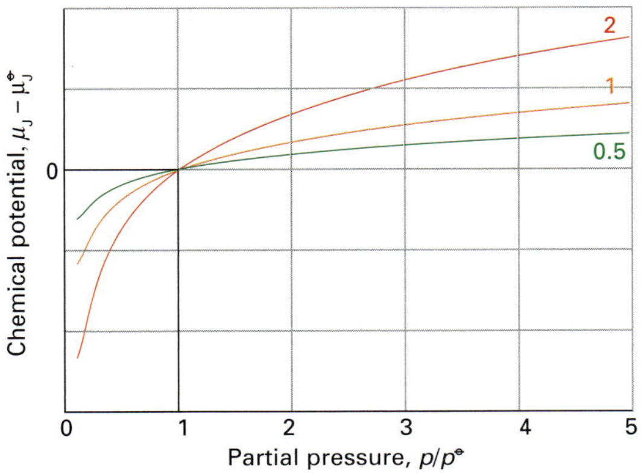

Fig. 6.2 The variation with partial pressure of the chemical potential of a perfect gas at three different temperatures (in the ratios 0.5:1:2). Note that the chemical potential increases with pressure and, at a given pressure, with temperature.

Figure 6.2 illustrates the pressure dependence of the chemical potential of a perfect gas predicted by this equation. The chemical potential becomes negatively infinite as the pressure tends to zero. As the pressure is increased from zero, the chemical potential rises to its standard value at 1 bar (because $\ln 1 = 0$), and then increases slowly (logarithmically, as $\ln p$) as the pressure is increased further.

As always, we can become familiar with an equation by listening to what it tells us. In this case:

- As p_J increases, so does $\ln p_J$. Therefore, eqn 6.4 tells us that the higher the partial pressure of a gas, the higher its chemical potential.

This conclusion is consistent with the interpretation of the chemical potential as an indication of the potential of a substance to be active chemically: the higher the partial pressure, the more active chemically the species. In this instance the chemical potential represents the tendency of the substance to react when it is in its standard state (the significance of the term $\mu^{\ominus}$) plus an additional tendency that reflects whether it is at a different pressure. For a given amount of substance, a higher partial pressure gives that substance more chemical 'punch', just like winding a spring gives a spring more physical punch (that is, enables it to do more work).

Self-test 6.4

Suppose that the partial pressure of a perfect gas falls from 100 kPa to 50 kPa as it is consumed in a reaction at 25°C. What is the change in chemical potential of the substance?

[*Answer:* −1.7 kJ mol^{-1}]

We saw in Section 5.1 that the molar Gibbs energy of a pure substance is the same in all the phases at equilibrium. We use the same argument in Derivation 6.2 to show that *a system is at equilibrium when the chemical potential of each substance has the same value in every phase in which it occurs*. We can think of the chemical potential as the pushing power of each substance, and equilibrium is reached only when each substance pushes with the same strength in any phase it occupies.

Derivation 6.2

The uniformity of chemical potential

Suppose a substance J occurs in different phases in different regions of a system. For instance, we might have a liquid mixture of ethanol and water and a mixture of their vapours. Let the substance J have chemical potential $\mu_J(\text{l})$ in the liquid mixture and $\mu_J(\text{g})$ in the vapour. We could imagine an infinitesimal amount, $\mathrm{d}n_J$, of J migrating from the liquid to the vapour. As a result, the Gibbs energy of the liquid phase falls by $\mu_J(\text{l})\mathrm{d}n_J$ and that of the vapour rises by $\mu_J(\text{g})\mathrm{d}n_J$. The net change in Gibbs energy is

$$\mathrm{d}G = \mu_J(\text{g})\mathrm{d}n_J - \mu_J(\text{l})\mathrm{d}n_J = \{\mu_J(\text{g}) - \mu_J(\text{l})\}\mathrm{d}n_J$$

There is no tendency for this migration (and the reverse process, migration from the vapour to the liquid) to occur if $\mathrm{d}G = 0$. The argument applies to each component of the system. Therefore, *for a substance to be at equilibrium throughout the system, its chemical potential must be the same everywhere.*

6.2 Spontaneous mixing

All gases mix spontaneously with one another because the molecules of one gas can mingle with the molecules of the other gas. But how can we show *thermodynamically* that mixing is spontaneous? At constant temperature and pressure, we need to show that $\Delta G < 0$. The first step is therefore to find an expression for ΔG when two gases mix, and then to decide whether it is negative. As we see in Derivation 6.3, when an amount n_A of A and n_B of B of two gases mingle at a temperature T,

$$\Delta G = nRT\{x_A \ln x_A + x_B \ln x_B\} \qquad (6.5)$$

with $n = n_A + n_B$ and the x_J the mole fractions of the components J in the mixture.

Derivation 6.3

The Gibbs energy of mixing

Suppose we have an amount n_A of a perfect gas A at a certain temperature T and pressure p, and an amount n_B of a perfect gas B at the same temperature and pressure. The two gases are in separate compartments initially (Fig. 6.3). The Gibbs energy of the system (the two unmixed gases) is the sum of their individual Gibbs energies:

(eqn 6.2b) (eqn 6.4b) (eqn 6.4b)

$$G_i = n_A\mu_A + n_B\mu_B$$
$$= n_A\{\mu_A^\ominus + RT\ln p\} + n_B\{\mu_B^\ominus + RT\ln p\}$$

The chemical potentials are those for the two gases, each at a pressure p. When the partition is removed, the total pressure remains the same, but according to Dalton's law (Section 1.3), the partial pressures fall to $p_A = x_A p$ and $p_B = x_B p$, where the x_J are the mole fractions of the two gases in the mixture ($x_J = n_J/n$, with $n = n_A + n_B$). The final Gibbs energy of the system is therefore

$$G_f = n_A\{\mu_A^\ominus + RT\ln p_A\} + n_B\{\mu_B^\ominus + RT\ln p_B\}$$
$$= n_A\{\mu_A^\ominus + RT\ln x_A p\} + n_B\{\mu_B^\ominus + RT\ln x_B p\}$$

The difference $G_f - G_i$ is the change in Gibbs energy that accompanies mixing. The standard chemical potentials cancel, and by making use of the relation

(Use $\ln a - \ln b = \ln(a/b)$)

$$\ln x_J p - \ln p = \ln \frac{x_J \cancel{p}}{\cancel{p}} = \ln x_J$$

for each gas, we obtain

($n_A = nx_A$) ($n_B = nx_B$)

$$\Delta G = RT\{n_A \ln x_A + n_B \ln x_B\} = nRT\{x_A \ln x_A + x_B \ln x_B\}$$

which is eqn 6.5.

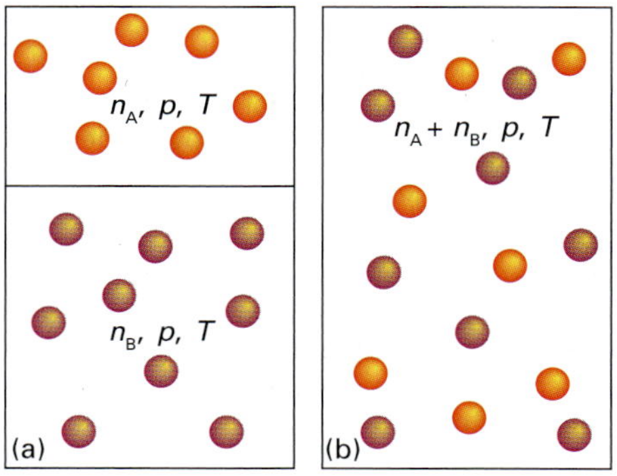

Fig. 6.3 The (a) initial and (b) final states of a system in which two perfect gases mix. The molecules do not interact, so the enthalpy of mixing is zero. However, because the final state is more disordered than the initial state, there is an increase in entropy.

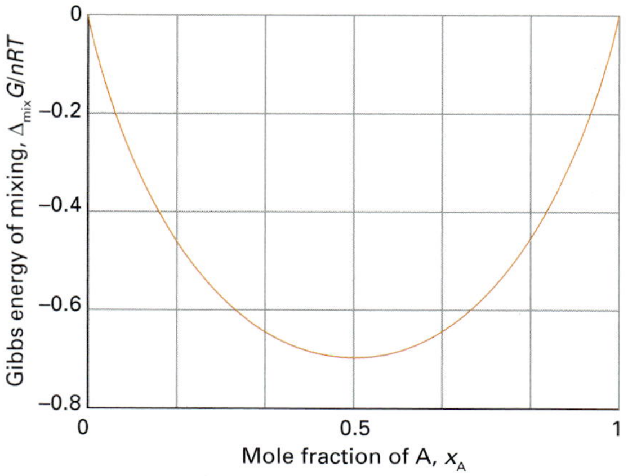

Fig. 6.4 The variation of the Gibbs energy of mixing with composition for two perfect gases at constant temperature and pressure. Note that $\Delta G < 0$ for all compositions, which indicates that two gases mix spontaneously in all proportions.

Equation 6.5 tells us the change in Gibbs energy when two gases mix at constant temperature and pressure (Fig. 6.4). The crucial feature is that because x_A and x_B are both less than 1, the two logarithms are negative ($\ln x < 0$ if $x < 1$), so $\Delta G < 0$ at all compositions. Therefore, *perfect gases mix spontaneously in all proportions*. Furthermore, if we compare eqn 6.5 with $\Delta G = \Delta H - T\Delta S$,

$$\Delta G = \Delta H - \cancel{T}\Delta S$$
$$\Delta G = \quad nR\cancel{T}\{x_A \ln x_A + x_B \ln x_B\}$$

we can conclude that:

$$\Delta H = 0 \qquad (6.6a)$$

$$\Delta S = -nR\{x_A \ln x_A + x_B \ln x_B\} \qquad (6.6b)$$

That is, there is no change in enthalpy when two perfect gases mix, which reflects the fact that there are no interactions between the molecules. There is an increase in entropy, because the mixed gas is more disordered than the unmixed gases (Fig. 6.5). The entropy of the surroundings is unchanged because the enthalpy of the system is constant, so no energy escapes as heat into the surroundings. It follows that the increase in entropy of the system is the 'driving force' of the mixing.

6.3 Ideal solutions

In chemistry we are concerned with liquids as well as gases, so we need an expression for the chemical potential of a substance in a liquid solution. We can anticipate that the chemical potential of a species ought to increase with concentration, because the higher its

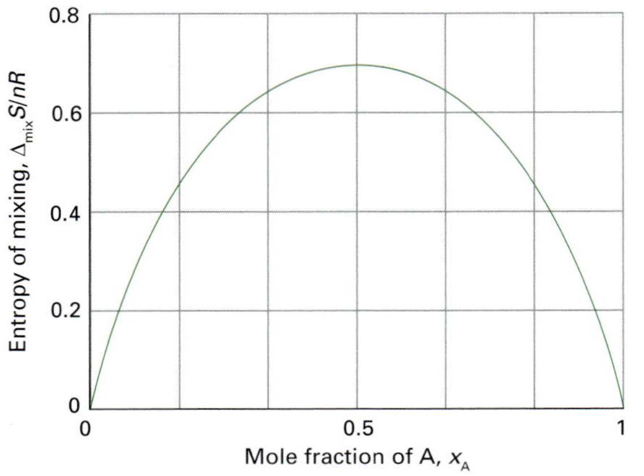

Fig. 6.5 The variation of the entropy of mixing with composition for two perfect gases at constant temperature and pressure.

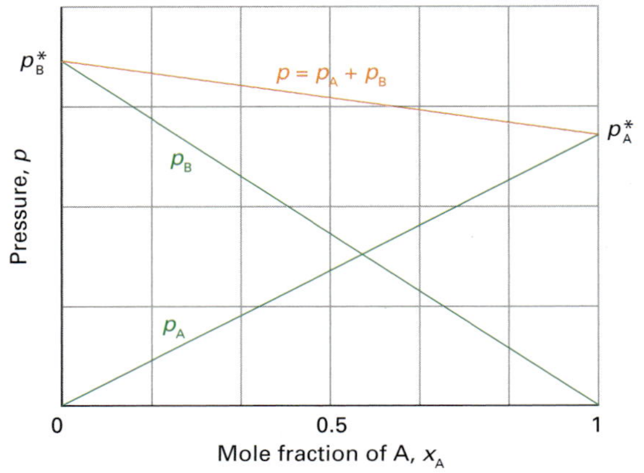

Fig. 6.6 The partial vapour pressures of the two components of an ideal binary mixture are proportional to the mole fractions of the components in the liquid. The total pressure of the vapour is the sum of the two partial vapour pressures.

concentration the greater its chemical 'punch'. In the following, we use the following notation:

J denotes a substance in general
A denotes a solvent
B denotes a solute

The key to setting up an expression for the chemical potential of a solute is the work done by the French chemist François Raoult (1830–1901), who spent most of his life measuring the vapour pressures of solutions. He measured the **partial vapour pressure**, p_J, of each component in the mixture, the partial pressure of the vapour of each component in dynamic equilibrium with the liquid mixture, and established what is now called **Raoult's law**:

The partial vapour pressure of a substance in a liquid mixture is proportional to its mole fraction in the mixture and its vapour pressure when pure:

$$p_J = x_J p_J^* \tag{6.7}$$

In this expression, p_J^* is the vapour pressure of the pure substance. For example, when the mole fraction of water in an aqueous solution is 0.90, then, provided Raoult's law is obeyed, the partial vapour pressure of the water in the solution is 90 per cent that of pure water. This conclusion is approximately true whatever the identity of the solute and the solvent (Fig. 6.6).

Self-test 6.5

A solution is prepared by dissolving 1.5 mol $C_{10}H_8$ (naphthalene) in 1.00 kg of benzene. The vapour pressure of pure benzene is 12.6 kPa at 25°C. What is the partial vapour pressure of benzene in the solution?

[*Answer:* 11.3 kPa]

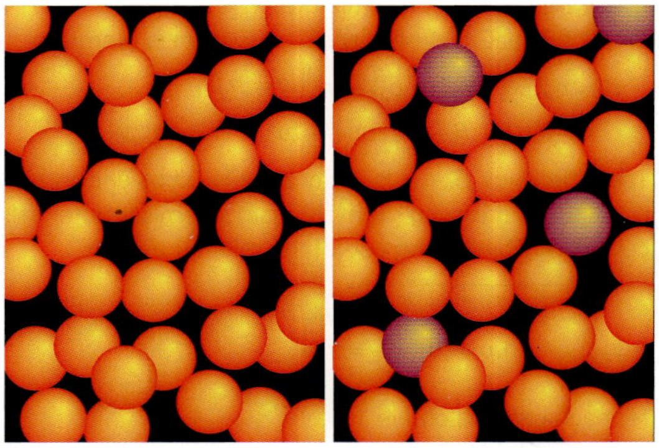

Fig. 6.7 (a) In a pure liquid, we can be confident that any molecule selected from the sample is a solvent molecule. (b) When a solute is present, we cannot be sure that blind selection will give a solvent molecule, so the entropy of the system is greater than in the absence of the solute.

The molecular origin of Raoult's law is the effect of the solute on the entropy of the solution. In the pure solvent, the molecules have a certain disorder and a corresponding entropy; the vapour pressure then represents the tendency of the system and its surroundings to reach a higher entropy. When a solute is present, the solution has a greater disorder than the pure solvent because we cannot be sure that a molecule chosen at random will be a solvent molecule (Fig. 6.7). Because the entropy of the solution is higher than that of the pure solvent, the solution has a lower tendency to acquire an even higher entropy by the solvent vaporizing. In other words, the vapour pressure of the solvent in the solution is lower than that of the pure solvent.

An **ideal solution** is a hypothetical solution of a solute B in a solvent A that obeys Raoult's law throughout the composition range from pure A to

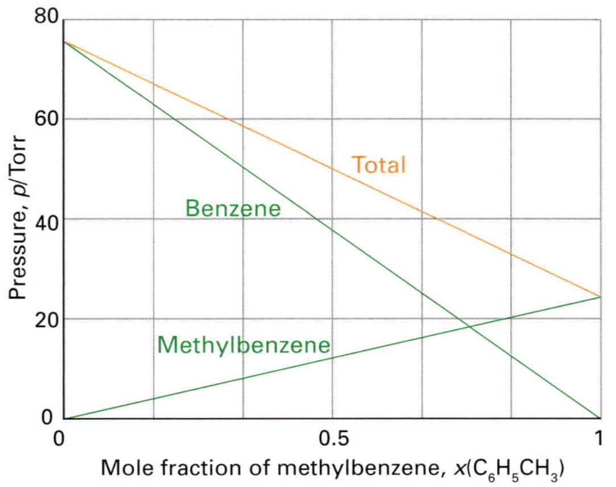

Fig. 6.8 Two similar substances, in this case benzene and methylbenzene (toluene) behave almost ideally and have vapour pressures that closely resemble those for the ideal case depicted in Fig. 6.6.

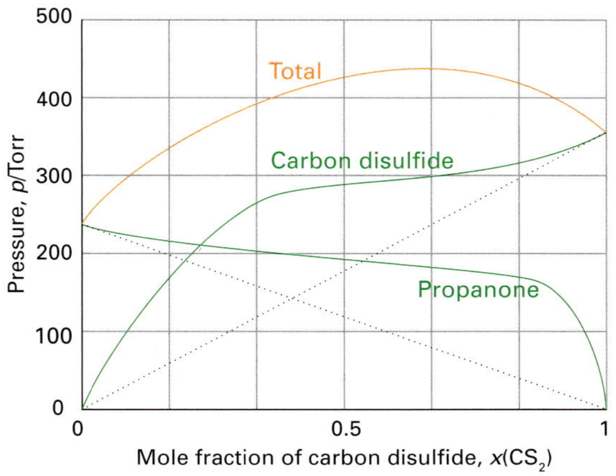

Fig. 6.9 Strong deviations from ideality are shown by dissimilar substances, in this case carbon disulfide and acetone (propanone). Note, however, that Raoult's law is obeyed by propanone when only a small amount of carbon disulfide is present (on the left) and by carbon disulfide when only a small amount of propanone is present (on the right).

pure B. The law is most reliable when the components of a mixture have similar molecular shapes and are held together in the liquid by similar types and strengths of intermolecular forces. An example is a mixture of two structurally similar hydrocarbons. A mixture of benzene and methylbenzene (toluene) is a good approximation to an ideal solution, for the partial vapour pressure of each component satisfies Raoult's law reasonably well throughout the composition range from pure benzene to pure methylbenzene (Fig. 6.8).

No mixture is perfectly ideal and all real mixtures show deviations from Raoult's law. However, the deviations are small for the component of the mixture that is in large excess (the solvent) and become smaller as the concentration of solute decreases (Fig. 6.9). We can usually be confident that Raoult's law is reliable for the solvent when the solution is very dilute. More formally, Raoult's law is a *limiting law* (like the perfect gas law), and is strictly valid only in the limit of zero concentration of solute.

The theoretical importance of Raoult's law is that, because it relates vapour pressure to composition, and we know how to relate pressure to chemical potential, we can use the law to relate chemical potential to the composition of a solution. As we show in Derivation 6.4, the chemical potential of a solvent A present in solution at a mole fraction x_A is

$$\mu_A = \mu_A^* + RT \ln x_A \qquad (6.8)$$

where μ_A^* is the chemical potential of pure A. This expression is valid throughout the concentration range for either component of a binary ideal solution.

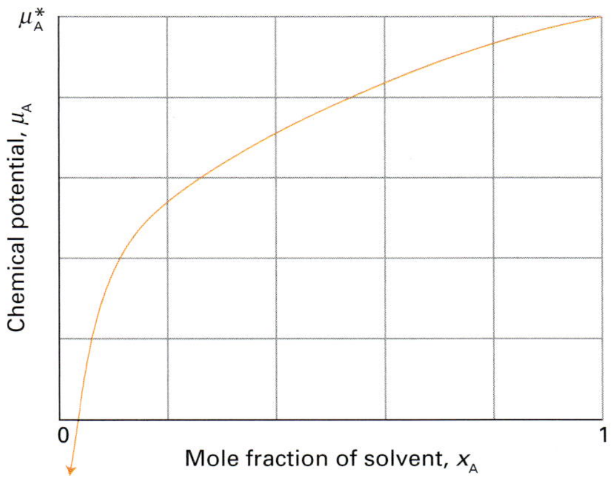

Fig. 6.10 The variation of the chemical potential of the solvent with the composition of the solution. Note that the chemical potential of the solvent is lower in the mixture than for the pure liquid (for an ideal system). This behaviour is likely to be shown by a dilute solution in which the solvent is almost pure (and obeys Raoult's law).

It is valid for the solvent of a real solution the closer the composition approaches pure solvent (pure A).

A note on good practice An asterisk (*) is used to denote a pure substance, but not one that is necessarily in its standard state. Only if the pressure is 1 bar would μ_A^* be the standard chemical potential of A, and it would then be written $\mu_A^\ominus$.

Figure 6.10 shows the variation of chemical potential of the solvent predicted by eqn 6.8. The essential feature is as follows:

higher values. In other words, the freezing point is depressed, the boiling point is elevated, and the liquid phase exists over a wider range of temperatures. Derivation 6.5 shows how to express these changes quantitatively.

Derivation 6.5

The modification of transition temperatures

To derive an expression for the elevation of boiling point, we note that at the normal boiling point, T_b^*, of the pure solvent A, the solvent vapour and liquid are in equilibrium at 1 atm, so their chemical potentials are equal:

$$\mu_A^*(\text{g, 1 atm}, T_b^*) = \mu_A^*(\text{l, 1 atm}, T_b^*)$$

This equality is depicted in Fig. 6.18a. In the presence of a solute B, the mole fraction of A is reduced from 1 to $x_A = 1 - x_B$ and the boiling point is T_b; then, because the solvent vapour and liquid remain in equilibrium under these new conditions,

Vapour remains pure | Solvent mole fraction is now x_A

$$\mu_A^*(\text{g, 1 atm}, T_b) = \mu_A(\text{l}, x_A, \text{1 atm}, T_b)$$

According to eqn 6.8, the chemical potential of the solvent in the solution is related to its mole fraction by

$$\mu_A(\text{l}, x_A, \text{1 atm}, T_b) = \mu_A^*(\text{l, 1 atm}, T_b) + RT_b \ln x_A$$

Therefore, the last equation becomes

$$\mu_A^*(\text{g, 1 atm}, T_b) = \mu_A^*(\text{l, 1 atm}, T_b) + RT_b \ln x_A$$

which rearranges into

$$\ln x_A = \frac{\mu_A^*(\text{g, 1 atm}, T_b) - \mu_A^*(\text{l, 1 atm}, T_b)}{RT_b}$$

The chemical potential of a pure substance is the same as the molar Gibbs energy of the substance, so this expression is the same as

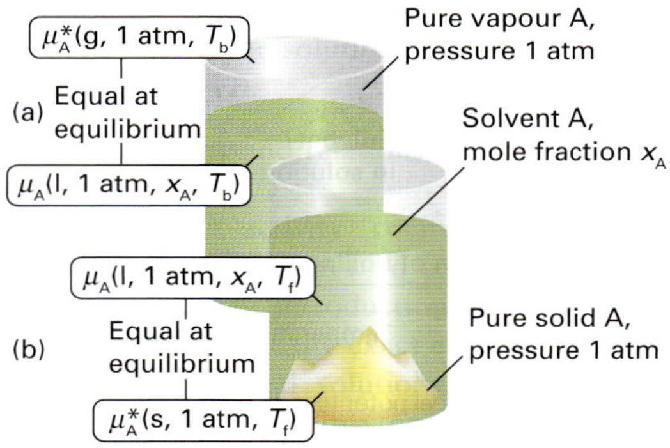

Fig. 6.18 The equilibria between phases and the corresponding relation between chemical potentials of the solvent in solution at (a) the normal boiling point and (b) the normal freezing point.

$$\ln x_A = \frac{G_m(\text{g, 1 atm}, T_b) - G_m(\text{l, 1 atm}, T_b)}{RT_b}$$

$$= \frac{\Delta_{vap}G(\text{1 atm}, T_b)}{RT_b}$$

When $x_A = 1$ (so $\ln x_A = 0$), the pure liquid solvent, the boiling point is T_b^*, and we can write

$$0 = \frac{\Delta_{vap}G(\text{1 atm}, T_b^*)}{RT_b^*}$$

The difference between these two equations is

$$\ln x_A = \frac{\Delta_{vap}G(T_b)}{RT_b} - \frac{\Delta_{vap}G(T_b^*)}{RT_b^*}$$

(For simplicity, we no longer specify the pressure as 1 atm, but remember that that is its value.) Now we use $\Delta G = \Delta H - T\Delta S$ and the approximate temperature independence of ΔH and ΔS to turn this equation into

Use $\Delta_{vap}G(T) = \Delta_{vap}H(T) - T\Delta_{vap}S(T)$ twice

$$\ln x_A = \frac{\Delta_{vap}H(T_b) - T_b\Delta_{vap}S(T_b)}{RT_b} - \frac{\Delta_{vap}H(T_b^*) - T_b^*\Delta_{vap}S(T_b^*)}{RT_b^*}$$

Assume $\Delta_{vap}H$ and $\Delta_{vap}S$ independent of temperature

$$= \frac{\Delta_{vap}H}{RT_b} - \frac{\Delta_{vap}S}{R} - \frac{\Delta_{vap}H}{RT_b^*} + \frac{\Delta_{vap}S}{R}$$

$$= \frac{\Delta_{vap}H}{R}\left(\frac{1}{T_b} - \frac{1}{T_b^*}\right)$$

Now we use $\ln x_A = \ln(1 - x_B) \approx -x_B$ to express this equation as

$$x_B \approx \frac{\Delta_{vap}H}{R}\left(\frac{1}{T_b^*} - \frac{1}{T_b}\right) = \frac{\Delta_{vap}H}{R}\left(\frac{T_b - T_b^*}{T_b^* T_b}\right)$$

A brief comment The series expansion of a natural logarithm is

$$\ln(1 - x) = -x - \tfrac{1}{2}x^2 - \tfrac{1}{3}x^3 \ldots$$

If $x \ll 1$, then the terms involving x raised to a power greater than 1 are much smaller than x, so $\ln(1 - x) \approx -x$. For example, $\ln(1 - 0.050) = \ln 0.950 = -0.051$, which is close to -0.050.

We are almost there. First, we note that the elevation of boiling point is $\Delta T_b = T_b - T_b^*$. Then we note that because the value of T_b is very close to T_b^*, little error is introduced by replacing $T_b^* T_b$ by T_b^{*2}. In this way we arrive at

$$x_B \approx \frac{\Delta_{vap}H}{R} \times \frac{\Delta T_b}{T_b^{*2}}$$

which we can rearrange into

$$\Delta T_b \approx \frac{RT_b^{*2}}{\Delta_{vap}H} \times x_B$$

At this point, we see that the elevation of boiling point is proportional to the mole fraction of solute B and independent of its identity (the $\Delta_{vap}H$ and T_b are properties of the solvent). The mole fraction of the solute is proportional to its molality, b_B, so the equation we have derived has the form $\Delta T_b = K_b b_B$, as in eqn 6.17.

The calculation of the depression of freezing point starts with the equilibrium condition depicted in Fig. 6.18b, which implies

$$\mu_A^*(s, 1\text{ atm}, T_f) = \mu_A(l, x_A, 1\text{ atm}, T_f)$$

The calculation then proceeds in exactly the same way, and we arrive at

$$\Delta T_f \approx \frac{RT_f^{*2}}{\Delta_{fus}H} \times x_B$$

with $\Delta T_f = T_f^* - T_f$, as in eqn 6.17.

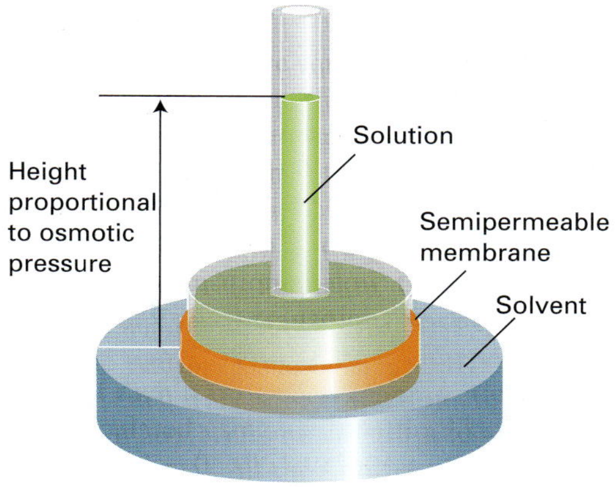

Fig. 6.19 In a simple osmosis experiment, a solution is separated from the pure solvent by a semipermeable membrane. Pure solvent passes through the membrane and the solution rises in the inner tube. The net flow ceases when the pressure exerted by the column of liquid is equal to the osmotic pressure of the solution.

The elevation of boiling point is too small to have any practical significance. A practical consequence of the lowering of the freezing point, and hence the lowering of the melting point of the pure solid, is its employment in organic chemistry to judge the purity of a sample, for any impurity lowers the melting point of a substance from its accepted value. The salt water of the oceans freezes at temperatures lower than that of fresh water, and salt is spread on highways to delay the onset of freezing. The addition of 'antifreeze' to car engines and, by natural processes, to arctic fish, is commonly held up as an example of the lowering of freezing point, but the concentrations are far too high for the arguments we have used here to be applicable. The 1,2-ethanediol ('glycol') used as antifreeze and the proteins present in fish body fluids probably simply interfere with bonding between water molecules.

6.7 Osmosis

The phenomenon of **osmosis** (from the Greek word for 'push') is the passage of a pure solvent into a solution separated from it by a semipermeable membrane. A **semipermeable membrane** is a membrane that is permeable to the solvent but not to the solute. The membrane might have microscopic holes that are large enough to allow water molecules to pass through, but not ions or carbohydrate molecules with their bulky coating of hydrating water molecules. The **osmotic pressure,** Π (uppercase pi), is the pressure that must be applied to the solution to stop the inward flow of solvent.

In the simple arrangement shown in Fig. 6.19, the pressure opposing the passage of solvent into the solution arises from the hydrostatic pressure of the column of solution that the osmosis itself produces. This column is formed when the pure solvent flows through the membrane into the solution and pushes the column of solution higher up the tube. Equilibrium is reached when the downward pressure exerted by the column of solution is equal to the upward osmotic pressure. A complication of this arrangement is that the entry of solvent into the solution results in dilution of the latter, so it is more difficult to treat mathematically than an arrangement in which an externally applied pressure opposes any flow of solvent into the solution.

The osmotic pressure of a solution is proportional to the concentration of solute. In fact, we show in Derivation 6.6 that the expression for the osmotic pressure of an ideal solution, which is called the **van 't Hoff equation,** bears an uncanny resemblance to the expression for the pressure of a perfect gas:

$$\Pi V \approx n_B RT \qquad (6.18a)$$

Because $n_B/V = [B]$, the molar concentration of the solute, a simpler form of this equation is

$$\Pi \approx [B]RT \qquad (6.18b)$$

This equation applies only to solutions that are sufficiently dilute to behave as ideal-dilute solutions.

Derivation 6.6

The van 't Hoff equation

The thermodynamic treatment of osmosis makes use of the fact that, at equilibrium, the chemical potential of the solvent A is the same on each side of the membrane (Fig. 6.20). The starting relation is therefore

$$\mu_A(\text{pure solvent at pressure } p) = \mu_A(\text{solvent in the solution at pressure } p+\Pi)$$

The pure solvent is at atmospheric pressure, p, and the solution is at a pressure $p + \Pi$ on account of the additional pressure, Π, that has to be exerted on the solution to establish equilibrium. We shall write the chemical potential of the pure solvent at the pressure p as $\mu_A^*(p)$. The chemical potential of the solvent in the solution is lowered by the solute but it is raised on account of the greater pressure, $p + \Pi$, acting on the solution. We denote this chemical potential by $\mu_A(x_A, p+\Pi)$. Our task is to find the extra pressure Π needed to balance the lowering of chemical potential caused by the solute.

The condition for equilibrium written above is

$$\mu_A^*(p) = \mu_A(x_A, p+\Pi)$$

We take the effect of the solute into account by using eqn 6.8:

Eqn 6.4b

$$\mu_A(x_A, p+\Pi) = \mu_A^*(p+\Pi) + RT\ln x_A$$

The effect of pressure on an (assumed incompressible) liquid is given by eqn 5.1 ($\Delta G_m = V_m\Delta p$) but now expressed in terms of the chemical potential and the partial molar volume of the solvent:

$$\mu_A^*(p+\Pi) = \mu_A^*(p) + V_A\Delta p$$

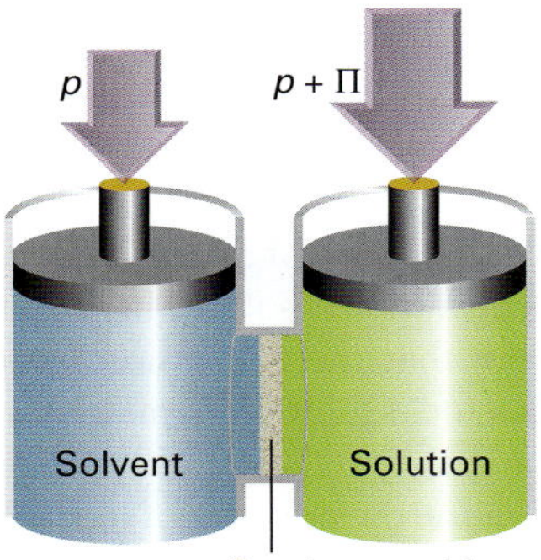

Fig. 6.20 The basis of the calculation of osmotic pressure. The presence of a solute lowers the chemical potential of the solvent in the right-hand compartment, but the application of pressure raises it. The osmotic pressure is the pressure needed to equalize the chemical potential of the solvent in the two compartments.

At this point we identify the difference in pressure Δp as Π, so

$$\mu_A^*(p+\Pi) = \mu_A^*(p) + V_A\Pi$$

When we combine this relation with $\mu_A^*(p) = \mu_A^*(p+\Pi) + RT\ln x_A$ we get

$$\cancel{\mu_A^*(p)} = \cancel{\mu_A^*(p)} + V_A\Pi + RT\ln x_A$$

and therefore

$$-RT\ln x_A = \Pi V_A$$

The mole fraction of the solvent is equal to $1 - x_B$, where x_B is the mole fraction of solute molecules. In dilute solution, $\ln(1 - x_B)$ is approximately equal to $-x_B$ (recall the comment in Derivation 6.5), so this equation becomes

$$RTx_B \approx \Pi V_A$$

When the solution is dilute, $x_B = n_B/n \approx n_B/n_A$, so

$n_AV_A = V$ the volume of solvent

$$RTn_B \approx n_A\Pi V_A = \Pi V$$

which is eqn 6.18.

Osmosis helps biological cells maintain their structure. Cell membranes are semipermeable and allow water, small molecules, and hydrated ions to pass, while blocking the passage of biopolymers synthesized inside the cell. The difference in concentrations of solutes inside and outside the cell gives rise to an osmotic pressure, and water passes into the more concentrated solution in the interior of the cell, carrying small nutrient molecules. The influx of water also keeps the cell swollen, whereas dehydration causes the cell to shrink.

One of the most common applications of osmosis is **osmometry**, the measurement of molar masses of proteins and synthetic polymers from the osmotic pressure of their solutions. As these huge molecules dissolve to produce solutions that are far from ideal, we assume that the van 't Hoff equation is only the first term of an expansion:

$$\Pi = [\mathrm{B}]RT\{1 + B[\mathrm{B}] + \cdots\} \qquad (6.19\text{a})$$

Exactly the same strategy was used in Section 1.12 to extend the perfect gas equation to real gases and there it led to the virial equation of state. The empirical parameter B in this expression is called the **osmotic virial coefficient.** To use eqn 6.19a, we rearrange it into a form that gives a straight line by dividing both sides by [B]:

y = intercept + slope × x

$$\frac{\Pi}{[\mathrm{B}]} = RT + BRT[\mathrm{B}] + \cdots \qquad (6.19\text{b})$$

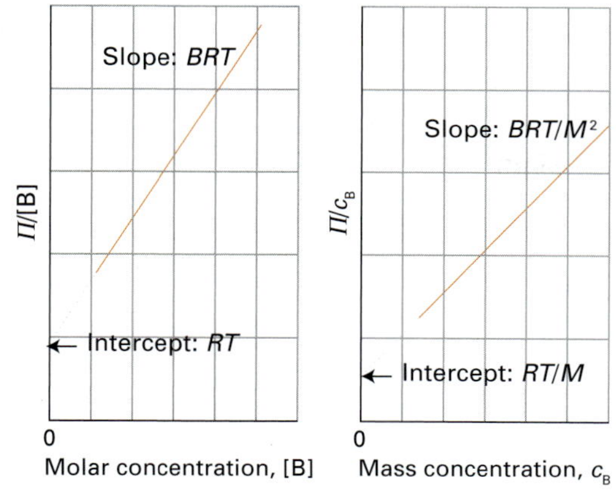

Fig. 6.21 The plot and extrapolation made to analyse the results of an osmometry experiment.

As we illustrate in Example 6.4, we can find the molar mass of the solute B by measuring the osmotic pressure at a series of mass concentrations and making a plot of Π/[B] against [B] (Fig. 6.21).

Example 6.4

Using osmometry to determine molar mass

The osmotic pressures of solutions of an enzyme in water at 298 K are given below. Determine the molar mass of the enzyme.

$c/(\mathrm{g\ dm^{-3}})$	1.00	2.00	4.00	7.00	9.00
Π/Pa	27	70	197	500	785

Strategy First, we need to express eqn 6.19b in terms of the mass concentration, c. The molar concentration [B] of the solute is related to the mass concentration $c_B = m_B/V$ by

$$c_B = \frac{m_B}{V} = \frac{m_B}{n_B} \times \frac{n_B}{V} = M \times [B]$$

where M is the molar mass of the solute (its mass, m_B, divided by its amount in moles, n_B), so $[B] = c_B/M$. With this substitution, eqn 6.19b becomes

$$\frac{M\Pi}{c_B} = RT + \frac{BRTc_B}{M} + \cdots$$

Division through by M gives

$$\underbrace{\frac{\Pi}{c_B}}_{y} = \underbrace{\frac{RT}{M}}_{\text{intercept}} + \underbrace{\left(\frac{RTB}{M^2}\right)}_{\text{slope}} \times \underbrace{c_B}_{x} + \cdots$$

It follows that, by plotting Π/c_B against c_B, the results should fall on a straight line with intercept RT/M on the vertical axis at $c_B = 0$. Therefore, by locating the intercept by extrapolation of the data to $c_B = 0$, we can find the molar mass of the solute.

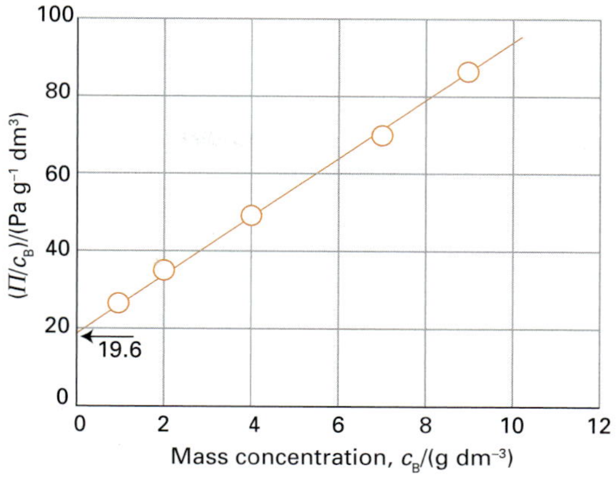

Fig. 6.22 The plot of the data in Example 6.5. The molar mass is determined from the intercept at $c = 0$.

Solution The following values of Π/c_B can be calculated from the data:

$c_B/(\mathrm{g\ dm^{-3}})$	1.00	2.00	4.00	7.00	9.00
$(\Pi/\mathrm{Pa})/(c_B/\mathrm{g\ dm^{-3}})$	27	35	49.2	71.4	87.2

The points are plotted in Fig. 6.22. The intercept with the vertical axis at $c_B = 0$ (which is best found by using linear regression and mathematical software, as we have done) is at

$$\frac{\Pi/\mathrm{Pa}}{c_B/(\mathrm{g\ dm^{-3}})} = 19.6$$

which we can rearrange into

$$(\Pi/c_B)_{\text{intercept}} = 19.6\ \mathrm{Pa\ g^{-1}\ dm^3}$$

$$= 19.6\ \frac{\mathrm{Pa\ dm^3}}{\mathrm{g}}$$

$$= 19.6\ \frac{\overbrace{\mathrm{kg\ m^{-1}\ s^{-2}}}^{\mathrm{Pa}} \times \overbrace{10^{-3}\ \mathrm{m^3}}^{\mathrm{dm^3}}}{\underbrace{10^{-3}\ \mathrm{kg}}_{\mathrm{g}}}$$

$$= 19.6\ \mathrm{m^2\ s^{-2}}$$

Therefore, because this intercept is equal to RT/M, so that $M = RT/\text{intercept}$, we can write

$$M = \frac{RT}{19.6\ \mathrm{m^2\ s^{-2}}}$$

It follows that

$$M = \frac{(8.314\,47\ \mathrm{J\ K^{-1}\ mol^{-1}}) \times (298\ \mathrm{K})}{19.6\ \mathrm{m^2\ s^{-2}}}$$

$$= \frac{8.314\,47 \times 298}{19.6}\ \frac{\mathrm{J\ mol^{-1}}}{\mathrm{m^2\ s^{-2}}}$$

J mol⁻¹ m⁻² s² = (kg m² s⁻²) mol⁻¹ m⁻² s² = kg mol⁻¹

$$= 126\ \mathrm{kg\ mol^{-1}}$$

The molar mass of the enzyme is therefore close to 130 kDa.

A note on good practice Graphs should be plotted on axes labelled with pure numbers. Note how the plotted quantities are divided by their units, so that $c_B/(\text{g dm}^{-3})$, for instance, is a dimensionless number. By carrying the units through every stage of the calculation, we end up with the correct units for M.

Self-test 6.11

The osmotic pressure of a solution of poly(vinyl chloride), PVC, in dioxane at 25°C were as follows:

$c/(\text{g dm}^{-3})$	0.50	1.00	1.50	2.00	2.50
$\Pi/c(\text{Pa g}^{-1}\text{ dm}^3)$	33.6	35.2	36.8	38.4	40.0

Determine the molar mass of the polymer.

[*Answer:* 77 kg mol^{-1}]

When pressure greater than the osmotic pressure is applied to the solution, there is a thermodynamic tendency for the solvent to flow out of the solution and into the pure solvent. This process is called **reverse osmosis.** Reverse osmosis is of great importance for the purification of sea water so that it is potable (drinkable) and can be used for irrigation, and many reverse osmosis plants are in operation around the world to supply fresh water to arid or water-deficient regions. The principal technical problem is to manufacture semipermeable membranes that are strong enough to withstand the high pressures required but still allow an economic flow.

Phase diagrams of mixtures

As in the discussion of pure substances (Chapter 5), the phase diagram of a mixture shows which phase is most stable for the given conditions. However, for mixtures composition is a variable in addition to the pressure and temperature.

It will be useful to keep in mind the implications of the phase rule ($F = C - P + 2$, Section 5.7). We shall consider only **binary mixtures,** which are mixtures of two components (such as ethanol and water) and may therefore set $C = 2$. Then $F = 4 - P$. For simplicity we keep the pressure constant (at 1 atm, for instance), which uses up one of the degrees of freedom, and write $F' = 3 - P$ for the number of degrees of freedom remaining. One of these degrees of freedom is the temperature, the other is the composition. Hence we should be able to depict the phase equilibria of the system on a **temperature–composition diagram** in which one axis is the temperature and the other axis is the mole fraction. In a region where there is only one phase, $F' = 2$ and both the temperature and the composition can be varied (Fig. 6.23). If two phases are present at equilibrium, $F' = 1$, and only one of the two variables may be changed at will. For example, if we change the composition, then to maintain equilibrium between the two phases we have to adjust the temperature too. Such two-phase equilibria therefore define a line in the phase diagram. If three phases are present, $F' = 0$ and there is no degree of freedom for the system. To establish equilibrium between three phases we must adopt a specific temperature and composition. Such a condition is therefore represented by a point on the phase diagram.

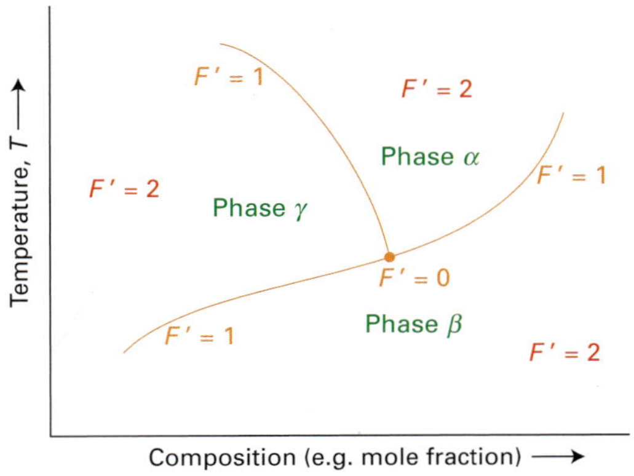

Fig. 6.23 The interpretation of a temperature–composition phase diagram at constant pressure. In a region, where only one phase is present, $F' = 2$ and both composition and temperature can be varied. On a phase boundary, where two phases are in equilibrium, $F' = 1$ and only one variable can be changed independently. At a point where three phases are present in equilibrium, $F' = 0$, and the temperature and composition are fixed.

6.8 Mixtures of volatile liquids

First, we consider the phase diagram of a binary mixture of two volatile components. This kind of system is important for understanding fractional distillation, which is a widely used technique in industry and the laboratory. Intuitively, we might expect the boiling point of a mixture of two volatile liquids to vary smoothly from the boiling point of one pure component when only that liquid is present to the boiling point of the other pure component when only that liquid is present. This expectation is often borne out in practice, and Fig. 6.24 shows a typical plot of boiling point against composition (the lower curve).

The vapour in equilibrium with the boiling mixture is also a mixture of the two components. We

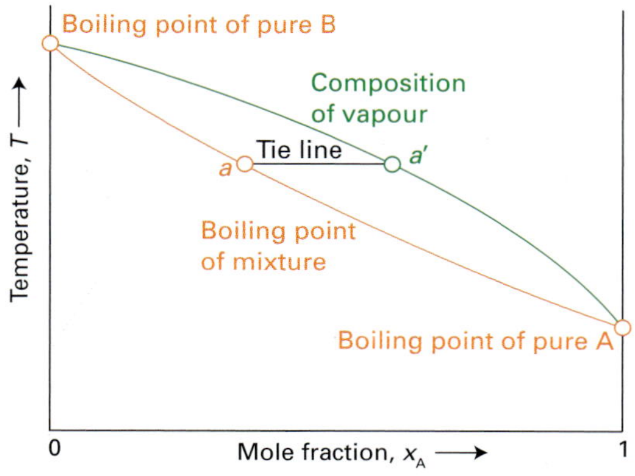

Fig. 6.24 A temperature–composition diagram for a binary mixture of volatile liquids. The tie-line connects the points that represent the compositions of liquid and vapour that are in equilibrium at each temperature. The lower curve is a plot of the boiling point of the mixture against composition.

should expect the vapour to be richer than the liquid mixture in the more volatile of the two substances. This difference is also often found in practice, and the upper curve in the illustration shows the composition of the vapour in equilibrium with the boiling liquid. To identify the composition of the vapour, we note the boiling point of the liquid mixture (point *a*, for instance) and draw a horizontal **tie-line**, a line joining two phases that are in equilibrium with each other, across to the upper curve. Its point of intersection (*a*′) gives the composition of the vapour. In this example, we see that the mole fraction of A in the vapour is about 0.6. As expected, the vapour is richer than the liquid in the more volatile component. Graphs like these are determined empirically, by measuring the boiling points of a series of mixtures (to plot the lower curve of boiling point against composition), and measuring the composition of the vapour in equilibrium with each boiling mixture (to plot the corresponding points of the vapour-composition curve).

We can follow the changes that occur during the fractional distillation of a mixture of volatile liquids by following what happens when a mixture of composition a_1 is heated (Fig. 6.25). The mixture boils at a_2 and its vapour has composition a_2'. This vapour condenses to a liquid of the same composition when it has risen to a cooler part of the 'fractionating column', a vertical column packed with glass rings or beads to give a large surface area. This condensate boils at the temperature corresponding to the point a_3 and yields a vapour of composition a_3'. This vapour is even richer in the more volatile component. That vapour condenses to a liquid that boils at the temperature corresponding to the point a_4. The cycle

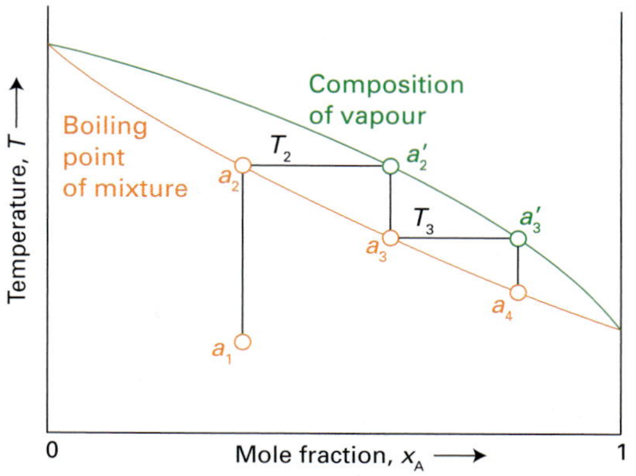

Fig. 6.25 The process of fractional distillation can be represented by a series of steps on a temperature–composition diagram like that in Fig. 6.24. The initial liquid mixture may be at a temperature and have a composition like that represented by point a_1. It boils at the temperature T_2, and the vapour in equilibrium with the boiling liquid has composition a_2'. If that vapour is condensed (to a_3 or below), the resulting condensate boils at T_3 and gives rise to a vapour of composition represented by a_3'. As the succession of vaporizations and condensations is continued, the composition of the distillate moves towards pure A (the more volatile component).

is repeated until almost pure A emerges from the top of the column.

Whereas many binary liquid mixtures do have temperature–composition diagrams resembling that shown in Fig. 6.25, in a number of important cases there are marked differences. For example, a maximum in the boiling point curve is sometimes found (Fig. 6.26). This behaviour is a sign that favourable

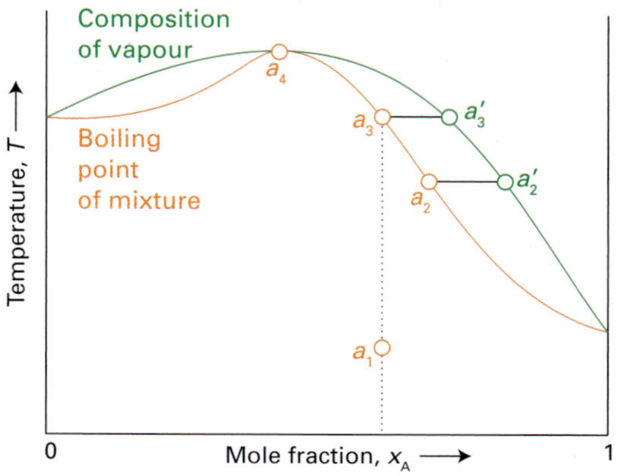

Fig. 6.26 The temperature–composition diagram for a high-boiling azeotrope. As fractional distillation proceeds, the composition of the remaining liquid moves towards a_4; however, once there, the vapour in equilibrium with that liquid has the same composition, so the mixture evaporates with an unchanged composition and no further separation can be achieved.

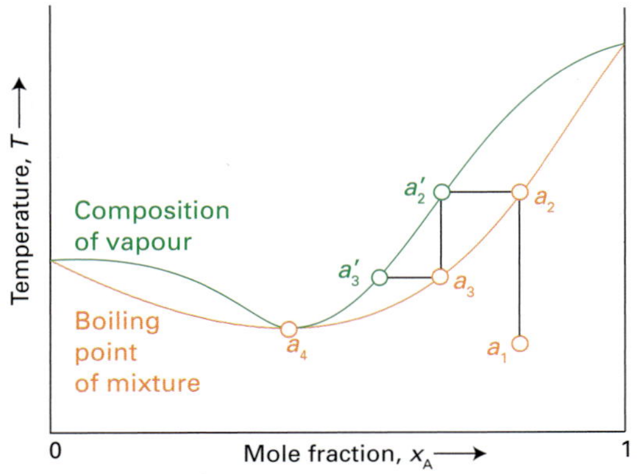

Fig. 6.27 The temperature–composition diagram for a low-boiling azeotrope. As fractional distillation proceeds, the composition of the vapour moves towards a_4; however, once there, the vapour in equilibrium with that liquid has the same composition, so no further separation of the distillate can be achieved.

interactions between the molecules of the two components reduce the vapour pressure of the mixture below the ideal value. Examples of this behaviour include trichloromethane/propanone and nitric acid/water mixtures. Temperature–composition curves are also found that pass through a minimum (Fig. 6.27). This behaviour indicates that the (A,B) interactions are unfavourable and hence that the mixture is more volatile than expected on the basis of simple mingling of the two species. Examples include dioxane/water and ethanol/water.

There are important consequences for distillation when the temperature–composition diagram has a maximum or a minimum. Consider a liquid of composition a_1 on the right of the maximum in Fig. 6.25. It boils at a_2 and its vapour (of composition a_2') is richer in the more volatile component A. If that vapour is removed, the composition of the remaining liquid moves towards a_3. The vapour in equilibrium with this boiling liquid has composition a_3': note that the two compositions are more similar than the original pair (a_3 and a_3' are closer together than a_2 and a_2'). If that vapour is removed, the composition of the boiling liquid shifts towards a_4 and the vapour of that boiling mixture has a composition identical to that of the liquid. At this stage, evaporation occurs without change of composition. The mixture is said to form an **azeotrope** (from the Greek words for 'boiling without changing'). When the azeotropic composition has been reached, distillation cannot separate the two liquids because the condensate retains the composition of the liquid. One example of azeotrope formation is hydrochloric acid/water, which is azeotropic at 80 per cent water (by mass) and boils unchanged at 108.6°C.

The system shown in Fig. 6.27 is also azeotropic, but shows this character in a different way. Suppose we start with a mixture of composition a_1 and follow the changes in the composition of the vapour that rises through a fractionating column. The mixture boils at a_2 to give a vapour of composition a_2'. This vapour condenses in the column to a liquid of the same composition (now marked a_3). That liquid reaches equilibrium with its vapour at a_3', which condenses higher up the tube to give a liquid of the same composition. The fractionation therefore shifts the vapour towards the azeotropic composition at a_4, but the composition cannot move beyond a_4 because now the vapour and the liquid have the same composition. Consequently, the azeotropic vapour emerges from the top of the column. An example is ethanol/water, which boils unchanged when the water content is 4 per cent and the temperature is 78°C.

6.9 Liquid–liquid phase diagrams

Partially miscible liquids are liquids that do not mix together in all proportions. An example is a mixture of hexane and nitrobenzene: when the two liquids are shaken together, the liquid consists of two liquid phases, one is a saturated solution of hexane in nitrobenzene and the other is a saturated solution of nitrobenzene in hexane. Because the two solubilities vary with temperature, the compositions and proportions of the two phases change as the temperature is changed. We can use a temperature–composition diagram to display the composition of the system at each temperature.

Suppose we add a small amount of nitrobenzene to hexane at a temperature T'. The nitrobenzene dissolves completely; however, as more nitrobenzene is added, a stage comes when no more dissolves. The sample now consists of two phases in equilibrium with each other, the more abundant one consisting of hexane saturated with nitrobenzene, the less abundant one a trace of nitrobenzene saturated with hexane. In the temperature–composition diagram drawn in Fig. 6.28, the composition of the former is represented by the point a' and that of the latter by the point a''. The relative abundances of the two phases are given by the **lever rule** (Fig. 6.29):

$$\frac{\text{Amount of phase of composition } a''}{\text{Amount of phase of composition } a'} = \frac{l'}{l''} \tag{6.20}$$

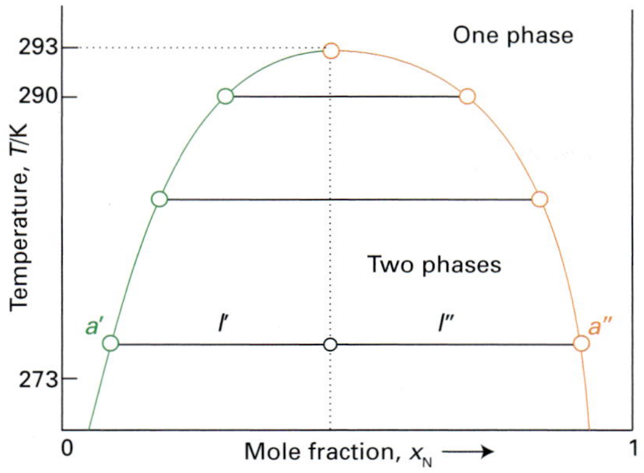

Fig. 6.28 The temperature–composition diagram for hexane and nitrobenzene at 1 atm. The upper critical solution temperature, T_{uc}, is the temperature above which no phase separation occurs. For this system it lies at 293 K (when the pressure is 1 atm).

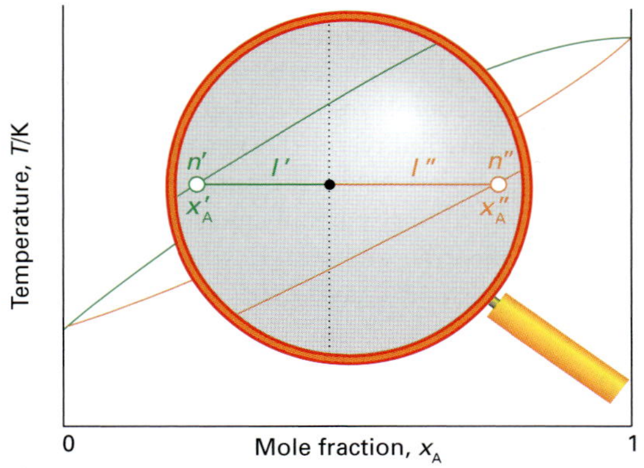

Fig. 6.29 The coordinates and compositions referred to by the lever rule.

Derivation 6.7

The lever rule

We write $n = n' + n''$, where n' is the total amount of molecules in the one phase, n'' is the total amount in the other phase, and n is the total amount of molecules in the sample. The total amount of A in the sample is nx_A, where x_A is the overall mole fraction of A in the sample (this is the quantity plotted along the horizontal axis). The overall amount of A is also the sum of its amounts in the two phases, where it has the mole fractions x_A' and x_A'', respectively:

Total amount of A

$$nx_A = n'x_A' + n''x_A''$$

Amount of A in one phase | Amount of A in the other phase

We can also multiply each side of the relation $n = n' + n''$ by x_A and obtain

$$nx_A = n'x_A + n''x_A$$

Then, by equating these two expressions we get first

$$n'x_A' + n''x_A'' = n'x_A + n''x_A$$

and then after a slight rearrangement

$$n'(x_A' - x_A) = n''(x_A - x_A'')$$

or (as can be seen by referring to Fig. 6.29)

$$n'l' = n''l''$$

which is eqn 6.20.

Example 6.5

Interpreting a liquid–liquid phase diagram

A mixture of 50 g (0.59 mol) of hexane and 50 g (0.41 mol) of nitrobenzene was prepared at 290 K. What are the compositions of the phases, and in what proportions do they occur? To what temperature must the sample be heated in order to obtain a single phase?

Strategy The answer is based on Fig. 6.28. First, we need to identify the tie-line corresponding to the temperature specified: the points at its two ends give the compositions of the two phases in equilibrium. Next, we identify the location on the horizontal axis corresponding to the overall composition of the system and draw a vertical line. Where that line cuts the tie-line it divides it into the two lengths needed to use the lever rule, eqn 6.20. For the final part, we note the temperature at which the same vertical line cuts through the phase boundary: at that temperature and above, the system consists of a single phase.

Solution We denote hexane by H and nitrobenzene by N. The horizontal tie-line at 290 K cuts the phase boundary at $x_N = 0.37$ and at $x_N = 0.83$, so those mole fractions are the compositions of the two phases. The overall composition of the system corresponds to $x_N = 0.41$, so we draw a vertical line at that mole fraction. The lever rule then gives the ratio of amounts of each phase as

$$\frac{l'}{l''} = \frac{0.41-0.37}{0.83-0.41} = \frac{0.04}{0.42} = 0.1$$

We conclude that the hexane-rich phase is ten times more abundant than the nitrobenzene-rich phase at this temperature. Heating the sample to 292 K takes it into the single-phase region.

Self-test 6.12

Repeat the problem for 50 g hexane and 100 g nitrobenzene at 273 K.

[*Answer:* $x_N = 0.07$ and 0.91 in the ratio 1:1.52; 292 K]

When more nitrobenzene is added to the two-phase mixture at the temperature T', hexane dissolves in it slightly. The overall composition moves to the right in the phase diagram, but the compositions of the two phases in equilibrium remain a' and a''. The difference is that the amount of the second phase increases at the expense of the first. A stage is reached when so much nitrobenzene is present that it can dissolve all the hexane, and the system reverts to a single phase. Now the point representing the overall composition and temperature lies to the right of the phase boundary in the illustration and the system is a single phase.

The **upper critical solution temperature,** T_{uc} (which is also called the *upper consolute temperature*), is the upper limit of temperatures at which phase separation occurs. Above the upper critical solution temperature the two components are fully miscible. In molecular terms, this temperature exists because the greater thermal motion of the molecules leads to greater miscibility of the two components. In thermodynamic terms, the Gibbs energy of mixing becomes negative above a certain temperature, regardless of the composition.

Some systems show a **lower critical solution temperature,** T_{lc} (which is also called the *lower consolute temperature*), below which they mix in all proportions and above which they form two phases. An example is water and triethylamine (Fig. 6.30). In this case, at low temperatures the two components are more miscible because they form a weak complex; at higher temperatures the complexes break up and the two components are less miscible.

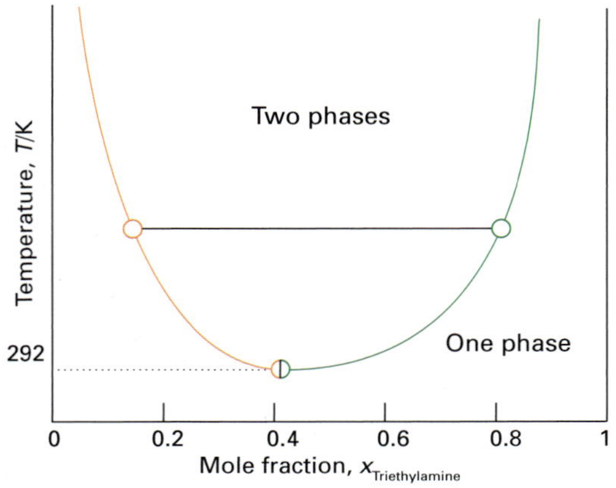

Fig. 6.30 The temperature–composition diagram for water and triethylamine. The lower critical solution temperature, T_{lc}, is the temperature below which no phase separation occurs. For this system it lies at 292 K (when the pressure is 1 atm).

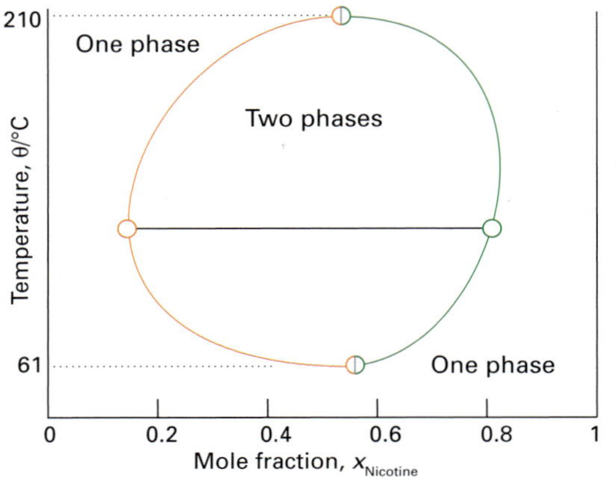

Fig. 6.31 The temperature–composition diagram for water and nicotine, which has both upper and lower critical solution temperatures. Note the high temperatures on the graph: the diagram corresponds to a sample under pressure.

A few systems have both upper and lower critical temperatures. The reason can be traced to the fact that after the weak complexes have been disrupted, leading to partial miscibility, the thermal motion at higher temperatures homogenizes the mixture again, just as in the case of ordinary partially miscible liquids. One example is nicotine and water, which are partially miscible between 61°C and 210°C (Fig. 6.31).

6.10 Liquid–solid phase diagrams

Phase diagrams are also used to show the regions of temperature and composition at which solids and liquids exist in binary systems. Such diagrams are useful for discussing the techniques that are used to prepare the high-purity materials used in the electronics industry and are also of great importance in metallurgy.

Figure 6.32 shows a simple phase diagram for an alloy of two metals that are miscible in all proportions. The **liquidus** is the line above which the entire sample is liquid; the **solidus** is the line below which the sample is entirely solid. When a sample of composition and temperature a_1 is cooled, at a_2 it initially deposits a solid of composition b_2. As the temperature is lowered, the equilibrium composition of the deposited solid moves towards b_3 and that of the remaining liquid moves towards a_3. Below the solidus, only solid of the original composition is present.

Phase diagrams such as these are constructed by monitoring the **cooling curve** at a series of compositions (Fig. 6.33). The different slopes of the liquid-phase and solid-phase cooling curves is due to their different heat capacities: the rate at which energy is

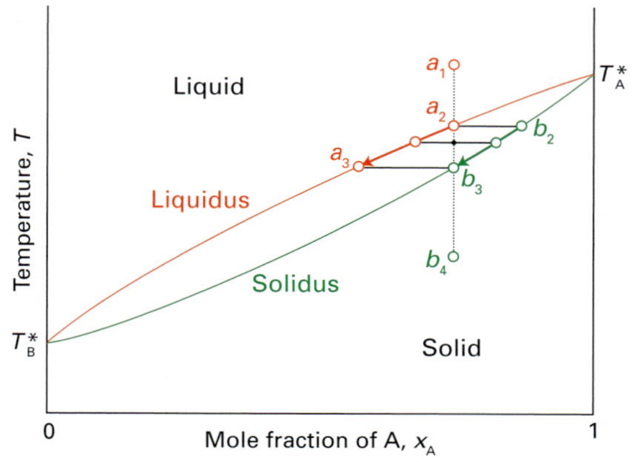

Fig. 6.32 The phase diagram for an alloy formed from two metals (with normal melting points T_A^* and T_B^* in their pure form) that are miscible in all proportions in the liquid and solid phases.

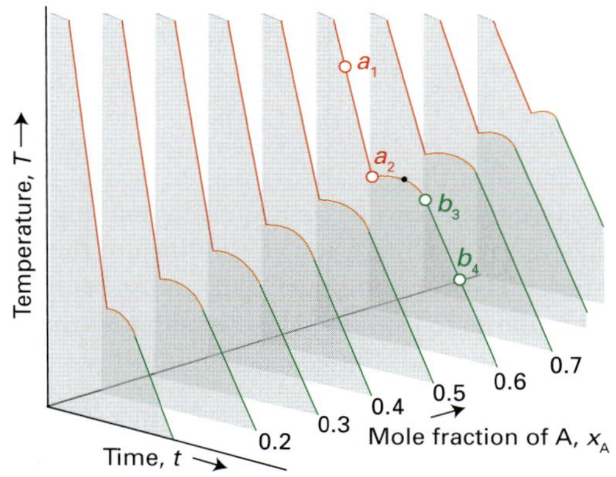

Fig. 6.33 The cooling curves for the alloy shown in Fig. 6.32; the labelled points correspond to those in that phase diagram.

lost as heat from a sample is proportional to the temperature difference between it and its surroundings, and the corresponding change in temperature of the sample will be large if the heat capacity is small (from $\Delta T = q/C$) and small if the heat capacity is high. Cooling to the temperature of the surroundings also slows if the surroundings are at a constant temperature. The changing slope between the temperatures corresponding to the liquidus and the solidus is due to the exothermic character of the phase transition: the progressive release of heat as the solid forms retards the cooling. Figure 6.34 shows a portrayal of the sequence of cooling curves plotted against the initial composition of the liquid but with the time dependence removed. By joining the points that terminate each cooling region the liquidus and solidus can be constructed.

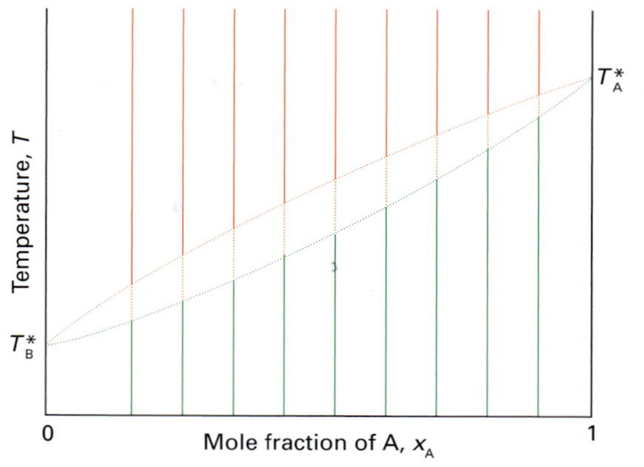

Fig. 6.34 When the time dependence of the cooling curves in Fig. 6.33 is removed and each curve is plotted vertically against the mole fraction of A, the liquidus and solidus can be identified and the phase diagram constructed.

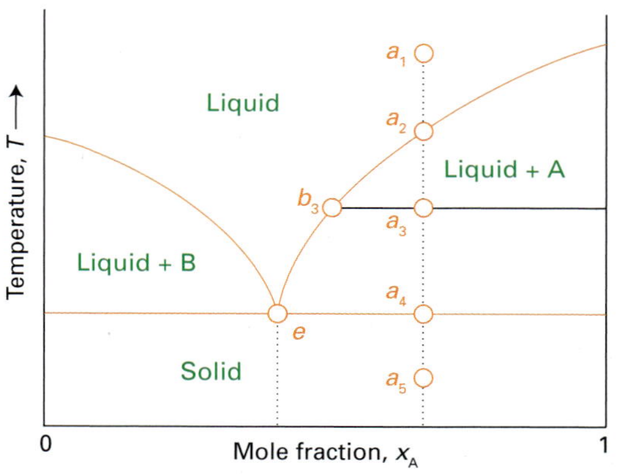

Fig. 6.35 The temperature–composition diagram for two almost immiscible solids and their completely immiscible liquids. The vertical line through *e* corresponds to the eutectic composition, the mixture with the lowest melting point.

Figure 6.35 shows the phase diagram for a system composed of two metals that are almost completely immiscible right up to their melting points (such as antimony and bismuth). Consider the molten liquid of composition a_1. When the liquid is cooled to a_2, the system enters the two-phase region labelled 'Liquid + A'. Almost pure solid A begins to come out of solution and the remaining liquid becomes richer in B. On cooling to a_3, more of the solid forms, and the relative amounts of the solid and liquid (which are in equilibrium) are given by the lever rule: at this stage there are nearly equal amounts of each. The liquid phase is richer in B than before (its composition is given by b_3) because A has been deposited. At a_4 there is less liquid than at a_3 and its composition is given by e. This liquid now freezes to give a two-phase system of almost pure A and almost pure B and cooling down to a_5 leads to no further change in composition.

The vertical line through e in Fig. 6.35 corresponds to the **eutectic composition** (from the Greek words for 'easily melted'). A solid with the eutectic composition melts, without change of composition, at the lowest temperature of any mixture. Solutions of composition to the right of e deposit A as they cool, and solutions to the left deposit B: only the eutectic mixture (apart from pure A or pure B) solidifies at a single definite temperature without gradually unloading one or other of the components from the liquid.

One technologically important eutectic is solder, which typically consists of about 67 per cent tin and 33 per cent lead by mass and melts at 183°C. Eutectic formation occurs in the great majority of binary alloy systems. It is of great importance for the microstructure of solid materials, for although a eutectic solid is a two-phase system, it crystallizes out in a nearly homogeneous mixture of microcrystals. The two microcrystalline phases can be distinguished by microscopy and structural techniques such as X-ray diffraction (Chapter 15).

Cooling curves are used to detect eutectics. We can see how it is used by considering the rate of cooling down the vertical line at a_1 in Fig. 6.35. The liquid cools steadily until it reaches a_2, when A begins to be deposited. Cooling is now slower because the solidification of A is exothermic and retards the cooling (Fig. 6.36). When the remaining liquid reaches the eutectic composition, the temperature remains constant until the whole sample has solidified: this pause in the decrease in temperature is known as the **eutectic halt**. If the liquid has the eutectic composition e initially, then the liquid cools steadily down to the freezing temperature of the eutectic, when there is a long eutectic halt as the entire sample solidifies just like the freezing of a pure liquid.

Phase diagrams are important for representing the process used to get ultrapure materials for use in the semiconductor industry (Box 6.2).

6.11 The Nernst distribution law

Suppose we shake a compound up with a mixture of two immiscible liquids, and allow the two phases to separate into layers. What can be said about the relative concentrations of the compound in the two layers?

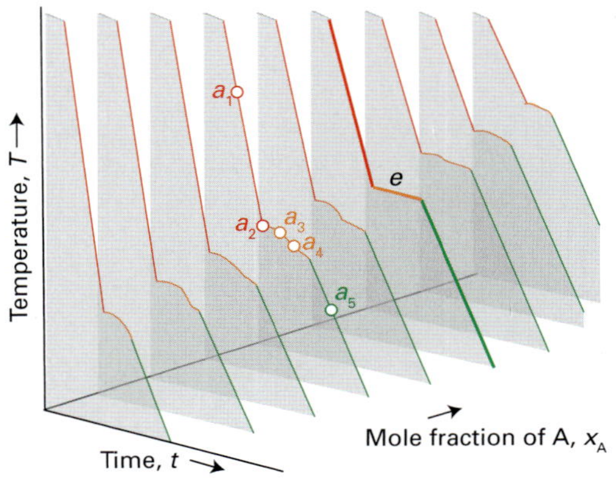

Fig. 6.36 The cooling curves for the system shown in Fig. 6.35. For a sample of composition represented by the vertical line through a_1 to a_5 in Fig. 6.35, the rate of cooling decreases at a_2 because solid A comes out of solution until the eutectic composition is reached, when there is a plateau. The cooling curve at the eutectic composition e has a complete halt at e when the eutectic solidifies without change of composition.

This question can be answered by making use of the chemical potential, for at equilibrium the chemical potential of the compound, μ_C, must be the same in each phase: $\mu_C(1) = \mu_C(2)$. If we suppose that each solution is an ideal-dilute solution, it follows that

$$\mu_C^{\ominus}(1) + RT \ln x_C(1) = \mu_C^{\ominus}(2) + RT \ln x_C(2)$$

The two standard states are different because we are using Henry's law to define the chemical potentials and the constants that occur in that law vary between solvents. It follows that

$$\ln \frac{x_C(2)}{x_C(1)} = \frac{\mu_C^{\ominus}(1) - \mu_C^{\ominus}(2)}{RT}$$

The right-hand side is a constant for a given pair of liquids, so we arrive at the **Nernst distribution law:**

$$\frac{x_C(2)}{x_C(1)} = \text{constant} \tag{6.21}$$

That is, regardless of the overall concentration (but providing the solutions are both behaving ideally), the ratio of mole fractions in the two phases is the same. For instance, suppose a certain amount of benzoic acid, C_6H_5COOH, is shaken up with a mixture of benzene and water; then the acid distributes itself between the two phases such that the ratio of mole fractions is equal to a constant. If twice that amount of acid is shaken up in the mixture, the ratio of mole fractions will be the same.

Box 6.2 Ultrapurity and controlled impurity

Advances in technology have called for materials of extreme purity. For example, semiconductor devices consist of almost perfectly pure silicon or germanium doped to a precisely controlled extent. For these materials to operate successfully, the impurity level must be kept down to less than 1 in 10^9. The technique of *zone refining* makes use of the nonequilibrium properties of mixtures. It relies on the impurities typically being more soluble in the molten sample than in the solid, and sweeps them up by passing a molten zone repeatedly from one end to the other along a sample (see the first illustration). In practice, a train of hot and cold zones are swept repeatedly from one end to the other. The zone at the end of the sample is the impurity dump: when the heater has gone by, it cools to a dirty solid that can be discarded.

We can use a phase diagram to discuss zone refining, but we have to allow for the fact that the molten zone moves along the sample and the sample is uniform in neither temperature nor composition. Consider a liquid (which represents the molten zone) on the vertical line at a_1 in the second illustration, and let it cool without the entire sample

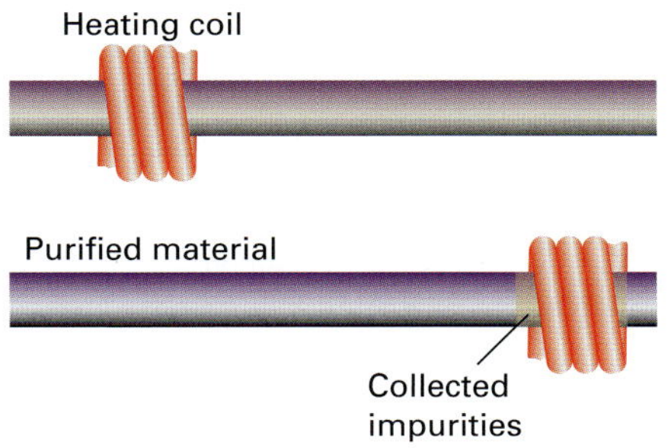

In the zone refining procedure, a heater is used to melt a small region of a long cylindrical sample of the impure solid, and that zone is swept to the other end of the rod. As it moves, it collects impurities. If a series of passes are made, the impurities accumulate at one end of the rod and can be discarded.

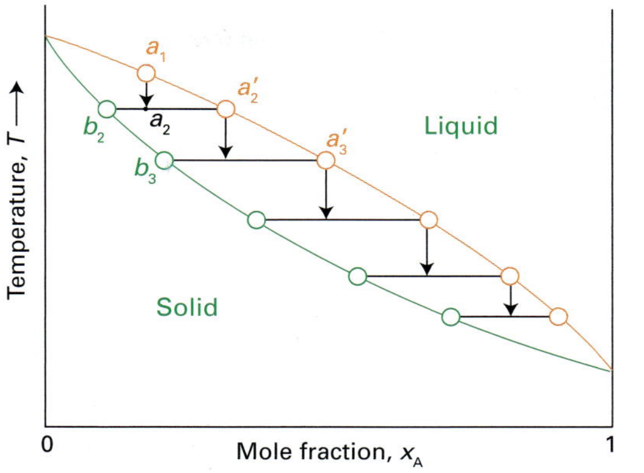

A binary temperature–composition diagram can be used to discuss zone refining, as explained in the text.

coming to overall equilibrium. If the temperature falls to a_2, a solid of composition b_2 is deposited and the remaining liquid (the zone where the heater has moved on) is at a_2'. Cooling that liquid down a vertical line passing through a_2' deposits solid of composition b_3 and leaves liquid at a_3'. The process continues until the last drop of liquid to solidify is heavily contaminated with A. There is plenty of everyday evidence that impure liquids freeze in this way. For example, an ice cube is clear near the surface but misty in the core. The water used to make ice normally contains dissolved air; freezing proceeds from the outside, and air is accumulated in the retreating liquid phase. The air cannot escape from the interior of the cube, so when that freezes the air is trapped in a mist of tiny bubbles.

A modification of zone refining is *zone levelling*. This technique is used to introduce controlled amounts of impurity (for example, of indium into germanium). A sample rich in the required dopant is put at the head of the main sample, and made molten. The zone is then dragged repeatedly in alternate directions through the sample, where it deposits a uniform distribution of the impurity.

Checklist of key ideas

You should now be familiar with the following concepts.

☐ **1** A partial molar quantity is the contribution of a component (per mole) to the overall property of a mixture.

☐ **2** The chemical potential of a component is the partial molar Gibbs energy of that component in a mixture.

☐ **3** An ideal solution is one in which both components obey Raoult's law over the entire composition range.

☐ **4** An ideal-dilute solution is one in which the solute obeys Henry's law.

☐ **5** The activity of a substance is an effective concentration; see Table 6.2.

☐ **6** A colligative property is a property that depends on the number of solute particles, not their chemical identity; they arise from the effect of a solute on the entropy of the solution.

☐ **7** Colligative properties include lowering of vapour pressure, depression of freezing point, elevation of boiling point, and osmotic pressure.

☐ **8** The osmotic pressure, Π, of an ideal solution is given by the van 't Hoff equation.

☐ **9** The equilibria between phases (at constant pressure) are represented by lines on a temperature–composition phase diagram, and the relative abundance of phases are obtained by using the lever rule.

☐ **10** An azeotrope is a mixture that vaporizes and condenses without a change of composition; a eutectic is a mixture that freezes and melts without change of composition.

Table of key equations

The following table summarizes the equations developed in this chapter.

Property	Equation	Comment
Total Gibbs energy of a mixture	$G = n_A\mu_A + n_B\mu_B$	
Chemical potential of a gas in a mixture	$\mu_J = \mu_J^{\ominus} + RT\ln p_J$	Perfect gas
Chemical potential of a solute	$\mu_J = \mu_J^* + RT\ln x_J$	Ideal solution (*denotes pure solute)
Raoult's law	$p_J = x_J p_J^*$	Ideal solution
Henry's law	$p_J = x_J K_J$ or $[J] = K_H p_J$	Ideal-dilute solution
Boiling point elevation	$\Delta T_B = K_b b_B$	Ideal solution
Freezing point depression	$\Delta T_f = K_f b_B$	Ideal solution
van 't Hoff equation	$\Pi V = n_B RT$	Ideal solution
lever rule	$n'l' = n''l''$	
Nernst distribution law	$x_C(2)/x_C(1) = \text{constant}$	Ideal solutions

Questions and exercises

Discussion questions

6.1 Explain the significance of a partial molar quantity and how it depends on composition.

6.2 Define and describe the applications of the chemical potential of a substance.

6.3 State and justify the thermodynamic criterion for solution–vapour equilibrium.

6.4 Justify Raoult's and Henry's laws in terms of the molecular interactions in a mixture.

6.5 Explain the origin of the colligative properties.

6.6 What is meant by the activity of a solute?

6.7 Explain the origin of osmosis in terms of the thermodynamic and molecular properties of a mixture.

6.8 Explain how osmotic pressure measurements can be used to determine the molar mass of a polymer.

Exercises

✓6.1 The partial molar volumes of propanone and trichloromethane in a mixture in which the mole fraction of $CHCl_3$ is 0.4693 are 74.166 $cm^3\ mol^{-1}$ and 80.235 $cm^3\ mol^{-1}$, respectively. What is the volume of a solution of total mass 1.000 kg?

6.2 Use Fig. 6.1 to estimate the total volume of a solution formed by mixing 50.0 cm^3 of ethanol with 50.0 cm^3 of water. The densities of the two liquids are 0.789 and 1.000 g cm^{-3}, respectively.

6.3 By how much does the chemical potential of carbon dioxide at 310 K and 2.0 bar differ from its standard value at that temperature?

6.4 The standard state of a substance used to be defined as 1 atm at the specified temperature. By how much does the current definition of standard chemical potential (at 1 bar) differ from its former value at 298.15 K?

6.5 Calculate (a) the (molar) Gibbs energy of mixing, (b) the (molar) entropy of mixing when the two major components of air (nitrogen and oxygen) are mixed to form air at 298 K. The mole fractions of N_2 and O_2 are 0.78 and 0.22, respectively. Is the mixing spontaneous?

6.6 Suppose now that argon is added to the mixture in Exercise 6.5 to bring the composition closer to real air, with mole fractions 0.780, 0.210, and 0.0096, respectively. What is the additional change in molar Gibbs energy and entropy at 298 K? Is the mixing spontaneous?

6.7 A solution is prepared by dissolving 2.33 g of C_{60} (buckminsterfullerene) in 100.0 g of toluene (methylbenzene). Given that the vapour pressure of pure toluene is 5.00 kPa at 30°C, what is the vapour pressure of toluene in the solution?

6.8 Estimate the vapour pressure of sea water at 20°C given that the vapour pressure of pure water is 2.338 kPa at that temperature and the solute is largely Na^+ and Cl^- ions, each present at about 0.50 mol dm^{-3}.

6.9 At 300 K, the vapour pressure of dilute solutions of HCl in liquid $GeCl_4$ are as follows:

x(HCl)	0.005	0.012	0.019
p/kPa	32.0	76.9	121.8

Show that the solution obeys Henry's law in this range of mole fractions and calculate the Henry's law constant at 300 K.

6.10 Calculate the concentration of carbon dioxide in fat given that the Henry's law constant is 8.6×10^4 Torr and the partial pressure of carbon dioxide is 55 kPa.

6.11 What partial pressure of hydrogen results in a molar concentration of 1.0 mmol dm^{-3} in water at 25°C?

6.12 The rise in atmospheric carbon dioxide results in higher concentrations of dissolved carbon dioxide in natural waters. Use Henry's law and the data in Table 6.1 to calculate the solubility of CO_2 in water at 25°C when its partial pressure is (a) 3.8 kPa, (b) 50.0 kPa.

6.13 The mole fractions of N_2 and O_2 in air at sea level are approximately 0.78 and 0.21, respectively. Calculate the molalities of the solution formed in an open flask of water at 25°C.

6.14 A water-carbonating plant is available for use in the home and operates by providing carbon dioxide at 1.0 atm. Estimate the molar concentration of the CO_2 in the soda water it produces.

6.15 At 90°C the vapour pressure of toluene (methylbenzene) is 53 kPa and that of *o*-xylene (1,2-dimethylbenzene) is 20 kPa. What is the composition of the liquid mixture that boils at 25°C when the pressure is 0.50 atm? What is the composition of the vapour produced?

6.16 The vapour pressures of the two components A and B of a binary mixture varied as follows:

$$p_A/\text{Torr} = 68x_A - 12x_A^2 + 643x_A^3 - 283x_A^4$$
$$p_B/\text{Torr} = 780 - 440x_A - 401x_A^2 + 92x_A^3$$

Confirm that the mixture conforms to Raoult's law for the component in large excess and to Henry's law for the minor component. Find the Henry's law constants.

6.17 The vapour pressures of the two components A and B of a binary mixture varied as follows:

x_A	0	0.20	0.40	0.60	0.80	1
p_A/Torr	0	127	246	357	457	539
p_B/Torr	701	631	526	394	234	38

Confirm that the mixture conforms to Raoult's law for the component in large excess and to Henry's law for the minor component. Find the Henry's law constants. Hint: proceed as in Example 6.3.

6.18 What is the change in chemical potential of glucose when its concentration in water at 20.0°C is changed from 0.10 mol dm^{-3} to 1.00 mol dm^{-3}?

6.19 The vapour pressure of a sample of benzene is 53.0 kPa at 60.6°C, but it fell to 51.2 kPa when 0.133 g of an organic compound was dissolved in 5.00 g of the solvent. Calculate the molar mass of the compound.

6.20 Estimate the freezing point of 200 cm^3 of water sweetened by the addition of 2.5 g of sucrose. Treat the solution as ideal.

6.21 Estimate the freezing point of 200 cm^3 of water to which 2.5 g of sodium chloride has been added. Treat the solution as ideal.

6.22 The addition of 28.0 g of a compound to 750 g of tetrachloromethane, CCl_4, lowered the freezing point of the solvent by 5.40 K. Calculate the molar mass of the compound.

6.23 A compound A existed in equilibrium with its dimer, A_2, in propanone solution. Derive an expression for the equilibrium constant $K = [A_2]/[A]^2$ in terms of the depression in vapour pressure caused by a given concentration of compound. (*Hint.* Suppose that a fraction *f* of the A molecules are present as the dimer. The depression of vapour pressure is proportional to the total concentration of A and A_2 molecules regardless of their chemical identities.)

6.24 The osmotic pressure of an aqueous solution of urea at 300 K is 150 kPa. Calculate the freezing point of the same solution.

6.25 The osmotic pressure of a solution of polystyrene in toluene (methylbenzene) was measured at 25°C with the following results:

$c/(\mathrm{g\ dm^{-3}})$	2.042	6.613	9.521	12.602
Π/Pa	58.3	188.2	270.8	354.6

Determine the molar mass of the polymer.

6.26 The molar mass of an enzyme was determined by dissolving it in water, measuring the osmotic pressure at 20°C and extrapolating the data to zero concentration. The following data were used:

$c/(\mathrm{mg\ cm^{-3}})$	3.221	4.618	5.112	6.722
h/cm	5.746	8.238	9.119	11.990

What is the molar mass of the enzyme? *Hint:* Begin by expressing eqn 6.19 in terms of the height of the solution, by using $\Pi = \rho g h$; take $\rho = 1.000\ \mathrm{g\ cm^{-3}}$.

6.27 The following temperature–composition data were obtained for a mixture of octane (O) and toluene (T) at 760 Torr, where x is the mole fraction in the liquid and y the mole fraction in the vapour at equilibrium.

$\theta/°\mathrm{C}$	110.9	112.0	114.0	115.8	117.3	119.0	120.0	123.0
x_T	0.908	0.795	0.615	0.527	0.408	0.300	0.203	0.097
y_T	0.923	0.836	0.698	0.624	0.527	0.410	0.297	0.164

The boiling points are 110.6°C for toluene and 125.6°C for octane. Plot the temperature–composition diagram of the mixture. What is the composition of the vapour in equilibrium with the liquid of composition (a) $x_T = 0.250$ and (b) $x_O = 0.250$?

6.28 Sketch the phase diagram of the system NH_3/N_2H_4 given that the two substances do not form a compound with each other, that NH_3 freezes at –78°C and N_2H_4 freezes at +2°C, and that a eutectic is formed when the mole fraction of N_2H_4 is 0.07 and that the eutectic melts at –80°C.

6.29 Figure 6.37 shows the phase diagram for two partially miscible liquids, which can be taken to be that for water (A) and 2-methyl-1-propanol (B). Describe what will be observed when a mixture of composition b_3 is heated, at each stage giving the number, composition, and relative amounts of the phases present.

6.30 Figure 6.38 is the phase diagram for silver/tin. Label the regions, and describe what will be observed when liquids of compositions *a* and *b* are cooled to 200°C.

6.31 Sketch the cooling curves for the compositions *a* and *b* in Fig. 6.38.

6.32 Use the phase diagram in Fig. 6.38 to determine (a) the solubility of silver in tin at 800°C, (b) the solubility of Ag_3Sn in silver at 460°C, and (c) the solubility of Ag_3Sn in silver at 300°C.

6.33 Figure 6.39 shows a part of the phase diagram of an alloy of copper and aluminium. Describe what you will

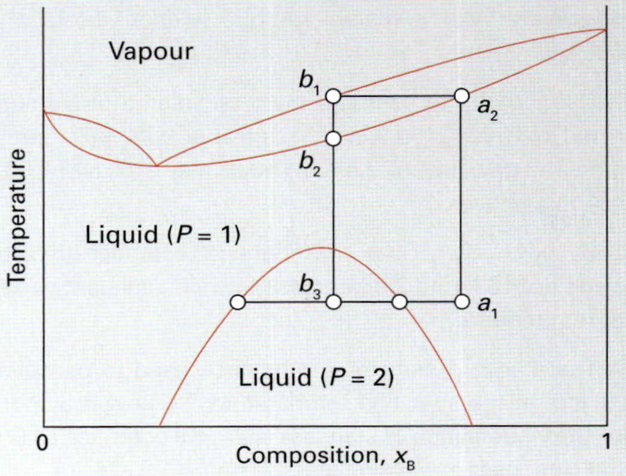

Fig. 6.37 A phase diagram for two partially miscible liquids.

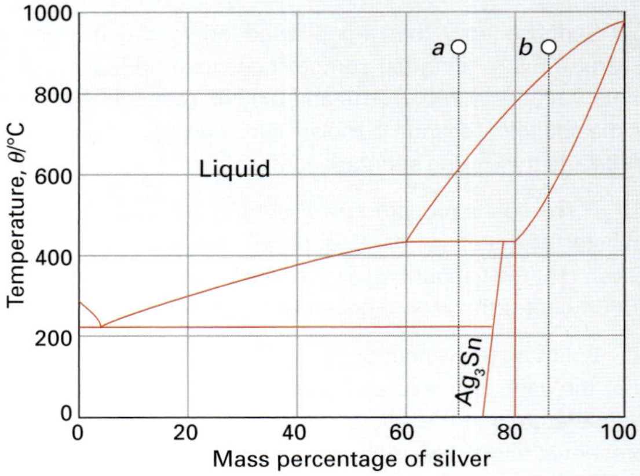

Fig. 6.38 The phase diagram for silver/tin.

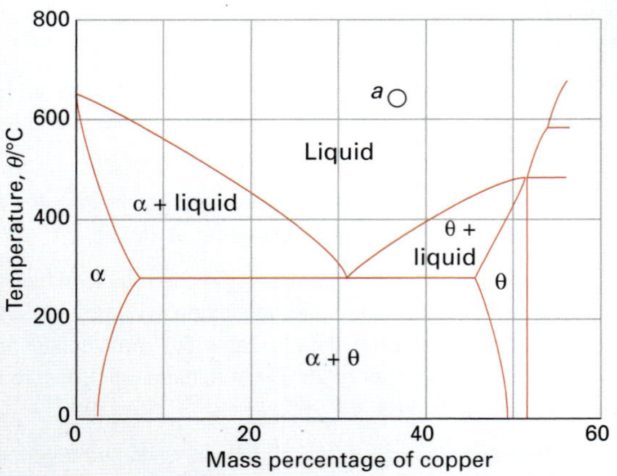

Fig. 6.39 A part of the phase diagram of an alloy of copper and aluminium.

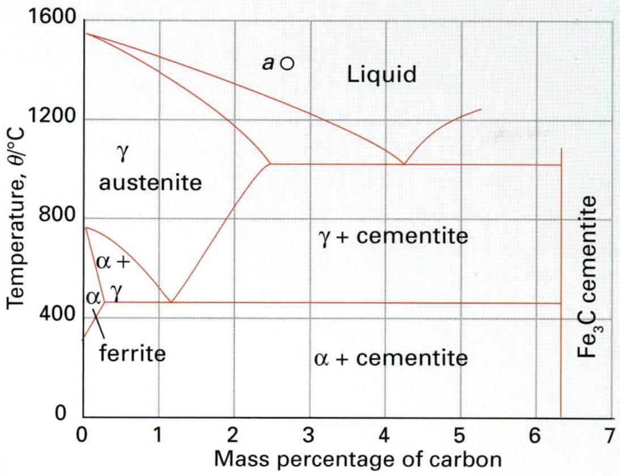

Fig. 6.40 A part of the phase diagram typical of a simple steel.

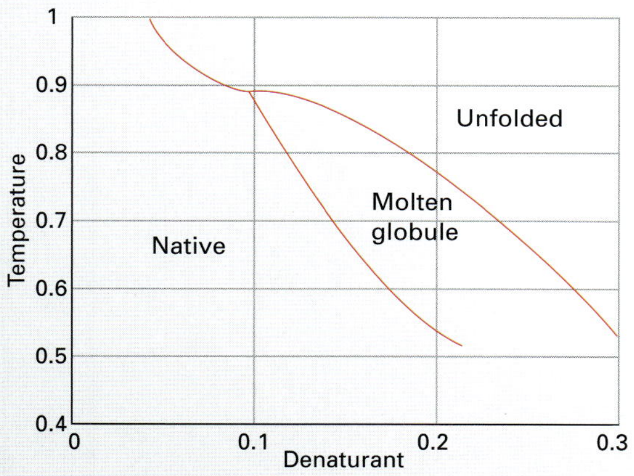

Fig. 6.41 A theoretical phase diagram for a model protein.

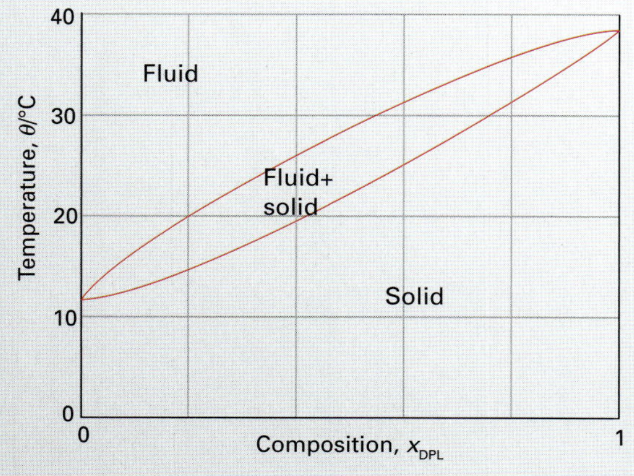

Fig. 6.42 The phase diagram for a membrane-forming two-component system.

observe as a melt of composition *a* is cooled. What is the solubility of copper in aluminium at 500°C?

6.34 Figure 6.40 shows a part of the phase diagram typical of a simple steel. Describe what you will observe as the melt at *a* is allowed to cool to room temperature.

6.35 Hexane and perfluorohexane (C_6F_{14}) show partial miscibility below 22.70°C. The critical concentration at the upper critical temperature is $x = 0.355$, where x is the mole fraction of C_6F_{14}. At 22.0°C the two solutions in equilibrium have $x = 0.24$ and $x = 0.48$, respectively, and at 21.5°C the mole fractions are 0.22 and 0.51. Sketch the phase diagram. Describe the phase changes that occur when perfluorohexane is added to a fixed amount of hexane at (a) 23°C, (b) 25°C.

6.36 In a theoretical study of protein-like polymers, the phase diagram shown in Fig. 6.41 was obtained. It shows three structural regions: the native form, the unfolded form, and a 'molten globule' form. (a) Is the molten-globule form ever stable when the denaturant concentration is below 0.1? (b) Describe what happens to the polymer as the native form is heated in the presence of denaturant at concentration 0.15.

6.37 In an experimental study of membrane-like assemblies of synthetic materials, a phase diagram like that shown in Fig. 6.42 was obtained. The two components are dielaidoylphosphatidylcholine (DEL) and dipalmitoylphosphatidylcholine (DPL). Explain what happens as a liquid mixture of composition $x_{DEL} = 0.5$ is cooled from 45°C.

6.38 When 2.0 g of aspirin was shaken up in a flask containing two immiscible liquids it was found that the mole factions in the two liquids were 0.11 and 0.18. When a further 1.0 g of aspirin was added, it was found that the mole fraction in the first liquid increased to 0.15. What would you expect the mole fraction the second liquid to become?

Projects

The symbol ‡ indicates that the exercise requires calculus.

6.39‡ (a) The partial molar volume of ethanol in a mixture at 25°C is $V_{ethanol}/(cm^3\ mol^{-1}) = 54.6664 - 0.727\,88b + 0.084\,768b^2$, where b is the numerical value of the molality of ethanol. Plot this quantity as a function of b and identify the composition at which the partial molar volume is a minimum. Express that composition as a mole fraction. (b) Use differentiation to identify the minium in part (a).

6.40‡ (a) The total volume of a water/ethanol mixture at 25°C fits the expression $V/cm^3 = 1002.93 + 54.6664b - 0.363\,94b^2 + 0.028\,256b^3$, where b is the numerical value of the molality of ethanol. With the information in Exercise 6.3, find an expression for the partial molar volume of water. Plot the curve. Show that the partial molar volume of water has a maximum value where the partial molar volume of ethanol is a minimum. (b) Use calculus to plot the partial molar volumes of ethanol and water from the data in part (a). *Hint*: Convert b to a mole fraction, then use $V_J = dV/dx_J$.

6.41 Haemoglobin, the red blood protein responsible for oxygen transport, binds about 1.34 cm^3 of oxygen per gram. Normal blood has a haemoglobin concentration of 150 g dm^{-3}. Haemoglobin in the lungs is about 97 per cent saturated with oxygen, but in the capillary is only about 75 per cent saturated. (a) What volume of oxygen is given up by 100 cm^3 of blood flowing from the lungs in the capillary? Breathing air at high pressures, such as in scuba diving, results in an increased concentration of dissolved nitrogen. The Henry's law constant in the form $c = Kp$ for the solubility of nitrogen is 0.18 μg/(g H_2O atm). (b) What mass of nitrogen is dissolved in 100 g of water saturated with air at 4.0 atm and 20°C? Compare your answer to that for 100 g of water saturated with air at 1.0 atm. (Air is 78.08 mole per cent N_2.) (c) If nitrogen is four times as soluble in fatty tissues as in water, what is the increase in nitrogen concentration in fatty tissue in going from 1 atm to 4 atm?

Chapter 7

Chemical equilibrium: the principles

Now we arrive at the point where real chemistry begins. Chemical thermodynamics is used to predict whether a mixture of reactants has a spontaneous tendency to change into products, to predict the composition of the reaction mixture at equilibrium, and to predict how that composition will be modified by changing the conditions. Although reactions in industry are rarely allowed to reach equilibrium, knowing whether equilibrium lies in favour of reactants or products under certain conditions is a good indication of the feasibility of a process. Much the same is true of biochemical reactions, where the avoidance of equilibrium is life and the attainment of equilibrium is death.

There is one word of warning that is essential to remember: *thermodynamics is silent about the rates of reaction*. All it can do is to identify whether a particular reaction mixture has a tendency to form products, it cannot say whether that tendency will ever be realized. Chapters 10 and 11 explore what determines the rates of chemical reactions.

Thermodynamic background

The thermodynamic criterion for spontaneous change at constant temperature and pressure is $\Delta G < 0$. The principal idea behind this chapter, therefore, is that, *at constant temperature and pressure, a reaction mixture tends to adjust its composition until its Gibbs energy is a minimum*. If the Gibbs energy of a mixture varies as shown in Fig. 7.1a, very little of the reactants convert into products before G has reached its minimum value and the reaction 'does not go'. If G varies as shown in Fig. 7.1c, then a high proportion of products must form before G reaches its minimum and the reaction 'goes'. In many cases, the equilibrium mixture contains almost no reactants or

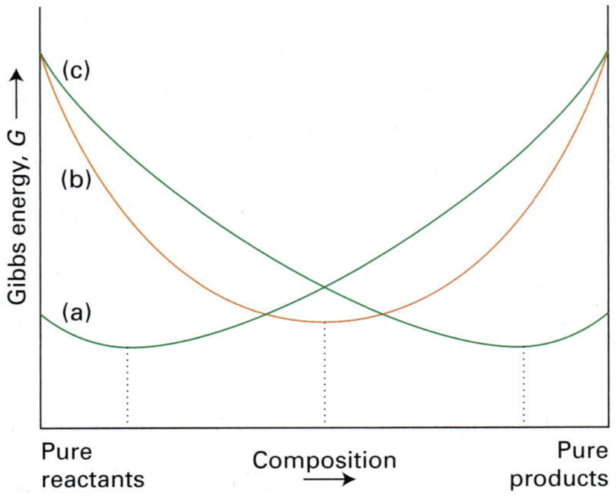

Fig. 7.1 The variation of Gibbs energy of a reaction mixture with progress of the reaction, pure reactants on the left and pure products on the right. (a) This reaction 'does not go': the minimum in the Gibbs energy occurs very close to reactants. (b) This reaction reaches equilibrium with approximately equal amounts of reactants and products present in the mixture. (c) This reaction goes almost to completion, as the minimum in Gibbs energy lies very close to pure products.

almost no products. Many reactions have a Gibbs energy that varies as shown in Fig. 7.1b, and at equilibrium the reaction mixture contains substantial amounts of both reactants and products. One of our tasks is to see how to use thermodynamic data to predict the equilibrium composition and to see how that composition depends on the conditions.

7.1 The reaction Gibbs energy

To keep our ideas in focus, we consider two important reactions. One is the isomerism of glucose-6-phosphate (**1**, G6P) to fructose-6-phosphate (**2**, F6P), which is an early step in the anaerobic breakdown of glucose:

$$\text{G6P(aq)} \rightleftharpoons \text{F6P(aq)} \quad \text{(A)}$$

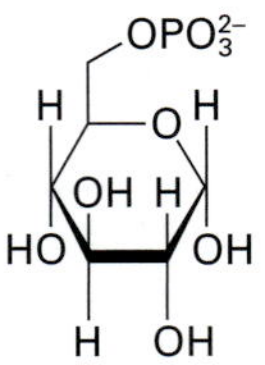

1 Glucose-6-phosphate

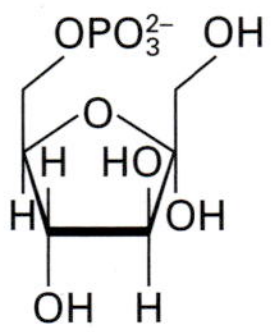

2 Fructose-6-phosphate

This reaction takes place in the aqueous environment of the cell. The second is the synthesis of ammonia, which is of crucial importance for industry and agriculture:

$$N_2(g) + 3\,H_2(g) \rightleftharpoons 2\,NH_3(g) \quad \text{(B)}$$

These two reactions are specific examples of a general reaction of the form

$$a\,\text{A} + b\,\text{B} \rightleftharpoons c\,\text{C} + d\,\text{D} \quad \text{(C)}$$

with arbitrary physical states.

First, consider reaction A. Suppose that in a short interval while the reaction is in progress, the amount of G6P changes by $-\Delta n$. As a result of this change in amount, the contribution of G6P to the total Gibbs energy of the system changes by $-\mu_{\text{G6P}}\Delta n$, where μ_{G6P} is the chemical potential (the partial molar Gibbs energy) of G6P in the reaction mixture. In the same interval, the amount of F6P changes by $+\Delta n$, so its contribution to the total Gibbs energy changes by $+\mu_{\text{F6P}}\Delta n$, where μ_{F6P} is the chemical potential of F6P. Provided Δn is small enough to leave the composition virtually unchanged, the net change in Gibbs energy of the system is

$$\Delta G = \mu_{\text{F6P}} \times \Delta n - \mu_{\text{G6P}} \times \Delta n$$

If we divide through by Δn, we obtain the **reaction Gibbs energy**, $\Delta_r G$:

$$\Delta_r G = \frac{\Delta G}{\Delta n} = \mu_{\text{F6P}} - \mu_{\text{G6P}} \quad (7.1a)$$

There are two ways to interpret $\Delta_r G$. First, it is the difference of the chemical potentials of the products and reactants *at the composition of the reaction mixture*. Second, because $\Delta_r G$ is the change in G divided by the change in composition, we can think of $\Delta_r G$ as being the slope of the graph of G plotted against the changing composition of the system (Fig. 7.2).

The synthesis of ammonia provides a slightly more complicated example. If the amount of N_2 changes by $-\Delta n$, then from the reaction stoichiometry we

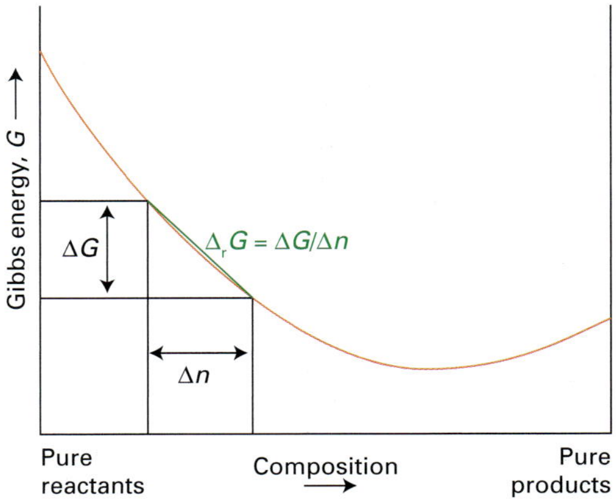

Fig. 7.2 The variation of Gibbs energy with progress of reaction showing how the reaction Gibbs energy, $\Delta_r G$, is related to the slope of the curve at a given composition.

know that the change in the amount of H_2 will be $-3\Delta n$ and the change in the amount of NH_3 will be $+2\Delta n$. Each change contributes to the change in the total Gibbs energy of the mixture, and the overall change is

$$\Delta G = \mu_{NH_3} \times 2\Delta n - \mu_{N_2} \times \Delta n - \mu_{H_2} \times 3\Delta n$$
$$= (2\mu_{NH_3} - \mu_{N_2} - 3\mu_{H_2})\Delta n$$

where the μ_J are the chemical potentials of the species in the reaction mixture. In this case, therefore, the reaction Gibbs energy is

$$\Delta_r G = \frac{\Delta G}{\Delta n} = 2\mu_{NH_3} - (\mu_{N_2} + 3\mu_{H_2}) \qquad (7.1b)$$

Note that each chemical potential is multiplied by the corresponding stoichiometric coefficient and that reactants are subtracted from products. For the general reaction C,

$$\Delta_r G = (c\mu_C + d\mu_D) - (a\mu_A + b\mu_B) \qquad (7.1c)$$

The chemical potential of a substance depends on the composition of the mixture in which it is present, and is high when its concentration or partial pressure is high. Therefore, $\Delta_r G$ changes as the composition changes (Fig. 7.3). Remember that $\Delta_r G$ is the *slope* of G plotted against composition. We see that $\Delta_r G < 0$ and the slope of G is negative (down from left to right) when the mixture is rich in the reactants A and B because μ_A and μ_B are then high. Conversely, $\Delta_r G > 0$ and the slope of G is positive (up from left to right) when the mixture is rich in the products C and D because μ_C and μ_D are then high.

At compositions corresponding to $\Delta_r G < 0$ the reaction tends to form more products. At compositions corresponding to $\Delta_r G > 0$, the *reverse* reaction is spontaneous, and the products tend to decompose into reactants. Where $\Delta_r G = 0$ (at the minimum of the graph where the slope is zero), the reaction has no tendency to form either products or reactants. In other words, the reaction is at equilibrium. That is, *the criterion for chemical equilibrium at constant temperature and pressure* is

$$\Delta_r G = 0 \qquad (7.2)$$

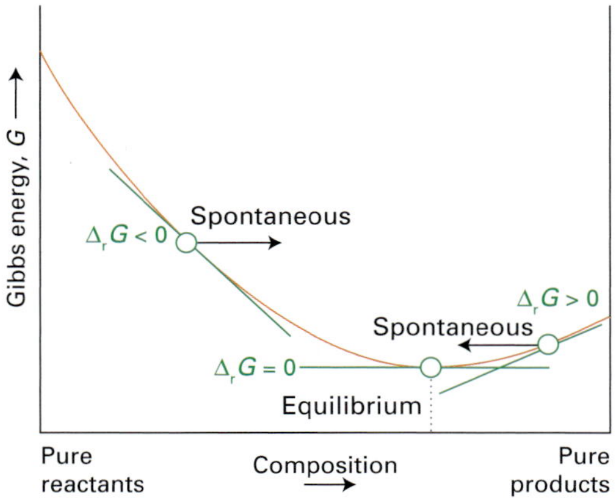

Fig. 7.3 At the minimum of the curve, corresponding to equilibrium, $\Delta_r G = 0$. To the left of the minimum, $\Delta_r G < 0$, and the forward reaction is spontaneous. To the right of the minimum, $\Delta_r G > 0$ and the reverse reaction is spontaneous.

7.2 The variation of $\Delta_r G$ with composition

Our next step is to find how $\Delta_r G$ varies with the composition of the system. Once we know that, we shall be able to identify the composition corresponding to $\Delta_r G = 0$. Our starting point is the general expression for the composition dependence of the chemical potential derived in Section 6.5:

$$\mu_J = \mu_J^\ominus + RT \ln a_J \qquad (7.3)$$

where a_J is the activity of the species J. When we are dealing with ideal systems, which will be the case in this chapter, we use the identifications given in Table 6.2:

For solutes in an ideal solution, $a_J = [J]/c^\ominus$, the molar concentration of J relative to the standard value $c^\ominus = 1$ mol dm^{-3}.

For perfect gases, $a_J = p_J/p^\ominus$, the partial pressure of J relative to the standard pressure $p^\ominus = 1$ bar.

For pure solids and liquids, $a_J = 1$.

As in Chapter 6, to simplify the appearance of expressions in what follows, we shall not write $c^\ominus$ and $p^\ominus$ explicitly.

Substitution of eqn 7.3 into eqn 7.1c gives

$$\Delta_r G = \{c(\mu_C^\ominus + RT \ln a_C) + d(\mu_D^\ominus + RT \ln a_D)\}$$
$$- \{a(\mu_A^\ominus + RT \ln a_A) + b(\mu_B^\ominus + RT \ln a_B)\}$$
$$= \{(c\mu_C^\ominus + d\mu_D^\ominus) - (a\mu_A^\ominus + b\mu_B^\ominus)\}$$
$$+ RT\{c \ln a_C + d \ln a_D - a \ln a_A - b \ln a_B\}$$

The first term on the right in the second equality is the **standard reaction Gibbs energy**, $\Delta_r G^\ominus$:

$$\Delta_r G^\ominus = \{c\mu_C^\ominus + d\mu_D^\ominus\} - \{a\mu_A^\ominus + b\mu_B^\ominus\} \qquad (7.4a)$$

Because the standard states refer to the pure materials, the standard chemical potentials in this expression are the standard molar Gibbs energies of the (pure) species. Therefore, eqn 7.4a is the same as

$$\Delta_r G^\ominus = \{cG_m^\ominus(C) + dG_m^\ominus(D)\} - \{aG_m^\ominus(A) + bG_m^\ominus(B)\} \qquad (7.4b)$$

We consider this important quantity in more detail shortly. At this stage, therefore, we know that

$$\Delta_r G = \Delta_r G^{\ominus} + RT\{c \ln a_C + d \ln a_D - a \ln a_A - b \ln a_B\}$$

and the expression for $\Delta_r G$ is beginning to look much simpler.

To make further progress, we rearrange the remaining terms on the right as follows:

$$c \ln a_C + d \ln a_D - a \ln a_A - b \ln a_B$$

Use $n \ln x = \ln x^n$

$$= \ln a_C^c + \ln a_D^d - \ln a_A^a - \ln a_B^b$$

Use $\ln x + \ln y = \ln xy$

$$= \ln a_C^c a_D^d - \ln a_A^a a_B^b$$

$$= \ln \frac{a_C^c a_D^d}{a_A^a a_B^b}$$

Use $\ln x - \ln y = \ln x/y$

At this point, we have deduced that

$$\Delta_r G = \Delta_r G^{\circ} + RT \ln \frac{a_C^c a_D^d}{a_A^a a_B^b}$$

To simplify the appearance of this expression still further we introduce the (dimensionless) **reaction quotient**, Q, for reaction C:

$$Q = \frac{a_C^c a_D^d}{a_A^a a_B^b} \qquad (7.5)$$

Note that Q has the form of products divided by reactants, with the activity of each species raised to a power equal to its stoichiometric coefficient in the reaction. We can now write the overall expression for the reaction Gibbs energy at any composition of the reaction mixture as

$$\Delta_r G = \Delta_r G^{\ominus} + RT \ln Q \qquad (7.6)$$

This simple but hugely important equation will occur several times in different disguises.

A brief illustration The reaction quotient for reaction A is

$$Q = \frac{a_{F6P}}{a_{G6P}} = \frac{[F6P]/c^{\ominus}}{[G6P]/c^{\ominus}}$$

However, our convention is not to write the standard concentration explicitly (in this example, it cancels anyway), so this expression simplifies to

$$Q = \frac{[F6P]}{[G6P]}$$

with [J] the numerical value of the molar concentration of J in moles per cubic decimetre (so, if [F6P] = 2.0 mmol dm^{-3}, corresponding to 2.0×10^{-3} mol dm^{-3}, we just write [F6P] = 2.0×10^{-3} when using this expression). For reaction B, the synthesis of ammonia, a gas-phase reaction, the reaction quotient is

$$Q = \frac{(p_{NH_3}/p^{\ominus})^2}{(p_{N_2}/p^{\ominus})(p_{H_2}/p^{\ominus})^3}$$

When we do not write the standard pressure explicitly, this expression simplifies to

$$Q = \frac{p_{NH_3}^2}{p_{N_2} p_{H_2}^3}$$

with p_J the numerical value of the partial pressure of J in bar (so, if p_{NH_3} = 2 bar, we just write p_{NH_3} = 2 when using this expression).

Self-test 7.1

Write the reaction quotient for the esterification reaction $CH_3COOH + C_2H_5OH \rightleftharpoons CH_3COOC_2H_5 + H_2O$. (All four components are present in the reaction mixture as liquids: the mixture is not an aqueous solution.)

[Answer: $Q \approx [CH_3COOC_2H_5][H_2O]/[CH_3COOH][C_2H_5OH]$]

7.3 Reactions at equilibrium

When the reaction has reached equilibrium, the composition has no further tendency to change because $\Delta_r G = 0$ and the reaction is spontaneous in neither direction. At equilibrium, the reaction quotient has a certain value called the **equilibrium constant**, K, of the reaction:

$$K = \left(\frac{a_C^c a_D^d}{a_A^a a_B^b}\right)_{\text{equilibrium}} \qquad (7.7)$$

We shall not normally write 'equilibrium'; the context will always make it clear that Q refers to an *arbitrary* stage of the reaction, whereas K, the value of Q at equilibrium, is calculated from the *equilibrium* composition. It now follows from eqn 7.6, that at equilibrium

$\Delta_r G = 0$ at equilibrium; $Q = K$ at equilibrium

$$0 = \Delta_r G^{\ominus} + RT \ln K$$

and therefore that

$$\Delta_r G^{\ominus} = -RT \ln K \qquad (7.8)$$

This is one of the most important equations in the whole of chemical thermodynamics. Its principal use is to predict the value of the equilibrium constant of any reaction from tables of thermodynamic data, like those in the Data section. Alternatively, we can use it to determine $\Delta_r G^{\ominus}$ by measuring the equilibrium constant of a reaction.

A brief illustration Suppose we know that $\Delta_r G^\ominus = +3.40\ \text{kJ mol}^{-1}$ for the reaction $H_2(g) + I_2(s) \rightarrow 2\ HI(g)$ at 25°C, then to calculate the equilibrium constant we write

$$\ln K = -\frac{\Delta_r G^\ominus}{RT} = -\frac{3.40 \times 10^3\ \text{J mol}^{-1}}{(8.314\ 47\ \text{J K}^{-1}\ \text{mol}^{-1}) \times (298\ \text{K})}$$
$$= -\frac{3.40 \times 10^3}{8.314\ 47 \times 298}$$

This expression evaluates to $\ln K = -1.37$, but to avoid rounding errors, we leave the evaluation until the next step, which is to use the relation $e^{\ln x} = x$ with $x = K$ to write.

$$K = e^{-\frac{3.40\times10^3}{8.314\ 47\times298}} = 0.25$$

A note on good practice All equilibrium constants and reaction quotients are dimensionless numbers.

Self-test 7.3

Calculate the equilibrium constant of the reaction $N_2(g) + 3\ H_2(g) \rightarrow 2\ NH_3(g)$ at 25°C, given that $\Delta_r G^\ominus = -32.90$ kJ mol^{-1}.

[*Answer:* 5.8×10^5]

An important feature of eqn 7.8 is that it tells us that $K > 1$ if $\Delta_r G^\ominus < 0$. Broadly speaking, $K > 1$ implies that products are dominant at equilibrium, so we can conclude that *a reaction is thermodynamically feasible if* $\Delta_r G^\ominus < 0$ (Fig. 7.4). Reactions for which $\Delta_r G^\ominus < 0$ are called **exergonic**. Conversely, because eqn 7.8 tells us that $K < 1$ if $\Delta_r G^\ominus > 0$, then we know that the reactants will be dominant in a reaction mixture at equilibrium if $\Delta_r G^\ominus > 0$. In other words, *a reaction with* $\Delta_r G^\ominus > 0$ *is not thermodynamically feasible.* Reactions for which $\Delta_r G^\ominus > 0$ are called **endergonic**. Some care must be exercised with these rules, however, because the products will be significantly more abundant than reactants only if $K \gg 1$ (more than about 10^3) and even a reaction with $K < 1$ may have a reasonable abundance of products at equilibrium.

Table 7.1 summarizes the conditions under which $\Delta_r G^\ominus < 0$ and $K > 1$. Because $\Delta_r G^\ominus = \Delta_r H^\ominus - T\Delta_r S^\ominus$, the standard reaction Gibbs energy is certainly negative if both $\Delta_r H^\ominus < 0$ (an exothermic reaction) and $\Delta_r S^\ominus > 0$ (a reaction system that becomes more disorderly, such as by forming a gas). The standard reaction Gibbs energy is also negative if the reaction is endothermic ($\Delta_r H^\ominus > 0$) and $T\Delta_r S^\ominus$ is sufficiently large and positive. Note that for an endothermic reaction to have $\Delta_r G^\ominus < 0$, its standard reaction entropy *must* be positive. Moreover, the temperature must be high enough for $T\Delta_r S^\ominus$ to be greater than $\Delta_r H^\ominus$ (Fig. 7.5). The switch of $\Delta_r G^\ominus$ from positive

Table 7.1
Thermodynamic criteria of spontaneity

(1)	If the reaction is exothermic ($\Delta_r H^\ominus < 0$) and $\Delta_r S^\ominus > 0$: $\Delta_r G^\ominus < 0$ and $K > 1$ at all temperatures
(2)	If the reaction is exothermic ($\Delta_r H^\ominus < 0$) and $\Delta_r S^\ominus < 0$: $\Delta_r G^\ominus < 0$ and $K > 1$ provided that $T < \Delta_r H^\ominus/\Delta_r S^\ominus$
(3)	If the reaction is endothermic ($\Delta_r H^\ominus > 0$) and $\Delta_r S^\ominus > 0$: $\Delta_r G^\ominus < 0$ and $K > 1$ provided that $T > \Delta_r H^\ominus/\Delta_r S^\ominus$
(4)	If the reaction is endothermic ($\Delta_r H^\ominus > 0$) and $\Delta_r S^\ominus < 0$: $\Delta_r G^\ominus < 0$ and $K > 1$ at no temperature

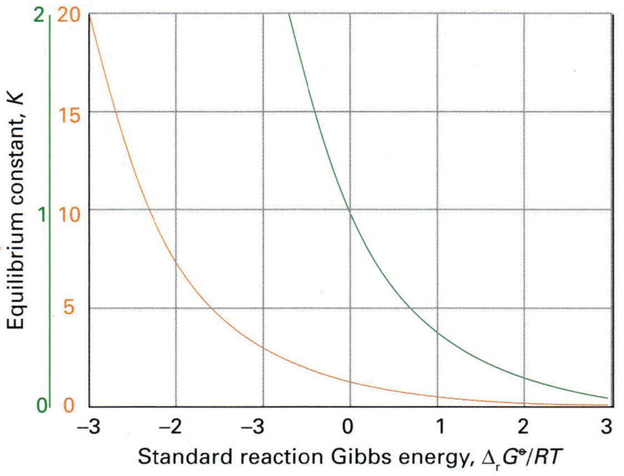

Fig. 7.4 The relation between standard reaction Gibbs energy and the equilibrium constant of the reaction. The pale curve is magnified by a factor of 10.

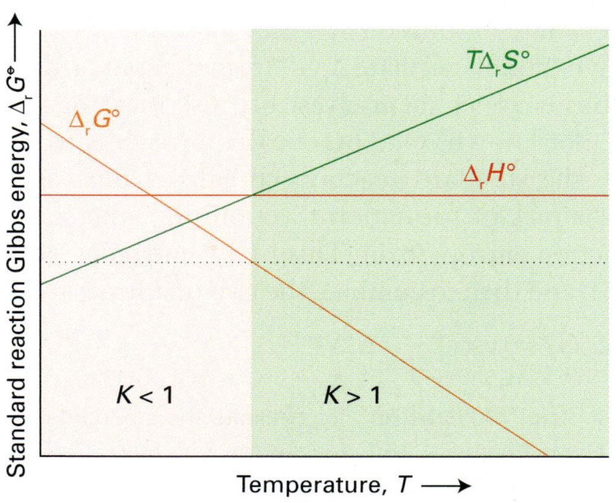

Fig. 7.5 An endothermic reaction may have $K > 1$ provided the temperature is high enough for $T\Delta_r S^\ominus$ to be large enough that, when subtracted from $\Delta_r H^\ominus$ the result is negative.

to negative, corresponding to the switch from $K < 1$ (the reaction 'does not go') to $K > 1$ (the reaction 'goes'), occurs at a temperature given by equating $\Delta_r H^\ominus - T\Delta_r S^\ominus$ to 0, which gives:

$$T = \frac{\Delta_r H^\ominus}{\Delta_r S^\ominus} \qquad (7.9)$$

A brief illustration Consider the (endothermic) thermal decomposition of calcium carbonate:

$$CaCO_3(s) \rightarrow CaO(s) + CO_2(g)$$

For this reaction $\Delta_r H^\ominus = +178$ kJ mol^{-1} and $\Delta_r S^\ominus = +161$ J K^{-1} mol^{-1}. The decomposition temperature, the temperature at which the decomposition becomes spontaneous, is

$$T = \frac{1.78 \times 10^5 \text{ J mol}^{-1}}{161 \text{ J K}^{-1}\text{ mol}^{-1}} = 1.11 \times 10^3 \text{ K}$$

or about 832°C. Because the entropy of decomposition is similar for all such reactions (they all involve the decomposition of a solid into a gas), we can conclude that the decomposition temperatures of solids increase as their enthalpy of decomposition increases.

7.4 The standard reaction Gibbs energy

The standard reaction Gibbs energy, $\Delta_r G^\ominus$, is central to the discussion of chemical equilibria and the calculation of equilibrium constants. We have seen that it is defined as the difference in standard molar Gibbs energies of the products and the reactants weighted by the stoichiometric coefficients, v, in the chemical equation:

$$\Delta_r G^\ominus = \sum v G_m^\ominus(\text{products}) - \sum v G_m^\ominus(\text{reactants}) \qquad (7.10)$$

For example, the standard reaction Gibbs energy for reaction A is the difference between the molar Gibbs energies of fructose-6-phosphate and glucose-6-phosphate in solution at 1 mol dm^{-3} and 1 bar.

We cannot calculate $\Delta_r G^\ominus$ from the standard molar Gibbs energies themselves, because these quantities are not known. One practical approach is to calculate the standard reaction enthalpy from standard enthalpies of formation (Section 3.5), the standard reaction entropy from Third-Law entropies (Section 4.6), and then to combine the two quantities by using

$$\Delta_r G^\ominus = \Delta_r H^\ominus - T\Delta_r S^\ominus \qquad (7.11)$$

A brief illustration To evaluate the standard reaction Gibbs energy at 25°C for the reaction $H_2(g) + \frac{1}{2}O_2(g) \rightarrow H_2O(l)$, we note that

$$\Delta_r H^\ominus = \Delta_f H^\ominus(H_2O, l) = -285.83 \text{ kJ mol}^{-1}$$

The standard reaction entropy, calculated in Chapter 4, is $\Delta_r S^\ominus = -163.34$ J K^{-1} mol^{-1}, which, because 163.34 J is the same as 0.163 34 kJ, corresponds to −0.163 34 kJ K^{-1} mol^{-1}. Therefore, from eqn 7.11,

$$\Delta_r G^\ominus = (-285.83 \text{ kJ mol}^{-1}) - (298.15 \text{ K}) \times (-0.163\,34 \text{ kJ K}^{-1}\text{ mol}^{-1}) = -237.13 \text{ kJ mol}^{-1}$$

Self-test 7.4

Use the information in the Data section to determine the standard reaction Gibbs energy for $3\,O_2(g) \rightarrow 2\,O_3(g)$ from standard enthalpies of formation and standard entropies.

[*Answer:* +326.4 kJ mol^{-1}]

We saw in Section 3.5 how to use standard enthalpies of formation of substances to calculate standard reaction enthalpies. We can use the same technique for standard reaction Gibbs energies. To do so, we list the **standard Gibbs energy of formation**, $\Delta_f G^\ominus$, of a substance, which is the standard reaction Gibbs energy (per mole of the species) for its formation from the elements in their reference states. The concept of reference state was introduced in Section 3.5; the temperature is arbitrary, but we shall almost always take it to be 25°C (298 K). For example, the standard Gibbs energy of formation of liquid water, $\Delta_f G^\ominus(H_2O, l)$, is the standard reaction Gibbs energy for

$$H_2(g) + \tfrac{1}{2}O_2(g) \rightarrow H_2O(l)$$

and is −237 kJ mol^{-1} at 298 K. Some standard Gibbs energies of formation are listed in Table 7.2 and more can be found in the Data section. It follows from the definition that the standard Gibbs energy of formation of an element in its reference state is zero because reactions such as

$$C(s, \text{graphite}) \rightarrow C(s, \text{graphite})$$

are null (that is, nothing happens). The standard Gibbs energy of formation of an element in a phase different from its reference state is nonzero:

$$C(s, \text{graphite}) \rightarrow C(s, \text{diamond}) \qquad \Delta_f G^\ominus(C, \text{diamond}) = +2.90 \text{ kJ mol}^{-1}$$

Many of the values in the tables have been compiled by combining the standard enthalpy of formation of the species with the standard entropies of the compound and the elements, as illustrated above, but there are other sources of data and we encounter some of them later.

Standard Gibbs energies of formation can be combined to obtain the standard Gibbs energy of almost any reaction. We use the now familiar expression

Table 7.2

*Standard Gibbs energies of formation at 298.15 K**

Substance	$\Delta_f G^\ominus$/(kJ mol^{-1})
Gases	
Ammonia, NH_3	−16.45
Carbon dioxide, CO_2	−394.36
Dinitrogen tetroxide, N_2O_4	+97.89
Hydrogen iodide, HI	+1.70
Nitrogen dioxide, NO_2	+51.31
Sulfur dioxide, SO_2	−300.19
Water, H_2O	−228.57
Liquids	
Benzene, C_6H_6	+124.3
Ethanol, CH_3CH_2OH	−174.78
Water, H_2O	−237.13
Solids	
Calcium carbonate, $CaCO_3$	−1128.8
Iron(III) oxide, Fe_2O_3	−742.2
Silver bromide, AgBr	−96.90
Silver chloride, AgCl	−109.79

* Additional values are given in the Data section and the text's website.

$$\Delta_r G^\ominus = \Sigma \nu \Delta_f G^\ominus(\text{products}) - \Sigma \nu \Delta_f G^\ominus(\text{reactants}) \quad (7.12)$$

A brief illustration To determine the standard reaction Gibbs energy for

$$2\,CO(g) + O_2(g) \rightarrow 2\,CO_2(g)$$

we carry out the following calculation:

$$\begin{aligned}\Delta_r G^\ominus &= 2\Delta_f G^\ominus(CO_2, g) - \{2\Delta_f G^\ominus(CO, g) + \Delta_f G^\ominus(O_2, g)\}\\ &= 2 \times (-394\ \text{kJ mol}^{-1}) - \{2 \times (-137\ \text{kJ mol}^{-1}) + 0\}\\ &= -514\ \text{kJ mol}^{-1}\end{aligned}$$

Self-test 7.5

Calculate the standard reaction Gibbs energy of the oxidation of ammonia to nitric oxide according to the equation $4\,NH_3(g) + 5\,O_2(g) \rightarrow 4\,NO(g) + 6\,H_2O(g)$.

[*Answer:* −959.42 kJ mol^{-1}]

Standard Gibbs energies of formation of compounds have their own significance as well as being useful in calculations of K. They are a measure of the 'thermodynamic altitude' of a compound above or

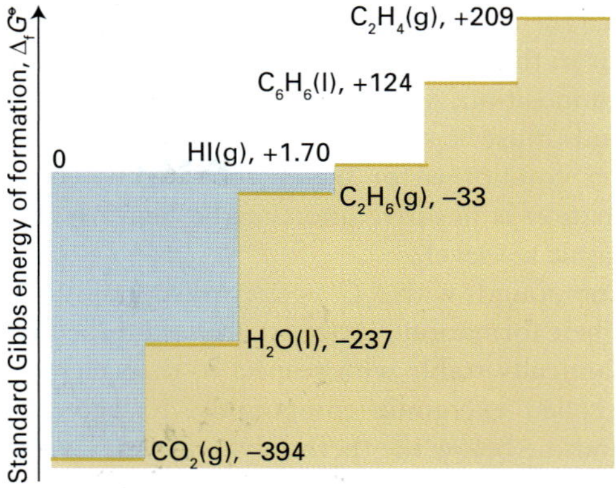

Fig. 7.6 The standard Gibbs energy of formation of a compound is like a measure of the compound's altitude above (or below) sea level: compounds that lie above sea level have a spontaneous tendency to decompose into the elements (and to revert to sea level). Compounds that lie below sea level are stable with respect to decomposition into the elements. The numerical values are in kilojoules per mole.

below a 'sea level' of stability represented by the elements in their reference states (Fig. 7.6). If the standard Gibbs energy of formation is positive and the compound lies above 'sea level', then the compound has a spontaneous tendency to sink towards thermodynamic sea level and decompose into the elements. That is, $K < 1$ for their formation reaction. We say that a compound with $\Delta_f G^\ominus > 0$ is **thermodynamically unstable** with respect to its elements or that it is an **endergonic compound**. Thus, ozone, for which $\Delta_f G^\ominus = +163$ kJ mol^{-1}, has a spontaneous tendency to decompose into oxygen under standard conditions at 25°C. More precisely, the equilibrium constant for the reaction $\frac{3}{2}\,O_2(g) \rightleftharpoons O_3(g)$ is less than 1 (much less in fact, for $K = 2.7 \times 10^{-29}$). However, although ozone is thermodynamically unstable, it can survive if the reactions that convert it into oxygen are slow. That is the case in the upper atmosphere, and the O_3 molecules in the ozone layer survive for long periods. Benzene ($\Delta_f G^\ominus = +124$ kJ mol^{-1}) is also thermodynamically unstable with respect to its elements ($K = 1.8 \times 10^{-22}$). However, the fact that bottles of benzene are everyday laboratory commodities also reminds us of the point made at the start of the chapter, that *spontaneity is a thermodynamic tendency that might not be realized at a significant rate in practice*.

Another useful point that can be made about standard Gibbs energies of formation is that there is no point in searching for *direct* syntheses of a thermodynamically unstable compound from its elements (under standard conditions, at the temperature to which the data apply), because the reaction does not occur in the required direction: the *reverse* reaction, decomposition, is spontaneous. Endergonic compounds must be synthesized by alternative routes or under conditions for which their Gibbs energy of formation is negative and they lie beneath thermodynamic sea level.

Compounds with $\Delta_f G^\ominus < 0$ (corresponding to $K > 1$ for their formation reactions) are said to be **thermodynamically stable** with respect to their elements or are called **exergonic compounds**. Exergonic compounds lie below the thermodynamic sea level of the elements (under standard conditions). An example is the exergonic compound ethane, with $\Delta_f G^\ominus = -33$ kJ mol^{-1}: the negative sign shows that the formation of ethane gas is spontaneous in the sense that $K > 1$ (in fact, $K = 7.1 \times 10^5$ at 25°C).

7.5 The equilibrium composition

The magnitude of an equilibrium constant is a good *qualitative* indication of the feasibility of a reaction regardless of whether the system is ideal or not. Broadly speaking, if $K \gg 1$ (typically $K > 10^3$, corresponding to $\Delta_r G^\ominus < -17$ kJ mol^{-1} at 25°C), then the reaction has a strong tendency to form products. If $K \ll 1$ (that is, for $K < 10^{-3}$, corresponding to $\Delta_r G^\ominus >$ +17 kJ mol^{-1} at 25°C), then the equilibrium composition will consist of largely unchanged reactants. If K is comparable to 1 (typically lying between 10^{-3} and 10^3), then significant amounts of both reactants and products will be present at equilibrium.

An equilibrium constant expresses the composition of an equilibrium mixture as a ratio of products of activities. Even if we confine our attention to ideal systems it is still necessary to do some work to extract the actual equilibrium concentrations or partial pressures of the reactants and products given their initial values.

Example 7.2

Calculating an equilibrium composition 1

Estimate the fraction f of F6P in a solution, where f is defined as

$$f = \frac{[\text{F6P}]}{[\text{F6P}] + [\text{G6P}]}$$

in which G6P and F6P are in equilibrium at 25°C given that $\Delta_r G^\ominus = +1.7$ kJ mol^{-1} at that temperature.

Strategy Express f in terms of K. To do so, recognize that if the numerator and denominator in the expression for f are both divided by [G6P], then the ratios [F6P]/[G6P] can be replaced by K. Calculate the value of K by using eqn 7.8.

Solution Division of the numerator and denominator by [G6P] gives

$$f = \frac{[\text{F6P}]/[\text{G6P}]}{[\text{F6P}]/[\text{G6P}] + 1} = \frac{K}{K+1}$$

We find the equilibrium constant by using $K = e^{\ln K}$ and rearranging eqn 7.8 into

$$K = e^{-\Delta_r G^\circ/RT}$$

First, note that because +1.7 kJ mol^{-1} is the same as $+1.7 \times 10^3$ J mol^{-1},

$$\frac{\Delta_r G^\circ}{RT} = \frac{1.7 \times 10^3 \text{ J mol}^{-1}}{(8.3145 \text{ J K}^{-1}\text{ mol}^{-1}) \times (298 \text{ K})} = \frac{1.7 \times 10^3}{8.3145 \times 298}$$

Therefore,

$$K = e^{-\frac{1.7 \times 10^3}{8.3145 \times 298}} = 0.50$$

and

$$f = \frac{0.50}{1 + 0.50} = 0.33$$

That is, at equilibrium, 33 per cent of the solute is F6P and 67 per cent is G6P.

Self-test 7.6

Estimate the composition of a solution in which two isomers A and B are in equilibrium (A ⇌ B) at 37°C and $\Delta_r G^\ominus = -2.2$ kJ mol^{-1}.

[*Answer:* The fraction of B at equilibrium is $f = 0.70$]

In more complicated cases it is best to organize the necessary work into a systematic procedure resembling a spreadsheet by constructing a table with columns headed by the species and, in successive rows:

1. The initial molar concentrations of solutes or partial pressures of gases.
2. The changes in these quantities that must take place for the system to reach equilibrium.
3. The resulting equilibrium values.

In most cases, we do not know the change that must occur for the system to reach equilibrium, so the change in the concentration or partial pressure of one species is written as x and the reaction stoichiometry is used to write the corresponding changes in the other species. When the values at equilibrium (the last row of the table) are substituted into the expression for the equilibrium constant, we obtain an equation for x in terms of K. This equation can be solved for x, and hence the concentrations of all the species at equilibrium may be found.

Example 7.3

Calculating an equilibrium composition 2

Suppose that in an industrial process, N_2 at 1.00 bar is mixed with H_2 at 3.00 bar and the two gases are allowed to come to equilibrium with the product ammonia in a reactor of constant volume (in the presence of a catalyst, so the reaction proceeds quickly). At the temperature of the reaction, it has been determined experimentally that $K = 977$. What are the equilibrium partial pressures of the three gases?

Strategy Proceed as set out above. Write down the chemical equation of the reaction and the expression for K. Set up the equilibrium table, express K in terms of x, and solve the equation for x. Because the volume of the reaction vessel is constant, each partial pressure is proportional to the amount of its molecules present ($p_J = n_J RT/V$), so the stoichiometric relations apply to the partial pressures directly. In general, solution of the equation for x results in several mathematically possible values of x. Select the chemically acceptable solution by considering the signs of the predicted concentrations or partial pressures: they must be positive. Confirm the accuracy of the calculation by substituting the calculated equilibrium partial pressures into the expression for the equilibrium constant to verify that the value so calculated is equal to the experimental value used in the calculation.

Solution The chemical equation is reaction B ($N_2(g) + 3\,H_2(g) \rightarrow 2\,NH_3(g)$), and the equilibrium constant is

$$K = \frac{p_{NH_3}^2}{p_{N_2} p_{H_2}^3}$$

with the partial pressures those at equilibrium (and, as usual, relative to $p^{\ominus}$). The equilibrium table is

Species	N_2	H_2	NH_3
Initial partial pressure/bar	1.00	3.00	0
Change/bar	$-x$	$-3x$	$+2x$
Equilibrium partial pressure/bar	$1.00 - x$	$3.00 - 3x$	$2x$

The equilibrium constant for the reaction is therefore

$$K = \frac{(2x)^2}{(1.00 - x)\times(3.00 - 3x)^3}$$

Our task is to solve this equation for x. Because $K = 977$, this equation rearranges first to

$$977 = \tfrac{4}{27}\left(\frac{x}{(1.00 - x)^2}\right)^2$$

and then, after multiplying both sides by $\tfrac{27}{4}$ and taking the square root, to

$$\sqrt{\tfrac{27}{4}\times 977} = \frac{x}{(1.00 - x)^2}$$

To keep the appearance of this equation simple, we write $g = (\tfrac{27}{4}\times 977)^{1/2}$, so it becomes

$$g = \frac{x}{(1.00 - x)^2} = \frac{x}{1.00 - 2.00x + x^2}$$

This expression can now be rearranged into

$$gx^2 - 2.00gx + 1.00g = x$$

and then into

$$gx^2 - (2.00g + 1)x + 1.00g = 0$$

This equation has the form $ax^2 + bx + c = 0$ (the quadratic equation) with $a = g$, $b = -(2.00g + 1)$, and $c = 1.00g$. Its solutions are given by the quadratic formula

$$x = \frac{-b \pm (b^2 - 4ac)^{1/2}}{2a}$$

and we find $x = 1.12$ and $x = 0.895$. Because p_{N_2} cannot be negative, and $p_{N_2} = 1.00 - x$ (from the equilibrium table), we know that x cannot be greater than 1.00; therefore, we select $x = 0.895$ as the acceptable solution. It then follows from the last line of the equilibrium table that (with the units bar restored):

$$p_{N_2} = 0.10\text{ bar} \qquad p_{H_2} = 0.31\text{ bar} \qquad p_{NH_3} = 1.8\text{ bar}$$

This is the composition of the reaction mixture at equilibrium. Note that, because K is large (of the order of 10^3), the products dominate. To verify the result, we calculate

$$\frac{p_{NH_3}^2}{p_{N_2} p_{H_2}^3} = \frac{(1.8)^2}{(0.10)\times(0.32)^3} = 9.9\times 10^2$$

The results is close to the experimental value (the discrepancy stems from rounding errors).

Self-test 7.7

In an experiment to study the formation of nitrogen oxides in jet exhausts, N_2 at 0.100 bar is mixed with O_2 at 0.200 bar and the two gases are allowed to come to equilibrium with the product NO in a reactor of constant volume. Take $K = 3.4 \times 10^{-21}$ at 800 K. What is the equilibrium partial pressure of NO?

[*Answer:* 8.2 pbar]

7.6 The equilibrium constant in terms of concentration

An important point to appreciate is that the equilibrium constant K calculated from thermodynamic data refers to activities. For gas-phase reactions, that means partial pressures (and explicitly, $p_J/p^{\ominus}$). This requirement is sometimes emphasized by writing K as K_p, but the practice is unnecessary if the thermodynamic origin of K is remembered. In practical applications, however, we might wish to discuss gas-phase reactions in terms of molar concentrations. The equilibrium constant is then denoted K_c, and for reaction B is

$$K_c = \frac{[NH_3]^2}{[N_2][H_2]^3}$$

with, as usual, the molar concentration [J] interpreted as $[J]/c^{\ominus}$ with $c^{\ominus} = 1\ \text{mol dm}^{-3}$. To obtain the value of K_c from thermodynamic data, we must first calculate K and then convert K to K_c by using, as shown in Derivation 7.1,

$$K = K_c \times \left(\frac{c^{\ominus}RT}{p^{\ominus}}\right)^{\Delta\nu_{\text{gas}}} \quad (7.13a)$$

In this expression, $\Delta\nu_{\text{gas}}$ is the difference in the stoichiometric coefficients of the gas-phase species, products – reactants. We get a very convenient form of this expression by substituting the values of $c^{\ominus}$, $p^{\ominus}$, and R, which gives

$$K = K_c \times \left(\frac{T}{12.027\ \text{K}}\right)^{\Delta\nu_{\text{gas}}} \quad (7.13b)$$

Derivation 7.1

The relation between K and K_c

In this derivation, we need to be fussy about units, and will write the equilibrium constants of reaction C in all their glory as

$$K = \frac{(p_C/p^{\ominus})^c(p_D/p^{\ominus})^d}{(p_A/p^{\ominus})^a(p_B/p^{\ominus})^b} \qquad K_c = \frac{([C]/c^{\ominus})^c([D]/c^{\ominus})^d}{([A]/c^{\ominus})^a([B]/c^{\ominus})^b}$$

The inclusion of $p^{\ominus}$ and $c^{\ominus}$ ensures that the two equilibrium constants are dimensionless. Now we use the perfect gas law to replace each partial pressure by

$$p_J = n_JRT/V = [J]RT$$

(because $[J] = n_J/V$). This substitution turns the expression for K into

$$K = \frac{([C]RT/p^{\ominus})^c([D]RT/p^{\ominus})^d}{([A]RT/p^{\ominus})^a([B]RT/p^{\ominus})^b} = \frac{[C]^c[D]^d}{[A]^a[B]^b} \times \left(\frac{RT}{p^{\ominus}}\right)^{(c+d)-(a+b)}$$

Next, we recognize that

$$K_c = \frac{[C]^c[D]^d}{[A]^a[B]^b} \times \left(\frac{1}{c^{\ominus}}\right)^{(c+d)-(a+b)}$$

and so conclude that

$$K = K_c \times \left(\frac{c^{\ominus}RT}{p^{\ominus}}\right)^{(c+d)-(a+b)}$$

We obtain eqn 7.13 by writing $(c + d) - (a + b) = \Delta\nu_{\text{gas}}$.

A brief illustration For reaction *b* we have $\Delta\nu_{\text{gas}} = 2 - (1 + 3) = -2$; therefore, from eqn 7.13b,

$$K_c = \frac{K}{(T/12.027\ \text{K})^{-2}} = K \times \left(\frac{T}{12.027\ \text{K}}\right)^2$$

At 298 K, $K = 5.8 \times 10^5$, so at this temperature

$$K_c = 5.8 \times 10^5 \times \left(\frac{298\ \text{K}}{12.027\ \text{K}}\right)^2 = 3.6 \times 10^8$$

The response of equilibria to the conditions

In introductory chemistry, we meet the empirical rule of thumb known as **Le Chatelier's principle:**

> When a system at equilibrium is subjected to a disturbance, the composition of the system adjusts so as to tend to minimize the effect of the disturbance.

For instance, if a system is compressed, then the equilibrium position can be expected to shift in the direction that leads to a reduction in the number of molecules in the gas phase, for that tends to minimize the effect of compression. Le Chatelier's principle, though, is only a rule of thumb, and to understand why reactions respond as they do, and to calculate the new equilibrium composition, we need to use thermodynamics. We need to keep in mind that some changes in conditions affect the value of $\Delta_rG^{\ominus}$ and therefore of K (temperature is the only instance) whereas others change the consequences of K having a particular fixed value without changing the value of K (the pressure, for instance).

7.7 The presence of a catalyst

A catalyst is a substance that accelerates a reaction without itself appearing in the overall chemical equation. Enzymes are biological versions of catalysts. We study the action of catalysts in Section 10.12, and at this stage do not need to know in detail how they work other than that they provide an alternative, faster route from reactants to products.

Although the new route from reactants to products is faster, the initial reactants and the final products are the same. The quantity $\Delta_rG^{\ominus}$ is defined as the difference of the standard molar Gibbs energies of the reactants and products, so it is independent of the path linking the two. It follows that an alternative pathway between reactants and products leaves $\Delta_rG^{\ominus}$ and therefore K unchanged. That is, *the presence of a catalyst does not change the equilibrium constant of a reaction.*

7.8 The effect of temperature

According to Le Chatelier's principle, we can expect a reaction to respond to a lowering of temperature by releasing heat and to respond to an increase of temperature by absorbing heat. That is:

> When the temperature is raised, the equilibrium composition of an exothermic reaction will tend to shift towards reactants; the equilibrium composition of an endothermic reaction will tend to shift towards products.

In each case, the response tends to minimize the effect of raising the temperature. But *why* do reactions at equilibrium respond in this way? Le Chatelier's principle is only a rule of thumb, and gives no clue to the reason for this behaviour. As we shall now see, the origin of the effect is the dependence of $\Delta_r G^\ominus$, and therefore of K, on the temperature.

First, we consider the effect of temperature on $\Delta_r G^\ominus$. We use the relation $\Delta_r G^\ominus = \Delta_r H^\ominus - T\Delta_r S^\ominus$ and make the assumption that neither the reaction enthalpy nor the reaction entropy varies much with temperature (over small ranges, at least). It follows that

$$\text{Change in } \Delta_r G^\ominus = -(\text{change in } T) \times \Delta_r S^\ominus \quad (7.14)$$

This expression is easy to apply when there is a consumption or formation of gas because, as we have seen (Section 4.6), gas formation dominates the sign of the reaction entropy.

A brief illustration Consider the three reactions

(i) $\frac{1}{2}C(s) + \frac{1}{2}O_2(g) \rightarrow \frac{1}{2}CO_2(g)$

(ii) $C(s) + \frac{1}{2}O_2(g) \rightarrow CO(g)$

(iii) $CO(g) + \frac{1}{2}O_2(g) \rightarrow CO_2(g)$

all of which are important in the discussion of the extraction of metals from their ores. In reaction (i), the amount of gas is constant, so the reaction entropy is small and $\Delta_r G^\ominus$ for this reaction changes only slightly with temperature. (Note, however, that K changes, because $-\Delta_r G^\ominus/RT$ becomes less negative as T increases, so K decreases). Because in reaction (ii) there is a net increase in the amount of gas molecules, from $\frac{1}{2}$ mol to 1 mol, the reaction entropy is large and positive; therefore, $\Delta_r G^\ominus$ for this reaction decreases sharply with increasing temperature. In reaction (iii), there is a similar net decrease in the amount of gas molecules, from $\frac{3}{2}$ mol to 1 mol, so $\Delta_r G^\ominus$ for this reaction increases sharply with increasing temperature. These remarks are summarized in Fig. 7.7.

Now consider the effect of temperature on K itself. At first, this problem looks troublesome, because both T and $\Delta_r G^\ominus$ appear in the expression for K. However, as we show in Derivation 7.2, the effect of temperature can be expressed very simply as the **van 't Hoff equation**.

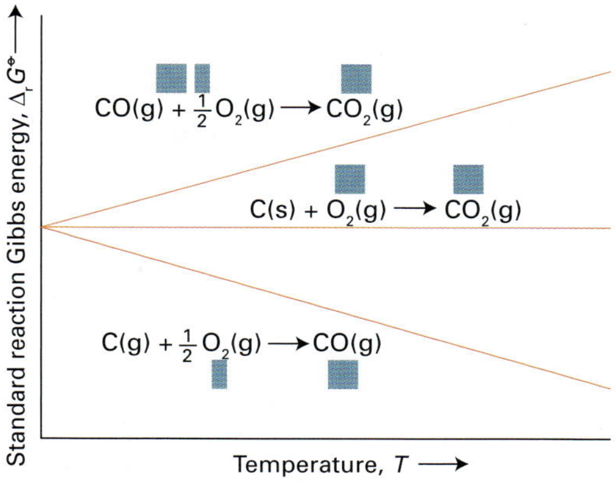

Fig. 7.7 The variation of reaction Gibbs energy with temperature depends on the reaction entropy and therefore on the net production or consumption of gas in a reaction (as indicated by the blue boxes, which show the relative amounts of gas on each side of the equations). The Gibbs energy of a reaction that produces gas decreases with increasing temperature. The Gibbs energy of a reaction that results in a net consumption of gas increases with temperature.

$$\ln K' - \ln K = \frac{\Delta_r H^\ominus}{R}\left(\frac{1}{T} - \frac{1}{T'}\right) \quad (7.15)$$

where K is the equilibrium constant at the temperature T and K' is its value when the temperature is T'. All we need to know to calculate the temperature dependence of an equilibrium constant, therefore, is the standard reaction enthalpy.

Derivation 7.2

The van 't Hoff equation

As before, we use the approximation that the standard reaction enthalpy and entropy are independent of temperature over the range of interest, so the entire temperature dependence of $\Delta_r G^\ominus$ stems from the T in $\Delta_r G^\ominus = \Delta_r H^\ominus - T\Delta_r S^\ominus$. At a temperature T,

$$\ln K = \underbrace{-\frac{\Delta_r G^\ominus}{RT}}_{\text{From } \Delta_r G^\ominus = -RT\ln K} = \underbrace{-\frac{\Delta_r H^\ominus}{RT} + \frac{\Delta_r S^\ominus}{R}}_{\text{From } \Delta_r G^\ominus = \Delta_r H^\ominus - T\Delta_r S^\ominus}$$

At another temperature T', when $\Delta_r G^{\ominus\prime} = \Delta_r H^\ominus - T'\Delta_r S^\ominus$ and the equilibrium constant is K', a similar expression holds:

$$\ln K' = -\frac{\Delta_r H^\ominus}{RT'} + \frac{\Delta_r S^\ominus}{R}$$

The difference between the two is

$$\ln K' - \ln K = \frac{\Delta_r H^\ominus}{R}\left(\frac{1}{T} - \frac{1}{T'}\right)$$

which is the van 't Hoff equation, eqn 7.15.

Box 7.1 Coupled reactions in biochemical processes

A reaction that is not spontaneous may be driven forward by coupling it to a reaction that is spontaneous. A simple mechanical analogy is a pair of weights joined by a string (see the illustration): the lighter of the pair of weights will be pulled up as the heavier weight falls. Although the lighter weight has a natural tendency to move downwards, its coupling to the heavier weight results in it being raised. The thermodynamic analogue is an endergonic reaction (the analogue of the lighter weight) being forced to occur by coupling it to an exergonic reaction (the analogue of the heavier weight falling to the ground). The overall reaction is spontaneous because the sum $\Delta_r G + \Delta_r G'$ is negative. The whole of life's activities depend on coupling of this kind, for the oxidation reactions of food act as the heavy weights that drive other reactions forward and result in the formation of proteins from amino acids, the actions of muscles for propulsion, and even the activities of the brain for reflection, learning, and imagination.

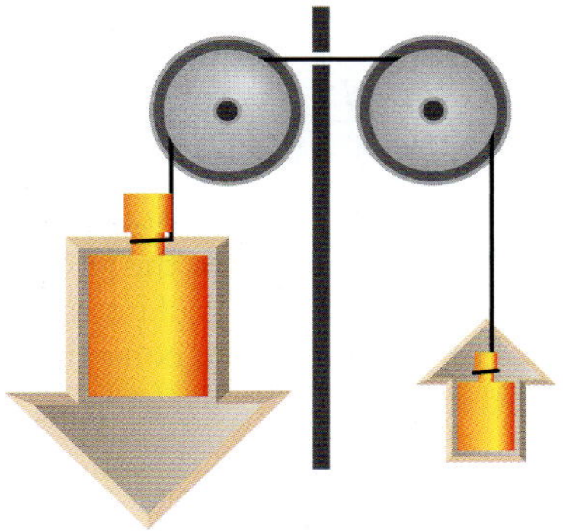

If two weights are coupled as shown here, then the heavier weight will move the lighter weight in its nonspontaneous direction: overall, the process is still spontaneous. The weights are the analogues of two chemical reactions: a reaction with a large negative ΔG can force another reaction with a smaller ΔG to run in its nonspontaneous direction.

The function of adenosine triphosphate, ATP (**3**), for instance, is to store the energy made available when food is oxidized and then to supply it on demand to a wide variety of processes, including muscular contraction, reproduction, and vision. The essence of ATP's action is its ability to lose its terminal phosphate group by hydrolysis and to form adenosine diphosphate, ADP (**4**):

$$\text{ATP(aq)} + \text{H}_2\text{O(l)} \rightarrow \text{ADP(aq)} + \text{P}_i^-\text{(aq)} + \text{H}^+\text{(aq)}$$

where P_i^- denotes an inorganic phosphate group, such as $H_2PO_4^-$. This reaction is exergonic under the conditions prevailing in cells and can drive an endergonic reaction forward if suitable enzymes are available to couple the reactions. For example, the endergonic phosphorylation of glucose (Example 7.1) is coupled to the hydrolysis of ATP in the cell, so the net reaction

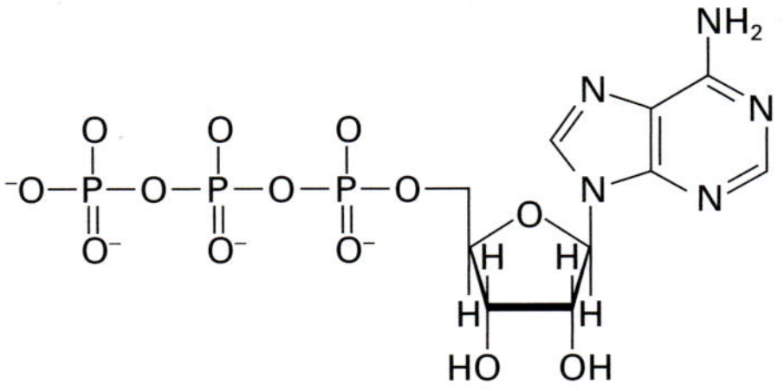

3 ATP

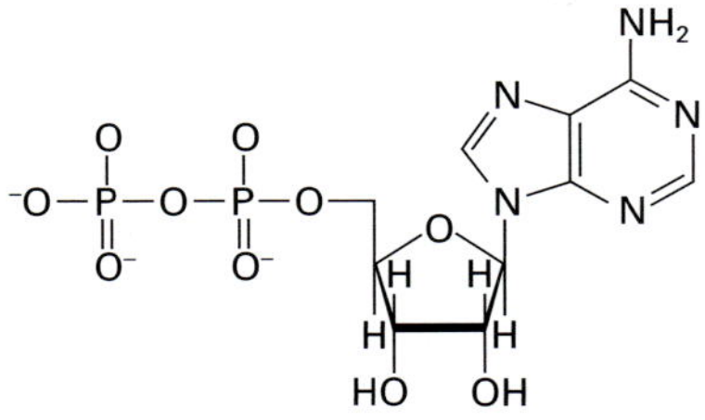

4 ADP

$$\text{glucose(aq)} + \text{ATP(aq)} \rightarrow \text{G6P(aq)} + \text{ADP(aq)}$$

is exergonic and initiates glycolysis.

Before discussing the hydrolysis of ATP quantitatively, we need to note that the conventional standard state of hydrogen ions ($a = 1$, corresponding to pH = 0, a strongly acidic solution; recall from introductory chemistry that $\text{pH} = -\log a_{H_3O^+} \approx -\log [H_3O^+]$) is not appropriate to normal biological conditions inside cells, where the $\text{pH} = \log a_{H_3O^+}$ is close to 7. Therefore, in biochemistry it is common to adopt the *biological standard state*, in which pH = 7, a neutral solution. We shall adopt this convention in this section, and label the corresponding standard quantities as $G^{\oplus}$, $H^{\oplus}$, and $S^{\oplus}$. Another convention to denote the biological standard state is to write $X^{\circ\prime}$ or $X^{\ominus\prime}$.

The biological standard values for the hydrolysis of ATP at 37°C (310 K, body temperature) are

$$\Delta_r G^{\oplus} = -31 \text{ kJ mol}^{-1} \qquad \Delta_r H^{\oplus} = -20 \text{ kJ mol}^{-1}$$
$$\Delta_r S^{\oplus} = +34 \text{ kJ mol}^{-1}$$

The hydrolysis is therefore exergonic ($\Delta_r G < 0$) under these conditions, and 31 kJ mol^{-1} is available for driving other reactions. On account of its exergonic character, the ADP–phosphate bond has been called a 'high-energy phosphate bond'. The name is intended to signify a high tendency to undergo reaction and should not be confused with 'strong' bond in its normal chemical sense (that of a high bond enthalpy). In fact, even in the biological sense it is not of very 'high energy'. The action of ATP depends on the bond being intermediate in strength. Thus ATP acts as a phosphate donor to a number of acceptors (such as glucose), but is recharged with a new phosphate group by more powerful phosphate donors in the phosphorylation steps in the respiration cycle.

Let's explore the information in the van 't Hoff equation. Consider the case when $T' > T$. Then the term in parentheses in eqn 7.15 is positive. If $\Delta_r H^\ominus > 0$, corresponding to an endothermic reaction, the entire term on the right is positive. In this case, therefore, $\ln K' > \ln K$. That being so, we conclude that $K' > K$ for an endothermic reaction. In general, *the equilibrium constant of an endothermic reaction increases with temperature*. The opposite is true when $\Delta_r H^\ominus < 0$, so we can conclude that *the equilibrium constant of an exothermic reaction decreases with an increase in temperature*.

The conclusions we have outlined are of considerable commercial and environmental significance. For example, the synthesis of ammonia is exothermic, so its equilibrium constant decreases as the temperature is increased; in fact, K falls below 1 when the temperature is raised to above 200°C. Unfortunately, the reaction is slow at low temperatures and is commercially feasible only if the temperature exceeds about 750°C even in the presence of a catalyst; but then K is very small. We shall see shortly how Fritz Haber, the inventor of the Haber process for the commercial synthesis of ammonia, was able to overcome this difficulty. Another example is the oxidation of nitrogen:

$$N_2(g) + O_2(g) \rightarrow 2\ NO(g)$$

This reaction is endothermic ($\Delta_r H^\ominus = +180\ \text{kJ mol}^{-1}$) largely as a consequence of the very high bond enthalpy of N_2, so its equilibrium constant increases with temperature. It is for this reason that nitrogen monoxide (nitric oxide) is formed in significant quantities in the hot exhausts of jet engines and in the hot exhaust manifolds of internal combustion engines, and then goes on to contribute to the problems caused by acid rain.

A final point in this connection is that to use the van 't Hoff equation for the temperature dependence of K_c, we first convert K_c to K by using eqn 7.13 at the temperature to which it applies, use eqn 7.15 to convert K to the new temperature, and then use eqn 7.13 again, but with the new temperature, to convert the new K to K_c. As you might appreciate, on the whole it is better to stick to using K.

7.9 The effect of compression

We have seen that Le Chatelier's principle suggests that the effect of compression (decrease in volume) on a gas-phase reaction at equilibrium is as follows:

> When a system at equilibrium is compressed, the composition of a gas-phase equilibrium adjusts so as to reduce the number of molecules in the gas phase.

For example, in the synthesis of ammonia, reaction B, four reactant molecules give two product molecules, so compression favours the formation of ammonia. Indeed, this is the key to resolving Haber's dilemma, for by working with highly compressed gases he was able to increase the yield of ammonia. Pressure plays an important role in governing the uptake and release of oxygen from oxygen transport and storage proteins (Box 7.2).

Box 7.2 Binding of oxygen to myoglobin and haemoglobin

The protein myoglobin (Mb) stores O_2 in muscle and the protein haemoglobin (Hb) transports O_2 in blood; haemoglobin is composed of four myoglobin-like molecules. In each protein, the O_2 molecule attaches to an iron ion in a haem group, and each myoglobin-like component of haemoglobin responds to the change in shape of the others when O_2 binds to them.

First, consider the equilibrium between Mb and O_2:

$$\text{Mb(aq)} + O_2(g) \rightleftharpoons \text{MbO}_2\text{(aq)} \qquad K = \frac{[\text{MbO}_2]}{p[\text{Mb}]}$$

where p is the numerical value of the partial pressure (in bar) of O_2 gas. It follows that the *fractional saturation*, s, the fraction of Mb molecules that are oxygenated, is

$$s = \frac{[\text{MbO}_2]}{[\text{Mb}]_{\text{total}}} = \frac{[\text{MbO}_2]}{[\text{Mb}]+[\text{MbO}_2]} = \frac{Kp}{1+Kp}$$

The dependence of s on p is shown in the illustration.

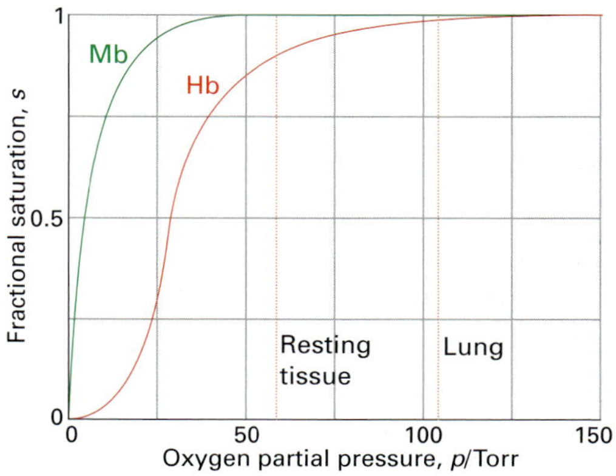

The variation of the fractional saturation of myoglobin and haemoglobin molecules with the partial pressure of oxygen. The different shapes of the curves account for the different biological functions of the two proteins.

Now consider the equilibria between Hb and O_2:

$$Hb(aq) + O_2(g) \rightleftharpoons HbO_2(aq) \qquad K_1 = \frac{[HbO_2]}{p[Hb]}$$

$$HbO_2(aq) + O_2(g) \rightleftharpoons Hb(O_2)_2(aq) \qquad K_2 = \frac{[Hb(O_2)_2]}{p[HbO_2]}$$

$$Hb(O_2)_2(aq) + O_2(g) \rightleftharpoons Hb(O_2)_3(aq) \qquad K_3 = \frac{[Hb(O_2)_3]}{p[Hb(O_2)_2]}$$

$$Hb(O_2)_3(aq) + O_2(g) \rightleftharpoons Hb(O_2)_4(aq) \qquad K_4 = \frac{[Hb(O_2)_4]}{p[Hb(O_2)_3]}$$

To develop an expression for *s*, we express $[Hb(O_2)_2]$ in terms of $[HbO_2]$ by using K_2, then express $[HbO_2]$ in terms of [Hb] by using K_1, and likewise for all the other concentrations of $Hb(O_2)_3$ and $Hb(O_2)_4$. It follows that

$$[HbO_2] = K_1p[Hb] \qquad [Hb(O_2)_2] = K_1K_2p^2[Hb]$$

$$[Hb(O_2)_3] = K_1K_2K_3p^3[Hb] \qquad [Hb(O_2)_4] = K_1K_2K_3K_4p^4[Hb]$$

The total concentration of bound O_2 is

$$\begin{aligned}[O_2]_{bound} &= [HbO_2] + 2[Hb(O_2)_2] + 3[Hb(O_2)_3] + 4[Hb(O_2)_4]\\ &= (1 + 2K_2p + 3K_2K_3p^2 + 4K_2K_3K_4p^3)K_1p[Hb]\end{aligned}$$

and the total concentration of haemoglobin is

$$[Hb]_{total} = (1 + K_1p + K_1K_2p^2 + K_1K_2K_3p^3 + K_1K_2K_3K_4p^4)[Hb]$$

Because each Hb molecule has four sites at which O_2 can attach, the fractional saturation is

$$s = \frac{[O_2]_{bound}}{4[Hb]_{total}} = \frac{(1+2K_2p+3K_2K_3p^2+4K_2K_3K_4p^3)K_1p}{4(1+K_1p+K_1K_2p^2+K_1K_2K_3p^3+K_1K_2K_3K_4p^4)}$$

A reasonable fit of the experimental data can be obtained with $K_1 = 0.01$, $K_2 = 0.02$, $K_3 = 0.04$, and $K_4 = 0.08$ when *p* is expressed in torr. The binding of O_2 to haemoglobin is an example of *cooperative binding*, in which the binding of a ligand (in this case O_2) to a biopolymer (in this case Hb) becomes more favourable thermodynamically (that is, the equilibrium constant increases) as the number of bound ligands increases up to the maximum number of binding sites. We see the effect of cooperativity in the illustration. Unlike the myoglobin saturation curve, the haemoglobin saturation curve is *sigmoidal* (S-shaped): the fractional saturation is small at low ligand concentrations, increases sharply at intermediate ligand concentrations, and then levels off at high ligand concentrations. Cooperative binding of O_2 by haemoglobin is explained by an *allosteric effect*, in which an adjustment of the conformation of a molecule when one substrate binds affects the ease with which a subsequent substrate molecule binds.

The differing shapes of the saturation curves for myoglobin and haemoglobin have important consequences for the way O_2 is made available in the body: in particular, the greater sharpness of the Hb saturation curve means that Hb can load O_2 more fully in the lungs and unload it more fully in different regions of the organism. In the lungs, where $p \approx 14$ kPa, $s \approx 0.98$, representing almost complete saturation. In resting muscular tissue, *p* is equivalent to about 5 kPa, corresponding to $s \approx 0.75$, implying that sufficient O_2 is still available should a sudden surge of activity take place. If the local partial pressure falls to 3 kPa, *s* falls to about 0.1. Note that the steepest part of the curve falls in the range of typical tissue oxygen partial pressure. Myoglobin, on the other hand, begins to release O_2 only when *p* has fallen below about 3 kPa, so it acts as a reserve to be drawn on only when the Hb oxygen has been used up.

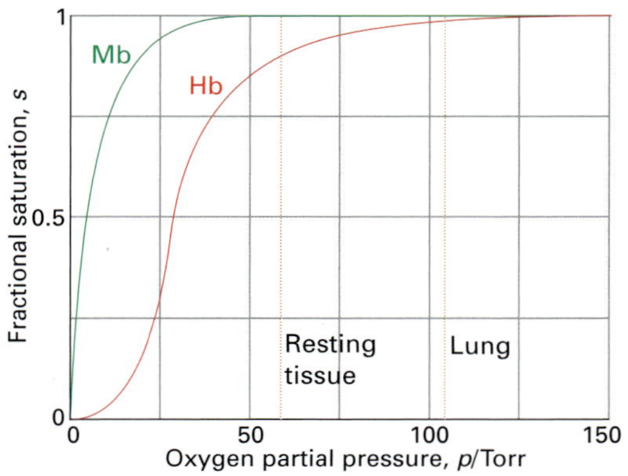

The variation of the fractional saturation of myoglobin and haemoglobin molecules with the partial pressure of oxygen. The different shapes of the curves account for the different biological functions of the two proteins.

Self-test 7.9

Is the formation of products in the reaction $4\ NH_3(g) + 5\ O_2(g) \rightleftharpoons 4\ NO(g) + 6\ H_2O(g)$ favoured by compression or expansion of the reaction vessel?

[*Answer:* expansion]

Let's explore the thermodynamic basis of this dependence. First, we note that $\Delta_rG^\ominus$ is defined as the difference between the Gibbs energies of substances in their standard states and therefore at 1 bar. It follows that $\Delta_rG^\ominus$ has the same value whatever the actual pressure used for the reaction. Therefore, because ln *K* is proportional to $\Delta_rG^\ominus$, *K is independent of the pressure at which the reaction is carried out.* Thus, if the reaction mixture in which ammonia is being synthesized is compressed isothermally, the equilibrium constant remains unchanged.

This rather startling conclusion should not be misinterpreted. The value of *K* is independent of the pressure to which the system is subjected, but because partial pressures occur in the expression for K in a rather complicated way, that does not mean that the *individual* partial pressures or concentrations are unchanged. Suppose, for example, the volume of the reaction vessel in which the reaction $H_2(g) + I_2(s) \rightleftharpoons 2\,HI(g)$ has reached equilibrium is reduced by a factor of 2 and the system is allowed to reach equilibrium again. If the partial pressures were simply to double (that is, there is no adjustment of composition by further reaction), the equilibrium constant would change from

$$K = \frac{p_{HI}^2}{p_{H_2}} \text{ to } K' = \frac{(2p_{HI})^2}{2p_{H_2}} = 2K$$

However, we have seen that compression leaves K unchanged. Therefore, the two partial pressures must adjust by different amounts. In this instance, K' will remain equal to K if the partial pressure of HI changes by a factor of less than 2 and the partial pressure of H_2 increases by more than a factor of 2. In other words, the equilibrium composition must shift in the direction of the reactants in order to preserve the equilibrium constant.

We can express this effect quantitatively by expressing the partial pressures in terms of the mole fractions and the total pressure. For the reaction above, we find

Use $p_J = x_J p$

$$K = \frac{p_{HI}^2}{p_{H_2}} = \frac{x_{HI}^2 p^2}{x_{H_2} p} = \frac{x_{HI}^2 p}{x_{H_2}}$$

For K to remain constant as the pressure is increases, the ratio of mole fractions must decrease, implying that the proportion of HI in the mixture must increase. Because $x_{HI} + x_{H_2} = 1$, the explicit dependence of the mole fractions can be found by substituting $x_{H_2} = 1 - x_{HI}$, which gives

$$K = \frac{x_{HI}^2 p}{1 - x_{HI}}$$

and solving the resulting quadratic equation

$$\underset{ax^2}{p x_{HI}^2} + \underset{bx}{K x_{HI}} \underset{+c}{- K} \underset{=0}{= 0}$$

for x_{HI}:

Use $x = \dfrac{-b + (b^2 - 4ac)^{1/2}}{2a}$

$$x_{HI} = \frac{-K \pm (K^2 + 4Kp)^{1/2}}{2p} = \left(\frac{K}{2p}\right)\left\{-1 \pm \left(1 + \frac{4p}{K}\right)^{1/2}\right\}$$

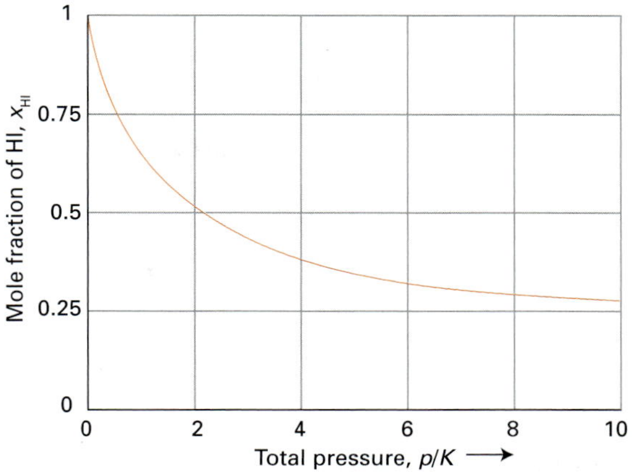

Fig. 7.8 The mole fraction of HI molecules in a gas-phase reaction mixture of H_2 and HI as a function of pressure (expressed as $4p/K$); the I_2 is present as a solid throughout.

Because a mole fraction must be positive, we select the following solution:

$$x_{HI} = \left(\frac{K}{2p}\right)\left\{\left(1 + \frac{4p}{K}\right)^{1/2} - 1\right\} \qquad (7.16)$$

The dependence of the mole fractions implied by this expression is shown in Fig. 7.8. Notice that as p becomes zero, x_{HI} approaches 1; this limit is derived in Derivation 7.3.

Derivation 7.3
Taking a limit

To show that x_{HI} approaches 1 as p becomes zero, we cannot simply substitute $p = 0$ in eqn 7.16, because the first factor gives infinity and the second factor gives zero, and infinity times zero is not defined. Instead, we have to allow p to become very small and use the following expansion (see Appendix 2):

$$(1 + x)^{1/2} = 1 + \tfrac{1}{2}x + \cdots.$$

In this case, with $x = 4p/K$,

$$x_{HI} = \left(\frac{K}{2p}\right)\left\{\left(1 + \frac{2p}{K} + \cdots\right) - 1\right\} = \left(\frac{K}{2p}\right)\left\{\frac{2p}{K} + \cdots\right\} = 1 + \cdots$$

and x_{HI} becomes 1 (the first of the unwritten terms is proportional to p, so all the other terms are zero when $p = 0$).

A note on good practice When one factor increases and another decreases, always evaluate the limit of an expression in this way: never rely on simply setting a term equal to zero.

Compression has no effect on the composition when the number of gas-phase molecules is the same in the reactants as in the products. An example is the synthesis of hydrogen iodide in which all three substances are present in the gas phase and the chemical equation is $H_2(g) + I_2(g) \rightleftharpoons 2\,HI(g)$.

A more subtle example is the effect of the addition of an inert gas to a reaction mixture contained inside a vessel of constant volume. The overall pressure increases as the gas (such as argon) is added, but the addition of a foreign gas does not affect the *partial* pressures of the other gases present: the partial pressure of a perfect gas (Section 1.3), the pressure a gas would exert if it alone occupied the vessel, is independent of the presence or absence of any other gases. Therefore, under these circumstances, not only does the equilibrium constant remain unchanged, but the partial pressures of the reactants and products remain the same whatever the stoichiometry of the reaction.

Checklist of key ideas

You should now be familiar with the following concepts.

☐ **1** The reaction Gibbs energy, $\Delta_r G$, is the slope of a plot of Gibbs energy against composition.

☐ **2** The condition of chemical equilibrium at constant temperature and pressure is $\Delta_r G = 0$.

☐ **3** The equilibrium constant is the value of the reaction quotient at equilibrium.

☐ **4** A compound is thermodynamically stable with respect to its elements if $\Delta_f G^\ominus < 0$.

☐ **5** The equilibrium constant of a reaction is independent of the presence of a catalysts and independent of the pressure.

☐ **6** The variation of an equilibrium constant with temperature is expressed by the van 't Hoff equation.

☐ **7** The equilibrium constant K increases with temperature if $\Delta_r H^\ominus > 0$ (an endothermic reaction), and decreases if $\Delta_r H^\ominus < 0$ (an exothermic reaction).

☐ **8** When a system at equilibrium is compressed, the composition of a gas-phase equilibrium adjusts so as to reduce the number of molecules in the gas phase.

Table of key equations

The following table summarizes the equations developed in this chapter.

Property	Equation	Comment
Gibbs energy of reaction	$\Delta_r G = (c\mu_C + d\mu_D) - (a\mu_A + b\mu_B)$	For the reaction $a\,A + b\,B \rightarrow c\,C + d\,D$
Gibbs energy of reaction and composition	$\Delta_r G = \Delta_r G^\ominus + RT \ln Q$, $Q = a_C^c a_D^d / a_A^a a_B^b$	For the reaction $a\,A + b\,B \rightarrow c\,C + d\,D$
Standard Gibbs energy of reaction	$\Delta_r G^\ominus = \sum \nu \Delta_f G^\ominus(\text{products}) - \sum \nu \Delta_f G^\ominus(\text{reactants})$	
Equilibrium constant	$\Delta_r G^\ominus = -RT \ln K$, $K = Q_{\text{equilibrium}}$	
Relation between equilibrium constants	$K = (c^\ominus RT/p^\ominus)^{\Delta \nu_{\text{gas}}} K_c$	Perfect gas
van 't Hoff equation	$\ln K' - \ln K = (\Delta_r H^\ominus/R)(1/T - 1/T')$	$\Delta_r H^\ominus$ constant in the temperature range

Questions and exercises

Discussion questions

7.1 Explain how the mixing of reactants and products affects the position of chemical equilibrium.

7.2 Explain how a reaction that is not spontaneous may be driven forward by coupling it to a spontaneous reaction.

7.3 State and explain Le Chatelier's principle in terms of thermodynamic quantities. Could there be exceptions to Le Chatelier's principle?

7.4 Suggest how the thermodynamic equilibrium constant and the equilibrium constant expressed in terms of partial pressures may respond differently to changes in pressure and temperature.

7.5 Identify and justify the approximations made in the derivation of the van 't Hoff equation, eqn 7.15.

Exercises

7.1 Write the reaction quotients for the following reactions making the approximation of replacing activities by molar concentrations or partial pressures:

(a) $2\ CH_3COCOOH(aq) + 5\ O_2(g) \rightarrow 6\ CO_2(g) + 4\ H_2O(l)$

(b) $Fe(s) + PbSO_4(aq) \rightleftharpoons FeSO_4(aq) + Pb(s)$

(c) $Hg_2Cl_2(s) + H_2(g) \rightleftharpoons 2\ HCl(aq) + 2\ Hg(l)$

(d) $2\ CuCl(aq) \rightleftharpoons Cu(s) + CuCl_2(aq)$

7.2 Write the expressions for the equilibrium constants of the following reactions:

(a) $CO(g) + Cl_2(g) \rightleftharpoons COCl(g) + Cl(g)$

(b) $2\ SO_2(g) + O_2(g) \rightleftharpoons 2\ SO_3(g)$

(c) $H_2(g) + Br_2(g) \rightleftharpoons 2\ HBr(g)$

(d) $2\ O_3(g) \rightleftharpoons 3\ O_2(g)$

7.3 One of the most extensively studied reactions of industrial chemistry is the synthesis of ammonia, as its successful operation helps to govern the efficiency of the entire economy. The standard Gibbs energy of formation of $NH_3(g)$ is -16.5 kJ mol^{-1} at 298 K. What is the reaction Gibbs energy when the partial pressure of the N_2, H_2, and NH_3 (treated as perfect gases) are 3.0 bar, 1.0 bar, and 4.0 bar, respectively? What is the spontaneous direction of the reaction in this case?

7.4 If the equilibrium constant for the reaction $A + B \rightleftharpoons C$ is reported as 0.432, what would be the equilibrium constant for the reaction written as $C \rightleftharpoons A + B$?

7.5 The equilibrium constant for the reaction $A + B \rightleftharpoons 2\ C$ is reported as 7.2×10^5. What would it be for the reaction written as (a) $2\ A + 2\ B \rightleftharpoons 4\ C$, (b) $\frac{1}{2}A + \frac{1}{2}B \rightleftharpoons C$?

7.6 The equilibrium constant for the isomerization of *cis*-2-butene to *trans*-2-butene is $K = 2.07$ at 400 K. Calculate the standard reaction Gibbs energy for the isomerization.

7.7 The standard reaction Gibbs energy of the isomerization of *cis*-2-pentene to *trans*-2-pentene at 400 K is -3.67 kJ mol^{-1}. Calculate the equilibrium constant of the isomerization.

7.8 One reaction has a standard Gibbs energy of -320 kJ mol^{-1} and a second reaction has a standard Gibbs energy of -55 kJ mol^{-1}. What is the ratio of their equilibrium constants at 300 K?

7.9 One enzyme-catalysed reaction in a biochemical cycle has an equilibrium constant that is 8.4 times the equilibrium constant of a second reaction. If the standard Gibbs energy of the former reaction is -250 kJ mol^{-1}, what is the standard reaction Gibbs energy of the second reaction?

7.10 What is the value of the equilibrium constant of a reaction for which $\Delta_r G^{\ominus} = 0$?

7.11 The standard reaction Gibbs energies (at pH = 7) for the hydrolysis of glucose-1-phosphate, glucose-6-phosphate, and glucose-3-phosphate are -21, -14, and -9.2 kJ mol^{-1}. Calculate the equilibrium constants for the hydrolyses at 37°C.

7.12 The standard Gibbs energy for the hydrolysis of ATP to ADP is -30.5 kJ mol^{-1}; what is the Gibbs energy of reaction in an environment at 37°C in which the ATP, ADP, and P_i concentrations are all (a) 1.0 mmol dm^{-3}, (b) 1.0 μmol dm^{-3}?

7.13 The standard reaction Gibbs energy for the hydrolysis of ATP (given in Exercise 7.12) is +10 kJ mol^{-1} at 298 K. What is the biological standard state value (see Box 7.1)?

7.14 The overall reaction for the glycolysis reaction is $C_6H_{12}O_6(aq) + 2\ NAD^+(aq) + 2\ ADP(aq) + 2\ P_i^-(aq) + 2\ H_2O(l) \rightarrow 2\ CH_3COCO_2^-(aq) + 2\ NADH(aq) + 2\ ATP(aq) + 2\ H_3O^+(aq)$. For this reaction, $\Delta_r G^{\oplus} = -80.6$ kJ mol^{-1} at 298 K. What is the value of $\Delta_r G^{\ominus}$? *Hint*: See Box 7.1. Use the relation $\Delta_r G = \Delta_r G^{\circ} + RT \ln Q$ with the appropriate value of Q for the presence of H_3O^+ at pH = 7.

7.15 The distribution of Na^+ ions across a typical biological membrane is 10 mmol dm^{-3} inside the cell and 140 mmol dm^{-3} outside the cell. At equilibrium the concentrations are equal. What is the Gibbs energy difference across the membrane at 37°C?

7.16 Use the information in the Data section to estimate the temperature at which (a) $CaCO_3$ decomposes spontaneously and (b) $CuSO_4{\cdot}5H_2O$ undergoes dehydration.

7.17 The standard reaction enthalpy of $Zn(s) + H_2O(g) \rightarrow ZnO(s) + H_2(g)$ is approximately constant at +224 kJ mol^{-1} from 920 K up to 1280 K. The standard reaction Gibbs energy is +33 kJ mol^{-1} at 1280 K. Assuming that both quantities remain constant, estimate the temperature at which the equilibrium constant becomes greater than 1.

7.18 The equilibrium constant for the reaction $I_2(g) \rightarrow 2\, I(g)$ is 0.26 at 1000 K. What is the corresponding value of K_c?

7.19 The second step in glycolysis is the isomerization of glucose-6-phosphate (G6P to fructose-6-phosphate (F6P). Example 7.2 considered the equilibrium between F6P and G6P. Draw a graph to show how the reaction Gibbs energy varies with the fraction *f* of F6P in solution. Label the regions of the graph that correspond to the formation of F6P and G6P being spontaneous, respectively.

7.20 Classify the following compounds as endergonic or exergonic: (a) glucose, (b) methylamine, (c) octane, (d) ethanol.

7.21 Combine the reaction entropies calculated in the following reactions with the reaction enthalpies and calculate the standard reaction Gibbs energies at 298 K:

(a) $HCl(g) + NH_3(g) \rightarrow NH_4Cl(s)$

(b) $2\, Al_2O_3(s) + 3\, Si(s) \rightarrow 3\, SiO_2(s) + 4\, Al(s)$

(c) $Fe(s) + H_2S(g) \rightarrow FeS(s) + H_2(g)$

(d) $FeS_2(s) + 2\, H_2(g) \rightarrow Fe(s) + 2\, H_2S(g)$

(e) $2\, H_2O_2(l) + H_2S(g) \rightarrow H_2SO_4(l) + 2\, H_2(g)$

7.22 Use the Gibbs energies of formation in the Data section to decide which of the following reactions have $K > 1$ at 298 K.

(a) $2\, CH_3CHO(g) + O_2(g) \rightleftharpoons 2\, CH_3COOH(l)$

(b) $2\, AgCl(s) + Br_2(l) \rightleftharpoons 2\, AgBr(s) + Cl_2(g)$

(c) $Hg(l) + Cl_2(g) \rightleftharpoons HgCl_2(s)$

(d) $Zn(s) + Cu^{2+}(aq) \rightleftharpoons Zn^{2+}(aq) + Cu(s)$

(e) $C_{12}H_{22}O_{11}(s) + 12\, O_2(g) \rightleftharpoons 12\, CO_2(g) + 11\, H_2O(l)$

7.23 Recall from Chapter 6 that the change in Gibbs energy can be identified with the maximum nonexpansion work that can be extracted from a process. What is the maximum energy that can be extracted as (a) heat, (b) nonexpansion work when 2.0 kg of natural gas (taken to be pure methane) is burned under standard conditions at 25°C? Take the reaction to be $CH_4(g) + 2\, O_2(g) \rightarrow CO_2(g) + 2\, H_2O(l)$.

7.24 In assessing metabolic processes we are usually more interested in the work that may be performed for the consumption of a given mass of compound than the heat it can produce (which merely keeps the body warm). What is the maximum energy that can be extracted as (a) heat, (b) nonexpansion work when 2.0 kg of glucose is burned under standard conditions at 25°C with the production of water vapour? The reaction is $C_6H_{12}O_6(s) + 6\, O_2(g) \rightarrow 6\, CO_2(g) + 6\, H_2O(g)$.

7.25 Is it more energy effective to ingest sucrose or glucose? Calculate the nonexpansion work, the expansion work, and the total work that can be obtained from the combustion of 2.0 kg of sucrose under standard conditions at 25°C when the product includes (a) water vapour, (b) liquid water.

7.26 The standard enthalpy of combustion of solid phenol, C_6H_5OH, is -3054 kJ mol^{-1} at 298 K and its standard molar entropy is 144.0 J K^{-1} mol^{-1}. Calculate the standard Gibbs energy of formation of phenol at 298 K.

7.27 Calculate the maximum nonexpansion work per mole that may be obtained from a fuel cell in which the chemical reaction is the combustion of methane at 298 K.

7.28 Calculate the standard biological Gibbs energy for the reaction

$$\text{Pyruvate}^-(aq) + NADH(aq) + H^+(aq) \rightarrow \text{lactate}^-(aq) + NAD^+(aq)$$

at 310 K given that $\Delta_r G^{\ominus} = -66.6$ kJ mol^{-1}. (NAD^+ is the oxidized form of nicotinamide dinucleotide.) This reaction occurs in muscle cells deprived of oxygen during strenuous exercise and can lead to cramp.

7.29 The standard biological reaction Gibbs energy for the removal of the phosphate group from adenosine monophosphate is -14 kJ mol^{-1} at 298 K. What is the value of the thermodynamic standard reaction Gibbs energy? *Hint*: See Box 7.1.

7.30 Show that if the logarithm of an equilibrium constant is plotted against the reciprocal of the temperature, then the standard reaction enthalpy may be determined.

7.31 Use the following data on the reaction $H_2(g) + Cl_2(g) \rightarrow 2\, HCl(g)$ to determine the standard reaction enthalpy:

T/K	300	500	1000
K	4.0×10^{31}	4.0×10^{18}	5.1×10^{8}

7.32 The equilibrium constant of the reaction $2\, C_3H_6(g) \rightleftharpoons C_2H_4(g) + C_4H_8(g)$ is found to fit the expression

$$\ln K = -1.04 - \frac{1088\ \text{K}}{T} + \frac{1.51 \times 10^5\ \text{K}^2}{T^2}$$

between 300 K and 600 K. Calculate the standard reaction enthalpy and standard reaction entropy at 400 K. *Hint*. Begin by calculating ln *K* at 390 K and 410 K; then use eqn 7.14.

7.33 Borneol is a pungent compound obtained from the camphorwood tree of Borneo and Sumatra. The standard reaction Gibbs energy of the isomerization of borneol (**5**) to isoborneol (**6**) in the gas phase at 503 K is +9.4 kJ mol^{-1}. Calculate the reaction Gibbs energy in a mixture consisting of 0.15 mol of borneol and 0.30 mol of isoborneol when the total pressure is 600 Torr.

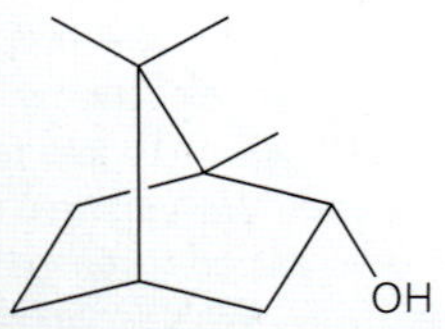

5 d-Borneol

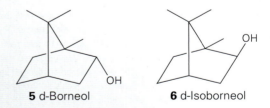

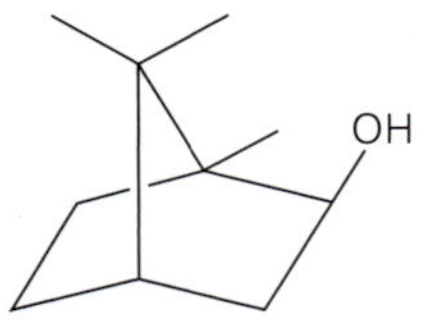

6 d-Isoborneol

7.34 The equilibrium constant for the gas-phase isomerization of borneol, $C_{10}H_{17}OH$, to isoborneol (see Exercise 7.33) at 503 K is 0.106. A mixture consisting of 6.70 g of borneol and 12.5 g of isoborneol in a container of volume 5.0 dm^3 is heated to 503 K and allowed to come to equilibrium. Calculate the mole fractions of the two substances at equilibrium.

7.35 Calculate the composition of a system in which nitrogen and hydrogen are mixed at partial pressures of 1.00 bar and 4.00 bar and allowed to reach equilibrium with their product, ammonia, under conditions when $K = 89.8$.

7.36 In a gas-phase equilibrium mixture of $SbCl_5$, $SbCl_3$, and Cl_2 at 500 K, $p_{SbCl_5} = 0.17$ bar and $p_{SbCl_3} = 0.22$ bar. Calculate the equilibrium partial pressure of Cl_2 given that $K = 3.5 \times 10^{-4}$ for the reaction $SbCl_5(g) \rightleftharpoons SbCl_3(g) + Cl_2(g)$.

7.37 The equilibrium constant $K = 0.36$ for the reaction $PCl_5(g) \rightleftharpoons PCl_3(g) + Cl_2(g)$ at 400 K. (a) Given that 1.5 g of PCl_5 was initially placed in a reaction vessel of volume 250 cm^3, determine the molar concentrations in the mixture at equilibrium. (b) What is the percentage of PCl_5 decomposed at 400 K?

7.38 In the Haber process for ammonia, $K = 0.036$ for the reaction $N_2(g) + 3\,H_2(g) \rightleftharpoons 2\,NH_3(g)$ at 500 K. If a reactor is charged with partial pressures of 0.020 bar of N_2 and 0.020 bar of H_2, what will be the equilibrium partial pressure of the components?

7.39 Express the equilibrium constant for $N_2O_4(g) \rightleftharpoons 2\,NO_2(g)$ in terms of the fraction α of N_2O_4 that has dissociated and the total pressure p of the reaction mixture, and show that when the extent of dissociation is small ($\alpha \ll 1$), α is inversely proportional to the square root of the total pressure ($\alpha \propto p^{-1/2}$).

7.40 The equilibrium pressure of H_2 over a mixture of solid uranium and solid uranium hydride at 500 K is 1.04 Torr. Calculate the standard Gibbs energy of formation of $UH_3(s)$ at 500 K.

7.41 What is the standard enthalpy of a reaction for which the equilibrium constant is (a) doubled, (b) halved when the temperature is increased by 10 K at 298 K?

7.42 The dissociation vapour pressure (the pressure of gaseous products in equilibrium with the solid reactant) of NH_4Cl at 427°C is 608 kPa but at 459°C it has risen to 1115 kPa. Calculate (a) the equilibrium constant, (b) the standard reaction Gibbs energy, (c) the standard enthalpy, (d) the standard entropy of dissociation, all at 427°C. Assume that the vapour behaves as a perfect gas and that $\Delta H^{\ominus}$ and $\Delta S^{\ominus}$ are independent of temperature in the range given.

Projects

The symbol ‡ indicates that calculus is required.

7.43‡ Here we explore the van't Hoff equation in more detail. (a) The bond in molecular iodine is weak, and hot iodine vapour contains a proportion of iodine atoms. When 1.00 g of I_2 is heated to 1000 K in a sealed container of volume 1.00 dm^3, the resulting equilibrium mixture contains 0.830 g of I_2. Calculate K for the dissociation equilibrium $I_2(g) \rightleftharpoons 2\,I(g)$. (b) The thermodynamically exact form of the van 't Hoff equation (eqn 7.15) is $d(\ln K)/dT = -\Delta_r H^{\ominus}/RT^2$. Use the data in part (a) to deduce an expression for the temperature dependence of the standard reaction enthalpy for the reaction treated there, and draw a graph to show the variation. (c) The van 't Hoff equation (eqn 7.15) applies to K, not to K_c. Find the corresponding expression for K_c.

7.44 The saturation curves shown in Box 7.2 may also be modelled mathematically by the equation

$$\log \frac{s}{1-s} = \mathbf{v} \log p - \mathbf{v} \log K$$

where s is the saturation, p is the partial pressure of O_2 (specifically, $p/p^{\ominus}$), K is a constant (not the binding constant for one ligand), and $\mathbf{v}$ is the *Hill coefficient*, which varies from 1, for no cooperativity, to N for all-or-none binding of N ligands ($N = 4$ in Hb). The Hill coefficient for myoglobin is 1, and for haemoglobin it is 2.8. (a) Determine the constant K for both Mb and Hb from the graph of fractional saturation (at $s = 0.5$) and then calculate the fractional saturation of Mb and Hb for the following values of p/kPa: 1.0, 1.5, 2.5, 4.0, 8.0. (b) Use the information from part (a) to calculate the value of s at the same p values assuming that $\mathbf{v}$ has the theoretical maximum value of 4.

Chapter 8

Chemical equilibrium: equilibria in solution

In this chapter we examine some consequences of dynamic chemical equilibria. We concentrate on the equilibria that exist in solutions of acids, bases, and their salts in water, where rapid proton transfer between species ensures that equilibrium is maintained at all times. The link between Chapter 7 and the discussion here is that, provided the temperature is held constant, *an equilibrium constant retains its value even though the individual activities may change*. So, if one substance is added to a mixture at equilibrium, the other substances adjust their abundances to restore the value of K.

Proton transfer equilibria

The reaction of acids and bases are central to chemistry and its applications, such as chemical analysis and synthesis. One particularly important application of proton transfer equilibrium is in living cells, for even small drifts in the equilibrium concentration of hydrogen ions can result in disease, cell damage, and death. Throughout this chapter, keep in mind that a free hydrogen ion (H^+, a proton) does not exist in water: it is always attached to a water molecule and exists as H_3O^+, a hydronium ion.

8.1 Brønsted–Lowry theory

According to the **Brønsted–Lowry theory** of acids and bases, an **acid** is a proton donor and a **base** is a proton acceptor. The proton, which in this context means a hydrogen ion, H^+, is highly mobile and acids and bases in water are always in equilibrium with their deprotonated and protonated counterparts and hydronium ions (H_3O^+). Thus, an acid HA, such as HCN, immediately establishes the equilibrium

$$\mathrm{HA(aq) + H_2O(l) \rightleftharpoons H_3O^+(aq) + A^-(aq)}$$

$$K = \frac{a_{\mathrm{H_3O^+}} a_{\mathrm{A^-}}}{a_{\mathrm{HA}} a_{\mathrm{H_2O}}} \qquad (8.1a)$$

A base B (such as NH_3) immediately establishes the equilibrium

$$\mathrm{B(aq) + H_2O(l) \rightleftharpoons BH^+(aq) + OH^-(aq)} \qquad K = \frac{a_{\mathrm{BH^+}} a_{\mathrm{OH^-}}}{a_{\mathrm{B}} a_{\mathrm{H_2O}}} \qquad (8.1b)$$

In these equilibria, A^- is the **conjugate base** of the acid HA and BH^+ is the **conjugate acid** of the base B. Even in the absence of added acids and bases, proton transfer occurs between water molecules and the **autoprotolysis equilibrium**

$$\mathrm{2\,H_2O(l) \rightleftharpoons H_3O^+(aq) + OH^-(aq)} \qquad K = \frac{a_{\mathrm{H_3O^+}} a_{\mathrm{OH^-}}}{a^2_{\mathrm{H_2O}}} \qquad (8.2)$$

is always present. Autoprotolysis is also called *autoionization.*

As will be familiar from introductory chemistry, the hydronium ion concentration is commonly expressed in terms of the pH, which is defined formally as

$$\mathrm{pH} = -\log a_{\mathrm{H_3O^+}} \qquad (8.3)$$

where the logarithm is to base 10. In elementary work, the hydronium ion activity is replaced by the numerical value of its molar concentration, $[H_3O^+]$, which is equivalent to setting the activity coefficient γ equal to 1 and writing $a_{\mathrm{H_3O^+}} = [\mathrm{H_3O^+}]/c^{\ominus}$ with $c^{\ominus} = 1\ \mathrm{mol\ dm^{-3}}$. For example, if the molar concentration of H_3O^+ is 2.0 mmol dm^{-3} (where 1 mmol = 10^{-3} mol), then

$$\mathrm{pH} \approx -\log(2.0 \times 10^{-3}) = 2.70$$

If the molar concentration were ten times less, at 0.20 mmol dm^{-3}, then the pH would be 3.70. Notice that *the higher the pH, the lower the concentration of hydronium ions in the solution* and that a change in pH by 1 unit corresponds to a 10-fold change in their molar concentration. However, it should never be forgotten that the replacement of activities by molar concentration is invariably hazardous. Because ions interact over long distances, the replacement is unreliable for all but the most dilute solutions (less than about 10^{-3} mol dm^{-3}).

Self-test 8.1

Death is likely if the pH of human blood plasma changes by more than ±0.4 from its normal value of 7.4. What is the approximate range of molar concentrations of hydrogen ions for which life can be sustained?

[*Answer:* 16 nmol dm^{-3} to 100 nmol dm^{-3} (1 nmol = 10^{-9} mol)]

8.2 Protonation and deprotonation

All the solutions we consider are so dilute that we can regard the water present as being a nearly pure liquid and therefore as having unit activity (see Table 6.2). When we set $a_{\mathrm{H_2O}} = 1$ for all the solutions we consider, the resulting equilibrium constant is called the **acidity constant,** K_a, of the acid HA:

$$K_a = \frac{a_{\mathrm{H_3O^+}} a_{\mathrm{A^-}}}{a_{\mathrm{HA}}} \approx \frac{([\mathrm{H_3O^+}]/c^{\ominus})([\mathrm{A^-}]/c^{\ominus})}{([\mathrm{HA}]/c^{\ominus})} \qquad (8.4a)$$

This rather cumbersome expression is normally written

$$K_a = \frac{[\mathrm{H_3O^+}][\mathrm{A^-}]}{[\mathrm{HA}]} \qquad (8.4b)$$

with '[J]' interpreted as $[\mathrm{J}]/c^{\ominus}$ (that is, as the numerical value of the molar concentration of J with the units mol dm^{-3} struck out). Acidity constants are also called *acid ionization constants* and, less appropriately (because deprotonation is not a simple fragmentation into atoms), *dissociation constants.* Data are widely reported in terms of the negative common logarithm of this quantity:

$$\mathrm{p}K_a = -\log K_a \qquad (8.5)$$

It follows from eqn 7.8 ($\Delta_r G^{\ominus} = -RT \ln K$) that $\mathrm{p}K_a$ is proportional to $\Delta_r G^{\ominus}$ for the proton transfer reaction. More explicitly, $\mathrm{p}K_a = \Delta_r G^{\ominus}/(RT \ln 10)$, with $\ln 10 = 2.303\ldots$. Therefore, manipulations of $\mathrm{p}K_a$ and related quantities are actually manipulations of standard reaction Gibbs energies in disguise.

Self-test 8.2

Show that $\mathrm{p}K_a = \Delta_r G^{\ominus}/(RT \ln 10)$. *Hint:* $\ln x = \ln 10 \times \log x$.

The value of the acidity constant indicates the extent to which proton transfer occurs at equilibrium in aqueous solution. The smaller the value of K_a, and therefore the larger the value of $\mathrm{p}K_a$, the lower is the concentration of deprotonated molecules. In short, the higher the value of $\mathrm{p}K_a$, the weaker the acid. Most acids have $K_a < 1$ (and usually much less than 1), with $\mathrm{p}K_a > 0$, indicating only a small extent of deprotonation in water. These acids are classified as **weak acids.** A few acids, most notably, in aqueous solution, HCl, HBr, HI, HNO_3, H_2SO_4 and $HClO_4$, are classified as **strong acids,** and are commonly regarded as being completely deprotonated in aqueous solution.

A brief comment Sulfuric acid, H_2SO_4, is strong with respect only to its first deprotonation; HSO_4^- is weak.

Example 8.1

Assessing the extent of deprotonation of a weak acid

Estimate the pH and the fraction of CH_3COOH molecules deprotonated in 0.15 M $CH_3COOH(aq)$.

Strategy The aim is to calculate the equilibrium composition of the solution. To do so, we use the technique illustrated in Example 7.3, with x the change in molar concentration of H_3O^+ ions required to reach equilibrium. We ignore the tiny concentration of hydronium ions present in pure water. Once x has been found, calculate $pH = -\log x$. Because we can anticipate that the extent of deprotonation is small (the acid is weak), use the approximation that x is very small to simplify the equations.

Solution We draw up the following equilibrium table:

Species	CH_3COOH	H_3O^+	$CH_3CO_2^-$
Initial concentration/ (mol dm^{-3})	0.15	0	0
Change to reach equilibrium/ (mol dm^{-3})	$-x$	$+x$	$+x$
Equilibrium concentration/ (mol dm^{-3})	$0.15 - x$	x	x

The value of x is found by inserting the equilibrium concentrations into the expression for the acidity constant:

$$K_a = \frac{[H_3O^+][CH_3CO_2^-]}{[CH_3COOH]} = \frac{x \times x}{0.15 - x}$$

We could arrange the expression into a quadratic equation and use the solution in Example 7.4. However, it is more instructive to make use of the smallness of x to replace $0.15 - x$ by 0.15 (this approximation is valid if $x \ll 0.15$, which is likely because the acid is weak, but should be verified at the end of the calculation once x has been calculated). Then the simplified equation $K_a = x^2/0.15$ rearranges first to $0.15 \times K_a = x^2$ and then to

$$x = (0.15 \times K_a)^{1/2} = (0.15 \times 1.8 \times 10^{-5})^{1/2} = 1.6 \times 10^{-3}$$

Therefore,

$$pH = -\log(1.6 \times 10^{-3}) = 2.80$$

Calculations of this kind are rarely accurate to more than one decimal place in the pH (and even that may be too optimistic) because the effects of ion–ion interactions have been ignored, so this answer would be reported as pH = 2.8. The fraction deprotonated, α, is

$$\alpha = \frac{[CH_3CO_2^-]_{equilibrium}}{F(CH_3COOH)} = \frac{x}{0.15} = \frac{1.6 \times 10^{-3}}{0.15} = 0.011$$

That is, only 1.1 per cent of the acetic acid molecules have donated a proton.

A note on good practice When an approximation has been assumed, verify at the end of the calculation that the approximation is consistent with the result obtained. In this case, we assumed that $x \ll 0.15$ and have found that $x = 0.011$, which is consistent.

Another note on good practice Acetic acid (ethanoic acid) is written CH_3COOH because the two O atoms are inequivalent; its conjugate base, the acetate ion (ethanoate ion) is written $CH_3CO_2^-$ because the two O atoms are now equivalent (by resonance).

Self-test 8.4

Estimate the pH of 0.010 M $CH_3CH(OH)COOH(aq)$ (lactic acid) from the data in Table 8.1. Before carrying out the numerical calculation, decide whether you expect the pH to be higher or lower than that calculated for the same concentration of acetic acid.

[*Answer:* 2.5]

Example 8.2

Assessing the extent of protonation of a weak base

The conjugate acid of the base quinoline (**1**) has $pK_a = 4.88$. Estimate the pH and the fraction of molecules protonated in a 0.010 M aqueous solution of quinoline.

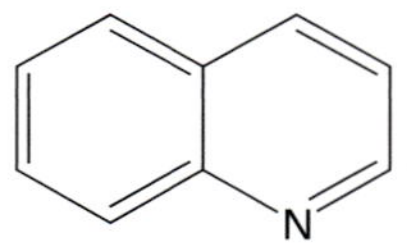

1 Quinoline

Strategy The calculation of the pH of a solution of a base involves one more step than that for the pH of a solution of an acid. The first step is to calculate the concentration of OH^- ions in the solution by using the equilibrium-table technique, and to express it as the pOH of the solution. The additional step is to convert that pOH into a pH by using the water autoprotolysis equilibrium, eqn 8.6, in the form $pH = pK_w - pOH$, with $pK_w = 14.00$ at 25°C. We also need to compute $pK_b = pK_w - pK_a$.

Solution First, we write

$$pK_b = 14.00 - 4.88 = 9.12, \text{ corresponding to}$$

$$K_b = 10^{-9.12} = 7.6 \times 10^{-10}$$

Now draw up the following equilibrium table, denoting quinoline by Q and its conjugate acid by QH^+:

Species	Q	OH^-	QH^+
Initial concentration/(mol dm^{-3})	0.010	0	0
Change to reach equilibrium/(mol dm^{-3})	$-x$	$+x$	$+x$
Equilibrium concentration/(mol dm^{-3})	$0.010 - x$	x	x

The value of x is found by inserting the equilibrium concentrations into the expression for the basicity constant:

$$K_b = \frac{[OH^-][QH^+]}{[Q]} = \frac{x \times x}{0.010 - x}$$

We suppose that $x \ll 0.010$. Then the simplified equation $K_b = x^2/0.010$ rearranges to

$$x = (0.010 \times K_b)^{1/2} = (0.010 \times 7.6 \times 10^{-10})^{1/2} = 2.8 \times 10^{-6}$$

This value is consistent with the assumption that $x \ll 0.010$. Therefore,

$$pOH = -\log(2.8 \times 10^{-6}) = 5.55$$

and consequently pH = 14.00 − 5.55 = 8.45, or about 8.4. The fraction protonated, α, is

$$\alpha = \frac{[QH^+]_{equilibrium}}{F(Q)} = \frac{x}{0.010} = \frac{2.8 \times 10^{-6}}{0.010} = 2.8 \times 10^{-4}$$

or 1 molecule in about 3500.

Self-test 8.5

The pK_a for the first protonation of nicotine (**2**) is 8.02. What is the pH and the fraction of molecules protonated in a 0.015 M aqueous solution of nicotine?

[*Answer:* 10.1; 1/120]

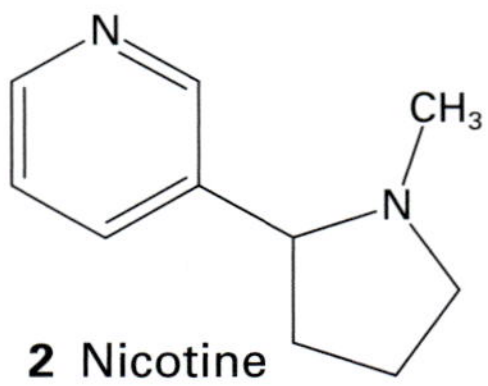

2 Nicotine

8.3 Polyprotic acids

A **polyprotic acid** is a molecular compound that can donate more than one proton. Two examples are sulfuric acid, H_2SO_4, which can donate up to two protons, and phosphoric acid, H_3PO_4, which can donate up to three. A polyprotic acid is best considered to be a molecular species that can give rise to a series of Brønsted acids as it donates its succession of protons. Thus, sulfuric acid is the parent of two Brønsted acids, H_2SO_4 itself and HSO_4^-, and phosphoric acid is the parent of three Brønsted acids, namely H_3PO_4, $H_2PO_4^-$, and HPO_4^{2-}.

For a species H_2A with two acidic protons (such as H_2SO_4), the successive equilibria we need to consider are

$$H_2A(aq) + H_2O(l) \rightleftharpoons H_3O^+(aq) + HA^-(aq)$$

$$K_{a1} = \frac{a_{H_3O^+}a_{HA^-}}{a_{H_2A}}$$

$$HA^-(aq) + H_2O(l) \rightleftharpoons H_3O^+(aq) + A^{2-}(aq)$$

$$K_{a2} = \frac{a_{H_3O^+}a_{A^{2-}}}{a_{HA^-}}$$

In the first of these equilibria, HA^- is the conjugate base of H_2A. In the second, HA^- acts as the acid and A^{2-} is its conjugate base. Values are given in Table 8.2. In all cases, K_{a2} is smaller than K_{a1}, typically by three orders of magnitude for small molecular species, because the second proton is more difficult to remove, partly on account of the negative charge on HA^-. Enzymes are polyprotic acids, for they possess many protons that can be donated to a substrate molecule or to the surrounding aqueous medium of the cell. For them, successive acidity constants vary much less because the molecules are so large that the loss of a proton from one part of the molecule has little effect on the ease with which another some distance away may be lost.

Table 8.2

Successive acidity constants of polyprotic acids at 298.15 K

Acid	K_{a1}	pK_{a1}	K_{a2}	pK_{a2}	K_{a3}	pK_{a3}
Carbonic acid, H_2CO_3	4.3×10^{-7}	6.37	5.6×10^{-11}	10.25		
Hydrosulfuric acid, H_2S	1.3×10^{-7}	6.88	7.1×10^{-15}	14.15		
Oxalic acid, $(COOH)_2$	5.9×10^{-2}	1.23	6.5×10^{-5}	4.19		
Phosphoric acid, H_3PO_4	7.6×10^{-3}	2.12	6.2×10^{-8}	7.21	2.1×10^{-13}	12.67
Phosphorous acid, H_2PO_3	1.0×10^{-2}	2.00	2.6×10^{-7}	6.59		
Sulfuric acid, H_2SO_4	Strong		1.2×10^{-2}	1.92		
Sulfurous acid, H_2SO_3	1.5×10^{-2}	1.81	1.2×10^{-7}	6.91		
Tartaric acid, $C_2H_4O_2(COOH)_2$	6.0×10^{-4}	3.22	1.5×10^{-5}	4.82		

Example 8.3

Calculating the concentration of carbonate ion in carbonic acid

Ground water contains dissolved carbon dioxide, carbonic acid, hydrogencarbonate ions, and a very low concentration of carbonate ions. Estimate the molar concentration of CO_3^{2-} ions in a solution in which water and CO_2(g) are in equilibrium.

Strategy We must be very cautious in the interpretation of calculations involving carbonic acid because equilibrium between dissolved CO_2 and H_2CO_3 is achieved only very slowly. In organisms, attainment of equilibrium is facilitated by the enzyme carbonic anhydrase. We start with the equilibrium that produces the ion of interest (such as A^{2-}) and write its activity in terms of the acidity constant for its formation (K_{a2}). That expression will contain the activity of the conjugate acid (HA^-), which we can express in terms of the activity of *its* conjugate acid (H_2A) by using the appropriate acidity constant (K_{a1}). This equilibrium dominates all the rest provided the molecule is small and there are marked differences between its acidity constants, so it may be possible to make an approximation at this stage.

Solution The CO_3^{2-} ion, the conjugate base of the acid HCO_3^-, is produced in the equilibrium

$$HCO_3^-(aq) + H_2O(l) \rightleftharpoons H_3O^+(aq) + CO_3^{2-}(aq)$$

$$K_{a2} = \frac{a_{H_3O^+}a_{CO_3^{2-}}}{a_{HCO_3^-}}$$

Hence,

$$a_{CO_3^{2-}} = \frac{a_{HCO_3^-}K_{a2}}{a_{H_3O^+}}$$

The HCO_3^- ions are produced in the equilibrium

$$H_2CO_3(aq) + H_2O(l) \rightleftharpoons H_3O^+(aq) + HCO_3^-(aq)$$

One H_3O^+ ion is produced for each HCO_3^- ion produced. These two concentrations are not exactly the same, because a little HCO_3^- is lost in the second deprotonation and the amount of H_3O^+ has been increased by it. Also, HCO_3^- is a weak base and abstracts a proton from water to generate H_2CO_3 (see Section 8.4). However, those secondary changes can safely be ignored in an approximate calculation. Because the molar concentrations of HCO_3^- and H_3O^+ are approximately the same, we can suppose that their activities are also approximately the same, and set $a_{HCO_3^-} \approx a_{H_3O^+}$. When this equality is substituted into the expression for $a_{CO_3^{2-}}$ and we make the approximation that $a_{CO_3^{2-}} = [CO_3^{2-}]/c^\ominus$, then we obtain

$$[CO_3^{2-}] \approx K_{a2}c^\ominus$$

Because we know from Table 8.2 that $pK_{a2} = 10.25$, it follows that $[CO_3^{2-}] = 5.6 \times 10^{-11}\ c^\ominus$, and therefore, if equilibrium has in fact been achieved, that the molar concentration of CO_3^{2-} ions is 5.6×10^{-11} mol dm^{-3} and (within the approximations we have made) independent of the concentration of H_2CO_3 present initially.

Self-test 8.6

Calculate the molar concentration of S^{2-} ions in H_2S(aq).

[*Answer:* 7.1×10^{-15} mol dm^{-3}]

Example 8.4

Calculating the fractional composition of a solution

Oxalic acid, $H_2C_2O_4$ (ethandioic acid, HOOC—COOH), exists in solution in equilibrium with $HC_2O_4^-$ and $C_2O_4^{2-}$. Show how the composition of an aqueous solution that contains 0.010 mol dm^{-3} of oxalic acid varies with pH.

Strategy We expect the fully protonated species ($H_2C_2O_4$) at low pH, the partially protonated species ($HC_2O_4^-$) at intermediate pH, and the fully deprotonated species ($C_2O_4^{2-}$) at high pH. Set up the expressions for the two acidity constants, treating $H_2C_2O_4$ as the parent acid, and an expression for the total concentration of oxalic acid. Solve the resulting expressions for the fraction of each species in terms of the hydronium ion concentration.

Solution The two acidity constants are

$$H_2C_2O_4(aq) + H_2O(l) \rightleftharpoons H_3O^+(aq) + HC_2O_4^-(aq)$$

$$K_{a1} = \frac{[H_3O^+][HC_2O_4^-]}{[H_2C_2O_4]}$$

$$HC_2O_4^-(aq) + H_2O(l) \rightleftharpoons H_3O^+(aq) + C_2O_4^{2-}(aq)$$

$$K_{a2} = \frac{[H_3O^+][C_2O_4^{2-}]}{[HC_2O_4^-]}$$

We also know that the total concentration of oxalic acid in all its forms, its formal concentration $F(H_2C_2O_4)$, is

$$[H_2C_2O_4] + [HC_2O_4^-] + [C_2O_4^{2-}] = F(H_2C_2O_4)$$

We now have three equations for three unknown concentrations. To solve the equations, we proceed systematically by using K_{a2} to express $[C_2O_4^{2-}]$ in terms of $[HC_2O_4^-]$, then K_{a1} to express $[HC_2O_4^-]$ in terms of $[H_2C_2O_4]$:

$$[HC_2O_4^-] = \frac{K_{a1}[H_2C_2O_4]}{[H_3O^+]}$$

$$[C_2O_4^{2-}] = \frac{K_{a2}[HC_2O_4^-]}{[H_3O^+]} = \frac{K_{a1}K_{a2}[H_2C_2O_4]}{[H_3O^+]^2}$$

The expression for the total concentration $F(H_2C_2O_4)$ can now be written in terms of $[H_2C_2O_4]$ and $[H_3O^+]$:

$$F(H_2C_2O_4) = [H_2C_2O_4] + \frac{K_{a1}[H_2C_2O_4]}{[H_3O^+]} + \frac{K_{a1}K_{a2}[H_2C_2O_4]}{[H_3O^+]^2}$$

$$= \left\{1 + \frac{K_{a1}}{[H_3O^+]} + \frac{K_{a1}K_{a2}}{[H_3O^+]^2}\right\}[H_2C_2O_4]$$

$$= \frac{1}{[H_3O^+]^2}\{[H_3O^+]^2 + [H_3O^+]K_{a1} + K_{a1}K_{a2}\}[H_2C_2O_4]$$

It follows that the fractions of each species present in the solution are

$$F(H_2C_2O_4) = \frac{[H_2C_2O_4]}{F_{total}}$$

$$= \frac{[H_2C_2O_4]}{(1/[H_3O^+])^2\{[H_3O^+]^2 + [H_3O^+]K_{a1} + K_{a1}K_{a2}\}[H_2C_2O_4]}$$

$$= \frac{[H_3O^+]^2}{[H_3O^+]^2 + [H_3O^+]K_{a1} + K_{a1}K_{a2}} \qquad (8.11a)$$

and similarly

$$\alpha(HC_2O_4^-) = \frac{[HC_2O_4^-]}{F(H_2C_2O_4)} = \frac{[H_3O^+]K_{a1}}{[H_3O^+]^2 + [H_3O^+]K_{a1} + K_{a1}K_{a2}} \qquad (8.11b)$$

$$\alpha(C_2O_4^{2-}) = \frac{[C_2O_4^{2-}]}{F(H_2C_2O_4)} = \frac{K_{a1}K_{a2}}{[H_3O^+]^2 + [H_3O^+]K_{a1} + K_{a1}K_{a2}} \qquad (8.11c)$$

These fractions are plotted against pH = $-\log[H_3O^+]$ in Fig. 8.1. Note how $H_2C_2O_4$ is dominant for pH < pK_{a1}, that $H_2C_2O_4$ and $HC_2O_4^-$ have the same concentration at pH = pK_{a1}, and that $HC_2O_4^-$ is dominant for pH > pK_{a1}, until $C_2O_4^{2-}$ becomes dominant.

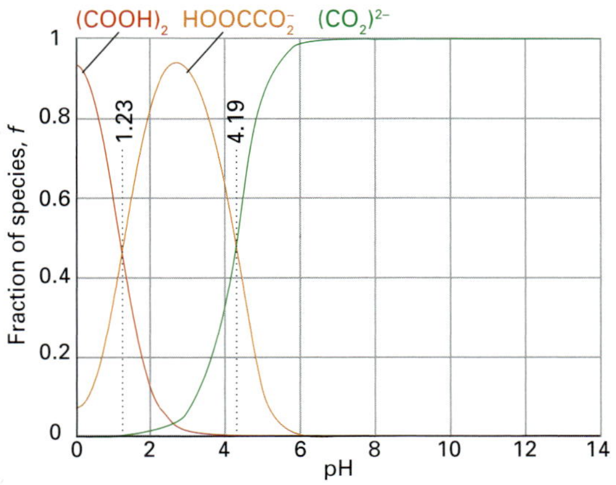

Fig. 8.1 The fractional composition of the protonated and deprotonated forms of oxalic acid in aqueous solution as a function of pH. Note that conjugate pairs are present at equal concentrations when the pH is equal to the pK_a of the acid member of the pair.

A note on good practice Be ready to take advantage of symmetries in the expressions: inspection of the three expressions for the fractions of the species present shows a symmetry in the appearance of $[H_3O^+]$ and the *K*s. By noting this symmetry, it is possible to write down the expression for the species present in a solution of a triprotic acid without further calculation (see Self-test 8.7).

Self-test 8.7

Construct the diagram for the fraction of protonated species in an aqueous solution of phosphoric acid.

[*Answer:* Fig. 8.2]

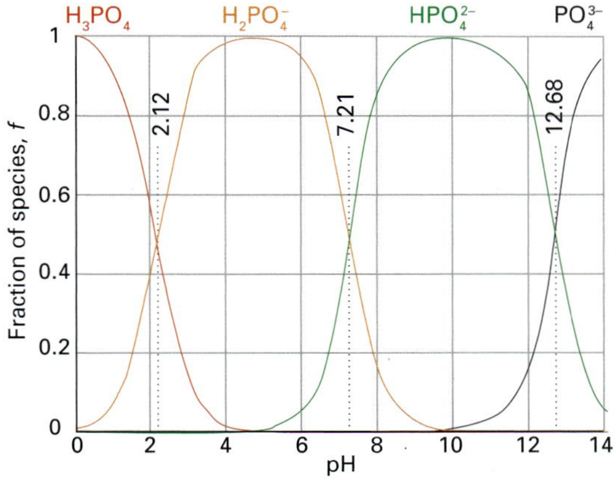

Fig. 8.2 The fractional composition of the protonated and deprotonated forms of phosphoric acid in aqueous solution as a function of pH.

We can summarize the behaviour found in Example 8.4 and illustrated in Figs. 8.1 and 8.2 as follows. Consider each conjugate acid–base pair, with acidity constant K_a; then:

- The acid form is dominant for pH < pK_a
- The conjugate pair have equal concentrations at pH = pK_a
- The base form is dominant for pH > pK_a

In each case, the other possible forms of a polyprotic system can be ignored, provided the pK_a values are not too close together.

8.4 Amphiprotic systems

An **amphiprotic** species is a molecule or ion that can both accept and donate protons. For instance, HCO_3^- can act as an acid (to form CO_3^{2-}) and as a base (to form H_2CO_3). The question we need to tackle is the pH of a solution of a salt with an amphiprotic anion, such as a solution of $NaHCO_3$. Is the solution acidic on account of the acid character of HCO_3^-, or is it basic on account of the anion's basic character? As we show in Derivation 8.1, the pH of such a solution is given by

$$\mathrm{pH} = \tfrac{1}{2}(\mathrm{p}K_{a1} + \mathrm{p}K_{a2}) \qquad (8.12)$$

This expression is valid provided the molar concentration of the salt is high in the sense (using the notation in Derivation 8.1) that $F/c^\ominus \gg K_w/K_{a2}$ and $F/c^\ominus \gg K_{a1}$, where F is the formal concentration of the salt. If these conditions are not satisfied, a much more complicated expression must be used (see the website).

A brief illustration If the dissolved salt is sodium hydrogencarbonate, we can immediately conclude that the pH of the solution *of any concentration* (provided the approximations remain valid) is

$$\mathrm{pH} = \tfrac{1}{2}(6.37 + 10.25) = 8.31$$

The solution is basic. This result is reliable provided $F/c^\ominus \gg 2 \times 10^{-4}$. We can treat a solution of potassium dihydrogenphosphate in the same way, taking into account only the second and third acidity constants of H_3PO_4 because protonation as far as H_3PO_4 is negligible:

$$\mathrm{pH} = \tfrac{1}{2}(7.21 + 12.67) = 9.94$$

This expression is reliable provided $F/c^\ominus \gg 0.05$.

Derivation 8.1

The pH of an amphiprotic salt solution

Let's suppose that we make up a solution of the salt MHA with formal concentration F, where HA^- is the amphiprotic anion (such as HCO_3^-) and M^+ is a cation (such as Na^+). The equilibrium table is as follows:

Species	H_2A	HA^-	A^{2-}	H_3O^+
Initial molar concentration/(mol dm^{-3})	0	F	0	0
Change to reach equilibrium/(mol dm^{-3})	$+x$	$-(x+y)$	$+y$	$+(y-x)$
Equilibrium concentration/(mol dm^{-3})	x	$F-x-y$	y	$y-x$

The two acidity constants are

$$K_{a1} = \frac{[H_3O^+][HA^-]}{[H_2A]} = \frac{(y-x)(F-x-y)}{x}$$

$$K_{a2} = \frac{[H_3O^+][A^{2-}]}{[HA^-]} = \frac{(y-x)y}{F-x-y}$$

Multiplication of these two expressions, noting from the equilibrium table that at equilibrium $y - x = [H_3O^+]$, gives

$$K_{a1}K_{a2} = \frac{(y-x)^2 y}{x} = [H_3O^+]^2 \times \frac{y}{x}$$

Next, we show that, to a good approximation, $y/x \approx 1$ and therefore that $[H_3O^+] = (K_{a1}K_{a2})^{1/2}$. For this step we rearrange the expression for K_{a1} as follows:

$$xK_{a1} = Fy - y^2 - Fx + x^2$$

Because xK_{a1}, x^2, and y^2 are all very small compared with terms that have F in them, and for typical (but by no means all) formal concentrations $F/c^\ominus \gg K_w/K_{a2}$ and $F/c^\ominus \gg K_{a1}$, this expression reduces to

$$0 \approx Fy - Fx$$

We conclude that $x \approx y$, and therefore that $y/x \approx 1$, as required. Equation 8.12 now follows by taking the negative common logarithm of both sides of $[H_3O^+] = (K_{a1}K_{a2})^{1/2}$.

Salts in water

The ions present when a salt is added to water may themselves be either acids or bases and consequently affect the pH of the solution. For example, when ammonium chloride is added to water, it provides both an acid (NH_4^+) and a base (Cl^-). The solution consists of a weak acid (NH_4^+) and a very weak base (Cl^-). The net effect is that the solution is acidic. Similarly, a solution of sodium acetate consists of a neutral ion (the Na^+ ion) and a base ($CH_3CO_2^-$). The net effect is that the solution is basic, and its pH is greater than 7.

Self-test 8.8

Is an aqueous solution of potassium lactate likely to be acidic or basic?

[*Answer:* basic]

To estimate the pH of the solution, we proceed in exactly the same way as for the addition of a 'conventional' acid or base, for in the Brønsted–Lowry theory, there is no distinction between 'conventional' acids like acetic acid and the conjugate acids of bases (like NH_4^+). For example, to calculate the pH of 0.010 M NH_4Cl(aq) at 25°C, we proceed exactly as in Example 8.1, taking the initial concentration of the acid (NH_4^+) to be 0.010 mol dm^{-3}. The K_a to use is the acidity constant of the acid NH_4^+, which is listed in Table 8.1. Alternatively, we use K_b for the conjugate base (NH_3) of the acid and convert that quantity to K_a by using eqn 8.6 ($K_aK_b = K_w$). We find pH = 5.63, which is on the acid side of neutral. Exactly the same procedure is used to find the pH of a solution of a salt of a weak acid, such as sodium acetate. The equilibrium table is set up by treating the anion $CH_3CO_2^-$ as a base (which it is), and using for K_b the value obtained from the value of K_a for its conjugate acid (CH_3COOH).

Self-test 8.9

Estimate the pH of 0.0025 M $NH(CH_3)_3Cl(aq)$ at 25°C.

[*Answer:* 6.2]

8.5 Acid–base titrations

Acidity constants play an important role in acid–base titrations, for we can use them to decide the value of the pH that signals the **stoichiometric point**, the stage at which a stoichiometrically equivalent amount of acid has been added to a given amount of base. The plot of the pH of the **analyte**, the solution being analysed, against the volume of **titrant**, the solution in the burette, added is called the **pH curve**. It shows a number of features that are still of interest even nowadays when many titrations are carried out in automatic titrators with the pH monitored electronically: automatic titration equipment is built to make use of the concepts we describe here.

A brief comment For historical reasons, the stoichiometric point is widely called the *equivalence point* of a titration. The meaning of *end point* is explained in Section 8.6.

First, consider the titration of a strong acid with a strong base, such as the titration of hydrochloric acid with aqueous sodium hydroxide. The reaction is

$$HCl(aq) + NaOH(aq) \rightarrow NaCl(aq) + H_2O(l)$$

Initially, the analyte (hydrochloric acid) has a low pH. The ions present at the stoichiometric point (the Na^+ ions from the strong base and the Cl^- ions from the strong acid) barely affect the pH, so the pH is that of almost pure water, namely pH = 7. After the stoichiometric point, when base is added to a neutral solution, the pH rises sharply to a high value. The pH curve for such a titration is shown in Fig. 8.3.

Figure 8.4 shows the pH curve for the titration of a weak acid (such as CH_3COOH) with a strong base (NaOH). At the stoichiometric point the solution contains $CH_3CO_2^-$ ions and Na^+ ions together with any ions stemming from autoprotolysis. The presence of the Brønsted base $CH_3CO_2^-$ in the solution means that we can expect pH > 7. In a titration of a weak

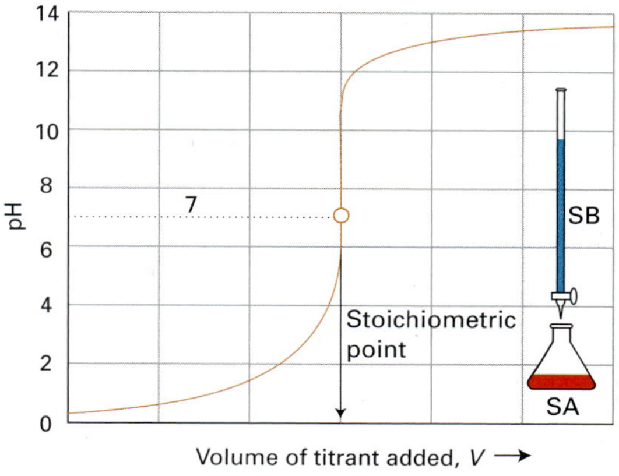

Fig. 8.3 The pH curve for the titration of a strong acid (SA, the analyte) with a strong base (SB, the titrant). There is an abrupt change in pH near the stoichiometric point at pH = 7. The final pH of the medium approaches that of the titrant.

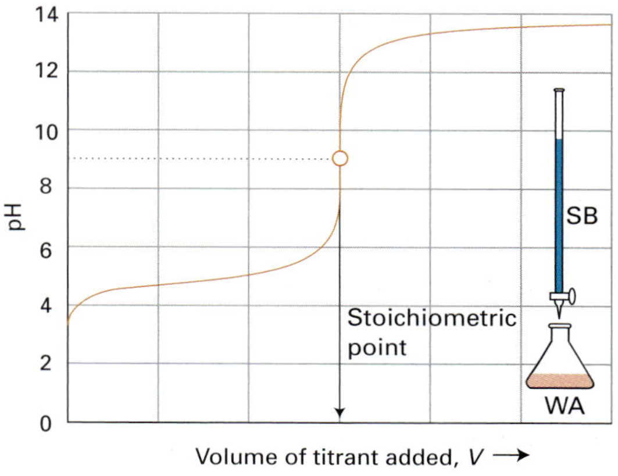

Fig. 8.4 The pH curve for the titration of a weak acid (WA, the analyte) with a strong base (WB, the titrant). Note that the stoichiometric point occurs at pH > 7 and that the change in pH near the stoichiometric point is less abrupt than in Fig. 8.3. The pK_a of the acid is equal to the pH half-way to the stoichiometric point.

base (such as NH_3) and a strong acid (HCl), the solution contains NH_4^+ ions and Cl^- ions at the stoichiometric point. Because Cl^- is only a very weak Brønsted base and NH_4^+ is a weak Brønsted acid the solution is acidic and its pH will be less than 7.

Now we consider the shape of the pH curve in Fig. 8.4 in terms of the acidity constants of the species involved. The approximations we make are based on the fact that the acid is weak, and therefore that HA is more abundant than any A^- ions in the solution. Furthermore, when HA is present, it provides so many H_3O^+ ions, even though it is a weak acid, that they greatly outnumber any H_3O^+ ions that come from the very feeble autoprotolysis of water. Finally, when excess base is present after the stoichiometric point has been passed, the OH^- ions it provides dominate any that come from the water autoprotolysis.

To be specific, let's suppose that we are titrating 25.00 cm^3 of 0.10 M $CH_3COOH(aq)$ with 0.20 M $NaOH(aq)$ at 25°C. We can calculate the pH at the start of a titration of a weak acid with a strong base as explained in Example 8.1, and find pH = 2.9. The addition of titrant converts some of the acid to its conjugate base in the reaction

$$CH_3COOH(aq) + OH^-(aq) \rightarrow H_2O(l) + CH_3CO_2^-(aq)$$

Suppose we add enough titrant to produce a concentration [base] of the conjugate base and simultaneously reduce the concentration of acid to [acid]. Then, because the acid and its conjugate base remain at equilibrium:

$$CH_3COOH(aq) + H_2O(l) \rightleftharpoons H_3O^+(aq) + CH_3CO_2^-(aq)$$

we can write

$$K_a = \frac{a_{H_3O^+}a_{CH_3CO_2^-}}{a_{CH_3COOH}} \approx \frac{a_{H_3O^+}[\text{base}]}{[\text{acid}]}$$

This expression rearranges first to

$$a_{H_3O^+} \approx \frac{K_a[\text{acid}]}{[\text{base}]}$$

and then, by taking negative common logarithms, we obtain

Use log xy = log x + log y

$$-\log a_{H_3O^+} \approx -\log\frac{K_a[\text{acid}]}{[\text{base}]} = -\log K_a - \log\frac{[\text{acid}]}{[\text{base}]}$$

which can be written as the **Henderson–Hasselbalch equation**

$$\text{pH} \approx \text{p}K_a - \log\frac{[\text{acid}]}{[\text{base}]} \qquad (8.13)$$

Example 8.5

Estimating the pH at an intermediate stage in a titration

Calculate the pH of the solution after the addition of 5.00 cm^3 of the titrant to the analyte in the titration described above.

Strategy The first step involves deciding the amount of OH^- ions added in the titrant, and then to use that amount to calculate the amount of CH_3COOH remaining. Notice that because the ratio of acid to base molar concentrations occurs in eqn 8.13, the volume of solution cancels, and we can equate the ratio of concentrations to the ratio of amounts:

$$\frac{[\text{acid}]}{[\text{base}]} = \frac{n_{\text{acid}}/V}{n_{\text{base}}/V} = \frac{n_{\text{acid}}}{n_{\text{base}}}$$

Solution The addition of 5.00 cm^3, or 5.00×10^{-3} dm^3 (because 1 cm^3 = 10^{-3} dm^3), of titrant corresponds to the addition of

$$n_{OH^-} = (5.00 \times 10^{-3}\ \text{dm}^3) \times (0.200\ \text{mol dm}^{-3}) = 1.00 \times 10^{-3}\ \text{mol}$$

This amount of OH^- (1.00 mmol) converts 1.00 mmol CH_3COOH to the base $CH_3CO_2^-$. The initial amount of CH_3COOH in the analyte is

$$n_{CH_3COOH} = (25.00 \times 10^{-3}\ \text{dm}^3) \times (0.200\ \text{mol dm}^{-3}) = 2.50 \times 10^{-3}\ \text{mol}$$

so the amount remaining after the addition of titrant is 1.50 mmol. It then follows from the Henderson–Hasselbalch equation that

$$\text{pH} \approx 4.75 - \log\frac{1.50 \times 10^{-3}}{1.00 \times 10^{-3}} = 4.6$$

As expected, the addition of base has resulted in an increase in pH from 2.9. You should not take the value 4.6 too seriously because we have already pointed out that such calculations are approximate. However, it is important to note that the pH has increased from its initial acidic value.

Self-test 8.10

Calculate the pH after the addition of a further 5.00 cm^3 of titrant.

[*Answer:* 5.4]

Half-way to the stoichiometric point, when enough base has been added to neutralize half the acid, the concentrations of acid and base are equal and because log 1 = 0 the Henderson–Hasselbalch equation gives

$$\text{pH} \approx \text{p}K_a \qquad (8.14)$$

In the present titration, we see that at this stage of the titration, pH ≈ 4.75. Note from the pH curve in Fig. 8.4 how much more slowly the pH is changing compared with initially: this point will prove important shortly. Equation 8.14 implies that we can determine the pK_a of the acid directly from the pH of the mixture. Indeed, an approximate value of the pK_a

may be calculated by recording the pH during a titration and then examining the record for the pH half-way to the stoichiometric point.

At the stoichiometric point, enough base has been added to convert all the acid to its base, so the solution consists—nominally—only of $CH_3CO_2^-$ ions. These ions are Brønsted bases, so we can expect the solution to be basic with a pH of well above 7. We have already seen how to estimate the pH of a solution of a weak base in terms of its concentration B (Example 8.2), so all that remains to be done is to calculate the concentration of $CH_3CO_2^-$ at the stoichiometric point.

A brief illustration Because the analyte initially contained 2.50 mmol CH_3COOH, the volume of titrant needed to neutralize it is the volume that contains the same amount of base:

$$V_{\text{base}} = \frac{2.50\times10^{-3}\,\text{mol}}{0.200\ \text{mol dm}^{-3}} = 1.25\times10^{-2}\,\text{dm}^3$$

or 12.5 cm^3. The total volume of the solution at this stage is therefore 37.5 cm^3, so the concentration of base is

$$[CH_3CO_2^-] = \frac{2.50\times10^{-3}\ \text{mol}}{37.5\times10^{-3}\ \text{dm}^3} = 6.67\times10^{-2}\ \text{mol dm}^{-3}$$

It then follows from a calculation similar to that in Example 8.2 (with $pK_b = 9.25$ for $CH_3CO_2^-$) that the pH of the solution at the stoichiometric point is 8.8.

It is very important to note that the pH at *the stoichiometric point of a weak-acid–strong-base titration is on the basic side of neutrality* (pH > 7). At the stoichiometric point, the solution consists of a weak base (the conjugate base of the weak acid, here the $CH_3CO_2^-$ ions) and neutral cations (the Na^+ ions from the titrant).

The general form of the pH curve suggested by these estimates throughout a weak-acid–strong-base titration is illustrated in Fig. 8.4. The pH rises slowly from its initial value, passing through the values given by the Henderson–Hasselbalch equation when the acid and its conjugate base are both present, until the stoichiometric point is approached. It then changes rapidly to and through the value characteristic of a solution of a salt, which takes into account the effect on the pH of a solution of a weak base, the conjugate base of the original acid. The pH then climbs less rapidly towards the value corresponding to a solution consisting of excess base, and finally approaches the pH of the original base solution when so much titrant has been added that the solution is virtually the same as the titrant itself. The stoichiometric point is detected by observing where the pH changes

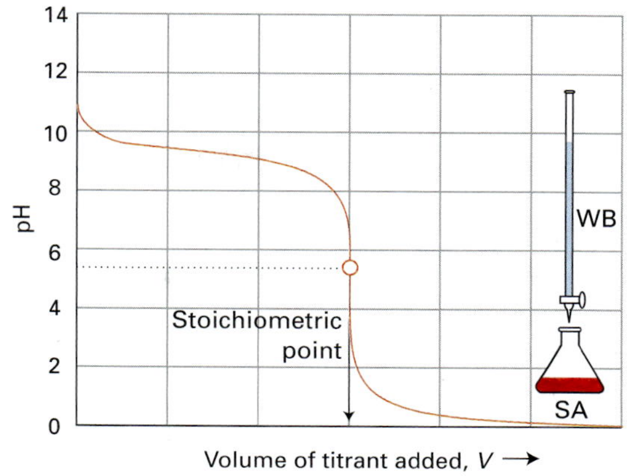

Fig. 8.5 The pH curve for the titration of a weak base (the analyte) with a strong acid (the titrant). The stoichiometric point occurs at pH < 7. The final pH of the solution approaches that of the titrant.

rapidly through the value calculated in the illustration at the beginning of this section.

A similar sequence of changes occurs when the analyte is a weak base (such as ammonia) and the titrant is a strong acid (such as hydrochloric acid). In this case the pH curve is like that shown in Fig. 8.5: the pH falls as acid is added, plunges through the pH corresponding to a solution of a weak acid (the conjugate acid of the original base, such as NH_4^+), and then slowly approaches the pH of the original strong acid. The pH of the stoichiometric point is that of a solution of a weak acid, and is calculated as illustrated in Example 8.2.

8.6 Buffer action

The slow variation of the pH when the concentrations of the conjugate acid and base are nearly equal, when pH ≈ pK_a, is the basis of **buffer action**, the ability of a solution to oppose changes in pH when small amounts of strong acids and bases are added (Fig. 8.6). An **acid buffer** solution, one that stabilizes the solution at a pH below 7, is typically prepared by making a solution of a weak acid (such as acetic acid) and a salt that supplies its conjugate base (such as sodium acetate). A **base buffer**, one that stabilizes a solution at a pH above 7, is prepared by making a solution of a weak base (such as ammonia) and a salt that supplies its conjugate acid (such as ammonium chloride). Physiological buffers are responsible for maintaining the pH of blood within a narrow range of 7.37 to 7.43, thereby stabilizing the active conformations of biological macromolecules and optimizing the rates of biochemical reactions (Box 8.1).

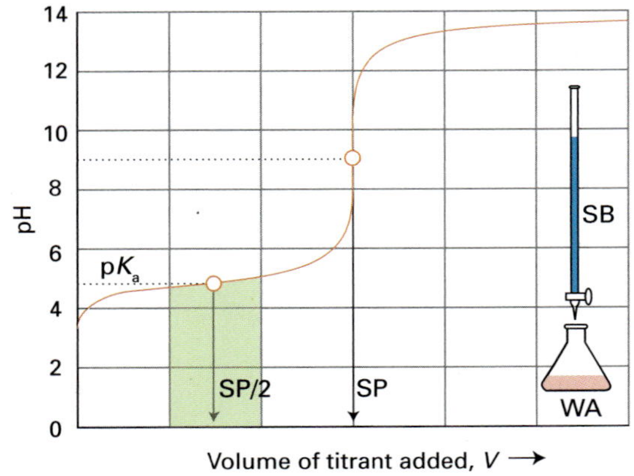

Fig. 8.6 The pH of a solution changes only slowly in the region of half-way to the stoichiometric point (SP). In this region the solution is buffered to a pH close to pK_a.

A brief illustration Suppose we need to estimate the pH of a buffer formed from equal amounts of $KH_2PO_4(aq)$ and $K_2HPO_4(aq)$. We note that the two anions present are $H_2PO_4^-$ and HPO_4^{2-}. The former is the conjugate acid of the latter:

$$H_2PO_4^-(aq) + H_2O(l) \rightleftharpoons H_3O^+(aq) + HPO_4^{2-}(aq)$$

so we need the pK_a of the acid form, $H_2PO_4^-$. In this case we can take it from Table 8.1, or recognize it as the pK_{a2} of phosphoric acid, and take it from Table 8.2 instead. In either case, $pK_a = 7.21$. Hence, the solution should buffer close to pH = 7.

Self-test 8.11

Calculate the pH of an aqueous buffer solution that contains equal amounts of NH_3 and NH_4Cl.

[*Answer:* 9.25; more realistically: 9]

Box 8.1 Buffer action in blood

The pH of blood in a healthy human being varies from 7.37 to 7.43. There are two buffer systems that help maintain the pH of blood relatively constant: one arising from a carbonic acid/bicarbonate (hydrogencarbonate) ion equilibrium and another involving protonated and deprotonated forms of haemoglobin, the protein responsible for the transport of O_2 in blood (Box 7.1).

Carbonic acid forms in blood from the reaction between water and CO_2 gas, which comes from inhaled air and is also a by-product of metabolism:

$$CO_2(g) + H_2O(l) \rightleftharpoons H_2CO_3(aq)$$

In red blood cells, this reaction is catalysed by the enzyme carbonic anhydrase. Aqueous carbonic acid then deprotonates to form a bicarbonate (hydrogencarbonate) ion:

$$H_2CO_3(aq) \rightleftharpoons H^+(aq) + HCO_3^-(aq)$$

The fact that the pH of normal blood is approximately 7.4 implies that $[HCO_3^-]/[H_2CO_3] \approx 20$. The body's control of the pH of blood is an example of *homeostasis*, the ability of an organism to counteract environmental changes with physiological responses. For instance, the concentration of carbonic acid can be controlled by respiration: exhaling air depletes the system of $CO_2(g)$ and $H_2CO_3(aq)$ so the pH of blood rises when air is exhaled. Conversely, inhalation increases the concentration of carbonic acid in blood and lowers its pH. The kidneys also play a role in the control of the concentration of hydronium ions. There, ammonia formed by the release of nitrogen from some aminoacids (such as glutamine) combines with excess hydronium ions and the ammonium ion is excreted through urine.

The condition known as *alkalosis* occurs when the pH of blood rises above about 7.45. *Respiratory alkalosis* is caused by hyperventilation, or excessive respiration. The simplest remedy consists of breathing into a paper bag in order to increase the levels of inhaled CO_2. *Metabolic alkalosis* may result from illness, poisoning, repeated vomiting, and overuse of diuretics. The body may compensate for the increase in the pH of blood by decreasing the rate of respiration.

Acidosis occurs when the pH of blood falls below about 7.35. In *respiratory acidosis*, impaired respiration increases the concentration of dissolved CO_2 and lowers the blood's pH. The condition is common in victims of smoke inhalation and patients with asthma, pneumonia, and emphysema. The most efficient treatment consists of placing the patient in a ventilator. *Metabolic acidosis* is caused by the release of large amounts of lactic acid or other acidic by-products of metabolism, which react with hydrogencarbonate ion to form carbonic acid, thus lowering the blood's pH. The condition is common in patients with diabetes and severe burns.

The concentration of hydronium ions in blood is also controlled by haemoglobin, which can exist in deprotonated (basic) or protonated (acidic) forms, depending on the state of protonation of several amino acid residues on the protein's surface. The carbonic acid/bicarbonate ion equilibrium and proton equilibria in haemoglobin also regulate the oxygenation of blood. The key to this regulatory mechanism is the *Bohr effect*, the observation that haemoglobin binds O_2 strongly when it is deprotonated and releases O_2 when it is protonated. It follows that when dissolved CO_2 levels are high and the pH of blood falls slightly, haemoglobin becomes protonated and releases bound O_2 to tissue. Conversely, when CO_2 is exhaled and the pH rises slightly, haemoglobin becomes deprotonated and binds O_2.

An acid buffer stabilizes the pH of a solution because the abundant supply of A^- ions (from the salt) can remove any H_3O^+ ions brought by additional acid; furthermore, the abundant supply of HA molecules can provide H_3O^+ ions to react with any base that is added. Similarly, in a base buffer the weak base B can accept protons when an acid is added and its conjugate acid BH^+ can supply protons if a base is added.

These abilities can be expressed quantitatively by considering the changes in equilibrium composition of hydronium ions in the presence of a buffer. This behaviour is best illustrated with a specific example.

Example 8.6

Illustrating the effect of a buffer

When 1 drop ($0.20\ cm^3$, say) of $1.0\ mol\ dm^{-3}$ HCl(aq) is added to $25\ cm^3$ of pure water, the resulting hydronium ion concentration rises to $0.0080\ mol\ dm^{-3}$ and so the pH changes from 7.0 to 2.1, a big change. Now suppose the drop is added to $25\ cm^3$ of an acetate buffer solution that is $0.040\ mol\ dm^{-3}$ $NaCH_3CO_2$(aq) and $0.080\ mol\ dm^{-3}$ CH_3COOH(aq). What will be the change in pH?

Strategy The presence of the acid tells us that this mixture will be an acid buffer. Use the Henderson–Hasselbalch equation (or, better, first principles), to estimate the initial pH. Then calculate the amount of H_3O^+ added in the drop and the consequent changes to the amounts of acetic acid and acetate ions in the solution. Use the Henderson–Hasselbalch equation to estimate the pH of the resulting solution.

Solution The initial pH of the buffer solution is

$$\text{pH} = 4.75 - \log\frac{0.080}{0.040} = 4.45$$

The drop of HCl(aq) contains

$$n(H_3O^+) = (0.20 \times 10^{-3}\ dm^3) \times (1.0\ mol\ dm^{-3}) = 0.20\ mmol$$

The buffer solution contains

$$n(CH_3CO_2^-) = (25 \times 10^{-3}\ dm^3) \times (0.040\ mol\ dm^{-3}) = 1.0\ mmol$$

$$n(CH_3COOH) = (25 \times 10^{-3}\ dm^3) \times (0.080\ mol\ dm^{-3}) = 2.0\ mmol$$

The acetate ion is protonated by the incoming acid, and so its amount is reduced to 0.8 mmol. As a result, the amount of CH_3COOH rises from 2.0 mmol to 2.2 mmol. The volume of the solution barely changes, so the two concentrations become $0.032\ mol\ dm^{-3}$ $NaCH_3CO_2$(aq) and $0.088\ mol\ dm^{-3}$ CH_3COOH(aq). It then follows from the Henderson–Hasselbalch equation that

$$\text{pH} = 4.75 - \log\frac{0.088}{0.032} = 4.31$$

The change in pH is from 4.45 to 4.31, far smaller than in the absence of the buffer.

Self-test 8.12

Estimate the change in pH when $0.20\ cm^3$ of $1.5\ mol\ dm^{-3}$ NaOH(aq) is added to $30\ cm^3$ of (a) pure water and (b) a phosphate buffer that is $0.20\ mol\ dm^{-3}$ KH_2PO_4(aq) and $0.30\ mol\ dm^{-3}$ K_2HPO_4(aq).

[*Answer:* (a) 7.00 to 9.00; (b) 7.39 to 4.42]

8.7 Indicators

The rapid change of pH near the stoichiometric point of an acid–base titration is the basis of indicator detection. An **acid–base indicator** is a water-soluble organic molecule with acid (HIn) and conjugate base (In^-) forms that differ in colour. The two forms are in equilibrium in solution:

$$\text{HIn(aq)} + \text{H}_2\text{O(l)} \rightleftharpoons \text{H}_3\text{O}^+\text{(aq)} + \text{In}^-\text{(aq)}$$

$$K_{In} = \frac{a_{H_3O^+}a_{In^-}}{a_{HIn}}$$

The pK_{In} of some indicators are listed in Table 8.3. The ratio of the concentrations of the conjugate acid and base forms of the indicator is

$$\frac{[In^-]}{[HIn]} \approx \frac{K_{In}}{a_{H_3O^+}}$$

This expression can be rearranged (after taking common logarithms) to

$$\log\frac{[In^-]}{[HIn]} \approx \log\frac{K_{In}}{a_{H_3O^+}} = \underbrace{\boxed{\log K_{In}}}_{-pK_{In}} - \underbrace{\boxed{\log a_{H_3O^+}}}_{-pH}$$

Use $\log(x/y) = \log x - \log y$

and written as

$$\log\frac{[In^-]}{[HIn]} \approx \text{pH} - pK_{In} \qquad (8.15)$$

We see that as the pH swings from higher than pK_{In} to lower than pK_{In} as acid is added to the solution, the ratio of In^- to HIn swings from well above 1 to

Table 8.3
Indicator colour changes

Indicator	Acid colour	pH range of colour change	pK_{In}	Base colour
Thymol blue	Red	1.2 to 2.8	1.7	Yellow
Methyl orange	Red	3.2 to 4.4	3.4	Yellow
Bromophenol blue	Yellow	3.0 to 4.6	3.9	Blue
Bromocresol green	Yellow	4.0 to 5.6	4.7	Blue
Methyl red	Red	4.8 to 6.0	5.0	Yellow
Bromothymol blue	Yellow	6.0 to 7.6	7.1	Blue
Litmus	Red	5.0 to 8.0	6.5	Blue
Phenol red	Yellow	6.6 to 8.0	7.9	Red
Thymol blue	Yellow	9.0 to 9.6	8.9	Blue
Phenolphthalein	Colorless	8.2 to 10.0	9.4	Pink
Alizarin yellow	Yellow	10.1 to 12.0	11.2	Red
Alizarin	Red	11.0 to 12.4	11.7	Purple

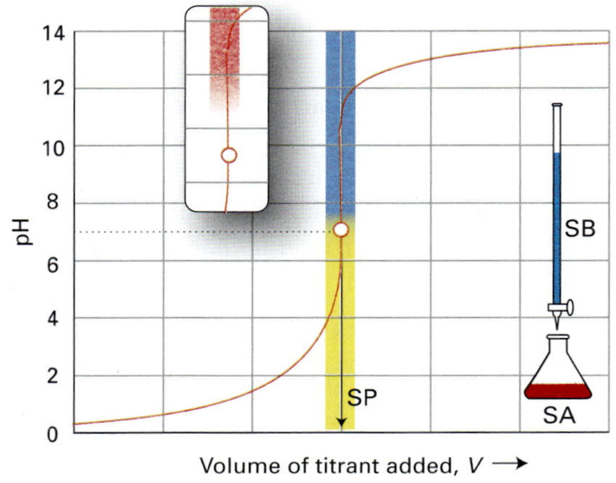

Fig. 8.7 The range of pH over which an indicator changes colour is depicted by the tinted band. For a strong acid–strong base titration, the stoichiometric point is indicated accurately by an indicator that changes colour at pH = 7 (such as bromothymol blue). However, the change in pH is so sharp that accurate results are also obtained even if the indicator changes colour in neighbouring values. Thus, phenolphthalein (which has $pK_{In} = 9.4$, see Table 8.3) is also often used.

well below 1 (Fig. 8.7). For instance, if $pH = pK_{In} - 1$, then $[In^-]/[HIn] = 10$, but if $pH = pK_{In} + 1$, then $[In^-]/[HIn] = 10^{-1}$, a decrease of two orders of magnitude.

Self-test 8.13

What is the ratio of the yellow and blue forms of bromocresol green in solution of pH (a) 3.7, (b) 4.7, and (c) 5.7?

[*Answer:* (a) 10:1, (b) 1:1, (c) 1:10]

At the stoichiometric point, the pH changes sharply through several pH units, so the molar concentration of H_3O^+ changes through several orders of magnitude. The indicator equilibrium changes so as to accommodate the change of pH, with HIn the dominant species on the acid side of the stoichiometric point, when H_3O^+ ions are abundant, and In^- dominant on the basic side, when the base can remove protons from HIn. The accompanying colour change signals the stoichiometric point of the titration. The colour in fact changes over a range of pH, typically from $pH \approx pK_{In} - 1$, when HIn is ten times as abundant as In^-, to $pH \approx pK_{In} + 1$, when In^- is ten times as abundant as HIn. The pH half-way through a colour change, when $pH \approx pK_{In}$ and the two forms, HIn and In^-, are in equal abundance, is the **end point** of the indicator. With a well-chosen indicator, the end point of the indicator coincides with the stoichiometric point of the titration.

Care must be taken to use an indicator that changes colour at the pH appropriate to the type of titration. Specifically, we need to match the end point to the stoichiometric point, and therefore select an indicator for which pK_{In} is close to the pH of the stoichiometric point. Thus, in a weak-acid–strong-base titration, the stoichiometric point lies at pH > 7, and we should select an indicator that changes at that pH (Fig. 8.8). Similarly, in a strong-acid–weak-base titration, we need to select an indicator with an end point at pH < 7. Qualitatively, we should choose an indicator with $pK_{In} \approx 7$ for strong-acid–strong-base titrations, one with $pK_{In} < 7$ for strong-acid–weak-base titrations, and one with $pK_{In} > 7$ for weak-acid–strong-base titrations.

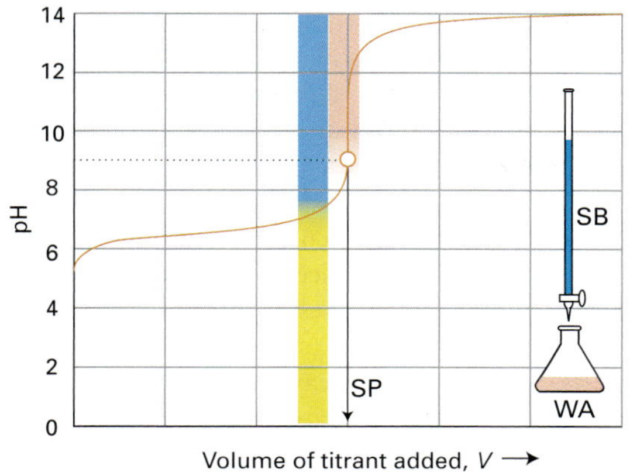

Fig. 8.8 In a weak acid–strong base titration, an indicator with pK_{In} ≈ 7 (the lower band, like bromothymol blue) would give a false indication of the stoichiometric point; it is necessary to use an indicator that changes colour close to the pH of the stoichiometric point. If that lies at about pH = 9, then phenolphthalein would be appropriate.

Self-test 8.14

Vitamin C is a weak acid (ascorbic acid), and the amount in a sample may be determined by titration with sodium hydroxide solution. Should you use methyl red or phenolphthalein as the indicator?

[*Answer:* phenolphthalein]

Solubility equilibria

A solid dissolves in a solvent until the solution and the solid solute are in equilibrium. At this stage, the solution is said to be **saturated**, and its molar concentration is the **molar solubility** of the solid. That the two phases—the solid solute and the solution—are in dynamic equilibrium implies that we can use equilibrium concepts to discuss the composition of the saturated solution. You should note that a solubility equilibrium is an example of a *heterogeneous* equilibrium in which the species are in different phases (the solid solute and the solution). The properties of aqueous solutions of electrolytes are commonly treated in terms of equilibrium constants, and in this section we shall confine our attention to them. We shall also limit our attention to **sparingly soluble** compounds, which are compounds that dissolve only slightly in water. This restriction is applied because the effects of ion–ion interactions are a complicating feature of more concentrated solutions and more advanced techniques are then needed before the calculations are reliable. Once again, we shall concentrate on general trends and properties rather than expecting to obtain numerically precise results and confine numerical calculations to systems at 298 K.

8.8 The solubility constant

The heterogeneous equilibrium between a sparingly soluble ionic compound, such as calcium hydroxide, $Ca(OH)_2$, and its ions in aqueous solution is

$$Ca(OH)_2(s) \rightleftharpoons Ca^{2+}(aq) + 2\,OH^-(aq)$$

$$K_s = \frac{a_{Ca^{2+}}a_{OH^-}^2}{\boxed{a_{Ca(OH)_2}}_{\to 1}} = a_{Ca^{2+}}a_{OH^-}^2$$

The equilibrium constant for an ionic equilibrium such as this, bearing in mind that the solid does not appear in the equilibrium expression because its activity is 1, is called the **solubility constant** (which is also called the *solubility product constant* or simply the *solubility product*). As usual, for very dilute solutions, we can replace the activity a_J of a species J by the numerical value of its molar concentration. Experimental values for solubility constants are given in Table 8.4.

We can interpret the solubility constant in terms of the numerical value of the **molar solubility**, S, of a sparingly soluble substance. For instance, it follows from the stoichiometry of the equilibrium equation written above that the molar concentration of Ca^{2+} ions in solution is equal to that of the $Ca(OH)_2$ dissolved in solution, so $S = [Ca^{2+}]$. Likewise, because the concentration of OH^- ions is twice that of $Ca(OH)_2$ formula units, it follows that $S = \frac{1}{2}[OH^-]$. Therefore, provided it is permissible to replace activities by molar concentrations,

$$K_s \approx S \times (2S)^2 = 4S^3$$

from which it follows that

$$S \approx (\tfrac{1}{4}K_s)^{1/3} \qquad (8.16)$$

This expression is only approximate because ion–ion interactions have been ignored. However, because the solid is sparingly soluble, the concentrations of the ions are low and the inaccuracy is moderately low. Thus, from Table 8.4, $K_s = 5.5 \times 10^{-6}$, so $S \approx 1 \times 10^{-2}$ and the molar solubility is 1×10^{-2} mol dm^{-3}. Solubility constants (which are determined by electrochemical measurements of the kind described in Chapter 9) provide a more accurate way of measuring solubilities of very sparingly soluble compounds than the direct measurement of the mass that dissolves.

Table 8.4

Solubility constants at 298.15 K

Compound	Formula	K_s
Aluminium hydroxide	$Al(OH)_3$	1.0×10^{-33}
Antimony sulfide	Sb_2S_3	1.7×10^{-93}
Barium carbonate	$BaCO_3$	8.1×10^{-9}
fluoride	BaF_2	1.7×10^{-6}
sulfate	$BaSO_4$	1.1×10^{-10}
Bismuth sulfide	Bi_2S_3	1.0×10^{-97}
Calcium carbonate	$CaCO_3$	8.7×10^{-9}
fluoride	CaF_2	4.0×10^{-11}
hydroxide	$Ca(OH)_2$	5.5×10^{-6}
sulfate	$CaSO_4$	2.4×10^{-5}
Copper(I) bromide	CuBr	4.2×10^{-8}
chloride	CuCl	1.0×10^{-6}
iodide	CuI	5.1×10^{-12}
sulfide	Cu_2S	2.0×10^{-47}
Copper(II) iodate	$Cu(IO_3)_2$	1.4×10^{-7}
oxalate	CuC_2O_4	2.9×10^{-8}
sulfide	CuS	8.5×10^{-45}
Iron(II) hydroxide	$Fe(OH)_2$	1.6×10^{-14}
sulfide	FeS	6.3×10^{-18}
Iron(III) hydroxide	$Fe(OH)_3$	2.0×10^{-39}
Lead(II) bromide	$PbBr_2$	7.9×10^{-5}
chloride	$PbCl_2$	1.6×10^{-5}
fluoride	PbF_2	3.7×10^{-8}
iodate	$Pb(IO_3)_2$	2.6×10^{-13}
iodide	PbI_2	1.4×10^{-8}
sulfate	$PbSO_4$	1.6×10^{-8}
sulfide	PbS	3.4×10^{-28}
Magnesium ammonium phosphate	$MgNH_4PO_4$	2.5×10^{-13}
carbonate	$MgCO_3$	1.0×10^{-5}
fluoride	MgF_2	6.4×10^{-9}
hydroxide	$Mg(OH)_2$	1.1×10^{-11}
Mercury(I) chloride	Hg_2Cl_2	1.3×10^{-18}
iodide	Hg_2I_2	1.2×10^{-28}
Mercury(II) sulfide	HgS black:	1.6×10^{-52}
	red:	1.4×10^{-53}
Nickel(II) hydroxide	$Ni(OH)_2$	6.5×10^{-18}
Silver bromide	AgBr	7.7×10^{-13}
carbonate	Ag_2CO_3	6.2×10^{-12}
chloride	AgCl	1.6×10^{-10}
hydroxide	AgOH	1.5×10^{-8}
iodide	AgI	1.5×10^{-16}
sulfide	Ag_2S	6.3×10^{-51}
Zinc hydroxide	$Zn(OH)_2$	2.0×10^{-17}
sulfide	ZnS	1.6×10^{-24}

Self-test 8.15

Copper occurs in many minerals, one of which is chalcocite, Cu_2S. What is the approximate solubility of this compound in water at 25°C? Use the data for Cu_2S in Table 8.4.

[*Answer:* 1.7×10^{-16} mol dm^{-3}]

Because solubility products are equilibrium constants, they may be calculated from thermodynamic data, particularly the standard Gibbs energies of formation of ions in solution and the relation $\Delta_r G^{\ominus} = -RT \ln K$). The direct determination of solubilities is very difficult for almost insoluble salts. Another application is to the discussion of qualitative analysis, where judicious choice of concentrations guided by the values of K_s can result in the successive precipitation of compounds (sulfides, for instance) and the recognition of the heavy elements (barium, for instance), present in a mixture.

8.9 The common-ion effect

The principle that an equilibrium constant remains unchanged whereas the individual concentrations of species may change is applicable to solubility constants, and may be used to assess the effect of the addition of species to solutions. An example of particular importance is the effect on the solubility of a compound of the presence of another freely soluble solute that provides an ion in common with the sparingly soluble compound already present. For example, we may consider the effect on the solubility of adding sodium chloride to a saturated solution of silver chloride, the common ion in this case being Cl^-.

We know what to expect from Le Chatelier's principle: when the concentration of the common ion is increased, we can expect the equilibrium to respond by tending to minimize that increase. As a result, the solubility of the original salt can be expected to decrease. To treat the effect quantitatively, we note that the molar solubility of silver chloride in pure water is related to its solubility constant by $S \approx K_s^{1/2}$. To assess the effect of the common ion, we suppose that Cl^- ions are added to a concentration C mol dm^{-3}, which greatly exceeds the concentration of the same ion that stems from the presence of the silver chloride. Therefore, we can write

$$K_s = a_{Ag^+}a_{Cl^-} \approx [Ag^+]C$$

It is very dangerous to neglect deviations from ideal behaviour in ionic solutions, so from now on the calculation will only be indicative of the kinds of changes that occur when a common ion is added to a solution of a sparingly soluble salt: the qualitative trends are reproduced, but the quantitative calculations are unreliable. With these remarks in mind, it follows that the solubility S' of silver chloride in the presence of added chloride ions is

$$S' \approx \frac{K_s}{C}$$

The solubility is greatly reduced by the presence of the common ion. For example, whereas the solubility of silver chloride in water is 1.3×10^{-5} mol dm^{-3}, in the presence of 0.10 M NaCl(aq) it is only 2×10^{-9} mol dm^{-3}, which is nearly ten thousand times less. The reduction of the solubility of a sparingly soluble salt by the presence of a common ion is called the **common-ion effect.**

Self-test 8.16

Estimate the molar solubility of calcium fluoride, CaF_2, in (a) water, (b) 0.010 M NaF(aq).

[*Answer:* (a) 2.2×10^{-4} mol dm^{-3}; (b) 4.0×10^{-7} mol dm^{-3}]

8.10 The effect of added salts on solubility

Even a salt that has no ion in common with the sparingly soluble salt can affect the latter's solubility. At low concentrations of the added salt, the solubility of the sparingly soluble salt is increased. The explanation lies in a phenomenon that will move to centre stage in Chapter 9, where we shall see that in aqueous solution, cations tend to be found near anions and anions tend to be found near cations. That is, each ion is in an environment, called an 'ionic atmosphere', of opposite charge. The charge imbalance is not great, because the ions are ceaselessly churned around by thermal motion, but it is enough to lower the energy of the central ion slightly.

When a soluble salt is added to the solution of a sparingly soluble salt, the abundant ions of the latter form ionic atmospheres around the ions of the sparingly soluble salt. As a result of the lowering of energy that results, the sparingly soluble salt has a greater tendency to go into solution. That is, the presence of the added salt raises the solubility of the sparingly soluble salt.

To estimate the effect of an added salt MX on a sparingly soluble salt AB, we write the solubility constant for AB in terms of activities:

$$K_s = a_A a_B = \gamma_A \gamma_B [A][B] = \gamma_A \gamma_B S^2$$

It follows that $S = (K_s/\gamma_A\gamma_B)^{1/2}$ and therefore that

$$\log S = \tfrac{1}{2}\log(K_s/\gamma_A\gamma_B) = \tfrac{1}{2}\log K_s - \tfrac{1}{2}\log \gamma_A\gamma_B$$

We shall see in Chapter 9 that the logarithm of the product of activity coefficients is proportional to the square root of the concentration, C, of the added salt; with $\log \gamma_A\gamma_B = -2AC^{1/2}$ where A is a constant that depends on the identity of the solvent and the temperature; for water at 25°C, $A = 0.51$. It follows that

$$\log S = \tfrac{1}{2}\log K_s + AC^{1/2} \qquad (8.17)$$

Note that $AC^{1/2}$ increases with increasing concentration of added salt, so log S, and therefore S itself, also increases, as we anticipated. The linear dependence of log S on $C^{1/2}$ is observed, but only for low concentrations of added salt.

A brief illustration The solubility of AgCl in water at 25°C is $S = K_s^{1/2}$; because $K_s = 1.6 \times 10^{-10}$, it follows that $S = 1.3 \times 10^{-5}$ mol dm^{-3}. In the presence of 0.10 mol dm^{-3} KNO_3(aq) its solubility increases to

$$\log S = \tfrac{1}{2} \times \log(1.6 \times 10^{-10}) + 0.51 \times (0.10)^{1/2} = -4.74$$

corresponding to $S = 1.8 \times 10^{-5}$ mol dm^{-3}.

Checklist of key ideas

You should now be familiar with the following concepts.

- ☐ **1** The strength of an acid HA is reported in terms of its acidity constant and that of a base B in terms of its basicity constant.
- ☐ **2** The acid form of a species is dominant if pH < pK_a and the base form is dominant if pH > pK_a.
- ☐ **3** The pH of a mixed solution of a weak acid and its conjugate base is given by the Henderson–Hasselbalch equation.
- ☐ **4** The pH of a buffer solution containing equal concentrations of a weak acid and its conjugate base is pH = pK_a.
- ☐ **5** The end-point of the colour change of an indicator occurs at pH = pK_{In}; in a titration, choose an indicator with an end-point that coincides with the stoichiometric point.
- ☐ **6** The common-ion effect is the reduction in solubility of a sparingly soluble salt by the presence of a common ion.

Table of key equations

The following table summarizes the equations developed in this chapter.

Property	Equation	Comment
Acidity constant	$K_a = a_{H_3O^+}a_{A^-}/a_{HA}$	Definition
Basicity constant	$K_b = a_{BH^+}a_{OH^-}/a_B$	Definition
Autoprotolysis constant of water	$K_w = a_{H_3O^+}a_{OH^-}$	Definition
Relation between pH and pOH	$pH + pOH = pK_w$	
Conjugate relation	$K_aK_b = K_w$ and $pK_a + pK_b = pK_w$	
pH of the solution of an amphiprotic salt of concentration *A*	$pH = \frac{1}{2}(pK_{a1} + pK_{a2})$	$F \gg K_w/K_{a2}$ and $F \gg K_{a1}$
Henderson–Hasselbalch equation	$pH = pK_a - \log([acid]/[base])$	Weak acid and base
Solubility constant	$K_s = a_{M^+}a_{A^-}$	Sparingly soluble salt M^+A^-

Questions and exercises

Discussion questions

8.1 Describe the changes in pH that take place during the titration of: (a) a weak acid with a strong base, (b) a weak base with a strong acid.

8.2 Describe the basis of buffer action and indicator detection.

8.3 Explain the difference between 'stoichiometric (equivalence) point' and 'end-point' in the context of a titration.

8.4 Outline the change in composition of a solution of the salt of a triprotic acid as the pH is changed from 1 to 14.

8.5 State the limits to the generality of the expression for estimating the pH of an amphiprotic salt solution. Suggest reasons for why these limitations exist.

8.6 Describe and justify the approximations used in the derivation of the Henderson–Hasselbalch equation.

8.7 Explain the common-ion effect.

Exercises

8.1 Write the proton-transfer equilibria for the following acids in aqueous solution and identify the conjugate acid–base pairs in each one: (a) H_2SO_4, (b) HF (hydrofluoric acid), (c) $C_6H_5NH_3^+$ (anilinium ion), (d) $H_2PO_4^-$ (dihydrogenphosphate ion), (e) HCOOH (formic acid), (f) $NH_2NH_3^+$ (hydrazinium ion).

8.2 Numerous acidic species are found in living systems. Write the proton-transfer equilibria for the following biochemically important acids in aqueous solution: (a) lactic acid ($CH_3CHOHCOOH$), (b) glutamic acid ($HOOCCH_2CH_2CH(NH_2)COOH$), (c) glycine (NH_2CH_2COOH), (d) oxalic acid (HOOCCOOH).

8.3 For biological and medical applications we often need to consider proton transfer equilibria at body temperature (37°C). The value of K_w for water at body temperature is 2.5×10^{-14}. (a) What is the value of $[H_3O^+]$ and the pH of neutral water at 37°C? (b) What is the molar concentration of OH^- ions and the pOH of neutral water at 37°C?

8.4 Suppose that something had gone wrong in the Big Bang, and instead of ordinary hydrogen there was an abundance of deuterium in the universe. There would be many subtle changes in equilibria, particularly the deuteron transfer equilibria of heavy atoms and bases. The K_w for D_2O, heavy water, at 25°C is 1.35×10^{-15}. (a) Write the chemical equation for the autoprotolysis (more precisely, autodeuterolysis) of D_2O. (b) Evaluate pK_w for D_2O at 25°C. (c) Calculate the molar concentrations of D_3O^+ and OD^- in neutral heavy water at 25°C. (d) Evaluate the pD and pOD of neutral heavy water at 25°C. (e) Formulate the relation between pD, pOD, and $pK_w(D_2O)$.

8.5 Estimate the pH of a solution of 0.50 M HCl(aq). The mean activity coefficient at this concentration is 0.769. What is a more reliable value of the pH?

8.6 Use the van 't Hoff equation (eqn 7.15) to derive an expression for the slope of a plot of pK_a against temperature.

8.7 The pK_w of water varies with temperature as follows:

θ/°C	10	15	20	25	30	35
pK_w	14.5346	14.3463	14.1669	13.9965	13.8330	13.6801

Determine the standard enthalpy of deprotonation of water.

8.8 The pK_b of ammonia in water varies with temperature as follows:

θ/°C	10	15	20	25	30	35
pK_b	4.804	4.782	4.767	4.751	4.740	4.733

Deduce as much information as you can from these values.

8.9 The pK_b of the organic base nicotine (denoted Nic) is 5.98. Write the corresponding protonation reaction, the deprotonation reaction of the conjugate acid, and the value of pK_a for nicotine.

8.10 The molar concentration of H_3O^+ ions in the following solutions was measured at 25°C. Calculate the pH and pOH of the solution: (a) 1.5×10^{-5} mol dm^{-3} (a sample of rain water), (b) 1.5 mmol dm^{-3}, (c) 5.1×10^{-14} mol dm^{-3}, (d) 5.01×10^{-5} mol dm^{-3}.

8.11 Calculate the molar concentration of H_3O^+ ions and the pH of the following solutions: (a) 25.0 cm^3 of 0.144 M HCl(aq) was added to 25.0 cm^3 of 0.125 M NaOH(aq), (b) 25.0 cm^3 of 0.15 M HCl(aq) was added to 35.0 cm^3 of 0.15 M KOH(aq), (c) 21.2 cm^3 of 0.22 M HNO_3(aq) was added to 10.0 cm^3 of 0.30 M NaOH(aq).

8.12 Determine whether aqueous solutions of the following salts have a pH equal to, greater than, or less than 7; if pH > 7 or pH < 7, write a chemical equation to justify your answer. (a) NH_4Br, (b) Na_2CO_3, (c) KF, (d) KBr, (e) $AlCl_3$, (f) $Co(NO_3)_2$.

8.13 Sodium acetate, $NaCH_3CO_2$, of mass 7.4 g is used to prepare 250 cm^3 of aqueous solution. What is the pH of the solution?

8.14 What is the pH of a solution when 2.75 g of ammonium chloride, NH_4Cl, is used to make 100 cm^3 of aqueous solution?

8.15 An aqueous solution of volume 1.0 dm^3 contains 10.0 g of potassium bromide. What is the percentage of Br^- ions that are protonated?

8.16 There are many organic acids and bases in our cells, and their presence modifies the pH of the fluids inside them. It is useful to be able to assess the pH of solutions of acids and bases and to make inferences from measured values of the pH. A solution of equal concentrations of lactic acid and sodium lactate was found to have pH = 3.08. (a) What are the values of pK_a and K_a of lactic acid? (b) What would the pH be if the acid had twice the concentration of the salt?

8.17 Sketch reasonably accurately the pH curve for the titration of 25.0 cm^3 of 0.15 M $Ba(OH)_2$(aq) with 0.22 M HCl(aq). Mark on the curve (a) the initial pH, (b) the pH at the stoichiometric point.

8.18 Determine the fraction of solute deprotonated or protonated in (a) 0.25 M C_6H_5COOH(aq), (b) 0.150 M NH_2NH_2(aq) (hydrazine), (c) 0.112 M $(CH_3)_3N$(aq) (trimethylamine).

8.19 Calculate the pH, pOH, and fraction of solute protonated or deprotonated in the following aqueous solutions: (a) 0.150 M $CH_3CH(OH)COOH$(aq) (lactic acid), (b) 2.4×10^{-4} M $CH_3CH(OH)COOH$(aq), (c) 0.25 M $C_6H_5SO_3H$(aq) (benzenesulfonic acid).

8.20 Show how the composition of an aqueous solution that contains 20 mmol dm^{-3} glycine varies with pH.

8.21 Show how the composition of an aqueous solution that contains 30 mmol dm^{-3} tyrosine varies with pH.

8.22 Estimate the pH of an aqueous solution of sodium hydrogenoxalate. Under what conditions is this estimate reasonably reliable?

8.23 Calculate the pH of the following acid solutions at 25°C; ignore second deprotonations only when that approximation is justified. (a) 1.0×10^{-4} M H_3BO_3(aq) (boric acid acts as a monoprotic acid), (b) 0.015 M H_3PO_4(aq), (c) 0.10 M H_2SO_3(aq).

8.24 The weak base colloquially known as Tris, and more precisely as tris(hydroxymethyl)aminomethane (**3**), has pK_a = 8.3 at 20°C and is commonly used to produce a buffer for biochemical applications. At what pH would you expect Tris to act as a buffer in a solution that has equal molar concentrations of Tris and its conjugate acid?

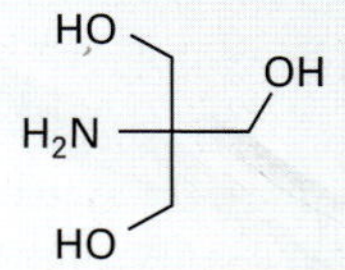

3 tris(hydroxymethyl)aminomethane

8.25 The amino acid tyrosine has pK_a = 2.20 for deprotonation of its carboxylic acid group. What are the relative concentrations of tyrosine and its conjugate base at a pH of (a) 7, (b) 2.2, (c) 1.5?

8.26 (a) Calculate the molar concentrations of $(COOH)_2$, $HOOCCO_2^-$, $(CO_2)_2^{2-}$, H_3O^+, and OH^- in 0.15 M $(COOH)_2$(aq). (b) Calculate the molar concentrations of H_2S, HS^-, S^{2-}, H_3O^+, and OH^- in 0.065 M H_2S(aq).

8.27 A sample of 0.10 M CH_3COOH(aq) of volume 25.0 cm^3 is titrated with 0.10 M NaOH(aq). The K_a for CH_3COOH is 1.8×10^{-5}. (a) What is the pH of 0.10 M CH_3COOH(aq)? (b) What is the pH after the addition of 10.0 cm^3 of 0.10 M NaOH(aq)? (c) What volume of 0.10 M NaOH(aq) is required to reach halfway to the stoichiometric point? (d) Calculate the pH at that half-way point. (e) What volume of 0.10 M NaOH(aq) is required to reach the stoichiometric point? (f) Calculate the pH at the stoichiometric point.

8.28 A buffer solution of volume 100 cm^3 consists of 0.10 M CH_3COOH(aq) and 0.10 M $Na(CH_3CO_2)$(aq). (a) What is its pH? (b) What is the pH after the addition of 3.3 mmol NaOH to the buffer solution? (c) What is the pH after the addition of 6.0 mmol HNO_3 to the initial buffer solution?

8.29 Predict the pH region in which each of the following buffers will be effective, assuming equal molar concentrations of the acid and its conjugate base: (a) sodium lactate and lactic acid, (b) sodium benzoate and benzoic acid, (c) potassium hydrogenphosphate and potassium phosphate, (d) potassium hydrogenphosphate and potassium dihydrogenphosphate, (e) hydroxylamine and hydroxylammonium chloride.

8.30 At the half-way point in the titration of a weak acid with a strong base the pH was measured as 5.16. What is the acidity constant and the pK_a of the acid? What is the pH of the solution that is 0.025 M in the acid?

8.31 Calculate the pH of (a) 0.10 M $NH_4Cl(aq)$, (b) 0.25 M $NaCH_3CO_2(aq)$, (c) 0.200 M $CH_3COOH(aq)$.

8.32 Calculate the pH at the stoichiometric point of the titration of 25.00 cm^3 of 0.150 M lactic acid with 0.188 M NaOH(aq).

8.33 Sketch the pH curve of a solution containing 0.10 M $NaCH_3CO_2(aq)$ and a variable amount of acetic acid.

8.34 From the information in Tables 8.1 and 8.2, select suitable buffers for (a) pH = 2.2 and (b) pH = 7.0.

8.35 Write the expression for the solubility constants of the following compounds: (a) AgI, (b) Hg_2S, (c) $Fe(OH)_3$, (d) Ag_2CrO_4.

8.36 Use the data in Table 8.4 to estimate the molar solubilities of (a) $BaSO_4$, (b) Ag_2CO_3, (c) $Fe(OH)_3$, (d) Hg_2Cl_2 in water.

8.37 Use the data in Table 8.4 to estimate the solubility in water of each sparingly soluble substance in its respective solution: (a) silver bromide in 1.4×10^{-3} M NaBr(aq), (b) magnesium carbonate in 1.1×10^{-5} M $Na_2CO_3(aq)$, (c) lead(II) sulfate in a 0.10 M $CaSO_4(aq)$, (d) nickel(II) hydroxide in 2.7×10^{-5} M $NiSO_4(aq)$.

8.38 The solubility of mercury(I) iodide is 5.5 fmol dm^{-3} (1 fmol = 10^{-15} mol) in water at 25°C. What is the standard Gibbs energy of dissolution of the salt?

8.39 Thermodynamic data can be used to predict the solubilities of compounds that would be very difficult to measure directly. Calculate the solubility of mercury(II) chloride in water at 25°C from standard Gibbs energies of formation.

8.40 (a) Derive an expression for the ratio of solubilities of AgCl at two different temperatures; assume that the standard enthalpy of solution of AgCl is independent of temperature in the range of interest. (b) Do you expect the solubility of AgCl to increase or decrease as the temperature is raised?

Projects

8.41 Deduce expressions for the fractions of each type of species present in an aqueous solution of lysine (**4**) as a function of pH and plot the appropriate speciation diagram. Use the following values of the acidity constants: $pK_a(H_3Lys^{2+}) = 2.18$, $pK_a(H_2Lys^{+}) = 8.95$, $pK_a(HLys) = 10.53$. *Hint:* Although it is instructive to rework Example 8.4 for a triprotic species, the expressions for the fraction can easily be written down by analogy with those in the example.

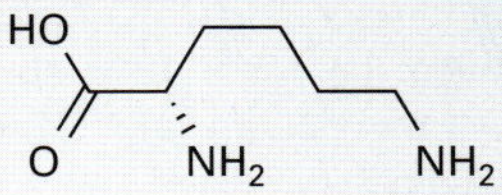

4 Lysine (Lys)

8.42 Using the insights gained through your work on Exercise 8.41, and *without doing a calculation*, sketch the speciation diagram for histidine (**5**) in water and label the axes with the significant values of pH. Use $pK_a(H_3His^{2+}) = 1.77$, $pK_a(H_2His^{+}) = 6.10$, $pK_a(HHis) = 9.18$.

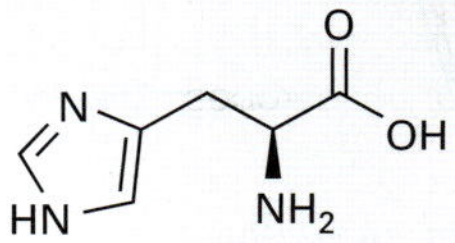

5 Histidine (His)

8.43 Here we explore buffer action in blood more quantitatively. (a) What are the values of the ratio $[HCO_3^-]/[H_2CO_3]$ at the onset of acidosis and alkalosis? (b) The Bohr effect may be understood in terms of a dependence on pH of the degree of cooperativity in the binding of O_2 by haemoglobin. Based on the description of the Bohr effect given here and the information provided in Exercise 7.44, does the Hill coefficient of haemoglobin increase or decrease with pH?

Chapter 9

Chemical equilibrium: electrochemistry

Such apparently unrelated processes as combustion, respiration, photosynthesis, and corrosion are actually all closely related, for in each of them an electron, sometimes accompanied by a group of atoms, is transferred from one species to another. Indeed, together with the proton transfer typical of acid–base reactions, processes in which electrons are transferred, the so-called **redox reactions**, account for many of the reactions encountered in chemistry. Redox reactions—the principal topic of this chapter—are of immense practical significance, not only because they underlie many biochemical and industrial processes, but also because they are the basis of the generation of electricity by chemical reactions and the investigation of reactions by making electrical measurements.

Measurements like the ones we describe in this chapter lead to a collection of data that are very useful for discussing the characteristics of electrolyte solutions and of a wide range of different types of equilibria in solution. They are also used throughout inorganic chemistry to assess the thermodynamic feasibility of reactions and the stabilities of compounds. They are used in physiology to discuss the details of the propagation of signals in neurons.

Before getting down to business, a word about notation. Throughout this chapter (and book) we use $\ln x$ for the natural logarithm of x (to the base e); this logarithm is sometimes written $\log_e x$. We use $\log x$ for the common logarithm of x (to the base 10); this logarithm is sometimes denoted $\log_{10} x$. The two logarithms are related by

$$\ln x = \ln 10 \times \log x \approx 2.303 \log x$$

Ions in solution

The most significant difference between the solution of an electrolyte and a nonelectrolyte is that there are long-range Coulombic interactions between the ions in the former. As a result, electrolyte solutions exhibit nonideal behaviour even at very low concentrations because the solute particles, the ions, do not move independently of one another. Some idea of the importance of ion–ion interactions is obtained by noting their average separations in solutions of different molar concentration c and, to appreciate the scale, the typical number of H_2O molecules that can fit between them:

$c/(\text{mol dm}^{-3})$	0.001	0.01	0.1	1	10
Separation/nm	90	40	20	9	4
Number of H_2O molecules	30	14	6	3	1

We see how to take the interactions between ions into account—which become very important for concentrations of 0.01 mol dm^{-3} and more—in the first part of this chapter. A second difference is that an ion in solution responds to the presence of an electric field, migrates through the solution, and carries charge from one location to another. Our bodies are electric conductors and some of the thoughts you are currently having as you read this sentence can be traced to the migration of ions through membranes in the enormously complex electrical circuits of your brain.

A brief comment The Coulomb interaction between two charges Q_1 and Q_2 separated by a distance r is described by the *Coulombic potential energy*:

$$E_p = \frac{Q_1 Q_2}{4\pi\varepsilon_0 r}$$

where $\varepsilon_0 = 8.854 \times 10^{-12}$ J^{-1} C^2 m^{-1} is the vacuum permittivity. Note that the interaction is attractive ($E_p < 0$) when Q_1 and Q_2 have opposite signs and repulsive ($E_p > 0$) when their signs are the same. The potential energy of a charge is zero when it is at an infinite distance from the other charge. Concepts related to electricity are reviewed in Appendix 3.

9.1 The Debye–Hückel theory

We have seen that the thermodynamic properties of solutes are expressed in terms of their activities, a_J, which is a kind of dimensionless effective concentration, and that activities are related to concentrations by multiplication by an activity coefficient, γ_J. There are various ways of expressing concentration; in the first part of this chapter we use the molality, b_J, and write

$$a_J = \gamma_J b_J / b^{\ominus} \qquad (9.1a)$$

where $b^{\ominus} = 1$ mol kg^{-1}. For notational simplicity, we shall replace $b_J/b^{\ominus}$ by b_J itself, treat b as the numerical value of the molality, and write

$$a_J = \gamma_J b_J \qquad (9.1b)$$

Because the solution becomes more ideal as the molality approaches zero, we know that $\gamma_J \to 1$ as $b_J \to 0$. Once we know the activity of the species J, we can write its chemical potential by using

$$\mu_J = \mu_J^{\ominus} + RT \ln a_J \qquad (9.2)$$

The thermodynamic properties of the solution—such as the equilibrium constants of reactions involving ions—can then be derived in the same way as for ideal solutions but with activities in place of concentrations. However, when we want to relate the results we derive to observations, we need to know how to relate activities to concentrations. We ignored that problem when discussing acids and bases, and simply assumed that all activity coefficients were 1. In this chapter, we see how to improve that approximation.

One problem that confronts us from the outset is that cations and anions always occur together in solution. Therefore, there is no experimental procedure for distinguishing the deviations from ideal behaviour due to the cations from those of the anions: we cannot measure the activity coefficients of cations and anions separately. The best we can do experimentally is to ascribe deviations from ideal behaviour equally to each kind of ion and to talk in terms of a **mean activity coefficient**, $\gamma_\pm$. For a salt MX, such as NaCl, we show in Derivation 9.1 that the mean activity coefficient is related to the activity coefficients of the individual ions as follows:

$$\gamma_\pm = (\gamma_+ \gamma_-)^{1/2} \qquad (9.3a)$$

For a salt M_pX_q, such as $Mg_3(PO_4)_2$ where $p = 3$ and $q = 2$, the mean activity coefficient is related to the activity coefficients of the individual ions as follows:

$$\gamma_\pm = (\gamma_+^p \gamma_-^q)^{1/s} \qquad s = p + q \qquad (9.3b)$$

Thus, for $Mg_3(PO_4)_2$, $s = 5$ and the mean activity coefficient for each type of ion is

$$\gamma_\pm = (\gamma_+^3 \gamma_-^2)^{1/5}$$

A brief illustration Suppose we found a way to calculate the actual activity coefficients of Na^+ and SO_4^{2-} ions in 0.010 m Na_2SO_4(aq) and found them to be 0.98 and 0.84, respectively (these values are invented), the mean activity coefficient would be

$$\gamma_\pm = \{(0.98)^2 \times (0.84)\}^{1/3} = 0.93$$

because $p = 2$ and $q = 1$ and $s = 3$. We would then write the activities of the two ions as

$$a_+ = \gamma_\pm b_+ = 0.93 \times (2 \times 0.010) = 0.019$$
$$a_- = \gamma_\pm b_- = 0.93 \times (0.010) = 0.0093$$

Derivation 9.1

Mean activity coefficients

In this derivation, we use the relation $\ln xy = \ln x + \ln y$ several times (sometimes as $\ln x + \ln y = \ln xy$), and its implication (by setting $y = x$) that $\ln x^2 = 2 \ln x$. For a salt MX that dissociates completely in solution, the molar Gibbs energy of the ions is

$$G_m = \mu_+ + \mu_-$$

where μ_+ and μ_- are the chemical potentials of the cations and anions, respectively. Each chemical potential can be expressed in terms of a molality b and an activity coefficient γ by using eqn 9.2 ($\mu = \mu^\ominus + RT \ln a$) and then eqn 9.1 ($a = \gamma b$) together with $\ln \gamma b = \ln \gamma + \ln b$, which gives

$$G_m = (\mu_+^\ominus + RT \ln \gamma_+ b_+) + (\mu_-^\ominus + RT \ln \gamma_- b_-)$$
$$= (\mu_+^\ominus + RT \ln \gamma_+ + RT \ln b_+) + (\mu_-^\ominus + RT \ln \gamma_- + RT \ln b_-)$$

We now use $\ln x + \ln y = \ln xy$ again to combine the two terms involving the activity coefficients as

$$G_m = (\mu_+^\ominus + RT \ln \gamma_+ + RT \ln b_+) + (\mu_-^\ominus + RT \ln \gamma_- + RT \ln b_-)$$
$$= (\mu_+^\ominus + RT \ln b_+) + (\mu_-^\ominus + RT \ln b_-) + RT \ln \gamma_+\gamma_-$$

We now write the term inside the logarithm as $\gamma_\pm^2$, and use $\ln x^2 = 2 \ln x$ to obtain

$$G_m = (\mu_+^\ominus + RT \ln b_+) + (\mu_-^\ominus + RT \ln b_-) + 2RT \ln \gamma_\pm$$
$$= (\mu_+^\ominus + RT \ln b_+ + RT \ln \gamma_\pm) + (\mu_-^\ominus + RT \ln b_- + RT \ln \gamma_\pm)$$
$$= (\mu_+^\ominus + RT \ln \gamma_\pm b_+) + (\mu_-^\ominus + RT \ln \gamma_\pm b_-)$$

We see that, with the mean activity coefficient defined as in eqn 9.3a, the deviation from ideal behaviour (as expressed by the activity coefficient) is now shared equally between the two types of ion. In exactly the same way, the Gibbs energy of a salt M_pX_q can be written

$$G_m = p(\mu_+^\ominus + RT \ln \gamma_\pm b_+) + q(\mu_-^\ominus + RT \ln \gamma_\pm b_-)$$

with the mean activity coefficient defined as in eqn 9.3b.[1]

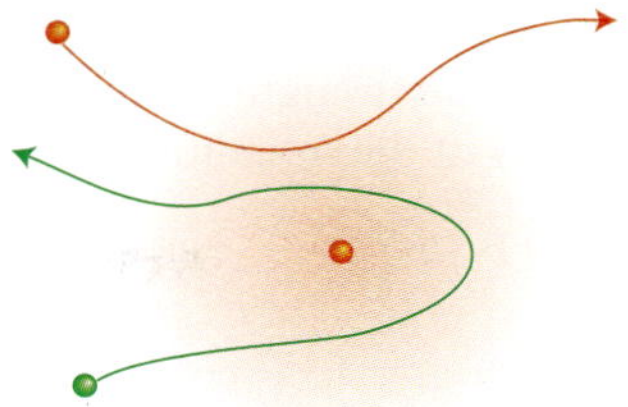

Fig. 9.1 The ionic atmosphere surrounding an ion consists of a slight excess of opposite charge as ions move through the vicinity of the central ion, with counterions lingering longer than ions of the same charge. The ionic atmosphere lowers the energy of the central ion.

The question still remains, however, about how the mean activity coefficients may be estimated. A theory that accounts for their values in very dilute solutions was developed by Peter Debye and Erich Hückel in 1923. They supposed that each ion in solution is surrounded by an **ionic atmosphere** of counter charge. This 'atmosphere' is actually the slight imbalance of charge arising from the competition between thermal motion, which tends to keep all the ions distributed uniformly throughout the solution, and the Coulombic interaction between ions, which tends to attract counterions (ions of opposite charge) into each other's vicinity and repel ions of like charge (Fig. 9.1). As a result of this competition, there is a slight excess of cations near any anion, giving a positively charged ionic atmosphere around the anion, and a slight excess of anions near any cation, giving a negatively charged ionic atmosphere around the cation. Because each ion is in an atmosphere of opposite charge, its energy is lower than in a uniform, ideal solution, and therefore its chemical potential is lower than in an ideal solution. A lowering of the chemical potential of an ion below its ideal solution value is equivalent to the activity coefficient of the ion being less than 1 (because $\ln \gamma$ is negative when $\gamma < 1$). Debye and Hückel were able to derive an expression that is a limiting law in the sense that it becomes increasingly valid as the concentration of ions approaches zero. The **Debye–Hückel limiting law**[2] is

$$\log \gamma_\pm = -A|z_+z_-|I^{1/2} \qquad (9.4)$$

(Note the common logarithm.) In this expression, A is a constant that for water at 25°C works out as 0.509. The z_J are the charge numbers of the ions (so $z_+ = +1$ for Na^+ and $z_- = -2$ for SO_4^{2-}); the vertical bars

[1] For the details of this general case, see our *Physical chemistry* (2006); see also Exercise 9.3.

[2] For a derivation of the Debye–Hückel limiting law, see our *Physical chemistry* (2006).

means that we ignore the sign of the product. The quantity I is the **ionic strength** of the solution, which is defined in terms of the molalities of the ions as

$$I = \tfrac{1}{2}(z_+^2 b_+ + z_-^2 b_-)/b^{\ominus} \tag{9.5a}$$

A brief illustration To estimate the mean activity coefficient for the ions in 0.0010 m Na_2SO_4(aq) at 25°C, we first evaluate the ionic strength of the solution from eqn 9.5 using $b_+/b^{\ominus} = 2 \times 0.0010$ and $z_+ = +1$ for Na^+ and $b_-/b^{\ominus} = 0.0010$ and z_- for SO_4^{2-}:

$$I = \tfrac{1}{2}\{(+1)^2 \times (2 \times 0.0010) + (-2)^2 \times (0.0010)\} = 0.0030$$

Then we use the Debye–Hückel limiting law, eqn 9.4, to write

$$\log \gamma_\pm = -0.509 \times |(+1)(-2)| \times (0.0030)^{1/2}$$
$$= -2 \times 0.509 \times (0.0030)^{1/2}$$

(This expression evaluates to −0.056.) On taking antilogarithms ($x = 10^{\log x}$), we conclude that $\gamma_\pm = 0.88$.

When using eqn 9.5, make sure to include all the ions present in the solution, not just those of interest. For instance, if you are calculating the ionic strength of a solution of silver chloride and potassium nitrate, there are contributions to the ionic strength from all four types of ion. When more than two ions contribute to the ionic strength, we write:

$$I = \tfrac{1}{2}\sum_i z_i^2 b_i / b^{\ominus} \tag{9.5b}$$

where the symbol Σ denotes a sum (in this case of all terms of the form $z_i^2 b_i$), z_i is the charge number of an ion i (positive for cations and negative for anions) and b_i is its molality.

As we have stressed, eqn 9.4 is a *limiting* law and is reliable only in very dilute solutions. For solutions more concentrated than about 10^{-3} mol dm^{-3} ion–ion interactions become even more important and it is better to use an empirical modification known as the **extended Debye–Hückel law:**

$$\log \gamma_\pm = -\frac{A|z_+ z_-| I^{1/2}}{1 + BI^{1/2}} + CI \tag{9.6}$$

where B and C are dimensionless constants (Fig. 9.2). Although B can be interpreted as a measure of the closest approach of the ions, it (like C) is best regarded as an adjustable empirical parameter.

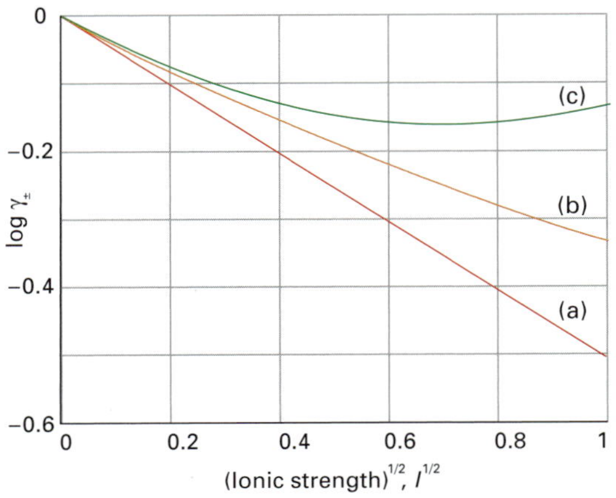

Fig. 9.2 The variation of the activity coefficient with ionic strength according to the extended Debye–Hückel theory. (a) The limiting law for a 1,1-electrolyte. (b) The extended law with $B = 0.5$. (c) The extended law, extended further by the addition of a term CI; in this case with $C = 0.2$. The last form of the law reproduces the observed behaviour reasonably well.

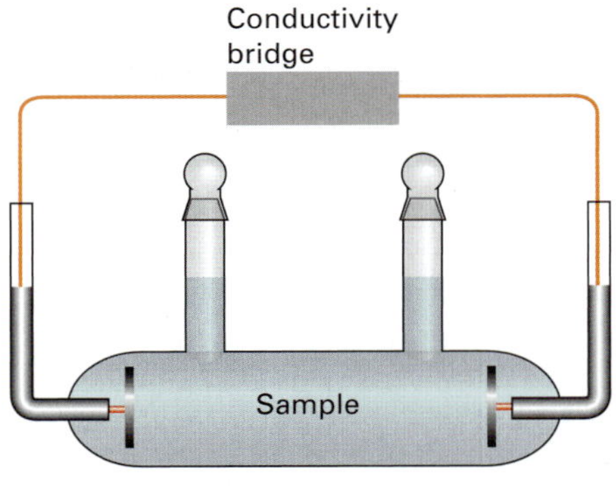

Fig. 9.3 A typical conductivity cell. The cell is made part of a 'bridge' and its resistance is measured. The conductivity is normally determined by comparison of its resistance to that of a solution of known conductivity. An alternating current is used to avoid the formation of decomposition products at the electrodes.

9.2 The migration of ions

Ions are mobile in solution, and the study of their motion down a potential gradient gives an indication of their size, the effect of solvation, and details of the type of motion they undergo. The migration of ions in solution is studied by measuring the electrical resistance of a solution of known concentration in a cell like that in Fig. 9.3. The resistance, R (in ohms, Ω), of the solution is related to the current, I (in amperes, A), that flows when a potential difference, $\mathcal{V}$ (in volts, V), is applied between the two electrodes, by Ohm's law: $\mathcal{V} = IR$ (see Appendix 3 for additional concepts of electrostatics). Certain technicalities must be dealt with in practice, such as using an alternating current to minimize the effects of electrolysis, but the essential point is the determination of R.

It is found empirically that the resistance of a sample is proportional to its length, L, and inversely proportional to its cross-sectional area, A. The constant of proportionality is called the **resistivity**, ρ (rho), and we write $R = \rho L/A$. The units of resistivity are ohm metre (Ω m). The reciprocal of the resistivity is called the **conductivity**, κ (kappa), which is expressed in Ω^{-1} m^{-1}. Reciprocal ohms appear so widely in electrochemistry that they are given their own name, siemens (S, $1\ S = 1\ \Omega^{-1}$); then conductivities are expressed in siemens per metre (S m^{-1}).

A brief comment The ampere (A) is one of the SI base units. Charge is reported in coulombs, with 1 C = 1 A s. For potential and potential difference the volt, V, is defined as $1\ V = 1\ J\ C^{-1}$ (or, in base units, $1\ V = 1\ kg\ m^2\ A^{-1}\ s^{-3}$) and for resistance the ohm (Ω) is defined as $1\ \Omega = 1\ V\ A^{-1}$ (in base units, $1\ \Omega = 1\ kg\ m^2\ A^{-2}\ s^{-3}$), so the siemens is $1\ S = 1\ A\ V^{-1}$. For most applications, it is possible to work in amperes, volts, and coulombs.

Once we have determined κ (in practice, by calibrating the cell with a solution of known conductivity), we find the **molar conductivity**, Λ_m (u.c. lambda), when the solute molar concentration is c by forming

$$\Lambda_m = \frac{\kappa}{c} \tag{9.7}$$

With molar concentration in moles per cubic decimetre, molar conductivity is expressed in siemens per metre per (moles per cubic decimetre), or S m^{-1} (mol $dm^{-3})^{-1}$. These awkward units are useful in practical applications but can be simplified to siemens metre-squared per mole (S m^2 mol^{-1}). Specifically, the relation between units is $1\ S\ m^{-1}\ (mol\ dm^{-3})^{-1} = 1\ mS\ m^2\ mol^{-1}$, where $1\ mS = 10^{-3}$ S.

The molar conductivity of a *strong* electrolyte (one that is fully dissociated into ions in solution, such as the solution of a salt) varies with molar concentration in accord with the empirical law discovered by Friedrich Kohlrausch in 1876:

$$\Lambda_m = \Lambda_m^\circ - \mathcal{K}c^{1/2} \tag{9.8}$$

The constant Λ_m°, the **limiting molar conductivity**, is the molar conductivity in the limit of such low concentration that the ions no longer interact with one another. The constant $\mathcal{K}$ takes into account the effect of these interactions when the concentration is nonzero. The fact that the interactions give rise to a square-root dependence on the concentration suggests that they arise from effects like those responsible for activity coefficients in the Debye–Hückel theory, and in particular the effect of an ionic atmosphere on the mobilities of ions. The fact that the molar conductivity *decreases* with increasing concentration can be traced to the retarding effect of the ions on the motion of one another. We shall concentrate on the limiting conductivity.

When the ions are so far apart that their interactions can be ignored, we can suspect that the molar conductivity is due to the independent migration of cations in one direction and of anions in the opposite direction, and write

$$\Lambda_m^\circ = \lambda_+ + \lambda_- \tag{9.9}$$

where λ_+ and λ_- are the **ionic conductivities** of the individual cations and anions (Table 9.1).

The molar conductivity of a *weak* electrolyte varies in a more complex way with concentration. This variation reflects the fact that the degree of ionization (or, in the case of weak acids and bases, the degree of deprotonation or protonation) varies with the concentration, with relatively more ions present at low concentrations than at high. Because we can use simple equilibrium-table techniques to relate the ion concentrations to the nominal (initial) concentration, we can use measurements of molar conductivity to determine acidity constants. The same kind of measurements can also be used to monitor the progress of reactions in solution, provided that they involve ions.

Table 9.1

*Ionic conductivities, λ/(mS m^2 mol^{-1})**

Cations		Anions	
H^+ (H_3O^+)	34.96	OH^-	19.91
Li^+	3.87	F^-	5.54
Na^+	5.01	Cl^-	7.64
K^+	7.35	Br^-	7.81
Rb^+	7.78	I^-	7.68
Cs^+	7.72	CO_3^{2-}	13.86
Mg^{2+}	10.60	NO_3^-	7.15
Ca^{2+}	11.90	SO_4^{2-}	16.00
Sr^{2+}	11.89	$CH_3CO_2^-$	4.09
NH_4^+	7.35	HCO_2^-	5.46
$[N(CH_3)_4]^+$	4.49		
$[N(CH_2CH_3)_4]^+$	3.26		

* The same numerical values apply when the units are S m^{-1} (mol $dm^{-3})^{-1}$.

Example 9.1

Determining the acidity constant from the conductivity of a weak acid

The molar conductivity of 0.010 M $CH_3COOH(aq)$ is 1.65 mS m^2 mol^{-1}. What is the acidity constant of the acid?

Strategy Because acetic acid is weak, it is only partly deprotonated in aqueous solution. Only the fraction of acid molecules present as ions contributes to the conduction, so we need to express Λ_m in terms of the fraction deprotonated. To do so, we set up an equilibrium table, find the molar concentration of H_3O^+ and $CH_3CO_2^-$ ions, and relate those concentrations to the observed molar conductivity.

Solution The equilibrium table for $CH_3COOH(aq) + H_2O(l) \rightleftharpoons H_3O^+(aq) + CH_3CO_2^-(aq)$ is

	Species		
	CH_3COOH	H_3O^+	$CH_3CO_2^-$
Initial molar concentration/(mol dm^{-3})	0.010	0	0
Change/(mol dm^{-3})	$-x$	$+x$	$+x$
Equilibrium molar concentration/(mol dm^{-3})	$0.010 - x$	x	x

The value of x is found by substituting the entries in the last line into the expression for K_a:

$$K_a = \frac{[H_3O^+][CH_3CO_2^-]}{[CH_3COOH]} = \frac{x^2}{0.010 - x}$$

On the assumption that x is small, we replace $0.010 - x$ by 0.010 and find that $x = (0.010K_a)^{1/2}$. The fraction, α, of CH_3COOH molecules present as ions is therefore $x/0.010$, or $\alpha = (K_a/0.010)^{1/2}$. The molar conductivity of the solution is therefore this fraction multiplied by the molar conductivity of acetic acid calculated on the assumption that deprotonation is complete:

$$\Lambda_m = \alpha\Lambda_m^\circ = \alpha(\lambda_{H_3O^+} + \lambda_{CH_3CO_2^-})$$

where $\Lambda_m^\circ = \lambda_{H_3O^+} + \lambda_{CH_3CO_2^-}$. Because

$$\lambda_{H_3O^+} + \lambda_{CH_3CO_2^-} = 34.96 \text{ mS m}^2\text{ mol}^{-1} + 4.09 \text{ mS m}^2\text{ mol}^{-1} = 39.05 \text{ mS m}^2\text{ mol}^{-1}$$

it follows that $\alpha = (1.65 \text{ mS m}^2\text{ mol}^{-1})/(39.05 \text{ mS m}^2\text{ mol}^{-1}) = 0.0423$. Therefore,

$$K_a = 0.010\alpha^2 = 0.010 \times (0.0423)^2 = 1.8 \times 10^{-5}$$

This value corresponds to $pK_a = 4.75$.

Self-test 9.1

The molar conductivity of 0.0250 M $HCOOH(aq)$ is 4.61 mS m^2 mol^{-1}. What is the pK_a of formic acid?

[*Answer:* 3.49]

The ability of an ion to conduct electricity depends on its ability to move through the solution. When an ion is subjected to an electric field $\mathcal{E}$, it accelerates. However, the faster it travels through the solution, the greater the retarding force it experiences from the viscosity of the medium. As a result, it settles down into a limiting velocity called its **drift velocity**, s, which is proportional to the strength of the applied field:

$$s = u\mathcal{E} \tag{9.10}$$

The **mobility**, u, depends on the radius, a, of the ion and the viscosity, η (eta), of the solution:

$$u = \frac{ez}{6\pi\eta a} \tag{9.11}$$

where ez is the charge of the moving ion.

Derivation 9.2

The ionic mobility

An *electric field* is an influence that accelerates a charged particle. An ion of charge ze in an electric field $\mathcal{E}$ (typically, in volts per metre, V m^{-1}) experiences a force of magnitude $ze\mathcal{E}$, which accelerates it. However, the ion experiences a frictional force due to its motion through the medium, which increases the faster the ion travels. The retarding force due to the viscosity on a spherical particle of radius a travelling at a speed s is given by 'Stokes' law':

$$F = 6\pi\eta as$$

When the particle has reached its drift speed, the accelerating and viscous retarding forces are equal, so we can write

$$ez\mathcal{E} = 6\pi\eta as$$

and solve this expression for s:

$$s = \frac{ez\mathcal{E}}{6\pi\eta a}$$

At this point we can compare this expression for the drift speed with eqn 9.10, and hence find the the expression for mobility given in eqn 9.11.

Equation 9.11 tells us that the mobility of an ion is high if it is highly charged, is small, and if it is in a solution with low viscosity. These features appear to contradict the trends in Table 9.2, which lists the mobilities of a number of ions. For instance, the mobilities of the Group 1 cations *increase* down the group despite their increasing radii. The explanation is that the radius to use in eqn 9.11 is the **hydrodynamic radius**, the *effective* radius for the

Table 9.2

Ionic mobilities in water at 298 K, $u/(10^{-8}\ m^2\ s^{-1}\ V^{-1})$

Cations		Anions	
H^+ (H_3O^+)	36.23	OH^-	20.64
Li^+	4.01	F^-	5.74
Na^+	5.19	Cl^-	7.92
K^+	7.62	Br^-	8.09
Rb^+	8.06	I^-	7.96
Cs^+	8.00	CO_3^{2-}	7.18
Mg^{2+}	5.50	NO_3^-	7.41
Ca^{2+}	6.17	SO_4^{2-}	8.29
Sr^{2+}	6.16		
NH_4^+	7.62		
$[N(CH_3)_4]^+$	4.65		
$[N(CH_2CH_3)_4]^+$	3.38		

migration of the ions taking into account the entire object that moves. When an ion migrates, it carries its hydrating water molecules with it, and as small ions are more extensively hydrated than large ions (because they give rise to a stronger electric field in their vicinity), ions of small radius actually have a large hydrodynamic radius. Thus, hydrodynamic radius *decreases* down Group 1 because the extent of hydration decreases with increasing ionic radius.

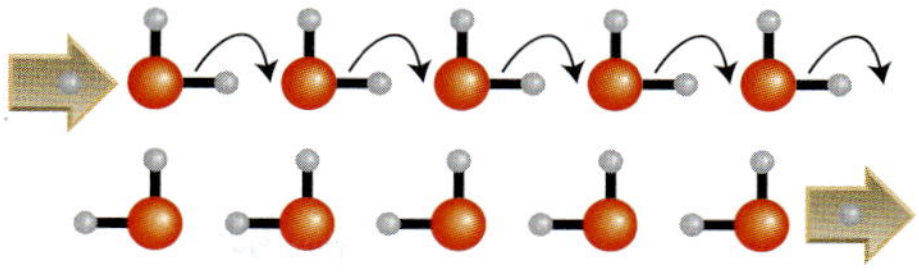

Fig. 9.4 A simplified version of the 'Grotthus mechanism' of proton conduction through water. The proton leaving the chain on the right is not the same as the proton entering the chain on the left.

See an animated version of this figure in the interactive ebook.

One significant deviation from this trend is the very high mobility of the proton in water. It is believed that this high mobility reflects an entirely different mechanism for conduction, the **Grotthus mechanism**, in which the proton on one H_2O molecule migrates to its neighbour, the proton on that H_2O molecule migrates to its neighbour, and so on along a chain (Fig. 9.4). The motion is therefore an *effective* motion of a proton, not the actual motion of a single proton.[3] The motion of protons and other ions across biological membranes is even more complicated and makes use of special proteins called *ion channels* and *ion pumps* (Box 9.1).

[3] For a detailed account of the modern version of this mechanism, see our *Physical chemistry* (2006).

Box 9.1 Ion channels and pumps

Controlled transport of molecules and ions across biological membranes is at the heart of a number of key cellular processes, such as the transmission of nerve impulses, the transfer of glucose into red blood cells, and the synthesis of ATP. Here, we examine in some detail the various ways in which ions cross the alien environment of the lipid bilayer.

Suppose that a membrane provides a barrier that slows down the transfer of molecules or ions into or out of the cell. The thermodynamic tendency to transport a species A through the membrane is partially determined by a concentration gradient (more precisely, an activity gradient) across the membrane, which results in a difference in molar Gibbs energy between the inside and the outside of the cell

$$\Delta G_m = G_{m,in} - G_{m,out} = RT \ln \frac{a_{in}}{a_{out}}$$

The equation implies that transport into the cell of either neutral or charged species is thermodynamically favourable if $a_{in} < a_{out}$ or, if we set the activity coefficients to 1, if $[A]_{in} < [A]_{out}$. If A is an ion, there is a second contribution to ΔG_m that is due to the different potential energy of the ions on each side of the bilayer, where the difference in electrostatic potential is $\Delta\phi = \phi_{in} - \phi_{out}$. The final expression for ΔG is then

$$\Delta G_m = RT \ln \frac{[A]_{in}}{[A]_{out}} + zF\Delta\phi$$

where z is the ion charge number and F is Faraday's constant. This equation implies that there is a tendency, called *passive transport*, for a species to move down concentration and membrane potential gradients. It is also possible to move a species against these gradients, but now the flow must be driven by an exergonic process, such as the hydrolysis of ATP. This process is called *active transport*.

The transport of ions into or out of a cell needs to be mediated (that is, facilitated by other species) because the hydrophobic environment of the membrane is inhospitable to ions. There are two mechanisms for ion transport:

mediation by a carrier molecule and transport through a *channel former*, a protein that creates a hydrophilic pore through which the ion can pass. An example of a channel former is the polypeptide gramicidin A, which increases the membrane permeability to cations such as H^+, K^+, and Na^+.

Ion channels are proteins that effect the movement of specific ions down a membrane potential gradient. They are highly selective, so there is a channel protein for Ca^{2+}, another for Cl^-, and so on. The opening of the gate may be triggered by potential differences between the two sides of the membrane or by the binding of an *effector* molecule to a specific receptor site on the channel.

The *patch clamp technique* can be used to measure the transport of ions across cell membranes. One of many possible experimental arrangements is shown in the illustration. With mild suction, a 'patch' of membrane from a whole cell or a small section of a broken cell can be attached tightly to the tip of a micropipette filled with an electrolyte solution and containing an electrode, the *patch electrode*. A potential difference (the 'clamp') is applied between the patch electrode and an intracellular electrode in contact with the cytosol of the cell. If the membrane is permeable to ions at the applied potential difference, a current flows through the completed circuit. Using sufficiently narrow micropipette tips with diameters of less than 1 μm, ion currents of a few picoamperes (1 pA = 10^{-12} A) have been measured across sections of membranes containing only one ion-channel protein.

A striking example of the importance of ion channels is their role in the propagation of impulses by neurons, the fundamental units of the nervous system. The cell membrane of a neuron is more permeable to K^+ ions than to either Na^+ or Cl^- ions. The key to the mechanism of action of a nerve cell is its use of Na^+ and K^+ channels to move ions across the membrane, modulating its potential. For example, the concentration of K^+ inside an inactive nerve cell is about 20 times that on the outside, whereas the concentration of Na^+ outside the cell is about 10 times that on the inside. The difference in concentrations of ions results in a transmembrane potential difference of about –62 mV, with the negative sign denoting that the inside has a lower potential. This potential difference is also called the *resting potential* of the cell membrane.

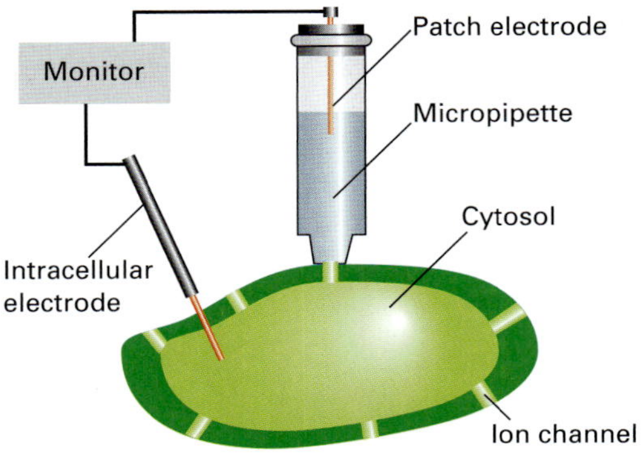

A representation of the patch clamp technique for the measurement of ionic currents through membranes in intact cells. A section of membrane containing an ion channel is in tight contact with the tip of a micropipette containing an electrolyte solution and the patch electrode. An intracellular electrode is inserted into the cytosol of the cell and the two electrodes are connected to a power supply and current-measuring device.

The transmembrane potential difference plays a particularly interesting role in the transmission of nerve impulses. Upon receiving an impulse, which is called an *action potential*, a site in the nerve cell membrane becomes transiently permeable to Na^+ and the transmembrane potential changes. To propagate along a nerve cell, the action potential must change the transmembrane potential by at least 20 mV, to values that are less negative than –40 mV. Propagation occurs when an action potential in one site of the membrane triggers an action potential in an adjacent site, with sites behind the moving action potential returning to the resting potential.

Ions such as H^+, Na^+, K^+, and Ca^{2+} are often transported actively across membranes by integral proteins called *ion pumps*. Ion pumps are molecular machines that work by adopting conformations that are permeable to one ion but not others, depending on the state of phosphorylation of the protein. Because protein phosphorylation requires dephosphorylation of ATP, the conformational change that opens or closes the pump is endergonic and requires the use of energy stored during metabolism.

Electrochemical cells

An **electrochemical cell** consists of two electronic conductors (metal or graphite, for instance) dipping into an electrolyte (an ionic conductor), which may be a solution, a liquid, or a solid. The electronic conductor and its surrounding electrolyte is an **electrode**. The physical structure containing them is called an **electrode compartment**. The two electrodes may share the same compartment (Fig. 9.5). If the electrolytes are different, then the two compartments may be joined by a **salt bridge**, which is an electrolyte solution that completes the electrical circuit by permitting ions to move between the compartments (Fig. 9.6). Alternatively, the two solutions may be in direct physical contact (for example, through a porous membrane) and form a **liquid junction**.

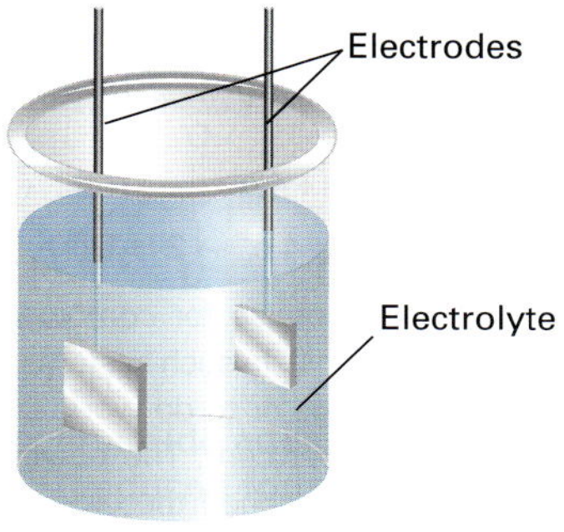

Fig. 9.5 The arrangement for an electrochemical cell in which the two electrodes share a common electrolyte.

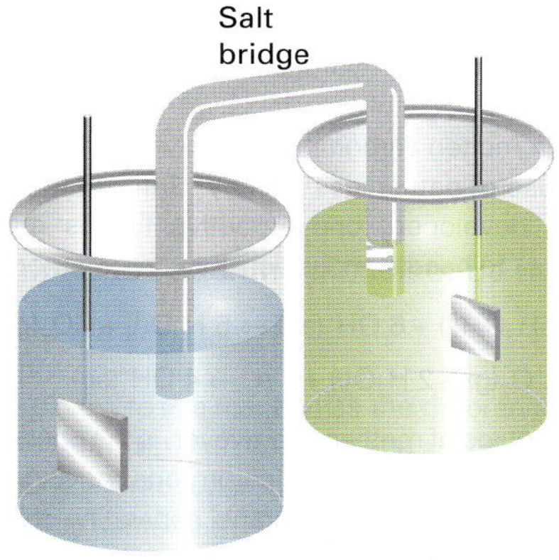

Fig. 9.6 When the electrolytes in the electrode compartments of a cell are different, they need to be joined so that ions can travel from one compartment to another. One device for joining the two compartments is a salt bridge.

However, a liquid junction introduces complications into the interpretation of measurements, and we shall not consider it further.

A **galvanic cell** (also called a *voltaic cell*) is an electrochemical cell that produces electricity as a result of the spontaneous reaction occurring inside it. An **electrolytic cell** is an electrochemical cell in which a nonspontaneous reaction is driven by an external source of direct current. The commercially available dry cells, mercury cells, nickel–cadmium ('nicad'), and lithium ion cells used to power electrical equipment are all galvanic cells and produce electricity as a result of the spontaneous chemical reaction between the substances built into them at manufacture. A **fuel cell** is a galvanic cell in which the reagents, such as hydrogen and oxygen or methane and oxygen, are supplied continuously from outside. Fuel cells are used on manned spacecraft, are beginning to be considered for use in automobiles, and gas supply companies hope that one day they may be used as a convenient, compact source of electricity in homes (Box 9.2). Electric eels and electric catfish are biological versions of fuel cells in which the fuel is food and the cells are adaptations of muscle cells. Electrolytic cells include the arrangement used to electrolyse water into hydrogen and oxygen and to obtain aluminium from its oxide in the *Hall–Hérault process*. Electrolysis is the only commercially viable means for the production of fluorine. The electron-transfer processes that occur in respiration and photosynthesis can be modelled by electrochemical cells in which electrons are transferred between proteins.

9.3 Half-reactions and electrodes

A redox reaction is the outcome of the loss of electrons, and perhaps atoms, from one species and their gain by another species. It will be familiar from introductory chemistry that we identify the loss of electrons (oxidation) by noting whether an element has undergone an increase in oxidation number (see Appendix 4 for a review of oxidation numbers). We identify the gain of electrons (reduction) by noting whether an element has undergone a decrease in oxidation number. The requirement to break and form covalent bonds in some redox reactions, as in the conversion of PCl_3 to PCl_5 or of NO_2^- to NO_3^-, is one of the reasons why redox reactions often achieve equilibrium quite slowly, often much more slowly than acid–base proton-transfer reactions.

Self-test 9.2

Identify the species that have undergone oxidation and reduction in the reaction $CuS(s) + O_2(g) \rightarrow Cu(s) + SO_2(g)$.

[*Answer:* Cu(+2) reduced to Cu(0), S(−2) oxidized to S(+4), O(0) reduced to O(−2)]

Any redox reaction may be expressed as the difference of two reduction **half-reactions**. Two examples are

Reduction of Cu^{2+}: $Cu^{2+}(aq) + 2\,e^- \rightarrow Cu(s)$

Reduction of Zn^{2+}: $Zn^{2+}(aq) + 2\,e^- \rightarrow Zn(s)$

Difference: $Cu^{2+}(aq) + Zn(s) \rightarrow Cu(s) + Zn^{2+}(aq)$ (A)

A half-reaction in which atom transfer accompanies electron transfer is

Reduction of MnO_4^-: $MnO_4^-(aq) + 8\,H^+(aq) + 5\,e^- \rightarrow Mn^{2+}(aq) + 4\,H_2O(l)$ (B)

where oxygen atoms are lost from $MnO_4^-(aq)$ and form $H_2O(l)$. In the discussion of redox reactions, the hydrogen ion is commonly denoted simply $H^+(aq)$ rather than treated as a hydronium ion, $H_3O^+(aq)$, as proton transfer is less of an issue and the chemical equations are simplified.

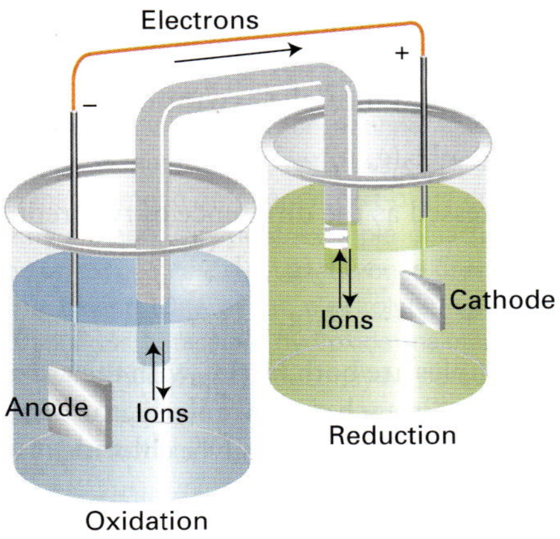

Fig. 9.7 The flow of electrons in the external circuit is from the anode of a galvanic cell, where they have been lost in the oxidation reaction, to the cathode, where they are used in the reduction reaction. Electrical neutrality is preserved in the electrolytes by the flow of cations and anions in opposite directions through the salt bridge.

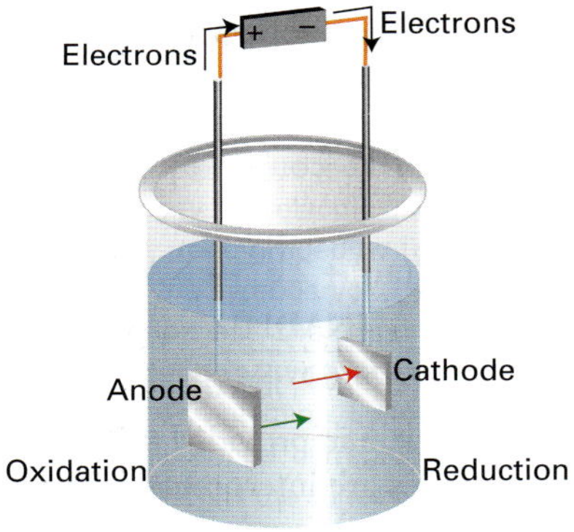

Fig. 9.8 The flow of electrons and ions in an electrolytic cell. An external supply forces electrons into the cathode, where they are used to bring about a reduction, and withdraws them from the anode, which results in an oxidation reaction at that electrode. Cations migrate towards the negatively charged cathode and anions migrate towards the positively charged anode. An electrolytic cell usually consists of a single compartment, but a number of industrial versions have two compartments.

be connected to the positive terminal of an external supply. Similarly, electrons must pass from the cathode to the species undergoing reduction, so the cathode must be connected to the negative terminal of a supply (Fig. 9.8).

In a **gas electrode** (Fig. 9.9), a gas is in equilibrium with a solution of its ions in the presence of an inert metal. The inert metal, which is often platinum, acts as a source or sink of electrons but takes no other part in the reaction except perhaps acting as a catalyst. One important example is the *hydrogen electrode*, in which

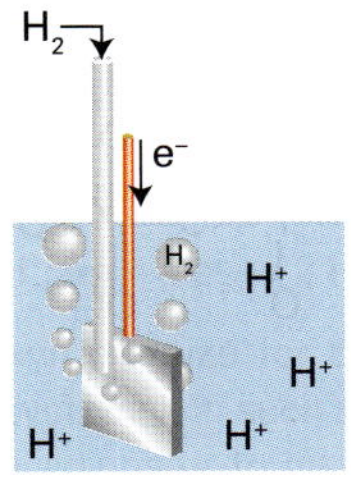

Fig. 9.9 The schematic structure of a hydrogen electrode, which is like other gas electrodes. Hydrogen is bubbled over a black (that is, finely divided) platinum surface that is in contact with a solution containing hydrogen ions. The platinum, as well as acting as a source or sink for electrons speeds the electrode reaction because hydrogen attaches to (adsorbs on) the surface as atoms.

hydrogen is bubbled through an aqueous solution of hydrogen ions and the redox couple is H^+/H_2. This electrode is denoted $Pt(s)|H_2(g)|H^+(aq)$. The vertical lines denote junctions between phases. In this electrode, the junctions are between the platinum and the gas and between the gas and the liquid containing its ions.

Example 9.3

Writing the half-reaction for a gas electrode

Write the half-reaction and the reaction quotient for the reduction of oxygen to water in acidic solution.

Strategy Write the chemical equation for the half-reaction. Then express the reaction quotient in terms of the activities and the corresponding stoichiometric coefficients, with products in the numerator and reactants in the denominator. Pure (and nearly pure) solids and liquids do not appear in Q; nor does the electron. The activity of a gas is set equal to the numerical value of its partial pressure in bar (more formally: $a_J = p_J/p^{\ominus}$).

Solution The equation for the reduction of O_2 in acidic solution is

$$O_2(g) + 4\,H^+(aq) + 4\,e^- \rightarrow 2\,H_2O(l)$$

The reaction quotient for the half-reaction is therefore

$$Q = \frac{1}{p_{O_2} a_{H^+}^4}$$

Note the very strong dependence of Q on the hydrogen ion activity.

Self-test 9.4

Write the half-reaction and the reaction quotient for a chlorine gas electrode.

[*Answer:* $Cl_2(g) + 2\,e^- \rightarrow 2\,Cl^-(aq)$, $Q = a_{Cl^-}^2/p_{Cl_2}$]

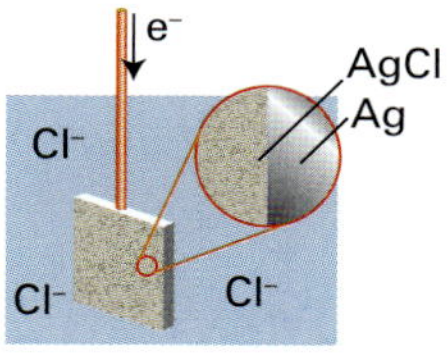

Fig. 9.10 The schematic structure of a silver–silver-chloride electrode (as an example of an insoluble-salt electrode). The electrode consists of metallic silver coated with a layer of silver chloride in contact with a solution containing Cl^- ions.

A **metal–insoluble-salt electrode** consists of a metal M covered by a porous layer of insoluble salt MX, the whole being immersed in a solution containing X^- ions (Fig. 9.10). The electrode is denoted $M|MX|X^-$, where the vertical line denotes a boundary across which electron transfer takes place. An example is the silver–silver-chloride electrode, $Ag(s)|AgCl(s)|Cl^-(aq)$, for which the reduction half-reaction is

$$AgCl(s) + e^- \rightarrow Ag(s) + Cl^-(aq) \qquad Q = a_{Cl^-} \approx [Cl^-]$$

The activities of both solids are 1. Note that the reaction quotient (and therefore, as we see later, the potential of the electrode) depends on the activity of chloride ions in the electrolyte solution.

Example 9.4

Writing the half-reaction for a metal–insoluble-salt electrode

Write the half-reaction and the reaction quotient for the lead–lead-sulfate electrode of the lead–acid battery, in which Pb(II), as lead(II) sulfate, is reduced to metallic lead in the presence of hydrogensulfate ions in the electrolyte.

Strategy Begin by identifying the species that is reduced, and writing the half-reaction. Balance that half-reaction by using H_2O molecules if O atoms are required, hydrogen ions (because the solution is acidic) if H atoms are needed, and electrons for the charge. Then write the reaction quotient in terms of the stoichiometric numbers and activities of the species present. Products appear in the numerator, reactants in the denominator.

Solution The electrode is

$$Pb(s)|PbSO_4(s)|HSO_4^-(aq)$$

in which Pb(II) is reduced to metallic lead. The equation for the reduction half-reaction is therefore

$$PbSO_4(s) + H^+(aq) + 2\,e^- \rightarrow Pb(s) + HSO_4^-(aq)$$

and the reaction quotient is

$$Q = \frac{a_{HSO_4^-}}{a_{H^+}}$$

Self-test 9.5

Write the half-reaction and the reaction quotient for the *calomel electrode*, $Hg(l)|Hg_2Cl_2(s)|Cl^-(aq)$, in which mercury(I) chloride (calomel) is reduced to mercury metal in the presence of chloride ions. This electrode is a component of instruments used to measure pH, as explained later.

[*Answer:* $Hg_2Cl_2(s) + 2\,e^- \rightarrow 2\,Hg(l) + 2\,Cl^-(aq)$, $Q = a_{Cl^-}^2$]

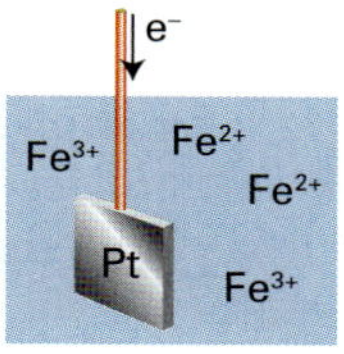

Fig. 9.11 The schematic structure of a redox electrode. The platinum metal acts as a source or sink for electrons required for the interconversion of (in this case) Fe^{2+} and Fe^{3+} ions in the surrounding solution.

The term **redox electrode** is normally reserved for an electrode in which the couple consists of the same element in two nonzero oxidation states (Fig. 9.11). An example is an electrode in which the couple is Fe^{3+}/Fe^{2+}. In general, the equilibrium is

$$Ox + \nu\,e^- \rightarrow Red \qquad Q = \frac{a_{Red}}{a_{Ox}}$$

A redox electrode is denoted M|Red,Ox, where M is an inert metal (typically platinum) making electrical contact with the solution. The electrode corresponding to the Fe^{3+}/Fe^{2+} couple is therefore denoted $Pt(s)|Fe^{2+}(aq),Fe^{3+}(aq)$ and the reduction half-reaction and reaction quotient are

$$Fe^{3+}(aq) + e^- \rightarrow Fe^{2+}(aq) \qquad Q = \frac{a_{Fe^{2+}}}{a_{Fe^{3+}}}$$

Another example of a similar kind is the electrode $Pt(s)|NADH(aq),NAD^+(aq),H^+(aq)$ used to study the $NAD^+/NADH$ couple.

9.5 Varieties of cell

The simplest type of galvanic cell has a single electrolyte common to both electrodes (as in Fig. 9.5). In

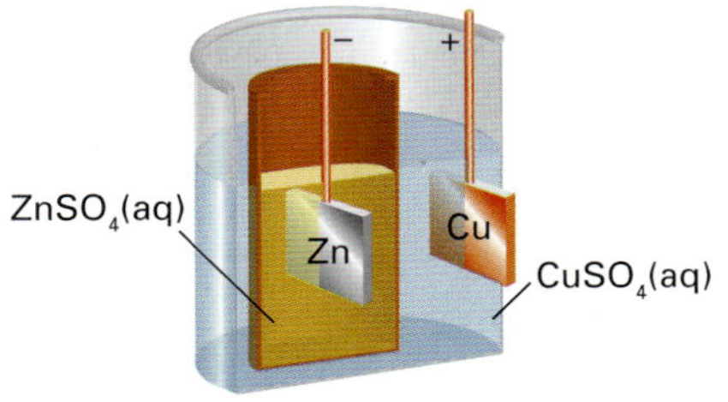

Fig. 9.12 A Daniell cell consists of copper in contact with copper(II) sulfate solution and zinc in contact with zinc sulfate solution; the two compartments are in contact through the porous pot that contains the zinc sulfate solution. The copper electrode is the cathode and the zinc electrode is the anode.

some cases it is necessary to immerse the electrodes in different electrolytes, as in the *Daniell cell* (Fig. 9.12), in which the redox couple at one electrode is Cu^{2+}/Cu and at the other is Zn^{2+}/Zn. In an **electrolyte concentration cell**, which would be constructed like the cell in Fig. 9.6, the electrode compartments are of identical composition except for the concentrations of the electrolytes. In an **electrode concentration cell** the electrodes themselves have different concentrations, either because they are gas electrodes operating at different pressures or because they are amalgams (solutions in mercury) with different concentrations.

In a cell with two different electrolyte solutions in contact, as in the Daniell cell or an electrolyte concentration cell, the **liquid junction potential**, E_j, the potential difference across the interface of the two electrolytes, contributes to the overall potential difference generated by the cell. The contribution of the liquid junction to the potential can be decreased (to about 1 to 2 mV) by joining the electrolyte compartments through a salt bridge consisting of a saturated electrolyte solution (usually KCl) in agar jelly (as in Fig. 9.6). The reason for the success of the salt bridge is that the mobilities of the K^+ and Cl^- ions are very similar and the liquid junctions at each end of the bridge are minimized.

In the notation for cells, an interface between phases is denoted by a vertical bar, |. For example, a cell in which the left-hand electrode is a hydrogen electrode and the right-hand electrode is a silver–silver-chloride electrode is denoted

$$Pt(s)|H_2(g)|HCl(aq)|AgCl(s)|Ag(s)$$

A double vertical line || denotes an interface for which the junction potential has been eliminated. Thus a cell in which the left-hand electrode, in an arrangement like that in Fig. 9.6, is zinc in contact with aqueous zinc sulfate and the right-hand electrode is copper in contact with aqueous copper(II) sulfate is denoted

$$Zn(s)|ZnSO_4(aq)||CuSO_4(aq)|Cu(s)$$

9.6 The cell reaction

The current produced by a galvanic cell arises from the spontaneous reaction taking place inside it. The **cell reaction** is the reaction in the cell written on the assumption that the right-hand electrode is the cathode, and hence that reduction is taking place in the right-hand compartment. Later we see how to predict if the right-hand electrode is in fact the cathode; if it is, then the cell reaction is spontaneous as written. If the left-hand electrode turns out to be the cathode, then the reverse of the cell reaction is spontaneous.

To write the cell reaction corresponding to the cell diagram, we first write the half-reactions at both electrodes as reductions, and then subtract the equation for the left-hand electrode from the equation for the right-hand electrode. Thus, we saw in Example 9.2 that for the cell used to study the reaction between NADH and O_2,

$$Pt(s)|NADH(aq),NAD^+(aq),H^+(aq)|| \\ H_2O_2(aq), H^+(aq)|O_2(g)|Pt(s)$$

the two reduction half-reactions are

$$\text{Right (R): } O_2(g) + 2\,H^+(aq) + 2\,e^- \rightarrow H_2O_2(aq)$$
$$\text{Left (L): } NAD^+(aq) + H^+(aq) + 2\,e^- \rightarrow NADH(aq)$$

The equation for the cell reaction is the difference:

$$\text{Overall (R − L): } NADH(aq) + O_2(g) + H^+(aq) \\ \rightarrow NAD^+(aq) + H_2O_2(aq)$$

In other cases, it may be necessary to match the numbers of electrons in the two half-reactions by multiplying one of the equations through by a numerical factor: there should be no spare electrons showing in the overall equation.

Self-test 9.6

Write the chemical equation for the reaction in the cell $Ag(s)|AgBr(s)|NaBr(aq)||NaCl(aq)|Cl_2(g)|Pt(s)$.

[*Answer:* $2\,Ag(s) + 2\,Br^-(aq) + Cl_2(g) \rightarrow 2\,AgBr(s) + 2\,Cl^-(aq)$]

9.7 The cell potential

A galvanic cell does electrical work as the reaction drives electrons through an external circuit. The work done by a given transfer of electrons depends on the potential difference between the two electrodes. This potential difference is measured in volts (V, where $1\ V = 1\ J\ C^{-1}$). When the potential difference is large (for instance, 2 V), a given number of electrons travelling between the electrodes can do a lot of electrical work. When the potential difference is small (such as 2 mV), the same number of electrons can do only a little work. A cell in which the reaction is at equilibrium can do no work and the potential difference between its electrodes is zero.

According to the discussion in Section 4.12, we know that the maximum nonexpansion work, w'_{max}, that a system (in this context, the cell) can do is given by the value of ΔG, and in particular that

At constant temperature and pressure: $$w'_{max} = \Delta G \quad (9.12)$$

Therefore, by measuring the potential difference and converting it to the electrical work done by the reaction, we have a means of determining a thermodynamic quantity, the reaction Gibbs energy. Conversely, if we know ΔG for a reaction, then we have a route to the prediction of the potential difference between the electrodes of a cell. However, to use eqn 9.12 we need to recall that maximum work is achieved only when a process occurs reversibly. We saw in Section 2.3 that the criterion of thermodynamic reversibility is the reversal of a process by an infinitesimal change in the external conditions. In the present context, reversibility means that the cell should be connected to an external source of potential difference that opposes and exactly matches the potential difference generated by the cell. Then an infinitesimal change of the external potential difference will allow the reaction to proceed in its spontaneous direction and an opposite infinitesimal change will drive the reaction in its reverse direction. The potential difference measured when a cell is balanced against an external source of potential is called the **cell potential** and denoted E_{cell} (Fig. 9.13). An alternative name for this quantity, which formerly was called the *electromotive force* (emf) of the cell, is the *zero-current cell potential*. In practice, all we need do is to measure the potential difference with a voltmeter that draws negligible current.

As we show in Derivation 9.3, the relation between the cell potential and the Gibbs energy of the cell reaction is

$$-\nu F E_{cell} = \Delta_r G \quad (9.13)$$

where F is **Faraday's constant**, the magnitude of electric charge per mole of electrons:

$$F = eN_A = 96.485\ \text{kC mol}^{-1}$$

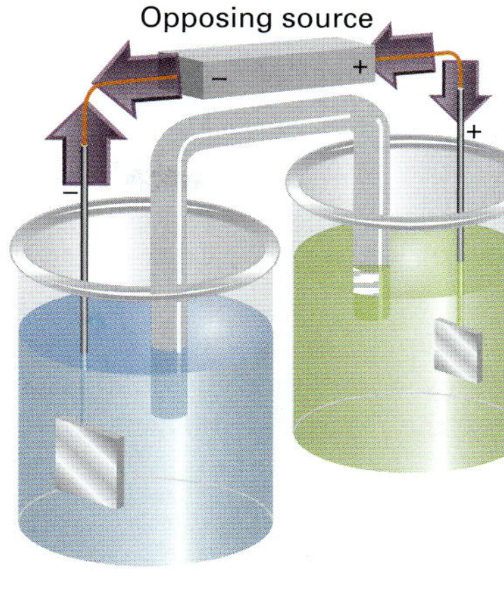

Fig. 9.13 The potential of a cell is measured by balancing the cell against an external potential that opposes the reaction in the cell. When there is no current flow, the external potential difference is equal to the cell potential.

Derivation 9.3

The cell potential

Suppose the cell reaction can be broken down into half-reactions of the form $A + \nu e^- \rightarrow B$. Then, when the reaction takes place, νN_A electrons are transferred from the reducing agent to the oxidizing agent per mole of reaction events, so the charge transferred between the electrodes is $\nu N_A \times (-e)$, or $-\nu F$. The electrical work w' done when this charge travels from the anode to the cathode is equal to the product of the charge and the potential difference E_{cell}:

$$w' = -\nu F \times E_{cell}$$

Provided the work is done reversibly at constant temperature and pressure, we can equate this electrical work to the reaction Gibbs energy and obtain eqn 9.13.

Equation 9.13 shows that the sign of the cell potential is opposite to that of the reaction Gibbs energy, which we should recall is the slope of a graph of G plotted against the composition of the reaction mixture (Section 7.1). When the reaction is spontaneous in the forward direction, $\Delta_r G < 0$ and $E_{cell} > 0$. When $\Delta_r G > 0$, the reverse reaction is spontaneous and $E_{cell} < 0$. At equilibrium $\Delta_r G = 0$ and therefore $E_{cell} = 0$ too.

Equation 9.13 provides an electrical method for measuring a reaction Gibbs energy at any composition of the reaction mixture: we simply measure the cell potential and convert it to $\Delta_r G$. Conversely, if we know the value of $\Delta_r G$ at a particular composition, then we can predict the cell potential.

A brief illustration Suppose $\Delta_r G \approx -1 \times 10^2$ kJ mol^{-1} and $\nu = 1$, then

$$E_{\text{cell}} = -\frac{\Delta_r G}{\nu F} = -\frac{(-1\times 10^5 \text{ J mol}^{-1})}{1\times(9.6485\times 10^4 \text{ C mol}^{-1})} = 1 \text{ V}$$

Most electrochemical cells bought commercially are indeed rated at between 1 and 2 V.

Our next step is to see how E_{cell} varies with composition by combining eqn 9.13 and eqn 7.6 showing how the reaction Gibbs energy varies with composition:

$$\Delta_r G = \Delta_r G^{\ominus} + RT \ln Q$$

In this expression, $\Delta_r G^{\ominus}$ is the standard reaction Gibbs energy and Q is the reaction quotient for the cell reaction. When we substitute this relation into eqn 9.13 written $E_{\text{cell}} = -\Delta_r G/\nu F$ we obtain the **Nernst equation:**

$$E_{\text{cell}} = E^{\ominus}_{\text{cell}} - \frac{RT}{\nu F}\ln Q \qquad (9.14)$$

$E^{\ominus}_{\text{cell}}$ is the **standard cell potential:**

$$E^{\ominus}_{\text{cell}} = -\frac{\Delta_r G^{\ominus}}{\nu F} \qquad (9.15)$$

The standard cell potential is often interpreted as the cell potential when all the reactants and products are in their standard states (unit activity for all solutes, pure gases and solids, a pressure of 1 bar). However, because such a cell is not in general attainable, it is better to regard $E^{\ominus}_{\text{cell}}$ simply as the standard Gibbs energy of the reaction expressed as a potential. Note that if all the stoichiometric coefficients in the equation for a cell reaction are multiplied by a factor, then $\Delta_r G^{\ominus}$ is increased by the same factor; but so too is ν, so the standard cell potential is unchanged. Likewise, Q is raised to a power equal to the factor (so if the factor is 2, Q is replaced by Q^2) and because $\ln Q^2 = 2 \ln Q$, and likewise for other factors, the second term on the right-hand side of the Nernst equation is also unchanged. That is, E_{cell} is independent of how we write the balanced equation for the cell reaction.

At 25.00°C,

$$\frac{RT}{F} = \frac{(8.314\,47 \text{ J K}^{-1} \text{ mol}^{-1})\times(298.15 \text{ K})}{9.6485\times 10^4 \text{ C mol}^{-1}}$$
$$= 2.5693 \times 10^{-2} \text{ J C}^{-1}$$

Because 1 J = 1 V C, 1 J C^{-1} = 1 V, and 10^{-3} V = 1 mV, we can write this result as

$$\frac{RT}{F} = 25.693 \text{ mV}$$

or approximately 25.7 mV. It follows from the Nernst equation that for a reaction in which $\nu = 1$, if Q is decreased by a factor of 10, then the cell potential becomes more positive by $(25.7 \text{ mV}) \times \ln 10 = 59.2$ mV. The reaction has a greater tendency to form products. If Q is increased by a factor of 10, then the cell potential falls by 59.2 mV and the reaction has a lower tendency to form products.

9.8 Cells at equilibrium

A special case of the Nernst equation has great importance in chemistry. Suppose the reaction has reached equilibrium; then $Q = K$, where K is the equilibrium constant of the cell reaction. However, because a chemical reaction at equilibrium cannot do work, it generates zero potential difference between the electrodes. Setting $Q = K$ and $E_{\text{cell}} = 0$ in the Nernst equation gives

$$0 = E^{\ominus}_{\text{cell}} - \frac{RT}{\nu F}\ln K$$

and therefore

$$\ln K = \frac{\nu F E^{\ominus}_{\text{cell}}}{RT} \qquad (9.16)$$

This very important equation lets us predict equilibrium constants from the standard cell potential. Equation 9.16, of course, is simply eqn 7.8 expressed electrochemically. Note that

If $E^{\ominus}_{\text{cell}} > 0$, then $K > 1$ and at equilibrium the cell reaction lies in favour of products.

If $E^{\ominus}_{\text{cell}} < 0$, then $K < 1$ and at equilibrium the cell reaction lies in favour of reactants.

A brief illustration Because the standard potential of the Daniell cell is +1.10 V, the equilibrium constant for the cell reaction (reaction A) is

$$\ln K = \frac{2\times(9.6485\times 10^4 \text{ C mol}^{-1})\times(1.10 \text{ V})}{(8.3145 \text{ J K}^{-1} \text{ mol}^{-1})\times(298.15 \text{ K})}$$
$$= \frac{2\times 9.6485\times 1.10\times 10^4}{8.3145\times 298.15}$$

(we have used 1 C V = 1 J to cancel units) and therefore $K = 1.5 \times 10^{37}$. Hence, the displacement of copper by zinc goes virtually to completion in the sense that the ratio of concentrations of Zn^{2+} ions to Cu^{2+} ions at equilibrium is about 10^{37}. This value is far too large to be measured by classical analytical techniques but its electrochemical measurement is straightforward. Note that a standard cell potential of +1 V corresponds to a very large equilibrium constant (and −1 V would correspond to a very small one).

9.9 Standard potentials

Each electrode in a galvanic cell makes a characteristic contribution to the overall cell potential. Although it is not possible to measure the contribution of a single electrode, one electrode can be assigned a value zero and the others assigned relative values on that basis. The specially selected electrode is the **standard hydrogen electrode** (SHE):

$$\mathrm{Pt(s)|H_2(g)|H^+(aq)} \qquad E^{\ominus} = 0 \text{ at all temperatures}$$

The **standard potential**, $E^{\ominus}$(Ox/Red), of a couple Ox/Red is then measured by constructing a cell in which the couple of interest forms the right-hand electrode and the standard hydrogen electrode is on the left. For example, the standard potential of the Ag^+/Ag couple is the standard potential of the cell

$$\mathrm{Pt(s)|H_2(g)|H^+(aq)||Ag^+(aq)|Ag(s)}$$

and is +0.80 V. Table 9.3 lists a selection of standard potentials; a longer list will be found in the Data section.

A brief comment Standard potentials are also called *standard electrode potentials* and *standard reduction potentials.* If in an older source of data you come across a 'standard oxidation potential', reverse its sign and use it as a standard reduction potential.

Table 9.3

Standard potentials at 25°C

Reduction half-reaction				$E^{\ominus}$/V
Oxidizing agent			Reducing agent	
Strongly oxidizing				
F_2	$+ 2 e^-$	→	$2 F^-$	+2.87
$S_2O_8^{2-}$	$+ 2 e^-$	→	$2 SO_4^{2-}$	+2.05
Au^+	$+ e^-$	→	Au	+1.69
Pb^{4+}	$+ 2 e^-$	→	Pb^{2+}	+1.67
Ce^{4+}	$+ e^-$	→	Ce^{3+}	+1.61
$MnO_4^- + 8 H^+$	$+ 5 e^-$	→	$Mn^{2+} + 4 H_2O$	+1.51
Cl_2	$+ 2 e^-$	→	$2 Cl^-$	+1.36
$Cr_2O_7^{2-} + 14 H^+$	$+ 6 e^-$	→	$2 Cr^{3+} + 7 H_2O$	+1.33
$O_2 + 4 H^+$	$+ 4 e^-$	→	$2 H_2O$	+1.23, +0.81 at pH = 7
Br_2	$+ 2 e^-$	→	$2 Br^-$	+1.09
Ag^+	$+ e^-$	→	Ag	+0.80
Hg_2^{2+}	$+ 2 e^-$	→	2 Hg	+0.79
Fe^{3+}	$+ e^-$	→	Fe^{2+}	+0.77
I_2	$+ e^-$	→	$2 I^-$	+0.54
$O_2 + 2 H_2O$	$+ 4 e^-$	→	$4 OH^-$	+0.40, +0.81 at pH = 7
Cu^{2+}	$+ 2 e^-$	→	Cu	+0.34
AgCl	$+ e^-$	→	$Ag + Cl^-$	+0.22
$2 H^+$	$+ 2 e^-$	→	H_2	0, by definition
Fe^{3+}	$+ 3 e^-$	→	Fe	−0.04
$O_2 + H_2O$	$+ 2 e^-$	→	$HO_2^- + OH^-$	−0.08
Pb^{2+}	$+ 2 e^-$	→	Pb	−0.13
Sn^{2+}	$+ 2 e^-$	→	Sn	−0.14
Fe^{2+}	$+ 2 e^-$	→	Fe	−0.44
Zn^{2+}	$+ 2 e^-$	→	Zn	−0.76
$2 H_2O$	$+ 2 e^-$	→	$H_2 + 2 OH^-$	−0.83, −0.42 at pH = 7
Al^{3+}	$+ 3 e^-$	→	Al	−1.66
Mg^{2+}	$+ 2 e^-$	→	Mg	−2.36
Na^+	$+ e^-$	→	Na	−2.71
Ca^{2+}	$+ 2 e^-$	→	Ca	−2.87
K^+	$+ e^-$	→	K	−2.93
Li^+	$+ e^-$	→	Li	−3.05
			Strongly reducing	

For a more extensive table, see the Data section.

To calculate the standard potential of a cell formed from any pair of electrodes we take the difference of their standard potentials:

$$E^{\ominus}_{\text{cell}} = E^{\ominus}_{\text{R}} - E^{\ominus}_{\text{L}} \tag{9.17}$$

Here, $E^{\ominus}_{\text{R}}$ is the standard potential of the right-hand electrode and $E^{\ominus}_{\text{L}}$ is that of the left. Once we have the numerical value of $E^{\ominus}_{\text{cell}}$ we can use it in eqn 9.16 to calculate the equilibrium constant of the cell reaction.

Example 9.5

Calculating an equilibrium constant

Calculate the equilibrium constant for the disproportionation reaction $2\,Cu^{+}(aq) \rightleftharpoons Cu(s) + Cu^{2+}(aq)$ at 298 K.

Strategy The aim is to find the values of $E^{\ominus}_{\text{cell}}$ and ν corresponding to the reaction, for then we can use eqn 9.16. To do so, we express the equation as the difference of two reduction half-reactions. The stoichiometric number of the electron in these matching half-reactions is the value of ν we require. We then look up the standard potentials for the couples corresponding to the half-reactions and calculate their difference to find $E^{\ominus}_{\text{cell}}$. Use $RT/F = 25.69$ mV (written as 2.569×10^{-2} V).

Solution The two half-reactions are

Right: $Cu^{+}(aq) + e^{-} \rightarrow Cu(aq)$ $\quad E^{\ominus}(Cu^{+},Cu) = +0.52$ V

Left: $Cu^{2+}(aq) + e^{-} \rightarrow Cu^{+}(aq)$ $\quad E^{\ominus}(Cu^{2+},Cu^{+}) = +0.15$ V

The difference is

$$E^{\ominus}_{\text{cell}} = E^{\ominus}_{\text{R}} - E^{\ominus}_{\text{L}} = (0.52\text{ V}) - (0.15\text{ V}) = +0.37\text{ V}$$

It then follows from eqn 9.16 with $\nu = 1$, that

$$\ln K = \frac{0.37\text{ V}}{2.569\times10^{-2}\text{ V}} = \frac{37}{2.569}$$

Therefore, because $K = e^{\ln K}$,

$$K = e^{37/2.569} = 1.8 \times 10^{6}$$

Because the value of K is so large, the equilibrium lies strongly in favour of products, and Cu^{+} disproportionates almost totally in aqueous solution.

A note on good practice Evaluate antilogarithms right at the end of the calculation, because e^{x} is very sensitive to the value of x and rounding an earlier numerical result can have a significant effect on the final answer.

Self-test 9.7

Calculate the equilibrium constant for the reaction $Sn^{2+}(aq) + Pb(s) \rightleftharpoons Sn(s) + Pb^{2+}(aq)$ at 298 K.

[*Answer:* 0.46]

9.10 The variation of potential with pH

The half-reactions of many redox couples involve hydrogen ions. For example, the fumaric acid/succinic acid couple (HOOCCH=CHCOOH/$HOOCCH_2CH_2COOH$), which plays a role in the aerobic breakdown of glucose in biological cells, is

$$\text{HOOCCH=CHCOOH(aq)} + 2\,\text{H}^{+}\text{(aq)} + 2\,\text{e}^{-} \rightarrow \text{HOOCCH}_2\text{CH}_2\text{COOH(aq)}$$

Half-reactions of this kind have potentials that depend on the pH of the medium. In this example, in which the hydrogen ions occur as reactants, an increase in pH, corresponding to a decrease in hydrogen ion activity, favours the formation of reactants, so the fumaric acid has a lower thermodynamic tendency to become reduced. We expect, therefore, that the potential of the fumaric/succinic acid couple should decrease as the pH is increased.

We can establish the quantitative variation of reduction potential with pH for a reaction by using the Nernst equation for the half-reaction and noting that (see the note at the beginning of the chapter pointing out the relation between $\ln x$ and $\log x$)

$$\ln a_{\text{H}^+} = \ln 10 \times \log a_{\text{H}^+} = -\ln 10 \times \text{pH}$$

with $\ln 10 = 2.303\ldots$. If we suppose that fumaric acid and succinic acid have fixed concentrations, the potential of the fumaric/succinic redox couple is

$$E = E^{\ominus} - \frac{RT}{2F}\ln\frac{a_{\text{suc}}}{a_{\text{fum}}a_{\text{H}^+}^2}$$

($\ln xy = \ln x + \ln y$)

$$= E^{\ominus} - \frac{RT}{2F}\ln\frac{a_{\text{suc}}}{a_{\text{fum}}} - \frac{RT}{2F}\ln\frac{1}{a_{\text{H}^+}^2}$$

($\ln 1/x^2 = -\ln x^2 = -2\ln x$)

$$= \underbrace{E^{\ominus} - \frac{RT}{2F}\ln\frac{a_{\text{suc}}}{a_{\text{fum}}}}_{E'} + \frac{RT}{F}\ln a_{\text{H}^+}$$

which, by using $\ln x = \ln 10 \times \log x$ and $\log a_{\text{H}^+} = -\text{pH}$, is easily rearranged into

$$E = E' - \frac{RT\ln 10}{F} \times \text{pH}$$

At 25°C,

$$E = E' - (59.2\text{ mV}) \times \text{pH}$$

We see that each increase of 1 unit in pH decreases the potential by 59.2 mV, which is in agreement with the remark above, that the reduction of fumaric acid is discouraged by an increase in pH.

We use the same approach to convert standard potentials to **biological standard potentials**, $E^{\oplus}$, which correspond to neutral solution (pH = 7) (see Box 7.1 for a discussion of biological standard states). If the hydrogen ions appear as reactants in the reduction half-reaction, then the potential is decreased below its standard value (for the fumaric/succinic couple, by 7 × 59.2 mV = 414 mV, or about 0.4 V). If the hydrogen ions appear as products, then the biological standard potential is higher than the thermodynamic standard potential. The precise change depends on the number of electrons and protons participating in the half-reaction.

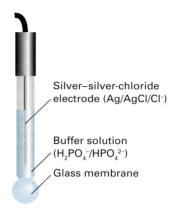

Example 9.6

Converting a standard potential to a biological standard value

Estimate the biological standard potential of the NAD⁺/NADH couple at 25°C (Example 9.2). The reduction half-reaction is

$$NAD^+(aq) + H^+(aq) + 2\,e^- \rightarrow NADH(aq) \qquad E^{\ominus}_{cell} = -0.11\ V$$

Strategy The Nernst equation applies not only to a complete cell but also to an individual electrode. Therefore, here too we write the Nernst equation for the potential, and express the reaction quotient in terms of the activities of the species. All species except H^+ are in their standard states, so their activities are all equal to 1. The remaining task is to express the hydrogen ion activity in terms of the pH, exactly as was done in the text, and set pH = 7.

Solution The Nernst equation for the half-reaction, with $v = 2$, is

$$E = E^{\ominus} - \frac{RT}{2F}\ln\frac{a_{NADH}}{a_{H^+}a_{NAD^+}} = E^{\ominus} + \frac{RT}{2F}\ln a_{H^+}$$

(ln 1/x = −ln x; a_{NADH} = 1, a_{NAD^+} = 1)

We rearrange this expression by using ln x = ln 10 × log x and log a_{H^+} = −pH into

$$E = E^{\ominus} + \frac{RT}{2F}\ln a_{H^+} = E^{\ominus} - \frac{RT\ln 10}{2F}\times \text{pH}$$

$$= E^{\ominus} - (29.58\ \text{mV}) \times \text{pH}$$

The biological standard potential (at pH = 7) is therefore

$$E^{\oplus} = (-0.11\ \text{V}) - (29.58 \times 10^{-3}\ \text{V}) \times 7 = -0.32\ \text{V}$$

Self-test 9.8

Calculate the biological standard potential of the half-reaction $O_2(g) + 4\,H^+(aq) + 4\,e^- \rightarrow 2\,H_2O(l)$ at 25 °C given its value +1.23 V under thermodynamic standard conditions.

[*Answer:* +0.82 V]

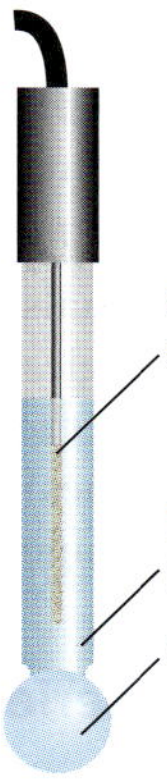

Fig. 9.14 A glass electrode has a potential that varies with the hydrogen ion concentration in the medium in which it is immersed. It consists of a thin glass membrane containing an electrolyte and a silver chloride electrode. The electrode is used in conjunction with a calomel (Hg_2Cl_2) electrode that makes contact with the test solution through a salt bridge; the electrodes are normally combined into a single unit.

9.11 The determination of pH

The potential of a hydrogen electrode is directly proportional to the pH of the solution. However, in practice, indirect methods are much more convenient to use than one based on the standard hydrogen electrode, and the hydrogen electrode is replaced by a *glass electrode* (Fig. 9.14). This electrode is sensitive to hydrogen ion activity and has a potential that depends linearly on the pH. It is filled with a phosphate buffer containing Cl^- ions, and conveniently has $E \approx 0$ when the external medium is at pH = 7. The glass electrode is much more convenient to handle than the gas electrode itself, and can be calibrated using solutions of known pH (for example, one of the buffer solutions described in Section 8.5).

Self-test 9.9

What range should a voltmeter have (in volts) to display changes of pH from 1 to 14 at 25°C if it is arranged to give a reading of zero when pH = 7?

[*Answer:* from −0.42 V to +0.35 V, a range of 0.77 V]

Finally, it should be noted that we now have a method for measuring the pK_a of an acid electrically. As we saw in Section 8.5, the pH of a solution containing equal amounts of the acid and its conjugate base is pH = pK_a. We now know how to determine pH and hence can determine pK_a in the same way.

Checklist of key ideas

You should now be familiar with the following concepts.

- ☐ **1** Deviations from ideal behaviour in ionic solutions are ascribed to the interaction of an ion with its ionic atmosphere.
- ☐ **2** The Debye–Hückel limiting law relates the mean activity of ions in a solution to the ionic strength.
- ☐ **3** The molar conductivity of a strong electrolyte follows the Kohlrausch law.
- ☐ **4** The rate at which an ion migrates through a solution is determined by its mobility, which depends on its charge, its hydrodynamic radius, and the viscosity of the solution.
- ☐ **5** Protons migrate by the Grotthus mechanism, Fig. 9.4.
- ☐ **6** A galvanic cell is an electrochemical cell in which a spontaneous chemical reaction produces a potential difference.
- ☐ **7** An electrolytic cell is an electrochemical cell in which an external source of current is used to drive a nonspontaneous chemical reaction.
- ☐ **8** A redox reaction is expressed as the difference of two reduction half-reactions.
- ☐ **9** A cathode is the site of reduction; an anode is the site of oxidation.
- ☐ **10** The cell potential is the potential difference it produces when operating reversibly.
- ☐ **11** The Nernst equation relates the cell potential to the composition of the reaction mixture.
- ☐ **12** The standard potential of a couple is the standard cell potential in which it forms the right-hand electrode and a hydrogen electrode is on the left.
- ☐ **13** The pH of a solution is determined by measuring the potential of a glass electrode.
- ☐ **14** A couple with a low standard potential has a thermodynamic tendency (in the sense $K > 1$) to reduce a couple with a high standard potential.
- ☐ **15** The entropy and enthalpy of a cell reaction are measured from the temperature dependence of the cell potential.

Table of key equations

The following table summarizes the equations developed in this chapter.

Property	Equation	Comment
Mean activity coefficient	$\gamma_\pm = (\gamma_+^p \gamma_-^q)^{1/s},\ s = p + q$	For fully dissociated salt M_pX_q
Debye–Hückel limiting law	$\log \gamma_\pm = -A\lvert z_+z_-\rvert I^{1/2}$	Concentration of ions approaches zero
Ionic strength	$I = \frac{1}{2}(z_+^2 b_+ + z_-^2 b_-)/b^\ominus$	Numerical values of b; I is dimensionless
Kohlrausch law	$\Lambda_m = \Lambda_m^\circ - \mathcal{K}c^{1/2}$	At low concentrations
Independent migration of ions	$\Lambda_m^\circ = \lambda_+ + \lambda_-$	Infinite dilution
Stokes law	$F = 6\pi\eta a$	Continuous fluid
Ionic mobility	$u = ez/6\pi\eta a$	Ion in a continuous fluid
Cell potential	$E_{cell} = -\Delta_r G/\nu F$	
Standard cell potential	$E_{cell}^\ominus = -\Delta_r G^\ominus/\nu F$	
Nernst equation	$E_{cell} = E_{cell}^\ominus - (RT/\nu F)\ln Q$	
Equilibrium constant of a cell reaction	$\ln K = \nu F E_{cell}^\ominus/RT$	
Standard cell potential from standard electrode potentials	$E_{cell}^\ominus = E_R^\ominus - E_L^\ominus$	Cell of the form L‖R
Entropy of a cell reaction	$\Delta_r S^\circ = \nu F(E_{cell}^\circ - E_{cell}^{\circ\prime})/(T - T')$	

Questions and exercises

Discussion questions

9.1 Describe the general features of the Debye–Hückel theory of electrolyte solutions. Which approximations limit its reliability to very low concentrations?

9.2 Describe the mechanism of proton conduction in water. Could a similar mechanism apply to proton conduction in liquid ammonia?

9.3 Distinguish between galvanic, electrolytic, and fuel cells. Explain why salt bridges are routinely used in electrochemical cell measurements.

9.4 Discuss how the electrochemical series can be used to determine if a redox reaction is spontaneous.

9.5 Describe an electrochemical method for the determination of thermodynamic properties of a chemical reaction.

Exercises

9.1 Calculate the ionic strength of a solution that is 0.15 mol kg^{-1} in KCl(aq) and 0.30 mol kg^{-1} in $CuSO_4$(aq).

9.2 Calculate the masses of (a) $Ca(NO_3)_2$ and, separately, (b) NaCl to add to a 0.150 mol kg^{-1} solution of KNO_3(aq) containing 500 g of solvent to raise its ionic strength to 0.250.

9.3 Express the mean activity coefficient of the ions in a solution of MgF_2 in terms of the activity coefficients of the individual ions.

9.4 Estimate the mean ionic activity coefficient and activity of a solution that is 0.015 mol kg^{-1} MgF_2(aq) and 0.025 mol kg^{-1} NaCl(aq).

9.5 The mean activity coefficients of HBr in three dilute aqueous solutions at 25°C are 0.930 (at 5.0 mmol kg^{-1}), 0.907 (at 10.0 mmol kg^{-1}), and 0.879 (at 20.0 mmol kg^{-1}). Estimate the value of B in the extended Debye–Hückel law.

9.6 The limiting molar conductivities of KCl, KNO_3, and $AgNO_3$ are 14.99 mS m^2 mol^{-1}, 14.50 mS m^2 mol^{-1} and 13.34 mS m^2 mol^{-1}, respectively (all at 25°C). What is the limiting molar conductivity of AgCl at this temperature?

9.7 The mobility of a chloride ion in aqueous solution at 25°C is 7.91×10^{-8} m^2 s^{-1} V^{-1}. Calculate its molar ionic conductivity.

9.8 The mobility of a Rb^+ ion in aqueous solution is 7.92×10^{-8} m^2 s^{-1} V^{-1} at 25°C. The potential difference between two electrodes placed in the solution is 35.0 V. If the electrodes are 8.00 mm apart, what is the drift speed of the Rb^+ ion?

9.9 The resistances of a series of aqueous NaCl solutions, formed by successive dilution of a sample, were measured in a cell with cell constant (the constant C in the relation $\kappa = C/R$) equal to 0.2063 cm^{-1}. The following values were found:

c/(mol dm^{-3})	0.000 50	0.0010	0.0050	0.010	0.020	0.050
R/Ω	3314	1669	342.1	174.1	89.08	37.14

(a) Verify that the molar conductivity follows the Kohlrausch law and find the limiting molar conductivity. (b) Determine the coefficient $\mathcal{K}$. (c) Use the value of $\mathcal{K}$ (which should depend only on the nature, not the identity of the ions) and the information that $\lambda(Na^+)$ = 5.01 mS m^2 mol^{-1} and $\lambda(I^-)$ = 7.68 mS m^2 mol^{-1} to predict (i) the molar conductivity, (ii) the conductivity, (iii) the resistance it would show in the cell, of 0.010 mol dm^{-3} NaI(aq) at 25°C.

9.10 After correction for the water conductivity, the conductivity of a saturated aqueous solution of AgCl at 25°C was found to be 0.1887 mS m^{-1}. What is the solubility of silver chloride at this temperature?

9.11 The molar conductivity of 0.020 M HCOOH(aq) is 3.83 mS m^2 mol^{-1}. What is the value of pK_a for formic acid?

9.12 The mobility of an ion depends on its charge and if a large molecule, such as a protein, can be contrived to have zero net charge, then it does not respond to an electric field. This 'isoelectric point' can be reached by varying the pH of the medium. The speed with which bovine serum albumin (BSA) moves through water under the influence of an electric field was monitored at several values of pH, and the data are listed below. What is the isoelectric point of the protein?

pH	4.20	4.56	5.20	5.65	6.30	7.00
Velocity/(μm s^{-1})	0.50	0.18	−0.25	−0.65	−0.90	−1.25

Hint: Use a plot of speed against pH to find the pH at which the speed is zero, which is the pH at which the molecule has zero net charge.

9.13 Express the oxidation of cysteine ($HSCH_2CH(NH_2)COOH$) to cystine ($HOOCCH(NH_2)CH_2SSCH_2CH(NH_2)COOH$) as the difference of two half-reactions, one of which is O_2(g) + 4 H^+(aq) + 4 e^- → 2 H_2O(l).

9.14 From the biological standard half-cell potentials $E^{\oplus}(O_2,H^+,H_2O)$ = +0.82 V and $E^{\oplus}(NADH^+,H^+,NADH)$ = −0.32 V, calculate the standard potential arising from the reaction in which NADH is oxidized to NAD^+ and the corresponding biological standard reaction Gibbs energy.

9.15 Consider a hydrogen electrode in HBr(aq) at 25°C operating at 1.45 bar. Estimate the change in the electrode potential when the solution is changed from 5.0 mmol dm^{-3} to 15.0 mmol dm^{-3}.

9.16 Devise a cell in which the cell reaction is Mn(s) + Cl_2(g) → $MnCl_2$(aq). Give the half-reactions for the electrodes and from the standard cell potential of +2.54 V deduce the standard potential of the Mn^{2+}/Mn couple.

9.17 Write the cell reactions and electrode half-reactions for the following cells:

(a) $Ag(s)|AgNO_3(aq,b_L)||AgNO_3(aq,b_R)|Ag(s)$

(b) $Pt(s)|H_2(g,p_L)|HCl(aq)|H_2(g,p_R)|Pt(s)$

(c) $Pt(s)|K_3[Fe(CN)_6](aq),K_4[Fe(CN)_6](aq)||Mn^{2+}(aq),H^+(aq)|MnO_2(s)|Pt(s)$

(d) $Pt(s)|Cl_2(g)|HCl(aq)||HBr(aq)|Br_2(l)|Pt(s)$

(e) $Pt(s)|Fe^{3+}(aq),Fe^{2+}(aq)||Sn^{4+}(aq),Sn^{2+}(aq)|Pt(s)$

(f) $Fe(s)|Fe^{2+}(aq)||Mn^{2+}(aq),H^+(aq)|MnO_2(s)|Pt(s)$

9.18 Write the Nernst equations for the cells in the preceding exercise.

9.19 Devise cells in which the following are the reactions. In each case state the value for ν to use in the Nernst equation.

(a) $Fe(s) + PbSO_4(aq) \rightarrow FeSO_4(aq) + Pb(s)$

(b) $Hg_2Cl_2(s) + H_2(g) \rightarrow 2\ HCl(aq) + 2\ Hg(l)$

(c) $2\ H_2(g) + O_2(g) \rightarrow 2\ H_2O(l)$

(d) $H_2(g) + O_2(g) \rightarrow H_2O_2(aq)$

(e) $H_2(g) + I_2(g) \rightarrow 2\ HI(aq)$

(f) $2\ CuCl(aq) \rightarrow Cu(s) + CuCl_2(aq)$

9.20 Use the standard potentials of the electrodes to calculate the standard potentials of the cells in Exercise 9.17.

9.21 Use the standard potentials of the electrodes to calculate the standard potentials of the cells devised in Exercise 9.19.

9.22 A fuel cell develops an electric potential from the chemical reaction between reagents supplied from an outside source. What is the potential of a cell fuelled by (a) hydrogen and oxygen, (b) the complete oxidation of benzene at 1.0 bar and 298 K?

9.23 A fuel cell is constructed in which both electrodes make use of the oxidation of methane. The left-hand electrode makes use of the complete oxidation of methane to carbon dioxide and water; the right-hand electrode makes use of the partial oxidation of methane to carbon monoxide and water. (a) Which electrode is the cathode? (b) What is the cell potential at 25°C when all gases are at 1 bar?

9.24 The permanganate ion is a common oxidizing agent. What is the standard potential of the $MnO_4^-,H^+/Mn^{2+}$ couple at (a) pH = 6.00, (b) general pH?

9.25 State what you would expect to happen to the cell potential when the following changes are made to the corresponding cells in Exercise 9.17. Confirm your prediction by using the Nernst equation in each case.

(a) The molar concentration of silver nitrate in the left-hand compartment is increased.

(b) The pressure of hydrogen in the left-hand compartment is increased.

(c) The pH of the right-hand compartment is decreased.

(d) The concentration of HCl is increased.

(e) Some iron(III) chloride is added to both compartments.

(f) Acid is added to both compartments.

9.26 State what you would expect to happen to the cell potential when the following changes are made to the corresponding cells devised in Exercise 9.19. Confirm your prediction by using the Nernst equation in each case.

(a) The molar concentration of $FeSO_4$ is increased.

(b) Some nitric acid is added to both cell compartments.

(c) The pressure of oxygen is increased.

(d) The pressure of hydrogen is increased.

(e) Some (i) hydrochloric acid, (ii) hydroiodic acid is added to both compartments.

(f) Hydrochloric acid is added to both compartments.

9.27 (a) Calculate the standard potential of the cell $Hg(l)|HgCl_2(aq)||TlNO_3(aq)|Tl(s)$ at 25°C. (b) Calculate the cell potential when the molar concentration of the Hg^{2+} ion is 0.230 mol dm^{-3} and that of the Tl^+ ion is 0.720 mol dm^{-3}.

9.28 (a) Can mercury produce zinc metal from aqueous zinc sulfate under standard conditions? (b) Can chlorine gas oxidize water to oxygen gas under standard conditions in basic solution?

9.29 Calculate the standard Gibbs energies at 25°C of the following reactions from the standard potential data in the Data section.

(a) $Ca(s) + 2\ H_2O(l) \rightarrow Ca(OH)_2(aq) + H_2(g)$

(b) $2\ Ca(s) + 4\ H_2O(l) \rightarrow 2\ Ca(OH)_2(aq) + 2\ H_2(g)$

(c) $Fe(s) + 2\ H_2O(l) \rightarrow Fe(OH)_2(aq) + H_2(g)$

(d) $Na_2S_2O_8(aq) + 2\ NaI(aq) \rightarrow I_2(s) + 2\ Na_2SO_4(aq)$

(e) $Na_2S_2O_8(aq) + 2\ KI(aq) \rightarrow I_2(s) + Na_2SO_4(aq) + K_2SO_4(aq)$

(f) $Pb(s) + Na_2CO_3(aq) \rightarrow PbCO_3(aq) + 2\ Na(s)$

9.30 Calculate the biological standard Gibbs energies of reactions of the following reactions and half-reactions:

(a) $2\ NADH(aq) + O_2(g) + 2\ H^+(aq) \rightarrow 2\ NAD^+(aq) + 2\ H_2O(l)$ $E^{\oplus} = +1.14$ V

(b) Malate(aq) + NAD^+(aq) → oxaloacetate(aq) + NADH(aq) + H^+(aq) $E^{\oplus} = -0.154$ V

(c) $O_2(g) + 4H^+(aq) + 4\ e^- \rightarrow 2\ H_2O(l)$ $E^{\oplus} = +0.81$ V

9.31 Tabulated thermodynamic data can be used to predict the standard potential of a cell even if it cannot be measured directly. The standard Gibbs energy of the reaction $K_2CrO_4(aq) + 2\ Ag(s) + 2\ FeCl_3(aq) \rightarrow Ag_2CrO_4(s) + 2\ FeCl_2(aq) + 2\ KCl(aq)$ is −62.5 kJ mol^{-1} at 298 K. (a) Calculate the standard potential of the corresponding galvanic cell and (b) the standard potential of the $Ag_2CrO_4/Ag,CrO_4^{2-}$ couple.

9.32 Estimate the potential at 25°C of the cell

$$Ag(s)|AgCl(s)|KCl(aq,\ 0.025\ mol\ kg^{-1})||AgNO_3(aq,\ 0.010\ mol\ kg^{-1})|Ag(s)$$

9.33 (a) Use the information in the Data section to calculate the standard potential of the cell $Ag(s)|AgNO_3(aq)||Cu(NO_3)_2(aq)|Cu(s)$ and the standard Gibbs energy and enthalpy of the cell reaction at 25°C. (b) Estimate the value of $\Delta_r G^\ominus$ at 35°C.

9.34 (a) Calculate the standard potential of the cell Pt(s)|cystine(aq), cysteine(aq)||H^+(aq)|O_2(g)|Pt(s) and the standard Gibbs energy and enthalpy of the cell reaction at 25°C. (b) Estimate the value of $\Delta_r G^\ominus$ at 35°C. Use $E^\ominus = -0.34$ V for the cysteine/cystine couple.

9.35 The biological standard potential of the couple pyruvic acid/lactic acid is −0.19 V at 25°C. What is the thermodynamic standard potential of the couple? Pyruvic acid is $CH_3COCOOH$ and lactic acid is $CH_3CH(OH)COOH$.

9.36 One ecologically important equilibrium is that between carbonate and hydrogencarbonate (bicarbonate) ions in natural water. (a) The standard Gibbs energies of formation of CO_3^{2-}(aq) and HCO_3^-(aq) are −527.81 kJ mol^{-1} and −586.77 kJ mol^{-1}, respectively. What is the standard potential of the $HCO_3^-/CO_3^{2-},H_2$ couple? (b) Calculate the standard potential of a cell in which the cell reaction is Na_2CO_3(aq) + H_2O(l) → $NaHCO_3$(aq) + NaOH(aq). (c) Write the Nernst equation for the cell, and (d) predict and calculate the change in potential when the pH is changed to 7.0. (e) Calculate the value of pK_a for HCO_3^-(aq).

9.37 Calculate the equilibrium constants of the following reactions at 25°C from standard potential data:

(a) Sn(s) + Sn^{4+}(aq) ⇌ 2 Sn^{2+}(aq)

(b) Sn(s) + 2 AgBr(s) ⇌ $SnBr_2$(aq) + 2 Ag(s)

(c) Fe(s) + $Hg(NO_3)_2$(aq) ⇌ Hg(l) + $Fe(NO_3)_2$(aq)

(d) Cd(s) + $CuSO_4$(aq) ⇌ Cu(s) + $CdSO_4$(aq)

(e) Cu^{2+}(aq) + Cu(s) ⇌ 2 Cu^+(aq)

(f) 3 Au^{2+}(aq) ⇌ Au(s) + 2 Au^{3+}(aq)

9.38 The dichromate ion in acidic solution is a common oxidizing agent for organic compounds. Derive an expression for the potential of an electrode for which the half-reaction is the reduction of $Cr_2O_7^{2-}$ ions to Cr^{3+} ions in acidic solution.

9.39 The molar solubilities of AgCl and $BaSO_4$ in water are 1.34×10^{-5} mol dm^{-3} and 9.51×10^{-4} mol dm^{-3}, respectively, at 25°C. Calculate their solubility constants from the appropriate standard potentials.

9.40 The potential of the cell Pt(s)|H_2(g)|HCl(aq)|AgCl(s)|Ag(s) is 0.312 V at 25°C. What is the pH of the electrolyte solution?

9.41 The molar solubility of AgBr is 2.6 μmol dm^{-3} at 25°C. What is the potential of the cell Ag(s)|AgBr(aq)|AgBr(s)|Ag(s) at that temperature?

9.42 The standard potential of the cell Ag(s)|AgI(s)|AgI(aq)|Ag(s) is +0.9509 V at 25°C. Calculate (a) the molar solubility of AgI and (b) its solubility constant.

Projects

9.43 Consider the Harned cell Pt(s)|H_2(g, 1 bar)|HCl(aq, b)|AgCl(s)|Ag(s). Show that the standard potential of the silver–silver-chloride electrode may be determined by plotting $E - (RT/F) \ln b$ against $b^{1/2}$. *Hint*: Express the cell potential in terms of activities, and use the Debye–Hückel law to estimate the mean activity coefficient. (b) Use the procedure you devised in part (a) and the following data at 25°C to determine the standard potential of the silver–silver-chloride electrode

$b/(10^{-3}\, b^\ominus)$	3.215	5.619	9.138	25.63
E/V	0.520 53	0.492 57	0.468 60	0.418 24

9.44 The standard potentials of proteins are not commonly measured by the methods described in this chapter because proteins often lose their native structure and their function when they react on the surfaces of electrodes. In an alternative method, the oxidized protein is allowed to react with an appropriate electron donor in solution. The standard potential of the protein is then determined from the Nernst equation, the equilibrium concentrations of all species in solution, and the known standard potential of the electron donor. We illustrate this method with the protein cytochrome *c*. (a) The one-electron reaction between cytochrome *c*, cyt, and 2,6-dichloroindophenol, D, can be written as

$$\text{cyt}_{ox} + \text{D}_{red} \rightleftharpoons \text{cyt}_{red} + \text{D}_{ox}$$

Consider $E^\ominus_{cyt}$ and $E^\ominus_D$ to be the standard potentials of cytochrome *c* and D, respectively. Show that, at equilibrium (eq), a plot of $\ln([\text{D}_{ox}]_{eq}/[\text{D}_{red}]_{eq})$ against $\ln([\text{cyt}_{ox}]_{eq}/[\text{cyt}_{red}]_{eq})$ is linear with slope of one and y-intercept $F(E^\ominus_{cyt} - E^\ominus_D)/RT$, where equilibrium activities are replaced by the numerical values of equilibrium molar concentrations. (b) The following data were obtained for the reaction between oxidized cytochrome *c* and reduced D at pH = 6.5 buffer and 298 K. The ratios $[\text{D}_{ox}]_{eq}/[\text{D}_{red}]_{eq}$ and $[\text{cyt}_{ox}]_{eq}/[\text{cyt}_{red}]_{eq}$ were adjusted by adding known volumes of a solution of sodium ascorbate, a reducing agent, to a solution containing oxidized cytochrome *c* and reduced D. From the data and the standard potential of D of 0.237 V, determine the standard potential of cytochrome *c* at pH = 6.5 and 298 K.

$[\text{D}_{ox}]_{eq}/[\text{D}_{red}]_{eq}$	0.002 79	0.008 43	0.0257	0.0497
$[\text{cyt}_{ox}]_{eq}/[\text{cyt}_{red}]_{eq}$	0.0106	0.0230	0.0894	0.197
$[\text{D}_{ox}]_{eq}/[\text{D}_{red}]_{eq}$	0.0748	0.238	0.534	
$[\text{cyt}_{ox}]_{eq}/[\text{cyt}_{red}]_{eq}$	0.335	0.809	1.39	

9.45 Here we explore ion channels in more quantitative detail. (a) Estimate the resting potential, the membrane potential at equilibrium, of a neuron at 298 K by using the fact that the concentration of K^+ inside an inactive nerve cell is about 20 times that on the outside. Now repeat the calculation, this time using the fact that the concentration of Na^+ outside the inactive cell is about 10 times that on the inside. Are the two values the same or different? How do each of the calculated values compare with the observed resting potential of −62 mV? (b) Your estimates of the resting potential from part (a) did not agree with the experimental value because the cell is never at equilibrium and ions continually cross the membrane, which is more permeable to some ions than others.

To take into account membrane permeability, we use the *Goldman equation* to calculate the resting potential:

$$\Delta\phi = \frac{RT}{F} \ln \frac{P_{M^+}[M^+]_{out} + P_{M'^+}[M'^+]_{out} + P_{X^-}[X^-]_{in} + P_{X'^-}[X'^-]_{in}}{P_{M^+}[M^+]_{in} + P_{M'^+}[M'^+]_{in} + P_{X^-}[X^-]_{out} + P_{X'^-}[X'^-]_{out}}$$

where the salts present on each side of the membrane are MX and M′X′ and the *P*s are the relative permeabilities of the ions. Consider an experiment in which $[Na^+]_{in} = 50$ mmol dm^{-3}, $[K^+]_{in} = 400$ mmol dm^{-3}, $[Cl^-]_{in} = 50$ mmol dm^{-3}, $[Na^+]_{out} = 440$ mmol dm^{-3}, $[K^+]_{out} = 20$ mmol dm^{-3}, and $[Cl^-]_{in} = 560$ mmol dm^{-3}. Use the Goldman equation and the relative permeabilities $P_{K^+} = 1.0$, $P_{Na^+} = 0.04$, and $P_{Cl^-} = 0.45$ to estimate the resting potential at 298 K under the stated conditions. How does your calculated value agree with the experimental value of −62 mV?

Chapter 10

Chemical kinetics: the rates of reactions

The branch of physical chemistry called **chemical kinetics** is concerned with the rates of chemical reactions. Chemical kinetics deals with how rapidly reactants are consumed and products formed, how reaction rates respond to changes in the conditions or the presence of a catalyst, and the identification of the steps by which a reaction takes place.

One reason for studying the rates of reactions is the practical importance of being able to predict how quickly a reaction mixture approaches equilibrium. The rate might depend on variables under our control, such as the pressure, the temperature, and the presence of a catalyst, and we might be able to optimize it by the appropriate choice of conditions. Another reason is that the study of reaction rates leads to an understanding of the **mechanism** of a reaction, its analysis into a sequence of elementary steps. For example, we might discover that the reaction of hydrogen and bromine to form hydrogen bromide proceeds by the dissociation of a Br_2 molecule, the attack of a Br atom on an H_2 molecule, and several subsequent steps. By analysing the rate of a biochemical reaction we may discover how an enzyme, a biological catalyst, acts. **Enzyme kinetics**, the study of the effect of enzymes on the rates of reactions, is also an important window on how these macromolecules work.

We need to cope with a wide variety of different rates and a process that appears to be slow may be the outcome of many faster steps. That is particularly true in the chemical reactions that underlie life. Photobiological processes like those responsible for photosynthesis and the slow growth of a plant may take place in about 1 ps. The binding of a neurotransmitter can have an effect after about 1 μs. Once a gene has been activated, a protein may emerge in about 100 s; but even that timescale incorporates many others, including the wriggling of a newly formed polypeptide chain into its working conformation,

special arrangements. Fast reactions are also studied by **pulse radiolysis** in which the flash of electromagnetic radiation is replaced by a short burst of high-velocity electrons.

In contrast to real-time analysis, **quenching methods** are based on stopping, or quenching, the reaction after it has been allowed to proceed for a certain time and the composition is analysed at leisure. The quenching (of the entire mixture or of a sample drawn from it) can be achieved either by cooling suddenly, by adding the mixture to a large volume of solvent, or by rapid neutralization of an acid reagent. This method is suitable only for reactions that are slow enough for there to be little reaction during the time it takes to quench the mixture.

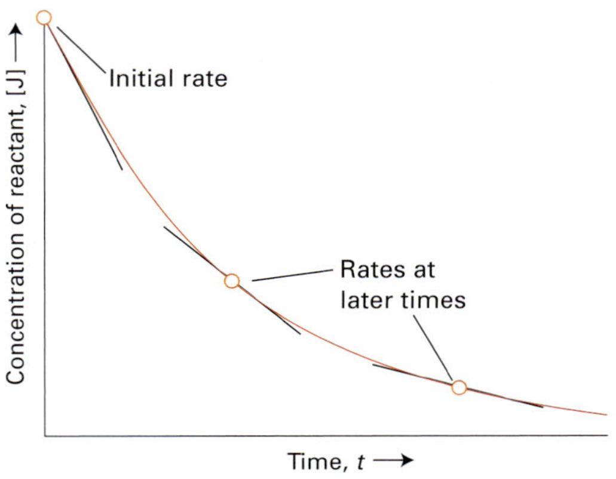

Fig. 10.4 The rate of a chemical reaction is the slope of the tangent to the curve showing the variation of concentration of a species with time. This graph is a plot of the concentration of a reactant, which is consumed as the reaction progresses. The rate of consumption decreases in the course of the reaction as the concentration of reactant decreases.

Reaction rates

The raw data from experiments to measure reaction rates are quantities (such as the absorbance of a sample) that are proportional to the concentrations or partial pressures of reactants and products at a series of times after the reaction is initiated. Ideally, information on any intermediates should also be obtained, but often they cannot be studied because their existence is so fleeting or their concentration so low. More information about the reaction can be extracted if data are obtained at a series of different temperatures. The next few sections look at these observations in more detail.

10.3 The definition of rate

The rate of a reaction taking place in a container of fixed volume is defined in terms of the rate of change of the concentration of a designated species:

Change in [J]

$$\text{Rate} = \frac{|\Delta[\mathrm{J}]|}{\Delta t} \qquad (10.5a)$$

|...| means ignore any negative sign

Time interval of interest

where $\Delta[\mathrm{J}]$ is the change in the molar concentration of the species J that occurs during the time interval Δt. We have put the change in concentration between modulus signs (|...|) to ensure that all rates are positive: if J is a reactant, its concentration will decrease and $\Delta[\mathrm{J}]$ will be negative, but $|\Delta[\mathrm{J}]|$ is positive.

Because the rates at which reactants are consumed and products are formed change in the course of a reaction, it is necessary to consider the **instantaneous rate** of the reaction, its rate at a specific instant. The instantaneous rate of consumption of a reactant is the slope of a graph of its molar concentration plotted against the time, with the slope evaluated as the tangent to the graph at the instant of interest (Fig. 10.4) and reported as a positive quantity. The instantaneous rate of formation of a product is also the slope if the tangent to the graph of its molar concentration plotted, and also reported as a positive quantity. The steeper the slope in either case, the greater the rate of the reaction. With the concentration measured in moles per cubic decimetre and the time in seconds, the reaction rate is reported in moles per cubic decimetre litre per second (mol dm^{-3} s^{-1}). From now on, we shall denote the instantaneous rate v (for 'velocity').

A brief comment The slope of the tangent to a plot of [J] against time curve at any instant is expressed mathematically as the magnitude of the derivative, $|d[\mathrm{J}]/dt|$, where the interval Δt (and consequently the change in concentration $\Delta[\mathrm{J}]$) in eqn 10.5a has been allowed to become infinitesimal. Therefore, the precise definition of reaction rate is

$$v = \frac{|d[\mathrm{J}]|}{dt} \qquad (10.5b)$$

The rate defined in eqn 10.5a is the average of this quantity over a range of times.

In general, the various reactants in a given reaction are consumed at different rates, and the various products are also formed at different rates. However, these rates are related by the stoichiometry of the reaction. For example, in the decomposition of urea, $(NH_2)_2CO$, in acidic solution

$$(NH_2)_2CO(aq) + 2\,H_2O(l) \rightarrow 2\,NH_4^+(aq) + CO_3^{2-}(aq)$$

the rate of formation of NH_4^+ is twice the rate of disappearance of $(NH_2)_2CO$, because for 1 mol $(NH_2)_2CO$ consumed, 2 mol NH_4^+ is formed. Once we know the rate of formation or consumption of one substance, we can use the reaction stoichiometry to deduce the rates of formation or consumption of the other participants in the reaction. In this example, for instance,

$$\text{Rate of formation of } NH_4^+ = 2 \times \text{rate of consumption of } (NH_2)_2CO$$

One consequence of this kind of relation is that we have to be careful to specify exactly what species we mean when we report a reaction rate.

A brief comment In terms of the preceding comment, the most sophisticated definition of rate is in terms of the stoichiometric numbers, ν_J, that appear in the chemical equation, the stoichiometric coefficients with sign: positive for products and negative for reactants. Then

$$v = \frac{1}{\nu_J}\frac{d[J]}{dt} \qquad (10.5c)$$

Self-test 10.1

The rate of formation of NH_3 in the reaction $N_2(g) + 3\,H_2(g) \rightarrow 2\,NH_3(g)$ was reported as 1.2 mmol dm^{-3} s^{-1} under a certain set of conditions. What is the rate of consumption of H_2?

[*Answer:* 1.8 mmol dm^{-3} s^{-1}]

There is a complication: if the reactants form a slowly decaying intermediate (we see examples later), then the products do not form at the same rate as the reactants turn into the intermediate. In such cases, we have to be very careful about the interpretation of the measured rate of reaction. This complication can be turned to advantage: the observation that the consumption and formation rates are not related by the reaction stoichiometry is a good sign that a long-lived intermediate is involved in the reaction.

10.4 Rate laws and rate constants

An empirical observation of the greatest importance is that *the rate of reaction is often found to be proportional to the molar concentrations of the reactants raised to a simple power*. For example, it may be found that the rate is directly proportional to the concentrations of the reactants A and B, so

$$v = k_r[A][B] \qquad (10.6)$$

The coefficient k_r, which is characteristic of the reaction being studied, is called the **rate constant** (or *rate coefficient*). The rate constant is independent of the concentrations of the species taking part in the reaction but depends on the temperature. An *experimentally determined* equation of this kind is called the 'rate law' of the reaction. More formally, a **rate law** is an equation that expresses the rate of reaction in terms of the molar concentrations (or partial pressures) of the species in the overall reaction (including, possibly, the products and any catalysts that might be present).

The units of k_r are always such as to convert the product of concentrations into a rate expressed as a change in concentration divided by time. For example, if the rate law is the one shown above, with concentrations expressed in mol dm^3, then the units of k_r will be dm^3 mol^{-1} s^{-1} because

$$\underbrace{\boxed{\text{dm}^3\ \text{mol}^{-1}\ \text{s}^{-1}}}_{k_r} \times \underbrace{\boxed{\text{mol dm}^{-3}}}_{[A]} \times \underbrace{\boxed{\text{mol dm}^{-3}}}_{[B]} = \underbrace{\boxed{\text{mol dm}^{-3}\ \text{s}^{-1}}}_{v}$$

In gas-phase studies, including studies of the processes taking place in the atmosphere, concentrations are commonly expressed in molecules cm^{-3}, so the rate constant for the reaction above would be expressed in cm^3 molecule^{-1} s^{-1}. We can use the approach just developed to determine the units of the rate constant from rate laws of any form. For example, the rate constant for a reaction with rate law of the form $k_r[A]$ is commonly expressed in s^{-1}.

A brief illustration The rate constant for the reaction $O(g) + O_3(g) \rightarrow 2\,O_2(g)$ is 8.0×10^{-15} cm^3 molecule^{-1} s^{-1} at 298 K. To express this rate constant in dm^3 mol^{-1} s^{-1}, we make use of

$$1\text{ cm} = 10^{-2}\text{ m} = 10^{-2} \times 10\text{ dm} = 10^{-1}\text{ dm} = \frac{1\text{ dm}}{10}$$

$$1\text{ mol} = 6.022 \times 10^{23},\text{ so } 1\text{ molecule} = \frac{1\text{ mol}}{6.022 \times 10^{23}}$$

It follows from the procedure described in Example 0.1 that

$$k_r = 8.0 \times 10^{-15}\,\boxed{\text{cm}^3}\,\boxed{\text{molecule}^{-1}}\,\text{s}^{-1}$$

$$= 8.0 \times 10^{-15}\left(\frac{1\text{ dm}}{10}\right)^3\left(\frac{1\text{ mol}}{6.022 \times 10^{23}}\right)^{-1}\text{s}^{-1}$$

$$= \frac{8.0 \times 10^{-15} \times 6.022 \times 10^{23}}{10^3}\text{ dm}^3\text{ mol}^{-1}\text{ s}^{-1}$$

$$= 4.8 \times 10^6\text{ dm}^3\text{ mol}^{-1}\text{ s}^{-1}$$

Note that, as should be expected (but is a good point to check), the rate per mole is much greater than the rate per molecule.

Self-test 10.2

A reaction has a rate law of the form $k_r[A]^2[B]$. What are the units of the rate constant k if the reaction rate was measured in mol dm^{-3} s^{-1}?

[*Answer:* dm^6 mol^{-2} s^{-1}]

Once we know the rate law and the rate constant of the reaction, we can predict the rate of the reaction for any given composition of the reaction mixture. We shall also see that we can use a rate law to predict the concentrations of the reactants and products at any time after the start of the reaction. Furthermore, a rate law is also an important guide to the mechanism of the reaction, for any proposed mechanism must be consistent with the observed rate law.

10.5 Reaction order

A rate law provides a basis for the classification of reactions according to their kinetics. The advantage of having such a classification is that reactions belonging to the same class have similar kinetic behaviour—their rates and the concentrations of the reactants and products vary with composition in a similar way. The classification of reactions is based on their **order**, the power to which the concentration of a species is raised in the rate law. For example, a reaction with the rate law in eqn 10.6 ($v = k_r[A][B]$) is *first-order* in A and first-order in B. A reaction with the rate law

$$v = k_r[A]^2 \qquad (10.7)$$

is *second-order* in A.

The **overall order** of a reaction with a rate law of the form $v = k_r[A]^a[B]^b[C]^c\cdots$ is the sum, $a + b + c + \cdots$, of the orders of all the components. The two rate laws just quoted both correspond to reactions that are *second-order* overall. An example of the first type of reaction is the reformation of a DNA double helix after the double helix has been separated into two strands by raising the temperature or the pH:

Strand + complementary strand → double helix

$v = k_r$[strand][complementary strand]

This reaction is first-order in each strand and second-order overall. An example of the second type is the reduction of nitrogen dioxide by carbon monoxide,

$$NO_2(g) + CO(g) \rightarrow NO(g) + CO_2(g) \qquad v = k_r[NO_2]^2$$

which is second-order in NO_2 and, because no other species occurs in the rate law, second-order overall. The rate of the latter reaction is independent of the concentration of CO provided that some CO is present. This independence of concentration is expressed by saying that the reaction is *zeroth-order* in CO, because a concentration raised to the power zero is 1 ($[CO]^0 = 1$, just as $x^0 = 1$ in algebra).

A reaction need not have an integral order, and many gas-phase reactions do not. For example, if a reaction is found to have the rate law

$$v = k_r[A]^{1/2}[B] \qquad (10.8)$$

then it is *half-order* in A, first-order in B, and three-halves-order overall.

If a rate law is not of the form $v = k_r[A]^a[B]^b[C]^c\ldots$ then the reaction does not have an overall order. Thus, the experimentally determined rate law for the gas-phase reaction $H_2(g) + Br_2(g) \rightarrow 2\,HBr(g)$ is

$$v = \frac{k_{r1}[H_2][Br_2]^{3/2}}{[Br_2] + k_{r2}[HBr]} \qquad (10.9)$$

Although the reaction is first-order in H_2, it has an indefinite order with respect to both Br_2 and HBr and an indefinite order overall. Similarly, a typical rate law for the action of an enzyme E on a substrate S is (see Chapter 11)

$$v = \frac{k_r[E][S]}{[S] + K_M} \qquad (10.10)$$

where K_M is a constant. This rate law is first-order in the enzyme but does not have a specific order with respect to the substrate.

Under certain circumstances a complicated rate law without an overall order may simplify into a law with a definite order. For example, if the substrate concentration in the enzyme-catalysed reaction is so low that $[S] \ll K_M$, then we can ignore [S] in the denominator of eqn 10.10, which simplifies to

$$v = \frac{k_r}{K_M}[S][E]$$

which is first-order in S, first-order in E, and second-order overall.

It is very important to note that *a rate law is established experimentally, and cannot in general be inferred from the chemical equation for the reaction.* The reaction of hydrogen and bromine, for example, has a very simple stoichiometry, but its rate law (eqn 10.9) is very complicated. In some cases, however, the rate law does happen to reflect the reaction stoichiometry. This is the case with the reaction of hydrogen and iodine, which has the same stoichiometry as the reaction of hydrogen with bromine but a much simpler rate law:

$$H_2(g) + I_2(g) \rightarrow 2\,HI(g) \qquad v = k_r[H_2][I_2]$$

10.6 The determination of the rate law

The determination of a rate law is simplified by the **isolation method**, in which all the reactants except one are present in large excess. We can find the dependence of the rate on each of the reactants by isolating each of them in turn—by having all the other substances present in large excess—and piecing together a picture of the overall rate law. For instance, we might use CH_3I in solution at a concentration of 0.2 mol dm^{-3} and an attacking nucleophile at only 0.01 mol dm^{-3}.

If a reactant B is in large excess, for example, it is a good approximation to take its concentration as constant throughout the reaction. Then, although the true rate law might be

$$v = k_r[A][B]^2$$

we can approximate [B] by its initial value $[B]_0$ (from which it hardly changes in the course of the reaction) and write

$$v = k_{r,eff}[A], \text{ with } k_{r,eff} = k_r[B]_0^2$$

Because the true rate law has been forced into first-order form by assuming a constant B concentration, the effective rate law is classified as **pseudofirst-order** and $k_{r,eff}$ is called the **effective rate constant** for a given, fixed concentration of B. If, instead, the concentration of A were in large excess, and hence effectively constant, then the rate law would simplify to

$$v = k_{r,eff}[B]^2, \text{ now with } k_{r,eff} = k_r[A]_0$$

This **pseudosecond-order rate law** is also much easier to analyse and identify than the complete law. Note that the order of the reaction and the form of the effective rate constant change according to whether A or B is in excess.

In a similar manner, a reaction may even appear to be zeroth-order. For instance, the oxidation of ethanol to acetaldehyde by NAD^+ in the liver in the presence of the enzyme liver alcohol dehydrogenase

$$CH_3CH_2OH(aq) + NAD^+(aq) + H_2O(l) \rightarrow CH_3CHO(aq) + NADH(aq) + H_3O^+(aq)$$

is zeroth-order overall as the ethanol is in excess and the concentration of the NAD^+ is maintained at a constant level by normal metabolic processes. Many reactions in aqueous solution that are reported as first- or second-order are actually pseudofirst- or pseudosecond-order: the solvent water participates in the reaction but it is in such large excess that its concentration remains constant.

In the method of **initial rates**, which is often used in conjunction with the isolation method, the instantaneous rate is measured at the beginning of the reaction for several different initial concentrations of reactants. For example, suppose the rate law for a reaction with A isolated is

$$v = k_r[A]^a$$

Then the initial rate of the reaction, v_0, is given by the initial concentration of A:

$$v_0 = k_r[A]_0^a$$

Taking logarithms gives

Use $\log xy = \log x + \log y$

$$\log v_0 = \log(k_{r,eff}[A]_0^a) = \log k_{r,eff} + \log[A]_0^a = \log k_{r,eff} + a\log[A]_0 \qquad (10.11)$$

Use $\log x^a = a \log x$

This equation has the form of the equation for a straight line:

y	=	intercept	+	slope x
$\log v_0$	=	$\log k_{r,eff}$	+	$a \log[A]_0$

It follows that, for a series of initial concentrations, a plot of the logarithms of the initial rates against the logarithms of the initial concentrations of A should be a straight line, and that the slope of the graph will be a, the order of the reaction with respect to the species A (Fig. 10.5).

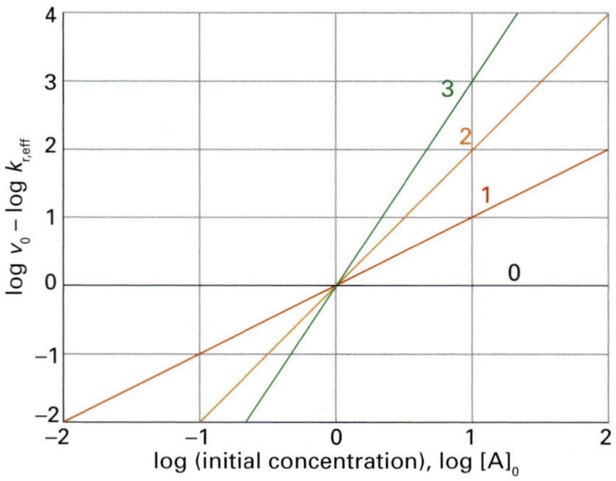

Fig. 10.5 The plot of $\log v_0$ (and, as shown here, of $\log v_0 - \log k_{r,eff}$) against $\log[A]_0$ gives straight lines with slopes equal to the order of the reaction.

Example 10.1

Using the method of initial rates

The recombination of I atoms in the gas phase in the presence of argon (which removes the energy released by the formation of an I–I bond, and so prevents the immediate dissociation of a newly formed I_2 molecule) was investigated and the order of the reaction was determined by the method of initial rates. The initial rates of reaction of 2 I(g) + Ar(g) → I_2(g) + Ar(g) were as follows:

$[I]_0/(10^{-5}$ mol dm$^{-3})$	1.0	2.0	4.0	6.0
$v_0/($mol dm^{-3} s$^{-1})$				
(a)	8.70×10^{-4}	3.48×10^{-3}	1.39×10^{-2}	3.13×10^{-2}
(b)	4.35×10^{-3}	1.74×10^{-2}	6.96×10^{-2}	1.57×10^{-1}
(c)	8.69×10^{-3}	3.47×10^{-2}	1.38×10^{-1}	3.13×10^{-1}

The Ar concentrations are (a) 1.0×10^{-3} mol dm^{-3}, (b) 5.0×10^{-3} mol dm^{-3}, and (c) 1.0×10^{-2} mol dm^{-3}. Find the orders of reaction with respect to I and Ar and the rate constant.

Strategy For constant $[Ar]_0$, the initial rate law has the form $v_0 = k_{r,eff}[I]_0^a$, with $k_{r,eff} = k_r[Ar]_0^b$, so

$$\log v_0 = \log k_{r,eff} + a \log [I]_0$$

We need to make a plot of $\log v_0$ against $\log [I]_0$ for a given $[Ar]_0$ and find the order from the slope and the value of $k_{r,eff}$ from the intercept at $\log [I]_0 = 0$. Then, because

$$\log k_{r,eff} = \log k_r + b \log [Ar]_0$$

plot $\log k_{r,eff}$ against $\log [Ar]_0$ to find $\log k_r$ from the intercept and b from the slope.

Solution The data give the following points for the graph:

$\log([I]_0/$mol dm$^{-3})$	−5.00	−4.70	−4.40	−4.22
$\log(v_0/$mol dm^{-3} s$^{-1})$				
(a)	−2.971	−2.458	−1.857	−1.504
(b)	−2.362	−1.760	−1.157	−0.804
(c)	−1.971	−1.460	−0.860	−0.504

The graph of the data is shown in Fig. 10.6. The slopes of the lines are 2 and the effective rate constants $k_{r,eff}$ are as follows:

$[Ar]_0/($mol dm$^{-3})$	1.0×10^{-3}	5.0×10^{-3}	1.0×10^{-2}
$\log([Ar]_0/$mol dm$^{-3})$	−3.00	−2.30	−2.00
$\log(k_{r,eff}/$mol^{-1} dm^3 s$^{-1})$	6.94	7.64	7.93

Figure 10.7 is the plot of $\log k_{r,eff}$ against $\log [Ar]_0$. The slope is 1, so $b = 1$. The intercept at $\log [Ar]_0 = 0$ is $\log k_{r,eff} = 9.91$, so $k = 8.6 \times 10^9$ mol^{-2} dm^6 s^{-1}. The overall (initial) rate law is

$$v = k_r[I]_0^2[Ar]_0$$

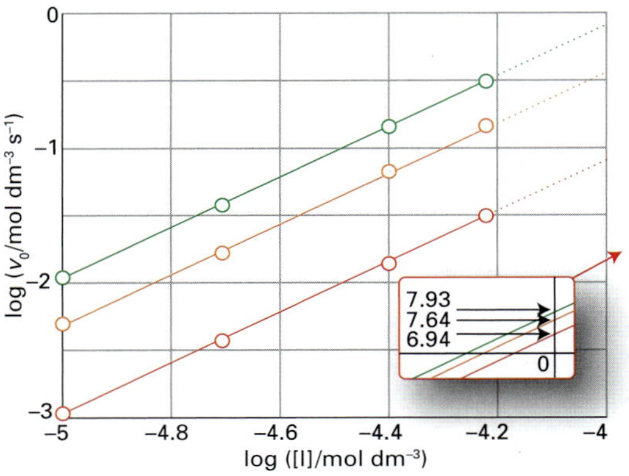

Fig. 10.6 The plots of the data in Example 10.1 for finding the order with respect to I.

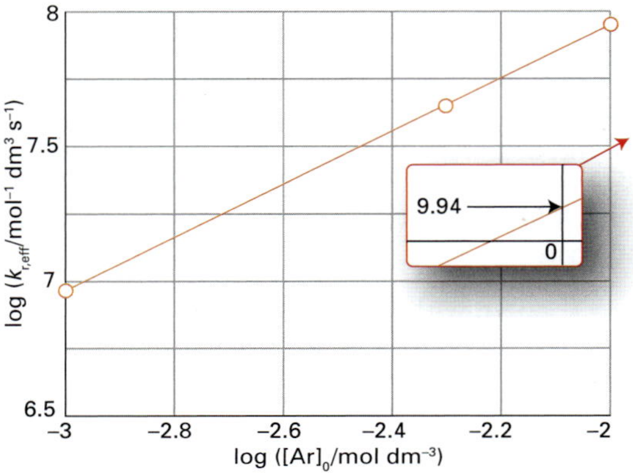

Fig. 10.7 The plots of the data in Example 10.1 for finding the order with respect to Ar.

A note on good practice When taking the common logarithm of a number of the form $x.xx \times 10^n$, there are *four* significant figures in the answer: the figure before the decimal point is simply the power of 10. Strictly, the logarithms are of the quantity divided by its units.

Self-test 10.3

The initial rate of a certain reaction depended on concentration of a substance J as follows:

$[J]_0/(10^{-3}$ mol dm$^{-3})$	5.0	10.2	17	30
$v_0/(10^{-7}$ mol dm^{-3} s$^{-1})$	3.6	9.6	41	130

Find the order of the reaction with respect to J and the rate constant.

[*Answer:* 2, 1.6×10^{-2} mol^{-1} dm^3 s^{-1}]

The method of initial rates might not reveal the entire rate law, as in a complex reaction the products themselves might affect the rate. That is the case for the synthesis of HBr, for eqn 10.4 shows that the rate law depends on the concentration of HBr, none of which is present initially.

10.7 Integrated rate laws

A rate law tells us the rate of the reaction at a given instant (when the reaction mixture has a particular composition). That is rather like being given the speed of a car at each point of its journey. For a car journey, we may want to know the distance that a car has travelled at a certain time given its varying speed. Similarly, for a chemical reaction, we may want to know the composition of the reaction mixture at a given time given the varying rate of the reaction. An **integrated rate law** is an expression that gives the concentration of a species as a function of the time.

Integrated rate laws have two principal uses. One is to predict the concentration of a species at any time after the start of the reaction. Another is to help find the rate constant and order of the reaction. Indeed, although we have introduced rate laws through a discussion of the determination of reaction rates, these rates are rarely measured directly because slopes are so difficult to determine accurately. Almost all experimental work in chemical kinetics deals with integrated rate laws; their great advantage being that they are expressed in terms of the experimental observables of concentration and time. Computers can be used to find the integrated form of even the most complex rate laws numerically and in some cases can be used to obtain closed, algebraic expressions. However, in a number of simple cases solutions can be obtained by elementary techniques and prove to be very useful.

For a chemical reaction and first-order rate law of the form

$$\text{A} \rightarrow \text{products,} \qquad \text{Rate of consumption of A} = k_r[\text{A}] \tag{10.12}$$

we show in Derivation 10.2 that the integrated rate law is

$$\ln \frac{[\text{A}]_0}{[\text{A}]} = k_r t \tag{10.13a}$$

where $[\text{A}]_0$ is the initial concentration of A. Two alternative forms of this expression are

$$\ln[\text{A}] = \ln[\text{A}]_0 - k_r t \tag{10.13b}$$

$$[\text{A}] = [\text{A}]_0 e^{-k_r t} \tag{10.13c}$$

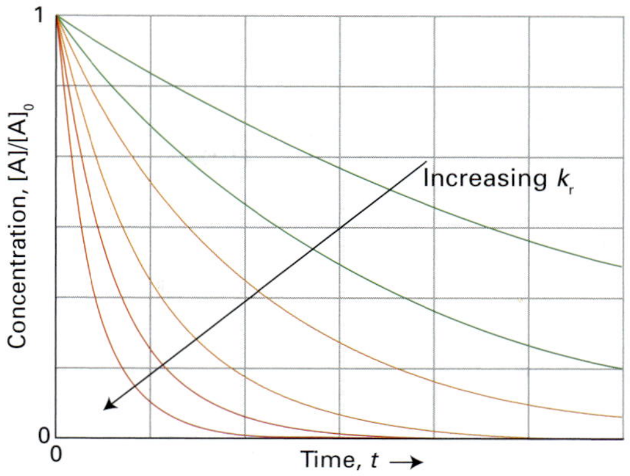

Fig. 10.8 The exponential decay of the reactant in a first-order reaction. The greater the rate constant, the more rapid is the decay.

Equation 10.13c has the form of an **exponential decay** (Fig. 10.8). A common feature of all first-order reactions, therefore, is that *the concentration of the reactant decays exponentially with time.*

Derivation 10.2

First-order integrated rate laws

Our first step is to express the rate of consumption of a reactant A mathematically. As remarked in Comment 10.1, the rate of a reaction is $|d[\text{A}]|/dt$. Because A is a reactant, the change d[A] is negative (the concentration of A decreases with time), so –d[A] is positive. We can therefore interpret the rate as $-d[\text{A}]/dt$. It follows that a first-order rate equation has the form

$$-\frac{d[\text{A}]}{dt} = k_r[\text{A}]$$

This expression is an example of a 'differential equation' (see Appendix 2 for a review of the concepts of calculus). Because the terms d[A] and d*t* may be manipulated like any algebraic quantity, we rearrange the differential equation into

$$\frac{d[\text{A}]}{[\text{A}]} = -k_r dt$$

and then integrate both sides. Integration from $t = 0$, when the concentration of A is $[\text{A}]_0$, to the time of interest, t, when the molar concentration of A is [A], is written as

$$\int_{[\text{A}]_0}^{[\text{A}]} \frac{d[\text{A}]}{[\text{A}]} = -k_r \int_0^t dt$$

We now use the standard integrals

$$\int \frac{dx}{x} = \ln x + \text{constant} \qquad \int dx = x + \text{constant}$$

and obtain the expression

$$\ln[\text{A}] - \ln[\text{A}]_0 = -k_r t$$

which rearranges into eqn 10.13a.

Table 10.2

Kinetic data for first-order reactions

Reaction	Phase	θ/°C	k_r/s^{-1}	$t_{1/2}$
$2\,N_2O_5 \rightarrow 4\,NO_2 + \mathbf{O_2}$	g	25	3.38×10^{-5}	2.85 h
$2\,N_2O_5 \rightarrow 4\,NO_2 + \mathbf{O_2}$	Br_2(l)	25	4.27×10^{-5}	2.25 h
$\mathbf{C_2H_6} \rightarrow 2\,CH_3$	g	700	5.46×10^{-4}	21.2 m
Cyclopropane → propene	g	500	6.17×10^{-4}	17.2 min

The rate constant is for the rate of formation or consumption of the species in bold type. The rate laws for the other species may be obtained from the reaction stoichiometry.

Equation 10.13c lets us predict the concentration of A at any time after the start of the reaction. Equation 10.13b shows that if we plot ln[A] against t, then we will get a straight line if the reaction is first-order. If the experimental data do not give a straight line when plotted in this way, then the reaction is not first-order. If the line is straight, then it follows from eqn 10.13b that its slope is $-k_r$, so we can also determine the rate constant from the graph. Some rate constants determined in this way are given in Table 10.2.

Example 10.2

Analysing a first-order reaction

The variation in the partial pressure p_A of azomethane with time was followed at 460 K, with the results given below. Confirm that the decomposition $CH_3N_2CH_3(g) \rightarrow CH_3CH_3(g) + N_2(g)$ is first-order in $CH_3N_2CH_3$, and find the rate constant at this temperature.

t/s	0	1000	2000	3000	4000
$p_A/(10^{-2}$ Torr)	10.20	5.72	3.99	2.78	1.94

Strategy The easiest procedure is to plot ln(p/Torr) against t/s and expect to obtain a straight line. If the graph is straight, then the slope is $-k_r$. To determine the slope use mathematical software, which automatically determines the best straight line through the data points and calculates the slope. (The text's website features interactive applets for data analysis.)

Solution We draw up the following table:

t/s	0	1000	2000	3000	4000
$p_A/(10^{-2}$ Torr)	10.20	5.72	3.99	2.78	1.94
ln(p_A/Torr)	−2.28	−2.86	−3.22	−3.58	−3.94

The graph of the data is shown in Fig. 10.9. The plot is straight, confirming a first-order reaction. Its least-squares best-fit slope is -4.04×10^{-4}, so $k_r = 4.04 \times 10^{-4}\ s^{-1}$.

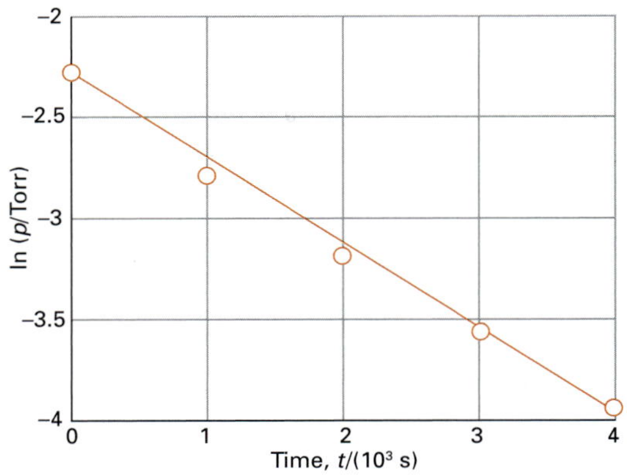

Fig. 10.9 The determination of the rate constant of a first-order reaction. A straight line is obtained when ln[A] (or ln p, where p is the partial pressure of the species of interest) is plotted against t; the slope is $-k_r$. The data are from Example 10.2.

Self-test 10.4

The concentration of N_2O_5 in liquid bromine varied with time as follows:

t/s	0	200	400	600	1000
$[N_2O_5]$/(mol dm^{-3})	0.110	0.073	0.048	0.032	0.014

Confirm that the reaction is first-order in N_2O_5 and determine the rate constant.

[*Answer:* $2.1 \times 10^{-3}\ s^{-1}$]

Now we need to see how the concentration varies with time for a reaction and second-order rate law if the form

$$A \rightarrow \text{products}, \quad \text{Rate of consumption of A} = k_r[A]^2 \qquad (10.14)$$

As before, we suppose that the concentration of A at $t = 0$ is $[A]_0$ and, as shown in Derivation 10.3, find that

$$\frac{1}{[A]} = \frac{1}{[A]_0} + k_r t \tag{10.15a}$$

that may also be written

$$[A] = \frac{[A]_0}{1 + k_r t[A]_0} \tag{10.15b}$$

Derivation 10.3

Second-order integrated rate laws

As before, the rate of consumption of the reactant A is $-d[A]/dt$, so the differential equation for the rate law is

$$-\frac{d[A]}{dt} = k_r[A]^2$$

To solve this equation, we rearrange it into

$$\frac{d[A]}{[A]^2} = -k_r dt$$

and integrate it between $t = 0$, when the concentration of A is $[A]_0$, and the time of interest t, when the concentration of A is $[A]$:

$$\int_{[A]_0}^{[A]} \frac{d[A]}{[A]^2} = -k_r \int_0^t dt$$

The term on the right is $-k_r t$. We evaluate the integral on the left by using the standard form

$$\int \frac{dx}{x^2} = -\frac{1}{x} + \text{constant}$$

which implies that

$$\int_a^b \frac{dx}{x^2} = \left\{-\frac{1}{x} + \text{constant}\right\}\Bigg|_b - \left\{-\frac{1}{x} + \text{constant}\right\}\Bigg|_a$$

$$= -\frac{1}{b} + \frac{1}{a}$$

and so obtain eqn 10.15a.

Equation 10.15a shows that to test for a second-order reaction we should plot $1/[A]$ against t and expect a straight line. If the line is straight, then the reaction is second-order in A and the slope of the line is equal to the rate constant (Fig. 10.10). Some rate constants determined in this way are given in Table 10.3. Equation 10.15b enables us to predict the concentration of A at any time after the start of the reaction (Fig. 10.11).

From plots of [A] against t, we see that the concentration of A approaches zero more slowly in a second-order reaction than in a first-order reaction with the same initial rate (Fig. 10.12). That is, reactants that decay by a second-order process die away more slowly at low concentrations than would be expected if the decay was first-order. A point of interest in this connection is that pollutants commonly disappear by second-order processes, so it takes a very long time for them to decline to acceptable levels.

Table 10.4 summarizes the integrated rate laws for a variety of simple reaction types.

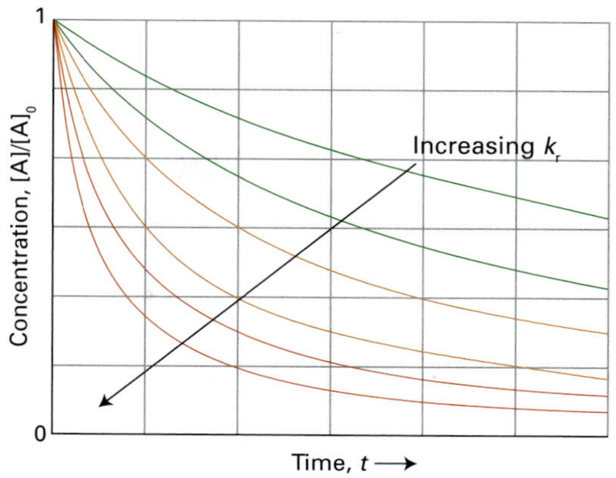

Fig. 10.10 The variation with time of the concentration of a reactant in a second-order reaction.

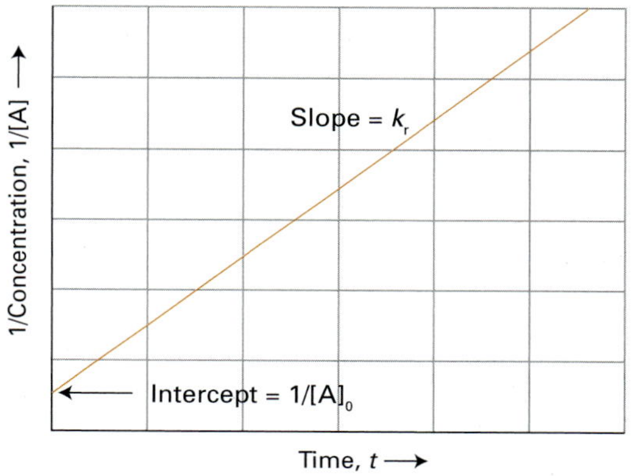

Fig. 10.11 The determination of the rate constant of a second-order reaction. A straight line is obtained when $1/[A]$ (or $1/p$, where p is the partial pressure of the species of interest) is plotted against t; the slope is k_r.

10.8 Half-lives and time constants

A useful indication of the rate of a first-order chemical reaction is the **half-life,** $t_{1/2}$, of a reactant, which is the time it takes for the concentration of the species to fall to half its initial value. We can find the half-life of a species A that decays in a first-order reaction

Table 10.3

Kinetic data for second-order reactions

Reaction	Phase	θ/°C	k_r/(dm^3 mol^{-1} s^{-1})
$2\,NOBr \rightarrow 2\,NO + \mathbf{Br_2}$	g	10	0.80
$2\,NO_2 \rightarrow 2\,NO + \mathbf{O_2}$	g	300	0.54
$\mathbf{H_2} + I_2 \rightarrow 2\,HI$	g	400	2.42×10^{-2}
$\mathbf{D_2} + HCl \rightarrow DH + DCl$	g	600	0.141
$2\,I \rightarrow \mathbf{I_2}$	g	23	7×10^{9}
	hexane	50	1.8×10^{10}
$\mathbf{CH_3Cl} + CH_3O^-$	$CH_3OH(l)$	20	2.29×10^{-6}
$\mathbf{CH_3Br} + CH_3O^-$	$CH_3OH(l)$	20	9.23×10^{-6}
$\mathbf{H^+} + OH^- \rightarrow H_2O$	water	25	1.5×10^{11}

The rate constant is for the rate of formation or consumption of the species in bold type. The rate laws for the other species may be obtained from the reaction stoichiometry.

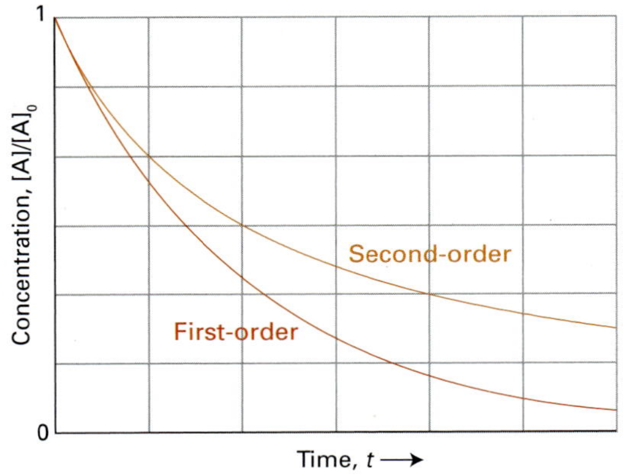

Fig. 10.12 Although the initial decay of a second-order reaction may be rapid, later the concentration approaches zero more slowly than in a first-order reaction with the same initial rate (compare Fig. 10.10).

(eqn 10.14) by substituting $[A] = \frac{1}{2}[A]_0$ and $t = t_{1/2}$ into eqn 10.15a:

Use ln(1/x) = −ln x

$$k_r t_{1/2} = -\ln \frac{\frac{1}{2}[A]_0}{[A]_0} = -\ln \tfrac{1}{2} = \ln 2$$

It follows that

$$t_{1/2} = \frac{\ln 2}{k_r} \qquad (10.16)$$

A brief illustration Because the rate constant for the first-order reaction

$$N_2O_5(g) \rightarrow 2\,NO_2(g) + \tfrac{1}{2}\,O_2(g)$$

$$\text{Rate of consumption of } N_2O_5 = k_r[N_2O_5]$$

is equal to 6.76×10^{-5} s^{-1} at 25°C, the half-life of N_2O_5 is 2.85 h. Hence, the concentration of N_2O_5 falls to half its

Table 10.4

Integrated rate laws

Order	Reaction type	Rate law	Integrated rate law
0	$A \rightarrow P$	$v = k_r$	$[P] = k_r t$ for $k_r t \leq [A]_0$
1	$A \rightarrow P$	$v = k_r[A]$	$[P] = [A]_0(1 - e^{-k_r t})$
2	$A \rightarrow P$	$v = k_r[A]^2$	$[P] = \dfrac{k_r t[A]_0^2}{1 + k_r t[A]_0}$
	$A + B \rightarrow P$	$v = k_r[A][B]$	$[P] = \dfrac{[A]_0[B]_0(1 - e^{([B]_0-[A]_0)k_r t})}{[A]_0 - [B]_0 e^{([B]_0-[A]_0)k_r t}}$

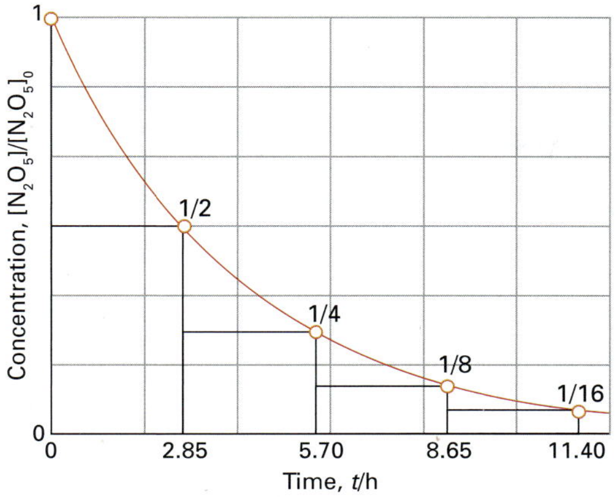

Fig. 10.13 The molar concentration of N_2O_5 after a succession of half-lives.

initial value in 2.85 h, and then to half that concentration again in a further 2.85 h, and so on (Fig. 10.13). This procedure is commonly used in reverse: the half-life is measured and then eqn 10.16 is used to determine k_r from $k_r = (\ln 2)/t_{1/2}$. ■

The main point to note about eqn 10.16 is that *for a first-order reaction, the half-life of a reactant is independent of its concentration.* It follows that if the concentration of A at some arbitrary stage of the reaction is [A], then the concentration will fall to $\frac{1}{2}$[A] after an interval of $(\ln 2)/k_r$ whatever the actual value of [A] (Fig. 10.14). Some half-lives are given in Table 10.2.

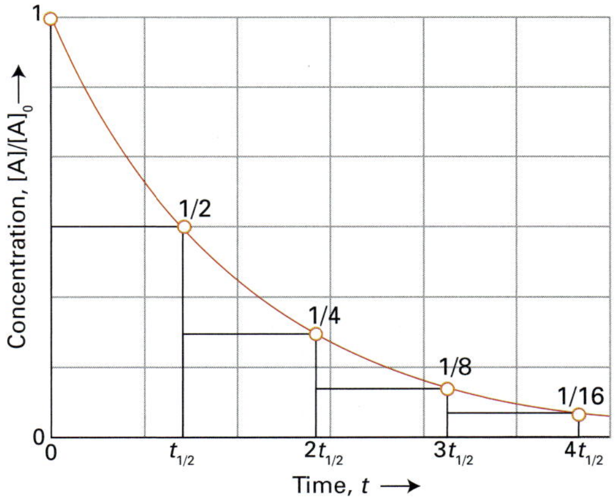

Fig. 10.14 In each successive period of duration $t_{1/2}$, the concentration of a reactant in a first-order reaction decays to half its value at the start of that period. After n such periods, the concentration is $(\frac{1}{2})^n$ of its initial concentration.

● **A brief illustration** In acidic solution, the disaccharide sucrose (cane sugar) is converted to a mixture of the monosaccharides glucose and fructose in a pseudofirst-order reaction. Under certain conditions of pH, the half-life of sucrose is 28.4 min. To calculate how long it takes for the concentration of a sample to fall from 8.0 mmol dm^{-3} to 1.0 mmol dm^{-3} we note that

Molar concentration/(mmol dm^{-3}):

$$8.0 \xrightarrow{28.4\ \text{min}} 4.0 \xrightarrow{28.4\ \text{min}} 2.0 \xrightarrow{28.4\ \text{min}} 1.0$$

The total time required is 3 × 28.4 min = 85.2 min. ■

Self-test 10.5

The half-life of a substrate in a certain enzyme-catalysed first-order reaction is 138 s. How long is required for the concentration of substrate to fall from 1.28 mmol dm^{-3} to 0.040 mmol dm^{-3}?

[*Answer:* 690 s]

Self-test 10.6

Derive an expression for the half-life of a second-order reaction in terms of the rate constant k.

[*Answer:* $t_{1/2} = 1/k_r[A]_0$]

In contrast to first-order reactions, the half-life of a second-order reaction does depend on the concentration of the reactant (see the answer to Self-test 10.6), and lengthens as the concentration of reactant falls. It is therefore not characteristic of the reaction itself, and for that reason is rarely used.

We can use the half-life of a substance to recognize first-order reactions. All we need do is inspect a set of data of composition against time. If we see that the initial concentration falls to half its value in a certain time, and that another concentration falls to half its value in the same time, then we can infer that the reaction is first-order. The first-order character can then be confirmed by plotting ln[A] against t and obtaining a straight line, as indicated earlier.

Another indication of the rate of a first-order reaction is the **time constant**, τ, the time required for the concentration of a reactant to fall to 1/e of its initial value. From eqn 10.15a it follows that

Set $[A] = [A]_0/e$

$$k_r\tau = -\ln\left(\frac{[A]_0/e}{[A]_0}\right) = -\ln\frac{1}{e} = \ln e = 1$$

Hence, the time constant is the reciprocal of the rate constant:

$$\tau = \frac{1}{k_r} \qquad (10.17)$$

The temperature dependence of reaction rates

The rates of most chemical reactions increase as the temperature is raised. Many organic reactions in solution lie somewhere in the range spanned by the hydrolysis of methyl ethanoate (for which the rate constant at 35°C is 1.8 times that at 25°C) and the hydrolysis of sucrose (for which the factor is 4.1). Enzyme-catalysed reactions may show a more complex temperature dependence because raising the temperature may provoke conformational changes that lower the effectiveness of the enzyme. Indeed, one of the reasons why we fight infection with a fever is to upset the balance of reaction rates in the infecting organism, and hence destroy it, by the increase in temperature. There is a fine line, though, between killing an invader and killing the invaded!

10.9 The Arrhenius parameters

As data on reaction rates were accumulated towards the end of the nineteenth century, the Swedish chemist Svante Arrhenius noted that almost all of them showed a similar dependence on the temperature. In particular, he noted that a graph of ln k_r, where k_r is the rate constant for the reaction, against $1/T$, where T is the (absolute) temperature at which k_r is measured, gives a straight line with a slope that is characteristic of the reaction (Fig. 10.15). The

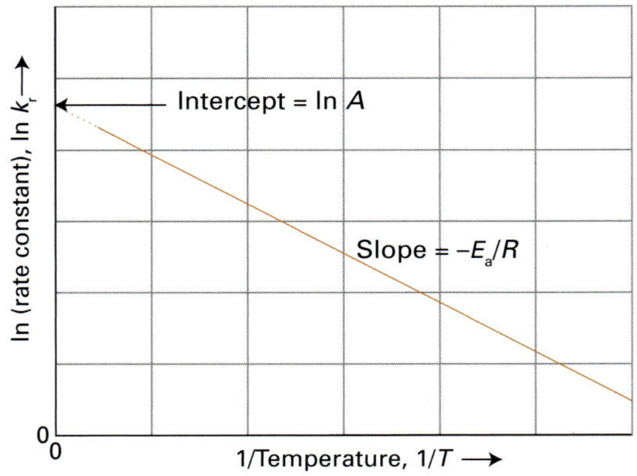

Fig. 10.15 The general form of an Arrhenius plot of ln k_r against $1/T$. The slope is equal to $-E_a/R$ and the intercept at $1/T = 0$ is equal to ln A.

Table 10.5
Arrhenius parameters

First-order reactions	A/s^{-1}	$E_a/(kJ\ mol^{-1})$
Cyclopropene → propane	1.58×10^{15}	272
$CH_3NC \rightarrow CH_3CN$	3.98×10^{13}	160
cis-CHD=CHD → *trans*-CHD=CHD	3.16×10^{12}	256
cyclobutane → $2\ C_2H_4$	3.98×10^{15}	261
$2\ N_2O_5 \rightarrow 4\ NO_2 + O_2$	4.94×10^{13}	103
$N_2O \rightarrow N_2 + O$	7.94×10^{11}	250
Second-order, gas phase	$A/(dm^3\ mol^{-1}\ s^{-1})$	$E_a/(kJ\ mol^{-1})$
$O + N_2 \rightarrow NO + H$	1×10^{11}	315
$OH + H_2 \rightarrow H_2 + H$	8×10^{10}	42
$Cl + H_2 \rightarrow HCl + H$	8×10^{10}	23
$CH_3 + CH_3 \rightarrow C_2H_6$	2×10^{10}	0
$NO + Cl_2 \rightarrow NOCl + Cl$	4×10^{9}	85
Second order, solution	$A/(dm^3\ mol^{-1}\ s^{-1})$	$E_a/(kJ\ mol^{-1})$
$NaC_2H_5O + CH_3I$ in ethanol	2.42×10^{11}	81.6
$C_2H_5Br + OH^-$ in water	4.30×10^{11}	89.5
$CH_3I + S_2O_3^{2-}$ in water	2.19×10^{12}	78.7
Sucrose + H_2O in acidic water	1.50×10^{15}	107.9

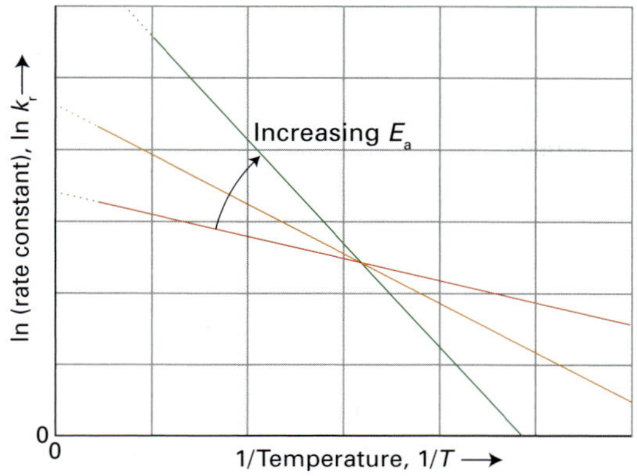

Fig. 10.16 These three Arrhenius plots correspond to three different activation energies. Note that the plot corresponding to the higher activation energy indicates that the rate of that reaction is more sensitive to temperature.

mathematical expression of this conclusion is that the rate constant varies with temperature as

$$\ln k_r = \text{intercept} + \text{slope} \times \frac{1}{T}$$

This expression is normally written as the **Arrhenius equation**

$$\ln k_r = \ln A - \frac{E_a}{RT} \qquad (10.18)$$

or alternatively (by using $\ln k_r - \ln A = \ln(k_r/A)$ and taking antilogarithms, e^x, of both sides) as

$$k_r = Ae^{-E_a/RT} \qquad (10.19)$$

The parameter A (which has the same units as k_r) is called the **pre-exponential factor**, and E_a (which is a molar energy and normally expressed as kilojoules per mole) is called the **activation energy**. Collectively, A and E_a are called the **Arrhenius parameters** of the reaction (Table 10.5).

A practical point to note from eqn 10.19 and illustrated in Fig. 10.16 is that a high activation energy corresponds to a reaction rate that is very sensitive to temperature (the Arrhenius plot has a steep slope). Conversely, a small activation energy indicates a reaction rate that varies only slightly with temperature (the slope is shallow). A reaction with zero activation energy, such as for some radical recombination reactions in the gas phase, has a rate that is largely independent of temperature.

Example 10.3
Determining the Arrhenius parameters

The rate of the second-order decomposition of acetaldehyde (ethanal, CH_3CHO) was measured over the range 700–1000 K, and the rate constants that were found are reported below. Determine the activation energy and the pre-exponential factor.

T/K	700	730	760	790
k_r/(mol^{-1} dm^3 s^{-1})	0.011	0.035	0.105	0.343
T/K	810	840	910	1000
k_r/(mol^{-1} dm^3 s^{-1})	0.789	2.17	20.0	145

Strategy We plot ln k against $1/T$ and expect a straight line. The slope is $-E_a/R$ and the intercept of the extrapolation to $1/T = 0$ is ln A. More precisely, we plot the dimensionless quantity $\ln(k_r/(\text{mol}^{-1}\ \text{dm}^3\ \text{s}^{-1}))$ against the dimensionless quantity T/K. Then from

$$\ln(k_r/\text{mol}^{-1}\ \text{dm}^3\ \text{s}^{-1}) = \ln(A/\text{mol}^{-1}\ \text{dm}^3\ \text{s}^{-1}) - \frac{E_a}{RT}$$

$$= \ln(A/\text{mol}^{-1}\ \text{dm}^3\ \text{s}^{-1}) - \frac{E_a}{R} \times \frac{1}{\text{K}} \times \frac{\text{K}}{T}$$

identify the dimensionless slope with $-(E_a/R)\ \text{K}^{-1}$, implying that $E_a = -R \times \text{slope} \times \text{K}$. It is best to use mathematical software to do a least-squares fit of the data to a straight line. Note (from eqn 10.19) that A has the same units as k_r.

Solution The Arrhenius plot is shown in Fig. 10.17. The least-squares best fit of the line has slope -2.265×10^4 and intercept at $1/T = 0$ (which is well off the graph) 27.7. Therefore,

$$E_a = -R \times \text{slope} \times \text{K}$$

$$= -(8.3145\ \text{J K}^{-1}\ \text{mol}^{-1}) \times (-2.265 \times 10^4\ \text{K}) = 188\ \text{kJ mol}^{-1}$$

and from $\ln(A/(\text{mol}^{-1}\ \text{dm}^3\ \text{s}^{-1})) = 27.7$,

$$A = e^{27.7}\ \text{mol}^{-1}\ \text{dm}^3\ \text{s}^{-1} = 1.1 \times 10^{12}\ \text{mol}^{-1}\ \text{dm}^3\ \text{s}^{-1}$$

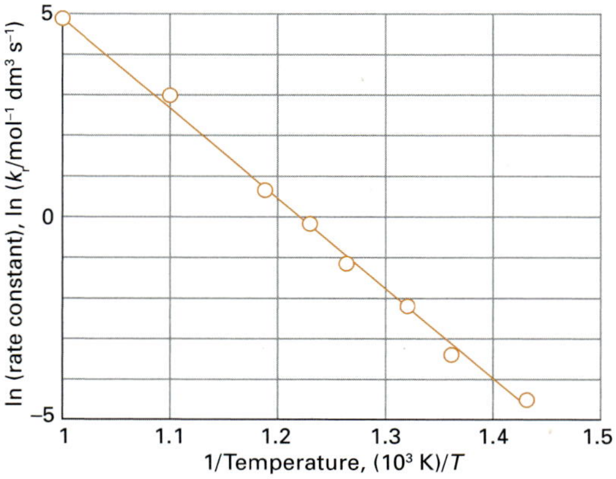

Fig. 10.17 The Arrhenius plot for the decomposition of CH_3CHO, and the best (least-squares) straight line fitted to the data points. The data are from Example 10.3.

Self-test 10.7

Determine A and E_a from the following data:

T/K	300	350	400
k_r/(mol^{-1} dm^3 s^{-1})	7.9×10^6	3.0×10^7	7.9×10^7
T/K	450	500	
k_r/(mol^{-1} dm^3 s^{-1})	1.7×10^8	3.2×10^8	

[*Answer:* 8×10^{10} mol^{-1} dm^3 s^{-1}, 23 kJ mol^{-1}]

Once the activation energy of a reaction is known, it is a simple matter to predict the value of a rate constant $k_r(T')$ at a temperature T' from its value $k_r(T)$ at another temperature T. To do so, we write

$$\ln k_r(T') = \ln A - \frac{E_a}{RT'} \quad \text{and} \quad \ln k_r(T) = \ln A - \frac{E_a}{RT}$$

and then subtract the second from the first to obtain

$$\ln k_r(T') - \ln k_r(T) = -\frac{E_a}{RT'} + \frac{E_a}{RT}$$

We can rearrange this expression to

$$\ln \frac{k_r(T')}{k_r(T)} = \frac{E_a}{R}\left(\frac{1}{T} - \frac{1}{T'}\right) \qquad (10.20)$$

A brief illustration For a reaction with an activation energy of 50 kJ mol^{-1}, an increase in the temperature from 25°C to 37°C (body temperature) corresponds to

$$\ln\frac{k_r(310\text{ K})}{k_r(298\text{ K})} = \frac{50\times10^3\text{ J mol}^{-1}}{8.3145\text{ J K}^{-1}\text{ mol}^{-1}}\left(\frac{1}{298\text{ K}} - \frac{1}{310\text{ K}}\right)$$
$$= \frac{50\times10^3}{8.3145}\left(\frac{1}{298} - \frac{1}{310}\right)$$

(The right-hand side evaluates to 0.7812… , but we take the next step before evaluating the answer.) By taking natural antilogarithms (that is, by forming e^x), $k_r(310\text{ K}) = 2.18k_r(298\text{ K})$. This result corresponds to slightly more than a doubling of the rate constant.

Self-test 10.8

The activation energy of one of the reactions in a biochemical process is 87 kJ mol^{-1}. What is the change in rate constant when the temperature falls from 37°C to 15°C?

[*Answer:* $k_r(15°\text{C}) = 0.076k_r(37°\text{C})$]

10.10 Collision theory

We can understand the origin of the Arrhenius parameters most simply by considering a class of gas-phase reactions in which reaction occurs when two molecules meet. That is, in the terminology to be introduced in Section 11.3, we are considering bimolecular gas-phase reactions. In this **collision theory** of reaction rates it is supposed that reaction occurs only if two molecules collide with a certain minimum kinetic energy along their line of approach (Fig. 10.18). In collision theory, a reaction resembles the collision of two defective billiard balls: the balls bounce apart if they collide with only a small energy, but might smash each other into fragments (products) if they collide with more than a certain minimum kinetic energy. This model of a reaction is a reasonable first approximation to the types of process that take place in planetary atmospheres and govern their compositions and temperature profiles.

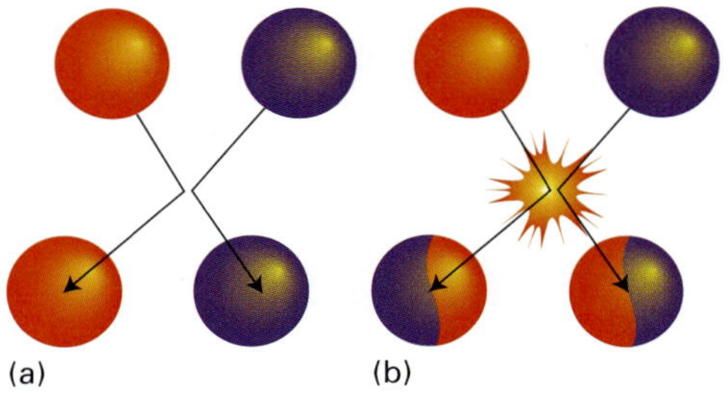

Fig. 10.18 In the collision theory of gas-phase chemical reactions, reaction occurs when two molecules collide, but only if the collision is sufficiently vigorous. (a) An insufficiently vigorous collision: the reactant molecules collide but bounce apart unchanged. (b) A sufficiently vigorous collision results in a reaction.

A **reaction profile** in collision theory is a graph showing the variation in potential energy as one reactant molecule approaches another and the products then separate (Fig. 10.19). On the left, the horizontal line represents the potential energy of the two reactant molecules that are far apart from one another. The potential energy rises from this value only when the separation of the molecules is so small that they are in contact, when it rises as bonds bend and start to break. The potential energy reaches a peak when the two molecules are highly distorted. Then it starts to decrease as new bonds are formed. At separations to the right of the maximum, the potential energy rapidly falls to a low value as the product molecules separate. For the reaction to be successful, the reactant molecules must approach with sufficient kinetic energy along their line of approach to carry them over the **activation barrier**, the peak in the reaction profile. As we shall see, we can identify the height of the activation barrier with the activation energy of the reaction.

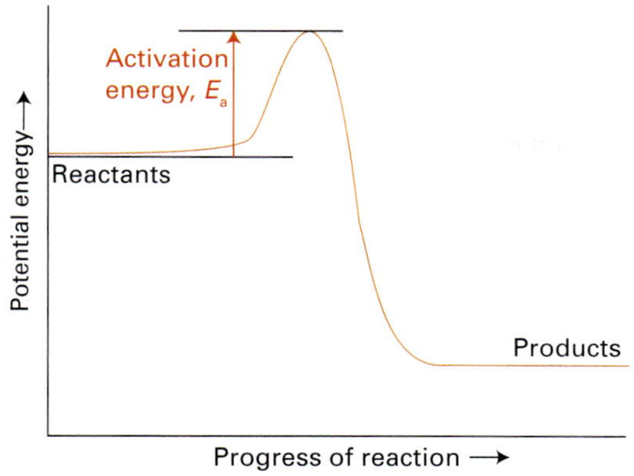

Fig. 10.19 A reaction profile. The graph depicts schematically the changing potential energy of two species that approach, collide, and then go on to form products. The activation energy is the height of the barrier above the potential energy of the reactants.

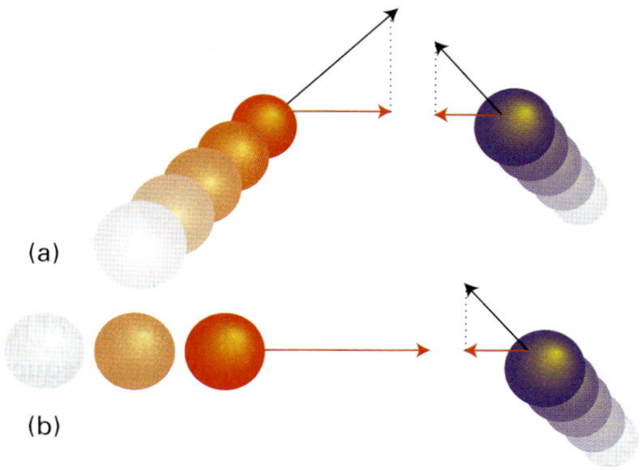

Fig. 10.20 The criterion for a successful collision is that the two reactant species should collide with a kinetic energy along their line of approach that exceeds a certain minimum value E_a that is characteristic of the reaction. The two molecules might also have components of velocity (and an associated kinetic energy) in directions other than those shown here (for example, the two molecules depicted here as (a) and (b) might be moving up the page as well as towards each other); but only the energy associated with their mutual approach can be used to overcome the activation energy.

With the reaction profile in mind, it is quite easy to establish that collision theory accounts for Arrhenius behaviour. Thus, **collision frequency**, the rate of collisions between species A and B, is proportional to both their concentrations: if the concentration of B is doubled, then the rate at which A molecules collide with B molecules is doubled, and if the concentration of A is doubled, then the rate at which B molecules collide with A molecules is also doubled. It follows that the collision frequency of A and B molecules is directly proportional to the concentrations of A and B, and we can write

Collision frequency $\propto$ [A][B]

Next, we need to multiply the collision frequency by a factor f that represents the fraction of collisions that occur with at least a kinetic energy E_a along the line of approach (Fig. 10.20), for only these collisions will lead to the formation of products. Molecules that approach with less than a kinetic energy E_a will behave like a ball that rolls toward the activation barrier, fails to surmount it, and rolls back. We saw in Section 1.6 that only small fractions of molecules in the gas phase have very high speeds and that the fraction with very high speeds increases sharply as the temperature is raised. Because the kinetic energy increases as the square of the speed (for a body of mass m moving at a speed v, the kinetic energy is $E_k = \frac{1}{2}mv^2$), we expect that, at higher temperatures, a larger fraction of molecules will have a speed and kinetic energy that exceed the minimum values required for collisions that lead to formation of products (Fig. 10.21). The fraction of collisions

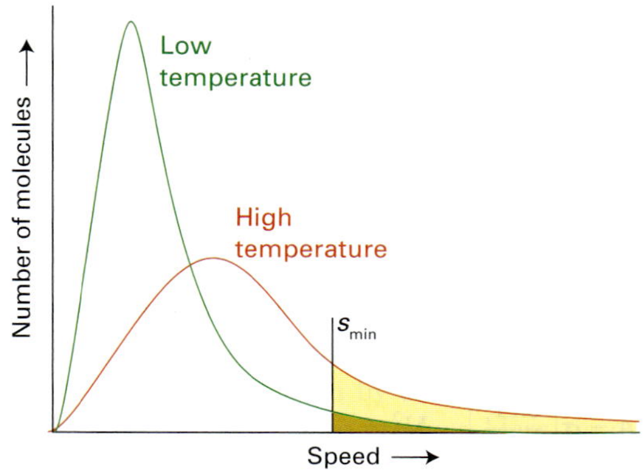

Fig. 10.21 According to the Maxwell distribution of speeds (Section 1.6), as the temperature increases, so does the fraction of gas-phase molecules with a speed that exceeds a minimum value s_{min}. Because the kinetic energy is proportional to the square of the speed, it follows that more molecules can collide with a minimum kinetic energy E_a (the activation energy) at higher temperatures.

that occur with at least a kinetic energy E_a can be calculated from general arguments developed in Chapter 21 concerning the probability that a molecule has a specified energy. The result is

$$f = e^{-E_a/RT} \tag{10.21}$$

This fraction increases from 0 when $T = 0$ to 1 when T is infinite.

Box 10.1 Femtochemistry

Until recently, activated complexes were not observed directly, as they have a very fleeting existence and often survive for only a few picoseconds. However, the development of femtosecond pulsed lasers (1 fs = 10^{-15} s) and their application to chemistry in the form of *femtochemistry* has made it possible to make observations on species that have such short lifetimes that in a number of respects they resemble activated complexes. Further developments have even brought attosecond investigations (1 as = 10^{-18} s) within reach.

In a typical experiment, energy from a femtosecond pulse is used to dissociate a molecule, and then a second femtosecond pulse is fired at an interval after the pulse. The frequency of the second pulse is set at an absorption of one of the free fragmentation products, so its absorption is a measure of the abundance of the dissociation product. For example, when ICN is dissociated by the first pulse, the emergence of CN can be monitored by watching the growth of the free CN absorption. In this way it has been found that the CN signal remains zero until the fragments have separated by about 600 pm, which takes about 205 fs.

Some sense of the progress that has been made in the study of the intimate mechanism of chemical reactions can be obtained by considering the decay of the ion pair Na^+I^-. Absorption of energy from the femtosecond laser by the ionic species leads to redistribution of electrons and forms a state that corresponds to a covalently bonded NaI molecule. The probe pulse examines the system at an absorption frequency either of the free Na atom or at a frequency at which the atom absorbs when it is a part of the complex. The latter frequency depends on the Na–I distance, so an absorption is obtained each time the vibration of the complex returns it to that separation.

A typical set of results is shown in the illustration. The bound Na absorption intensity shows up as a series of pulses that recur in about 1 ps, showing that the complex vibrates with about that period. The decline in intensity shows the rate at which the complex can dissociate as the two atoms swing away from each other. The free Na absorption also grows in an oscillating manner, showing the periodicity of the vibration of the complex, each swing of which gives it a chance to dissociate. The precise period of the oscillation in NaI is 1.25 ps. The complex survives for about ten oscillations. In contrast, although the oscillation frequency of NaBr is similar, it barely survives one oscillation.

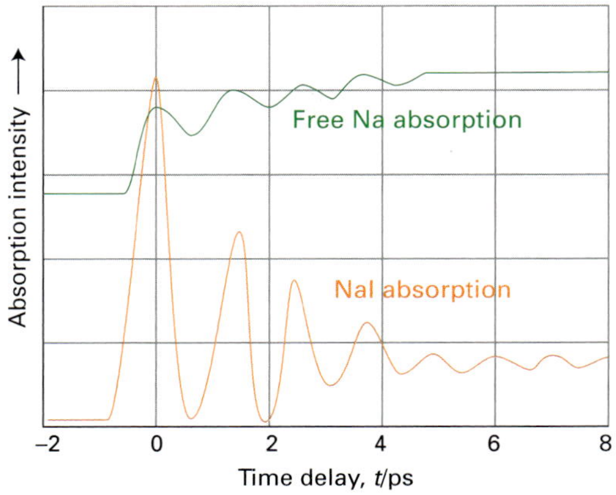

The absorption spectra of the species NaI and Na immediately after a femtosecond flash. The oscillations show how the species incipiently form then reform their precursors before finally forming products. (Adapted from A. H. Zewail, *Science*, **242**, 1645 (1988).)

Femtochemistry techniques have also been used to examine analogues of the activated complex involved in bimolecular reactions. As an example, consider the weakly bound complex (also called a 'van der Waals molecule'), IH···OCO. The HI bond can be dissociated by a femtosecond pulse, and the H atom is ejected towards the O atom of the neighbouring CO_2 molecule to form HOCO. Hence, the complex is a source of a species that resembles the activated complex of the reaction

$$H + CO_2 \rightarrow [HOCO]^{\ddagger} \rightarrow HO + CO$$

The probe pulse is tuned to the OH radical, which enables the evolution of $[HOCO]^{\ddagger}$ to be studied in real time. Femtosecond techniques have also been used to study more complex reactions, such as the Diels–Alder reaction, nucleophilic substitution reactions, and pericyclic addition and cleavage reactions. Biological processes that are open to study by femtochemistry include the photostimulated processes of vision (Box 20.1) and the energy-converting processes of photosynthesis (Box 20.2). In other experiments, the photoejection of carbon monoxide from myoglobin and the attachment of O_2 to the exposed site have been studied to obtain rate constants for the two processes.

between A, B, and $C^{\ddagger}$ and the rate of successful passage of $C^{\ddagger}$ through the transition state, it is possible to derive the **Eyring equation** for the rate constant:

$$k_r = \kappa \times \frac{kT}{h} \times K^{\ddagger} \qquad (10.23)$$

where $k = R/N_A = 1.381 \times 10^{-23}$ J K^{-1} is Boltzmann's constant and $h = 6.626 \times 10^{-34}$ J s is Planck's constant (which we meet in Chapter 12). The factor κ is the **transmission coefficient**, which takes into account the fact that the activated complex does not always pass through to the transition state. In the absence of information to the contrary, κ is assumed to be about 1.

A note on good practice Be very careful to distinguish the Boltzmann constant k from the symbol for a rate constant, k_r. In some expositions, you will see Boltzmann's constant denoted k_B to emphasize its significance (and sometimes, confusingly, the rate constant denoted k).

The term kT/h in eqn 10.23 (which has the dimensions of a frequency, as kT is an energy, and division by Planck's constant turns an energy into a frequency; with kT in joules, kT/h has the units s^{-1}) arises from consideration of the motions of atoms that lead to the decay of $C^{\ddagger}$ into products, as specific bonds are broken and formed. It follows that one way in which an increase in temperature enhances the rate is by causing more vigorous motion in the activated complex, facilitating the rearrangement of atoms and the formation of new bonds.

Calculation of the equilibrium constant $K^{\ddagger}$ is very difficult, except in certain simple model cases. For example, if we suppose that the reactants are two structureless atoms and that the activated complex is a diatomic molecule of bond length R, then k_r turns out to be the same as for collision theory provided we interpret the collision cross-section in eqn 10.23 as πR^2.

It is more useful to express the Eyring equation in terms of thermodynamic parameters and to discuss reactions in terms of their empirical values. Thus, we saw in Section 7.3 that an equilibrium constant may be expressed in terms of the standard reaction Gibbs energy ($-RT \ln K = \Delta_r G^{\ominus}$). In this context, the Gibbs energy is called the **activation Gibbs energy**, and written $\Delta^{\ddagger}G$. It follows that

$$\Delta^{\ddagger}G = -RT \ln K^{\ddagger} \quad \text{and} \quad K^{\ddagger} = e^{-\Delta^{\ddagger}G/RT}$$

Therefore, by writing

$$\Delta^{\ddagger}G = \Delta^{\ddagger}H - T\Delta^{\ddagger}S \qquad (10.24)$$

we conclude that (with $\kappa = 1$)

$$k_r = \frac{kT}{h} \underbrace{e^{-(\Delta^{\ddagger}H - T\Delta^{\ddagger}S)/RT}}_{\Delta^{\ddagger}G} = \left(\frac{kT}{h} e^{\Delta^{\ddagger}S/R}\right) e^{-\Delta^{\ddagger}H/RT} \qquad (10.25)$$

Use $e^{x+y} = e^x e^y$

This expression has the form of the Arrhenius expression, eqn 10.19, if we identify the **enthalpy of activation**, $\Delta^{\ddagger}H$, with the activation energy and the term in parentheses, which depends on the **entropy of activation**, $\Delta^{\ddagger}S$, with the pre-exponential factor.

The advantage of transition state theory over collision theory is that it is applicable to reactions in solution as well as in the gas phase. It also gives some clue to the calculation of the steric factor P, for the orientation requirements are carried in the entropy of activation. Thus, if there are strict orientation requirements (for example, in the approach of a substrate molecule to an enzyme), then the entropy of activation will be strongly negative (representing a decrease in disorder when the activated complex forms), and the pre-exponential factor will be small. In practice, it is occasionally possible to estimate the sign and magnitude of the entropy of activation and hence to estimate the rate constant. The general importance of transition state theory is that it shows that even a complex series of events—not only a collisional encounter in the gas phase—displays Arrhenius-like behaviour, and that the concept of activation energy is applicable.

Self-test 10.10

In a particular reaction in water, it is proposed that two ions of opposite charge come together to form an electrically neutral activated complex. Is the contribution of the solvent to the entropy of activation likely to be positive or negative?

[*Answer:* positive, as H_2O is less organized around the neutral species]

10.15 Carbonic anhydrase is a zinc-based enzyme that catalyses the conversion of carbon dioxide to carbonic acid. In an experiment to study its effect, it was found that the molar concentration of carbon dioxide in solution decreased from 220 mmol dm^{-3} to 56.0 mmol dm^{-3} in 1.22×10^4 s. What is the rate constant of the reaction?

10.16 The formation of NOCl from NO in the presence of a large excess of chlorine is pseudosecond-order in NO. In an experiment to study the reaction, the partial pressure of NOCl increased from zero to 100 Pa in 522 s. What is the rate constant of the reaction?

10.17 A number of reactions that take place on the surfaces of catalysts are zero-order in the reactant. One example is the decomposition of ammonia on hot tungsten. In one experiment, the partial pressure of ammonia decreased from 21 kPa to 10 kPa in 770 s. (a) What is the rate constant for the zero-order reaction? (b) How long will it take for all the ammonia to disappear?

10.18 The following kinetic data (v_0 is the initial rate) were obtained for the reaction $2\ ICl(g) + H_2(g) \rightarrow I_2(g) + 2\ HCl(g)$:

Experiment	$[ICl]_0$/(mmol dm^{-3})	$[H_2]_0$/(mmol dm^{-3})	v_0/(mol dm^{-3} s^{-1})
1	1.5	1.5	3.7×10^{-7}
2	3.0	1.5	7.4×10^{-7}
3	3.0	4.5	22×10^{-7}
4	4.7	2.7	?

(a) Write the rate law for the reaction. (b) From the data, determine the value of the rate constant. (c) Use the data to predict the reaction rate for Experiment 4.

10.19 The variation in the partial pressure p of mercury dimethyl with time was followed at 800 K, with the results given below. Confirm that the decomposition $Hg(CH_3)_2(g) \rightarrow Hg(g) + 2\ CH_3(g)$ is first-order in $Hg(CH_3)_2$ and find the rate constant at this temperature.

t/s	0	1.0	2.0	3.0	4.0
p/kPa	15.1	11.8	9.21	7.2	5.6

10.20 The following data were collected for the reaction $2\ HI(g) \rightarrow H_2(g) + I_2(g)$ at 580 K:

t/s	0	1000	2000	3000	4000
[HI]/(mol dm^{-3})	1.00	0.112	0.061	0.041	0.031

(a) Plot the data in an appropriate fashion to determine the order of the reaction. (b) From the graph, determine the rate constant.

10.21 The following data were collected for the reaction $H_2(g) + I_2(g) \rightarrow 2\ HI(g)$ at 780 K:

t/s	0	1	2	3	4
[HI]/(mol dm^{-3})	1	0.43	0.27	0.2	0.16

(a) Plot the data in an appropriate fashion to determine the order of the reaction. (b) From the graph, determine the rate constant.

10.22 Laser flash photolysis is often used to measure the binding rate of CO to haem proteins, such as myoglobin (Mb), because CO dissociates from the bound state relatively easily upon absorption of energy from an intense and narrow pulse of light. The reaction is usually run under pseudofirst-order conditions. For a reaction in which $[Mb]_0 = 10$ mmol dm^{-3}, $[CO] = 400$ mmol dm^{-3}, and the rate constant is 5.8×10^5 dm^3 mol^{-1} s^{-1}, plot a curve of [Mb] against time. The observed reaction is $Mb + CO \rightarrow MbCO$.

10.23 The integrated rate law of a second-order reaction of the form $3\ A \rightarrow B$ is $[A] = [A]_0/(1 + k_r t[A]_0)$. How does the concentration of B change with time?

10.24 The composition of a liquid-phase reaction $2\ A \rightarrow B$ was followed spectrophotometrically with the following results:

t/min	0	10	20	30	40	∞
[B]/(mol dm^{-3})	0	0.372	0.426	0.448	0.460	0.500

Determine the order of the reaction and its rate constant (written in the form of eqn 10.5c).

10.25 Example 10.2 provided data on a first-order gas-phase reaction. How does the total pressure of the sample change with time?

10.26 The half-life of pyruvic acid in the presence of an aminotransferase enzyme (which converts it to alanine) was found to be 221 s. How long will it take for the concentration of pyruvic acid to fall to $\frac{1}{64}$ of its initial value in this first-order reaction?

10.27 The half-life for the (first-order) radioactive decay of ^{14}C is 5730 a (1 a is the SI unit 1 annum, for 1 year; the nuclide emits β particles, high-energy electrons, with an energy of 0.16 MeV). An archaeological sample contained wood that had only 69 per cent of the ^{14}C found in living trees. What is its age?

10.28 One of the hazards of nuclear explosions is the generation of ^{90}Sr and its subsequent incorporation in place of calcium in bones. This nuclide emits β particles of energy 0.55 MeV, and has a half-life of 28.1 a (1 a is the SI unit annum, for 1 year). Suppose 1.00 μg was absorbed by a newly born child. How much will remain after (a) 19 a, (b) 75 a, if none is lost metabolically?

10.29 The half-life of a first-order reaction was found to be 439 s; what is the time constant for the reaction?

10.30 The second-order rate constant for the reaction $CH_3COOC_2H_5(aq) + OH^-(aq) \rightarrow CH_3CO_2^-(aq) + CH_3CH_2OH(aq)$ is 0.11 dm^3 mol^{-1} s^{-1}. What is the concentration of ester after (a) 15 s, (b) 15 min, when ethyl acetate is added to sodium hydroxide so that the initial concentrations are [NaOH] = 0.055 mol dm^{-3} and $[CH_3COOC_2H_5] = 0.150$ mol dm^{-3}?

10.31 A reaction 2 A → P has a second-order rate law with k_r = 1.44 dm^3 mol^{-1} s^{-1}. Calculate the time required for the concentration of A to change from 0.460 mol dm^{-3} to 0.046 mol dm^{-3}.

10.32 The Arrhenius parameters for the reaction $C_4H_8(g) \rightarrow 2\ C_2H_4(g)$, where C_4H_8 is *cyclo*-butane are $\log(A/s^{-1}) = 15.6$ and E_a = 261 kJ mol^{-1}. What is the half-life of *cyclo*-butane at (a) 20°C, (b) 500°C?

10.33 A rate constant is 2.78 × 10^{-4} dm^3 mol^{-1} s^{-1} at 19°C and 3.38 × 10^{-3} dm^3 mol^{-1} s^{-1} at 37°C. Evaluate the Arrhenius parameters of the reaction.

10.34 The activation energy for the decomposition of benzene diazonium chloride is 99.1 kJ mol^{-1}. At what temperature will the rate be 10 per cent greater than its rate at 25°C?

10.35 Which reaction responds more strongly to changes of temperature, one with an activation energy of 52 kJ mol^{-1} or one with an activation energy of 25 kJ mol^{-1}?

10.36 The rate constant of a reaction increases by a factor of 1.41 when the temperature is increased from 20°C to 27°C. What is the activation energy of the reaction?

10.37 Make an appropriate Arrhenius plot of the following data for the conversion of *cyclo*-propane to propene and calculate the activation energy for the reaction.

T/K	750	800	850	900
k_r/s^{-1}	1.8 × 10^{-4}	2.7 × 10^{-3}	3.0 × 10^{-2}	0.26

10.38 Food rots about 40 times more rapidly at 25°C than when it is stored at 4°C. Estimate the overall activation energy for the processes responsible for its decomposition.

10.39 Suppose that the rate constant of a reaction *decreases* by a factor of 1.23 when the temperature is increased from 20°C to 27°C. How should you report the activation energy of the reaction?

10.40 The enzyme urease catalyses the reaction in which urea is hydrolysed to ammonia and carbon dioxide. The half-life of urea in the pseudofirst-order reaction for a certain amount of urease doubles when the temperature is lowered from 20°C to 10°C and the Michaelis constant is largely unchanged. What is the activation energy of the reaction?

10.41 What proportion of the collisions between NO_2 molecules have enough energy to result in reaction when the temperature is (a) 20°C, (b) 200°C, given that the activation energy for the reaction $2\ NO_2(g) \rightarrow 2\ NO(g) + O_2(g)$ is 111 kJ mol^{-1}?

10.42 Use collision theory to estimate the pre-exponential factor for the reaction in the preceding exercise. The experimental value is 2 × 10^9 dm^3 mol^{-1} s^{-1}. Suggest a reason for any discrepancy.

10.43 Suppose an electronegative reactant needs to come to within 500 pm of a reactant with low ionization energy before an electron can flip across from one to the other (as in the harpoon mechanism). Estimate the reaction cross-section.

10.44 Estimate the activation Gibbs energy for the decomposition of urea in the reaction $CO(NH_2)_2(aq) + 2\ H_2O(l) \rightarrow 2\ NH_4^+(aq) + CO_3^{2-}(aq)$ for which the pseudofirst-order rate constant is 1.2 × 10^{-7} s^{-1} at 60°C and 4.6 × 10^{-7} s^{-1} at 70°C.

10.45 Calculate the entropy of activation of the reaction in Exercise 10.44 at the two temperatures.

Projects

The symbol ‡ indicates that calculus is required.

10.46‡ Here we explore integrated rate laws in more detail. (a) Establish the integrated form of a third-order rate law of the form $v = k_r[A]^3$. What would be appropriate to plot to confirm that a reaction is third-order? (b) Establish the integrated form of a second-order rate law of the form $v = k_r[A][B]$ for a reaction A + B → products (i) with different initial concentrations of A and B, (ii) with the same concentrations of the two reactants. *Hints:* Note that when the concentration of A falls to $[A]_0 - x$, the concentration of B falls to $[B]_0 - x$. Use these relations to show that the rate law may be written as

$$\frac{dx}{dt} = k_r([A]_0 - x)([B]_0 - x)$$

To make progress with integration of the rate law, use the form:

$$\int \frac{dx}{(a-x)(b-x)} = \frac{1}{b-a}\left(\ln\frac{1}{a-x} - \ln\frac{1}{b-x}\right) + \text{constant}$$

10.47 Prebiotic reactions are reactions that might have occurred under the conditions prevalent on the Earth before the first living creatures emerged and that can lead to analogues of molecules necessary for life as we now know it. To qualify, a reaction must proceed with a favourable rate and have a reasonable value for the equilibrium constant. An example of a prebiotic reaction is the formation of 5-hydroxymethyluracil (HMU) from uracil and formaldehyde (HCHO). Amino acid analogues can be formed from HMU under prebiotic conditions by reaction with various nucleophiles, such as H_2S, HCN, indole, imidazole, etc. For the synthesis of HMU at pH = 7, the temperature dependence of the rate constant is given by

$$\log k_r/(dm^3\ mol^{-1}\ s^{-1}) = 11.75 - 5488/(T/K)$$

and the temperature dependence of the equilibrium constant is given by

$$\log K = -1.36 + 1794/(T/K)$$

(a) Calculate the rate constants and equilibrium constants over a range of temperatures corresponding to possible prebiotic conditions, such as 0–50°C, and plot them against temperature. (b) Calculate the activation energy and the standard reaction Gibbs energy and enthalpy at 25°C. (c) Prebiotic conditions are not likely to be standard conditions. Speculate about how the actual values of the reaction Gibbs energy and enthalpy might differ from the standard values. Do you expect that the reaction would still be favourable?

Chapter 11

Chemical kinetics: accounting for the rate laws

Even quite simple rate laws can give rise to complicated behaviour. The fact that the heart maintains a steady pulse throughout a lifetime, but may break into fibrillation during a heart attack, is one sign of that complexity. On a less personal scale, reaction intermediates come and go, and all reactions approach equilibrium. However, the complexity of the behaviour of reaction rates means that the study of reaction rates can give deep insight into the way that reactions actually take place. As remarked in Chapter 10, a rate law is a window on to the mechanism, the sequence of elementary molecular events that lead from the reactants to the products, of the reaction it summarizes.

Reaction schemes

So far, we have considered very simple rate laws, in which reactants are consumed or products formed. However, all reactions actually proceed towards a state of equilibrium in which the reverse reaction becomes increasingly important. Moreover, many reactions proceed to products through a series of intermediates. In industry, one of the intermediates may be of crucial importance and the ultimate products may represent waste.

11.1 The approach to equilibrium

All forward reactions are accompanied by their reverse reactions. At the start of a reaction, when little or no product is present, the rate of the reverse reaction is negligible. However, as the concentration of products increases, the rate at which they decompose into reactants becomes greater. At equilibrium, the reverse rate matches the forward rate and the reactants and products are present in abundances given by the equilibrium constant for the reaction.

We can analyse this behaviour by thinking of a very simple reaction of the form

Forward: $A \rightarrow B$ Rate of formation of $B = k_r[A]$

Reverse: $B \rightarrow A$ Rate of decomposition of $B = k_r'[B]$

A brief comment Throughout this chapter we write k_r for the rate constant of a general forward reaction and k_r' for the rate constant of the corresponding reverse reaction. When there are several steps *a*, *b*, ... in a mechanism, we write the forward and reverse rate constants k_a, k_b, ... and k_a', k_b', ..., respectively.

For instance, we could envisage this scheme as the interconversion of coiled (A) and uncoiled (B) DNA molecules. The *net* rate of formation of B, the difference of its rates of formation and decomposition, is

$$\text{Net rate of formation of B} = k_r[A] - k_r'[B]$$

When the reaction has reached equilibrium the concentrations of A and B are $[A]_{eq}$ and $[B]_{eq}$ and there is no net formation of either substance. It follows that

$$k_r[A]_{eq} = k_r'[B]_{eq}$$

and therefore that the equilibrium constant for the reaction is related to the rate constants by

$$K = \frac{[B]_{eq}}{[A]_{eq}} = \frac{k_r}{k_r'} \qquad (11.1)$$

If the forward rate constant is much larger than the reverse rate constant, then $K \gg 1$. If the opposite is true, then $K \ll 1$. This result is a crucial connection between the kinetics of a reaction and its equilibrium properties. It is also very useful in practice, for we may be able to measure the equilibrium constant and one of the rate constants, and can then calculate the missing rate constant from eqn 11.1. Alternatively, we can use the relation to calculate the equilibrium constant from kinetic measurements. This relation is valid even if the forward and reverse reactions have different orders. In that case we need to be careful with units. For instance, if the reaction $A + B \rightarrow C$ is second-order forward and first-order in reverse, the condition for equilibrium is $k_r[A]_{eq}[B]_{eq} = k_r'[C]_{eq}$ and the dimensionless equilibrium constant in full dress is

$$K = \frac{[C]_{eq}/c^{\ominus}}{([A]_{eq}/c^{\ominus})([B]_{eq}/c^{\ominus})} = \left(\frac{[C]}{[A][B]}\right)_{eq} c^{\ominus} = \frac{k_r}{k_r'} \times c^{\ominus}$$

The presence of $c^{\ominus} = 1\ \text{mol dm}^{-3}$ in the last term ensures that the ratio of a second-order to first-order rate constants, with their different units, is turned into a dimensionless quantity.

A brief illustration The rates of the forward and reverse reactions for the dimerization of an antibacterial agent were found to be $8.1 \times 10^8\ \text{dm}^3\ \text{mol}^{-1}\ \text{s}^{-1}$ (second-order) and $2.0 \times 10^6\ \text{s}^{-1}$ (first-order), respectively. The equilibrium constant for the dimerization is therefore

$$K = \frac{8.1 \times 10^8\ \text{dm}^3\ \text{mol}^{-1}\ \text{s}^{-1}}{2.0 \times 10^6\ \text{s}^{-1}} \times 1\ \text{mol dm}^{-3} = 4.0 \times 10^2$$

Equation 11.1 also gives us insight into the temperature dependence of equilibrium constants. First, we suppose that both the forward and reverse reactions show Arrhenius behaviour (Section 10.9). As we see from Fig. 11.1, for an exothermic reaction the activation energy of the forward reaction is smaller than that of the reverse reaction. Therefore, the forward rate constant increases less sharply with temperature than the reverse reaction does. Consequently, when we increase the temperature of a system at equilibrium, k_r' increases more steeply than k_r does, and the ratio k_r/k_r', and therefore K, decreases. This is exactly the conclusion we drew from the van 't Hoff equation (eqn 7.15), which was based on thermodynamic arguments.

Equation 11.1 tells us the ratio of concentrations after a long time has passed and the reaction has reached equilibrium. To find the concentrations at an intermediate stage, we need the integrated rate equation. If no B is present initially, we show in Derivation 11.1 that

$$[A] = \frac{(k_r' + k_r e^{-(k_r + k_r')t})[A]_0}{k_r + k_r'} \qquad (11.2a)$$

$$[B] = \frac{k_r(1 - e^{-(k_r + k_r')t})[A]_0}{k_r + k_r'} \qquad (11.2b)$$

where $[A]_0$ is the initial concentration of A.

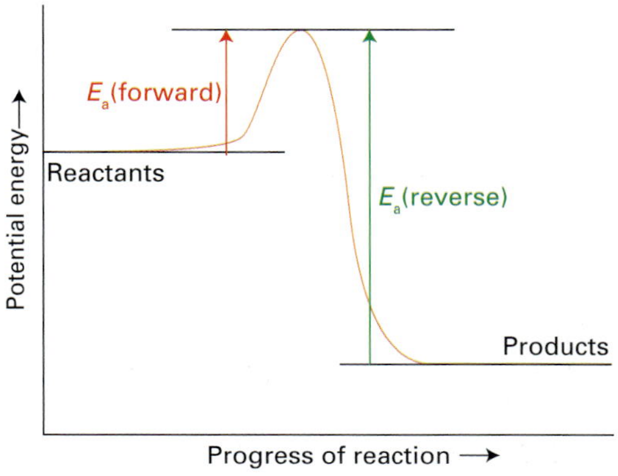

Fig. 11.1 The reaction profile for an exothermic reaction. The activation energy is greater for the reverse reaction than for the forward reaction, so the rate of the forward reaction increases less sharply with temperature. As a result, the equilibrium constant shifts in favour of the products as the temperature is raised.

Derivation 11.1

The approach to equilibrium

The concentration of A is reduced by the forward reaction (at a rate k_r[A]) but it is increased by the reverse reaction (at a rate k_r'[B]). Therefore, the net rate of change is

$$\underbrace{\frac{d[A]}{dt}}_{\text{Net rate of formation of A}} = \underbrace{-k_r[A]}_{\text{Rate of loss of A}} + \underbrace{k_r'[B]}_{\text{Rate of formation of A}}$$

If the initial concentration of A is $[A]_0$, and no B is present initially, then at all times $[A] + [B] = [A]_0$. Therefore, by replacing [B] with $[A]_0 - [A]$, we obtain

$$\frac{d[A]}{dt} = -k_r[A] + k_r'([A]_0 - [A]) = -(k_r + k_r')[A] + k_r'[A]_0$$

The solution of this differential equation is eqn 11.2a. To verify the result, differentiate eqn 11.2a by using the general relation

$$\frac{d}{dx}e^{-ax} = -ae^{-ax}$$

To obtain eqn 11.2b, we use eqn 11.2a and $[B] = [A]_0 - [A]$ again.

As we see in Fig. 11.2, the concentrations start from their initial values and move gradually towards their final equilibrium values as t approaches infinity. We find the latter by setting t equal to infinity in eqn 11.2 and using $e^{-x} = 0$ at $x = \infty$:

$$[B]_{eq} = \frac{k_r[A]_0}{k_r + k_r'} \qquad [A]_{eq} = \frac{k_r'[A]_0}{k_r + k_r'} \qquad (11.3)$$

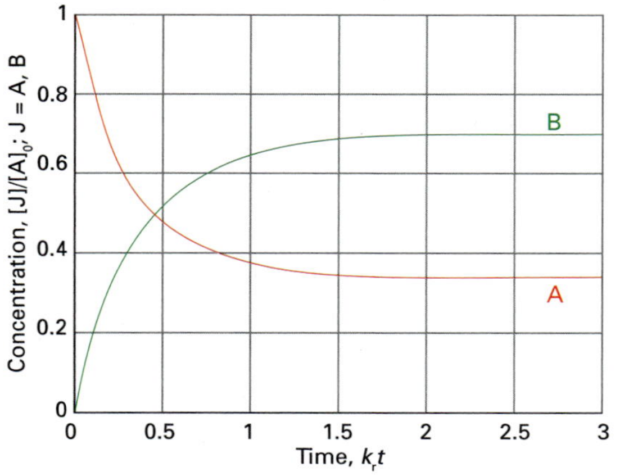

Fig. 11.2 The approach to equilibrium of a reaction that is first-order in both directions. Here we have taken $k_r = 2k_r'$. Note how, at equilibrium, the ratio of concentrations is 2:1, corresponding to $K = 2$.

As may be verified, the ratio of these two expressions is the equilibrium constant in eqn 11.1 (that is, $[B]_{eq}/[A]_{eq} = k_r/k_r'$).

11.2 Relaxation methods

The term **relaxation** denotes the return of a system to equilibrium. It is used in chemical kinetics to indicate that an externally applied influence has shifted the equilibrium position of a reaction, normally abruptly, and that the reaction is adjusting to the equilibrium composition characteristic of the new conditions (Fig. 11.3). We shall consider the response of reaction rates to a **temperature jump**, a sudden change in temperature. We know from Section 7.8 that the equilibrium composition of a reaction depends on the temperature (provided $\Delta_r H^{\ominus}$ is nonzero), so a shift in temperature acts as a perturbation on the system. One way of achieving a temperature jump is to discharge a capacitor through a sample made conducting by the addition of ions, but laser or microwave discharges can also be used. Temperature jumps of between 5 and 10 K can be achieved in about 1 μs with electrical discharges. The high energy output of pulsed lasers (Section 20.8) is sufficient to generate temperature jumps of between 10 and 30 K within nanoseconds in aqueous samples, making the technique suitable for the study of the events involved in protein folding (Box 11.1). Equilibria for which there is a change in volume between reactants and products are also sensitive to pressure, and **pressure-jump techniques** may then also be used.

We show in Derivation 11.2 that when a sudden temperature increase is applied to a simple A ⇌ B

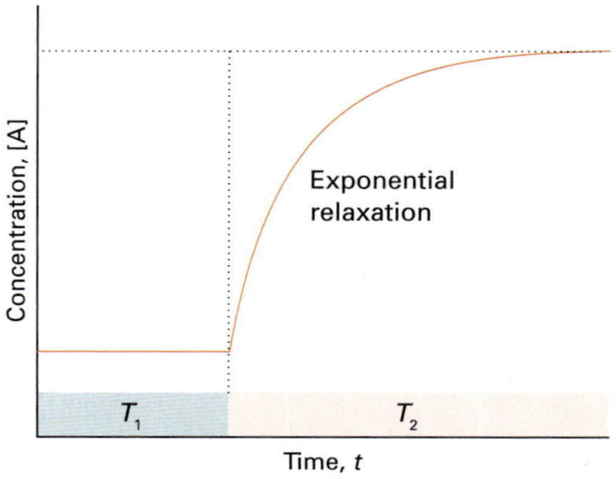

Fig. 11.3 The relaxation to the new equilibrium composition when a reaction initially at equilibrium at a temperature T_1 is subjected to a sudden change of temperature, which takes it to T_2.

Box 11.1 Kinetics of protein unfolding

Proteins are polymers that attain well-defined three-dimensional structures both in solution and in biological cells. They are *polypeptides* formed from different amino acids strung together by the *peptide link*, —CONH—. Hydrogen bonds between amino acids of a polypeptide give rise to stable helical or sheet structures, which may collapse into a random coil when certain conditions are changed. For example, the synthetic polypeptide poly-γ-benzyl-glutamate is helical in a nonhydrogen-bonding solvent, but in a hydrogen-bonding solvent it forms a random coil. The unwinding, or denaturation, of a helix into a random coil is a *cooperative transition*, in which the polymer becomes increasingly more susceptible to structural changes once the process has begun. Denaturation can be brought about by changes in temperature or pH, and by reaction with certain compounds, such as urea or guanidinium hydrochloride, known as *denaturants*. Here we examine the kinetics of the helix–coil transition, focusing primarily on experimental strategies and some recent results.

Earlier work on folding and unfolding of small polypeptides and large proteins relied primarily on rapid mixing and stopped-flow techniques. In a typical stopped-flow experiment, a sample of the protein with a high concentration of a chemical denaturant is mixed with a solution containing a much lower concentration of the same denaturant. Upon entering the mixing chamber, the denaturant is diluted and the protein refolds. Unfolding is observed by mixing a sample of folded protein with a solution containing a high concentration of denaturant. These experiments are ideal for sorting out events in the millisecond timescale, such as the formation of contacts between helical segments in a large protein. However, the available data also indicate that, in a number of proteins, a significant portion of the folding process occurs in less than 1 ms, a time range not accessible by the stopped-flow technique. More recent temperature-jump and flash-photolysis experiments have uncovered faster events. For example, at room temperature the formation of a loop between helical or sheet segments may be as fast as 1 μs and the formation of tightly packed cores with significant tertiary structure occurs in 10–100 μs. Among the fastest events are the formation of helices and sheets from fully unfolded peptide chains and we examine how the laser-induced temperature-jump technique has been used in the study of the helix–coil transition.

The laser-induced temperature-jump technique takes advantage of the fact that proteins unfold, or 'melt', at high temperatures and each protein has a characteristic melting temperature. Proteins also lose their native structures at very low temperatures, a process known as *cold denaturation*, and refold when the temperature in increased but kept significantly below the melting temperature. Hence, a temperature-jump experiment can be configured to monitor either folding or unfolding of a polypeptide, depending on the initial and final temperatures of the sample. The challenge of using melting or cold denaturation as the basis of kinetic measurements lies in increasing the temperature of the sample very quickly so fast relaxation processes can be monitored. A number of clever strategies have been employed. In one, a pulsed laser excites dissolved dye molecules that discard the extra energy largely by heat transfer to the solution. Another variation makes use of direct excitation of H_2O or D_2O with a pulsed infrared laser. The latter strategy leads to temperature jumps in a small irradiated volume of about 20 K in less than 100 ps. Relaxation of the sample can then be probed by a variety of spectroscopic techniques.

Much of the kinetic work on the helix–coil transition has been conducted in small synthetic polypeptides rich in alanine, an aminoacid that is known to stabilize helical structures. Both experimental and theoretical results suggest that the mechanism of unfolding consists of at least two steps: a very fast step in which aminoacids at either end of a helical segment undergo transitions to coil regions and a slower rate-determining step that corresponds to the cooperative melting of the rest of the chain and loss of helical content. Using *h* and *c* to denote an amino acid residue belonging to a helical and coil region, respectively, the mechanism may be summarized as follows:

hhhh... → *chhh*...	very fast
chhh... → *cccc*...	rate-determining step

The rate-determining step is thought to account for the relaxation time of 160 ns measured with a laser-induced temperature jump between 282.5 K and 300.6 K in an alanine-rich polypeptide containing 21 aminoacids. It is thought that the limitation on the rate of the helix–coil transition in this peptide arises from an activation energy barrier of 1.7 kJ mol^{-1} associated with initial events of the form... *hhhh*... → ... *hhch*... in the middle of the chain. Therefore, initiation is not only thermodynamically unfavourable but also kinetically slow. Theoretical models also suggest that a *hhhh*... → *chhh*... transition at either end of a helical segment has a significantly lower activation energy on account of the converting amino acid not being flanked by *h* regions.

The time constant for the helix–coil transition has also been measured in proteins. In apomyoglobin (myoglobin lacking the haem cofactor), the unfolding of the helices appears to have a relaxation time of about 50 ns, even shorter than in synthetic peptides. It is difficult to interpret these results because we do not yet know how the amino acid sequence or interactions between helices in a folded protein affect the helix–coil relaxation time.

equilibrium that is first-order in each direction, the composition relaxes exponentially to the new equilibrium composition:

$$x = x_0 e^{-t/\tau} \qquad \frac{1}{\tau} = k_r + k_r' \tag{11.4}$$

where x is the departure from equilibrium at the new temperature, x_0 is the departure from equilibrium immediately after the temperature jump, and τ is the **relaxation time.**

Derivation 11.2

Relaxation to equilibrium

We need to keep track of the fact that rate constants depend on temperature. At the initial temperature, when the rate constants are k_{initial} and k'_{initial}, the net rate of change of [A] is

$$\frac{d[A]}{dt} = -k_{r,\text{initial}}[A] + k'_{r,\text{initial}}[B]$$

At equilibrium under these conditions, we write the concentrations as $[A]_{\text{eq,initial}}$ and $[B]_{\text{eq,initial}}$ and because d[A]/dt is then zero,

$$k_{r,\text{initial}}[A]_{\text{eq,initial}} = k'_{r,\text{initial}}[B]_{\text{eq,initial}}$$

When the temperature is increased suddenly, the rate constants change to k_r and k_r', but the concentrations of A and B remain for an instant at their old equilibrium values. As the system is no longer at equilibrium, it readjusts to the new equilibrium concentrations, which are now given by

$$k_r[A]_{eq} = k_r'[B]_{eq}$$

and it does so at a rate that depends on the new rate constants.

We write the deviation of [A] from its new equilibrium value as x, so $[A] = x + [A]_{eq}$ and $[B] = [B]_{eq} - x$. The concentration of A then changes as follows:

$$\frac{d[A]}{dt} = -k_r[A] + k_r'[B]$$

[A] = x + $[A]_{eq}$ and [B] = $[B]_{eq} - x$

$$= -k_r(x + [A]_{eq}) + k_r'(-x + [B]_{eq})$$

Use $k_r[A]_{eq} = k_r'[B]_{eq}$

$$= -k_r x + k_r[A]_{eq} - k_r' x + k_r'[B]_{eq}$$

$$= -(k_r + k_r')x$$

From $[A] = x + [A]_{eq}$ it follows that d[A]/dt = dx/dt and therefore that

$$\frac{dx}{dt} = -(k_r + k_r')x$$

To solve this equation we divide both sides by x and multiply by dt:

$$\frac{dx}{x} = -(k_r + k_r')dt$$

Now integrate both sides. When $t = 0$, $x = x_0$, its initial value, so the integrated equation has the form

$$\overbrace{\int_{x_0}^{x} \frac{dx}{x}}^{\ln(x/x_0)} = -(k_r + k_r')\overbrace{\int_0^t dt}^{t}$$

The integrated equation is therefore

$$\ln \frac{x}{x_0} = -(k_r + k_r')t$$

When antilogarithms are taken of both sides, the result is eqn 11.4.

Equation 11.4 shows that the concentrations of A and B relax into the new equilibrium at a rate determined by the *sum* of the two new rate constants. Because the equilibrium constant under the new conditions is $K \approx k_r/k_r'$, its value may be combined with the relaxation time measurement to find the individual k_r and k_r'.

11.3 Consecutive reactions

It is commonly the case that a reactant produces an intermediate, which subsequently decays into a product. Radioactive decay is often of this type, with one nuclide decaying into another, and then that nuclide decaying into a third:

$$^{239}\text{U} \xrightarrow{2.35\ \text{min}} {}^{239}\text{Np} \xrightarrow{2.35\ \text{d}} {}^{239}\text{Pu}$$

The times are half-lives. Biochemical processes are often elaborate versions of this simple model. For instance, the restriction enzyme EcoRI catalyses the cleavage of DNA and brings about the sequence of reactions

Supercoiled DNA → open-circle DNA → linear DNA

To illustrate the kinds of considerations involved, we suppose that the reaction takes place in two steps. First, the intermediate I (the open-circle DNA, for instance) is formed from the reactant A (the supercoiled DNA) in a first-order reaction. Then I decays in a first-order reaction to form the product P (the linear DNA):

$$\text{A} \rightarrow \text{I} \qquad \text{Rate of formation of I} = k_a[\text{A}]$$

$$\text{I} \rightarrow \text{P} \qquad \text{Rate of formation of P} = k_b[\text{I}]$$

For simplicity, we are ignoring the reverse reactions, which is valid if they are slow. The first of these rate laws implies that the decay of A is first-order, and therefore that

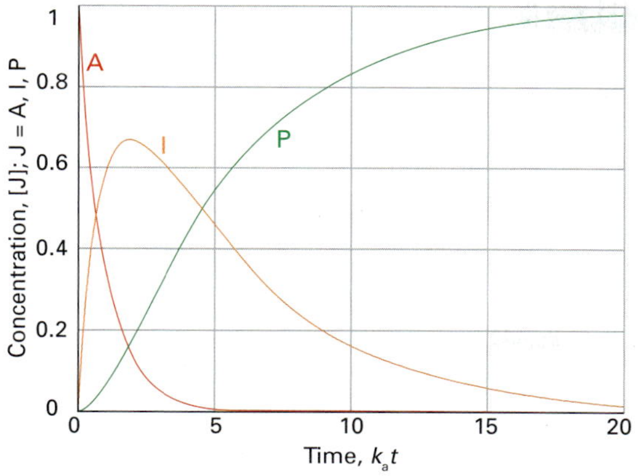

Fig. 11.4 The concentrations of the substances involved in a consecutive reaction of the form A → I → P, where I is an intermediate and P a product. We have used $k_a = 5k_b$. Note how, at each time the sum of the three concentrations is a constant.

$$[A] = [A]_0 e^{-k_a t} \quad (11.5a)$$

The net rate of formation of I is the difference between its rate of formation and its rate of consumption, so we can write

$$\text{Net rate of formation of I} = k_a[A] - k_b[I]$$

with [A] given by eqn 11.5a. This equation is more difficult to solve, but it is a standard form with the following solution:

$$[I] = \frac{k_a}{k_b - k_a}(e^{-k_a t} - e^{-k_b t})[A]_0 \quad (11.5b)$$

Finally, because [A] + [I] + [P] = $[A]_0$ at all stages of the reaction, the concentration of P is

$$[P] = \left(1 + \frac{k_a e^{-k_b t} - k_b e^{-k_a t}}{k_b - k_a}\right)[A]_0 \quad (11.5c)$$

These solutions are illustrated in Fig. 11.4. We see that the intermediate grows in concentration initially, then decays as A is exhausted. Meanwhile, the concentration of P rises smoothly to its final value. As we see in Derivation 11.3, the intermediate reaches its maximum concentration at

$$t = \frac{1}{k_a - k_b} \ln \frac{k_a}{k_b} \quad (11.6)$$

This is the optimum time for a manufacturer trying to make the intermediate in a batch process to extract it. For instance, if $k_a = 0.120\ \text{h}^{-1}$ and $k_b = 0.012\ \text{h}^{-1}$, then the intermediate is at a maximum at t = 21 h after the start of the process.

Derivation 11.3

The time of maximum concentration

To find the time corresponding to the maximum concentration of intermediate, we differentiate eqn 11.5b and look for the time at which d[I]/dt = 0. First, because $de^{at}/dt = ae^{at}$, we obtain

$$\frac{d[I]}{dt} = \frac{k_a}{k_b - k_a}(-k_a e^{-k_a t} + k_b e^{-k_b t})[A]_0 = 0$$

This equation is satisfied if

$$k_a e^{-k_a t} = k_b e^{-k_b t}$$

Because $e^x e^y = e^{x+y}$, this relation becomes

$$\frac{k_a}{k_b} = e^{(k_a - k_b)t}$$

Taking logarithms of both sides leads to eqn 11.6.

Reaction mechanisms

We have seen how two simple types of reaction—approach to equilibrium and consecutive reactions—result in a characteristic dependence of the concentration on the time. We can suspect that other variations with time will act as the signatures of other reaction mechanisms.

11.4 Elementary reactions

Many reactions occur in a series of steps called **elementary reactions**, each of which involves only one or two molecules. We shall denote an elementary reaction by writing its chemical equation without displaying the physical state of the species, as in

$$H + Br_2 \rightarrow HBr + Br$$

We have already used this convention without comment in some of the reactions discussed in Chapter 10. This equation signifies that a specific H atom attacks a specific Br_2 molecule to produce a molecule of HBr and a Br atom. Ordinary chemical equations summarize the overall stoichiometry of the reaction and do not imply any specific mechanism.

The **molecularity** of an elementary reaction is the number of molecules coming together to react. In a **unimolecular reaction** a single molecule shakes itself apart or its atoms into a new arrangement (Fig. 11.5). An example is the isomerization of cyclopropane into propene. The radioactive decay of nuclei (for example, the emission of a β particle from the nucleus of a tritium atom, which is used in mechanistic studies

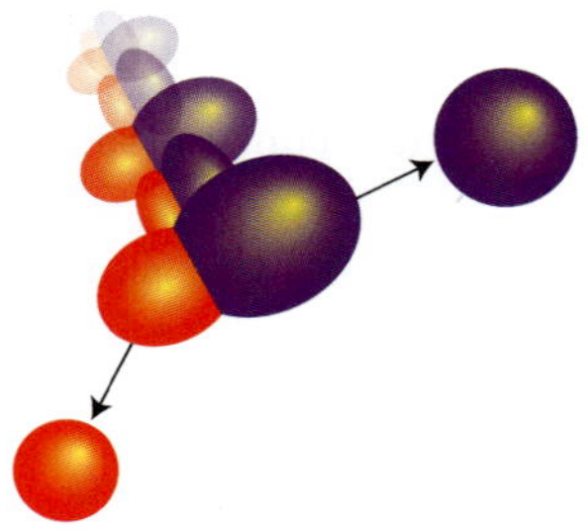

Fig. 11.5 In a unimolecular elementary reaction, an energetically excited species decomposes into products: it simply shakes itself apart.

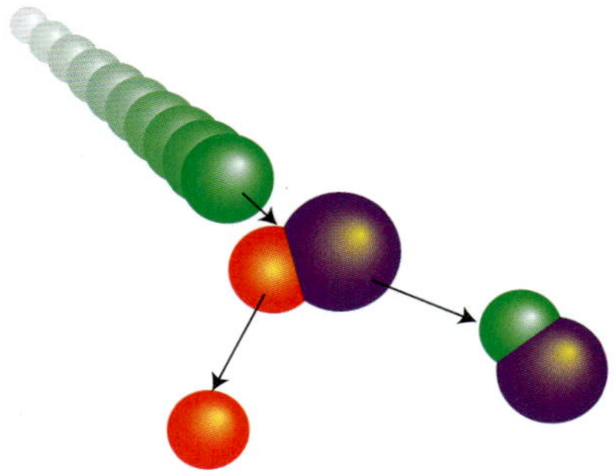

Fig. 11.6 In a bimolecular elementary reaction, two species are involved in the process.

to follow the course of particular groups of atoms) is 'unimolecular' in the sense that a single nucleus shakes itself apart. In a **bimolecular reaction**, two molecules collide and exchange energy, atoms, or groups of atoms, or undergo some other kind of change, as in the reaction between H and F_2 or between H and Br_2 (Fig. 11.6).

It is important to distinguish molecularity from order: the *order* of a reaction is an empirical quantity, and is obtained by inspection of the experimentally determined rate law; the *molecularity* of a reaction refers to an individual elementary reaction that has been postulated as a step in a proposed mechanism. Many substitution reactions in organic chemistry (for instance, S_N2 nucleophilic substitutions) are bimolecular and involve an activated complex that is formed from two reactant species. Enzyme-catalysed reactions (Section 11.13) can be regarded, to a good approximation, as bimolecular in the sense that they depend on the encounter of a substrate molecule and an enzyme molecule.

We can write down the rate law of an elementary reaction from its chemical equation. First, consider a unimolecular reaction. In a given interval, ten times as many A molecules decay when there are initially 1000 A molecules as when there are only 100 A molecules present. Therefore, the rate of decomposition of A is proportional to its concentration and we can conclude that *a unimolecular reaction is first-order*:

$$\mathrm{A} \rightarrow \text{products} \qquad v = k_r[\mathrm{A}] \tag{11.7}$$

The rate of a bimolecular reaction is proportional to the rate at which the reactants meet, which in turn is proportional to both their concentrations. Therefore, the rate of the reaction is proportional to the product of the two concentrations and *an elementary bimolecular reaction is second-order overall*:

$$\mathrm{A} + \mathrm{B} \rightarrow \text{products} \qquad v = k_r[\mathrm{A}][\mathrm{B}] \tag{11.8}$$

We must now explore how to string simple steps together into a mechanism and how to arrive at the corresponding overall rate law. For the present we emphasize that if the reaction is an elementary bimolecular process, then it has second-order kinetics; however, if the kinetics are second-order, then the reaction could be bimolecular but might be complex.

11.5 The formulation of rate laws

Suppose we propose that a particular reaction is the outcome of a sequence of elementary steps. How do we arrive at the rate law implied by the mechanism? We introduce the technique by considering the rate law for the gas-phase oxidation of nitric oxide (nitrogen monoxide, NO):

$$2\,\mathrm{NO(g)} + \mathrm{O_2(g)} \rightarrow 2\,\mathrm{NO_2(g)}$$

Nitric oxide is a very important component of polluted atmospheres. It is formed in the hot exhausts of vehicles and the jet engines of aircraft and its oxidation is a step in the formation of acid rain. The compound is also a neurotransmitter involved in the physiological changes taking place during sexual arousal. Experimentally, the reaction is found to be third-order overall:

$$v = k_r[\mathrm{NO}]^2[\mathrm{O_2}]$$

One explanation of the observed reaction order might be that the reaction is a single termolecular (three-molecule) elementary step involving the simultaneous collision of two NO molecules and one O_2 molecule. However, such collisions occur very infrequently. Therefore, although termolecular collisions may contribute, the rate of reaction by this mechanism is so slow that another mechanism usually dominates.

The following mechanism has been proposed:

Step 1. Two NO molecules combine to form a dimer:

(a) $NO + NO \rightarrow N_2O_2$
Rate of formation of $N_2O_2 = k_a[NO]^2$

This step is plausible, because NO is an odd-electron species, a radical, and two radicals can pair their electrons and form a covalent bond when they meet. That the N_2O_2 dimer is also known in the solid makes the suggestion plausible: it is often a good strategy to decide whether a proposed intermediate is the analogue of a known compound.

Step 2. The N_2O_2 dimer decomposes into NO molecules:

$N_2O_2 \rightarrow NO + NO,$
Rate of decomposition of $N_2O_2 = k'_a[N_2O_2]$

This step, the reverse of Step 1, is a unimolecular decay: the dimer shakes itself apart. We adopt the convention in which the rate constant of a reverse reaction is marked with a prime (as in k_a for the forward reaction and k'_a for its reverse).

Step 3. Alternatively, an O_2 molecule collides with the dimer and results in the formation of NO_2:

$N_2O_2 + O_2 \rightarrow NO_2 + NO_2,$
Rate of consumption of $N_2O_2 = k_b[N_2O_2][O_2]$

Now we proceed to derive the rate law on the basis of this proposed mechanism. The rate of formation of product comes directly from Step 3:

Rate of formation of $NO_2 = 2k_b[N_2O_2][O_2]$

The 2 appears in the rate law because two NO_2 molecules are formed in each reaction event, so the concentration of NO_2 increases at twice the rate that the concentration of N_2O_2 decays. However, this expression is not an acceptable overall rate law because it is expressed in terms of the concentration of the intermediate N_2O_2: an acceptable rate law for an overall reaction is expressed solely in terms of the species that appear in the overall reaction. Therefore, we need to find an expression for the concentration of N_2O_2. To do so, we consider the net rate of formation of the intermediate, the difference between its rates of formation and decay. Because N_2O_2 is formed by Step 1 but decays by Steps 2 and 3, its net rate of formation is

Net rate of formation of N_2O_2
$= k_a[NO]^2 - k'_a[N_2O_2] - k_b[N_2O_2][O_2]$

Notice that formation terms occur with a positive sign and decay terms occur with a negative sign because they reduce the net rate of formation.

If we could solve this equation for the concentration of N_2O_2 in terms of the concentrations of NO and O_2, we could substitute the result into the preceding expression and obtain the overall rate law. However, this involves solving a very difficult differential equation, and will give an enormously complex expression. In fact, even in this relatively simple case, we can obtain only a numerical solution using a computer. To make progress towards obtaining a simple formula, we must make an approximation.

11.6 The steady-state approximation

It is common at this stage of formulating a rate law to introduce the **steady-state approximation**, in which we suppose that *the concentrations of all intermediates remain constant and small throughout the reaction* (except right at the beginning and right at the end). An **intermediate** is any species that does not appear in the overall reaction but that has been invoked in the mechanism. For our mechanism the intermediate is N_2O_2, so we write

Net rate of formation of $N_2O_2 = 0$

which implies that

$$k_a[NO]^2 - k'_a[N_2O_2] - k_b[N_2O_2][O_2] = 0$$

We can rearrange this equation into an expression for the concentration of N_2O_2:

$$[N_2O_2] = \frac{k_a[NO]^2}{k'_a + k_b[O_2]}$$

It follows that the rate of formation of NO_2 is

From step 3

$$\text{Rate of formation of } NO_2 = 2k_b[N_2O_2][O_2] = \frac{2k_ak_b[NO]^2[O_2]}{k'_a + k_b[O_2]} \quad (11.9)$$

Substitute the expression for $[N_2O_2]$

At this stage, the rate law is more complex than the observed law, but the numerator resembles it. The two expressions become identical if we suppose that the rate of decomposition of the dimer is much greater than its rate of reaction with oxygen, for then $k'_a[N_2O_2] \gg k_b[N_2O_2][O_2]$, or, after cancelling the $[N_2O_2]$, $k'_a \gg k_b[O_2]$. When this condition is satisfied, we can approximate the denominator in the overall rate law by k'_a alone and conclude that

Derivation 11.4

The Lindemann mechanism

First, we write down the expression for the net rate of formation of A*, and set this rate equal to zero:

Net rate of formation of A*

Formation of A* | Deactivation of A* | Decay of A* into P

$$= k_a[A]^2 - k'_a[A^*][A] - k_b[A^*] = 0$$

Steady-state assumption

The solution of this equation is

$$[A^*] = \frac{k_a[A]^2}{k_b + k'_a[A]}$$

It follows that the rate law for the formation of products is

$$\text{Rate of formation of P} = k_b[A^*] = \frac{k_a k_b[A]^2}{k_b + k'_a[A]}$$

At this stage the rate law is not first-order in A. However, we can suppose that the rate of deactivation of A* by (A*,A) collisions is much greater than the rate of unimolecular decay of A* to products. That is, we suppose that the unimolecular decay of A* is the rate-determining step. Then $k'_a[A^*][A] \gg k_b[A^*]$, which corresponds to $k'_a[A] \gg k_b$. If that is the case, we can neglect k_b in the denominator of the rate law and obtain eqn 11.14.

Self-test 11.3

Suppose that an inert gas M is present and dominates the excitation of A and de-excitation of A*. Devise the rate law for the formation of products.

[*Answer:* $v = k_a k_b[A][M]/(k_b + k'_a[M])$]

Reactions in solution

We now turn specifically to reactions in solution, where the reactant molecules do not fly freely through a gaseous medium and collide with each other, but wriggle past their closely packed neighbours as gaps open up in the structure.

11.10 Activation control and diffusion control

The concept of the rate-determining step plays an important role for reactions in solution where it leads to the distinction between 'diffusion control' and 'activation control'. To develop this point, let's suppose that a reaction between two solute molecules A and B occurs by the following mechanism. First, we assume that A and B drift into each other's vicinity by the process of diffusion, and form an **encounter pair**, AB, at a rate proportional to each of their concentrations:

$$A + B \rightarrow AB \qquad \text{Rate of formation of AB} = k_{r,d}[A][B]$$

The 'd' subscript reminds us that this process is diffusional. The encounter pair may persist for some time as a result of the **cage effect**, the trapping of A and B near each other by their inability to escape rapidly through the surrounding solvent molecules. However, the encounter pair can break up when A and B have the opportunity to diffuse apart, so we must allow for the following process:

$$AB \rightarrow A + B \qquad \text{Rate of loss of AB} = k'_{r,d}[AB]$$

We suppose that this process is first-order in AB. Competing with this process is the reaction between A and B while they exist as an encounter pair. This process depends on their ability to acquire sufficient energy to react. That energy might come from the jostling of the thermal motion of the solvent molecules. We assume that this step is first-order in AB, but if the solvent molecules are involved it is more accurate to regard it as pseudofirst-order with the solvent molecules in great and constant excess. In any event, we can suppose that the reaction is

$$AB \rightarrow \text{products} \qquad \text{Rate of reactive loss of AB} = k_{r,a}[AB]$$

The 'a' subscript reminds us that this process is activated in the sense that it depends on the acquisition by AB of at least a minimum energy.

Now we use the steady-state approximation to set up the rate law for the formation of products. As shown in Derivation 11.5, we find

$$\text{Rate of formation of products} = k_r[A][B]$$

$$k_r = \frac{k_{r,a}k_{r,d}}{k_{r,a} + k'_{r,d}} \qquad (11.15)$$

Derivation 11.5

Diffusion control

The net rate of formation of AB is

$$\text{Net rate of formation of AB} = k_{r,d}[A][B] - k'_{r,d}[AB] - k_{r,a}[AB]$$

In a steady state, this rate is zero, so we can write

$$k_{r,d}[A][B] - k'_{r,d}[AB] - k_{r,a}[AB] = 0$$

which we can rearrange to find [AB]:

$$[AB] = \frac{k_{r,d}[A][B]}{k_{r,a} + k'_{r,d}}$$

The rate of formation of products (which is the same as the rate of reactive loss of AB) is therefore

$$\text{Rate of formation of products} = k_a[AB] = \frac{k_{r,a}k_{r,d}[A][B]}{k_{r,a} + k'_{r,d}}$$

which is eqn 11.15.

Now we distinguish two limits. Suppose the rate of reaction is much faster than the rate at which the encounter pair breaks up. In this case, $k_{r,a} \gg k'_{r,d}$ and we can neglect k'_d in the denominator of the expression for k_r in eqn 11.15. The $k_{r,a}$ in the numerator and denominator then cancel and we are left with

$$\text{Rate of formation of products} = k_{r,d}[A][B]$$

In this **diffusion-controlled limit**, the rate of the reaction is controlled by the rate at which the reactants diffuse together (as expressed by $k_{r,d}$), for once they have encountered the reaction is so fast that they will certainly go on to form products rather than diffuse apart before reacting. Alternatively, we may suppose that the rate at which the encounter pair accumulated enough energy to react is so low that it is highly likely that the pair will break up. In this case, we can set $k_{r,a} \ll k'_{r,d}$ in the expression for k_r, and obtain

$$\text{Rate of formation of products} = \frac{k_a k_d}{k'_d}[A][B] \quad (11.16)$$

In this **activation-controlled limit**, the reaction rate depends on the rate at which energy accumulates in the encounter pair (as expressed by $k_{r,a}$).

A lesson to learn from this analysis is that the concept of the rate-determining stage is rather subtle. Thus, in the diffusion-controlled limit, the condition for the encounter rate to be rate determining is not that it is the slowest step, but that the reaction rate of the encounter pair is much greater than the rate at which the pair breaks up. In the activation-controlled limit, the condition for the rate of energy accumulation to be rate determining is likewise a competition between the rate of reaction of the pair and the rate at which it breaks up, and all three rate constants control the overall rate. The best way to analyse competing rates is to do as we have done here: to set up the overall rate law, and then to analyse how it simplifies as we allow particular elementary processes to dominate others.

A detailed analysis of the rates of diffusion of molecules in liquids shows that the rate constant $k_{r,d}$ is related to the **coefficient of viscosity**, η (eta), of the medium by

$$k_{r,d} = \frac{8RT}{3\eta} \quad (11.17)$$

We see that the higher the viscosity, then the smaller the diffusional rate constant, and therefore the slower the rate of a diffusion-controlled reaction.

A brief illustration For a diffusion-controlled reaction in water, for which $\eta = 8.9 \times 10^{-4}$ kg m^{-1} s^{-1} at 25°C, we find

$$k_{r,d} = \frac{8 \times (8.3145 \text{ J K}^{-1}\text{ mol}^{-1}) \times (298 \text{ K})}{3 \times (8.9 \times 10^{-4} \text{ kg m}^{-1}\text{ s}^{-1})}$$

$$= \frac{8 \times 8.3145 \times 298}{3 \times 8.9 \times 10^{-4}} \frac{\text{J mol}^{-1}}{\text{kg m}^{-1}\text{ s}^{-1}}$$

$$= 7.4 \times 10^{6} \frac{\text{kg m}^2\text{ s}^{-2}\text{ mol}^{-1}}{\text{kg m}^{-1}\text{ s}^{-1}}$$

$$= 7.4 \times 10^{6} \text{ m}^3\text{ s}^{-1}\text{ mol}^{-1}$$

Because 1 m^3 = 10^3 dm^3, this result can be written $k_{r,d}$ = 7.4×10^9 dm^3 mol^{-1} s^{-1}, which is a useful approximate estimate to keep in mind for such reactions.

11.11 Diffusion

Diffusion plays such a central role in the processes involved in reactions in solution that we need to examine it more closely. The picture to hold in mind is that a molecule in a liquid is surrounded by other molecules and can move only a fraction of a diameter, perhaps because its neighbours move aside momentarily, before colliding. Molecular motion in liquids is a series of short steps, with incessantly changing directions, like people in an aimless, milling crowd.

The process of migration by means of a random jostling motion through a fluid (a gas as well as a liquid; even atoms in solids can migrate very slowly) is called **diffusion**. We can think of the motion of the molecule as a series of short jumps in random directions, a so-called **random walk**. If there is an initial concentration gradient in the liquid—for instance, a solution may have a high concentration of solute in one region—then the rate at which the molecules spread out is proportional to the concentration gradient, $\Delta c/\Delta x$, and we write

$$\text{Rate of diffusion} \propto \text{concentration gradient}$$

To express this relation mathematically, we introduce the **flux**, J, which is the number of particles passing through an imaginary window in a given time interval, divided by the area of the window and the duration of the interval:

$$J = \frac{\text{number of particles passing through window}}{\text{area of window} \times \text{time interval}} \quad (11.18a)$$

Then,

$$J = -D \times \text{concentration gradient} \quad (11.18b)$$

A brief comment Equation 11.18b is a verbal interpretation of the equation

$$J = -D\frac{dc}{dx}$$

where dc/dx is the gradient of the number concentration c.

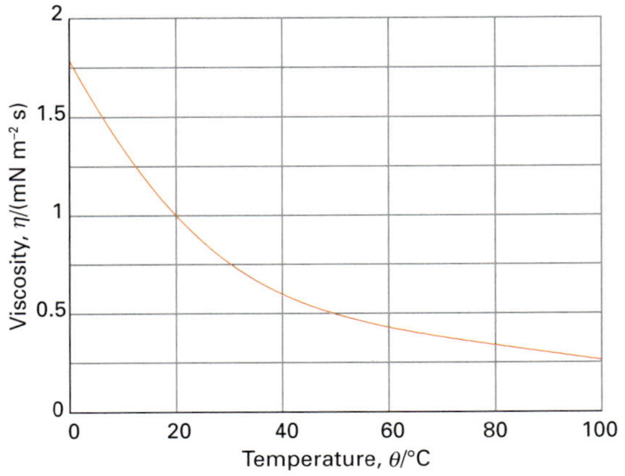

Fig. 11.11 The experimental temperature dependence of the viscosity of water. As the temperature is increased, more molecules are able to escape from the potential wells provided by their neighbours, so the liquid becomes more fluid.

where a is the radius of the molecule. It follows that

$$\eta = \eta_0 e^{E_a/RT} \tag{11.24}$$

Note the change in sign of the exponent: viscosity *decreases* as the temperature is raised. We are supposing that the strong temperature dependence of the exponential term dominates the weak linear dependence on T in the numerator of eqn 11.23. The temperature dependence described by eqn 11.24 is observed, at least over reasonably small temperature ranges (Fig. 11.11). The forces acting between the molecules govern the magnitude of E_a, but the problem of calculating it is immensely difficult and still largely unsolved.

Self-test 11.6

Estimate the activation energy for the viscosity of water from the graph in Fig. 11.11, by using the viscosities at 40°C and 80°C. *Hint:* Use an equation like eqn 11.24 to formulate an expression for the logarithm of the ratio of the two viscosities.

[*Answer:* 13 kJ mol^{-1}]

Catalysis

A **catalyst** is a substance that accelerates a reaction but undergoes no net chemical change. The catalyst lowers the activation energy of the reaction by providing an alternative path that avoids the slow, rate-determining step of the uncatalysed reaction (Fig. 11.12). Catalysts can be very effective; for instance, the activation energy for the decomposition of hydrogen peroxide in solution is 76 kJ mol^{-1}, and the reaction is slow at room temperature. When a little iodide ion is added, the activation energy falls to 57 kJ mol^{-1} and the rate constant increases by a factor of 2000. **Enzymes**, which are biological catalysts, are very selective and can have a dramatic effect on the reactions they control. For example, the enzyme catalase reduces the activation energy for the decomposition of hydrogen peroxide to 8 kJ mol^{-1}, corresponding to an acceleration of the reaction by a factor of 10^{15} at 298 K.

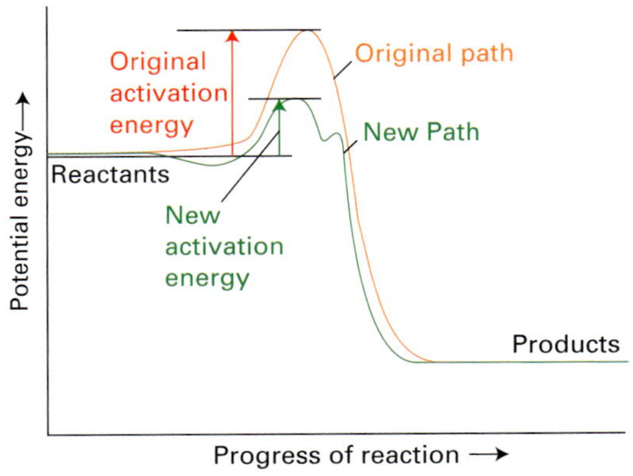

Fig. 11.12 A catalyst acts by providing a new reaction pathway between reactants and products, with a lower activation energy than the original pathway.

11.12 Homogeneous catalysis

A **homogeneous catalyst** is a catalyst in the same phase as the reaction mixture. For example, the decomposition of hydrogen peroxide in aqueous solution is catalysed by bromide ion or catalase. A **heterogeneous catalyst** is a catalyst in a different phase from the reaction mixture. For example, the hydrogenation of ethene to ethane, a gas-phase reaction, is accelerated in the presence of a solid catalyst such as palladium, platinum, or nickel. The metal provides a surface upon which the reactants bind; this binding facilitates encounters between reactants and increases the rate of the reaction. We examine heterogeneous catalysis in Chapter 16 and consider only homogeneous catalysis here.

In **acid catalysis** the crucial step is the transfer of a proton to the substrate:

$$X + HA \rightarrow HX^+ + A^- \qquad HX^+ \rightarrow \text{products}$$

Acid catalysis is the primary process in keto–enol tautomerism:

$$CH_3COCH_2CH_3 \xrightleftharpoons{H^+} CH_3C(OH){=}CHCH_3$$

In **base catalysis**, a proton is transferred from the substrate to a base, as in the hydrolysis of esters:

$$CH_3COOCH_2CH_3 + H_2O \xrightarrow{OH^-} CH_3COOH\ (\text{as } CH_3CO_2^-) + CH_3CH_2OH$$

11.13 Enzymes

One of the earliest descriptions of the action of enzymes is the **Michaelis–Menten mechanism.** The proposed mechanism, with all species in an aqueous environment, is as follows.

Step 1: The bimolecular formation of a combination, ES, of the enzyme E and the substrate S:

$$E + S \rightarrow ES \quad \text{Rate of formation of ES} = k_a[E][S]$$

Step 2: The unimolecular decomposition of the complex:

$$ES \rightarrow E + S \quad \text{Rate of decomposition of ES} = k_a'[ES]$$

Step 3: The unimolecular formation of products P and the release of the enzyme from its combination with the substrate:

$$ES \rightarrow P + E \quad \text{Rate of formation of P} = k_b[ES]$$

$$\text{Rate of consumption of ES} = k_b[ES]$$

As shown in Derivation 11.6, the rate law for the rate of formation of product in terms of the concentrations of enzyme and substrate turns out to be

$$\text{Rate of formation of P} = k_r[E]_0,$$

$$\text{with } k_r = \frac{k_b[S]}{[S] + K_M} \tag{11.25}$$

where the **Michaelis constant**, K_M (which has the dimensions of a concentration), is

$$K_M = \frac{k_a' + k_b}{k_a} \tag{11.26}$$

and $[E]_0$ is the total concentration of enzyme (both bound and unbound).

Derivation 11.6

The Michaelis–Menten rate law

Michaelis and Menten derived their rate law in 1913 in a more restrictive way, by assuming a rapid pre-equilibrium. The approach we take is a generalization using the steady-state approximation made by Briggs and Haldane in 1925.

The product is formed (irreversibly) in Step 3, so we begin by writing

$$\text{Rate of formation of P} = k_b[ES]$$

To calculate the concentration [ES] we set up an expression for the net rate of formation of ES allowing for its formation in Step 1 and its removal in Steps 2 and 3. Then we set that net rate equal to zero:

$$\text{Net rate of formation of ES} = k_a[E][S] - k_a'[ES] - k_b[ES] = 0$$

It follows that

$$[ES] = \frac{k_a[E][S]}{k_a' + k_b}$$

However, [E] and [S] are the molar concentrations of the *free* enzyme and *free* substrate. If $[E]_0$ is the total concentration of enzyme, then $[E] + [ES] = [E]_0$ and we can replace [E] in this expression by $[E]_0 - [ES]$. Therefore,

$$[ES] = \frac{k_a([E]_0 - [ES])[S]}{k_a' + k_b}$$

Multiplication by $k_a' + k_b$ gives first

$$k_a'[ES] + k_b[ES] = k_a[E]_0[S] - k_a[ES][S]$$

and then

$$(k_a' + k_b + k_a[S])[ES] = k_a[E]_0[S]$$

Division by k_a turns this expression into

$$\left(\frac{k_a' + k_b}{k_a} + [S]\right)[ES] = [E]_0[S]$$

We recognize the first term inside the parentheses as K_M, so this expression rearranges to

$$[ES] = \frac{[E]_0[S]}{[S] + K_M}$$

It follows from the first equation in this derivation that the rate of formation of product is given by eqn 11.25.

According to eqn 11.25, the rate of enzymolysis is first-order in the added enzyme concentration, but the effective rate constant k depends on the concentration of substrate. We can infer from eqn 11.25 that:

- When $[S] \ll K_M$, the effective rate constant is equal to $k_b[S]/K_M$. Therefore, the rate increases linearly with [S] at low concentrations.
- When $[S] \gg K_M$, the effective rate constant is equal to k_b, and the rate law in eqn 11.25 reduces to

$$\text{Rate of formation of P} = k_b[E]_0 \tag{11.27}$$

When $[S] \gg K_M$, the rate is independent of the concentration of S because there is so much substrate present that it remains at effectively the same concentration even though products are being formed. Under these conditions, the rate of formation of product is a maximum, and $k_b[E]_0$ is called the **maximum velocity**, v_{max}, of the enzymolysis:

$$v_{max} = k_b[E]_0 \tag{11.28}$$

the recombination of H atoms to form H_2 and the combination of H and Br atoms; however, it turns out that only Br atom recombination is important.

Now we establish the rate law for the reaction. The experimental rate law is expressed in terms of the rate of formation of product, HBr, so we start by writing an expression for its net rate of formation. Because HBr is formed in Step 2 (by both reactions) and consumed in Step 3,

$$\text{Net rate of formation of HBr} = k_b[\text{Br}][\text{H}_2] + k_c[\text{H}][\text{Br}_2] - k_d[\text{H}][\text{HBr}] \quad (11.34)$$

To make progress, we need the concentrations of the intermediates Br and H. Therefore, we set up the expressions for their net rate of formation and apply the steady-state assumption to both:

$$\text{Net rate of formation of H} = k_b[\text{Br}][\text{H}_2] - k_c[\text{H}][\text{Br}_2] - k_d[\text{H}][\text{HBr}] = 0$$

$$\text{Net rate of formation of Br} = 2k_a[\text{Br}_2] - k_b[\text{Br}][\text{H}_2] + k_c[\text{H}][\text{Br}_2] + k_d[\text{H}][\text{HBr}] - 2k_e[\text{Br}]^2 = 0$$

The steady-state concentrations of the intermediates are found by solving these two equations and are

$$[\text{Br}] = \left(\frac{k_a[\text{Br}_2]}{k_e}\right)^{1/2} \qquad [\text{H}] = \frac{k_b(k_a/k_e)^{1/2}[\text{H}_2][\text{Br}_2]^{1/2}}{k_c[\text{Br}_2] + k_d[\text{HBr}]}$$

When we substitute these concentrations into eqn 11.34 we obtain

$$\text{Rate of formation of HBr} = \frac{2k_b(k_a/k_e)^{1/2}[\text{H}_2][\text{Br}_2]^{3/2}}{[\text{Br}_2] + (k_d/k_c)[\text{HBr}]} \quad (11.35)$$

This equation has the same form as the empirical rate law, and we can identify the two empirical rate coefficients as

$$k_{r1} = 2k_b\left(\frac{k_a}{k_e}\right)^{1/2} \qquad k_{r2} = \frac{k_d}{k_c} \quad (11.36)$$

We can conclude that the proposed mechanism is at least consistent with the observed rate law. Additional support for the mechanism would come from the detection of the proposed intermediates (by spectroscopy), and the measurement of individual rate constants for the elementary steps and confirming that they correctly reproduced the observed composite rate constants.

Checklist of key ideas

You should now be familiar with the following concepts.

☐ **1** In relaxation methods of kinetic analysis, the equilibrium position of a reaction is first shifted suddenly and then allowed to readjust the equilibrium composition characteristic of the new conditions.

☐ **2** The molecularity of an elementary reaction is the number of molecules coming together to react.

☐ **3** An elementary unimolecular reaction has first-order kinetics; and an elementary bimolecular reaction has second-order kinetics.

☐ **4** In the steady-state approximation, it is assumed that the concentrations of all reaction intermediates remain constant and small throughout the reaction.

☐ **5** The rate-determining step is the slowest step in a reaction mechanism that controls the rate of the overall reaction.

☐ **6** Provided a reaction has not reached equilibrium, the products of competing reactions are controlled by kinetics.

☐ **7** The Lindemann mechanism of 'unimolecular' reactions is a theory that accounts for the first-order kinetics of gas-phase reactions.

☐ **8** A reaction in solution may be diffusion-controlled or activation-controlled.

☐ **9** Diffusion takes place in a random walk.

☐ **10** Catalysts are substances that accelerate reactions but undergo no net chemical change.

☐ **11** A homogeneous catalyst is a catalyst in the same phase as the reaction mixture.

☐ **12** Enzymes are homogeneous, biological catalysts.

☐ **13** The Michaelis–Menten mechanism of enzyme kinetics accounts for the dependence of rate on the concentration of the substrate.

☐ **14** In a chain reaction, an intermediate (the chain carrier) produced in one step generates a reactive intermediate in a subsequent step.

Table of key equations

The following table summarizes the equations developed in this chapter.

Property	Equation	Comment
Relation between rate constants and equilibrium constants	$K = [B]_{eq}/[A]_{eq} = k_r/k_r'$	First-order forward and reverse reactions (but applies generally with inclusion of $c^{\ominus}$)
Relaxation time of a temperature jump applied to a reaction A ⇌ B at equilibrium	$1/\tau = k_r + k_r'$, with $K = k_r/k_r'$	First-order forward and reverse reactions
Ratio of product concentrations for a reaction under kinetic control	$[P_2]/[P_1] = k_{r,2}/k_{r,1}$	Overall second-order reactions leading to products P1 and P_2
Relation between the rate constant of a diffusion-controlled reaction and the viscosity	$k_{r,d} = 8RT/3\eta$	Stokes law applies
Fick's first law of diffusion	$J = -D \times$ concentration gradient	
Fick's second law of diffusion	Rate of change of concentration in a region = $D \times$ (curvature of the concentration in the region)	
Root mean square distance d travelled by a diffusing molecule in a time t	$d = (2Dt)^{1/2}$	Molecular random walk
Einstein–Smoluchowski equation for the diffusion coefficient	$D = \lambda^2/2\tau$	Random walk with λ the length of each step and τ the time between steps
Einstein relation for the temperature dependence of the diffusion coefficient	$D = kT/6\pi\eta a$ with $\eta = \eta_0 e^{E_a/RT}$	Diffusion with activation energy E_a
Rate of enzymolysis according to the Michaelis–Menten mechanism	$v = [S]v_{max}/([S] + K_M)$	$v_{max} = k_b[E]_0$ and $K_M = (k_a' + k_b)/k_a$
Equation for the analysis of enzyme kinetics with a Lineweaver–Burk plot	$1/v = 1/v_{max} + (K_M/v_{max})/[S]$	The Michaelis–Menten mechanism applies
Turnover frequency or catalytic constant of an enzyme	$k_{cat} = v_{max}/[E]_0$	The Michaelis–Menten mechanism applies
Catalytic efficiency of an enzyme	$\eta = k_{cat}/K_M$	The Michaelis–Menten mechanism applies

Further information 11.1

Fick's laws of diffusion

1. *Fick's first law of diffusion.* Consider the arrangement in Fig. 11.18. Let's suppose that in an interval Δt the number of molecules passing through the window of area A from the left is proportional to the number in the slab of thickness l and area A, and therefore volume lA, just to the left of the window where the average (number) concentration is $c(x - \frac{1}{2}l)$ and to the length of the interval Δt:

$$\text{Number coming from left} \propto c(x - \tfrac{1}{2}l)lA\Delta t$$

Likewise, the number coming from the right in the same interval is

$$\text{Number coming from right} \propto c(x + \tfrac{1}{2}l)lA\Delta t$$

The net flux is therefore the difference in these numbers divided by the area and the time interval:

$$J \propto \frac{c(x - \frac{1}{2}l)lA\Delta t - c(x + \frac{1}{2}l)lA\Delta t}{A\Delta t}$$

$$= \{c(x - \tfrac{1}{2}l) - c(x + \tfrac{1}{2}l)\}l$$

Chapter 12

Quantum theory

The phenomena of chemistry cannot be understood thoroughly without a firm understanding of the principal concepts of quantum mechanics, the most fundamental description of matter that we currently possess. The same is true of virtually all the spectroscopic techniques that are now so central to investigations of composition and structure. Present-day techniques for studying chemical reactions have progressed to the point where the information is so detailed that quantum mechanics has to be used in its interpretation. And, of course, the very currency of chemistry—the electronic structures of atoms and molecules—cannot be discussed without making use of quantum-mechanical concepts.

The role—indeed, the existence—of quantum mechanics was appreciated only during the twentieth century. Until then it was thought that the motion of atomic and subatomic particles could be expressed in terms of the laws of classical mechanics introduced in the seventeenth century by Isaac Newton (see Appendix 3), as these laws were very successful at explaining the motion of planets and everyday objects such as pendulums and projectiles. However, towards the end of the nineteenth century, experimental evidence accumulated showing that classical mechanics failed when it was applied to very small particles, such as individual atoms, nuclei, and electrons, and when the transfers of energy were very small. It took until 1926 to identify the appropriate concepts and equations for describing them.

Three crucial experiments

Quantum theory emerged from a series of observations made during the late nineteenth century. As far as we are concerned, there are three crucially important experiments. One shows—contrary to what had been supposed for two centuries—that energy can be transferred between systems only in discrete amounts. Another showed that electromagnetic radiation (light), which had long been considered to be a wave, in fact behaved like a stream of particles. A third showed that electrons, which since their discovery in 1897 had been supposed to be particles, in fact behaved like waves. In this section we review these three experiments and establish the properties that a valid system of mechanics must accommodate.

12.1 Atomic and molecular spectra: discrete energies

A **spectrum** is a display of the frequencies or wavelengths (which are related by $\lambda = c/\nu$) of electromagnetic radiation that are absorbed or emitted by an atom or molecule. Figure 12.1 shows a typical atomic emission spectrum and Fig. 12.2 shows a typical molecular absorption spectrum. The obvious

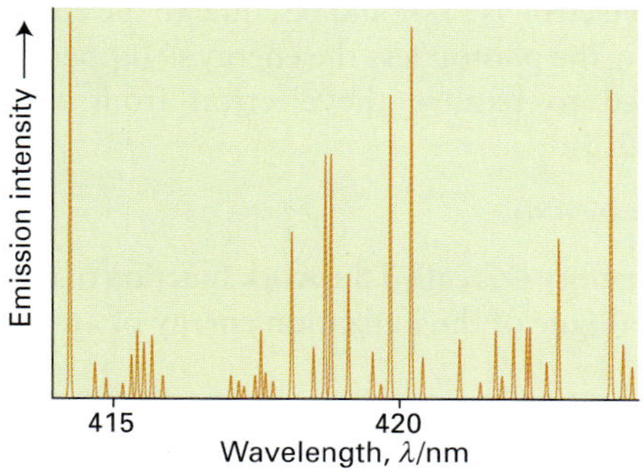

Fig. 12.1 A region of the spectrum of radiation emitted by excited iron atoms consists of radiation at a series of discrete wavelengths (or frequencies).

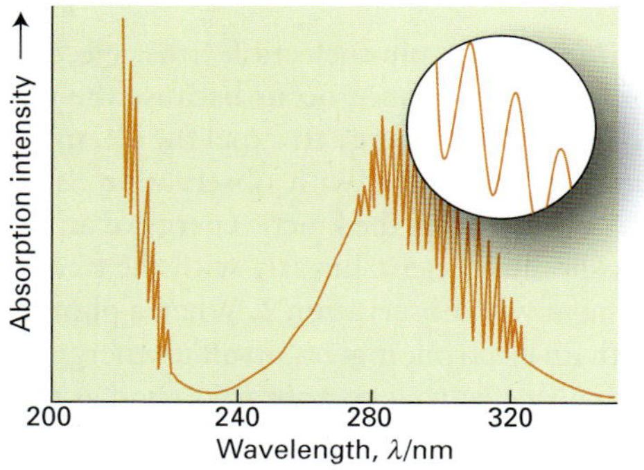

Fig. 12.2 When a molecule changes its state, it does so by absorbing radiation at definite frequencies. This spectrum is part of that due to sulfur dioxide (SO_2) molecules. This observation suggests that molecules can possess only discrete energies, not a continuously variable energy. Later we shall see that the shape of this curve is due to a combination of electronic and vibrational transitions of the molecule.

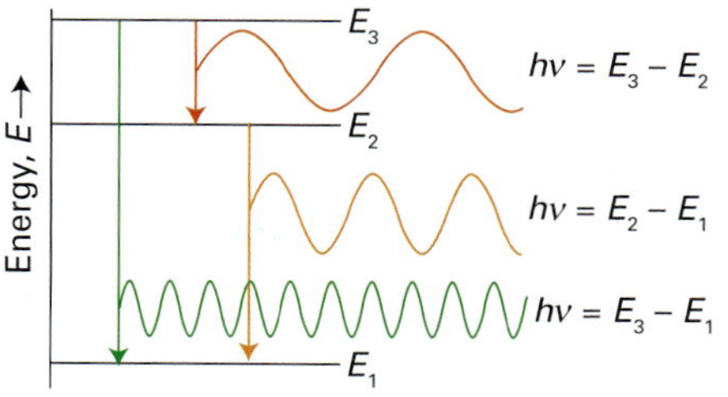

Fig. 12.3 Spectral lines can be accounted for if we assume that a molecule emits a photon as it changes between discrete energy levels. High-frequency radiation is emitted when the two states involved in the transition are widely separated in energy; low-frequency radiation is emitted when the two states are close in energy.

feature of both is that *radiation is absorbed or emitted at a series of discrete frequencies*. The emission of light at discrete frequencies can be understood if we suppose that

- The energy of the atoms or molecules is confined to discrete values, as then energy can be discarded or absorbed only in packets as the atom or molecule jumps between its allowed states (Fig. 12.3).
- The frequency of the radiation is related to the energy difference between the initial and final states.

The simplest assumption is the **Bohr frequency relation**, that the frequency ν (nu) is directly proportional to the difference in energy ΔE, and that we can write

$$\Delta E = h\nu \qquad (12.1)$$

where h is the constant of proportionality. The additional evidence that we describe below confirms this simple relation and gives the value $h = 6.626 \times 10^{-34}$ J s. This constant is now known as **Planck's constant,** for it arose in a context that had been suggested by the German physicist Max Planck.

A brief illustration The bright yellow light emitted by sodium atoms in some street lamps has wavelength 590 nm. Wavelength and frequency are related by $\nu = c/\lambda$, so the light is emitted when an atom loses an energy $\Delta E = hc/\lambda$. In this case,

$$\Delta E = \frac{(6.626\times10^{-34}\ \text{J s})\times(2.998\times10^{8}\ \text{m s}^{-1})}{5.9\times10^{-7}\ \text{m}}$$

$$= 3.4\times10^{-19}\ \text{J}$$

or 0.34 aJ (corresponding to 2.1 eV).

At this point we can conclude that one feature of nature that any system of mechanics must accommodate is that the internal modes of atoms and molecules can possess only certain energies; that is, these modes are **quantized.**

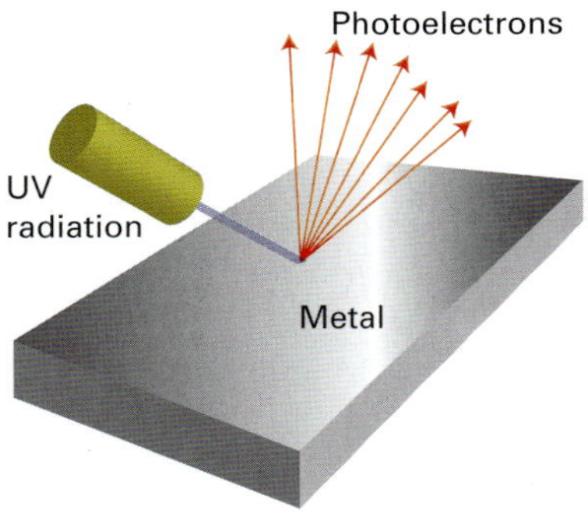

Fig. 12.4 The experimental arrangement to demonstrate the photoelectric effect. A beam of ultraviolet radiation is used to irradiate a patch of the surface of a metal, and electrons are ejected from the surface if the frequency of the radiation is above a threshold value that depends on the metal.

12.2 The photoelectric effect: light as particles

By the middle of the nineteenth century, the generally acceptable view was that electromagnetic radiation is a wave (see Appendix 3). There was a great deal of compelling information that supported this view, specifically that light underwent **diffraction**, the interference between waves caused by an object in their path, and that results in a series of bright and dark fringes where the waves are detected. However, evidence emerged that suggested that radiation can be interpreted as a stream of particles. The crucial experimental information came from the **photoelectric effect**, the ejection of electrons from metals when they are exposed to ultraviolet radiation (Fig. 12.4). The characteristics of the photoelectric effect are as follows:

1. No electrons are ejected, regardless of the intensity of the radiation, unless the frequency exceeds a threshold value characteristic of the metal.
2. The kinetic energy of the ejected electrons varies linearly with the frequency of the incident radiation but is independent of its intensity.
3. Even at low light intensities, electrons are ejected immediately if the frequency is above the threshold value.

A brief comment We say that y *varies linearly* with x if the relation between them is $y = a + bx$; we say that y is *proportional* to x if the relation is $y = bx$.

These observations strongly suggest an interpretation of the photoelectric effect in which an electron is ejected in a collision with a particle-like projectile, provided the projectile carries enough energy to expel the electron from the metal. If we suppose that the projectile is a **photon** of energy $h\nu$, where ν is the frequency of the radiation, then the conservation of energy requires that the kinetic energy, E_k, of the electron (which is equal to $\frac{1}{2}m_e v^2$, when the speed of the electron is v) should be equal to the energy supplied by the photon less the energy Φ (uppercase phi) required to remove the electron from the metal (Fig. 12.5):

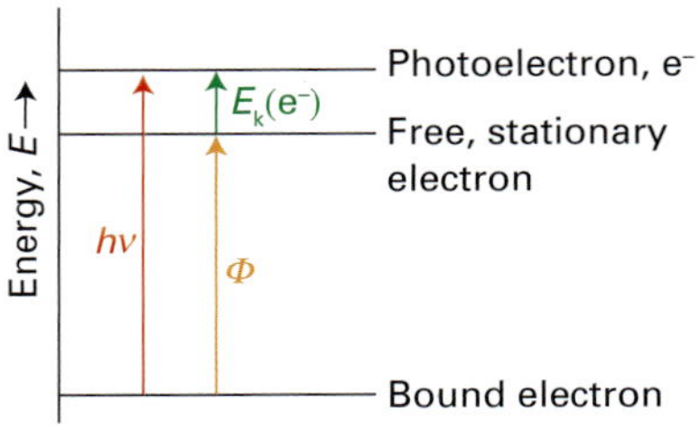

Fig. 12.5 In the photoelectric effect, an incoming photon brings a definite quantity of energy, $h\nu$. It collides with an electron close to the surface of the metal target, and transfers its energy to it. The difference between the work function, Φ, and the energy $h\nu$ appears as the kinetic energy of the ejected electron.

$$E_k = h\nu - \Phi \qquad (12.2)$$

The quantity Φ is called the **work function** of the metal, the analogue of the ionization energy of an atom.

Self-test 12.1

The work function of rubidium is 2.09 eV (1 eV = 1.60×10^{-19} J). Can blue (470 nm) light eject electrons from the metal?

[*Answer:* yes]

When $h\nu < \Phi$, photoejection (the ejection of electrons by light) cannot occur because the photon supplies insufficient energy to expel the electron: this conclusion is consistent with observation 1. Equation 12.2 predicts that the kinetic energy of an ejected electron should increase linearly with the frequency, in agreement with observation 2. When a photon collides with an electron, it gives up all its energy, so we should expect electrons to appear as soon as the collisions begin, provided the photons carry sufficient energy: this conclusion agrees with observation 3.

Thus, the photoelectric effect is strong evidence for the particle-like nature of light and the existence of photons. Moreover, it provides a route to the determination of h, for a plot of E_k against ν is a straight line of slope h.

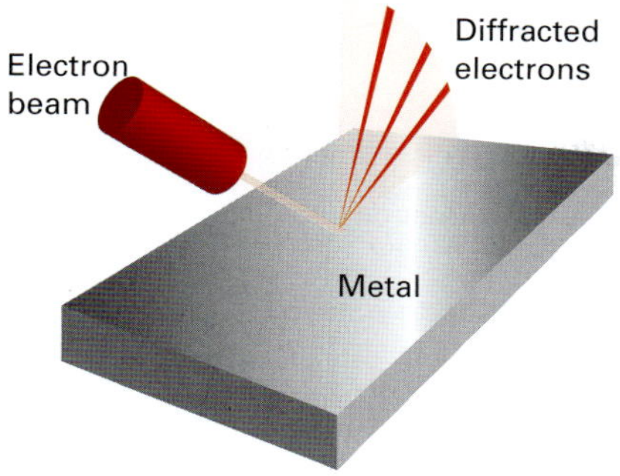

Fig. 12.6 In the Davisson–Germer experiment, a beam of electrons was directed on a single crystal of nickel, and the scattered electrons showed a variation in intensity with angle that corresponded to the pattern that would be expected if the electrons had a wave character and were diffracted by the layers of atoms in the solid.

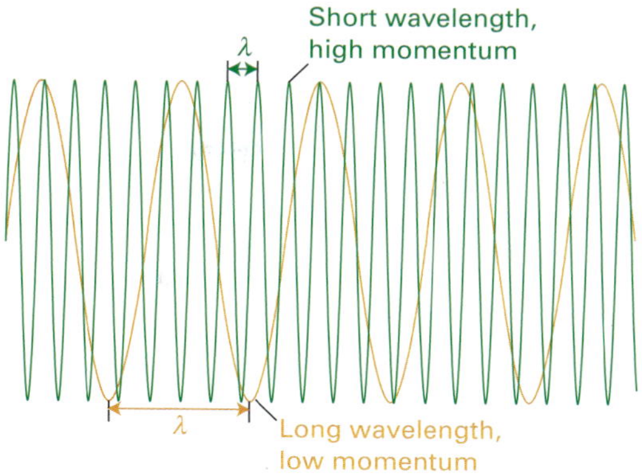

Fig. 12.7 According to the de Broglie relation, a particle with low momentum has a long wavelength, whereas a particle with high momentum has a short wavelength. A high momentum can result either from a high mass or from a high velocity (because $p = mv$). Macroscopic objects have such large masses that, even if they are travelling very slowly, their wavelengths are undetectably short.

12.3 Electron diffraction: electrons as waves

The photoelectric effect shows that light has certain properties of particles. Although contrary to the long-established wave theory of light, a similar view had been held before, but discarded. No significant scientist, however, had taken the view that matter is wave-like. Nevertheless, experiments carried out in the early 1920s forced people to question even that conclusion. The crucial experiment was performed by the American physicists Clinton Davisson and Lester Germer, who observed the diffraction of electrons by a crystal (Fig. 12.6).

There was an understandable confusion—which continues to this day—about how to combine both aspects of matter into a single description. Some progress was made by Louis de Broglie when, in 1924, he suggested that any particle travelling with a linear momentum, $p = mv$, should have (in some sense) a wavelength λ given by what we now call the **de Broglie relation:**

$$\lambda = \frac{h}{p} \qquad (12.3)$$

The wave corresponding to this wavelength, what de Broglie called a 'matter wave', has the mathematical form $\sin(2\pi x/\lambda)$. The de Broglie relation implies that the wavelength of a 'matter wave' should decrease as the particle's speed increases (Fig. 12.7). Equation 12.3 was confirmed by the Davisson–Germer experiment, as the wavelength it predicts for the electrons they used in their experiment agrees with the details of the diffraction pattern they observed.

Example 12.1

Estimating the de Broglie wavelength

Estimate the wavelength of electrons that have been accelerated from rest through a potential difference of 1.00 kV.

Strategy We need to establish a string of relations: from the potential difference we can deduce the kinetic energy acquired by the accelerated electron; then we need to find the electron's linear momentum from its kinetic energy; finally, we use that linear momentum in the de Broglie relation to calculate the wavelength.

Solution The kinetic energy acquired by an electron of charge $-e$ accelerated from rest by falling through a potential difference V is

$$E_k = eV$$

Because $E_k = \frac{1}{2}m_e v^2$ and $p = m_e v$ the linear momentum is related to the kinetic energy by $p = (2m_e E_k)^{1/2}$ and therefore

$$p = (2m_e eV)^{1/2}$$

This is the expression we use in the de Broglie relation, which becomes

$$\lambda = \frac{h}{(2m_e eV)^{1/2}}$$

At this stage, all we need do is to substitute the data and use the relations 1 C V = 1 J and 1 J = 1 kg m^2 s^{-2}:

$$\lambda = \frac{6.626 \times 10^{-34}\ \text{J s}}{\{2 \times (9.110 \times 10^{-31}\ \text{kg}) \times (1.602 \times 10^{-19}\ \text{C}) \times (1.00 \times 10^{3}\ \text{V})\}^{1/2}}$$

$$= \frac{6.626 \times 10^{-34}}{\{2 \times (9.110 \times 10^{-31}) \times (1.602 \times 10^{-19}) \times (1.00 \times 10^{3})\}^{1/2}} \frac{\text{J s}}{(\text{kg C V})^{1/2}}$$

$$= 3.88 \times 10^{-11}\ \text{m}$$

The wavelength of 38.8 pm is comparable to typical bond lengths in molecules (about 100 pm). Electrons accelerated in this way are used in the technique of *electron diffraction*, in which the diffraction pattern generated by interference when a beam of electrons passes through a sample is interpreted in terms of the locations of the atoms.

Self-test 12.2

Calculate the wavelength of an electron in a 10 MeV particle accelerator (1 MeV = 10^6 eV; 1 eV (electronvolt) = 1.602×10^{-19} J; energy units are described in Appendix 1).

[*Answer:* 0.39 pm]

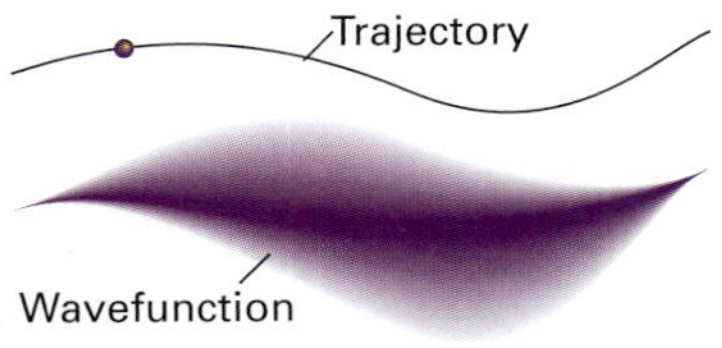

Fig. 12.8 According to classical mechanics, a particle may have a well-defined trajectory, with a precisely specified position and momentum at each instant (as represented by the precise path in the diagram). According to quantum mechanics, a particle cannot have a precise trajectory; instead, there is only a probability that it may be found at a specific location at any instant. The wavefunction that determines its probability distribution is a kind of blurred version of the trajectory. Here, the wavefunction is represented by areas of shading: the darker the area, the greater the probability of finding the particle there.

The Davisson–Germer experiment, which has since been repeated with other particles (including molecular hydrogen and C_{60}), shows clearly that 'particles' have wave-like properties. We have also seen that 'waves' have particle-like properties. Thus, we are brought to the heart of modern physics. When examined on an atomic scale, the concepts of particle and wave melt together, particles taking on the characteristics of waves, and waves the characteristics of particles. This joint wave–particle character of matter and radiation is called **wave–particle duality**. It will be central to all that follows.

The dynamics of microscopic systems

How can we accommodate the fact that atoms and molecules exist with only certain energies, waves exhibit the properties of particles, and particles exhibit the properties of waves?

We shall take the de Broglie relation as our starting point, and abandon the classical concept of particles moving along 'trajectories', precise paths at definite speeds. From now on, we adopt the quantum-mechanical view that *a particle is spread through space like a wave*. To describe this distribution, we introduce the concept of a **wavefunction**, ψ (psi), in place of the precise path, and then set up a scheme for calculating and interpreting ψ. A 'wavefunction' is the modern term for de Broglie's 'matter wave'. To a very crude first approximation, we can visualize a wavefunction as a blurred version of a path (Fig. 12.8); however, we refine this picture considerably in the following sections.

12.4 The Schrödinger equation

In 1926, the Austrian physicist Erwin Schrödinger proposed an equation for calculating wavefunctions. The **Schrödinger equation**, specifically the *time-independent* Schrödinger equation, for a single particle of mass m moving with energy E in one dimension is

$$-\frac{\hbar^2}{2m}\frac{d^2\psi}{dx^2} + V(x)\psi = E\psi \qquad (12.4a)$$

In this expression $V(x)$ is the potential energy; $\hbar$ (which is read h-bar) is a convenient modification of Planck's constant:

$$\hbar = \frac{h}{2\pi} = 1.055 \times 10^{-34} \text{ J s}$$

The term proportional to $d^2\psi/dx^2$ is closely related to the kinetic energy (so that its sum with V is the total energy, E). Mathematically, it can be interpreted as the way of measuring the curvature of the wavefunction at each point. Thus, if the wavefunction is sharply curved, then $d^2\psi/dx^2$ is large; if it is only slightly curved, then $d^2\psi/dx^2$ is small. We shall develop this interpretation later: just keep it in mind for now.

You will often see eqn 12.4 written in the very compact form

$$\hat{H}\psi = E\psi \qquad (12.4b)$$

where '$\hat{H}\psi$' stands for everything on the left of eqn 12.4a. The quantity $\hat{H}$ is called the **hamiltonian** of the system after the mathematician William Hamilton who had formulated a version of classical mechanics that used the concept. It is written with a ^ to signify that it is an 'operator', something that acts in a particular way on ψ rather than just multiplying it (as E

multiplies ψ in $E\psi$); see Derivation 12.1. You should be aware that a lot of quantum mechanics is formulated in terms of various operators, but we shall not encounter them again in this text.[1]

For a justification of the form of the Schrödinger equation, see Derivation 12.1. The fact that the Schrödinger equation is a 'differential equation', an equation in terms of the derivatives of a function, should not cause too much consternation for we shall simply quote the solutions and not go into the details of how they are found. The rare cases where we need to see the explicit forms of its solution will involve very simple functions.

A brief illustration Three simple but important cases, but not putting in various constants are as follows:

- The wavefunction for a freely moving particle is sin x, exactly as for de Broglie's matter wave.
- The wavefunction for a particle free to oscillate to-and-fro near a point is e^{-x^2}, where x is the displacement from the point.
- The wavefunction for an electron in the lowest energy state of a hydrogen atom is e^{-r}, where r is the distance from the nucleus.

As can be seen, none of these wavefunctions is particularly complicated mathematically.

Derivation 12.1

A justification of the Schrödinger equation

We can justify the form of the Schrödinger equation to a certain extent by showing that it implies the de Broglie relation for a freely moving particle. By free motion we mean motion in a region where the potential energy is zero ($V = 0$ everywhere). Then, eqn 12.4a simplifies to

$$-\frac{\hbar^2}{2m}\frac{d^2\psi}{dx^2} = E\psi \qquad (12.5a)$$

A solution of this equation is

$$\psi = \sin(kx) \qquad k = \frac{(2mE)^{1/2}}{\hbar}$$

as may be verified by substitution of the solution into both sides of the equation and using

$$\frac{d}{dx}\sin(kx) = k\cos(kx) \qquad \frac{d}{dx}\cos(kx) = -k\sin(kx)$$

$$\frac{d^2}{dx^2}\sin(kx) = -k^2\sin(kx)$$

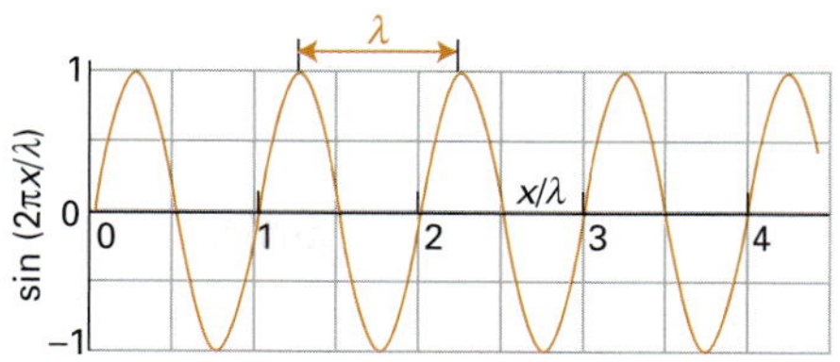

Fig. 12.9 The wavelength of a harmonic wave of the form $\sin(2\pi x/\lambda)$. The amplitude of the wave is the maximum height above the centre line.

Thus:

$$-\frac{\hbar^2}{2m}\frac{d^2\psi}{dx^2} = -\frac{\hbar^2}{2m}\frac{d^2\sin(kx)}{dx^2}$$

$$= -\frac{\hbar^2}{2m}(-k^2\sin(kx)) = \frac{k^2\hbar^2}{2m}\psi$$

The final term is equal (according to the Schrödinger equation) to $E\psi$, so we can recognize that $E = k^2\hbar^2/2m$ and therefore that $k = (2mE)^{1/2}/\hbar$.

The function $\sin(kx)$ is a wave of wavelength $\lambda = 2\pi/k$, as we can see by comparing $\sin(kx)$ with $\sin(2\pi x/\lambda)$, the standard form of a harmonic wave with wavelength λ (Fig. 12.9). Next, we note that the energy of the particle is entirely kinetic (because $V = 0$ everywhere), so the total energy of the particle is just its kinetic energy:

$$E = E_k = \frac{p^2}{2m}$$

Because E is related to k by $E = k^2\hbar^2/2m$, it follows from a comparison of the two equations that $p = k\hbar$. Therefore, the linear momentum is related to the wavelength of the wavefunction by

$$p = \frac{2\pi}{\lambda}\times\frac{h}{2\pi} = \frac{h}{\lambda}$$

which is the de Broglie relation. We see, in the case of a freely moving particle, that the Schrödinger equation has led to an experimentally verified conclusion.

12.5 The Born interpretation

Before going any further, it will be helpful to understand the physical significance of a wavefunction. The interpretation that is widely used is based on a suggestion made by the German physicist Max Born. He made use of an analogy with the wave theory of light, in which the square of the amplitude of an electromagnetic wave is interpreted as its intensity and therefore (in quantum terms) as the number of photons present. He argued that, by analogy, the square of a wavefunction gives an indication of the probability of finding a particle in a particular region of space. To be precise, the **Born interpretation** asserts that:

[1] See, for instance, our *Physical chemistry* (2006).

The probability of finding a particle in a small region of space of volume δV is proportional to $\psi^2\delta V$, where ψ is the value of the wavefunction in the region.

In other words, ψ^2 is a **probability density**. As for other kinds of density, such as mass density (ordinary 'density'), we get the probability itself by multiplying the probability density ψ^2 by the volume δV of the region of interest.

A note on good practice The symbol δ is used to indicate a small (and, in the limit, infinitesimal) change in a parameter, as in x changing to $x + \delta x$. The symbol Δ is used to indicate a finite (measurable) difference between two quantities, as in $\Delta X = X_{\text{final}} - X_{\text{initial}}$.

A brief comment We are supposing throughout that ψ is a real function (that is, one that does not depend on i, the square-root of −1). In general, ψ is complex (has both real and imaginary components); in such cases ψ^2 is replaced by $\psi^*\psi$, where ψ^* is the complex conjugate of ψ. We do not consider complex functions in this book.[2]

For a small 'inspection volume' δV of given size, the Born interpretation implies that wherever ψ^2 is large, there is a high probability of finding the particle. Wherever ψ^2 is small, there is only a small chance of finding the particle. The density of shading in Fig. 12.10 represents this **probabilistic interpretation,** an interpretation that accepts that we can make predictions only about the probability of finding a particle somewhere. This interpretation is in contrast to classical physics, which claims to be able to predict precisely that a particle will be at a given point on its path at a given instant.

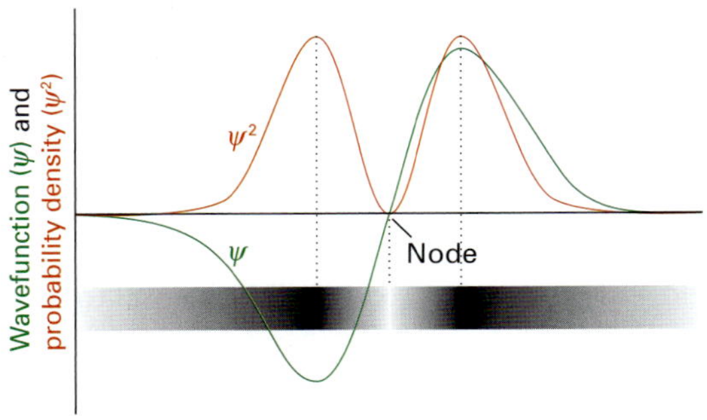

Fig. 12.10 (a) A wavefunction does not have a direct physical interpretation. However, (b) its square tells us the probability of finding a particle at each point. The probability density implied by the wavefunction shown here is depicted by the density of shading in (c).

[2] For the role, properties, and interpretation of complex wavefunctions, see our *Physical chemistry* (2006).

Example 12.2

Interpreting a wavefunction

The wavefunction of an electron in the lowest energy state of a hydrogen atom is proportional to e^{-r/a_0}, with a_0 = 52.9 pm and r the distance from the nucleus (Fig. 12.11). Calculate the relative probabilities of finding the electron inside a small cubic volume located at (a) the nucleus, (b) a distance a_0 from the nucleus.

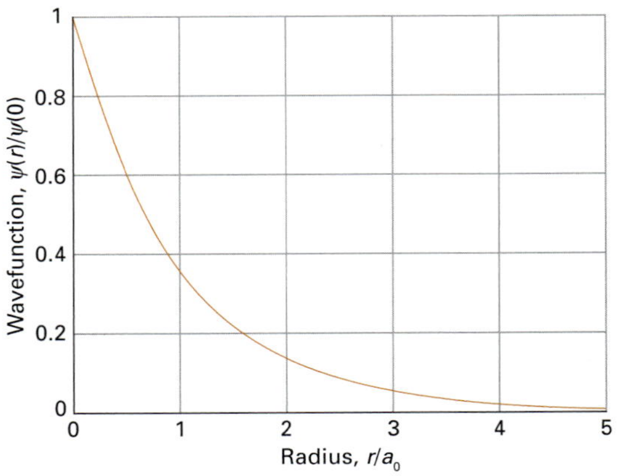

Fig. 12.11 The wavefunction for an electron in the ground state of a hydrogen atom is an exponentially decaying function of the form e^{-r/a_0}, where a_0 is the Bohr radius.

Strategy The probability is proportional to $\psi^2\delta V$ evaluated at the specified location. The volume of interest is so small (even on the scale of the atom) that we can ignore the variation of ψ within it and write

$$\text{Probability} \propto \psi^2\delta V$$

with ψ evaluated at the point in question.

Solution (a) At the nucleus, $r = 0$, so there $\psi^2 \propto 1.0$ (because $e^0 = 1$) and the probability is proportional to $1.0 \times \delta V$. (b) At a distance $r = a_0$ in an arbitrary direction, $\psi^2 \propto e^{-2} \times \delta V = 0.14 \times \delta V$. Therefore, the ratio of probabilities is 1.0/0.14 = 7.1. It is more probable (by a factor of 7.1) that the electron will be found at the nucleus than in the same tiny volume located at a distance a_0 from the nucleus.

Self-test 12.3

The wavefunction for the lowest energy state in the ion He^+ is proportional to e^{-2r/a_0}. Repeat the calculation for this ion. Any comment?

[*Answer:* 55; a more compact wavefunction on account of the higher nuclear charge]

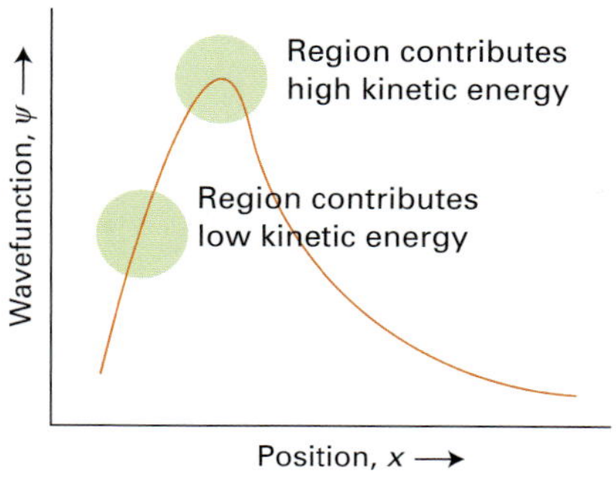

Fig. 12.12 The observed kinetic energy of a particle is the average of contributions from the entire space covered by the wavefunction. Sharply curved regions contribute a high kinetic energy to the average; slightly curved regions contribute only a small kinetic energy.

There is more information embedded in ψ than the probability that a particle will be found at a location. We saw a hint of that in the discussion of eqn 12.4 when we identified the first term as an indication of the relation between the kinetic energy of the particle and the curvature of the wavefunction: if the wavefunction is sharply curved, then the particle it describes has a high kinetic energy; if the wavefunction has only a low curvature, then the particle has only a low kinetic energy. This interpretation is consistent with the de Broglie relation, as a short wavelength corresponds to both a sharply curved wavefunction and a high linear momentum and therefore a high kinetic energy (Fig. 12.12). For more complicated wavefunctions, the curvature changes from point to point, and the total contribution to the kinetic energy is an average over the entire region of space.

The central point to remember is that *the wavefunction contains all the dynamical information about the particle it describes*. By 'dynamical' we mean all aspects of the particle's motion. Its amplitude at any point tells us the probability density of the particle at that point and other details of its shape tells us all that it is possible to know about other aspects of its motion, such as its momentum and its kinetic energy.

The Born interpretation has a further important implication: it helps us identify the conditions that a wavefunction must satisfy for it to be acceptable:

1. It must be single valued (that is, have only a single value at each point): there cannot be more than one probability density at each point.
2. It cannot become infinite over a finite region of space: the total probability of finding a particle in a region cannot exceed 1.

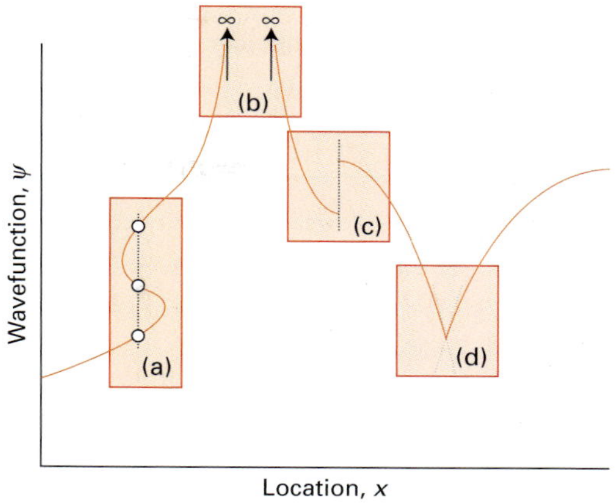

Fig. 12.13 These wavefunctions are unacceptable because (a) it is not single-valued, (b) it is infinite over a finite range, (c) it is not continuous, (d) its slope is not continuous.

These conditions turn out to be satisfied if the wavefunction takes on particular values at various points, such as at a nucleus, at the edge of a region, or at infinity. That is, the wavefunction must satisfy certain **boundary conditions**, values that the wavefunction must adopt at certain positions. We shall see plenty of examples later. Two further conditions stem from the Schrödinger equation itself, which could not be written unless:

3. The wavefunction is continuous everywhere.
4. It has a continuous slope everywhere.

These last two conditions mean that the 'curvature' term, the first term in eqn 12.4, is well defined everywhere. All four conditions are summarized in Fig. 12.13.

These requirements have a profound implication. One feature of the solution of any given Schrödinger equation, a feature common to all differential equations, is that an infinite number of possible solutions are allowed mathematically. For instance, if $\sin x$ is a solution of the equation, then so too is $a \sin(bx)$, where a and b are arbitrary constants, with each solution corresponding to a particular value of E. However, it turns out that only some of these solutions fulfill the requirements stated above. Suddenly, we are at the heart of quantum mechanics: *the fact that only some solutions are acceptable, together with the fact that each solution corresponds to a characteristic value of E, implies that only certain values of the energy are acceptable*. That is, *when the Schrödinger equation is solved subject to the boundary conditions that the solutions must satisfy, we find that the energy of the system is quantized* (Fig. 12.14).

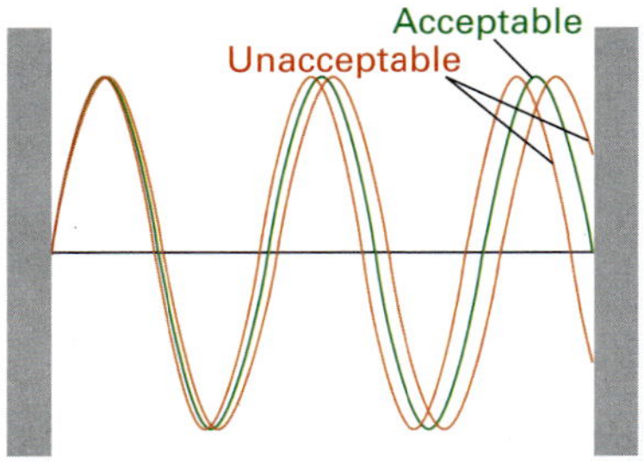

Fig. 12.14 Although an infinite number of solutions of the Schrödinger equation exist, not all of them are physically acceptable. In the example shown here, where the particle is confined between two impenetrable walls, the only acceptable wavefunctions are those that fit between the walls (like the vibrations of a stretched string). Because each wavefunction corresponds to a characteristic energy, and the boundary conditions rule out many solutions, only certain energies are permissible.

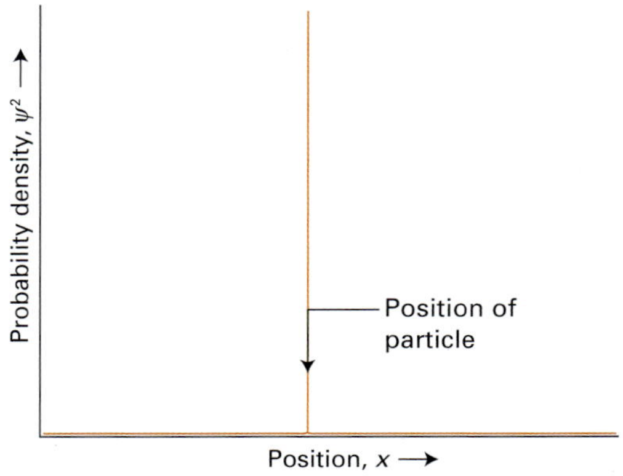

Fig. 12.15 The wavefunction for a particle with a well-defined position is a sharply spiked function that has zero amplitude everywhere except at the particle's position.

12.6 The uncertainty principle

We have seen that, according to the de Broglie relation, a wave of constant wavelength, the wavefunction $\sin(2\pi x/\lambda)$, corresponds to a particle with a definite linear momentum $p = h/\lambda$. However, a wave does not have a definite location at a single point in space, so we cannot speak of the precise position of the particle if it has a definite momentum. Indeed, because a sine wave spreads throughout the whole of space we cannot say anything about the location of the particle: because the wave spreads everywhere, the particle may be found anywhere in the whole of space. This statement is one half of the **uncertainty principle** proposed by Werner Heisenberg in 1927, in one of the most celebrated results of quantum mechanics:

> It is impossible to specify simultaneously, with arbitrary precision, both the momentum and the position of a particle.

More precisely, this is the *position–momentum uncertainty principle*: there are many other pairs of observables with simultaneous values that are restricted in a similar way; we meet some later.

Before discussing the principle further, we must establish the other half: that if we know the position of a particle exactly, then we can say nothing about its momentum. If the particle is at a definite location, then its wavefunction must be nonzero there and zero everywhere else (Fig. 12.15). We can simulate such a wavefunction by forming a **superposition** of many wavefunctions; that is, by adding together the amplitudes of a large number of sine functions (Fig. 12.16). This procedure is successful because the amplitudes of the waves add together at one location to give a nonzero total amplitude, but cancel everywhere else. In other words, we can create a sharply localized wavefunction by adding together wavefunctions corresponding to many different wavelengths, and therefore, by the de Broglie relation, of many different linear momenta.

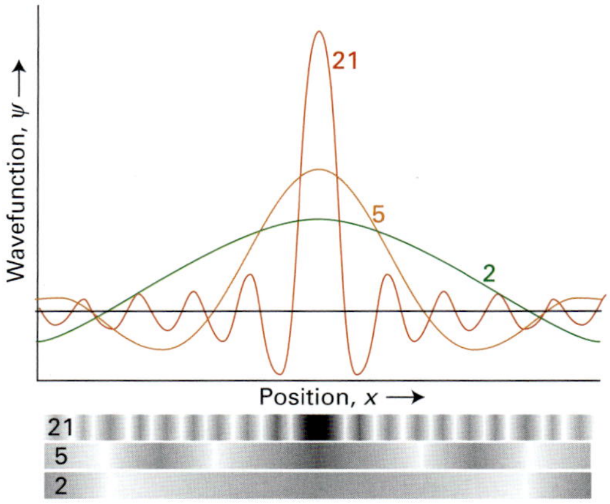

Fig. 12.16 The wavefunction for a particle with an ill-defined location can be regarded as the sum (superposition) of several wavefunctions of different wavelengths that interfere constructively in one place but destructively elsewhere. As more waves are used in the superposition, the location becomes more precise at the expense of greater uncertainty in the particle's momentum. An infinite number of waves are needed to construct the wavefunction of a perfectly localized particle. The numbers against each curve are the number of sine waves used in the superpositions.

The superposition of a few sine functions gives a broad, ill-defined wavefunction. As the number of functions increases, the wavefunction becomes sharper

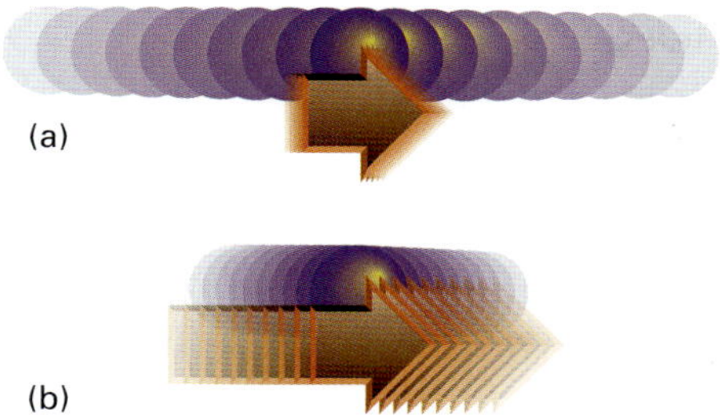

Fig. 12.17 A representation of the content of the uncertainty principle. The range of locations of a particle is shown by the circles, and the range of momenta by the arrows. In (a), the position is quite uncertain, and the range of momenta is small. In (b), the location is much better defined, and now the momentum of the particle is quite uncertain.

because of the more complete interference between the positive and negative regions of the components. When an infinite number of components are used, the wavefunction is a sharp, infinitely narrow spike like that in Fig. 12.15, which corresponds to perfect localization of the particle. Now the particle is perfectly localized, but at the expense of discarding all information about its momentum.

The quantitative version of the position–momentum uncertainty relation is

$$\Delta p \Delta x \geq \tfrac{1}{2}\hbar \qquad (12.6)$$

The quantity Δp is the 'uncertainty' in the linear momentum and Δx is the uncertainty in position (which is proportional to the width of the peak in Fig. 12.16). Equation 12.6 expresses quantitatively the fact that the more closely the location of a particle is specified (the smaller the value of Δx), then the greater the uncertainty in its momentum (the larger the value of Δp) parallel to that coordinate, and vice versa (Fig. 12.17). The position–momentum uncertainty principle applies to location and momentum *along the same axis*. It does not limit our ability to specify location on one axis and momentum along a perpendicular axis.

Example 12.3

Using the uncertainty principle

The speed of a certain projectile of mass 1.0 g is known to within 1.0 μm s^{-1}. What is the minimum uncertainty in its position along its line of flight?

Strategy Estimate Δp from $m\Delta v$, where Δv is the uncertainty in the speed; then use eqn 12.6 to estimate the minimum uncertainty in position, Δx, where x is the direction in which the projectile is travelling.

Solution From $\Delta p\Delta x \geq \tfrac{1}{2}\hbar$, the uncertainty in position is

$$\Delta x \geq \frac{\hbar}{2\Delta p} = \frac{1.054\times10^{-34}\text{ J s}}{2\times(1.0\times10^{-3}\text{ kg})\times(1.0\times10^{-6}\text{ m s}^{-1})}$$
$$= 5.3\times10^{-26}\text{ m}$$

This degree of uncertainty is completely negligible for all practical purposes, which is why the need for quantum mechanics was not recognized for over 200 years after Newton had proposed his system of mechanics and why in daily life we are completely unaware of the restrictions it implies. However, when the mass is that of an electron, the same uncertainty in speed implies an uncertainty in position far larger than the diameter of an atom, so the concept of a trajectory—the simultaneous possession of a precise position and momentum—is untenable.

Self-test 12.4

Estimate the minimum uncertainty in the speed of an electron in a hydrogen atom (taking its diameter as 100 pm).

[*Answer:* 580 km s^{-1}]

The uncertainty principle captures one of the principal differences between classical and quantum mechanics. Classical mechanics supposed, falsely as we now know, that the position and momentum of a particle can be specified simultaneously with arbitrary precision. However, quantum mechanics shows that position and momentum are **complementary**, that is, not simultaneously specifiable. Quantum mechanics requires us to make a choice: we can specify position at the expense of momentum, or momentum at the expense of position. As we shall see, there are many other complementary observables, and if any one is known precisely, the other is completely unknown.

The uncertainty principle has profound implications for the description of electrons in atoms and molecules and therefore for chemistry as a whole. When the nuclear model of the atom was first proposed it was supposed that the motion of an electron around the nucleus could be described by classical mechanics and that it would move in some kind of orbit. But to specify an orbit, we need to specify the position and momentum of the electron at each point of its path. The possibility of doing so is ruled out by the uncertainty principle. The properties of electrons in atoms, and therefore the foundations of chemistry, have had to be formulated (as we shall see) in a completely different way.

Applications of quantum mechanics

To prepare for applying quantum mechanics to chemistry we need to understand three basic types of motion: translation (motion through space), rotation, and vibration. It turns out that the wavefunctions for free translational and rotational motion in a plane can be constructed directly from the de Broglie relation, without solving the Schrödinger equation itself, and we shall take that simple route. That is not possible for rotation in three dimensions and vibrational motion where the motion is more complicated, so there we shall have to use the Schrödinger equation to find the wavefunctions.

12.7 Translational motion

The simplest type of motion is translation in one dimension. When the motion is confined between two infinitely high walls, the appropriate boundary conditions imply that only certain wavefunctions and their corresponding energies are acceptable. That is, the motion is quantized. When the walls are of finite height, the solutions of the Schrödinger equation reveal surprising features of particles, especially their ability to penetrate into and through regions where classical physics would forbid them to be found.

(a) Motion in one dimension

First, we consider the translational motion of a 'particle in a box', a particle of mass m that can travel in a straight line in one dimension (along the x-axis) but is confined between two walls separated by a distance L. The potential energy of the particle is zero inside the box but rises abruptly to infinity at the walls (Fig. 12.18). The particle might be a bead free to slide along a horizontal wire between two stops. Although this problem is very elementary, there has been a resurgence of research interest in it now that nanometre-scale structures are used to trap electrons in cavities resembling square wells.

The boundary conditions for this system are the requirement that each acceptable wavefunction of the particle must fit inside the box exactly, like the vibrations of a violin string (as in Fig. 12.10). It follows that the wavelength, λ, of the permitted wavefunctions must be one of the values

$$\lambda = 2L, L, \tfrac{2}{3}L, \ldots \quad \text{or} \quad \lambda = \frac{2L}{n}, \text{ with } n = 1, 2, 3, \ldots$$

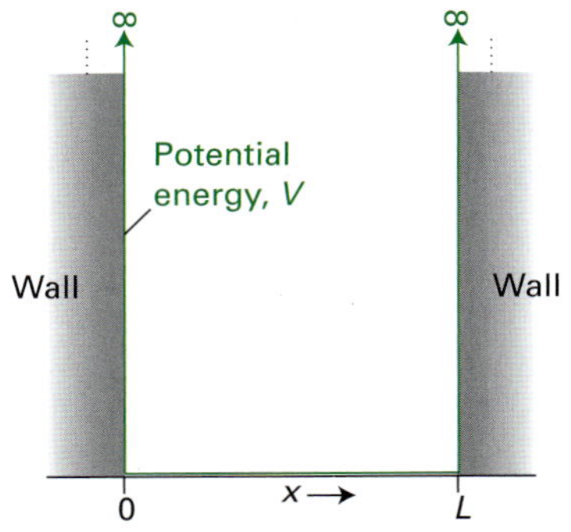

Fig. 12.18 A particle in a one-dimensional region with impenetrable walls at either end. Its potential energy is zero between $x = 0$ and $x = L$ and rises abruptly to infinity as soon as the particle touches either wall.

A brief comment More precisely, the boundary conditions stem from the requirement that the wavefunction is continuous everywhere: because the wavefunction is zero outside the box, it must therefore be zero at $x = 0$ and at $x = L$. This requirement rules out $n = 0$, which would be a line of constant, zero amplitude. Wavelengths are positive, so negative values of n do not exist.

Each wavefunction is a sine wave with one of these wavelengths; therefore, because a sine wave of wavelength λ has the form $\sin(2\pi x/\lambda)$, the permitted wavefunctions are

$$\psi_n = N \sin\left(\frac{n\pi x}{L}\right) \quad n = 1, 2, \ldots \qquad (12.7)$$

The constant N is called the **normalization constant.** It is chosen so that the total probability of finding the particle inside the box is 1, and as we show in Derivation 12.2, has the value $N = (2/L)^{1/2}$.

Derivation 12.2

The normalization constant

According to the Born interpretation, the probability of finding a particle in the infinitesimal region of length dx at the point x given that its normalized wavefunction has the value ψ at that point, is equal to $\psi^2 dx$. Therefore, the total probability of finding the particle between $x = 0$ and $x = L$ is the sum (integral) of all the probabilities of its being in each infinitesimal region. That total probability is 1 (the particle is certainly in the range somewhere), so we know that

$$\int_0^L \psi^2 dx = 1$$

Substitution of the form of the wavefunction turns this expression into

$$N^2\int_0^L \sin^2\left(\frac{n\pi x}{L}\right)dx = 1$$

Our task is to find N. To do so, we use the standard integral

$$\int \sin^2(ax)dx = \tfrac{1}{2}x - \frac{\sin(2ax)}{4a} + \text{constant}$$

It follows that, because the sine term in this expression is zero at $x = 0$ and $x = L$,

$$\int_0^L \sin^2\left(\frac{n\pi x}{L}\right)dx = \tfrac{1}{2}L$$

Therefore,

$$N^2 \times \tfrac{1}{2}L = 1$$

and hence $N = (2/L)^{1/2}$. Note that, in this case but not in general, the same normalization factor applies to all the wavefunctions regardless of the value of n.

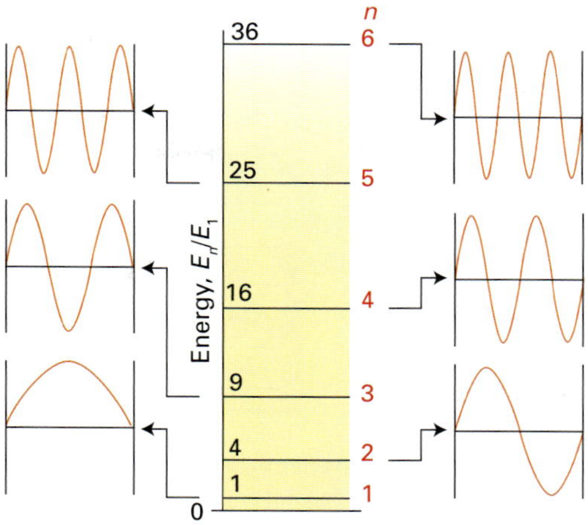

Fig. 12.19 The allowed energy levels and the corresponding (sine wave) wavefunctions for a particle in a box. Note that the energy levels increase as n^2, and so their spacing increases as n increases. Each wavefunction is a standing wave, and successive functions possess one more half-wave and a correspondingly shorter wavelength.

It is now a simple matter to find the permitted energy levels because the only contribution to the energy is the kinetic energy of the particle: the potential energy is zero everywhere inside the box, and the particle is never outside the box. First, we note that it follows from the de Broglie relation that the only acceptable values of the linear momentum are

$$p = \frac{h}{\lambda} \overset{\lambda = 2L/n}{=} \frac{nh}{2L}, \quad n = 1, 2, \ldots$$

Then, because the kinetic energy of a particle of momentum p and mass m is $E = p^2/2m$, it follows that the permitted energies of the particle are

$$E_n = \frac{n^2h^2}{8mL^2}, \quad n = 1, 2, \ldots \qquad (12.8)$$

As we see in eqns 12.7 and 12.8, the wavefunctions and energies of a particle in a box are labelled with the number n. A **quantum number**, of which n is an example, is an integer (or in certain cases, as we shall see in Chapter 13, a half-integer) that labels the state of the system. As well as acting as a label, a quantum number specifies certain physical properties of the system: in the present example, n specifies the energy of the particle through eqn 12.8.

The permitted energies of the particle are shown in Fig. 12.19 together with the shapes of the wavefunctions for $n = 1$ to 6. All the wavefunctions except the one of lowest energy ($n = 1$) possess points called **nodes** where the function passes through zero. Passing *through* zero is an essential part of the definition: just becoming zero is not sufficient. The points at the edges of the box where $\psi = 0$ are not nodes, because the wavefunction does not pass through zero there. The number of nodes in the wavefunctions shown in the illustration increases from 0 (for $n = 1$) to 5 (for $n = 6$), and is $n - 1$ for a particle in a box in general. It is a general feature of quantum mechanics that the wavefunction corresponding to the state of lowest energy has no nodes, and as the number of nodes in the wavefunctions increases, the energy increases too.

The solutions of a particle in a box introduce another important general feature of quantum mechanics. Because the quantum number n cannot be zero (for this system), the lowest energy that the particle may possess is not zero, as would be allowed by classical mechanics, but $h^2/8mL^2$ (the energy when $n = 1$). This lowest, irremovable energy is called the **zero-point energy**. The existence of a zero-point energy is consistent with the uncertainty principle. If a particle is confined to a finite region, its location is not completely indefinite; consequently its momentum cannot be specified precisely as zero, and therefore its kinetic energy cannot be precisely zero either. The zero-point energy is not a special, mysterious kind of energy. It is simply the last remnant of energy that a particle cannot give up. For a particle in a box it can be interpreted as the energy arising from a ceaseless fluctuating motion of the particle between the two confining walls of the box.

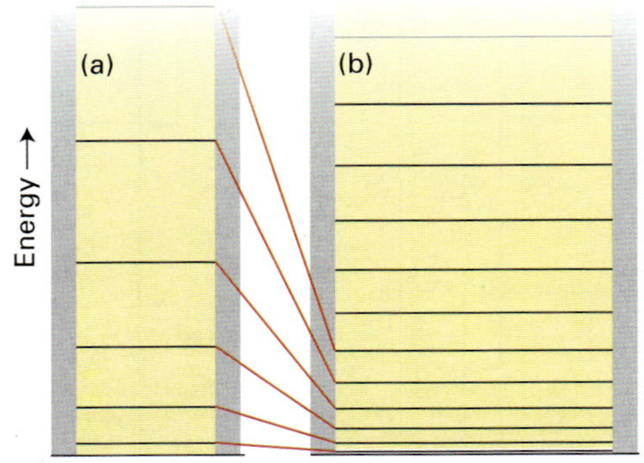

Fig. 12.20 (a) A narrow box has widely spaced energy levels; (b) a wide box has closely spaced energy levels. (In each case, the separations depend on the mass of the particle too.)

The energy difference between adjacent levels is

$$\Delta E = E_{n+1} - E_n = \overbrace{(n+1)^2}^{n^2+2n+1}\frac{h^2}{8mL^2} - n^2\frac{h^2}{8mL^2}$$

$$= (2n+1)\frac{h^2}{8mL^2} \qquad (12.9)$$

This expression shows that the difference decreases as the length L of the box increases, and that it becomes zero when the walls are infinitely far apart (Fig. 12.20). Atoms and molecules free to move in laboratory-sized vessels may therefore be treated as though their translational energy is not quantized, because L is so large. The expression also shows that the separation decreases as the mass of the particle increases. Particles of macroscopic mass (like balls sand planets, and even minute specks of dust) behave as though their translational motion is unquantized. Both the following conclusions are true in general:

- The greater the extent of the confining region, the less important are the effects of quantization. Quantization is very important for highly confining regions.
- The greater the mass of the particle, the less important are the effects of quantization. Quantization is very important for particles of very small mass.

This chapter opened with the remark that the correct description of Nature must account for the observation of transitions at discrete frequencies. This is exactly what is predicted for a system that can be modelled as a particle in a box, as it follows that when a particle makes a transition from a state with quantum number n_{initial} to one with quantum number n_{final}, the change in energy is

$$\Delta E = E_{n_{\text{final}}} - E_{n_{\text{initial}}} = (n_{\text{final}}^2 - n_{\text{initial}}^2)\frac{h^2}{8mL^2} \qquad (12.10)$$

Because the two quantum numbers can take only integer values, only certain energy changes are allowed, and therefore, through $\nu = \Delta E/h$, only certain frequencies will appear in the spectrum of transitions.

A brief illustration Suppose we can treat the π electrons of a long polyene, such as β-carotene (**1**), as a collection of electrons in a box of length 2.94 nm. Then for an electron to be excited from the level with $n = 11$ to the next higher level requires light of frequency

$$\nu = (n_{\text{final}}^2 - n_{\text{initial}}^2)\frac{h}{8m_eL^2}$$

$$= (12^2 - 11^2) \times \frac{6.626\times10^{-34}\ \text{J s}}{8\times(9.110\times10^{-31}\ \text{kg})\times(2.94\times10^{-9}\ \text{m})^2}$$

$$= 2.42\times10^{14}\ \text{s}^{-1}$$

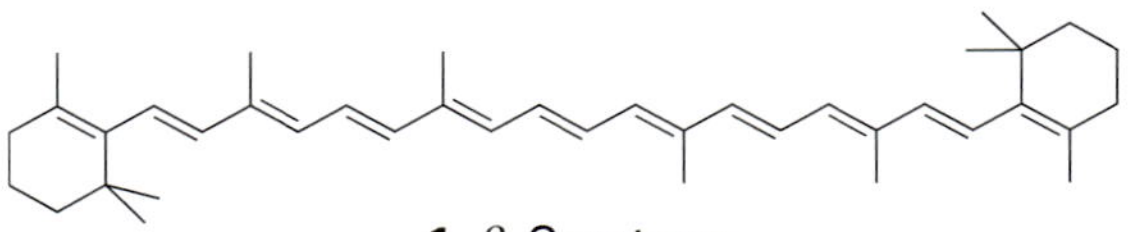

1 β-Carotene

This frequency (which we could report as 242 THz) corresponds to a wavelength of 1240 nm. The first absorption of β-carotene actually occurs at 497 nm, so although the numerical result of this very crude model is unreliable, the order-of-magnitude agreement is satisfactory. Why did we set $n = 11$? You should recall from introductory chemistry that only two electrons can occupy any state (the Pauli exclusion principle, Section 13.9); then, because each of the 22 carbon atoms in the polyene provides one π electron, the uppermost occupied state is the one with $n = 11$. The excitation of lowest energy is then from this state to the one above.

A note on good practice The ability to make such quick 'back-of-the-envelope' estimates of orders of magnitude of physical properties should be a part of every scientist's toolkit.

(b) Tunnelling

If the potential energy of a particle does not rise to infinity when it is in the walls of the container, and $E < V$ (so that the total energy is less than the potential energy and classically the particle cannot escape

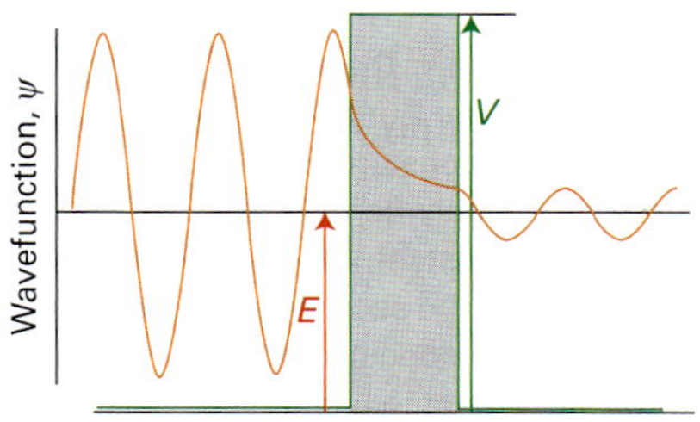

Fig. 12.21 A particle incident on a barrier from the left has an oscillating wavefunction, but inside the barrier there are no oscillations (for $E < V$). If the barrier is not too thick, the wavefunction is nonzero at its opposite face, and so oscillation begins again there.

from the container), the wavefunction does not decay abruptly to zero. The wavefunction oscillates inside the box (eqn 12.6), decays exponentially inside the region representing the wall, and oscillates again on the other side of the wall outside the box (Fig. 12.21). Hence, if the walls are so thin and the particle is so light that the exponential decay of the wavefunction has not brought it to zero by the time it emerges on the right, the particle might be found on the outside of a container even though according to classical mechanics it has insufficient energy to escape. Such leakage by penetration into or through classically forbidden zones is called **tunnelling**.

The Schrödinger equation can be used to determine the probability of tunnelling of a particle incident on a barrier.[3] It turns out that the tunnelling probability decreases sharply with the thickness of the wall and with the mass of the particle. Hence, tunnelling is very important for electrons, moderately important for protons, and less important for heavier particles. The very rapid equilibration of proton-transfer reactions (Chapter 8) is also a manifestation of the ability of protons to tunnel through barriers and transfer quickly from an acid to a base. Tunnelling of protons between acidic and basic groups is also an important feature of the mechanism of some enzyme-catalysed reactions. Electron tunnelling is one of the factors that determine the rates of electron-transfer reactions at electrodes in electrochemical cells and in biological systems, and is of the greatest importance in the semiconductor industry. The important technique of 'scanning tunnelling microscopy' relies on the dependence of electron tunnelling on the thickness of the region between a point and a surface (Section 18.2).

[3] For details of the calculation, see our *Physical chemistry* (2006).

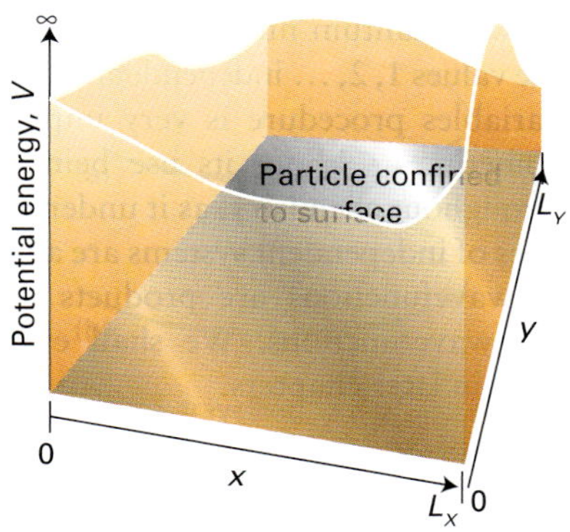

Fig. 12.22 A two-dimensional square well. The particle is confined to a rectangular plane bounded by impenetrable walls. As soon as the particle touches a wall, its potential energy rises to infinity.

(c) Motion in two dimensions

Once we have dealt with translation in one dimension it is quite easy to step into higher dimension. In doing so, we encounter two very important features of quantum mechanics that will occur many times in what follows. One feature is the simplification of the Schrödinger equation by the technique known as 'separation of variables'; the other is the existence of 'degeneracy'.

The arrangement we shall consider is like a particle —a marble—confined to the floor of a rectangular box (Fig. 12.22). The box is of side L_X in the x-direction and L_Y in the y-direction. The wavefunction varies from place to place across the floor of the box, so it is a function of both the x- and y-coordinates; we write it $\psi(x,y)$. We show in Derivation 12.3 that for this problem, according to the **separation of variables procedure**, the wavefunction can be expressed as a product of wavefunctions for each direction:

$$\psi(x,y) = X(x)Y(y) \tag{12.11}$$

with each wavefunction satisfying its 'own' Schrödinger equation like that in eqn 12.5, and that the solutions are

$$\psi_{n_X,n_Y}(x,y) = X_{n_X}(x)Y_{n_Y}(y) = \left(\frac{4}{L_X L_Y}\right)^{1/2} \sin\left(\frac{n_X \pi x}{L_X}\right) \sin\left(\frac{n_Y \pi y}{L_Y}\right) \tag{12.12a}$$

with energies

$$E_{n_X n_Y} = E_{n_X} + E_{n_Y} = \frac{n_X^2 h^2}{8mL_X^2} + \frac{n_Y^2 h^2}{8mL_Y^2} = \left(\frac{n_X^2}{L_X^2} + \frac{n_Y^2}{L_Y^2}\right)\frac{h^2}{8m} \tag{12.12b}$$

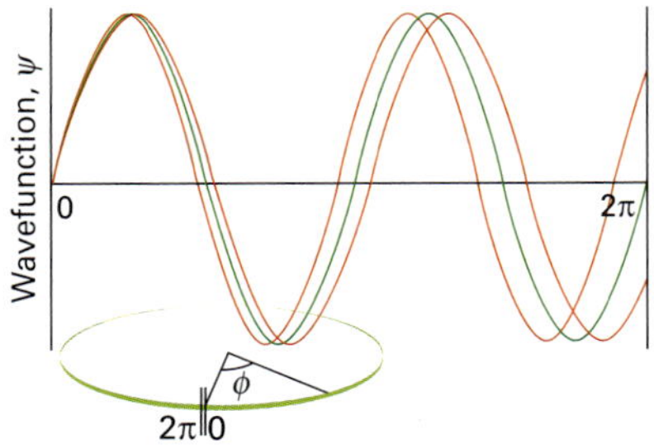

Fig. 12.25 Three solutions of the Schrödinger equation for a particle on a ring. The circumference has been opened out into a straight line; the points at $\phi = 0$ and 2π are identical. The waves shown in red are unacceptable because they have different values after each circuit and so interfere destructively with themselves. The solution shown in green is acceptable because it reproduces itself on successive circuits.

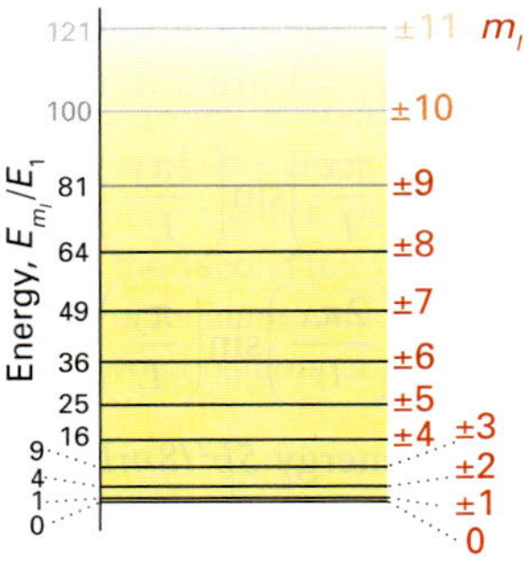

Fig. 12.26 The energy levels of a particle that can move on a circular path. Classical physics allowed the particle to travel with any energy (as represented by the continuous tinted band); quantum mechanics, however, allows only discrete energies. Each energy level, other than the one with $m_l = 0$, is doubly degenerate, because the particle may rotate either clockwise or counterclockwise with the same energy.

When the angle increases beyond 2π (that is, beyond 360°), the wavefunction continues to change on its next circuit. For an arbitrary wavelength it gives rise to a different amplitude at each point and the wavefunction will not be single-valued (a requirement for acceptable wavefunctions, Section 12.5). Thus, this particular arbitrary wave is not acceptable. An acceptable solution is obtained if the wavefunction reproduces itself on successive circuits in the sense that the wavefunction at $\phi = 2\pi$ (after a complete revolution) must be the same as the wavefunction at $\phi = 0$: we say that the wavefunction must satisfy **cyclic boundary conditions**. Specifically, the acceptable wavefunctions that match after each circuit have wavelengths that are given by the expression

$$\lambda = \frac{2\pi r}{n} \qquad n = 0, 1, \ldots$$

where the value $n = 0$, which gives an infinite wavelength, corresponds to a uniform nonzero amplitude. It follows that the permitted energies are

$$E_n = \overbrace{\frac{(hr/\lambda)^2}{2I}}^{E = J_z^2/2I \text{ and } J_z = hr/\lambda} \underbrace{=}_{\lambda = 2\pi r/n} \frac{(nh/2\pi)^2}{2I} \overbrace{=}^{h/2\pi = \hbar} \frac{n^2\hbar^2}{2I}$$

with $n = 0, \pm1, \pm2, \ldots$

In the discussion of rotational motion it is conventional—for reasons that will become clear—to denote the quantum number by m_l in place of n. Therefore, the final expression for the energy levels is

$$E_{m_l} = \frac{m_l^2\hbar^2}{2I} \qquad m_l = 0, \pm1, \ldots \qquad (12.16)$$

These energy levels are drawn in Fig. 12.26. The occurrence of m_l^2 in the expression for the energy means that two states of motion with opposite values of m_l, such as those with $m_l = +1$ and $m_l = -1$, correspond to the same energy. This degeneracy arises from the fact that the energy is independent of the direction of travel. The state with $m_l = 0$ is nondegenerate. A further point is that the particle does not have a zero-point energy: m_l may take the value 0, and $E_0 = 0$.

An important additional conclusion is that *the angular momentum of the particle is quantized*. We can use the relation between angular momentum and linear momentum ($J_z = pr$), and between linear momentum and the allowed wavelengths of the particle ($\lambda = 2\pi r/m_l$), to conclude that the angular momentum of a particle around the z-axis is confined to the values

$$J_z = pr = \frac{hr}{\lambda} = \frac{hr}{2\pi r/m_l} = m_l \times \frac{h}{2\pi}$$

That is, the angular momentum of the particle around the axis is confined to the values

$$J_z = m_l\hbar \qquad (12.17)$$

with $m_l = 0, \pm1, \pm2, \ldots$ Positive values of m_l correspond to clockwise rotation (as seen from below) and negative values correspond to counterclockwise rotation (Fig. 12.27). The quantized motion can be thought of in terms of the rotation of a bicycle wheel that can rotate only with a discrete series of angular momenta, so that as the wheel is accelerated, the angular momentum jerks from the values 0 (when the wheel is stationary) to $\hbar$, $2\hbar$, … but can have no intermediate value.

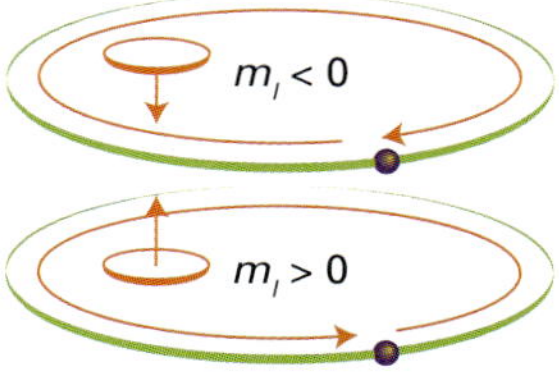

Fig. 12.27 The significance of the sign of m_l. When $m_l < 0$, the particle travels in a counterclockwise direction as viewed from below; then $m_l > 0$, the motion is clockwise.

Self-test 12.5

Consider an electron that is part of a cyclic, aromatic molecule (such as benzene). Treat the molecule as a ring of diameter 280 pm and the electron as a particle that moves only along the perimeter of the ring. What is the energy in electronvolts (1 eV = 1.602×10^{-19} J) required to excite the electron from the level with $m_l = \pm 1$ (according to the Pauli exclusion principle, one of the uppermost filled levels for this six-electron system) to the next higher level?

[*Answer:* 5.83 eV, corresponding to $\lambda = 220$ nm (the first absorption in fact lies close to 260 nm)]

(b) Rotation in three dimensions

Rotational motion in three dimensions includes the motion of electrons around nuclei in atoms. Consequently, understanding rotational motion in three dimensions is crucial to understanding the electronic structures of atoms. Gas-phase molecules also rotate freely in three dimensions and by studying their allowed energies (using the spectroscopic techniques described in Chapter 19) we can infer bond lengths, bond angles, and dipole moments.

Just as the location of a city on the surface of the Earth is specified by giving its latitude and longitude, the location of a particle free to move at a constant distance from a point is specified by two angles, the **colatitude** θ (theta) and the **azimuth** ϕ (phi) (Fig. 12.28). The wavefunction for the particle is therefore a function of both angles and is written $\psi(\theta,\phi)$. It turns out that this wavefunction factorizes by the separation of variables procedure into the product of a function of θ and a function of ϕ, and that the latter are exactly the same as those we have already found for a particle on a ring. In other words, motion of a particle over the surface of a sphere is like the motion of the particle over a stack of rings, with the additional freedom to migrate between rings.

There are two sets of cyclic boundary conditions that limit the selection of solutions of the Schrödinger equation. One is that the wavefunctions must match

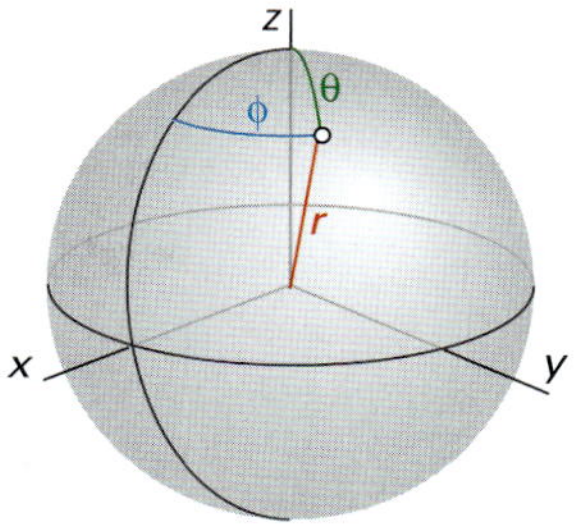

Fig. 12.28 The spherical polar coordinates r (the radius), θ (the colatitude), and ϕ (the azimuth).

as we travel round the equator (just like the particle on a ring); as we have seen, that boundary condition introduces the quantum number m_l. The other condition is that the wavefunction must match as we travel over the poles. This constraint introduces a second quantum number, which is called the **orbital angular momentum quantum number** and denoted l. We shall not go into the details of the solution, but just quote the results. It turns out that the quantum numbers are allowed the following values:

$$l = 0, 1, 2, \ldots \qquad m_l = l, l-1, \ldots, -l$$

Note that there are $2l + 1$ values of m_l for a given value of l. The energy of the particle is given by the expression

$$E_l = l(l+1)\frac{\hbar^2}{2mr^2} \qquad (12.18)$$

where r is the radius of the surface of the sphere on which the particle moves. Note that, for reasons that will become clear in a moment, the energy depends on l and is independent of the value of m_l. The wavefunctions appear in a number of applications, and are called **spherical harmonics**. They are commonly denoted $Y_{l,ml}(\theta,\phi)$ and can be imagined as wave-like distortions of a spherical shell (Fig. 12.29).

We can draw a very important additional conclusion by comparing the expression for the energy in eqn 12.18 with the classical expression for the energy:

Classical	Quantum mechanical
$E = \dfrac{J^2}{2mr^2}$	$E_l = \dfrac{l(l+1)\hbar^2}{2mr^2}$

where J is the magnitude of the angular momentum of the particle. We can conclude that the magnitude of the angular momentum is quantized and limited to the values

$$J = \{l(l+1)\}^{1/2}\hbar \qquad (12.19)$$

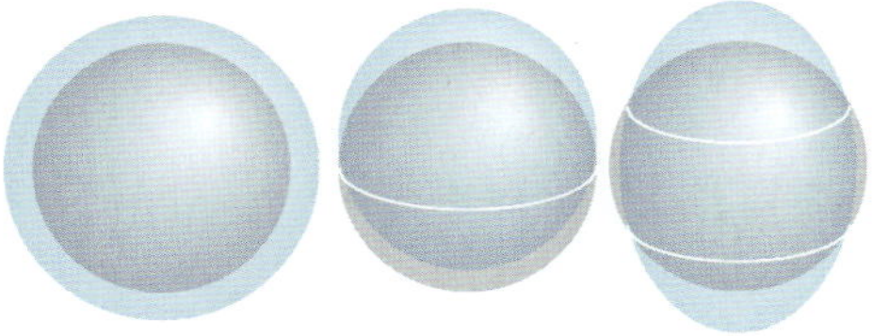

Fig. 12.29 The wavefunctions of a particle on a sphere can be imagined as having the shapes that the surface would have when the sphere is distorted. Three of these 'spherical harmonics' are shown here: amplitudes above the surface of the sphere represent positive regions of the functions and amplitudes below the surface represent negative regions.

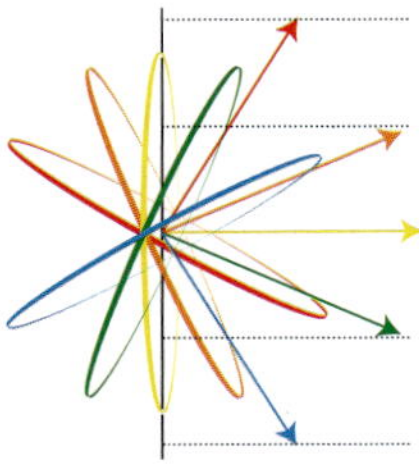

Fig. 12.30 The significance of the quantum numbers l and m_l shown for $l = 2$: l determines the magnitude of the angular momentum (as represented by the length of the arrow), and m_l the component of that angular momentum about the z-axis.

Thus, the allowed values of the magnitude of the angular momentum are 0, $2^{1/2}\hbar$, $6^{1/2}\hbar$, ... We have already seen that m_l tells us the value, as $m_l\hbar$, of the angular momentum around the z-axis (the polar axis of a sphere). In summary:

- The orbital angular momentum quantum number l can have the non-negative integral values 0, 1, 2, ...; it tells us (through eqn 12.19) the magnitude of the orbital angular momentum of the particle.
- The magnetic quantum number m_l is limited to the $2l + 1$ values $l, l - 1, \ldots, -l$; it tells us, through $m_l\hbar$, the z-component of the orbital angular momentum.

Several features now fall into place. First, we can now see why m_l is confined to a *range* of values that depend on l: the angular momentum around a single axis (as expressed by m_l) cannot exceed the magnitude of the angular momentum (as expressed by l). Second, for a given magnitude to correspond to different values of the angular momentum around the z-axis, the angular momentum must lie at different angles (Fig. 12.30). The value of m_l therefore indicates the angle to the z-axis of the motion of the particle.

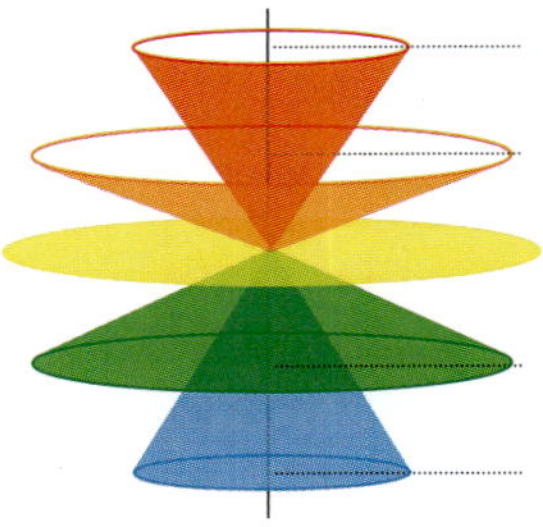

Fig. 12.31 The vector model of angular momentum acknowledges that nothing can be said about the x- and y-components of angular momentum if the z-component is known, by representing the states of angular momentum by cones.

Providing the particle has a given amount of angular momentum, its kinetic energy (its only source of energy) is independent of the orientation of its path: hence, the energy is independent of m_l, as asserted above.

What can we say about the component of angular momentum about the x- and y-axes? Almost nothing. We know that these components cannot exceed the magnitude of the angular momentum, but there is no quantum number that tells us their precise values. In fact, J_x, J_y, and J_z, the three components of angular momentum, are complementary observables in the sense described in Section 12.8 in connection with the uncertainty principle, and if one is known exactly (the value of J_z, for instance, as $m_l\hbar$), then the values of the other two cannot be specified. For this reason, the angular momentum is often represented as lying anywhere on a cone with a given z-component (indicating the value of m_l) and side (indicating the value of $\{l(l+1)\}^{1/2}$, but with indefinite projection on the x- and y-axes (Fig. 12.31). This **vector model** of angular momentum is intended to be only a representation of the quantum-mechanical aspects of angular momentum, expressing the fact that the magnitude is well defined, one component is well defined, and the two other components are indeterminate.

A brief illustration Suppose that a particle is in a state with $l = 3$. We would know that the magnitude of its angular momentum is $12^{1/2}\hbar$ (or 3.65×10^{-34} J s). The angular momentum could have any of seven orientations with z-components $m_l\hbar$, with $m_l = +3, +2, +1, 0, -1, -2$, or -3. The kinetic energy of rotation in any of these states is $12\hbar^2/mr^2$.

12.9 Vibrational motion

One very important type of motion of a molecule is the vibration of its atoms—bonds stretching, compressing, and bending. A molecule is not just a frozen, static array of atoms: all of them are in constant motion relative to one another. In the type of vibrational motion known as **harmonic oscillation**, a particle vibrates backwards and forwards restrained by a spring that obeys **Hooke's law** of force. Hooke's law states that the restoring force is proportional to the displacement, x:

$$\text{Restoring force} = -kx \qquad (12.20a)$$

The constant of proportionality k is called the **force constant:** a stiff spring has a high force constant (the restoring force is strong even for a small displacement) and a weak spring has a low force constant. The units of k are newtons per metre (N m^{-1}). The negative sign in eqn 12.20a is included because a displacement to the right (to positive x) corresponds to a force directed to the left (towards negative x). The potential energy of a particle subjected to this force increases as the square of the displacement, and specifically

$$V(x) = \tfrac{1}{2}kx^2 \qquad (12.20b)$$

A brief comment This result is easy to verify, because force is the negative gradient of the potential energy ($F = -\mathrm{d}V/\mathrm{d}x$), and differentiating $V(x)$ with respect to x gives eqn 12.20a.

The variation of V with x is shown in Fig. 12.32: it has the shape of a parabola (a curve of the form $y = ax^2$), and we say that a particle undergoing harmonic motion has a 'parabolic potential energy'.

Unlike the earlier cases we considered, the potential energy varies with position in the regions where the particle may be found, so we have to use $V(x)$ in the

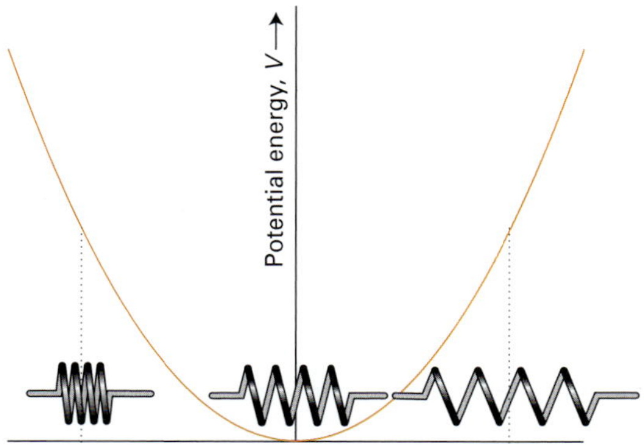

Fig. 12.32 The parabolic potential energy characteristic of an harmonic oscillator. Positive displacements correspond to extension of the spring; negative displacements correspond to compression of the spring.

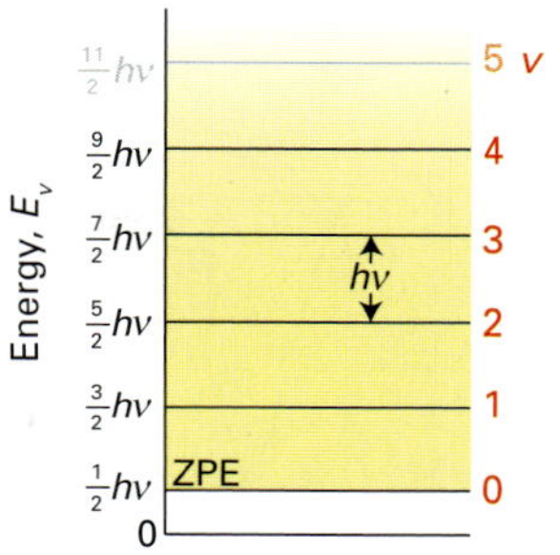

Fig. 12.33 The array of energy levels of an harmonic oscillator (the levels continue upwards to infinity). The separation depends on the mass and the force constant. Note the zero-point energy (ZPE).

Schrödinger equation. Then we have to select the solutions that satisfy the boundary equations, which in this case means that they must fit into the parabola representing the potential energy. More precisely, the wavefunctions must all go to zero for large displacements in either direction from $x = 0$: they do not have to go abruptly to zero at the edges of the parabola.

The solutions of the equation are quite hard to find, but once found they turn out to be very simple. For instance, the energies of the solutions that satisfy the boundary conditions are

$$E_v = (v + \tfrac{1}{2})h\nu \quad v = 0, 1, 2, \ldots \quad \nu = \frac{1}{2\pi}\left(\frac{k}{m}\right)^{1/2} \qquad (12.21)$$

where m is the mass of the particle and v is the **vibrational quantum number**. These energies form a uniform ladder of values separated by $h\nu$ (Fig. 12.33). The quantity ν is a frequency (in cycles per second, or hertz, Hz), and is in fact the frequency that a classical oscillator of mass m and force constant k would be calculated to have. In quantum mechanics, though, ν tells us (through $h\nu$) the separation of any pair of adjacent energy levels. The separation is large for stiff springs and low masses.

A brief illustration The force constant for an H—Cl bond is 516 N m^{-1}, where the newton (N) is the SI unit of force (1 N = 1 kg m s^{-2}). If we suppose that, because the chlorine atom is relatively very heavy, only the hydrogen atom moves, we take m as the mass of the H atom (1.67×10^{-27} kg for ^{1}H). We find

$$\nu = \frac{1}{2\pi}\left(\frac{k}{m}\right)^{1/2} = \frac{1}{2\pi}\left(\frac{516\ \mathrm{N\,m^{-1}}}{1.67\times10^{-27}\ \mathrm{kg}}\right)^{1/2} = 8.85\times10^{13}\ \mathrm{Hz}$$

The separation between adjacent levels is h times this frequency, or 5.86×10^{-20} J (58.6 zJ). Be very careful to distinguish the quantum number v (italic vee) from the frequency ν (Greek nu).

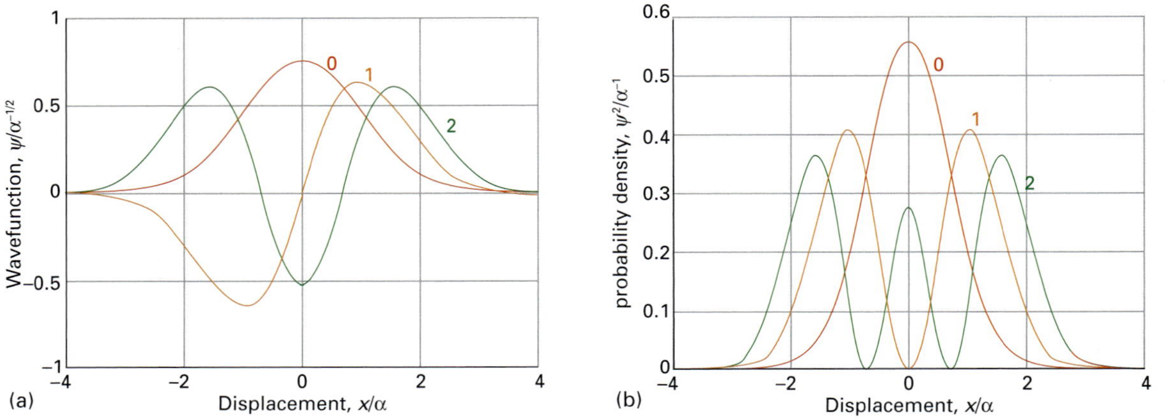

Fig. 12.34 (a) The wavefunctions and (b) the probability densities of the first three states of an harmonic oscillator. Note how the probability of finding the oscillator at large displacements increases as the state of excitation increases. The wavefunctions and displacements are expressed in terms of the parameter $\alpha = (\hbar^2/mk)^{1/4}$.

Figure 12.34 shows the shapes of the first few wavefunctions of a harmonic oscillator. The ground-state wavefunction (corresponding to $v = 0$ and having the zero-point energy $\frac{1}{2}h\nu$) is a bell-shaped curve, a curve of the form e^{-x^2} (a Gaussian function; see Section 1.6), with no nodes. This shape shows that the particle is most likely to be found at $x = 0$ (zero displacement), but may be found at greater displacements with decreasing probability. The first excited wavefunction has a node at $x = 0$ and positive and negative peaks on either side. Therefore, in this state, the particle will be found most probably with the 'spring' stretched or compressed to the same amount. However, the wavefunctions extend beyond the limits of motion of a classical oscillator (Fig. 12.35), which is another example of quantum-mechanical tunnelling.

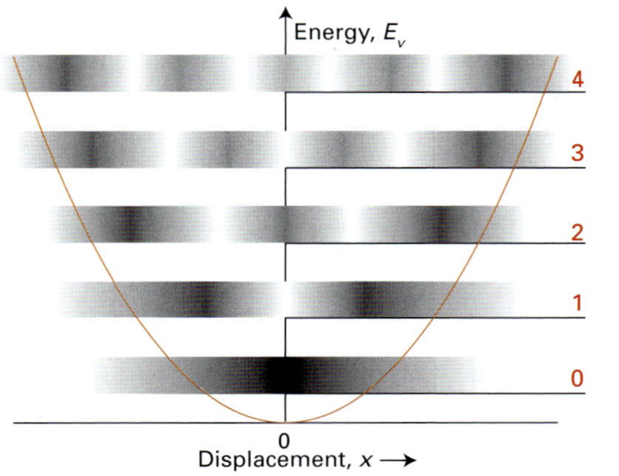

Fig. 12.35 A schematic illustration of the probability density for finding an harmonic oscillator at a given displacement. Classically, the oscillator cannot be found at displacements at which its total energy is less than its potential energy (because the kinetic energy cannot be negative). A quantum oscillator, though, may tunnel into regions that are classically forbidden.

Checklist of key ideas

You should now be familiar with the following concepts.

☐ **1** Atomic and molecular spectra show that the energies of atoms and molecules are quantized.

☐ **2** The photoelectric effect is the ejection of electrons when radiation of greater than a threshold frequency is incident on a metal.

☐ **3** The wave-like character of electrons was demonstrated by the Davisson–Germer diffraction experiment.

☐ **4** The joint wave–particle character of matter and radiation is called wave–particle duality.

☐ **5** A wavefunction, ψ, contains all the dynamical information about a system and is found by solving the appropriate Schrödinger equation subject to the constraints on the solutions known as boundary conditions.

☐ **6** According to the Born interpretation, the probability of finding a particle in a small region of space of volume δV is proportional to $\psi^2\delta V$, where ψ is the value of the wavefunction in the region.

☐ **7** According to the Heisenberg uncertainty principle, it is impossible to specify simultaneously, with arbitrary precision, both the momentum and the position of a particle.

☐ **8** The energy levels of a particle of mass m in a box of length L are quantized and the wavefunctions are sine functions (see the following table).

☐ **9** The zero-point energy is the lowest permissible energy of a system.

☐ **10** Different states with the same energy are said to be degenerate.

☐ **11** Because wavefunctions do not, in general, decay abruptly to zero, particles may tunnel into classically forbidden regions.

☐ **12** The angular momentum and the kinetic energy of a particle free to move on a circular ring are quantized; the quantum number is denoted m_l.

☐ **13** A particle on a ring and on a sphere must satisfy cyclic boundary conditions (the wavefunctions must repeat on successive cycles).

☐ **14** The angular momentum and the kinetic energy of a particle on a sphere are quantized with values determined by the quantum numbers l and m_l (see the following table); the wavefunctions are the spherical harmonics.

☐ **15** A particle undergoes harmonic motion if it is subjected to a Hooke's-law restoring force (a force proportional to the displacement).

☐ **16** The energy levels of a harmonic oscillator are equally spaced and specified by the quantum number $v = 0, 1, 2, \ldots$.

Table of key equations

The following table summarizes the equations developed in this chapter.

Property	Equation	Comment
Relation between the energy change and the frequency of radiation	$\Delta E = h\nu$	
Photoelectric effect	$E_k = h\nu - \Phi$	
de Broglie relation	$\lambda = h/p$	
Schrödinger equation	$\hat{H}\psi = E\psi$	
Heisenberg uncertainty relation	$\Delta p \Delta x \geq \frac{1}{2}\hbar$.	
Particle in a box energies	$E_n = n^2h^2/8mL^2$	$n = 1, 2, \ldots$
Particle in a box wavefunctions	$\psi_n(x) = (2/L)^{1/2}\sin(2\pi nx/L)$	
Energy of a particle on a ring	$E_{m_l} = m_l^2\hbar^2/2I,\ I = mr^2$	$m_l = 0, \pm1, \pm2 \ldots$
Angular momentum of a particle on a ring	$J_z = m_l\hbar$	$m_l = 0, \pm1, \pm2 \ldots$
Energy of a particle on a sphere	$E_l = l(l+1)\hbar^2/2I$	$l = 0, 1, 2, \ldots$
Magnitude of angular momentum of a particle on a sphere	$J = \{l(l+1)\}^{1/2}\hbar$	$l = 0, 1, 2, \ldots$
z-Component of angular momentum	$J_z = m_l\hbar$	$m_l = l, l-1, \ldots, -l$
Hooke's law	$F = -kx$	
Potential energy of a particle undergoing harmonic motion	$V = \frac{1}{2}kx^2$	
Energy of a harmonic oscillator	$E_v = (v + \frac{1}{2})h\nu$, with $\nu = (\frac{1}{2\pi})(k/m)^{1/2}$	$v = 0, 1, 2, \ldots$

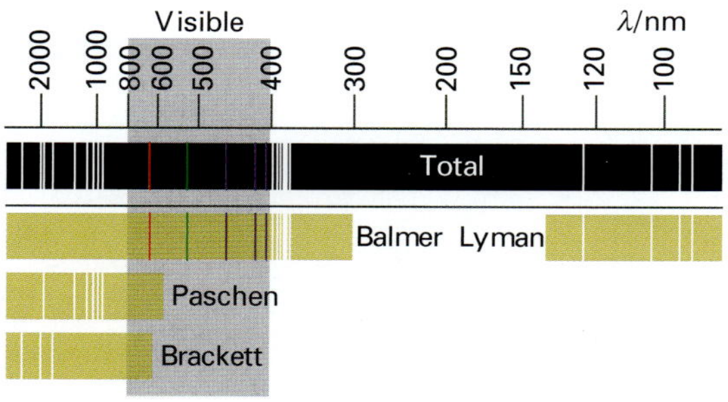

Fig. 13.1 The spectrum of atomic hydrogen. The spectrum is shown at the top, and is analysed into overlapping series below. The Balmer series lies largely in the visible region.

form, the radiation was detected photographically as a series of lines (the focused image of the slit that the light was sampled through), and the components of radiation present in a spectrum are still widely referred to as spectroscopic 'lines'.

13.1 The spectra of hydrogenic atoms

The first important contribution to understanding the spectrum of atomic hydrogen, which is observed when an electric discharge is passed through hydrogen gas, was made by the Swiss schoolteacher Johann Balmer. In 1885 he pointed out that (in modern terms) the wavenumbers of the light in the visible region of the electromagnetic spectrum fit the expression

$$\tilde{\nu} \propto \frac{1}{2^2} - \frac{1}{n^2}$$

with $n = 3, 4, \ldots$. The lines described by this formula are now called the **Balmer series** of the spectrum. Later, another set of lines was discovered in the ultraviolet region of the spectrum, and is called the **Lyman series**. Yet another set was discovered in the infrared region when detectors became available for that region, and is called the **Paschen series**. With this additional information available, the Swedish spectroscopist Johannes Rydberg noted (in 1890) that all the lines are described by the expression

$$\tilde{\nu} = R_{\mathrm{H}}\left(\frac{1}{n_1^2} - \frac{1}{n_2^2}\right) \qquad (13.1)$$

with $n_1 = 1, 2, \ldots, n_2 = n_1 + 1, n_1 + 2, \ldots$, and $R_{\mathrm{H}} =$ 109 677 cm^{-1}. The constant R_{H} is now called the **Rydberg constant** for hydrogen. The first five series of lines then correspond to n_1 taking the values 1 (Lyman), 2 (Balmer), 3 (Paschen), 4 (Brackett), and 5 (Pfund).

As we saw in Section 12.1, the existence of discrete spectroscopic lines strongly suggests that the energy of atoms is quantized and that when an atom changes its energy by ΔE, this difference is carried away as a photon of frequency ν related to ΔE by the **Bohr frequency condition** (eqn 12.1):

$$\Delta E = h\nu \qquad (13.2)$$

In terms of the wavenumber $\tilde{\nu}$ of the radiation the Bohr frequency condition is $\Delta E = hc\tilde{\nu}$. It follows that we can expect to observe discrete lines if an electron in an atom can exist only in certain energy states and electromagnetic radiation induces transitions between them.

13.2 The permitted energies of hydrogenic atoms

The quantum-mechanical description of the structure of a hydrogenic atom is based on Rutherford's **nuclear model**, in which the atom is pictured as consisting of an electron outside a central nucleus of charge Ze. To derive the details of the structure of this type of atom, we have to set up and solve the Schrödinger equation in which the potential energy, V, is the Coulomb potential energy for the interaction between the nucleus of charge $+Ze$ and the electron of charge $-e$. In general the **Coulombic potential energy** of a charge Q_1 at a distance r from another charge Q_2 is:

$$V = \frac{Q_1 Q_2}{4\pi\varepsilon_0 r} \qquad (13.3a)$$

The fundamental constant $\varepsilon_0 = 8.854 \times 10^{-12}\ \mathrm{J^{-1}\ C^2\ m^{-1}}$ is called the **vacuum permittivity**. When the charges are expressed in coulombs (C) and their separation in metres (m), the energy is expressed in joules. Note that according to this expression, the potential energy of a charge is zero when it is at an infinite distance from the other charge. On setting $Q_1 = +Ze$ and $Q_2 = -e$

$$V = -\frac{Ze^2}{4\pi\varepsilon_0 r} \qquad (13.3b)$$

The negative sign indicates that the potential energy falls (becomes more negative) as the distance between the nucleus and the electron decreases. It follows that the Schrödinger equation for the hydrogen atom has the following form:

$$-\frac{\hbar^2}{2\mu}\nabla^2\psi - \frac{Ze^2}{4\pi\varepsilon_0 r}\psi = E\psi \qquad \mu = \frac{m_{\mathrm{e}} m_{\mathrm{N}}}{m_{\mathrm{e}} + m_{\mathrm{N}}}$$

where the symbol ∇^2 is the three-dimensional version of the quantity $\mathrm{d}^2/\mathrm{d}x^2$ that we encountered in our first encounter with the Schrödinger equation (eqn 12.4) and μ (mu) is the **reduced mass.** For all except the most precise considerations, the mass of the nucleus is so much greater than the mass of the electron that the latter may be neglected in the denominator of μ, and then $\mu \approx m_e$.

A brief comment The explicit form of ∇^2 is the sum of three terms like d^2/dx^2, with one for each dimension:

$$\nabla^2 = \frac{\partial^2}{\partial x^2} + \frac{\partial^2}{\partial y^2} + \frac{\partial^2}{\partial z^2}$$

We have used the notation of partial derivatives. You can think of the expression $\nabla^2\psi$ as an indication of the total curvature in all three dimensions of the wavefunction ψ.

We also need to identify the appropriate conditions that the wavefunctions must satisfy in order to be acceptable. For the hydrogen atom, these conditions are that the wavefunction must not become infinite anywhere and that it must repeat itself (just like the particle on the surface of a sphere) on circling the nucleus either over the poles or round the equator. We should expect that, with three conditions to satisfy, three quantum numbers will emerge.

With a lot of work, the Schrödinger equation with this potential energy and these conditions can be solved, and we shall summarize the results. As usual, the need to satisfy conditions leads to the conclusion that the electron can have only certain energies, which is qualitatively in accord with the spectroscopic evidence. Schrödinger himself found that for a hydrogenic atom of atomic number Z with a nucleus of mass m_N, the allowed energy levels are given by the expression

$$E_n = -\frac{hcRZ^2}{n^2} \qquad (13.4a)$$

where

$$hcR = \frac{\mu e^4}{32\pi^2\varepsilon_0^2\hbar^2} \qquad \mu = \frac{m_e m_N}{m_e + m_N} \qquad (13.4b)$$

and $n = 1, 2, \ldots$. The constant R (not the gas constant!) is numerically identical to the experimental Rydberg constant R_H when m_N is set equal to the mass of the proton. Schrödinger must have been thrilled to find that when he calculated R_H, the value he obtained was in almost exact agreement with the experimental value.

Here we shall focus on eqn 13.4a, and unpack its significance. We shall examine (1) the role of n, (2) the significance of the negative sign, and (3) the appearance in the equation of Z^2.

The quantum number n is called the **principal quantum number.** We use it to calculate the energy of the electron in the atom by substituting its value into eqn 13.4a. The resulting energy levels are depicted in Fig. 13.2. Note how they are widely separated at low values of n, but then converge as n increases. At low values of n the electron is confined close to the nucleus by the attraction of opposite charges and the energy levels are widely spaced like those of a particle in a narrow box. At high values of n, when the electron has such a high energy that it can travel out to large distances, the energy levels are close together, like those of a particle in a large box.

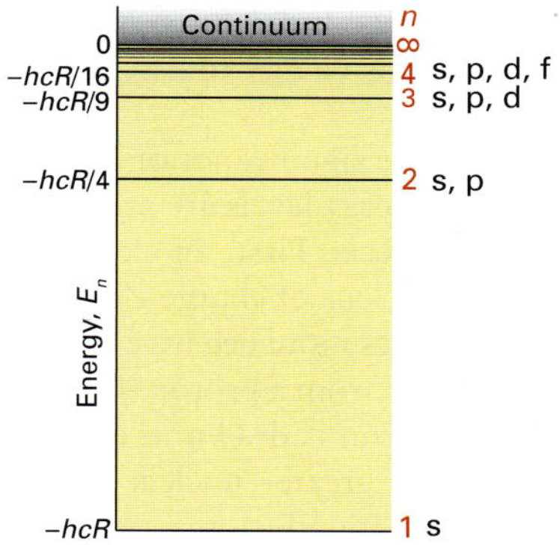

Fig. 13.2 The energy levels of the hydrogen atom. The energies are relative to a proton and an infinitely distant, stationary electron.

Now for the sign in eqn 13.4a. All the energies are negative, which signifies that an electron in an atom has a lower energy than when it is free. The zero of energy (which occurs at $n = \infty$) corresponds to the infinitely widely separated (so that the Coulomb potential energy is zero) and stationary (so that the kinetic energy is zero) electron and nucleus. The state of lowest, most negative, energy, the ground state of the atom, is the one with $n = 1$ (the lowest permitted value of n and hence the most negative value of the energy). The energy of this state is $E_1 = -hcRZ^2$: the negative sign means that the ground state lies $hcRZ^2$ *below* the energy of the infinitely separated stationary electron and nucleus. The first excited state of the atom, the state with $n = 2$, lies at $E_2 = -\frac{1}{4}hcRZ^2$. This energy level is $\frac{3}{4}hcRZ^2$ above the ground state.

These results allow us to explain the empirical expression for the spectroscopic lines observed in the spectrum of atomic hydrogen (for which $R = R_H$ and $Z = 1$). In a transition, an electron jumps from an energy level with one quantum number (n_2) to a level with a lower energy (with quantum number n_1). As a result, its energy changes by

$$\Delta E = \frac{hcR_H}{n_1^2} - \frac{hcR_H}{n_2^2} = hcR_H\left(\frac{1}{n_1^2} - \frac{1}{n_2^2}\right)$$

This energy is carried away by a photon of energy $hc\tilde{\nu}$. By equating this energy to ΔE, we immediately obtain eqn 13.1.

Now consider the significance of Z^2 in eqn 13.4a. The fact that the energy levels are proportional to Z^2 stems from two effects. First, an electron at a given distance from a nucleus of charge Ze has a potential energy that is Z times more negative than an electron at the same distance from a proton (for which $Z = 1$). However, the electron is drawn in to the vicinity of the nucleus by the greater nuclear charge, so it is more likely to be found closer to the nucleus of charge Z than the proton. This effect is also proportional to Z, so overall the energy of an electron can be expected to be proportional to the square of Z, one factor representing the Z times greater strength of the nuclear field and the second factor representing the fact that the electron is Z times more likely to be found closer to the nucleus.

Self-test 13.1

The shortest wavelength transition in the Paschen series in hydrogen occurs at 821 nm; at what wavelength does it occur in Li^{2+}? *Hint:* Think about the variation of energies with atomic number Z.

[*Answer:* $\frac{1}{9} \times 821$ nm = 91.2 nm]

The minimum energy needed to remove an electron completely from an atom is called the **ionization energy**, I. For a hydrogen atom, the ionization energy is the energy required to raise the electron from the ground state (with $n = 1$ and energy $E_1 = -hcR_H$) to the state corresponding to complete removal of the electron (the state with $n = \infty$ and zero energy). Therefore, the energy that must be supplied is $I = hcR_H = 2.180 \times 10^{-18}$ J, which corresponds to 1312 kJ mol^{-1} or 13.59 eV.

Self-test 13.2

Predict the ionization energy of He^+ given that the ionization energy of H is 13.59 eV. *Hint:* Decide how the energy of the ground state varies with Z.

[*Answer:* $I_{He^+} = 4I_H = 54.36$ eV]

13.3 Quantum numbers

The wavefunction of the electron in a hydrogenic atom is called an **atomic orbital.** The name is intended to express something less definite than the 'orbit' of classical mechanics. An electron that is described by a particular wavefunction is said to 'occupy' that orbital. So, in the ground state of the atom, the electron occupies the orbital of lowest energy (that with $n = 1$).

We have remarked that there are three mathematical conditions on the orbitals: that the wavefunctions must decay to zero as they extend to infinity, that they must match as we encircle the equator, and that they must match as we encircle the poles. Each condition gives rise to a quantum number, so each orbital is specified by three quantum numbers that act as a kind of 'address' of the electron in the atom. We can suspect that the values allowed to the three quantum numbers are linked because, as we saw in the discussion of a particle on a sphere, to get the right shape on a polar journey we also have to note how the wavefunction changes shape as we travel round the equator. It turns out that the relations between the allowed values are very simple.

We saw in Chapter 12 that in certain cases a wavefunction can be separated into factors that depend on different coordinates and that the Schrödinger equation separates into simpler versions for each variable. As may be expected for a system like a hydrogen atom, by using the separation of variables procedure, its Schrödinger equation separates into one equation for the electron moving around the nucleus (the analogue of the particle on a sphere treated in Section 12.10) and an equation for the radial dependence. The wavefunction correspondingly factorizes, and is written

$$\psi_{n,l,m_l}(r,\theta,\phi) = R_{n,l}(r)Y_{l,m_l}(\theta,\phi) \qquad (13.5)$$

The factor $R_{n,l}(r)$ is called the **radial wavefunction** and the factor $Y_{l,m_l}(\theta,\phi)$ is called the **angular wavefunction**; the latter is exactly the wavefunction we found for a particle on a sphere. As can be seen from this expression, the wavefunction is specified by three quantum numbers, all of which we have already met in different guises (Section 13.2 and Chapter 12):

Quantum number	Name	Allowed values	Determines
n	principal	1, 2, . . .,	Energy, through $E_n = -hcRZ^2/n^2$
l	orbital angular momentum	0, 1, . . . $n - 1$	Orbital angular momentum, through $J = \{l(l + 1)\}^{1/2}\hbar$
m_l	magnetic	l, $l - 1$, $l - 2$, . . ., $-l$	z-Component of orbital angular momentum, through $J_z = m_l\hbar$

Note that the radial wavefunction $R_{nl}(r)$ depends only on n and l, so all wavefunctions of a given n and l have the same radial shape regardless of the value of m_l. Similarly, the angular wavefunction $Y_{l,m_l}(\theta,\phi)$ depends only on l and m_l, so all wavefunctions of a given l and m_l have the same angular shape regardless of the value of n.

A brief illustration It follows from the restrictions on the values of the quantum numbers that there is only one orbital with $n = 1$, because when $n = 1$ the only value that l can have is 0, and that in turn implies that m_l can have only the value 0. Likewise, there are four orbitals with $n = 2$, because l can take the values 0 and 1, and in the latter case m_l can have the three values +1, 0, and −1. In general, there are n^2 orbitals with a given value of n.

A note on good practice Always give the sign of m_l, even when it is positive. So, write $m_l = +1$, not $m_l = 1$.

Although we need all three quantum numbers to specify a given orbital, eqn 13.4 reveals that for hydrogenic atoms—and, as we shall see, *only* for hydrogenic atoms—the energy depends only on the principal quantum number, n. Therefore, in hydrogenic atoms, and only in hydrogenic atoms, *all orbitals of the same value of n but different values of l and m_l have the same energy*. Recall from Section 12.9 that when we have more than one wavefunction corresponding to the same energy, we say that the wavefunctions are 'degenerate'; so, now we can say that in hydrogenic atoms all orbitals with the same value of n are degenerate.

The degeneracy of all orbitals with the same value of n (remember from the preceding *illustration* that there are n^2 of them) and, as we shall see, their similar mean radii, is the basis of saying that they all belong to the same **shell** of the atom. It is common to refer to successive shells by letters:

n	1	2	3	4 ...
	K	L	M	N ...

Thus, all four orbitals of the shell with $n = 2$ form the L shell of the atom.

Orbitals with the same value of n but different values of l belong to different **subshells** of a given shell. These subshells are denoted by the letters s, p, ... using the following correspondence:

l	0	1	2	3 ...
	s	p	d	f ...

Only these four types of subshell are important in practice. For the shell with $n = 1$, there is only one subshell, the one with $l = 0$. For the shell with $n = 2$

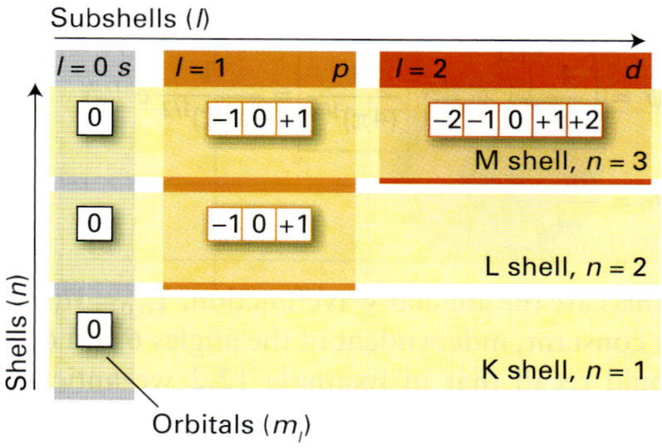

Fig. 13.3 The structures of atoms are described in terms of shells of electrons that are labelled by the principal quantum number n, and a series of n subshells of these shells, with each subshell of a shell being labelled by the quantum number l. Each subshell consists of $2l + 1$ orbitals.

(which allows $l = 0, 1$), there are two subshells, namely the 2s subshell (with $l = 0$) and the 2p subshell (with $l = 1$). The general pattern of the first three shells and their subshells is shown in Fig. 13.3. In a hydrogenic atom, all the subshells of a given shell correspond to the same energy (because, as we have seen, the energy depends on n and not on l).

We have seen that if the orbital angular momentum quantum number is l, then m_l can take the $2l + 1$ values $m_l = 0, \pm 1, \ldots, \pm l$. Therefore, each subshell contains $2l + 1$ individual orbitals (corresponding to the $2l + 1$ values of m_l for each value of l). It follows that in any given subshell, the number of orbitals is

s	p	d	f ...
1	3	5	7 ...

An orbital with $l = 0$ (and necessarily $m_l = 0$) is called an **s orbital**. A p subshell ($l = 1$) consists of three **p orbitals** (corresponding to $m_l = +1, 0, -1$). An electron that occupies an s orbital is called an **s electron**. Similarly, we can speak of p, d, ... electrons according to the orbitals they occupy.

Self-test 13.3

How many orbitals are there in a shell with $n = 5$ and what is their designation?

[*Answer:* 25; one s, three p, five d, seven f, nine g]

13.4 The wavefunctions: s orbitals

The mathematical form of a 1s orbital (the wavefunction with $n = 1$, $l = 0$, and $m_l = 0$) for a hydrogen atom is

Derivation 13.1

The radial distribution function

Consider two spherical shells centred on the nucleus, one of radius r and the other of radius $r + \delta r$. The probability of finding the electron at a radius r regardless of its direction is equal to the probability of finding it between these two spherical surfaces. The volume of the region of space between the surfaces is equal to the surface area of the inner shell, $4\pi r^2$, multiplied by the thickness, δr, of the region, and is therefore $4\pi r^2\delta r$. According to the Born interpretation, the probability of finding an electron inside a small volume of magnitude δV is given, for a normalized wavefunction that is constant throughout the region, by the value of $\psi^2\delta V$. An s orbital has the same value at all angles at a given distance from the nucleus, so it is constant throughout the shell (provided δr is very small). Therefore, interpreting δV as the volume of the shell, we obtain

$$\text{Probability} = \psi^2 \times (4\pi r^2\delta r)$$

as in eqn 13.8a. The result we have derived applies only to s orbitals.

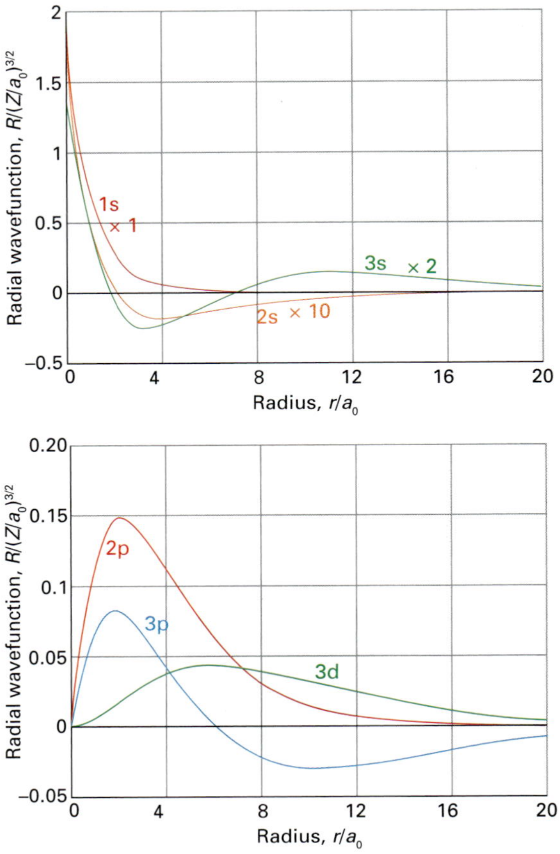

Fig. 13.9 The radial wavefunctions of some hydrogenic s, p, and d orbitals. Note that the s orbitals have a nonzero and finite value at the nucleus. The vertical scales are different in each case.

The radial distribution function tells us the probability of finding an electron at a distance r from the nucleus regardless of its direction. Because r^2 increases from 0 as r increases but ψ^2 decreases towards 0 exponentially, P starts at 0, goes through a maximum, and declines to 0 again. The location of the maximum marks the most probable *radius* (not point) at which the electron will be found. For a 1s orbital of hydrogen, the maximum occurs at a_0, the Bohr radius. An analogy that might help to fix the significance of the radial distribution function for an electron is the corresponding distribution for the population of the Earth regarded as a perfect sphere. The radial distribution function is zero at the centre of the Earth and for the next 6400 km (to the surface of the planet), when it peaks sharply and then rapidly decays again to zero. It remains virtually zero for all radii more than about 10 km above the surface. Almost all the population will be found very close to $r = 6400$ km, and it is not relevant that people are dispersed nonuniformly over a very wide range of latitudes and longitudes. The small probabilities of finding people above and below 6400 km anywhere in the world corresponds to the population that happens to be down mines or living in places as high as Denver or Tibet at the time.

A 2s orbital (an orbital with $n = 2$, $l = 0$, and $m_l = 0$) is also spherical, so its boundary surface is a sphere. Because a 2s orbital spreads further out from the nucleus than a 1s orbital—because the electron it describes has more energy to climb away from the nucleus—its boundary surface is a sphere of larger radius. The orbital also differs from a 1s orbital in its radial dependence (Fig. 13.9), for although the wavefunction has a nonzero value at the nucleus (like all s orbitals), it passes through zero before commencing its exponential decay towards zero at large distances. We summarize the fact that the wavefunction passes through zero everywhere at a certain radius by saying that the orbital has a **radial node.** A 3s orbital has two radial nodes, a 4s orbital has three radial nodes. In general, an ns orbital has $n - 1$ radial nodes.

A general feature of orbitals is that their mean radii increase with n, as more radial nodes have to be fitted into the wavefunction, with the result that it spreads out to greater radii. All orbitals of the same principal quantum number have similar mean radii, which reinforces the notion of the shell structure of the atom. Mean radii decrease with increasing Z,

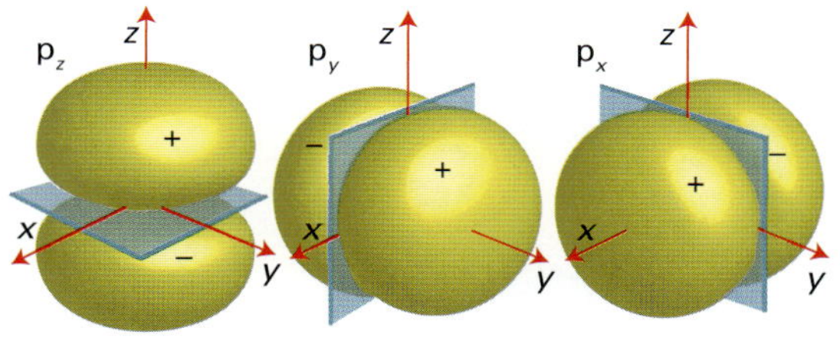

Fig. 13.10 The boundary surfaces of p orbitals. A nodal plane passes through the nucleus and separates the two lobes of each orbital. The light and dark tones denote regions of opposite sign of the wavefunction.

because the increased nuclear charge attracts the electron more strongly and it is confined more closely to the nucleus.

13.5 The wavefunctions: p and d orbitals

All p orbitals (orbitals with l = 1) have a double-lobed appearance like that shown in Fig. 13.10. The two lobes are separated by a **nodal plane** that cuts through the nucleus and arises from the angular wavefunction $Y(\theta,\phi)$. There is zero probability density for an electron on this plane. Here, for instance, is the explicit form of the $2p_z$ orbital:

$$\psi = \underbrace{\frac{1}{2}\left(\frac{1}{6a_0^3}\right)^{1/2}\frac{r}{a_0}e^{-r/2a_0}}_{R_{2,1}(r)} \times \underbrace{\left(\frac{3}{4\pi}\right)^{1/2}\cos\theta}_{Y_{1,0}(\theta,\phi)}$$

$$= \left(\frac{1}{32\pi a_0^5}\right)^{1/2} r\cos\theta\, e^{-r/2a_0}$$

Note that because ψ is proportional to r, it is zero at the nucleus, so there is zero probability density of the electron at the nucleus. The orbital is also zero everywhere on the plane with $\cos\theta = 0$, corresponding to $\theta = 90°$. The p_x and p_y orbitals are similar, but have nodal planes perpendicular to this one.

The exclusion of the electron from the nucleus is a common feature of all atomic orbitals except s orbitals. To understand its origin, we need to note that the value of the quantum number l tells us the magnitude of the angular momentum of the electron around the nucleus (in classical terms, how rapidly it is circulating around the nucleus). For an s orbital, the orbital angular momentum is zero (because $l = 0$), and in classical terms the electron does not circulate around the nucleus. Because $l = 1$ for a p orbital, the magnitude of the angular momentum of a p electron is $2^{1/2}\hbar$. As a result, a p electron—in classical terms—is flung away from the nucleus by the centrifugal force arising from its motion, but an s electron is not. The same centrifugal effect appears in all orbitals with angular momentum (those for which $l > 0$), such as d orbitals and f orbitals, and all such orbitals have nodal planes that cut through the nucleus.

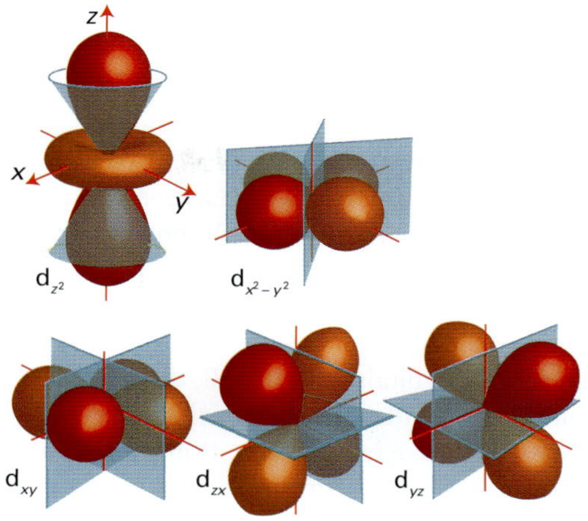

Fig. 13.11 The boundary surfaces of d orbitals. Two nodal planes in each orbital intersect at the nucleus and separate the four lobes of each orbital. (For a d_{z^2} orbital the planes are replaced by conical surfaces.) The light and dark tones denote regions of opposite sign of the wavefunction.

Each p subshell consists of three orbitals (m_l = +1, 0, −1). The three orbitals are normally represented by their boundary surfaces, as depicted in Fig. 13.10. The p_x orbital has a symmetrical double-lobed shape directed along the x-axis, and similarly the p_y and p_z orbitals are directed along the y- and z-axes, respectively. As n increases, the p orbitals become bigger (for the same reason as s orbitals) and have $n - 2$ radial nodes. However, their boundary surfaces retain the double-lobed shape shown in the illustration. Each d subshell consists of five orbitals (m_l = +2, +1, 0, −1, −2). These five orbitals are normally represented by the boundary surfaces shown in Fig. 13.11 and labelled as shown there.

A brief comment The radial wavefunction is zero at $r = 0$, but that is not a radial node because the wavefunction does not pass *through* zero there because r does not extend to negative values.

The quantum number m_l indicates, through the expression $m_l\hbar$, the component of the electron's orbital angular momentum around an arbitrary axis passing through the nucleus. As explained in Section 12.10, positive values of m_l correspond to clockwise motion seen from below and negative values correspond to anticlockwise motion. An s electron has $m_l = 0$, and has no orbital angular momentum about any axis. A p electron can circulate clockwise about an axis as seen from below ($m_l = +1$). Of its total orbital angular momentum of $2^{1/2}\hbar = 1.414\hbar$, an amount $\hbar$ is due to motion around the selected axis (the rest is due to motion around the other two axes). A p electron can also circulate counterclockwise as seen from below ($m_l = -1$), or not at all ($m_l = 0$) about that selected axis. An electron in the d subshell can circulate with five different amounts of orbital angular momentum about an arbitrary axis ($+2\hbar$, $+\hbar$, 0, $-\hbar$, $-2\hbar$).

Except for orbitals with $m_l = 0$, there is not a one-to-one correspondence between the value of m_l and the orbitals shown in the illustrations: we cannot say, for instance, that a p_x orbital has $m_l = +1$. For technical reasons, the orbitals we draw are combinations of orbitals with opposite values of m_l (p_x, for instance, is the sum—a superposition—of the orbitals with $m_l = +1$ and -1).

13.6 Electron spin

To complete the description of the state of a hydrogenic atom, we need to introduce one more concept, that of electron spin. The **spin** of an electron is an *intrinsic* angular momentum that every electron possesses and that cannot be changed or eliminated (just like its mass or its charge). The name 'spin' is evocative of a ball spinning on its axis, and (so long as it is treated with caution) this classical interpretation can be used to help to visualize the motion. However, in fact spin is a purely quantum-mechanical phenomenon and has no classical counterpart, so the analogy must be used with care.

We shall make use of two properties of electron spin (Fig. 13.12):

1. Electron spin is described by a **spin quantum number**, s (the analogue of l for orbital angular momentum), with s fixed at the single (positive) value of $\frac{1}{2}$ for all electrons at all times.
2. The spin can be clockwise or anticlockwise; these two states are distinguished by the **spin magnetic quantum number**, m_s, which can take the values $+\frac{1}{2}$ or $-\frac{1}{2}$ but no other values.

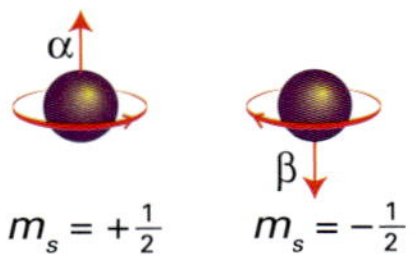

Fig. 13.12 A classical representation of the two allowed spin states of an electron. The magnitude of the spin angular momentum is $(3^{1/2}/2)\hbar$ in each case, but the directions of spin are opposite.

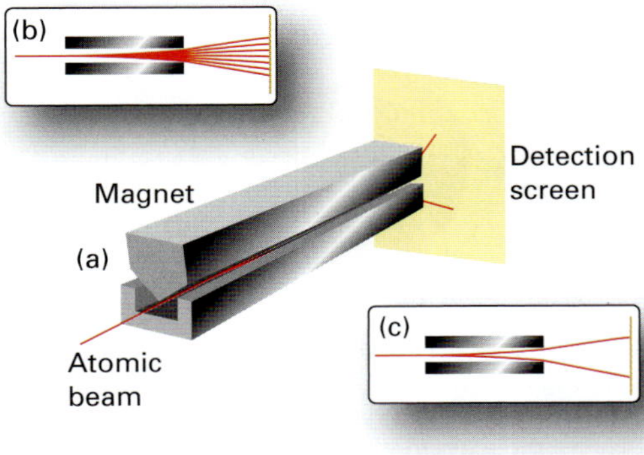

Fig. 13.13 (a) The experimental arrangement for the Stern–Gerlach experiment: the magnet is the source of an inhomogeneous field. (b) The classically expected result, when the orientations of the electron spins can take all angles. (c) The observed outcome using silver atoms, when the electron spins can adopt only two orientations (↑ and ↓).

An electron with $m_s = +\frac{1}{2}$ is called an **α electron** and commonly denoted α or ↑; an electron with $m_s = -\frac{1}{2}$ is called a **β electron** and denoted β or ↓.

A note on good practice The quantum number s is equal to $\frac{1}{2}$ for electrons. You will occasionally see its value written incorrectly as $s = +\frac{1}{2}$ or $s = -\frac{1}{2}$. For the projection, use m_s.

The existence of the electron spin was demonstrated by an experiment performed by Otto Stern and Walther Gerlach in 1921, who shot a beam of silver atoms through a strong, inhomogeneous magnetic field (Fig. 13.13). A silver atom has 47 electrons, and (as will be familiar from introductory chemistry and will be reviewed later in this chapter) 23 of the spins are ↑ and 23 spins are ↓; the one remaining spin may be either ↑ or ↓. Because the spin angular momenta of the ↑ and ↓ electrons cancel each other, the atom behaves as if it had the spin of a single electron. The idea behind the Stern–Gerlach experiment was that a rotating, charged body—in this case an electron—behaves like a magnet and interacts with the applied field. Because the magnetic field pushes or pulls the electron according to the orientation of the electron's spin, the initial beam of atoms should split into two beams, one corresponding to atoms with ↑ spin and the other to atoms with ↓ spin. This result was observed.

Other fundamental particles also have characteristic spins. For example, protons and neutrons are **spin-$\frac{1}{2}$ particles** (that is, for them $s = \frac{1}{2}$) so invariably spin with a single, irremovable angular momentum. Because the masses of a proton and a neutron are so much greater than the mass of an electron, yet they all have the same spin angular momentum, the classical picture of proton and neutron spin would be of particles spinning much more slowly than an electron. Some elementary particles have $s = 1$ and therefore have a higher intrinsic angular momentum than an electron. For our purposes the most important **spin-1 particle** is the photon. It is a very deep feature of nature, that the fundamental particles from which matter is built have half-integral spin (such as electrons and quarks, all of which have $s = \frac{1}{2}$). The particles that transmit forces between these particles, so binding them together into entities like nuclei, atoms, and planets, all have integral spin (such as $s = 1$ for the photon, which transmits the electromagnetic interaction between charged particles). Fundamental particles with half-integral spin are called **fermions**; those with integral spin are called **bosons**. Matter therefore consists of fermions bound together by bosons.

13.7 Spectral transitions and selection rules

We can think of the sudden change in the distribution of the electron as it changes its spatial distribution from one orbital to another orbital as jolting the electromagnetic field into oscillation, and that oscillation corresponds to the generation of a photon of light. It turns out, however, that not all transitions between all available orbitals are possible. For example, it is not possible for an electron in a 3d orbital to make a transition to a 1s orbital. Transitions are classified as either **allowed**, if they can contribute to the spectrum, or **forbidden**, if they cannot. The allowed or forbidden character of a transition can be traced to the role of the photon spin, which we mentioned above. When a photon, with its one unit of angular momentum, is generated in a transition, the angular momentum of the electron must change by one unit to compensate for the angular momentum carried away by the photon. That is, the angular momentum must be conserved—neither created nor destroyed—just as linear momentum is conserved in collisions. Thus, an electron in a d orbital (with $l = 2$) cannot make a transition into an s orbital (with $l = 0$) because the photon cannot carry away enough angular momentum. Similarly, an s electron cannot make a transition to another s orbital, because then there is no change in the electron's angular momentum to make up for the angular momentum carried away by the photon.

A **selection rule** is a statement about which spectroscopic transitions are allowed. They are derived (for atoms) by identifying the transitions that conserve angular momentum when a photon is emitted or absorbed. The selection rules for hydrogenic atoms are

$$\Delta l = \pm 1 \qquad \Delta m_l = 0, \pm 1$$

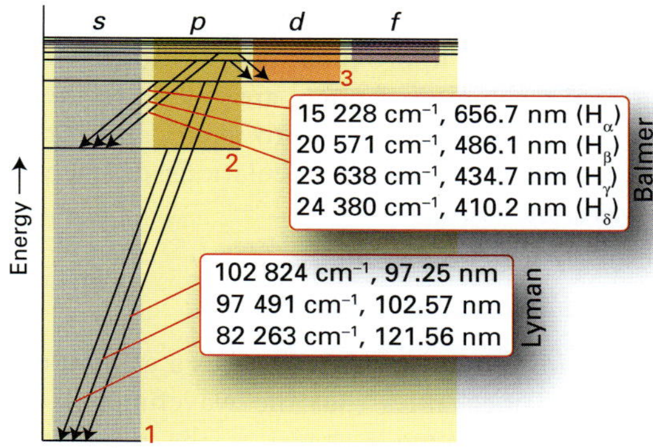

Fig. 13.14 A Grotrian diagram that summarizes the appearance and analysis of the spectrum of atomic hydrogen.

The principal quantum number n can change by any amount consistent with the Δl for the transition because it does not relate directly to the angular momentum.

A brief illustration To identify the orbitals to which an electron in a 4d orbital may make spectroscopic transitions we apply the selection rules, principally the rule concerning l. Because $l = 2$, the final orbital must have $l = 1$ or 3. Thus, an electron may make a transition from a 4d orbital to any np orbital (subject to $\Delta m_l = 0, \pm 1$) and to any nf orbital (subject to the same rule). However, it cannot undergo a transition to any other orbital, so a transition to any ns orbital or another nd orbital is forbidden.

Self-test 13.5

To what orbitals may a 4s electron make spectroscopic transitions?

[*Answer:* np orbitals only]

Selection rules enable us to construct a **Grotrian diagram** (Fig. 13.14), which is a diagram that summarizes the energies of the states and the allowed transitions between them. The thickness of a transition line in the diagram is sometimes used to indicate in a general way its relative intensity in the spectrum.

The structures of many-electron atoms

The Schrödinger equation for a many-electron atom is highly complicated because all the electrons interact with one another. Even for a He atom, with its two electrons, no mathematical expression for the orbitals and energies can be given and we are forced to make approximations. Modern computational techniques, though, are able to refine the approximations we are about to make, and permit highly accurate numerical calculations of energies and wavefunctions.

13.8 The orbital approximation

We show in Derivation 13.2 that it is a general rule in quantum mechanics that the wavefunction for several noninteracting particles is the product of the wavefunctions for each particle. This rule justifies the **orbital approximation**, in which we suppose that a reasonable first approximation to the exact wavefunction is obtained by letting each electron occupy its 'own' orbital, and writing

$$\psi = \psi(1)\psi(2)\ldots \qquad (13.8)$$

where $\psi(1)$ is the wavefunction of electron 1, $\psi(2)$ that of electron 2, and so on.

Derivation 13.2

Many-particle wavefunctions

Consider a two-particle system. If the particles do not interact with one another, the total hamiltonian that appears in the Schrödinger equation is the sum of contributions from each particle, and the equation itself is

$$\{\hat{H}(1)+\hat{H}(2)\}\psi(1,2)=E\psi(1,2)$$

We need to verify that $\psi(1,2) = \psi(1)\psi(2)$ is a solution, where each individual wavefunction is a solution of its 'own' Schrödinger equation:

$$\hat{H}(1)\psi(1)=E(1)\psi(1) \qquad \hat{H}(2)\psi(2)=E(2)\psi(2)$$

To do so, we substitute $\psi(1,2) = \psi(1)\psi(2)$ into the full equation, then let $\hat{H}(1)$ operate on $\psi(1)$ and $\hat{H}(2)$ operate on $\psi(2)$:

$$\begin{aligned}\{\hat{H}(1)+\hat{H}(2)\}\psi(1)\psi(2) &= \hat{H}(1)\psi(1)\psi(2)+\hat{H}(2)\psi(1)\psi(2)\\ &= \psi(2)\boxed{\hat{H}(1)\psi(2)}+\psi(1)\boxed{\hat{H}(2)\psi(2)}\\ &= \psi(2)\boxed{E(1)\psi(2)}+\psi(1)\boxed{E(2)\psi(2)}\\ &= \{E(1)+E(2)\}\psi(1)\psi(2)\end{aligned}$$

This expression has the form of the original Schrödinger equation, so $\psi(1,2) = \psi(1)\psi(2)$ is indeed a solution, and we can identify the total energy as $E = E(1) + E(2)$. Note that this argument fails if the particles interact with one another, because then there is an additonal term in the hamiltonian and the variables cannot be separated. For electrons, therefore, writing $\psi(1,2) = \psi(1)\psi(2)$ is an approximation.

We can think of the individual orbitals as resembling the hydrogenic orbitals, but with nuclear charges that are modified by the presence of all the other electrons in the atom. This description is only approximate, but it is a useful model for discussing the properties of atoms, and is the starting point for more sophisticated descriptions of atomic structure.

A brief illustration If both electrons occupy the same 1s orbital, the wavefunction for each electron in helium is $\psi = (8/\pi a_0^3)^{1/2}e^{-2r/a_0}$. If electron 1 is at a radius r_1 and electron 2 is at a radius r_2 (and at any angle), then the overall wavefunction for the two-electron atom is

$$\psi = \psi(1)\psi(2) = \left(\frac{8}{\pi a_0^3}\right)^{1/2} e^{-2r_1/a_0} \times \left(\frac{8}{\pi a_0^3}\right)^{1/2} e^{-2r_2/a_0}$$

$$\overset{\text{Use } e^xe^y=e^{x+y}}{=} \left(\frac{8}{\pi a_0^3}\right) e^{-2(r_1+r_2)/a_0}$$

The orbital approximation allows us to express the electronic structure of an atom by reporting its **configuration**, a statement of the orbitals that are occupied (usually, but not necessarily, in its ground state). For example, because the ground state of a hydrogen atom consists of a single electron in a 1s orbital, we report its configuration as $1s^1$ (read 'one s one'). A helium atom has two electrons. We can imagine forming the atom by adding the electrons in succession to the orbitals of the bare nucleus (of charge $2e$). The first electron occupies a hydrogenic 1s orbital, but because $Z = 2$, the orbital is more compact than in H itself. The second electron joins the first in the same 1s orbital, and so the electron configuration of the ground state of He is $1s^2$ (read 'one s two').

13.9 The Pauli principle

Lithium, with $Z = 3$, has three electrons. Two of its electrons occupy a 1s orbital drawn even more closely than in He around the more highly charged nucleus. The third electron, however, does not join the first two in the 1s orbital because a $1s^3$ configuration is forbidden by a fundamental feature of nature summarized by the **Pauli exclusion principle:**

> No more than two electrons may occupy any given orbital, and if two electrons do occupy one orbital, then their spins must be paired.

Electrons with **paired spins**, denoted ↑↓, have zero net spin angular momentum because the spin angular momentum of one electron is cancelled by the spin of the other. The exclusion principle is the key to understanding the structures of complex atoms, to chemical periodicity, and to molecular structure. It was proposed by the Austrian Wolfgang Pauli in 1924 when he was trying to account for the absence of some lines in the spectrum of helium. In Further information 13.1 we see that the exclusion principle is a consequence of an even deeper statement about wavefunctions.

Lithium's third electron cannot enter the 1s orbital because that orbital is already full: we say the K shell is **complete** and that the two electrons form a **closed shell.** Because a similar closed shell occurs in the He atom, we denote it [He]. The third electron is excluded from the K shell (n = 1) and must occupy the next available orbital, which is one with n = 2 and hence belonging to the L shell. However, we now have to decide whether the next available orbital is the 2s orbital or a 2p orbital, and therefore whether the lowest energy configuration of the atom is [He]$2s^1$ or [He]$2p^1$.

13.10 Penetration and shielding

Unlike in hydrogenic atoms, in many-electron atoms the 2s and 2p orbitals (and, in general, all the subshells of a given shell) are not degenerate. For reasons we shall now explain, s electrons generally lie lower in energy than p electrons of a given shell, and p electrons lie lower than d electrons.

An electron in a many-electron atom experiences a Coulombic repulsion from all the other electrons present. When the electron is at a distance r from the nucleus, the repulsion it experiences from the other electrons can be modelled by a point negative charge located on the nucleus and having a magnitude equal to the charge of the electrons within a sphere of radius r (Fig. 13.15). The effect of the point negative charge is to lower the full charge of the nucleus from Ze to

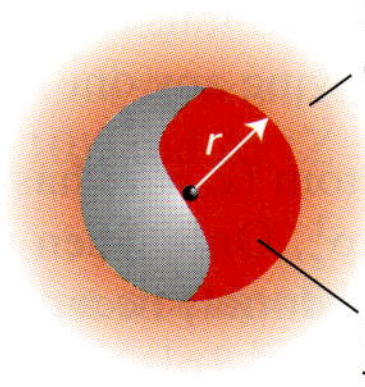

Fig. 13.15 An electron at a distance r from the nucleus experiences a Coulombic repulsion from all the electrons within a sphere of radius r and that is equivalent to a point negative charge located on the nucleus. The effect of the point charge is to reduce the apparent nuclear charge of the nucleus from Ze to $Z_{eff}e$.

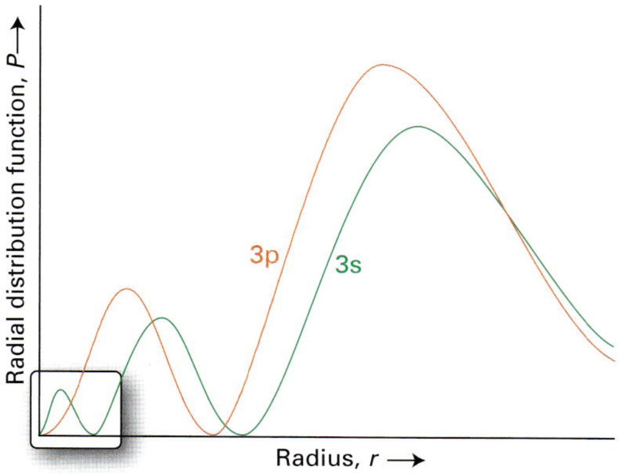

Fig. 13.16 An electron in an s orbital (here a 3s orbital) is more likely to be found close to the nucleus than an electron in a p orbital of the same shell. Hence it experiences less shielding and is more tightly bound.

$Z_{eff}e$, the **effective nuclear charge.** To express the fact that an electron experiences a nuclear charge that has been modified by the other electrons present, we say that the electron experiences a **shielded nuclear charge.** The electrons do not actually 'block' the full Coulombic attraction of the nucleus: the effective charge is simply a way of expressing the net outcome of the nuclear attraction and the electronic repulsions in terms of a single equivalent charge at the centre of the atom.

A note on good practice Commonly, Z_{eff} itself is referred to as the 'effective nuclear charge', although strictly that quantity is $Z_{eff}e$.

The effective nuclear charges experienced by s and p electrons are different because the electrons have different wavefunctions and therefore different distributions around the nucleus (Fig. 13.16). An s electron has a greater **penetration** through inner shells than a p electron of the same shell in the sense that an s electron is more likely to be found close to the nucleus than a p electron of the same shell (Fig. 13.17). As a result of this greater penetration, an s electron experiences less shielding than a p electron of the same shell and therefore experiences a larger Z_{eff}. Consequently, by the combined effects of penetration and shielding, an s electron is more tightly bound than a p electron of the same shell. Similarly, a d electron penetrates less than a p electron of the same shell, and it therefore experiences more shielding and an even smaller Z_{eff}.

The consequence of penetration and shielding is that, in general, the energies of orbitals in the same shell of a many-electron atom lie in the order s < p < d < f. The individual orbitals of a given subshell

inorganic (and bioinorganic) d-metal chemistry. A similar intrusion of the f orbitals in Periods 6 and 7 accounts for the existence of the **f block** of the periodic table (the lanthanoids and actinoids).

13.13 The configurations of cations and anions

The configurations of cations of elements in the s, p, and d blocks of the periodic table are derived by removing electrons from the ground-state configuration of the neutral atom in a specific order. First, we remove any valence p electrons, then the valence s electrons, and then as many d electrons as are necessary to achieve the stated charge. For instance, because the configuration of Fe is [Ar]$3d^6 4s^2$, an Fe^{3+} cation has the configuration [Ar]$3d^5$.

The configurations of anions are derived by continuing the building-up procedure and adding electrons to the neutral atom until the configuration of the next noble gas has been reached. Thus, the configuration of an O^{2-} ion is achieved by adding two electrons to [He]$2s^2 2p^4$, giving [He]$2s^2 2p^6$, the configuration of Ne.

Self-test 13.7

Predict the electron configurations of (a) a Cu^{2+} ion and (b) an S^{2-} ion.

[*Answer:* (a) [Ar]$3d^9$, (b) [Ne]$3s^2 3p^6$]

13.14 Self-consistent field orbitals

The treatment we have given to the electronic configuration of many-electron species is only approximate because it is hopeless to expect to find exact solutions of a Schrödinger equation that takes into account the interaction of all the electrons with one another. However, computational techniques are available that give very detailed and reliable approximate solutions for the wavefunctions and energies. The techniques were originally introduced by D.R. Hartree (before computers were available) and then modified by V. Fock to take into account the Pauli principle correctly. In broad outline, the **Hartree–Fock self-consistent field** (HF-SCF) procedure is as follows.

Imagine that we have a rough idea of the structure of the atom. In the Ne atom, for instance, the orbital approximation suggests the configuration $1s^2 2s^2 2p^6$ with the orbitals approximated by hydrogenic atomic orbitals. Now consider one of the 2p electrons. A Schrödinger equation can be written for this electron by ascribing to it a potential energy due to the nuclear attraction and the repulsion from the other electrons. Although the equation is for the 2p orbital, it depends on the wavefunctions of all the other occupied orbitals in the atom. To solve the equation, we guess an approximate form of the wavefunctions of all the orbitals except 2p and then solve the Schrödinger equation for the 2p orbital. The procedure is then repeated for the 1s and 2s orbitals. This sequence of calculations gives the form of the 2p, 2s, and 1s orbitals, and in general they will differ from the set used initially to start the calculation. These improved orbitals can be used in another cycle of calculation, and a second improved set of orbitals and a better energy are obtained. The recycling continues until the orbitals and energies obtained are insignificantly different from those used at the start of the current cycle. The solutions are then self-consistent and accepted as solutions of the problem.

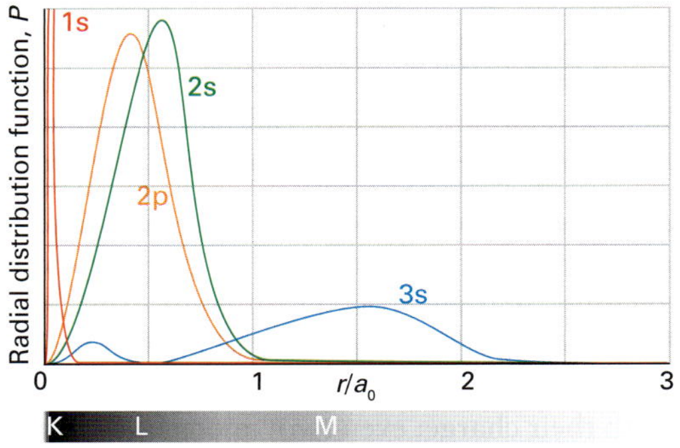

Fig. 13.19 The radial distribution functions for the orbitals of Na based on SCF calculations. Note the shell-like structure, with the 3s orbital outside the inner K and L shells.

Figure 13.19 shows plots of some of the HF-SCF radial distribution functions for sodium. They show the grouping of electron density into shells, as was anticipated by the early chemists, and the differences of penetration as discussed above. These SCF calculations therefore support the qualitative discussions that are used to explain chemical periodicity. They also considerably extend that discussion by providing detailed wavefunctions and precise energies.

Periodic trends in atomic properties

The periodic recurrence of analogous ground-state electron configurations as the atomic number increases accounts for the periodic variation in the

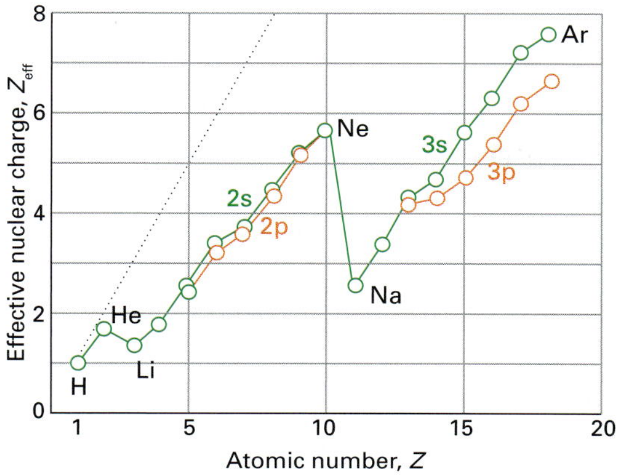

Fig. 13.20 The variation of the effective atomic number with actual atomic number for the elements of the first three periods. The value of Z_{eff} depends on the identity of the orbital occupied by the electron: we show the values only for the valence electrons.

Table 13.1

Atomic radii of main-group elements, r/pm

Li	Be	B	C	N	O	F
157	112	88	77	74	66	64
Na	Mg	Al	Si	P	S	Cl
191	160	143	118	110	104	99
K	Ca	Ga	Ge	As	Se	Br
235	197	153	122	121	117	114
Rb	Sr	In	Sn	Sb	Te	I
250	215	167	158	141	137	133
Cs	Ba	Tl	Pb	Bi	Po	
272	224	171	175	182	167	

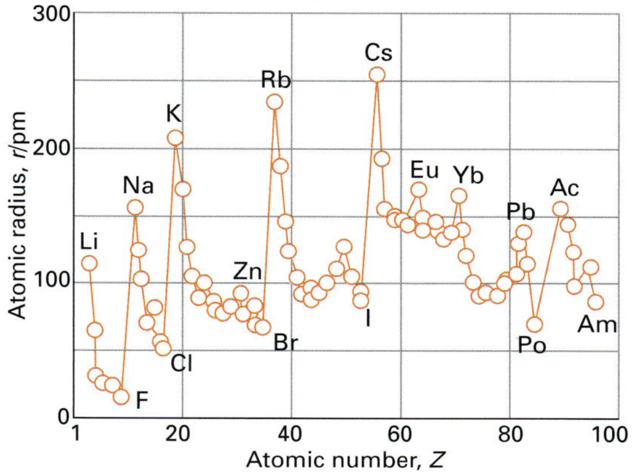

Fig. 13.21 The variation of atomic radius through the periodic table. Note the contraction of radius following the lanthanoids in Period 6 (following Yb, ytterbium).

properties of atoms. Here we concentrate on two aspects of atomic periodicity: atomic radius and ionization energy. Both can be correlated with the effective nuclear charge, and Fig. 13.20 shows how this quantity varies through the first three periods.

13.15 Atomic radius

The **atomic radius** of an element is half the distance between the centres of neighbouring atoms in a solid (such as Cu) or, for nonmetals, in a homonuclear molecule (such as H_2 or S_8). It is of great significance in chemistry, for the size of an atom is one of the most important controls on the number of chemical bonds the atom can form. Moreover, the size and shape of a molecule depend on the sizes of the atoms of which it is composed, and molecular shape and size are crucial aspects of a molecule's biological function. Atomic radius also has an important technological aspect, because the similarity of the atomic radii of the d-block elements is the main reason why they can be blended together to form so many different alloys, particularly varieties of steel.

In general, atomic radii decrease from left to right across a period and increase down each group (Table 13.1 and Fig. 13.21). The decrease across a period can be traced to the increase in nuclear charge, which draws the electrons in closer to the nucleus. The increase in nuclear charge is partly cancelled by the increase in the number of electrons, but because electrons are spread over a region of space, one electron does not fully shield one nuclear charge, so the increase in nuclear charge dominates. The increase in atomic radius down a group (despite the increase in nuclear charge) is explained by the fact that the valence shells of successive periods correspond to higher principal quantum numbers. That is, successive periods correspond to the start and then completion of successive (and more distant) shells of the atom that surround each other like the successive layers of an onion. The need to occupy a more distant shell leads to a larger atom despite the increased nuclear charge.

A modification of the increase down a group is encountered in Period 6, for the radii of the atoms late in the d block and in the following regions of the p block are not as large as would be expected by simple extrapolation down the group. The reason can be traced to the fact that in Period 6 the f orbitals are in the process of being occupied. An f electron is a very inefficient shielder of nuclear charge (for reasons connected with its radial extension), and as the atomic number increases from La to Yb, there is a considerable contraction in radius. By the time the d block resumes (at lutetium, Lu), the poorly shielded but considerably increased nuclear charge has drawn in the surrounding electrons, and the atoms are compact. They are so compact, that the metals in this region of the periodic table (iridium to lead) are very dense. The reduction in radius below that expected by extrapolation from preceding periods is called the **lanthanide contraction.**

13.16 Ionization energy and electron affinity

The minimum energy necessary to remove an electron from a many-electron atom is its **first ionization energy**, I_1. The **second ionization energy**, I_2, is the minimum energy needed to remove a second electron (from the singly charged cation):

$$X(g) \rightarrow X^+(g) + e^-(g) \qquad I_1 = E(X^+) - E(X) \quad (13.9)$$

$$X^+(g) \rightarrow X^{2+}(g) + e^-(g) \qquad I_2 = E(X^{2+}) - E(X^+)$$

The variation of the first ionization energy through the periodic table is shown in Fig. 13.22 and some numerical values are given in Table 13.2. The ionization energy of an element plays a central role in determining the ability of its atoms to participate in bond formation (for bond formation, as we shall see in Chapter 14, is a consequence of the relocation of electrons from one atom to another). After atomic radius, it is the most important property for determining an element's chemical characteristics.

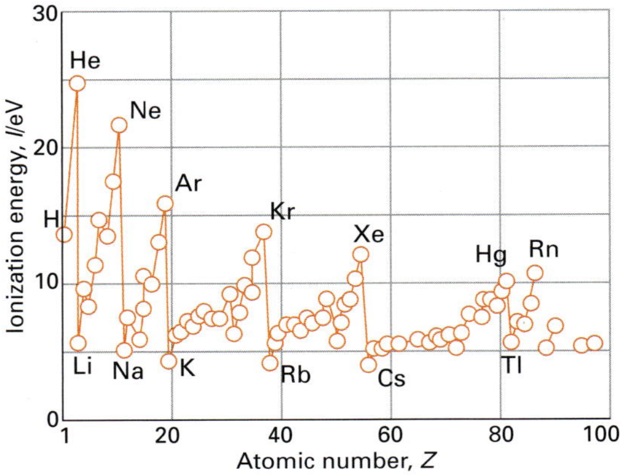

Fig. 13.22 The periodic variation of the first ionization energies of the elements.

The following trends are noteworthy:

- Lithium has a low first ionization energy: its outermost electron is well shielded from the weakly charged nucleus by the core ($Z_{eff} = 1.3$ compared with $Z = 3$) and it is easily removed.
- Beryllium has a higher nuclear charge than lithium, and its outermost electron (one of the two 2s electrons) is more difficult to remove: its ionization energy is larger.
- The ionization energy decreases between beryllium and boron because in the latter the outermost electron occupies a 2p orbital and is less strongly bound than if it had been a 2s electron.
- The ionization energy increases between boron and carbon because the latter's outermost

Table 13.2

*First ionization energies of main-group elements, I/eV**

H							He
13.60							24.59
Li	Be	B	C	N	O	F	Ne
5.32	9.32	8.30	11.26	14.53	13.62	17.42	21.56
Na	Mg	Al	Si	P	S	Cl	Ar
5.14	7.65	5.98	8.15	10.49	10.36	12.97	15.76
K	Ca	Ga	Ge	As	Se	Br	Kr
4.34	6.11	6.00	7.90	9.81	9.75	11.81	14.00
Rb	Sr	In	Sn	Sb	Te	I	Xe
4.18	5.70	5.79	7.34	8.64	9.01	10.45	12.13
Cs	Ba	Tl	Pb	Bi	Po	At	Rn
3.89	5.21	6.11	7.42	7.29	8.42	9.64	10.78

* 1 eV = 96.485 kJ mol^{-1}. See also Table 3.2.

Table 13.3

*Electron affinities of main-group elements, E_{ea}/eV**

H							He
+0.75							<0†
Li	Be	B	C	N	O	F	Ne
+0.62	−0.19	+0.28	+1.26	−0.07	+1.46	+3.40	−0.30†
Na	Mg	Al	Si	P	S	Cl	Ar
+0.55	−0.22	+0.46	+1.38	+0.46	+2.08	+3.62	−0.36†
K	Ca	Ga	Ge	As	Se	Br	Kr
+0.50	−1.99	+0.3	+1.20	+0.81	+2.02	+3.37	−0.40†
Rb	Sr	In	Sn	Sb	Te	I	Xe
+0.49	+1.51	+0.3	+1.20	+1.05	+1.97	+3.06	−0.42†
Cs	Ba	Tl	Pb	Bi	Po	At	Rn
+0.47	−0.48	+0.2	+0.36	+0.95	+1.90	+2.80	−0.42†

* 1 eV = 96.485 kJ mol^{-1}. See also Table 3.3.

† Calculated.

electron is also 2p and the nuclear charge has increased.

- Nitrogen has a still higher ionization energy because of the further increase in nuclear charge.
- There is now a kink in the curve because the ionization energy of oxygen is lower than would be expected by simple extrapolation.
- At oxygen, a 2p orbital must become doubly occupied, and the electron–electron repulsions are increased above what would be expected by simple extrapolation along the row. (The kink is less pronounced in the next row, between phosphorus and sulfur, because their orbitals are more diffuse.)
- The values for oxygen, fluorine, and neon fall roughly on the same line, the increase of their ionization energies reflecting the increasing attraction of the nucleus for the outermost electrons.
- The outermost electron in sodium is 3s. It is far from the nucleus, and the latter's charge is shielded by the compact, complete neon-like core. As a result, the ionization energy of sodium is substantially lower than that of neon.
- The periodic cycle starts again along this row, and the variation of the ionization energy can be traced to similar reasons.

The **electron affinity**, E_{ea}, is the difference in energy between a neutral atom and its anion. It is the energy *released* in the process

$$\begin{aligned} &X(g) + e^-(g) \rightarrow X^-(g) \\ &E_{ea} = E(X) - E(X^-) \end{aligned} \qquad (13.10a)$$

The electron affinity is positive if the anion has a lower energy than the neutral atom. Care should be taken to distinguish the electron affinity from the electron-gain enthalpy (Section 3.2): they have very similar numerical values but differ in sign:

$$\begin{aligned} &X(g) + e^-(g) \rightarrow X^-(g) \\ &\Delta_{ea}H^{\ominus} = H_m^{\ominus}(X^-) - H_m^{\ominus}(X) \end{aligned} \qquad (13.10b)$$

Electron affinities (Table 13.3) vary much less systematically through the periodic table than ionization energies. However, the following general observations are important:

- Broadly speaking the highest electron affinities are found close to fluorine. In the halogens, the incoming electron enters the valence shell and experiences a strong attraction from the nucleus.
- The electron affinities of the noble gases are negative—which means that the anion has a higher energy than the neutral atom—because the incoming electron occupies an orbital outside the closed valence shell. It is then far from the nucleus and repelled by the electrons of the closed shells.
- The first electron affinity of oxygen is positive for the same reason as for the halogens. However, the second electron affinity (for the formation of O^{2-} from O^-) is strongly negative because although the incoming electron enters the valence shell, it experiences a strong repulsion from the net negative charge of the O^- ion.

The spectra of complex atoms

The spectra of many-electron atoms can be very complicated, yet that complexity contains a great deal of detailed information about the interactions between electrons. Here we consider the notation used to specify the states of atoms. Chemists need to know how to designate the states of atoms when they are describing photochemical events in the atmosphere and the chemical composition of stars (Box 13.1). We also describe an interaction that has important consequences in molecular spectroscopy and magnetism.

13.17 Term symbols

For historical reasons, the energy level of an atom is called a **term** and the notation used to specify the term is called a **term symbol**. A term symbol looks like $^3D_2$, with each component (the 3, the D, and the 2) telling us something about the angular momentum of the electrons in the atom. The scheme we shall use to arrive at the term symbols of atoms is called **Russell–Saunders coupling** and is based on the notion that the orbital and spin angular momenta of each electron couple together, and then all these resultants couple together to give an overall total angular momentum. The calculations focus on the valence

Box 13.1 Spectroscopy of stars

The bulk of stellar material consists of neutral and ionized forms of hydrogen and helium atoms, with helium being the product of 'hydrogen burning' by nuclear fusion. However, nuclear fusion also makes heavier elements. It is generally accepted that the outer layers of stars are composed of lighter elements, such as H, He, C, N, O, and Ne in both neutral and ionized forms. Heavier elements, including neutral and ionized forms of Si, Mg, Ca, S, and Ar, are found closer to the stellar core. The core itself contains the heaviest elements and ^{56}Fe is particularly abundant because it is very stable. All of these elements are in the gas phase on account of the very high temperatures in stellar interiors. For example, the temperature is estimated to be 3.6 MK (1 MK = 10^6 K) half-way to the centre of the Sun.

Astronomers use spectroscopic techniques to determine the chemical composition of stars because each element, and indeed each isotope of an element, has a characteristic spectral signature that is transmitted through space by the star's light. To understand the spectra of stars, we must first know why they shine. Nuclear reactions in the dense stellar interior generate radiation that travels to less dense outer layers. Absorption and re-emission of photons by the atoms and ions in the interior give rise to a quasicontinuum of radiation energy that is emitted into space by a thin layer of gas called the *photosphere*. To a good approximation, the distribution of energy emitted from a star's photosphere resembles that for a very hot object. For example, the energy distribution of our Sun's photosphere is like that of an object heated to a temperature of 5800 K. Superimposed on the radiation continuum are sharp absorption and emission lines from neutral atoms and ions present in the photosphere. Analysis of stellar radiation with a spectrometer mounted onto a telescope, such as the Hubble Space Telescope, yields the chemical composition of the star's photosphere by comparison with known spectra of the elements. The data can also reveal the presence of small molecules, such as CN, C_2, TiO, and ZrO, in certain 'cold' stars, which are stars with relatively low effective temperatures.

The two outermost layers of a star are the *chromosphere*, a region just above the photosphere, and the *corona*, a region above the chromosphere that can be seen (with proper care) during eclipses. The photosphere, chromosphere, and corona comprise a star's 'atmosphere'. Our Sun's chromosphere is much less dense than its photosphere and its temperature is much higher, rising to about 10 kK (1 kK = 10^3 K). The reasons for this increase in temperature are not fully understood. The temperature of our Sun's corona is very high, rising up to 1.5 MK, so black-body emission is strong from the X-ray to the radio-frequency region of the spectrum. The spectrum of the Sun's corona is dominated by emission lines from electronically excited species, such as neutral atoms and a number of highly ionized species. The most intense emission lines in the visible range are from the Fe^{13+} ion at 530.3 nm, the Fe^{9+} ion at 637.4 nm, and the Ca^{4+} ion at 569.4 nm.

Because our telescopes detect only light from the outer layers of stars, their overall chemical composition must be inferred from theoretical work on their interiors and from spectral analysis of their atmospheres. Data on our Sun indicate that it is 92 per cent hydrogen and 7.8 per cent helium. The remaining 0.2 per cent are due to heavier elements, among which C, N, O, Ne, and Fe are the most abundant. More advanced analysis of spectra also permit the determination of other properties of stars, such as their relative speeds and their effective temperatures.

electrons, for core electrons do not contribute to the overall angular momentum of an atom.

The letter (D, for instance) tells us the total orbital angular momentum of the electrons in the atom. To find it, we work out the **total orbital angular momentum quantum number**, L, in the manner described below, and then we use the following code:

L	0	1	2	3...
	S	P	D	F...

Note that the code is the same as for orbitals, but we use upper-case Roman letters. To find L we identify the orbital angular momentum quantum numbers (l_1 and l_2 for instance) of the electrons in the valence shell of the atom, and then form the following series:

$$L = l_1 + l_2, l_1 + l_2 - 1, \ldots, |l_1 - l_2|$$

This and the analogous series introduced later are called **Clebsch–Gordan series.** The modulus signs (|...|) simply mean that the series terminates at a positive value. The highest total orbital angular momentum occurs when the two electrons are orbiting in the same direction (in classical terms, like the planets round the Sun); the lowest occurs when they are orbiting in opposite directions.

A brief illustration Suppose we are considering the excited-state configuration of carbon [He]$2s^2 2p^1 3p^1$ in which a 2p electron has been promoted to a 3p orbital (though not by electromagnetic radiation; see Section 13.19). We concentrate on the p electrons because the s electrons have no orbital angular momentum. For each electron $l = 1$ (that is, $l_1 = 1$ and $l_2 = 1$ for the two electrons we are considering). It follows that

$$L = 1 + 1, 1 + 1 - 1, \ldots |1 - 1| = 2, 1, 0$$

This result is shown pictorially in Fig. 13.23. It follows that the configuration gives rise to D, P, and S terms, corresponding to the three allowed values of the total orbital angular momentum.

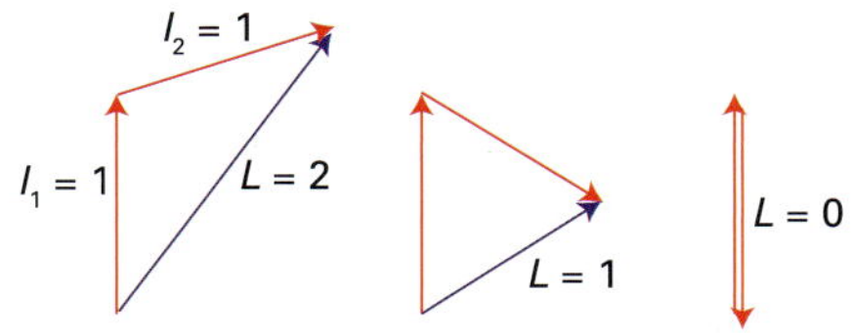

Fig. 13.23 A depiction of the rules for coupling two angular momenta into a resultant. In this case, $l_1 = l_2 = 1$ to give resultants with $L = 2$, 1, and 0. The lengths of the vectors are proportional to $\{l(l + 1)\}^{1/2}$ in each case.

Another brief illustration Consider the configuration p^3. When there are more than two electrons to couple together, we use two Clebsch–Gordan series in succession: first we couple two electrons, and then we couple the third to each combined state, and so on. Coupling two electrons with orbital angular momentum quantum numbers $l_1 = l_2 = 1$ gives a minimum value of $|1 - 1| = 0$. Therefore, using L' to denote the total orbital angular momentum quantum number for these two electrons only, we obtain

$$L' = 1 + 1, 1 + 1 - 1, \ldots, 0 = 2, 1, 0$$

Now we go on to calculate the values of L, the total orbital angular momentum quantum number for the three-electron system. We couple l_3 with $L' = 2$, to give $L = 3$, 2, 1; with $L' = 1$, to give $L = 2$, 1, 0; and with $L' = 0$, to give $L = 1$. The overall result is

$$L = 3, 2, 2, 1, 1, 1, 0$$

giving one F, two D, three P, and one S terms.

Next, we consider the total **spin angular momentum quantum number**, S. This quantum number is obtained in the same way as L, by adding together the individual spin angular momentum quantum numbers:

$$S = s_1 + s_2, s_1 + s_2 - 1, \ldots, |s_1 - s_2|$$

For electrons, $s = \frac{1}{2}$, so for two electrons

$$S = \tfrac{1}{2} + \tfrac{1}{2}, \tfrac{1}{2} + \tfrac{1}{2} - 1, \ldots, |\tfrac{1}{2} - \tfrac{1}{2}| = 1, 0$$

Then the value of S is represented in the term symbol by writing the **multiplicity** of the term, the value of $2S + 1$, as a left superscript. The higher the multiplicity of a term, the more electrons there are in the atom that are spinning in the same direction.

A brief illustration For the excited configuration of carbon, [He]$2s^2 2p^1 3p^1$, that we are considering, the two p electrons each have $s = \frac{1}{2}$, so $S = 1,0$. The corresponding multiplicities are $2 \times 1 + 1 = 3$ (a 'triplet' term) and $2 \times 0 + 1 = 1$ (a 'singlet' term). The corresponding term symbols are

Triplet terms: ^3D, ^3P, ^3S Singlet terms: ^1D, ^1P, ^1S

A note on good practice Except in casual conversation, the name 'state' should not be used in place of 'term'. As we shall see, in general a term consists of a number of different states.

Finally, we come to the right subscript. This label is the **total angular momentum quantum number**, J, the total angular momentum being the sum of the orbital angular momentum and the spin angular momentum. We find J (a positive number) by forming the series

$$J = L + S, L + S - 1, \ldots, |L - S|$$

If there are many electrons having spins in the same direction as their orbital motion, then J is large. If the spins are aligned against the orbital motion, then J is small. Each value of J corresponds to a particular **level** of a term.

A brief illustration The levels that occur in the ^{3}D term are found by setting $L = 2$ and $S = 1$; then

$$J = 2 + 1, 2 + 1 - 1, \ldots, |2 - 1| = 3, 2, 1$$

That is, the levels of the ^{3}D term are $^3D_3$, $^3D_2$, and $^3D_1$ (note that there are three levels for this triplet term). In the $^3D_3$ level, not only are the two p electrons orbiting in the same sense, the two spins are spinning in the same direction as each other, and the total spin is in the same direction as the orbital angular momentum. In $^3D_1$ the total spin is aligned oppositely to the total orbital momentum and the overall total angular momentum is relatively low.

A note on good practice 'Levels' are still not 'states'. Each level with a quantum number J consists of $2J + 1$ individual states distinguished by the quantum number M_J.

Self-test 13.8

What terms and levels can arise from the configuration ...$4p^1 3d^1$?

[*Answer:* $^1F_3$, $^1D_2$, $^1P_1$, $^3F_{4,3,2}$, $^3D_{3,2,1}$, $^3P_{2,1,0}$]

The terms of a configuration in general have different energies because they correspond to the occupation of different orbitals and to different numbers of electrons with parallel spins. Typically, Hund's rule enables us to identify the term of lowest energy as the one with the greatest number of parallel spins (Section 13.11), for parallel spins tend to stay apart and that allows the atom to shrink slightly. In other words,

> The term with the greatest multiplicity lies lowest in energy.

In the excited configuration of carbon in the *illustration* above, the triplet terms lie lower than the singlet terms, so one of the terms ^{3}D, ^{3}P, ^{3}S lies lowest. It is also commonly found that, having sorted the terms by multiplicity,

> The term with the greatest orbital angular momentum lies lowest in energy.

Classically, we can think of the term with the greatest orbital angular momentum as having electrons circulating in the same direction, like cars on a traffic circle, and therefore being able to stay far apart. We therefore predict that the ^{3}D term will lie lowest of all the terms arising from the [He]$2s^2 2p^1 3p^1$ configuration. To find which of the three *levels* of this term lies lowest, we need another concept.

13.18 Spin–orbit coupling

An electron is a charged particle, so its orbital angular momentum gives rise to a magnetic field, just as an electric current in a loop gives rise to a magnetic field in an electromagnet. That is, an electron with orbital angular momentum acts like a tiny bar magnet. An electron also has a spin angular momentum, and this intrinsic 'spinning motion' means that it also acts as a tiny bar magnet. The magnet arising from the spin interacts with the magnet arising from the orbital motion and gives rise to the interaction called **spin–orbit coupling.**

The two magnets have a higher energy when they are parallel than when they are antiparallel (Fig. 13.24). Therefore, because the relative orientation of the magnets reflects the relative orientation of the orbital and spin angular momenta, the energy of the atom depends on the total angular momentum quantum number J (because its value also reflects the relative orientation of the two kinds of momentum). A low energy is obtained when the angular momenta, and therefore the bar magnets, are antiparallel to each other. That arrangement of angular momenta corresponds to a low value of J. Therefore, we can predict that the level with the lowest value of J will lie lowest in energy. In our current example, we predict that the lowest level of the ^{3}D term is $^3D_1$. A more

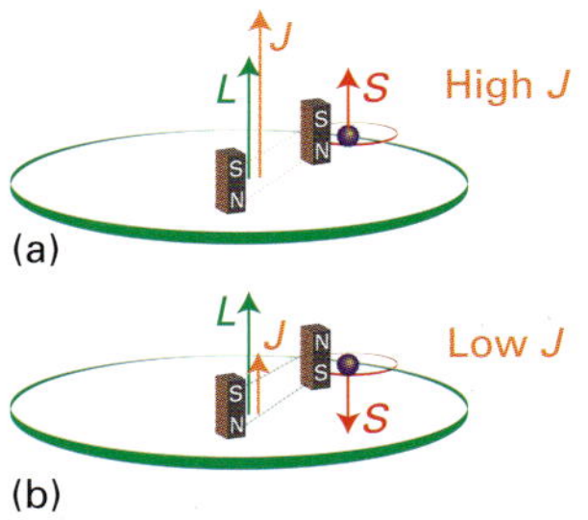

Fig. 13.24 The magnetic interaction responsible for spin–orbit coupling. (a) A high total angular momentum corresponds to a parallel arrangements of magnetic moments (represented by the bar magnets), and hence a high energy. (b) A low total angular momentum corresponds to an antiparallel arrangements of magnetic moments, and hence a low energy. Note that the difference in energy is not due *directly* to the differences in total angular momentum: the total simply tells us the relative orientations of the two magnetic moments.

general statement, which applies to many-electron systems, is as follows:

> For atoms with shells that are less than half full, the level with lowest J lies lowest in energy; for atoms with shells that are more than half full, the level with highest J lies lowest.

The strength of the spin–orbit coupling increases sharply with atomic number. In Period 2 atoms it gives rise to splittings between levels of the order of 10^2 cm^{-1}, but in Period 3 the difference approaches 10^3 cm^{-1}. We can understand this increase by thinking about the source of the orbital magnetic field. To do so, imagine that we are riding on the electron as it orbits the nucleus. From our viewpoint, the nucleus appears to orbit around us (rather as the pre-Copernicans thought the Sun revolved around the Earth). If the nucleus has a high atomic number it will have a high charge, we shall be at the centre of a strong electric current, and we experience a strong magnetic field. If the nucleus has a low atomic number, we experience a feeble magnetic field arising from the low current that encircles us.

Spin–orbit coupling has important consequences in photochemistry and in particular for the existence of the property of 'phosphorescence', which we discuss in Chapter 18.

13.19 Selection rules

Now that we have described the energy levels of complex atoms, we can decide which spectroscopic transitions are allowed or forbidden. We have seen that spectroscopic selection rules arise from the conservation of angular momentum during a transition and from the fact that a photon has a spin of 1. They can therefore be expressed in terms of the term symbols, because the latter carry information about angular momentum. A detailed analysis leads to the following rules:

$$\Delta S = 0 \qquad \Delta L = 0, \pm 1 \qquad \Delta l = \pm 1$$
$$\Delta J = 0, \pm 1, \text{ but } J = 0 \leftrightarrow J = 0 \text{ is forbidden}$$

The rule about ΔS (no change of overall spin) stems from the fact that the light does not affect the spin directly. The rules about ΔL and Δl express the fact that the orbital angular momentum of an individual electron must change (so $\Delta l = \pm 1$), but whether or not this results in an overall change of orbital momentum depends on the coupling of angular momenta. These selection rules apply strictly for relatively light atoms, those near the top of the periodic table. As the atomic number increases, the rules progressively fail on account of significant spin–orbit coupling. So, for example, transitions between singlet and triplet states are allowed in heavy atoms.

Checklist of key ideas

You should now be familiar with the following concepts.

☐ **1** Hydrogenic atoms are atoms with a single electron.

☐ **2** The wavefunctions of hydrogenic atoms are labelled with three quantum numbers, the principal quantum number $n = 1, 2, \ldots$, the orbital angular momentum quantum number $l = 0, 1, \ldots, n-1$, and the magnetic quantum number $m_l = l, l-1, \ldots, -l$.

☐ **3** s Orbitals are spherically symmetrical and have nonzero amplitude at the nucleus.

☐ **4** A radial distribution function, $P(r)$, is the probability density for finding an electron at a radius r; the probability of finding the electron between r and $r + \delta r$ is $P(r)\delta r$.

☐ **5** The magnitude of the orbital angular momentum of an electron is $\{l(l+1)\}^{1/2}\hbar$ and the component of angular momentum about an axis is $m_l\hbar$.

☐ **6** An electron possesses an intrinsic angular momentum, its spin, which is described by the quantum numbers $s = \frac{1}{2}$ and $m_s = \pm\frac{1}{2}$.

☐ **7** A selection rule is a statement about which spectroscopic transitions are allowed.

☐ **8** In the orbital approximation, each electron in a many-electron atom is supposed to occupy its own orbital.

☐ **9** The Pauli exclusion principle states that no more than two electrons may occupy any given orbital and if two electrons do occupy one orbital, then their spins must be paired.

☐ **10** In a many-electron atom, the orbitals of a given shell lie in the order s < p < d < f as a result of the effects of penetration and shielding.

☐ **11** Atomic radii decrease from left to right across a period and increase down a group.

☐ **12** Ionization energies increase from left to right across a period and decrease down a group.

☐ **13** Electron affinities are highest towards the top right of the periodic table (near fluorine).

☐ **14** A term symbol has the form $^{2S+1}\{L\}_J$, where $2S+1$ is the multiplicity and $\{L\}$ is a letter denoting the value of L, the total orbital angular momentum quantum number.

☐ **15** For a given configuration (and most reliably for the ground-state configuration) the term with the greatest multiplicity lies lowest in energy, that with the highest value of L lies lowest, and for atoms with shells that are less than half-full, the level with the lowest J lies lowest.

☐ **16** Different levels of a term have different energies on account of spin–orbit coupling, and the strength of spin–orbit coupling increases sharply with increasing atomic number.

Table of key equations

The following table summarizes the equations developed in this chapter.

Property	Equation	Comment
Wavenumbers of spectroscopic transitions for the hydrogen atom	$\tilde{\nu} = R_H\left(\frac{1}{n_1^2} - \frac{1}{n_2^2}\right)$	$n_1 = 1, 2, \ldots$ and $n_2 = n_1 + 1, n_1 + 2, \ldots$
Energies of hydrogenic atoms	$E_n = -hcRZ^2/n^2$	$n = 1, 2, \ldots$
Atomic orbitals	$\psi_{n,l,m_l}(r,\theta,\phi) = R_{n,l}(r)Y_{l,m_l}(\theta,\phi)$	$R_{n,l}(r)$ and $Y_{l,m_l}(\theta,\phi)$ and are the radial and angular wavefunctions, respectively
Radial distribution function	$P(r) = 4\pi r^2\psi^2$ $P(r) = r^2R(r)^2$	s Orbitals General form
Clebsch–Gordan series	$J = j_1 + j_2, j_1 + j_2 - 1, \ldots, \lvert j_1 - j_2\rvert$	Russell–Saunders coupling scheme; $j = l$ or s
Selection rules for atomic spectra	$\Delta l = \pm 1$ and $\Delta m_l = 0, \pm 1$ $\Delta S = 0, \Delta L = 0, \pm 1, \Delta l = \pm 1,$ $\Delta J = 0, \pm 1$, but $J = 0 \leftrightarrow J = 0$ is forbidden	Hydrogenic atoms Many-electron atoms

Further information 13.1

The Pauli principle

The Pauli exclusion principle is a special case of a general statement called the *Pauli principle*:

> When the labels of any two identical fermions are exchanged, the total wavefunction changes sign. When the labels of any two identical bosons are exchanged, the total wavefunction retains the same sign.

As remarked in the text, a *fermion* is a particle with half-integral spin (such as electrons, protons, and neutrons); a *boson* is a particle with integral spin (such as photons, which have spin 1). The Pauli *exclusion* principle applies only to fermions. By 'total wavefunction' is meant the entire wavefunction, including the spin of the particles.

Consider the wavefunction for two electrons $\Psi(1,2)$. The Pauli principle implies that it is a fact of nature that the wavefunction must change sign if we interchange the labels 1 and 2 wherever they occur in the function: $\Psi(2,1) = -\Psi(1,2)$. Suppose the two electrons in an atom occupy an orbital ψ, then in the orbital approximation the overall wavefunction is $\psi(1)\psi(2)$. To apply the Pauli principle, we must deal with the total wavefunction, the wavefunction including spin. There are four possibilities for two spins:

$$\alpha(1)\alpha(2) \qquad \alpha(1)\beta(2) \qquad \beta(1)\alpha(2) \qquad \beta(1)\beta(2)$$

Let's consider two of these possibilities: the state $\alpha(1)\alpha(2)$ corresponds to parallel spins, whereas (for technical reasons related to the cancellation of each spin's angular momentum by the other) the combination $\alpha(1)\beta(2) - \beta(1)\alpha(2)$ corresponds to paired spins. The total wavefunction of the system is one of the following:

Parallel spins: $\psi(1)\psi(2)\alpha(1)\alpha(2)$

Paired spins: $\psi(1)\psi(2)\{\alpha(1)\beta(2) - \beta(1)\alpha(2)\}$

The Pauli principle, however, asserts that for a wavefunction to be acceptable (for electrons), it must change sign when the electrons are exchanged. In each case, exchanging the labels 1 and 2 converts the factor $\psi(1)\psi(2)$ into $\psi(2)\psi(1)$, which is the same, because the order of multiplying the functions does not change the value of the product. The same is true of $\alpha(1)\alpha(2)$. Therefore, the first combination is not allowed, because it does not change sign. The second combination, however, changes to

$$\psi(2)\psi(1)\{\alpha(1)\beta(2) - \beta(1)\alpha(2)$$
$$= -\psi(1)\psi(2)\{\alpha(1)\beta(2) - \beta(1)\alpha(2)\}$$

This combination does change sign (it is 'antisymmetric'), and is therefore acceptable.

Now we see that the only possible state of two electrons in the same orbital allowed by the Pauli principle is the one that has paired spins. This is the content of the Pauli exclusion principle. The exclusion principle is irrelevant when the orbitals occupied by the electrons are different, and both electrons may then have (but need not have) the same spin state. Nevertheless, even then the overall wavefunction must still be antisymmetric overall, and must still satisfy the Pauli principle itself.

Questions and exercises

Discussion questions

13.1 List and describe the significance of the quantum numbers needed to specify the internal state of a hydrogenic atom.

13.2 Explain the significance of (a) a boundary surface and (b) the radial distribution function for hydrogenic orbitals.

13.3 Describe the orbital approximation for the wavefunction of a many-electron atom. What are the limitations of the approximation?

13.4 Discuss the relationship between the location of a many-electron atom in the periodic table and its electron configuration.

13.5 Describe and account for the variation of first ionization energies along Period 2 of the periodic table. Would you expect the same variation in Period 3?

13.6 Explain the origin of spin–orbit coupling and how it affects the appearance of a spectrum.

13.7 Specify and account for the selections rules for spectroscopic transitions in (a) hydrogenic atoms and (b) many-electron atoms.

Exercises

13.1 Calculate the wavelength of the line with $n = 6$ in the Balmer series of the spectrum of atomic hydrogen.

13.2 The frequency of one of the lines in the Paschen series of the spectrum of atomic hydrogen is 2.7415×10^{15} Hz. Identify the principal quantum number of the upper state in the transition.

13.3 One of the terms of the H atom is at 27 414 cm^{-1}. What is (a) the wavenumber, (b) the energy of the term with which it combines to produce light of wavelength 486.1 nm?

13.4 The Rydberg constant, eqn 13.4, depends on the mass of the nucleus. What is the difference in wavenumbers of the 3p → 1s transition in hydrogen and deuterium?

13.5 What transition in He^+ has the same frequency (disregarding mass differences) as the 2p → 1s transition in H?

13.6 Hydrogen is the most abundant element in all stars. However, neither absorption nor emission lines due to neutral hydrogen are found in the spectra of stars with effective temperatures higher than 25 000 K. Account for this observation.

13.7 The distribution of isotopes of an element may yield clues about the nuclear reactions that occur in the interior of a star. Show that it is possible to use spectroscopy to confirm the presence of both $^4He^+$ and $^3He^+$ in a star by calculating the wavenumbers of the $n = 3 \rightarrow n = 2$ and of the $n = 2 \rightarrow n = 1$ transitions for each isotope.

13.8 Predict the ionization energy of Li^{2+} given that the ionization energy of He^+ is 54.36 eV.

13.9 How many orbitals are present in the N shell of an atom?

13.10 The 'Humphreys series' is another group of lines in the spectrum of atomic hydrogen. It begins at 12 368 nm and has been traced to 3281.4 nm. (a) What are the transitions involved? (b) What are the wavelengths of the intermediate transitions?

13.11 At what wavelength would you expect the longest-wavelength transition of the Humphreys series to occur in He^+? *Hint*: The energy levels of hydrogenic atoms and ions are proportional to Z^2.

13.12 A series of lines in the spectrum of atomic hydrogen lies at 656.46, 486.27, 434.17, and 410.29 nm. (a) What is the wavelength of the next line in the series? (b) What is the ionization energy of the atom when it is in the lower state of the transitions?

13.13 The Li^{2+} ion is hydrogenic and has a Lyman series of lines at 740 747 cm^{-1}, 877 924 cm^{-1}, 925 933 cm^{-1}, and beyond. (a) Show that the energy levels are of the form $-hcR_{Li}/n^2$ and find the value of R_{Li} for this ion. (b) Go on to predict the wavenumbers of the two longest-wavelength transitions of the Balmer series of the ion and (c) find the ionization energy of the ion.

13.14 At what radius does the probability of finding an electron in a small volume located at a point in the ground state of an H atom fall to 30 per cent of its maximum value?

13.15 At what radius in the H atom does the radial distribution function of the ground state have (a) 30 per cent, (b) 5 per cent of its maximum value?

13.16 What is the probability of finding an electron anywhere in one lobe of a p orbital given that it occupies the orbital?

13.17 What is the probability of finding the electron in a volume of 6.5 pm^3 centred on the nucleus in (a) a hydrogen atom, (b) a He^+ ion?

13.18 Locate the radial nodes in (a) the 3s orbital, (b) the 4s orbital of an H atom.

13.19 The wavefunction of one of the d orbitals is proportional to $\sin\theta\cos\theta$. At what angles does it have nodal planes?

13.20 What is the orbital angular momentum (as multiples of $\hbar$) of an electron in the orbitals (a) 1s, (b) 3s, (c) 3d, (d) 2p, (e) 3p? Give the numbers of angular and radial nodes in each case.

13.21 State the orbital degeneracy of the levels in the hydrogen atom that have energy (a) $-hcR_H$, (b) $-\frac{1}{9}hcR_H$, and (c) $-\frac{1}{49}hcR_H$.

13.22 How many electrons can occupy subshells with the following values of l: (a) 0, (b) 3, (c) 5?

13.23 How is the ionization energy of an anion related to the electron affinity of the parent atom?

13.24 When ultraviolet radiation of wavelength 58.4 nm from a helium lamp is directed on to a sample of krypton, electrons are ejected with a speed of 1.59×10^6 m s^{-1}. Calculate the ionization energy of krypton.

13.25 One important function of atomic and ionic radius is in regulating the uptake of oxygen by haemoglobin, for the change in ionic radius that accompanies the conversion of Fe(II) to Fe(III) when O_2 attaches triggers a conformational change in the protein. Which do you expect to be larger: Fe^{2+} or Fe^{3+}? Why?

13.26 What terms (expressed as S, D, etc.) can arise from the $[He]2s^22p^13d^1$ excited configuration of carbon?

13.27 What are the total spin angular momenta (reported as the value of S) that can arise from four electrons? *Hint*: Use the Clebsch–Gordan series successively.

13.28 What levels can the following terms possess: (a) ^1S, (b) ^3F, (c) ^5S, (d) ^5P?

13.29 The ground configuration of a Ti^{2+} ion is $[Ar]3d^2$. (a) What is the term of lowest energy and which level of that term lies lowest? (b) How many states belong to that lowest level?

13.30 Which of the following transitions are allowed in the normal electronic emission spectrum of an atom: (a) 2s → 1s; (b) 2p → 1s; (c) 3d → 2p; (d) 5d → 2s; (e) 5p → 3s; (f) 6f → 4p?

13.31 To what orbitals may a 5f electron make spectroscopic transitions?

13.32 Which of the following transitions between terms are allowed in the normal electronic emission spectrum of a many-electron atom: (a) $^3D_2 \rightarrow {}^3P_1$; (b) $^3P_2 \rightarrow {}^1S_0$; (c) $^3F_4 \rightarrow {}^3D_3$?

Projects

The symbol ‡ indicates that calculus is required.

13.33‡ Here we explore hydrogenic wavefunctions in more quantitative detail. (a) What is the most probable distance of an electron from the nucleus in a hydrogen atom in its ground sate? *Hint*: Look for a maximum in the radial distribution function of a hydrogenic 1s electron). (b) The (normalized) wavefunction for a 2s orbital in hydrogen is

$$\psi = \left(\frac{1}{32\pi a_0^3}\right)^{1/2}\left(2 - \frac{r}{a_0}\right)e^{-r/2a_0}$$

Calculate the probability of finding an electron that is described by this wavefunction in a volume of 1.0 pm^3 (i) centred on the nucleus, (ii) at the Bohr radius, (iii) at twice the Bohr radius. (c) Construct an expression for the radial distribution function of a hydrogenic 2s electron (see part (b) for the form of the orbital), and plot the function against r. What is the most probable radius at which the electron will be found? (d) For a more accurate determination of the most probable radius at which an electron will be found in an H2s orbital, differentiate the radial distribution function to find where it is a maximum.

13.34 Thallium, a neurotoxin, is the heaviest member of Group 13 of the periodic table and is found most usually in the +1 oxidation state. Aluminium, which causes anaemia and dementia, is also a member of the group but its chemical properties are dominated by the +3 oxidation state. Examine this issue by plotting the first, second, and third ionization energies for the Group 13 elements against atomic number. Explain the trends you observe. *Hints*: The third ionization energy, I_3, is the minimum energy needed to remove an electron from the doubly charged cation: $E^{2+}(g) \rightarrow E^{3+}(g) + e^-(g)$, $I_3 = E(E^{3+}) - E(E^{2+})$. For data, see the links to databases of atomic properties provided in the text's web site.

13.35 The spectrum of a star is used to measure its *radial velocity* with respect to the Sun, the component of the star's velocity vector that is parallel to a vector connecting the star's centre to the centre of the Sun. The measurement relies on the Doppler effect, in which radiation is shifted in frequency when the source is moving towards or away from the observer. When a star emitting electromagnetic radiation of frequency ν moves with a speed s relative to an observer, the observer detects radiation of frequency $\nu_{receding} = \nu f$ or $\nu_{approaching} = \nu/f$, where $f = \{(1 - s/c)/(1 + s/c)\}^{1/2}$ and c is the speed of light. (a) Three Fe I lines of the star HDE 271 182, which belongs to the Large Magellanic Cloud, occur at 438.882 nm, 441.000 nm, and 442.020 nm. The same lines occur at 438.392 nm, 440.510 nm, and 441.510 nm in the spectrum of an Earth-bound iron arc. Determine whether HDE 271 182 is receding from or approaching the Earth and estimate the star's radial speed with respect to the Earth. (b) What additional information would you need to calculate the radial velocity of HDE 271 182 with respect to the Sun?

Chapter 14

Quantum chemistry: the chemical bond

The **chemical bond**, a link between atoms, is central to all aspects of chemistry. Reactions make them and break them, and the structures of solids and individual molecules depend on them. The physical properties of individual molecules and of bulk samples of matter also stem in large part from the shifts in electron density that take place when atoms form bonds. The theory of the origin of the numbers, strengths, and three-dimensional arrangements of chemical bonds between atoms is called **valence theory**. (The name comes from a Latin word for strength.)

Valence theory is an attempt to explain the properties of molecules ranging from the smallest to the largest. For instance, it explains why N_2 is so inert that it dilutes the aggressive oxidizing power of atmospheric oxygen. At the other end of the scale, valence theory deals with the structural origins of the function of protein molecules and the molecular biology of DNA. The description of chemical bonding has become highly developed through the use of computers, and it is now possible to compute details of the electron distribution in molecules of almost any complexity. However, much can also be achieved in terms of a simple qualitative understanding of bond formation, and that is the initial focus of this chapter.

There are two major approaches to the calculation of molecular structure, **valence bond theory** (VB theory) and **molecular orbital theory** (MO theory). Almost all modern computational work makes use of MO theory, and we concentrate on it in this chapter. Valence bond theory, though, has left its imprint on the language of chemistry, and it is important to know the significance of terms that chemists use every day. The structure of this chapter is therefore as follows. First, we set out a few concepts common to all levels of description. Then we present the concepts of VB theory that continue to be used in chemistry (such as hybridization and resonance). Next, we present the basic ideas of MO theory, and finally we see how computational techniques pervade all current discussions of molecular structure.

Introductory concepts

Certain ideas of valence theory will be well known from introductory chemistry. This section reviews this background.

14.1 The classification of bonds

We distinguish between two types of bond:

An **ionic bond** is formed by the transfer of electrons from one atom to another and the consequent attraction between the ions so formed.

A **covalent bond** is formed when two atoms share a pair of electrons.

The character of a covalent bond, on which we concentrate in this chapter, was identified by G. N. Lewis in 1916, before quantum mechanics was fully developed. We shall assume that Lewis's ideas are familiar, but for convenience they are reviewed in Appendix 4.2. In this chapter we develop the modern theory of chemical bond formation in terms of the quantum-mechanical properties of electrons and set Lewis's ideas in a modern context. We shall see that ionic and covalent bonds are two extremes of a single type of bond. However, because there are certain aspects of ionic solids that require special attention, we treat them separately in Chapter 15.

Lewis's original theory was unable to account for the shapes adopted by molecules. The most elementary but qualitatively quite successful explanation of the shapes adopted by molecules is the **valence-shell electron pair repulsion model** (VSEPR model) in which we suppose that the shape of a molecule is determined by the repulsions between electron pairs in the valence shell. This model is fully discussed in introductory chemistry texts, but we give a brief review of it in Appendix 4.3. Once again, the purpose of this chapter is to extend these elementary arguments and to indicate some of the contributions that quantum theory has made to understanding why a molecule adopts its characteristic shape.

14.2 Potential-energy curves

All theories of molecular structure adopt the **Born–Oppenheimer approximation.** In this approximation, it is supposed that the nuclei, being so much heavier than an electron, move relatively slowly and may be treated as stationary while the electrons move around them. We can therefore think of the nuclei as being fixed at arbitrary locations, and then solve the Schrödinger equation for the electrons alone. The approximation is quite good for molecules in their electronic ground states, for calculations suggest that (in classical terms) the nuclei in H_2 move through only about 1 pm while the electron speeds through 1000 pm.

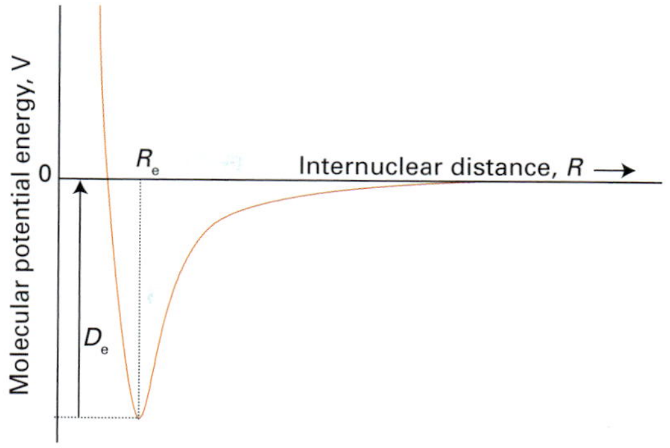

Fig. 14.1 A molecular potential energy curve. The equilibrium bond length R_e corresponds to the energy minimum (at $-D_e$).

By invoking the Born–Oppenheimer approximation, we can select an internuclear separation in a diatomic molecule and solve the Schrödinger equation for the electrons for that nuclear separation. Then we can choose a different separation and repeat the calculation, and so on. In this way we can explore how the energy of the molecule varies with bond length and obtain a **molecular potential energy curve**, a graph showing how the molecular energy depends on the internuclear separation (Fig. 14.1). The graph is called a *potential* energy curve because the nuclei are stationary and contribute no kinetic energy. Once the curve has been calculated, we can identify the **equilibrium bond length,** R_e, the internuclear separation at the minimum of the curve, and D_e, the depth of the minimum below the energy of the infinitely widely separated atoms. In Chapter 19 we shall also see that the narrowness of the potential well is an indication of the stiffness of the bond. Similar considerations apply to polyatomic molecules, where bond angles as well as bond lengths may be varied.

Valence bond theory

In valence bond theory, a bond is regarded as forming when an electron in an atomic orbital on one atom pairs its spin with that of an electron in an atomic

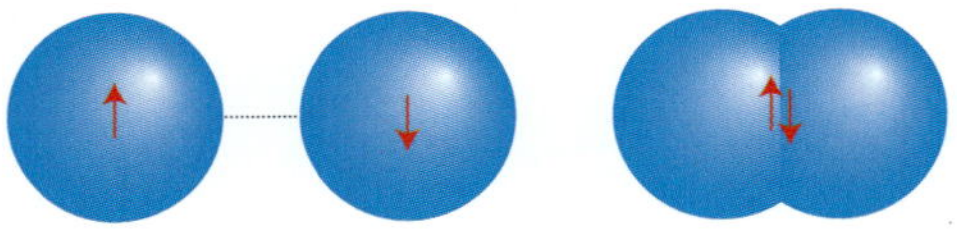

Fig. 14.2 In the valence bond theory, a σ bond is formed when two electrons in orbitals on neighbouring atoms, as in (a), pair and the orbitals merge to form a cylindrical electron cloud, as in (b).

orbital on another atom (Fig. 14.2). To understand why this pairing leads to bonding, we have to examine the wavefunction for the two electrons that form the bond.

14.3 Diatomic molecules

We begin by considering the simplest possible chemical bond, the one in molecular hydrogen, H—H. When the two ground-state H atoms are far apart, we can be confident that electron 1 is in the 1s orbital of atom A that we denote $\psi_A(1)$, and electron 2 is the 1s orbital of atom B, which we denote $\psi_B(2)$. We saw in Section 13.1 that it is a general rule in quantum mechanics that the wavefunction for several non-interacting particles is the product of the wavefunctions for each particle (this is the separation of variables argument), so providing we can ignore the interactions between the electrons we can write

$$\psi(1,2) = \psi_A(1)\psi_B(2)$$

When the two atoms are at their bonding distance, it may still be true that electron 1 is on A and electron 2 is on B. However, an equally likely arrangement is for electron 1 to escape from A and be found on B and for electron 2 to be on A. In this case the wavefunction is

$$\psi(1,2) = \psi_A(2)\psi_B(1)$$

Whenever two outcomes are equally likely, the rules of quantum mechanics tell us to add together, formally to **superimpose**, the two corresponding wavefunctions. Therefore, the (unnormalized) wavefunction for the two electrons in a hydrogen molecule is

$$\psi_{H-H}(1,2) = \psi_A(1)\psi_B(2) + \psi_A(2)\psi_B(1) \qquad (14.1)$$

This expression is the VB wavefunction for the bond in molecular hydrogen. It expresses the idea that we cannot keep track of either electron and their distributions blend together. The wavefunction is only an approximation, because when the two atoms are close together it is not true that the electrons do not interact. However, this approximate wavefunction is a reasonable starting point for all discussions of the VB theory of bonding.

We show in Derivation 14.1 that for technical reasons related to the Pauli exclusion principle, the wavefunction in eqn 14.1 can exist only if the two electrons it describes have opposite spins. It follows that the merging of orbitals that gives rise to a bond is accompanied by the pairing of the two electrons that contribute to it. Bonds do not form *because* electrons tend to pair: bonds are *allowed* to form by the electrons pairing their spins.

Derivation 14.1

The Pauli principle and bond formation

The VB wavefunction in eqn 14.1 does not change sign when the labels 1 and 2 are interchanged. To formulate a wavefunction that obeys the Pauli principle (Further information 13.1) and does change sign when the labels 1 and 2 are interchanged, we must combine this symmetric spatial wavefunction $\psi(1)\psi(2)$ with the antisymmetric spin function $\alpha(1)\beta(2) - \beta(1)\alpha(2)$ and write

$$\psi_{A-B}(2,1) = \{\psi_A(1)\psi_B(2) + \psi_A(2)\psi_B(1)\} \times \{\alpha(1)\beta(2) - \beta(1)\alpha(2)\}$$

This is the only permitted combination of space and spin functions. Because $\alpha(1)\beta(2) - \beta(1)\alpha(2)$ represents a spin-paired state of the two electrons, we see that the Pauli principle requires the two electrons in the bond to be paired (↑↓).

Because ψ is built from the merging of H1s orbitals, we can expect the overall distribution of the electrons in the molecule to be sausage-shaped (as in Fig. 14.2). A VB wavefunction with cylindrical symmetry around the internuclear axis is called a **σ bond**. The bond is so called because, when viewed along the internuclear axis it resembles a pair of electrons in an s orbital (and σ, sigma, is the Greek equivalent of s). All VB wavefunctions are constructed in a similar way, by using the atomic orbitals available on the participating atoms. In general, therefore, the (unnormalized) VB wavefunction for an A—B bond is

$$\psi_{A-B}(1,2) = \psi_A(1)\psi_B(2) + \psi_A(2)\psi_B(1) \qquad (14.2)$$

To calculate the energy of a molecule for a series of internuclear separations R, we substitute the VB wavefunction into the Schrödinger equation for the molecule and carry out the necessary mathematical manipulations to calculate the corresponding values of the energy. When this energy is plotted against R,

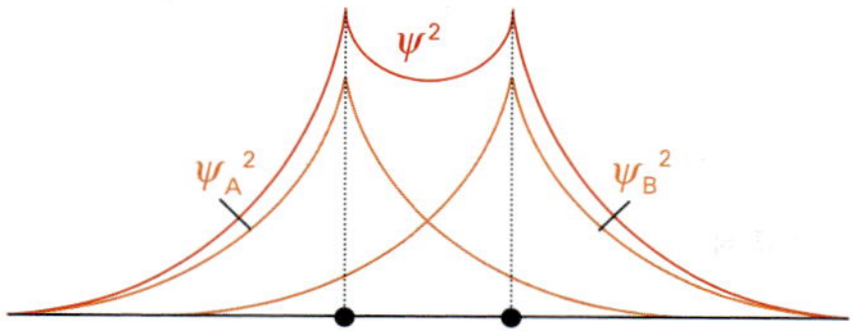

Fig. 14.3 The (normalized) electron density in H_2 according to the valence-bond model of the chemical bond and the electron densities corresponding to the contributing atomic orbitals. The nuclei are denoted by large dots on the horizontal line. Note the accumulation of electron density in the internuclear region.

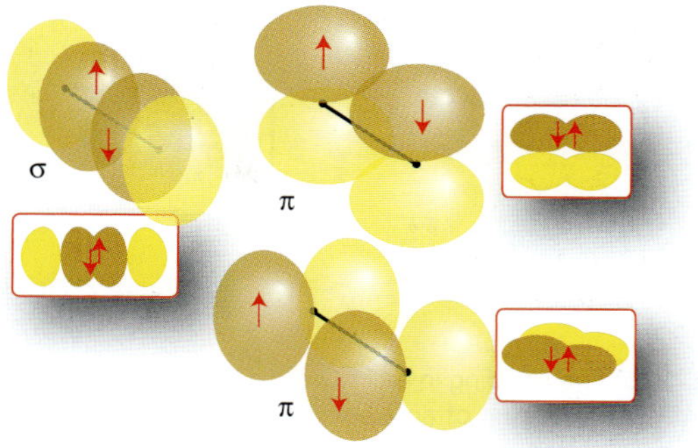

Fig. 14.4 The bonds in N_2 are built by allowing the electrons in the N2p orbitals to pair. However, only one orbital on each atom can form a σ bond: the orbitals perpendicular to the axis form π-bonds.

we get a curve like that shown in Fig. 14.1. As R decreases from infinity, the energy falls below that of two separated H atoms as each electron becomes free to migrate to the other atom. This decrease in energy is the outcome of several effects:

- As the two atoms approach each other, there is an accumulation of electron density between the two nuclei (Fig. 14.3). The electrons attract the two nuclei, and the potential energy is lowered.
- This accumulation between the nuclei is at the expense of removing electron density from close to the nuclei, which contributes an increase in potential energy.
- The freedom of the electrons to migrate between the atoms is like the transfer of an electron from a small box to a bigger box, which (as we saw in the discussion of a particle in a box) results in a lowering of their kinetic energy.

In H_2 the last is the dominant effect, but the relative importance of changes in potential and kinetic energy is still unclear in more complex molecules.

The overall decrease in energy due to the redistribution of electrons is counteracted by an increase in energy from the Coulombic repulsion between the two positively charged nuclei of charges $Z_A e$ and $Z_B e$, which has the form

$$V_{\text{nuc,nuc}} = \frac{Z_A Z_B e^2}{4\pi\varepsilon_0 R} \qquad (14.3)$$

(For H_2, $Z_A = Z_B = 1$.) This positive contribution to the energy becomes large as R becomes small (and the decrease in electronic kinetic energy becomes less significant as the 'big box' is no longer much bigger than the initial two 'little boxes'). As a result, the total energy curve passes through a minimum and then climbs to a strongly positive value as the two nuclei are pressed together.

A similar description is used for molecules built from atoms that contribute more than one electron to the bonding. For example, to construct the VB description of N_2, we consider the valence-electron configuration of each atom, which is $2s^2 2p_x^1 2p_y^1 2p_z^1$. It is conventional to take the z-axis to be the internuclear axis, so we can imagine each atom as having a $2p_z$ orbital pointing towards a $2p_z$ orbital on the other atom, with the $2p_x$ and $2p_y$ orbitals perpendicular to the axis (Fig. 14.4). Each of these p orbitals is occupied by one electron, so we can think of bonds as being formed by the merging of matching orbitals on neighbouring atoms and the pairing of the electrons that occupy them. We get a cylindrically symmetric σ bond from the merging of the two $2p_z$ orbitals and the pairing of the electrons they contain. However, the remaining p orbitals cannot merge to give σ bonds because they do not have cylindrical symmetry around the internuclear axis. Instead, the $2p_x$ orbitals merge and the two electrons pair to form a **π bond.** A π bond is so called because, viewed along the internuclear axis, it resembles a pair of electrons in a p orbital (and π is the Greek equivalent of p). Similarly, the $2p_y$ orbitals merge and their electrons pair to form another π bond. In general, a π bond arises from the merging of two p orbitals that approach side-by-side and the pairing of the electrons that they contain. It follows that the overall bonding pattern in N_2 is a σ bond plus two π bonds, which is consistent with the Lewis structure :N≡N: in which the atoms are linked by a triple bond.

Self-test 14.1

Describe the VB ground state of a Cl_2 molecule.

[*Answer:* one σ(Cl3p_z,Cl3p_z) bond]

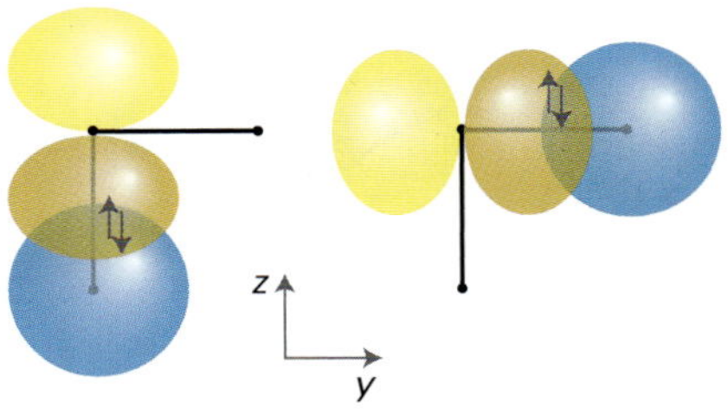

Fig. 14.5 The bonding in an H_2O molecule can be pictured in terms of the pairing of an electron belonging to one H atom with an electron in an O2p orbital; the other bond is formed likewise, but using a perpendicular O2p orbital. The predicted bond angle is 90°, which is in poor agreement with the experimental bond angle (104°).

14.4 Polyatomic molecules

Each σ bond in a polyatomic molecule is formed by the merging of orbitals with cylindrical symmetry about the relevant internuclear axis and the pairing of the spins of the electrons they contain. Likewise, π bonds are formed by pairing electrons that occupy atomic orbitals of the appropriate symmetry (broadly speaking, of the appropriate shape).

A brief illustration The valence electron configuration of an O atom is $2s^22p_x^22p_y^12p_z^1$. The two unpaired electrons in the O2p orbitals can each pair with an electron in a H1s orbital, and each combination results in the formation of a σ bond (each bond has cylindrical symmetry about the respective O—H internuclear distance). Because the $2p_y$ and $2p_z$ orbitals lie at 90° to each other, the two σ bonds they form also lie at 90° to each other (Fig. 14.5). We predict, therefore, that H_2O should be an angular molecule, which it is. However, the model predicts a bond angle of 90°, whereas the actual bond angle is 104°.

Self-test 14.2

Give a VB description of NH_3, and predict the bond angle of the molecule on the basis of this description.

[*Answer:* three σ(N2p,H1s) bonds; 90°; the experimental bond angle is 107°.]

While broadly correct, VB theory seems to have two deficiencies. One is the poor estimate it provides for the bond angle in H_2O and other molecules, such as NH_3. Indeed, the theory appears to make worse predictions than the qualitative VSEPR model, which predicts HOH and HNH bond angles of slightly less than 109° in H_2O and NH_3, respectively. The second major deficiency is the apparent inability of VB theory to account for the number of bonds that atoms can form, and in particular the tetravalence of carbon. To appreciate the latter problem, we note that the ground-state valence configuration of a carbon atom is $2s^22p_x^12p_y^1$, which suggests that it should be capable of forming only two bonds, not four.

14.5 Promotion and hybridization

Two modifications solve all these problems. First, we allow a valence electron to be **promoted** from a full atomic orbital to an empty atomic orbital as a bond is formed: that results in two unpaired electrons instead of two paired electrons, and each unpaired electron can participate in bond formation. In carbon, for example, the promotion of a 2s electron to a 2p orbital leads to the configuration $2s^12p_x^12p_y^12p_z^1$, with four unpaired electrons in separate orbitals. These electrons may pair with four electrons in orbitals provided by four other atoms (such as four H1s orbitals if the molecule is CH_4), and as a result the atom can form four σ bonds. Promotion is worthwhile if the energy it requires can be more than recovered in the greater strength or number of bonds that can be formed.

We can now see why tetravalent carbon is so common. The promotion energy of carbon is small because the promoted electron leaves a doubly occupied 2s orbital and enters a vacant 2p orbital, hence significantly relieving the electron–electron repulsion it experiences in the former. Furthermore, the energy required for promotion is more than recovered by the atom's ability to form four bonds in place of the two bonds of the unpromoted atom.

Promotion, however, appears to imply the presence of three σ bonds of one type (in CH_4, from the merging of H1s and C2p orbitals) and a fourth σ bond of a distinctly different type (formed from the merging of H1s and C2s). It is well known, however, that all four bonds in methane are exactly equivalent both in terms of their chemical properties and their physical properties (their lengths, strengths, and stiffnesses).

This problem is overcome in VB theory by drawing on another technical feature of quantum mechanics that allows the same electron distribution to be described in different ways. In this case, we can describe the electron distribution in the promoted atom either as arising from four electrons in one s and three p orbitals, or as arising from four electrons in four different *mixtures* of these orbitals. Mixtures (more formally, linear combinations) of atomic orbitals on the same atom are called **hybrid orbitals**. These wavefunctions interfere destructively or constructively in different regions and give rise to four

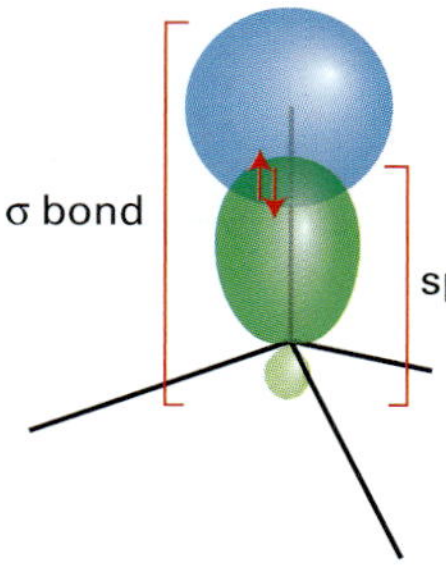

Fig. 14.6 The 2s and three 2p orbitals of a carbon atom hybridize, and the resulting hybrid orbitals point towards the corners of a regular tetrahedron. Each σ bond is formed by the pairing of an electron in an H1s orbital with an electron in one of the hybrid orbitals. The resulting molecule is regular tetrahedral.

new shapes (we saw in Section 12.3 that interference is a characteristic of waves). The specific linear combinations that give rise to four equivalent hybrid orbitals are

$$h_1 = s + p_x + p_y + p_z \quad h_2 = s - p_x - p_y + p_z$$
$$h_3 = s - p_x + p_y - p_z \quad h_4 = s + p_x - p_y - p_z \qquad (14.4)$$

A brief comment In general, a linear combination of two functions f and g is $c_1f + c_2g$, where c_1 and c_2 are numerical coefficients, so a linear combination is a more general term than 'sum'. In a sum, $c_1 = c_2 = 1$.

As a result of the constructive and destructive interference between the positive and negative regions of the component orbitals, each hybrid orbital has a large lobe pointing towards one corner of a regular tetrahedron (Fig. 14.6). Because each hybrid is built from one s orbital and three p orbitals, it is called an **sp^3 hybrid orbital.**

It is now easy to see how the valence bond description of the methane molecule leads to a tetrahedral molecule containing four equivalent C—H bonds. It is energetically favourable (in the end, after bonding has been taken into account) for the carbon atom to undergo promotion. The promoted configuration has a distribution of electrons that is equivalent to one electron occupying each of four tetrahedral hybrid orbitals. Each hybrid orbital of the promoted atom contains a single unpaired electron; a hydrogen 1s electron can pair with each one, giving rise to a σ bond pointing in a tetrahedral direction. Because each sp^3 hybrid orbital has the same composition, all four σ bonds are identical apart from their orientation in space.

Hybridization is also used in the VB description of alkenes. An ethene molecule is planar, with HCH and HCC bond angles close to 120°. To reproduce

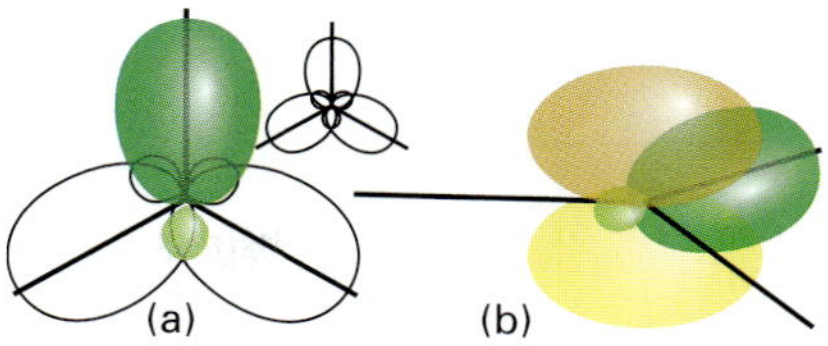

Fig. 14.7 (a) Trigonal planar hybridization is obtained when an s and two p orbitals are hybridized. The three lobes lie in a plane and make an angle of 120° to each other. (b) The remaining p orbital in the valence shell of an sp^2-hybridized atom lies perpendicular to the plane of the three hybrids.

this σ-bonding structure, we think of each C atom as being promoted to a $2s^1 2p_x^1 2p_y^1 2p_z^1$ configuration. However, instead of using all four orbitals to form hybrids, we form **sp^2 hybrid orbitals** by allowing the s orbital and two of the p orbitals to interfere. As shown in Fig. 14.7, the three hybrid orbitals

$$h_1 = s + 2^{1/2}\, p_x$$
$$h_2 = s + (\tfrac{3}{2})^{1/2} p_x - (\tfrac{1}{2})^{1/2} p_y \qquad (14.5)$$
$$h_3 = s - (\tfrac{3}{2})^{1/2} p_x - (\tfrac{1}{2})^{1/2} p_y$$

lie in a plane and point towards the corners of an equilateral triangle. The third 2p orbital ($2p_z$) is not included in the hybridization, and its axis is perpendicular to the plane in which the hybrids lie. The coefficients $2^{1/2}$, etc in the hybrids have been chosen to give the correct directional properties of the hybrids. The *squares* of the coefficients give the proportion of each atomic orbital in the hybrid. All three hybrids have s and p orbitals in the ratio 1:2, as indicated by the label sp^2.

The sp^2-hybridized C atoms each form three σ bonds with either the h_1 hybrid of the other C atom or with the H1s orbitals. The σ framework therefore consists of bonds at 120° to each other. Moreover, provided the two CH_2 groups lie in the same plane, the two electrons in the unhybridized $C2p_z$ orbitals can pair and form a π bond (Fig. 14.8). The formation of this π bond locks the framework into the planar arrangement, for any rotation of one CH_2 group relative to the other leads to a weakening of the π bond (and consequently an increase in energy of the molecule).

A similar description applies to a linear ethyne (acetylene) molecule, H—C≡C—H. Now the carbon atoms are **sp hybridized**, and the σ bonds are built from hybrid atomic orbitals of the form

$$h_1 = s + p_z \qquad h_2 = s - p_z \qquad (14.6)$$

Note that the s and p orbitals contribute in equal proportions. The two hybrids lie along the z-axis.

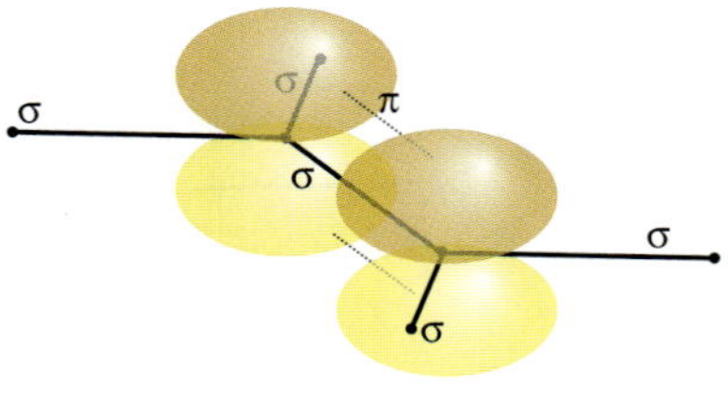

Fig. 14.8 The valence-bond description of the structure of a carbon–carbon double bond, as in ethene. The electrons in the two sp^2 hybrids that point towards each other pair and form a σ bond. Electrons in the two p orbitals that are perpendicular to the plane of the hybrids pair, and form a π bond. The electrons in the remaining hybrid orbitals are used to form bonds to other atoms (in ethene itelf, to H atoms).

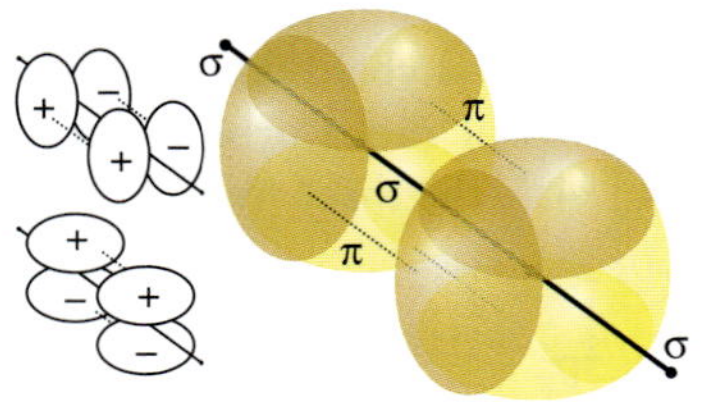

Fig. 14.9 The electronic structure of ethyne (acetylene). The electrons in the two sp hybrids on each atom pair to form σ bonds either with the other C atom or with an H atom. The remaining two unhybridized 2p orbitals on each atom are perpendicular to the axis: the electrons in corresponding orbitals on each atom pair to form two π bonds. The overall electron distribution is cylindrical.

The electrons in them pair either with an electron in the corresponding hybrid orbital on the other C atom or with an electron in the H1s orbitals. Electrons in the two remaining p orbitals on each atom, which are perpendicular to the molecular axis, pair to form two perpendicular π bonds (as in Fig. 14.9).

Other hybridization schemes, particularly those involving d orbitals, are often invoked to account for (or at least be consistent with) other molecular geometries (Table 14.1). An important point to note is that *the hybridization of N atomic orbitals always results in the formation of N hybrid orbitals*. For example, sp^3d^2 hybridization results in six equivalent hybrid orbitals pointing towards the corners of a regular octahedron. This octahedral hybridization scheme is sometimes invoked to account for the structure of octahedral molecules, such as SF_6.

Self-test 14.3

Describe the bonding in a PCl_5 molecule in VB terms.

[*Answer:* Five σ bonds formed from sp^3d hybrids on the central P atom.]

Table 14.1
Hybrid orbitals

Number	Shape	Hybridization*
2	Linear	sp
3	Trigonal planar	sp^2
4	Tetrahedral	sp^3
5	Trigonal bipyramidal	sp^3d
6	Octahedral	sp^3d^2

* Other combinations are possible.

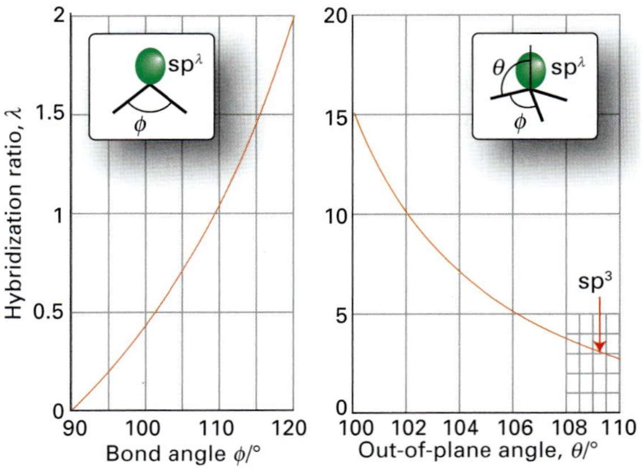

Fig. 14.10 The variation of hybridization with bond angle in (a) angular, (b) trigonal pyramidal molecules. The vertical axis gives the ratio of p to s character, so high values indicate mostly p character.

The 'pure' schemes in Table 14.1 are not the only possibilities: it is possible to form hybrid orbitals with intermediate proportions of atomic orbitals. For example, as more p-orbital character is included in an sp-hybridization scheme, the hybridization changes towards sp^2 and the angle between the hybrids changes continuously from 180° for pure sp hybridization to 120° for pure sp^2 hybridization. If the proportion of p character continues to be increased (by reducing the proportion of s orbital), then the hybrids eventually become pure p orbitals at an angle of 90° to each other (Fig. 14.10). Figure 14.11 shows contour plots of hybrid orbitals as the ratio of 2p character to 2s character increases. Now we can account for the structure of H_2O, with its bond angle of 104°. Each O—H σ bond is formed from an O atom hybrid orbital with a composition that lies between pure p

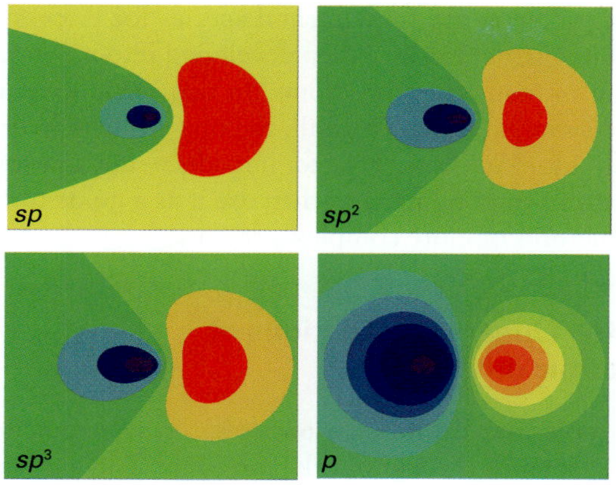

Fig. 14.11 Contour plots showing the amplitudes of sp^n hybrid orbitals. To construct these plots, we have used hydrogenic 2s and 2p orbitals.

(which would lead to a bond angle of 90°) and pure sp^2 (which would lead to a bond angle of 120°). The actual bond angle and hybridization adopted are found by calculating the energy of the molecule as the bond angle is varied, and looking for the angle at which the energy is a minimum.

14.6 Resonance

Another term introduced into chemistry by VB theory is **resonance**, the superposition of the wavefunctions representing different electron distributions in the same nuclear framework. To understand what this means, consider the VB description of a purely covalently bonded HCl molecule, which could be written

$$\psi_{H-Cl}(1,2) = \psi_H(1)\psi_{Cl}(2) + \psi_H(2)\psi_{Cl}(1)$$

We have supposed that the bond is formed by the spin pairing of electrons in the H1s orbital, ψ_H, and the Cl2p_z orbital, ψ_{Cl}. However, there is something wrong with this description: it allows electron 1 to be on the H atom when electron 2 is on the Cl atom, and vice versa, but it does not allow for unequal sharing of electron density between the atoms. On physical grounds, we should expect the purely covalent character of HCl to be an incomplete description of the molecule: because the Cl atom has higher ionization energy and electron affinity than the H atom, we can expect the ionic form H^+Cl^- to play a role. The wavefunction for this ionic structure, in which both electrons are in the Cl2p_z orbital, is

$$\psi_{H^+Cl^-}(1,2) = \psi_{Cl}(1)\psi_{Cl}(2)$$

However, this wavefunction alone is unrealistic, because HCl is not an ionic species. A better description of the wavefunction for the molecule is as a superposition of the covalent and ionic descriptions, and we write (with a slightly simplified notation)

$$\psi_{HCl} = \psi_{H-Cl} + \lambda\psi_{H^+Cl^-}$$

with λ (lambda) some numerical coefficient. In general, we write

$$\psi = \psi_{covalent} + \lambda\psi_{ionic} \qquad (14.7)$$

where $\psi_{covalent}$ is the wavefunction for the purely covalent form of the bond and ψ_{ionic} is the wavefunction for the ionic form of the bond. According to the general rules of quantum mechanics, in which probabilities are related to squares of wavefunctions, we interpret the square of λ as the relative proportion of the ionic contribution. If λ^2 is very small, the covalent description is dominant. If λ^2 is very large, the ionic description is dominant.

We find the numerical value of λ by using the **variation theorem**. First, we write down a plausible wavefunction, a **trial wavefunction**, for the molecule, such as the wavefunction in eqn 14.7 where λ is a variable parameter. The variation theorem then states that:

> The energy of a trial wavefunction is never less than the true energy.

The theorem implies that if we vary λ until we achieve the lowest energy, then the wavefunction with that value of λ is the best available of that particular kind.

The approach summarized by eqn 14.7, in which we express a wavefunction as the superposition of wavefunctions corresponding to a variety of structures *with the nuclei in the same locations*, is called **resonance**. In this case, where one structure is pure covalent and the other pure ionic, it is called **ionic–covalent resonance**. The interpretation of the wavefunction, which is called a **resonance hybrid**, is that if we were to inspect the molecule, then the probability that it would be found with an ionic structure is proportional to λ^2. For instance, we might find that the lowest energy is reached when $\lambda = 0.1$, so the best description of the bond in the molecule in terms of a wavefunction like that in eqn 14.7 is a resonance structure described by the wavefunction $\psi = \psi_{covalent} + 0.1\psi_{ionic}$. This wavefunction implies that the probabilities of finding the molecule in its covalent and ionic forms are in the ratio 100:1 (because $0.1^2 = 0.01$).

One of the most famous examples of resonance is in the VB description of benzene, where the wavefunction of the molecule is written as a superposition of the wavefunctions of the two covalent Kekulé structures (**1**) and (**2**):

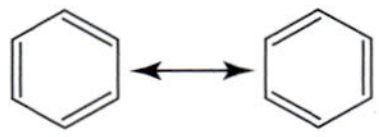

1 **2**

$$\psi = \psi_{\text{Kek1}} + \psi_{\text{Kek2}} \tag{14.8}$$

The two contributing structures have identical energies, so they contribute equally to the superposition. The effect of resonance (which is represented by a double-headed arrow) in this case is to distribute double-bond character around the ring and to make the lengths and strengths of all the carbon–carbon bonds identical. The wavefunction is improved by allowing resonance because it allows for a more accurate description of the location of the electrons, and in particular the distribution can adjust into a state of lower energy. This lowering is called the **resonance stabilization** of the molecule and, in the context of VB theory, is largely responsible for the unusual stability of aromatic rings. Resonance always lowers the energy, and the lowering is greatest when the contributing structures have similar energies. The wavefunction of benzene is improved still further, and the calculated energy of the molecule is lowered further still, if we allow ionic–covalent resonance too, by allowing a small admixture of structures such as that shown in (**3**).

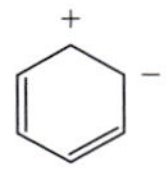

3

Resonance is not a flickering between the contributing states: it is a blending of their characteristics, much as a mule is a blend of a horse and a donkey. It is only a mathematical device for achieving a closer approximation to the true wavefunction of the molecule than that represented by any single contributing structure alone.

Molecular orbitals

In molecular orbital theory, electrons are treated as spreading throughout the entire molecule: every electron contributes to the strength of every bond. This theory has been more fully developed than valence bond theory and provides the language that is widely used in modern discussions of bonding in small inorganic molecules, d-metal complexes, and solids. To introduce it, we follow the same strategy as in Chapter 13, where the one-electron hydrogen atom was taken as the fundamental species for discussing atomic structure, and then developed into a description of many-electron atoms. In this section we use the simplest molecule of all, the one-electron hydrogen molecule–ion, H_2^+, to introduce the essential features of bonding, and then use it as a guide to the structures of more complex systems.

14.7 Linear combinations of atomic orbitals

A **molecular orbital** is a one-electron wavefunction that spreads throughout the molecule. The mathematical forms of such orbitals are highly complicated, even for such a simple species as H_2^+, and they are unknown in general. All modern work builds approximations to the true molecular orbital by formulating models based on linear combinations of the atomic orbitals on the atoms in the molecule.

First, we recall the general principle of quantum mechanics—which we used earlier to construct VB wavefunctions—that if there are several possible outcomes, then we superimpose—add together—the wavefunctions that represent those outcomes. In H_2^+, there are two possible outcomes: because an electron spreads throughout the molecule, it may be found either in an atomic orbital centred on A, ψ_A, or in an orbital centred on B, ψ_B. Therefore, we write

$$\psi = c_A\psi_A + c_B\psi_B \tag{14.9a}$$

where c_A and c_B are numerical coefficients. A wavefunction constructed in this way is called a **linear combination of atomic orbitals** (LCAO) and the corresponding molecular orbital is called an LCAO-MO. The squares of the coefficients tell us the relative proportions of the atomic orbitals contributing to the molecular orbital. In a homonuclear diatomic molecule an electron can be found with equal probability in orbital A or orbital B, so the *squares* of the coefficients must be equal, which implies that $c_B = \pm c_A$. The two possible (unnormalized) wavefunctions are therefore

$$\psi = \psi_A \pm \psi_B \tag{14.9b}$$

First, we consider the LCAO with the plus sign, $\psi = \psi_A + \psi_B$, as this molecular orbital will turn out to have the lower energy of the two. The form of this orbital is shown in Fig. 14.12. It is called a **σ orbital** because it resembles an s orbital when viewed along the axis. More precisely, it is so called because an electron that occupies a σ orbital has zero orbital angular momentum around the internuclear axis, just

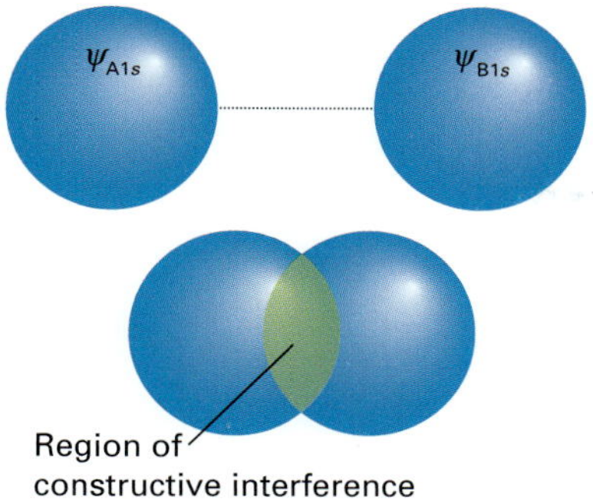

Fig. 14.12 The formation of a bonding molecular orbital (a σ orbital). (a) Two H1s orbitals come together. (b) The atomic orbitals overlap, interfere constructively, and give rise to an enhanced amplitude in the internuclear region. The resulting orbital has cylindrical symmetry about the internuclear axis. When it is occupied by two paired electrons, to give the configuration σ^2, we have a σ bond.

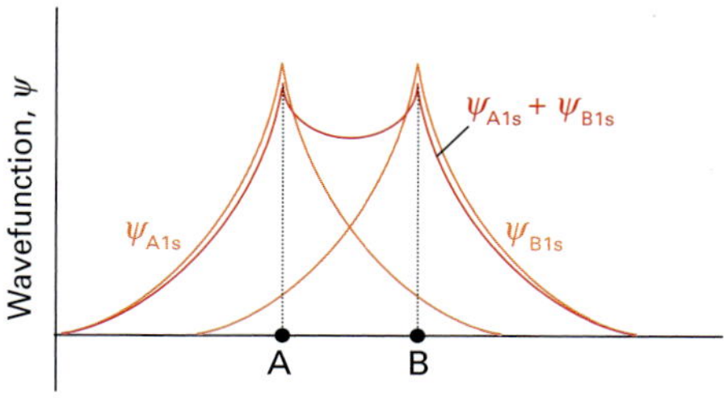

Fig. 14.13 The (normalized) bonding molecular orbital wavefunction along the internuclear axis. Note that there is an enhancement of amplitude between the nuclei, so there is an increased probability of finding the bonding electrons in that region.

as an s electron has zero orbital angular momentum around an axis passing through the nucleus. Because, as we shall see, it is the σ orbital of lowest energy, it is labelled 1σ. An electron that occupies a σ orbital is called a **σ electron**. In the ground state of the H_2^+ ion, there is a single 1σ electron, so we report the ground-state configuration of H_2^+ as $1\sigma^1$.

We can see the origin of the lowering of energy that is responsible for the formation of the bond by examining the LCAO-MO. The two atomic orbitals are like waves centred on adjacent nuclei. In the internuclear region, the amplitudes interfere constructively and the wavefunction has an enhanced amplitude there (Fig. 14.13). The three contributions that we listed for bonding in VB theory (Section 14.3) apply here too: there is an accumulation of electron density between the two nuclei, a removal of electron density from close to the nuclei, and a lowering of

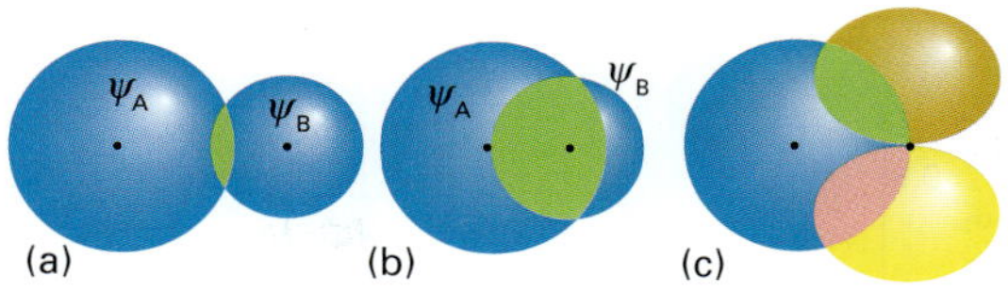

Fig. 14.14 A schematic representation of the contributions to the overlap integral. (a) $S \approx 0$ because the orbitals are far apart and their product is always small. (b) S is large (but less than 1) because the product $\psi_A\psi_B$ is large over a substantial region. (c) $S = 0$ because the positive region of overlap is exactly cancelled by the negative region.

kinetic energy as a result of the electron spreading over both nuclei.

The accumulation of probability density in the internuclear region is measured by the **overlap integral**, S. As we show in Derivation 14.2, when $S = 1$, there is perfect overlap between two atomic orbitals; when $S = 0$, there is no overlap at all (Fig. 14.14). Broadly speaking, the greater the overlap integral, the stronger is the bonding effect of electrons in the molecular orbital they form.

Derivation 14.2

Overlap integrals

An overlap integral is calculated by dividing space up into a large number of small regions, multiplying together the values of ψ_A and ψ_B in each region, then adding together (integrating) the resulting products for all the regions. Formally, we express this rule by writing

$$S = \int \psi_A \psi_B \mathrm{d}\tau \qquad (14.10)$$

where $\mathrm{d}\tau$ (dee tau) is an infinitesimal volume element (for instance, $\mathrm{d}\tau = \mathrm{d}x\mathrm{d}y\mathrm{d}z$ in three-dimensional Cartesian coordinates). If ψ_B is small wherever ψ_A is large, and vice versa (such as when two hydrogen nuclei are far apart), the products are all small and the integral is also small: this corresponds to a small value of S. At typical bonding distances, ψ_A and ψ_B are both large in the internuclear region, so their products there are large, and the integral is also large: this corresponds to a value of S approaching 1 (typically, about 0.4). If the two nuclei are coincident, the two atomic orbitals have identical values everywhere, and the integral of their products gives $S = 1$.

It is possible, but not easy, to evaluate the overlap integral for hydrogenic orbitals, and for two 1s-orbitals on hydrogen nuclei separated by a distance R, the result is

$$S = \left\{1 + \frac{R}{a_0} + \frac{R^2}{3a_0^2}\right\} e^{-R/a_0} \qquad (14.11)$$

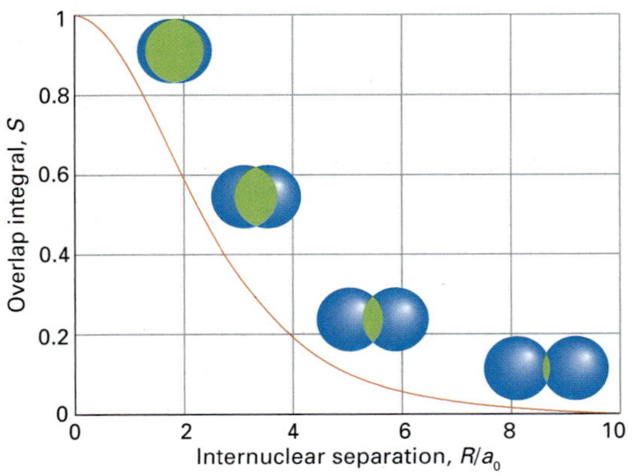

Fig. 14.15 The variation of the overlap integral with internuclear distance for two H1s orbitals.

This function is plotted in Fig. 14.15. The exponential factor guarantees that the overlap integral goes to zero at large separations.

14.8 Bonding and antibonding orbitals

A 1σ orbital is an example of a **bonding orbital**, a molecular orbital that, if occupied, contributes to the strength of a bond between two atoms. As in VB theory, we can substitute the wavefunction $\psi = \psi_A + \psi_B$ into the Schrödinger equation for the molecule–ion with the nuclei at a fixed separation R and solve the equation for the energy. The molecular potential energy curve obtained by plotting the energy against R is very similar to the one drawn in Fig. 14.1. The energy of the molecule falls as R is decreased from large values because the electron is increasingly likely to be found in the internuclear region as the two atomic orbitals interfere more effectively. However, at small separations, there is too little space between the nuclei for significant accumulation of electron density there. In addition, the nucleus–nucleus repulsion $V_{nuc,nuc}$ (given in eqn 14.3) becomes large and the kinetic energy of the electron is not lowered by very much. As a result, after an initial decrease, at small internuclear separations the potential energy curve passes through a minimum and then rises sharply to high values. Calculations on H_2^+ give the equilibrium bond length as 130 pm and the bond dissociation energy as 171 kJ mol^{-1}; the experimental values are 106 pm and 250 kJ mol^{-1}, so this simple LCAO-MO description of the molecule, while inaccurate, is not absurdly wrong.

Now consider the alternative LCAO, the one with a minus sign: $\psi = \psi_A - \psi_B$. Because this wavefunction

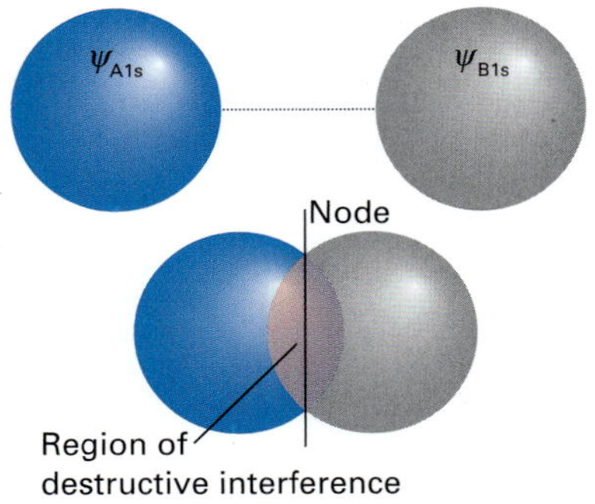

Fig. 14.16 The formation of an antibonding molecular orbital (a σ* orbital). (a) Two H1s orbitals come together. (b) The atomic orbtials overlap with opposite signs (as depicted by different shades of grey), interfere destructively, and give rise to a decreased amplitude in the internuclear region. There is a nodal plane exactly half-way between the nuclei, on which any electrons that occupy the orbital will not be found.

is also cylindrically symmetrical around the internuclear axis it is also a σ orbital, which we denote 1σ* (Fig. 14.16). When substituted into the Schrödinger equation, we find that it has a higher energy than the bonding 1σ orbital and, indeed, it has a higher energy than either of the two atomic orbitals.

Self-test 14.4

Show that the molecular orbital written above is zero on a plane cutting through the internuclear axis at its mid-point. Take each atomic orbital to be of the form e^{-r/a_0}, with r_A measured from nucleus A and r_B measured from nucleus B.

[*Answer:* The atomic orbitals cancel for values equidistant from the two nuclei.]

We can trace the origin of the high energy of the 1σ* orbital to the existence of a **nodal plane**, a plane on which the wavefunction passes through zero. This plane lies half-way between the nuclei and cuts through the internuclear axis. The two atomic orbitals cancel on this plane as a result of their destructive interference, because they have opposite signs. In drawings like those in Figs 14.13 and 14.16, we represent overlap of orbitals with the same sign (as in the formation of 1σ) by shading of the same tint; the overlap of orbitals of opposite sign (as in the formation of 1σ*) is represented by one orbital of a light tint and another orbital of a dark tint.

The 1σ* orbital is an example of an **antibonding orbital**, an orbital that, if occupied, decreases the strength of a bond between two atoms. The antibonding character of the 1σ* orbital is partly a result of the exclusion of the electron from the internuclear

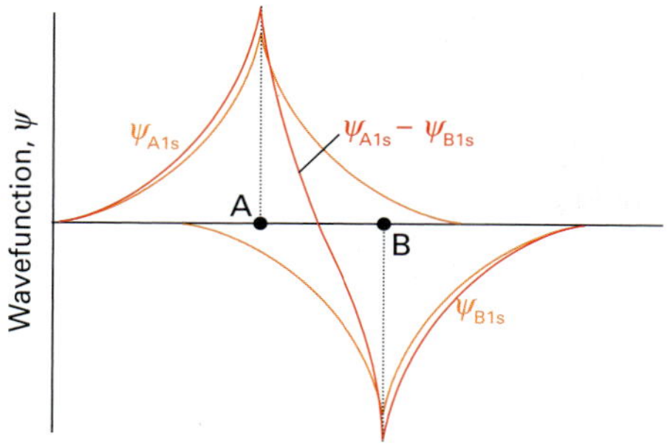

Fig. 14.17 The (normalized) antibonding molecular orbital wavefunction along the internuclear axis. Note that there is a decrease in amplitude between the nuclei, so there is a decreased probability of finding the bonding electrons in that region.

region and its relocation outside the bonding region where it helps to pull the nuclei apart rather than pulling them together (Fig. 14.17). An antibonding orbital is often slightly more strongly antibonding than the corresponding bonding orbital is bonding: although the 'gluing' effect of a bonding electron and the 'antigluing' effect of an antibonding electron are similar, the nuclei repel each other in both cases, and this repulsion pushes both levels up in energy.

We need to be aware of a few points regarding notation. For homonuclear diatomic molecules, it is helpful to identify the **inversion symmetry** of a molecular orbital, especially when discussing electronic transitions (Chapter 18). By 'inversion symmetry' is meant the behaviour of a wavefunction when it is inverted through the centre (more formally, the centre of inversion) of the molecule. Thus, if we consider any point of the σ bonding orbital, and then project it through the centre of the molecule and out an equal distance on the other side, then we arrive at an identical value of the wavefunction (Fig. 14.18). This so-called ***gerade* symmetry** (from the German word for 'even') is denoted by a subscript g, as in σ_g. On the other hand, the same procedure applied to the antibonding σ* orbital results in the same size but opposite sign of the wavefunction. This ***ungerade* symmetry** ('odd symmetry') is denoted by a subscript u, as in σ_u. This inversion symmetry classification (or 'parity') is not applicable to heteronuclear diatomic molecules (like CO) as they do not have a centre of inversion.

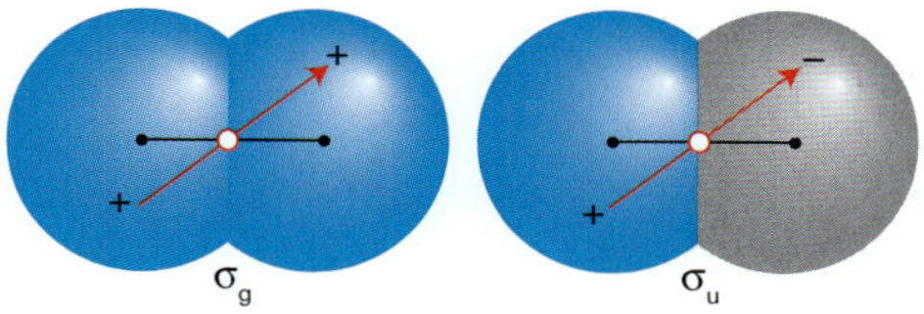

Fig. 14.18 The gerade/ungerade character of σ bonding and antibonding orbitals.

14.9 The structures of diatomic molecules

In Chapter 13 we used the hydrogenic atomic orbitals and the building-up principle to deduce the ground electronic configurations of many-electron atoms. Here we use the same procedure for many-electron diatomic molecules (such as H_2 with two electrons and even Br_2 with 70), but using the H_2^+ molecular orbitals as a basis. The general procedure is as follows:

1. Construct molecular orbitals by forming linear combinations of all suitable valence atomic orbitals supplied by the atoms (the meaning of 'suitable' will be explained shortly); N atomic orbitals result in N molecular orbitals.
2. Accommodate the valence electrons supplied by the atoms so as to achieve the lowest overall energy subject to the constraint of the Pauli exclusion principle, that no more than two electrons may occupy a single orbital (and then must be paired).
3. If more than one molecular orbital of the same energy is available, add the electrons to each individual orbital before doubly occupying any one orbital (because that minimizes electron–electron repulsions).
4. Take note of Hund's rule (Section 13.11), that if electrons occupy different degenerate orbitals, then they do so with parallel spins.

The following sections show how these rules are used in practice.

Self-test 14.5

How many molecular orbitals can be built from the valence shell orbitals in O_2?

[*Answer:* 8]

14.10 Hydrogen and helium molecules

The first step in the discussion of H_2, the simplest many-electron diatomic molecule, is to build the molecular orbitals. Because each H atom of H_2 contributes a 1s orbital (as in H_2^+), we can form the 1σ (more precisely, $1\sigma_g$) and 1σ* (that is, $1\sigma_u$) bonding and antibonding orbitals from them, as we have seen already. At the equilibrium internuclear separation these orbitals will have the energies represented by the horizontal lines in Fig. 14.19.

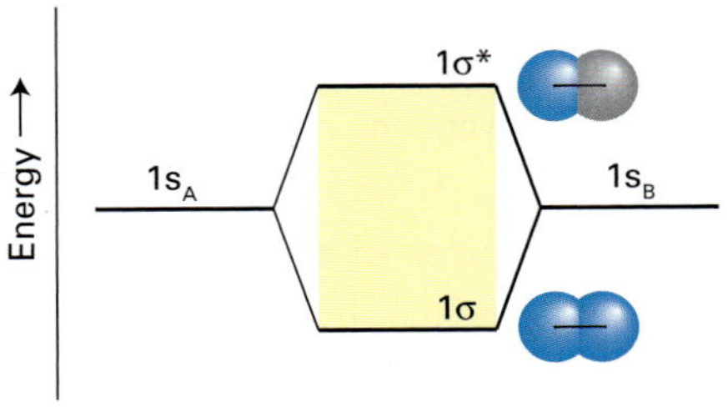

Fig. 14.19 A molecular orbital energy level diagram for orbitals constructed from (1s,1s)-overlap, the separation of the levels corresponding to the equilibrium bond length.

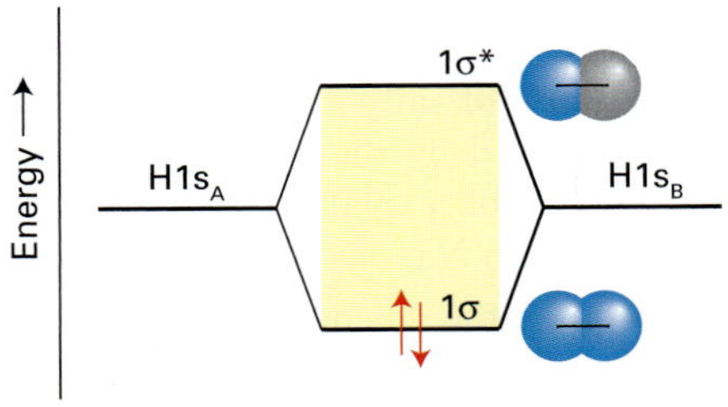

Fig. 14.20 The ground electronic configuration of H_2 is obtained by accommodating the two electrons in the lowest available orbital (the bonding orbital).

There are two electrons to accommodate (one from each atom). Both can enter the $1\sigma_g$ orbital by pairing their spins (Fig. 14.20). The ground-state configuration is therefore $1\sigma_g^2$, and the atoms are joined by a bond consisting of an electron pair in a bonding σ orbital. These two electrons bind the two nuclei together more strongly and closely than the single electron in H_2^+, and the bond length is reduced from 106 pm to 74 pm. A pair of electrons in a σ orbital is called a **σ bond**, and is very similar to the σ bond of VB theory. The two differ in certain details of the electron distribution between the two atoms joined by the bond, but both have an accumulation of density between the nuclei.

We can conclude that *the importance of an electron pair in bonding stems from the fact that two is the maximum number of electrons that can enter each bonding molecular orbital*. Electrons do not 'want' to pair: they pair because in that way they are able to occupy a low-energy orbital.

A similar argument shows why helium is a monatomic gas. Consider a hypothetical He_2 molecule. Each He atom contributes a 1s orbital to the linear combination used to form the molecular orbitals, and so we can construct $1\sigma_g$ and $1\sigma_u$ molecular orbitals. They differ in detail from those in H_2 because the He1s orbitals are more compact, but the general shape is the same, and for qualitative discussions we can use the same molecular orbital energy-level diagram as for H_2. Because each atom provides two electrons, there are four electrons to accommodate. Two can enter the $1\sigma_g$ orbital, but then it is full (by the Pauli exclusion principle). The next two electrons must enter the antibonding $1\sigma_u$ orbital (Fig. 14.21). The ground electronic configuration of He_2 is therefore $1\sigma_g^2 1\sigma_u^2$. Because an antibonding orbital is slightly more antibonding than a bonding orbital is bonding, the He_2 molecule has a higher energy than the separated atoms and is unstable. Hence, two ground-state He atoms do not form bonds to each other, and helium is a monatomic gas.

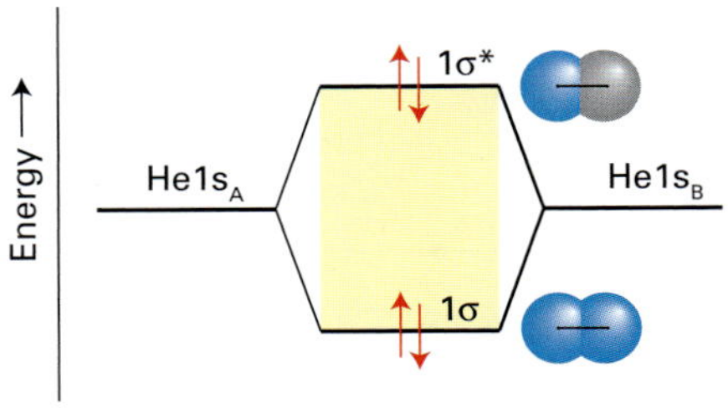

Fig. 14.21 The ground electronic configuration of the four-electron molecule He_2 has two bonding electrons and two antibonding electrons. It has a higher energy than the separated atoms, and so He_2 is unstable relative to two He atoms.

Example 14.1

Judging the stability of diatomic molecules

Decide whether Li_2 is likely to exist on the assumption that only the valence s orbitals contribute to its molecular orbitals.

Strategy Decide what molecular orbitals can be formed from the available valence orbitals, rank them in order of energy, then feed in the electrons supplied by the valence orbitals of the atoms. Judge whether there is a net bonding or net antibonding effect between the atoms.

Solution Each molecular orbital is built from 2s atomic orbitals, which give one bonding and one antibonding combination ($1\sigma_g$ and $1\sigma_u$, respectively). Each Li atom supplies one valence electron; the two electrons fill the $1\sigma_g$ orbital, to give the configuration $1\sigma_g^2$, which is bonding.

Self-test 14.6

Is LiH likely to exist if the Li atom uses only its 2s orbital for bonding

[*Answer:* Yes, σ(Li2s,H1s)^2]

14.11 Period 2 diatomic molecules

We shall now see how the concepts we have introduced apply to other homonuclear diatomic molecules, such as N_2 and Cl_2, and diatomic ions such as O_2^{2-}. In line with the building-up procedure, we first consider the molecular orbitals that may be formed from the valence orbitals and do not (at this stage) trouble about how many electrons are available.

In Period 2, the valence orbitals are 2s and 2p. Suppose first that we consider these two types of orbital separately. Then the 2s orbitals on each atom overlap to form bonding and antibonding combinations that we denote $1\sigma_g$ and $1\sigma_u$, respectively. Likewise, the two $2p_z$ orbitals (by convention, the internuclear axis is the z-axis) have cylindrical symmetry around the internuclear axis. They may therefore participate in σ-orbital formation to give the bonding and antibonding combinations $2\sigma_g$ and $2\sigma_u$, respectively (Fig. 14.22). The resulting energy levels of the σ orbitals are shown in the MO energy level diagram in Fig. 14.23. Note that we number the σ_g orbitals in sequence ($1\sigma_g$, $2\sigma_g$, ...) and the σ_u orbitals likewise.

Strictly, we should not consider the 2s and $2p_z$ orbitals separately, because both of them can contribute to the formation of σ orbitals. Therefore, in a more advanced treatment, we should combine all four orbitals together to form four σ molecular orbitals, each one of the form

$$\psi = c_1\psi_{A2s} + c_2\psi_{B2s} + c_3\psi_{A2p_z} + c_4\psi_{B2p_z}$$

We find the four coefficients, which represent the different contributions that each atomic orbital makes to the overall molecular orbital, by using the variation theorem. However, in practice, the two lowest-energy combinations of this kind are very similar to the combination $1\sigma_g$ and $1\sigma_u$ of 2s orbitals that we have described, and the two highest energy combinations are very similar to the $2\sigma_g$ and $2\sigma_u$ combinations of $2p_z$ orbitals. In each case there will be small differences: the $1\sigma_g$ orbital, for instance, will be contaminated by some $2p_z$ character and the $2\sigma_g$ orbital will be contaminated by some 2s character, and their energies will be slightly shifted from where they would be if we considered only the 'pure' combinations. Nevertheless, the changes are not great, and we can continue to think of 1σ and 1σ* as being one bonding and antibonding pair, and of $2\sigma_g$ and $2\sigma_u$ as being another pair. The four orbitals are shown in the centre column of Fig. 14.24. There is no guarantee that $1\sigma_u$ and $2\sigma_g$ will be in the exact

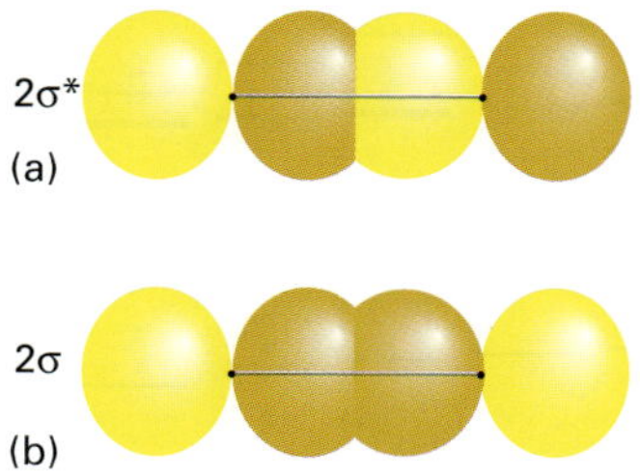

Fig. 14.22 (a) The interference leading to the formation of a σ bonding orbital and (b) the corresponding antibonding orbital when two p orbitals overlap along an internuclear axis.

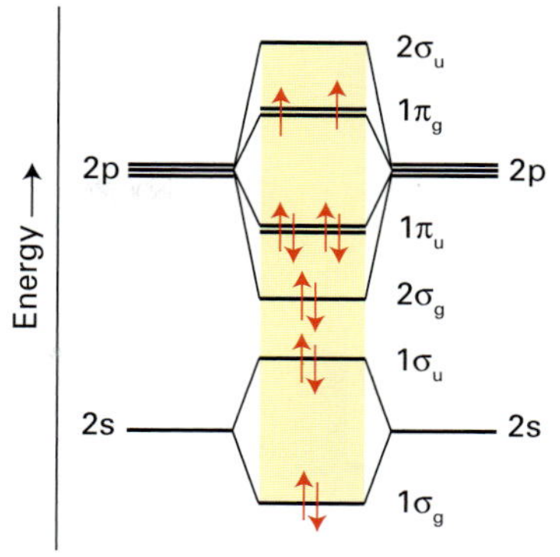

Fig. 14.23 A typical molecular orbital energy level diagram for Period 2 homonuclear diatomic molecules. the valence atomic orbitals are drawn in the columns on the left and the right; the molecular orbitals are shown in the middle. Note that the π orbitals form doubly degenerate pairs (the closely spaced lines denote orbitals lying at exactly the same energy). The sloping lines joining the molecular orbitals to the atomic orbitals show the principal composition of the molecular orbitals. This diagram is suitable for O_2 and F_2; the configuration of O_2 is shown.

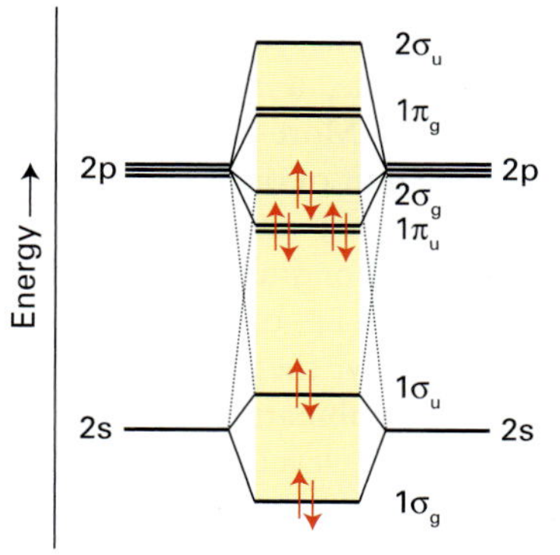

Fig. 14.24 A typical molecular orbital energy level diagram for Period 2 homonuclear diatomic molecules up to and including N_2.

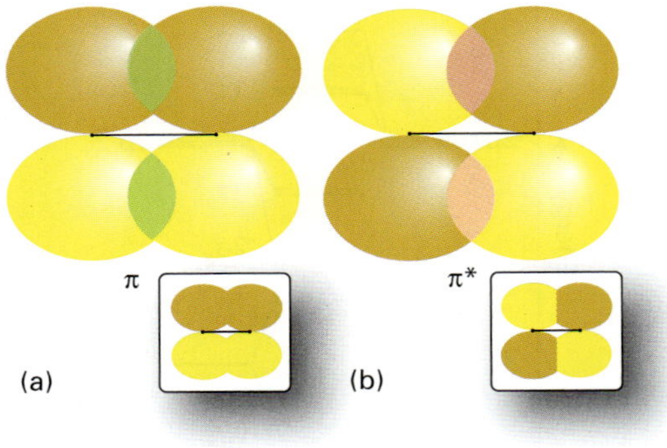

Fig. 14.25 (a) The interference leading to the formation of a π bonding orbital and (b) the corresponding antibonding orbital.

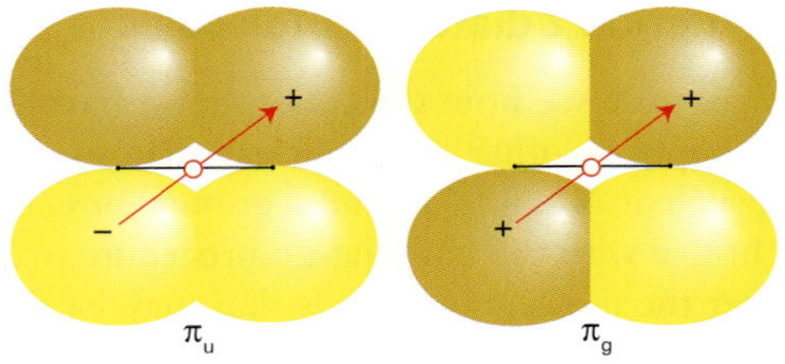

Fig. 14.26 The gerade/ungerade character of π bonding and antibonding orbitals.

location shown in the illustration and the locations shown in Fig. 14.23 are found in some molecules (see below).

There is one further point in this connection. As soon as we allow all four atomic orbitals to contribute to an LCAO it is no longer clear—except by appealing to the form of the simple pairwise LCAOs that each one resembles—whether a particular combination is bonding or antibonding: all we can say is that the four linear combinations have successively increasing energies. However, the parity classification is unaffected, and the orbitals can still be classified as g or u; in homonuclear diatomic molecules, inversion symmetry is a more fundamental classification scheme than bonding and antibonding.

Now consider the $2p_x$ and $2p_y$ orbitals of each atom, which are perpendicular to the internuclear axis and may overlap side-by-side. This overlap may be constructive or destructive and results in a bonding and an antibonding **π orbital**, which initially we label 1π and 1π*, respectively. The notation π is the analogue of p in atoms, for when viewed along the axis of the molecule, a π orbital looks like a p orbital (Fig. 14.25). More precisely, an electron in a π orbital has one unit of orbital angular momentum about the internuclear axis. The two $2p_x$ orbitals overlap to give a bonding and an antibonding π orbital, as do the two $2p_y$ orbitals too. The two bonding combinations have the same energy; likewise, the two antibonding combinations have the same energy. Hence, each π energy level is doubly degenerate and consists of two distinct orbitals. Typically (but not universally) the bonding effect of electrons in a π orbital is less than for a σ orbital in the same molecule because the electron density it represents does not lie between the nuclei so completely. Likewise, the antibonding effect of electrons in a π* orbital is typically less than when they occupy a σ* orbital in the same molecule. Two electrons in a π orbital constitute a **π bond:** such a bond resembles a π bond of valence bond theory, but the details of the electron distribution are slightly different.

The inversion-symmetry classification also applies to π orbitals. As we see from Fig. 14.26, a bonding π orbital changes sign on inversion, and is therefore classified as u. On the other hand, the antibonding π* orbital does not change sign, and is therefore g. The bonding and antibonding combinations will henceforth be denoted $1\pi_u$ and $1\pi_g$. The relative order of the σ and π orbitals in a molecule cannot be predicted without detailed calculation and varies with the energy separation between the 2s and 2p orbitals of the atoms; in some molecules the order shown in Fig. 14.23 applies, whereas others have the order shown in Fig. 14.24. The change in order can be seen in Fig. 14.27, which shows the calculated energy levels for the Period 2 homonuclear diatomic molecules. A useful rule is that, for neutral molecules, the

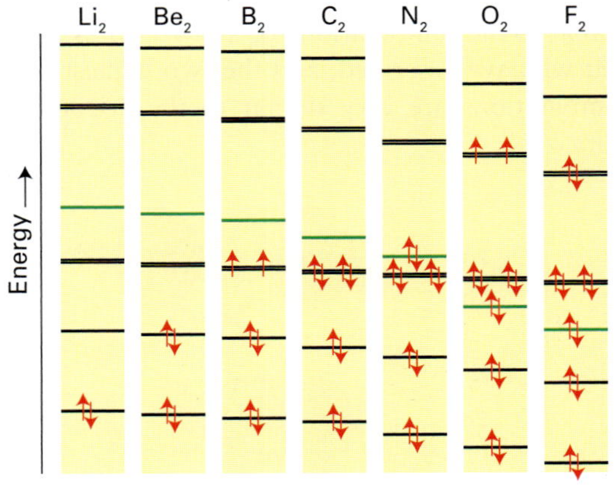

Fig. 14.27 The variation of the orbital energies of Period 2 homonuclear diatomic molecules. Only the valence-shell orbitals are shown.

order shown in Fig. 14.23 is valid for O_2 and F_2, whereas the order shown in Fig. 14.24 is valid for the preceding elements of the period.

14.12 Symmetry and overlap

One central feature of molecular orbital theory can now be addressed. We have seen that s and p_z orbitals may contribute to the formation of σ orbitals, and that p_x and p_y orbitals may contribute to π orbitals. However, we never have to consider orbitals formed by the overlap of s and p_x orbitals (or p_y orbitals). When building molecular orbitals, *we need consider linear combinations only of atomic orbitals of the same symmetry with respect to the internuclear axis.* Because an s orbital has cylindrical symmetry around the internuclear axis, but a p_x orbital does not, the two atomic orbitals cannot contribute to the same molecular orbital. The reason for this distinction based on symmetry can be understood by considering the interference between an s orbital and a p_x orbital (Fig. 14.28): although there is constructive interference between the two orbitals on one side of the axis, there is an exactly compensating amount of destructive interference on the other side of the axis, and the net bonding or antibonding effect is zero.

Consistent with this interpretation, the overlap of a 1s orbital on one atom and a $2p_x$ orbital on another atom (with z the internuclear axis) is zero. In terms of the discussion in Derivation 14.2, we see in Fig. 14.28 that at some point the product $\psi_A\psi_B$ may be large. However, there is a matching point in the lower half of the figure point where $\psi_A\psi_B$ has exactly the same magnitude but an opposite sign. When the integral is evaluated, these two contributions are added together and cancel. For every point in the upper half of the diagram, there is a point in the lower half that cancels it, so $S = 0$. Therefore, for symmetry reasons, there is no net overlap between the s and p orbitals in this arrangement.

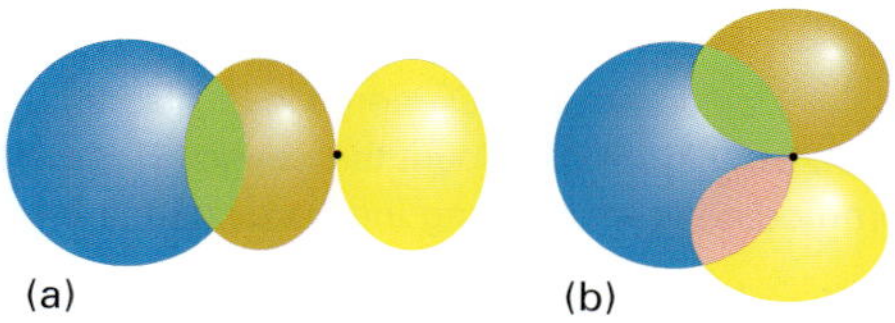

Fig. 14.28 Overlapping s and p orbitals. (a) End-on overlap leads to nonzero overlap and to the formation of an axially symmetric σ orbital. (b) Broadside overlap leads to no net accumulation or reduction of electron density and does not contribute to bonding.

We now have the criteria for selecting atomic orbitals from which molecular orbitals are to be built:

1. Use all available valence orbitals from both atoms.
2. Classify the atomic orbitals as having σ and π symmetry with respect to the internuclear axis, and build σ and π orbitals from all atomic orbitals of a given symmetry.
3. From N_σ atomic orbitals of σ symmetry, N_σ σ orbitals can be built with progressively higher energy from strongly bonding to strongly antibonding.
4. From N_π atomic orbitals of π symmetry, N_π π orbitals can be built with progressively higher energy from strongly bonding to strongly antibonding. The π orbitals occur in doubly degenerate pairs.

As a general rule, the energy of each type of orbital (σ or π) increases with the number of internuclear nodes. The lowest energy orbital of a given species has no internuclear nodes and the highest energy orbital has a nodal plane between each pair of adjacent atoms (Fig. 14.29).

A brief illustration A d_{z^2} orbital has cylindrical symmetry around z and so can contribute to σ orbitals. The d_{zx} and d_{yz} orbitals have π symmetry with respect to the axis (Fig. 14.30), so they can contribute to π orbitals.

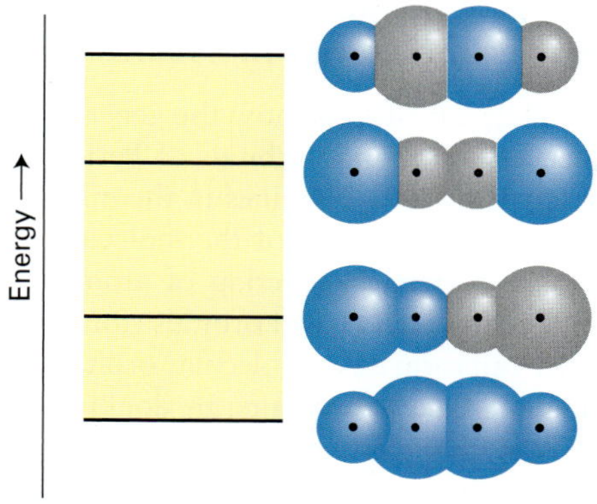

Fig. 14.29 A schematic representation of the four molecular orbitals that can be formed from four s orbitals in a chain of four atoms. The lowest energy combination (the bottom diagram) is formed from atomic orbitals with the same sign, and there are no internuclear nodes. The next higher orbital has one node (at the centre of the molecule. The next higher orbital has two internuclear nodes, and the uppermost, highest energy orbital, has three internuclear nodes, one between each neighbouring pair of atoms, and is fully antibonding. The sizes of the spheres reflect the contributions of each atom to the molecular orbital; the shading represents different signs.

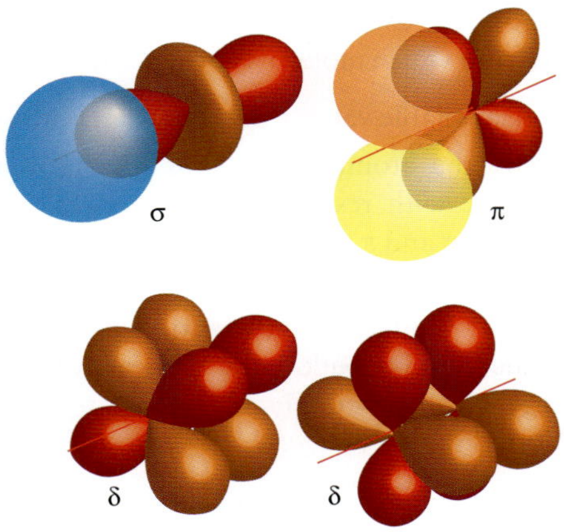

Fig. 14.30 The types of molecular orbital to which d orbitals can contribute. The σ and π combinations can be formed with s, p, and d orbitals of the appropriate symmetry, but the δ orbitals can be formed only by the d orbitals of the two atoms.

Self-test 14.7

Sketch the 'δ orbitals' (orbitals that resemble four-lobed d orbitals when viewed along the internuclear axis) that may be formed by the remaining two d orbitals (and which contribute to bonding in some d-metal cluster compounds). Give their inversion-symmetry classification.

[*Answer:* see Fig. 14.30: bonding are g, antibonding are u]

14.13 The electronic structures of homonuclear diatomic molecules

Figures 14.23 and 14.24 show the general layout of the valence-shell atomic orbitals of Period 2 atoms on the left and right. The lines in the middle are an indication of the energies of the molecular orbitals that can be formed by overlap of atomic orbitals. From the eight valence shell orbitals (four from each atom), we can form eight molecular orbitals: four are σ orbitals and four, in two pairs, are doubly degenerate π orbitals. With the orbitals established, we derive the ground-state electron configurations of the molecules by adding the appropriate number of electrons to the orbitals and following the building-up rules. Charged species (such as the peroxide ion, O_2^{2-}, and C_2^+) need either more or fewer electrons (for anions and cations, respectively) than the neutral molecules.

We illustrate the procedure with N_2, which has ten valence electrons; for this molecule we use Fig. 14.24. The first two electrons pair, enter, and fill the $1\sigma_g$ orbital. The next two electrons enter and fill the $1\sigma_u$ orbital. Six electrons remain. There are two $1\pi_u$ orbitals, so four electrons can be accommodated in them. The two remaining electrons enter the $2\sigma_g$ orbital. The ground-state configuration of N_2 is therefore $1\sigma_g^2 1\sigma_u^2 1\pi_u^4 2\sigma_g^2$. This configuration is also depicted in Fig. 14.24.

The strength of a bond in a molecule is the net outcome of the bonding and antibonding effects of the electrons in the orbitals. The **bond order**, b, in a diatomic molecule is defined as

$$b = \tfrac{1}{2}(N - N^*) \qquad (14.12)$$

where N is the number of electrons in bonding orbitals and N^* is the number of electrons in antibonding orbitals (as judged by their resemblance to the simple pairwise LCAOs). Each electron pair in a bonding orbital increases the bond order by 1 and each pair in an antibonding orbital decreases it by 1. For H_2, $b = 1$, corresponding to a single bond between the two atoms: this bond order is consistent with the Lewis structure H—H for the molecule. In He_2, which has equal numbers of bonding and antibonding electrons (with $N = 2$ and $N^* = 2$), the bond order is $b = 0$, and there is no bond. In N_2, $1\sigma_g$, $2\sigma_g$, and $1\pi_u$ are bonding orbitals, and $n = 2 + 2 + 4 = 8$; however, $1\sigma_u$ (the antibonding partner of $1\sigma_g$) is antibonding, so $N^* = 2$ and the bond order of N_2 is $b = \tfrac{1}{2}(8 - 2) = 3$. This value is consistent with the Lewis structure :N≡N:, in which there is a triple bond between the two atoms.

The bond order is a useful parameter for discussing the characteristics of bonds, because it correlates with bond length, and the greater the bond order between atoms of a given pair of atoms, the shorter the bond. The bond order also correlates with bond strength, and the greater the bond order, the greater the strength. The high bond order of N_2 is consistent with its high dissociation energy (942 kJ mol^{-1}).

Example 14.3

Writing the electron configuration of a diatomic molecule

Write the ground-state electron configuration of O_2 and calculate the bond order.

Strategy Decide which MO energy level diagram to use (Fig. 14.23 or Fig. 14.24). Count the valence electrons and accommodate them by using the building-up principle.

Solution Figure 14.23 is appropriate for oxygen. There are 12 valence electrons to accommodate. The first 10 electrons recreate the N_2 configuration (with a reversal of the order of the $2\sigma_g$ and $1\pi_u$ orbitals); the remaining two electrons must occupy the $1\pi_g$ orbitals. The configuration and bond order are therefore $1\sigma_g^2 1\sigma_u^2 2\sigma_g^2 1\pi_u^4 1\pi_g^2$.

This configuration is also depicted in Fig. 14.23. Because $1\sigma_g$, $2\sigma_g$, and $1\pi_u$ are regarded as bonding and $1\sigma_u$ and $1\pi_g$ as antibonding, the bond order is $b=\frac{1}{2}(8-4)=2$. This bond order accords with the classical view that oxygen has a double bond.

Self-test 14.8

Write the electron configuration of F_2 and deduce its bond order.

[*Answer:* $1\sigma_g^2 1\sigma_u^2 2\sigma_g^2 1\pi_u^4 1\pi_g^4$, $b=1$]

We see from Example 14.3 that the electron configuration of O_2 is $1\sigma_g^2 1\sigma_u^2 2\sigma_g^2 1\pi_u^4 1\pi_g^2$. According to the building-up principle, the two $1\pi_g$ electrons in O_2 will occupy different orbitals. One enters the $1\pi_g$ orbital formed by overlap of $2p_x$. The other enters its degenerate partner, the $1\pi_g$ orbital formed from overlap of the $2p_y$ orbitals. Because the two electrons occupy different orbitals, by Hund's rule they will have parallel spins (↑↑). Consequently, an O_2 molecule is sometimes regarded as a biradical, a species with two unpaired electrons. (A true biradical has two electron spins with random relative orientations; in O_2 the two spins are parallel.) Molecular orbital theory therefore suggests—correctly—that O_2 is a reactive component of the Earth's atmosphere; its most important biological role is as an oxidizing agent. By contrast, N_2, the major component of the air we breathe, is so unreactive that nitrogen fixation, the reduction of atmospheric N_2 to NH_3 by certain microorganisms, is among the most thermodynamically demanding of biological processes in the sense that it requires a great deal of energy derived from metabolic processes.

The electronic configuration of O_2 also suggests that it will be magnetic because the magnetic fields generated by the two unpaired spins do not cancel. Specifically, O_2 is predicted to be a **paramagnetic** substance, a substance that is drawn into a magnetic field. Most substances (those with paired electron spins) are **diamagnetic**, and are pushed out of a magnetic field. That O_2 is in fact a paramagnetic gas is a striking confirmation of the superiority of the molecular orbital description of the molecule over the Lewis and VB descriptions (which require all the electrons to be paired). The property of paramagnetism is utilized to monitor the oxygen content of incubators by measuring the magnetism of the gases they contain.

An F_2 molecule has two more electrons than an O_2 molecule, so its configuration is $1\sigma_g^2 1\sigma_u^2 2\sigma_g^2 1\pi_u^4 1\pi_g^4$ and its bond order is 1. We conclude that F_2 is a singly bonded molecule, in agreement with its Lewis structure :F̤—F̤:. The low bond order is consistent with the low dissociation energy of F_2 (154 kJ mol^{-1}). A hypothetical Ne_2 molecule would have two further electrons: its configuration would be $1\sigma_g^2 1\sigma_u^2 2\sigma_g^2 1\pi_u^4 1\pi_g^4 2\sigma_u^2$ and its bond order 0. The bond order of zero—which implies that two neon atoms do not bond together—is consistent with the monatomic character of neon.

Example 14.4

Judging the relative bond strengths of molecules and ions

The superoxide ion, O_2^-, plays an important role in the ageing processes that take place in organisms. Judge whether O_2^- is likely to have a larger or smaller dissociation energy than O_2.

Strategy Because a species with the larger bond order is likely to have the larger dissociation energy, we should compare their electronic configurations, and assess their bond orders.

Solution From Fig. 14.23,

O_2 $1\sigma_g^2 1\sigma_u^2 2\sigma_g^2 1\pi_u^4 1\pi_g^2$ $b=2$

O_2^- $1\sigma_g^2 1\sigma_u^2 2\sigma_g^2 1\pi_u^4 1\pi_g^3$ $b=1.5$

Because the anion has the smaller bond order, we expect it to have the smaller dissociation energy.

Self-test 14.9

Which can be expected to have the higher dissociation energy, F_2 or F_2^+?

[*Answer:* F_2^+]

14.14 Heteronuclear diatomic molecules

A **heteronuclear diatomic molecule** is a diatomic molecule formed from atoms of two different elements; two examples are CO and HCl. The electron distribution in the covalent bond between the atoms is not symmetrical between the atoms because it is energetically favourable for a bonding electron pair to be found closer to one atom rather than the other. This imbalance results in a **polar bond**, which is a covalent bond in which the electron pair is shared unequally by the two atoms. The **electronegativity**, χ (chi), of an element is the power of its atoms to draw electrons to itself when it is part of a compound, so we can

Table 14.2

Electronegativities of the main-group elements

H 2.1						
Li 1.01	Be 1.5	B 2.0	C 2.5	N 3.0	O 3.5	F 4.0
Na 0.9	Mg 1.2	Al 1.5	Si 1.8	P 2.1	S 2.5	Cl 3.0
K 0.8	Ca 1.0	Ga 1.6	Ge 1.8	As 2.0	Se 2.4	Br 2.8
Rb 0.8	Sr 1.0	In 1.7	Sn 1.8	Sb 1.9	Te 2.1	I 2.5
Cs 0.7	Ba 0.9	Tl 1.8	Pb 1.8	Bi 1.9	Po 2.0	

expect the polarity of a bond to depend on the relative electronegativities of the elements.

Linus Pauling formulated a numerical scale of electronegativity based on considerations of bond dissociation energies, $E(\text{A—B})$:

$$|\chi_A - \chi_B| = 0.102 \times (\Delta E/\text{kJ mol}^{-1})^{1/2} \qquad (14.13a)$$

with

$$\Delta E = E(\text{A—B}) - \tfrac{1}{2}\{E(\text{A—A}) + E(\text{B—B})\} \qquad (14.13b)$$

Table 14.2 lists values for the main-group elements. Robert Mulliken proposed an alternative definition in terms of the ionization energy, I, and the electron affinity, E_{ea}, of the element expressed in electronvolts:

$$\chi = \tfrac{1}{2}(I + E_{ea}) \qquad (14.14)$$

This relation is plausible, because an atom that has a high electronegativity is likely to be one that has a high ionization energy (so that it is unlikely to lose electrons to another atom in the molecule) and a high electron affinity (so that it is energetically favourable for an electron to move towards it). The Mulliken electronegativities are broadly in line with the Pauling electronegativities. Electronegativities show a periodicity, and the elements with the highest electronegativities are those close to fluorine in the periodic table.

The location of the bonding electron pair close to one atom in a heteronuclear molecule results in that atom having a net negative charge, which is called a **partial negative charge** and denoted δ–. There is a compensating **partial positive charge**, δ+, on the other atom. In a typical heteronuclear diatomic molecule, the more electronegative element has the partial negative charge and the more electropositive element has the partial positive charge.

Self-test 14.10

Predict the signs of the charge distribution of a C—H bond.

[Answer: $^{\delta-}$C—H$^{\delta+}$]

Molecular orbital theory takes polar bonds into its stride. A polar bond consists of two electrons in an orbital of the form

$$\psi = c_A\psi_A + c_B\psi_B \qquad (14.15)$$

with c_B^2 no longer equal to c_A^2. If $c_B^2 > c_A^2$, the electrons have a greater probability of being found on B than on A and the molecule is polar in the sense $^{\delta+}$A—B$^{\delta-}$. A nonpolar bond, a covalent bond in which the electron pair is shared equally between the two atoms and there are zero partial charges on each atom, has $c_A^2 = c_B^2$. A pure ionic bond, in which one atom has obtained virtually sole possession of the electron pair (as in Cs^+F^-, to a first approximation), has one coefficient zero (so that A^+B^- would have $c_A^2 = 0$ and $c_B^2 = 1$).

A general feature of molecular orbitals between dissimilar atoms is that the atomic orbital with the lower energy (that belonging to the more electronegative atom) makes the larger contribution to the lowest-energy molecular orbital. The opposite is true of the highest (most antibonding) orbital, for which the principal contribution comes from the atomic orbital with higher energy (the less electronegative atom):

Bonding orbitals: for $\chi_A > \chi_B$, $c_A^2 > c_B^2$

Antibonding orbitals: for $\chi_A > \chi_B$, $c_A^2 < c_B^2$

Figure 14.31 shows a schematic representation of this point.

These features of polar bonds can be illustrated by considering HF. The general form of the molecular orbitals of HF is $\psi = c_H\psi_H + c_F\psi_F$, where ψ_H is an H1s orbital and ψ_F is an F2p$_z$ orbital. Because the ionization energy of a hydrogen atom is 13.6 eV, we know that the energy of the H1s orbital is –13.6 eV. As usual, the zero of energy is the infinitely separated electron and proton (Fig. 14.32). Similarly, from the ionization energy of fluorine, which is 18.6 eV, we know that the energy of the F2p$_z$ orbital is approximately –18.6 eV, about 5 eV lower than the H1s orbital. It follows that the bonding σ orbital in HF is mainly F2p$_z$ and the antibonding σ orbital is mainly H1s orbital in character. The two electrons in the bonding orbital are most likely to be found in the F2p$_z$ orbital, so there is a partial negative charge on the F atom and a partial positive charge on the H atom.

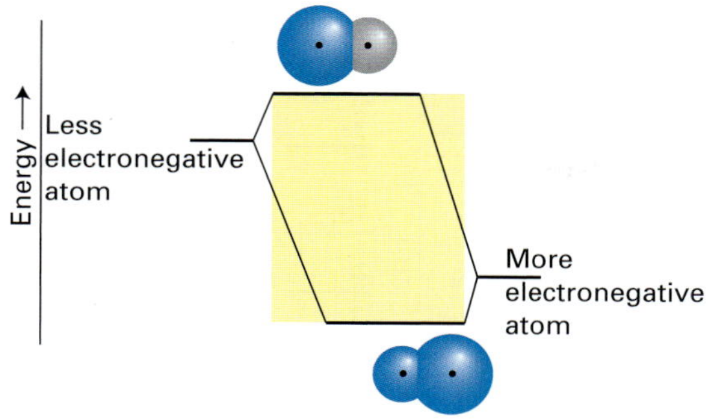

Fig. 14.31 A schematic representation of the relative contributions of atoms of different electronegativities to bonding and antibonding molecular orbitals. In the bonding orbital, the more electronegative atom makes the greater contribution (represented by the larger sphere), and the electrons of the bond are more likely to be found on that atom. The opposite is true of an antibonding orbital. A part of the reason why an antibonding orbital is of high energy is that the electrons that occupy it are likely to be found on the more electropositive atom.

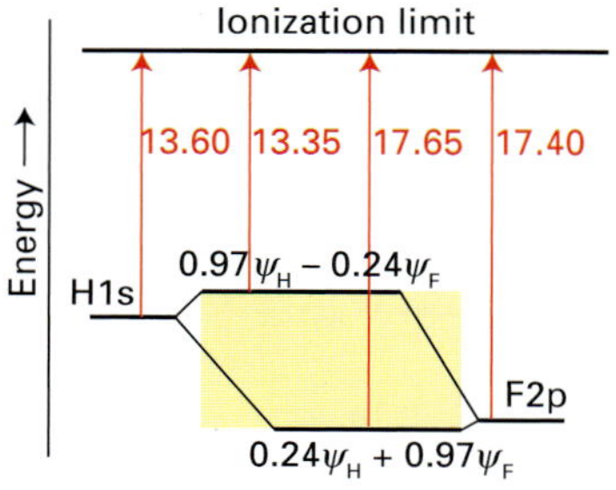

Fig. 14.32 The atomic orbital energy levels of H and F atoms and the molecular orbitals they form. The bonding orbital has predominantly F atom character and the antibonding orbital has predominantly H atom character. Energies are in electronvolts.

A systematic way of finding the coefficients in the linear combinations is to use the variation theorem and to look for the values of the coefficients that result in the lowest energy (Section 14.2). For example, when the variation principle is applied to an H_2 molecule, the calculated energy is lowest when the two H1s orbitals contribute equally to a bonding orbital. However, when we apply the principle to HF, the lowest energy is obtained for the orbital $\psi = 0.24\psi_H + 0.97\psi_F$. We see that indeed the F2p$_z$ orbital does make the greater contribution to the bonding σ orbital. An even lower energy is obtained—with a lot more calculation—if even more orbitals are included in the linear combination (such as F2s and F3p$_z$ orbitals) but the principal lowering of energy is achieved from atomic orbitals of similar energies.

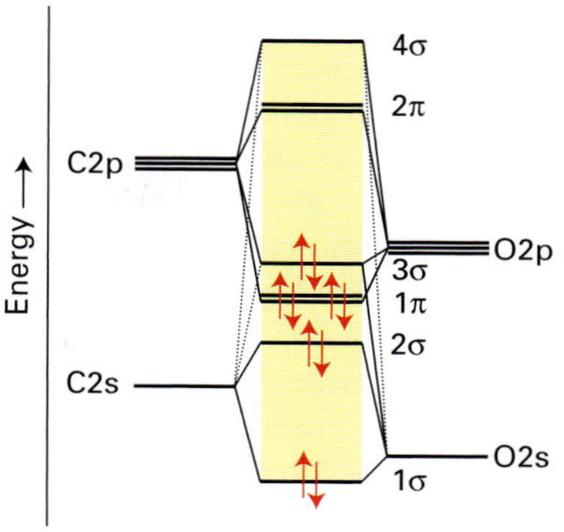

Fig. 14.33 The molecular orbital energy-level diagram for CO.

Self-test 14.11

What is the probability of finding a σ electron in HF in a F2p$_z$ orbital?

[*Answer:* 88 per cent = $(0.94)^2 \times 100$ per cent)]

Figure 14.33 shows the bonding scheme in CO and illustrates a number of points we have made. The ground configuration is $1\sigma^2 2\sigma^2 1\pi^4 3\sigma^2$. (The g,u designation is inapplicable because the molecule is heteronuclear and the σ orbitals are simply numbered in sequence, 1σ, 2σ, ..., and the π orbitals likewise.) The lowest-energy orbitals are predominantly of O character as that is the more electronegative element. The **highest occupied molecular orbital** (HOMO) is 3σ, which is a largely nonbonding orbital centred on C, so the two electrons that occupy it can be regarded as a lone pair on the C atom. The **lowest unoccupied molecular orbital** (LUMO) is 2π, which is largely a doubly degenerate orbital of 2p character on carbon. This combination of a lone-pair orbital on C and a pair of empty π orbitals also largely on C is at the root of the importance of carbon monoxide in d-block chemistry, because it enables it to form an extensive series of carbonyl complexes by a combination of electron donation from the 3σ orbital and electron acceptance into the 2π orbitals. The HOMO and the LUMO jointly form the **frontier orbitals** of the molecule, and are of great importance for assessing its reactions.

14.15 The structures of polyatomic molecules

The bonds in polyatomic molecules are built in the same way as in diatomic molecules, the only difference being that we use more atomic orbitals to construct the molecular orbitals, and these molecular orbitals spread over the entire molecule, not just the adjacent atoms of the bond. In general, a molecular orbital is a linear combination of all the atomic orbitals of all the atoms in the molecule. In H_2O, for instance, the atomic orbitals are the two H1s orbitals, the O2s orbital, and the three O2p orbitals (if we consider only the valence shell). From these six atomic orbitals we can construct six molecular orbitals that spread over all three atoms. The molecular orbitals differ in energy. The lowest energy, most strongly bonding orbital has the least number of nodes between adjacent atoms. The highest energy, most strongly antibonding orbital has the greatest numbers of nodes between neighbouring atoms.

According to MO theory, the bonding influence of a single electron pair is distributed over all the atoms, and each electron pair (the maximum number of electrons that can occupy any single molecular orbital) helps to bind all the atoms together. In the LCAO approximation, each molecular orbital is modelled as a sum of atomic orbitals, with atomic orbitals contributed by all the atoms in the molecule. Thus, a typical molecular orbital in H_2O constructed from H1s orbitals (denoted ψ_A and ψ_B) and O2s and O2p$_z$ orbitals (denoted ψ_{Os} and ψ_{Op_z}) will have the composition

$$\psi = c_1\psi_A + c_2\psi_{Os} + c_3\psi_{Op_z} + c_4\psi_B \qquad (14.16)$$

Because four atomic orbitals are being used to form the LCAO, there will be four possible molecular orbitals of this kind: the lowest energy (most bonding) orbital will have no internuclear nodes and the highest energy (most antibonding) orbital will have a node between each pair of neighbouring nuclei (Fig. 14.34).

An important example of the application of MO theory is to the orbitals that may be formed from the p orbitals perpendicular to the molecular plane of benzene, C_6H_6. Because there are six such atomic orbitals, it is possible to form six molecular orbitals of the form

$$\psi = c_1\psi_1 + c_2\psi_2 + c_3\psi_3 + c_4\psi_4 + c_5\psi_5 + c_6\psi_6 \qquad (14.17)$$

The lowest energy, most strongly bonding orbital has no internuclear nodes, and has the form

$$\psi = \psi_1 + \psi_2 + \psi_3 + \psi_4 + \psi_5 + \psi_6$$

A brief comment We are ignoring normalization factors, for clarity. In this and the following case it would be $1/6^{1/2}$ if we ignore overlap.

This orbital is illustrated at the bottom of Fig. 14.35. It is strongly bonding because the constructive interference between neighbouring p orbitals results in a good accumulation of electron density between the nuclei (but slightly off the internuclear axis, as in the π bonds of diatomic molecules). The most antibonding orbital has the form

$$\psi = \psi_1 - \psi_2 + \psi_3 - \psi_4 + \psi_5 - \psi_6$$

The alternation of signs in the linear combination results in destructive interference between neighbours, and the molecular orbital has a nodal plane between each pair of neighbours, as shown in the illustration. The remaining four molecular orbitals are more

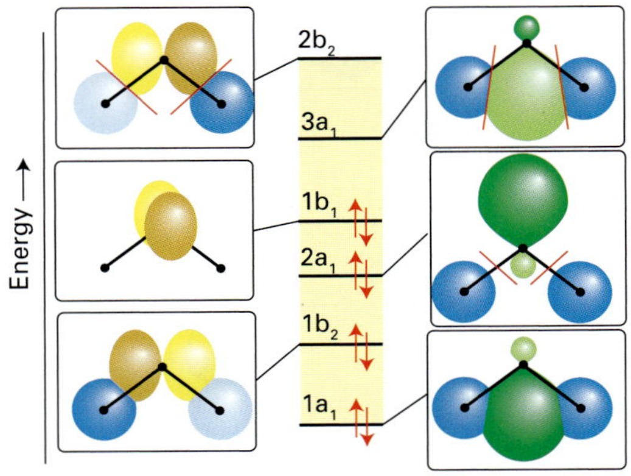

Fig. 14.34 Schematic form of the molecular orbitals of H_2O. The blue denotes s orbitals, the yellow p orbitals, and the green hybrids of s and p character. Dark and light tones denote positive and negative phases, respectively.

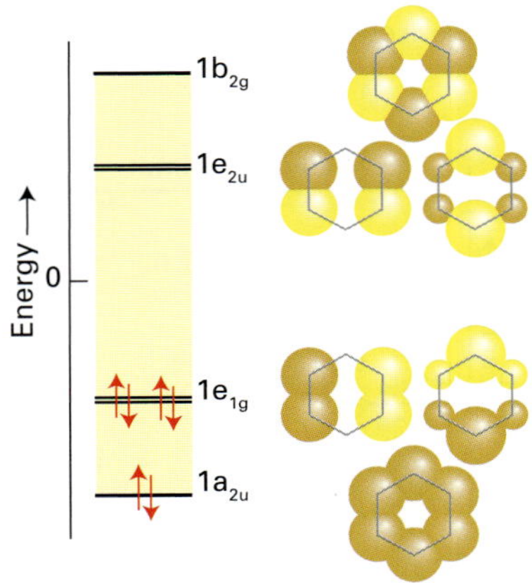

Fig. 14.35 The π orbitals and the π molecular orbital energy level diagram of benzene. The lowest-energy orbital is fully bonding between neighbouring atoms but the uppermost orbital is fully antibonding. The two pairs of doubly degenerate molecular orbitals have an intermediate number of internuclear nodes. As usual, light and dark shading represents different signs of the wavefunction. The orbitals have opposite signs below the plane of the ring. The symmetry designations are those appropriate to a hexagonal molecule.

difficult to establish by qualitative arguments, but they have the form shown in Fig. 14.35, and lie in energy between the most bonding and most antibonding orbitals. Note that the four intermediate orbitals form two doubly degenerate pairs, one net bonding and the other net antibonding.

We find the energies of the six π molecular orbitals in benzene by solving the Schrödinger equation (see the following section); they are also shown in the molecular orbital energy-level diagram. There are six electrons to be accommodated (one is supplied by each C atom), and they occupy the lowest three orbitals. The resulting electron distribution is like a double doughnut. It is an important feature of the configuration that the only molecular orbitals occupied have a net bonding character, as this is one contribution to the stability (in the sense of low energy) of the benzene molecule. It may be helpful to note the similarity between the molecular orbital energy-level diagram for benzene and that for N_2 (see Fig. 14.24): the strong bonding in benzene is echoed in the strong bonding in nitrogen.

A feature of the molecular orbital description of benzene is that each molecular orbital spreads either all round or partially round the C_6 ring. That is, π bonding is **delocalized**, and each electron pair helps to bind together several or all of the C atoms. The delocalization of bonding influence is a primary feature of molecular orbital theory and we shall encounter it in its extreme form when we come to consider the electronic structures of solids.

A final point is that the σ, π classification of molecular orbitals is not strictly applicable to nonlinear polyatomic molecules. Instead, a classification scheme based on the actual symmetry of the molecule is used, and you will see symbols such as a_1, e, and t_{2g} in place of σ and π[1]. However, the σ, π classification is relevant *locally*, in the sense that we can speak of the σ bond between the O and an H atom in H_2O and the π orbitals between a pair of C atoms in benzene. In benzene, and other planar molecules, the term 'π orbital' is extended to mean the delocalized molecular orbitals constructed from the p orbitals perpendicular to the ring.

14.16 The Hückel method

One of the earliest and simplest attempts to express these concepts quantitatively was introduced by Erich Hückel in 1931. A point to bear in mind throughout is that a linear combination of N atomic orbitals results in N molecular orbitals: the Hückel method provides an approximate procedure for putting these molecular orbitals in order of increasing energy. We shall introduce the method by considering ethene, $CH_2{=}CH_2$, and then consider larger molecules.

The first step that Hückel took was to ignore the σ-bonding framework and focus solely on the π electrons. That is, he assumed that the atoms had taken up the positions they have in the actual molecule, then calculated the properties of the π orbitals that matched that framework. The partial justification for this approximation is that atomic orbitals that contribute to σ molecular orbitals have a different symmetry from the atomic orbitals that contribute to π molecular orbitals and have zero net overlap with them. For ethene that meant that he wrote $\psi = c_A\psi_A + c_B\psi_B$, where ψ_A and ψ_B are the C2p_x orbitals (with x perpendicular to the molecular plane). We show in Derivation 14.3 that the Schrödinger equation for the orbitals, $\hat{H}\psi = E\psi$, then becomes the following pair of simultaneous equations for the coefficients:

$$\begin{aligned}(H_{AA} - E)c_A + (H_{AB} - ES)c_B &= 0\\ (H_{BA} - ES)c_A + (H_{BB} - E)c_B &= 0\end{aligned} \qquad (14.18)$$

with the various terms defined and interpreted in the derivation. These equations are called the **secular equations.**

Derivation 14.3

The secular equations

When $\psi = c_A\psi_A + c_B\psi_B$ is substituted into $\hat{H}\psi = E\psi$, we get

$$c_A\hat{H}\psi_A + c_B\hat{H}\psi_B = c_A E\psi_A + c_B E\psi_B$$

Now multiply through by ψ_A:

$$c_A\psi_A\hat{H}\psi_A + c_B\psi_A\hat{H}\psi_B = c_A E\psi_A\psi_A + c_B E\psi_A\psi_B$$

A common 'trick' in quantum chemistry is to turn products of functions (such as $\psi_A\psi_B$) into quantities with a particular numerical value by integrating them over all space ($\psi_A\psi_B$, for instance, then becomes the overlap integral S with a numerical value such as 0.2). When we integrate all four terms in this expression we get

$$c_A\underbrace{\int \psi_A\hat{H}\psi_A\,d\tau}_{H_{AA}} + c_B\underbrace{\int \psi_A\hat{H}\psi_B\,d\tau}_{H_{AB}}$$

$$= c_A E\underbrace{\int \psi_A^2\,d\tau}_{1} + c_B E\underbrace{\int \psi_A\psi_B\,d\tau}_{S_{AB}}$$

[1] The symmetry classification of orbitals in polyatomic molecules is described in our *Physical chemistry* (2006).

That is,

$$c_A H_{AA} + c_B H_{AB} = c_A E + c_B E S_{AB}$$

which is the first of the two secular equations. The second equation is generated similarly but by multiplying with ψ_B instead of ψ_A and then integrating.

In anticipation of the next step, we can interpret some of the integrals. The overlap integral S will be familiar. The integral we have called H_{AA} is the 'expectation value' of the hamiltonian calculated by using the wavefunction ψ_A centred on atom A: that means it is a kind of average of the hamiltonian, with the contribution to the total energy weighted by the probability, $\psi_A^2 d\tau$, that the electron is at each point; thus it can be interpreted as the energy of the electron in that orbital. The same is true of H_{BB}, which we interpret as the energy of an electron in the orbital ψ_B centred on atom B. If both atoms are the same, these integrals are the same, and we denote them α. The integral H_{AB} depends on both ψ_A and ψ_B, and we can interpret it as the contribution to the energy due to the accumulation of electron density where the two atomic orbitals overlap, including, for instance, the Coulombic attraction between the extra accumulation of electron density and both nuclei. This term also includes the contribution to the lowering of kinetic energy that stems from the spread of the electron over both nuclei. This integral (and H_{BA}, to which it is equal) is denoted β.

Hückel then made further approximations. First, he neglected all overlap integrals and set $S = 0$ wherever it appears. This approximation turns the secular equations into

$$\begin{aligned}(H_{AA} - E)c_A + H_{AB}c_B &= 0\\ H_{BA}c_A + (H_{BB} - E)c_B &= 0\end{aligned} \qquad (14.19)$$

Then, as explained in the preceding derivation, like him we suppose that the terms H_{AA} and H_{BB}, which represent the energy of an electron when it is on atom A and B, respectively, are equal (that is true in ethene and any molecule with equivalent carbon atoms) and replaced them by an empirical constant α. This parameter is approximately equal to (the negative of) the ionization energy of the atom. At this point, the equations are

$$\begin{aligned}(\alpha - E)c_A + H_{AB}c_B &= 0\\ H_{BA}c_A + (\alpha - E)c_B &= 0\end{aligned} \qquad (14.20)$$

Finally, as explained in the derivation, he also supposed that the terms H_{AB} and H_{BA}, which represent the energy of the interaction of the two nuclei with the accumulation of electron density in the internuclear region and the effect on the kinetic energy of the electrons of spreading over both nuclei can be replaced by the empirical parameter β (which, like α, is also a negative quantity as it represents a lowering of energy):

$$\begin{aligned}(\alpha - E)c_A + \beta c_B &= 0\\ \beta c_A + (\alpha - E)c_B &= 0\end{aligned} \qquad (14.21)$$

These are the **Hückel equations** for ethene. They are two simultaneous equations for the coefficients c_A and c_B.

We solve the Hückel equations like solving any pair of simultaneous equations. In this case we multiply the first by β and the second by $\alpha - E$:

$$\begin{aligned}\beta(\alpha - E)c_A + \beta^2 c_B &= 0\\ \beta(\alpha - E)c_A + (\alpha - E)^2 c_B &= 0\end{aligned}$$

and then subtract the second from the first, to obtain

$$\{\beta^2 - (\alpha - E)^2\}c_B = 0$$

We know that the coefficient c_B cannot be zero (for if it were, there would be no overlap and bonding), so the expression multiplying c_B must be zero:

$$\beta^2 - (\alpha - E)^2 = 0 \qquad (14.22)$$

The parameters α and β are fixed, so this is an equation for E. We rearrange it into $(\alpha - E)^2 = \beta^2$ and then take the square root of each side (allowing for two possible signs), to obtain $\alpha - E = \pm\beta$. That is, the energies of the orbitals are given by

$$E = \alpha \pm \beta \qquad (14.23)$$

This is essentially the end of the calculation, for we see that there are two energy levels, one at $\alpha + \beta$ and the other at $\alpha - \beta$, corresponding, respectively (because β is negative), to the bonding and antibonding orbitals (Fig. 14.36). Because two electrons can occupy the bonding orbital, the **π-electron binding energy**, E_π, of the molecule, the energy due to the π electrons, is

$$E_\pi = 2\alpha + 2\beta \qquad (14.24a)$$

If one electron is excited into the antibonding orbital, the π-electron energy changes to

$$E'_\pi = \alpha + \beta + (\alpha - \beta) = 2\alpha \qquad (14.24b)$$

The difference $E'_\pi - E_\pi = -2\beta$ is therefore the energy needed to excite a π electron in ethene. From spectroscopy it is known that that excitation energy is equal to about 7.7 eV, so we now know that β is approximately −3.8 eV (corresponding to −370 kJ mol^{-1}).

The Hückel method is relatively easy to extend to more complicated hydrocarbons, but more advanced techniques—now universally implemented on mathematical software—are needed to solve the secular

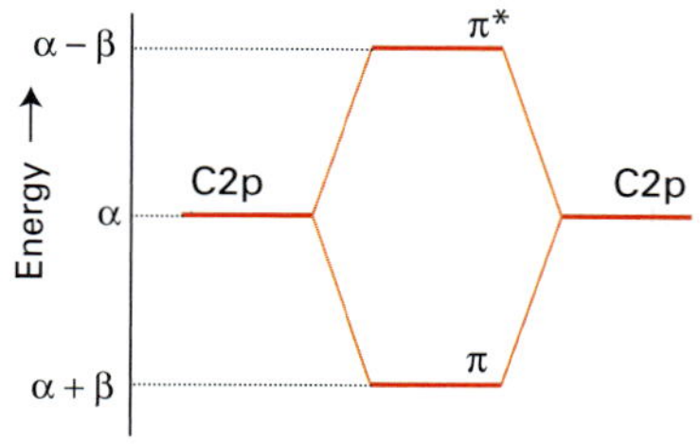

Fig. 14.36 The results of a Hückel calculation of the π orbital energies of ethene.

equations for the energy levels (and the coefficients of the orbitals). In addition to the approximations already mentioned, the method also supposes that the only interactions between immediate neighbours are included. In summary, the method is as follows:

- Focus solely on the π orbitals of the system.
- Set all overlap integrals equal to zero.
- Set all terms of the form H_{AA} equal to α (a negative quantity)
- Set all terms of the form H_{AB} equal to β (a negative quantity) if the atoms are neighbours and to zero otherwise.
- Solve the Hückel equations for the energies.

In practice, this means that for N carbon atoms there are N Hückel equations; the coefficients are multiplied by $\alpha - E$, β, or zero depending on whether they refer to a single atom, involve neighbours, or involve atoms that are not neighbours, respectively.

A brief illustration The Hückel equations for butadiene (**1**) are

$$(\alpha - E)c_A + \beta c_B = 0$$
$$\beta c_A + (\alpha - E)c_B + \beta c_C = 0$$
$$\beta c_B + (\alpha - E)c_C + \beta c_D = 0$$
$$\beta c_C + (\alpha - E)c_D = 0$$

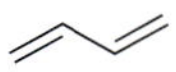

4 Butadiene

The solutions (which are best found by using mathematical software but in this simple case can also be found manually) are

$$E = \alpha \pm 1.62\beta,\ \alpha \pm 0.62\beta$$

Because α and β are both negative quantities, the order of increasing energy, from most bonding to most antibonding, is $\alpha + 1.62\beta$, $\alpha + 0.62\beta$, $\alpha - 0.62\beta$, $\alpha - 1.62\beta$.

Self-test 14.12

Calculate the π-electron energy of butadiene in its ground state.

[*Answer:* $4\alpha + 4.48\beta$]

The π-electron binding energy of butadiene is $E_\pi = 4\alpha + 4.48\beta$. If we had treated the molecule as two isolated π bonds between atoms A and B and atoms C and D, we would have calculated $E_\pi = 4\alpha + 4\beta$ (as for two ethene molecules). The difference, 0.48β (about -180 kJ mol^{-1}) is called the **delocalization energy**, E_{deloc}. The delocalization energy is the additional lowering of energy of the molecule due to the spreading of the π electrons throughout the molecule instead of being localized in discrete bonding regions.

A brief illustration The π electron energy levels of benzene (see Problem 14.36) according to the Hückel method are

$$E = \alpha \pm 2\beta,\ \alpha \pm \beta,\ \alpha \pm \beta$$

(Note the double degeneracy of levels with energy $\alpha \pm \beta$.) These levels are illustrated in Fig. 14.35. The π-electron binding energy of benzene is

$$E_\pi = 2(\alpha + 2\beta) + 2(\alpha + \beta) + 2(\alpha + \beta) = 6\alpha + 8\beta$$

The π-electron binding energy of three localized ethene molecules is $3(2\alpha + 2\beta) = 6\alpha + 6\beta$. Therefore, the delocalization energy of benzene is

$$E_{deloc} = 6\alpha + 8\beta - (6\alpha + 6\beta) = 2\beta$$

or about 740 kJ mol^{-1}. This considerable lowering of energy due to delocalization is a major contribution to the stability of the benzene ring and of aromatic compounds in general.

Computational chemistry

Computational chemistry is now a standard part of chemical research. One major application is in pharmaceutical chemistry, where the likely pharmacological activity of a molecule can be assessed computationally from its shape and electron density distribution before expensive *in-vivo* trials are started. Commercial software is now widely available for calculating the electronic structures of molecules and displaying the results graphically. All such calculations work within the Born–Oppenheimer approximation and express the molecular orbitals as linear combinations of atomic orbitals.

14.17 Techniques

There are two principal approaches to solving the Schrödinger equation for many-electron polyatomic molecules. In the **semiempirical methods**, certain expressions that occur in the Schrödinger equation are set equal to parameters that have been chosen to lead to the best fit to experimental quantities, such as enthalpies of formation. Semiempirical methods are applicable to a wide range of molecules with a virtually limitless number of atoms, and are widely popular. In the more fundamental ***abinitio* methods**, an attempt is made to calculate structures from first principles, using only the atomic numbers of the atoms present. Such an approach is intrinsically more reliable than a semiempirical procedure but is much more demanding computationally.

Both types of procedure typically adopt a **self-consistent field** (SCF) procedure, in which an initial guess about the composition of the LCAO is successively refined until the solution remains unchanged in a cycle of calculation. First, we guess the values of the coefficients in the LCAO used to build the molecular orbitals—and solve the Schrödinger equation for the coefficients of one LCAO on the basis of that guess for the coefficients of all the other occupied orbitals. Now we have a first approximation to the coefficients of one LCAO. We then repeat the procedure for all the other occupied molecular orbitals. At the end of that stage we have a new set of LCAO coefficients that differ from our first guess, and we also have an estimate of the energy of the molecule. We use that refined set of coefficients to repeat the calculation and calculate a new set of coefficients and a new energy. In general, these will differ from the new starting point. However, there comes a stage when repetition of the calculation leaves the coefficients and energy unchanged. The orbitals are now said to be 'self-consistent', and we accept them as a description of the molecule.

The severe approximations of the Hückel method have been removed over the years in a succession of better approximations. Each has given rise to an acronym, such as CNDO ('complete neglect of differential overlap'), INDO ('intermediate neglect of differential overlap'), MINDO ('modified neglect of differential overlap'), and AM1 ('Austin Model 1', version 2 of MINDO). Software for all these procedures are now readily available, and reasonably sophisticated calculations can now be run even on hand-held computers. A semiepirical technique that has gained considerable ground in recent years to become one of the most widely used techniques for the calculation of molecular structure is **density functional theory** (DFT). Its advantages include less demanding computational effort, less computer time, and—in some cases, particularly d-metal complexes—better agreement with experimental values than is obtained from other procedures.

The *abinitio* methods also simplify the calculations, but they do so by setting up the problem in a different manner, avoiding the need to estimate parameters by appeal to experimental data. In these methods, sophisticated techniques are used to solve the Schrödinger equation numerically. The difficulty with this procedure is the enormous time it takes to carry out the detailed calculation. That time can be reduced by replacing the hydrogenic atomic orbitals used to form the LCAO by a **gaussian-type orbital** (GTO) in which the exponential function e^{-r} characteristic of actual orbitals is replaced by a sum of gaussian functions of the form e^{-r^2}.

14.18 Graphical output

One of the most significant developments in computational chemistry has been the introduction of graphical representations of molecular orbitals and electron densities. The raw output of a molecular structure calculation is a list of the coefficients of the atomic orbitals in each molecular orbital and the energies of these orbitals. The graphical representation of a molecular orbital uses stylized shapes to represent the basis set, and then scales their size to indicate the value of the coefficient in the LCAO. Different signs of the wavefunctions are represented by different colours (Fig. 14.37).

Once the coefficients are known, we can build up a representation of the electron density in the molecule by noting which orbitals are occupied and then forming the squares of those orbitals. The total electron density at any point is then the sum of the squares of the wavefunctions evaluated at that point. The outcome is commonly represented by an **isodensity surface**, a surface of constant total electron density (Fig. 14.38). There are several styles of representing an isodensity surface, as a solid form, as a transparent form with a ball-and-stick representation of the molecule within, or as a mesh.

One of the most important aspects of a molecule other than its geometrical shape is the distribution of electric potential over its surface. A common procedure begins with calculation of the net potential at each point on an isodensity surface by subtracting the potential due to the electron density at that point from the potential due to the nuclei. The result is an

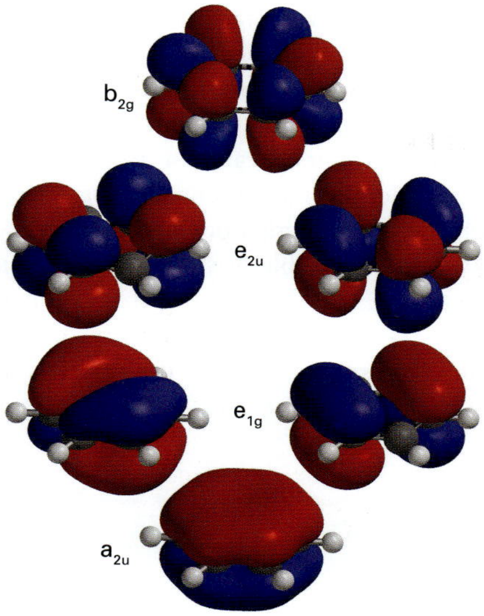

Fig. 14.37 The output of a computation of the π orbitals of benzene: opposite signs of the wavefunctions are represented by different colours. Compare these molecular orbitals with the more diagrammatic representation in Fig. 14.35.

Fig. 14.38 The isodensity surface of benzene obtained by using the same software as in Fig. 14.37.

electrostatic potential surface (an 'elpot surface') in which net positive potential is shown in one colour and net negative potential is shown in another, with intermediate gradations of colour (Fig. 14.39).

14.19 Applications

One goal of computational chemistry—at least when applied to large molecules—is to gain insight into trends in molecular properties without necessarily striving for ultimate accuracy. We have already encountered one example of this approach in Section 3.6, where we saw that they may be used to estimate enthalpies of formation of molecules. Molecular orbital calculations may also be used to predict trends in electrochemical properties, such as standard potentials (Chapter 9). Several experimental and computational studies of aromatic hydrocarbons indicate that

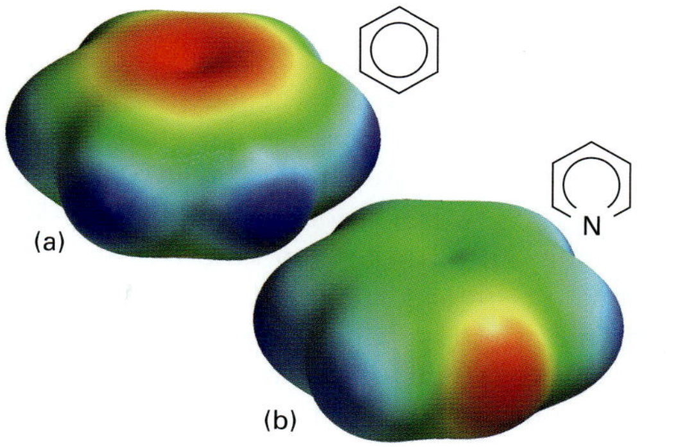

Fig. 14.39 The electrostatic potential surfaces of (a) benzene and (b) pyridine. Note the accumulation of electron density on the nitrogen atom of pyridine at the expense of the other atoms.

Table 14.3

Summary of ab-initio *calculations and spectroscopic data for four linear polyenes*

	$\Delta E_{\text{HOMO–LUMO}}$/eV	$\lambda_{\text{transition}}$/nm
=	18.1	163
=–=	14.5	217
=–=–=	12.7	252
=–=–=–=	11.6	304

1 eV = 1.602 × 10^{-19} J

decreasing the energy of the LUMO enhances the ability of a molecule to accept an electron into the LUMO, with an attendant increase in the value of the molecule's standard potential.

We remarked in Chapter 12 that a molecule can absorb or emit a photon of energy hc/λ, resulting in a transition between two quantized molecular energy levels. The transition of lowest energy (and longest wavelength) occurs between the HOMO and LUMO. We can use calculations to correlate the HOMO–LUMO energy gap with the wavelength of absorption. For example, consider the linear polyenes shown in Table 14.3, all of which absorb in the ultraviolet region of the spectrum. The table also shows that, as expected, the wavelength of the lowest-energy electronic transition decreases as the energy separation between the HOMO and LUMO increases. We also see that the smallest HOMO–LUMO gap and longest wavelength of absorption correspond to octatetraene, the longest polyene in the group. It follows that the wavelength of the transition increases with increasing number of conjugated double bonds in linear polyenes. Extrapolation of the trend suggests that a sufficiently long linear polyene should absorb light in the visible region of the electromagnetic spectrum. This is indeed the case for β-carotene (structure **1** in Section 12.9), which absorbs light with $\lambda \approx 450$ nm. The ability of β-carotene to absorb visible light is part of the strategy employed by plants to harvest solar energy for use in photosynthesis (Box 20.2).

There are several ways in which molecular orbital calculations lend insight into reactivity. For example, electrostatic potential surfaces may be used to reveal an electron-poor region of a molecule, a region that is susceptible to association with or chemical attack by an electron-rich region of another molecule. Such considerations are important for assessing the pharmacological activity of potential drugs. Computational chemistry may also be used to model species that may be too unstable or short-lived to be studied experimentally. For this reason, it is often used to study the transition state, with an eye toward describing factors that increase the reaction rate.

Checklist of key ideas

You should now be familiar with the following concepts.

- ☐ **1** The classification of bonds as covalent and ionic.
- ☐ **2** The Born–Oppenheimer approximation and molecular potential energy curves.
- ☐ **3** Valence bond theory and the concepts of σ and π bonds, promotion, hybridization, and resonance.
- ☐ **4** Molecular orbital theory and the construction of molecular orbitals as linear combinations of atomic orbitals.
- ☐ **5** Bonding and antibonding atomic orbitals and inversion (g,u) symmetry.
- ☐ **6** The building-up principle for constructing the electron configuration of molecules on the basis of their molecular orbital energy-level diagram.
- ☐ **7** The concepts of σ and π orbitals and the role of symmetry and the similarity of energy in the construction of molecular orbitals.
- ☐ **8** The concept of electronegativity and the Pauling and Mulliken definitions.
- ☐ **9** The concept of self-consistent field and the distinction between semiempirical and *ab-initio* methods of computation.
- ☐ **10** The Hückel method for the estimation of the energies of molecular orbitals.
- ☐ **11** Applications of molecular orbital calculations to the prediction of reactivity and thermochemical, electrochemical, and spectroscopic properties.

Table of key equations

The following table summarizes the equations developed in this chapter.

Property	Equation	Comment
Bond order	$b = \frac{1}{2}(N - N^*)$	Definition
Mulliken electronegativity	$\chi = \frac{1}{2}(I + E_{ea})$	Definition
Molecular orbital	$\psi = c_A\psi_A + c_B\psi_B$	LCAO approximation
Valence bond wavefunction	$\psi_{A-B}(1,2) = \psi_A(1)\psi_B(2) + \psi_A(2)\psi_B(1)$	
Ionic–covalent resonance	$\psi = \psi_{\text{covalent}} + \lambda\psi_{\text{ionic}}$	
Overlap integral	$S = \int \psi_A\psi_B \mathrm{d}\tau$	Definition

Questions and exercises

Discussion questions

14.1 Compare the approximations built into valence bond theory and molecular orbital theory.

14.2 Discuss the steps involved in the construction of sp^3, sp^2, and sp hybrid orbitals.

14.3 Describe how molecular orbital theory accommodates all the conventional types of bonding.

14.4 Why is the electron pair such a central concept in theories of the chemical bond?

14.5 Distinguish between the Pauling and Mulliken electronegativity scales.

14.6 Why is orbital overlap a guide to assessing the strengths of chemical bonds? Why, sometimes, is it not?

14.7 Identify and justify the approximations used in the Hückel theory of hydrocarbons.

14.8 Explain the differences between semiempirical and *ab-initio* methods of electronic structure determination.

Exercises

14.1 Give the valence-bond description of a C—H group in a molecule.

14.2 Give the valence-bond description of a P_2 molecule. Why is P_4 a more stable form of molecular phosphorus?

14.3 Write down the valence-bond wavefunction for a nitrogen molecule.

14.4 Calculate the molar energy of repulsion between two hydrogen nuclei at the separation in H_2 (74.1 pm). The result is the energy that must be overcome by the attraction from the electrons that form the bond.

14.5 Write down the valence-bond description of CH_4 based on hybrid orbitals h on the carbon atom.

14.6 The structure of the visual pigment retinal is shown in (**5**). Label each atom with its state of hybridization and specify the composition of each of the different type of bond.

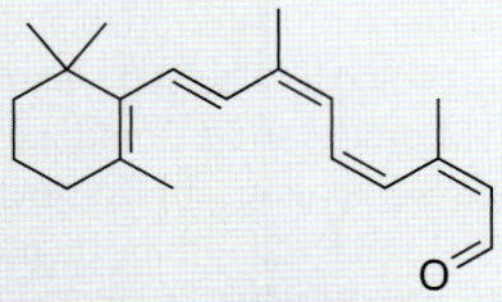

5 11-cis-Retinal

14.7 Write down three contributions to the resonance structure of naphthalene, $C_{10}H_8$.

14.8 A normalized valence-bond wavefunction turned out to have the form $\psi = 0.889\psi_{cov} + 0.458\psi_{ion}$. What is the chance that, in 1000 inspections of the molecule, both electrons of the bond will be found on one atom?

14.9 Benzene is commonly regarded as a resonance hybrid of the two Kekulé structures, but other possible structures can also contribute. Draw three other structures in which there are only covalent π bonds (allowing for bonding between some non-adjacent C atoms) and two structures in which there is one ionic bond. Why may these structures be ignored in simple descriptions of the molecule?

14.10‡ Show, if overlap is ignored, (a) that any molecular orbital expressed as a linear combination of two atomic orbitals may be written in the form $\psi = \psi_A \cos\theta + \psi_B \sin\theta$, where θ is a parameter that varies between 0 and $\frac{1}{2}\pi$, and (b) that if ψ_A and ψ_B are orthogonal and normalized to 1, then ψ is also normalized to 1. (c) To what values of θ do the bonding and antibonding orbitals in a homonuclear diatomic molecule correspond?

14.11 Draw diagrams to show the various orientations in which a p orbital and a d orbital on adjacent atoms may form bonding and antibonding molecular orbitals.

14.12 Give the ground-state electron configurations of (a) Li_2, (b) Be_2, and (c) C_2.

14.13 Give the ground-state electron configurations of (a) H_2^-, (b) N_2, and (c) O_2.

14.14 Three biologically important diatomic species, either because they promote or inhibit life, are (a) CO, (b) NO, and (c) CN^-. The first binds to haemoglobin, the second is a neurotransmitter, and the third interrupts the respiratory electron-transfer chain. Their biochemical action is a reflection of their orbital structure. Deduce their ground-state electron configurations. (For heteronuclear diatomic molecules, a good first approximation is that the energy-level diagram is much the same as for homonuclear diatomic molecules.)

14.15 From the ground-state electron configurations of B_2 and C_2, predict which molecule should have the greater dissociation energy.

14.16 Some chemical reactions proceed by the initial loss or transfer of an electron to a diatomic species. Which of the molecules N_2, NO, O_2, C_2, F_2, and CN would you expect to be stabilized by (a) the addition of an electron to form AB^-, (b) the removal of an electron to form AB^+?

14.17 From the ground-state electron configurations of B_2 and C_2, predict which molecule should have the greater bond dissociation energy.

14.18 The existence of compounds of the noble gases was once a great surprise and stimulated a great deal of theoretical work. Sketch the molecular orbital energy level diagram for XeF and deduce its ground-state electron configurations. Is XeF likely to have a shorter bond length than XeF^+?

14.19 Where it is appropriate, give the parity of (a) $1\pi^*$ in F_2, (b) 2σ in NO, (c) 1δ in Tl_2, (d) $1\delta^*$ in Fe_2.

14.20 Give the (g,u) parities of the first four levels of a particle-in-a-box wavefunction.

14.21 (a) Give the parities of the wavefunctions for the first four levels of a harmonic oscillator. (b) How may the parity be expressed in terms of the quantum number v?

14.22 State the parities of the six π-orbitals of benzene (see Fig. 14.35).

14.23 Two important diatomic molecules for the welfare of humanity are NO and N_2: the former is both a pollutant and a neurotransmitter, and the latter is the ultimate source of the nitrogen of proteins and other biomolecules. Use the electron configurations of NO and N_2 to predict which is likely to have the shorter bond length.

14.24 Put the following species in order of increasing bond length: F_2^-, F_2, F_2^+.

14.25 Identify the bond order of each of the species in the preceding exercise.

14.26 Arrange the species O_2^+, O_2, O_2^-, O_2^{2-} in order of increasing bond length.

14.27 Identify the bond order of each of the species in the preceding exercise.

14.28 Construct the molecular orbital energy level diagrams of (a) ethene and (b) ethyne on the basis that the molecules are formed from the appropriately hybridized CH_2 or CH fragments.

14.29 Try to anticipate the form that the bonding and antibonding 'ϕ orbitals' that could be constructed from two neighbouring f orbitals. What are their parities?

14.30 Predict the polarities of the bonds (a) P—H, (b) B—H.

14.31 Predict the electronic configurations of (a) the benzene anion, (b) the benzene cation. Estimate the π-bond energy in each case.

14.32 Many of the colours of vegetation are due to electronic transitions in conjugated π-electron systems. In the *free-electron molecular orbital* (FEMO) theory, the electrons in a conjugated molecule are treated as independent particles in a box of length L. Sketch the form of the two occupied orbitals in butadiene predicted by this model and predict the minium excitation energy of the molecule. The tetraene $CH_2{=}CHCH{=}CHCH{=}CHCH{=}CH_2$ can be treated as a box of length $8R$, where $R = 140$ pm (as in this case, an extra half bond length is often added at each end of the box). Calculate the minimum excitation energy of the molecule and sketch the HOMO and LUMO.

14.33 Suppose that the π-electron molecular orbitals of naphthalene can be represented by the wavefunctions of a particle in a rectangular box. What are the parities of the occupied orbitals?

14.34 How many molecular orbitals can be constructed from a diatomic molecule in which s, p, d, and f orbitals are all important for bonding?

14.35 The FEMO theory (Exercise 14.32) of conjugated molecules is rather crude and better results are obtained with simple Hückel theory. (a) For a linear conjugated polyene with each of N carbon atoms contributing an electron in a 2p orbital, the energies E_k of the resulting π molecular orbitals are given by:

$$E_k = \alpha + 2\beta\cos\frac{k\pi}{N+1} \qquad k = 1, 2, 3, \ldots, N$$

Use this expression to determine a reasonable empirical estimate of the resonance integral β for the series consisting of ethene, butadiene, hexatriene, and octatetraene given that π←π ultraviolet absorptions from the HOMO to the LUMO occur at 61 500, 46 080, 39 750, and 32 900 cm^{-1}, respectively. (b) Calculate the π-electron delocalization energy, $E_{\text{deloc}} = E_\pi - n(\alpha + \beta)$, of octatetraene, where E_π is the total π-electron binding energy and n is the total number of π-electrons.

14.36 For monocyclic conjugated polyenes (such as cyclobutadiene and benzene) with each of N carbon atoms contributing an electron in a 2p orbital, simple Hückel theory gives the following expression for the energies E_k of the resulting π molecular orbitals:

$$E_k = \alpha + 2\beta\cos\frac{2k\pi}{N+1} \qquad k = 0, \pm1, \pm2, \ldots, \pm N/2 \text{ (even } N\text{)}$$

$$k = 0, \pm1, \pm2, \ldots, \pm(N-1)/2 \text{ (odd } N\text{)}$$

(a) Calculate the energies of the π molecular orbitals of benzene and cyclooctaene. Comment on the presence or absence of degenerate energy levels. (b) Calculate and compare the delocalization energies of benzene (using the expression above) and hexatriene (see Exercise 14.35). What do you conclude from your results? (c) Calculate and compare the delocalization energies of cyclooctaene and octatetraene. Are your conclusions for this pair of molecules the same as for the pair of molecules investigated in part (b)?

Projects

The symbol ‡ indicates that calculus is required.

14.37‡ Here we explore hydrid orbitals in more quantitative detail. (a) Show that the orbitals $h_1 = s + p_x + p_y + p_z$ and $h_2 = s - p_x - p_y + p_z$ are orthogonal. *Hint:* Each atomic orbital individually normalized to 1. Also, note that: (i) s and p orbitals are orthogonal and, (ii) p orbitals with perpendicular orientations are orthogonal. (b) Show that the sp^2 hybrid orbital $(s + 2^{1/2}p)/3^{1/2}$ is normalized to 1 if the s and p orbitals are each normalized to 1. (c) Find another sp^2 hybrid orbital that is orthogonal to the hybrid orbital in par (b).

14.38‡ Now we explore orbital overlap and overlap integrals in detail. (a) Without doing a calculation, sketch how the overlap between an H1s orbital and a 2p orbital can be expected to depend on their separation. (b) The overlap integral between an H1s orbital and a H2p orbital on nuclei separated by a distance R is $S = (R/a_0)\{1 + (R/a_0) + \frac{1}{3}(R/a_0)^2\}e^{-R/a_0}$. Plot this function, and find the separation for which the overlap is a maximum. (c) Suppose that a molecular orbital has the form $N(0.245A + 0.644B)$. Find a linear combination of the orbitals A and B that does not overlap with (that is, is orthogonal to) this combination. (d) Normalize the wavefunction $\psi = \psi_{\text{cov}} + \lambda\psi_{\text{ion}}$ in terms of the parameter λ and the overlap integral S between the covalent and ionic wavefunctions.

Chapter 15

Molecular interactions

Atoms and molecules with complete valence shells are still able to interact with one another even though all their valences are satisfied. They attract one another over the range of several atomic diameters and they repel one another when pressed together. These residual interactions are highly important. They account, for instance, for the condensation of gases to liquids and the structures of molecular solids. All organic liquids and solids, ranging from small molecules like benzene to virtually infinite cellulose and the polymers from which fabrics are made, are bound together by the cohesive interactions we explore in this chapter. These interactions are also responsible for the structural organization of biological macromolecules, for they pin molecular building blocks—such as polypeptides, polynucleotides, and lipids—together in the arrangement essential to their proper physiological function.

In this chapter we present the basic theory of molecular interactions and then explore how they play a role in the properties of liquids. In the following chapter we explore how the same interactions contribute to the properties of macromolecules and molecular aggregates.

van der Waals interactions

The interactions between or within molecules (for example, within macromolecules) include the attractive and repulsive interactions involving partial electric charges and electron clouds of polar and nonpolar molecules or functional groups and the repulsive interactions that prevent the complete collapse of matter to densities as high as those characteristic of atomic nuclei. The repulsive interactions arise from the exclusion of electrons from regions of space where the orbitals of closed-shell species overlap. These interactions are called **van der Waals interactions;** the term excludes interactions that result in the formation of covalent or ionic bonds. We shall see that the *potential energy* arising from an attractive van der Waals interaction is commonly proportional to the inverse sixth power of the separation between molecules or functional groups. The intermolecular *force* depends inversely on one higher power of the separation, so a van der Waals interaction for which the potential energy is proportional to the inverse sixth power of the separation corresponds to a force that is proportional to the inverse seventh power.

A brief comment If the potential energy is denoted V, then the force is $-\mathrm{d}V/\mathrm{d}r$ (see Appendix 3). So, if $V = -C/r^6$, the magnitude of the force is

Use $\mathrm{d}(1/x^n)/\mathrm{d}x = -n(1/x^{n+1})$

$$\frac{\mathrm{d}}{\mathrm{d}r}\left(-\frac{C}{r^6}\right) = \frac{6C}{r^7}$$

15.1 Interactions between partial charges

Atoms in molecules in general have partial charges. Table 15.1 gives the partial charges typically found on the atoms in peptides. If these charges were separated by a vacuum, they would attract or repel each other in accord with Coulomb's law (see the Introduction) and we would write

$$V = \frac{Q_1 Q_2}{4\pi\varepsilon_0 r} \quad (15.1a)$$

where Q_1 and Q_2 are the partial charges and r is their separation. However, we need to take into account the possibility that other parts of the molecule, or other molecules, lie between the charges, and decrease the strength of the interaction. The simplest procedure for taking into account these very complicated effects is to treat the medium as a uniform substance and to write

Table 15.1
Partial charges in polypeptides

Atom	Partial charge/e
C(=O)	+0.45
C(—CO)	+0.06
H(—C)	+0.02
H(—N)	+0.18
H(—O)	+0.42
N	−0.36
O	−0.38

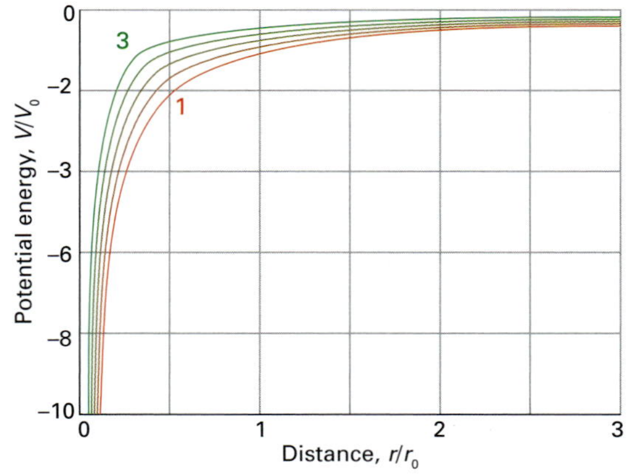

Fig. 15.1 The Coulomb potential for two charges and its dependence on their separation. The curves correspond to different relative permittivities (1 for a vacuum, 3 for a typical organic fluid).

$$V = \frac{Q_1 Q_2}{4\pi\varepsilon r} \quad (15.1b)$$

where ε is the **permittivity** of the medium: a high permittivity means that the medium reduces the strength of the interaction between the two charges. The permittivity is usually expressed as a multiple of the vacuum permittivity by writing $\varepsilon = \varepsilon_r\varepsilon_0$, where the dimensionless quantity ε_r is the **relative permittivity** (which is still widely called the *dielectric constant*). The effect of the medium can be very large: for water at 25°C, $\varepsilon_r = 78$, so the potential energy of two charges separated by bulk water is reduced by nearly two orders of magnitude compared to the value it would have if the charges were separated by a vacuum (Fig. 15.1). The problem is made worse in calculations on polypeptides and nucleic acids by the fact that two partial charges may have water and a biopolymer chain lying between them. Various models have been proposed to take this awkward effect into account, the simplest being to set $\varepsilon_r = 3.5$ and to hope for the best.

15.2 Electric dipole moments

When the molecules or groups that we are considering are widely separated, it turns out to be simpler to express the principal features of their interaction in terms of the dipole moments associated with the charge distributions rather than with each individual partial charge. At its simplest, an **electric dipole**

Table 15.2

Dipole moments and mean polarizability volumes

	μ/D	$\alpha'/(10^{-30}\ m^3)$
Ar	0	1.85
CCl_4	0	11.7
C_6H_6	0	11.6
H_2	0	0.911
H_2O	1.85	1.65
NH_3	1.47	2.47
HCl	1.08	2.93
HBr	0.80	4.01
HI	0.42	6.06

consists of two charges Q and $-Q$ separated by a distance l. The product Ql is called the **electric dipole moment**, μ. We represent dipole moments by an arrow with a length proportional to μ and pointing from the negative charge to the positive charge (**1**). (Be careful with this convention: for historical reasons the opposite convention is still widely adopted.) Because a dipole moment is the product of a charge (in coulombs, C) and a length (in metres, m), the SI unit of dipole moment is the coulomb metre (C m). However, it is often much more convenient to report a dipole moment in **debye**, D, where

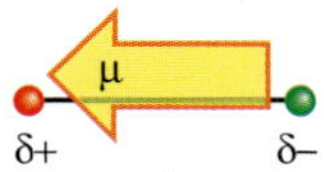

1

$$1\ \text{D} = 3.335\ 64 \times 10^{-30}\ \text{C m}$$

because then experimental values for molecules are close to 1 D (Table 15.2). The unit is named after Peter Debye, the Dutch pioneer of the study of dipole moments of molecules. The dipole moment of charges e and $-e$ separated by 100 pm is 1.6×10^{-29} C m, corresponding to 4.8 D. Dipole moments of small molecules are typically smaller than that, at about 1 D, confirming that the charge separation in simple molecules is only partial.

A **polar molecule** has a permanent electric dipole moment arising from the partial charges on its atoms (Section 14.14). A **nonpolar molecule** has no permanent electric dipole moment. All heteronuclear diatomic molecules are polar because the difference in electronegativities of their two atoms results in nonzero partial charges. Typical dipole moments are 1.08 D for HCl and 0.42 D for HI (Table 15.2). A very approximate relation between the dipole moment and the difference $\Delta\chi$ of Pauling electronegativities (Table 14.2) χ_A and χ_B of two atoms A and B, is

$$\mu/\text{D} \approx \chi_A - \chi_B = \Delta\chi \qquad (15.2)$$

A brief illustration The electronegativities of hydrogen and bromine are 2.1 and 2.8, respectively. The difference is 0.7, so we predict an electric dipole moment of about 0.7 D for HBr. The experimental value is 0.80 D.

Because it attracts the electrons more strongly, the more electronegative atom is usually the negative end of the dipole. However, there are exceptions, particularly when antibonding orbitals are occupied. Thus, the dipole moment of CO is very small (0.12 D) but the negative end of the dipole is on the C atom even though the O atom is more electronegative. This apparent paradox is resolved as soon as we realize that antibonding orbitals are occupied in CO (see Fig. 14.33) and, because electrons in antibonding orbitals tend to be found closer to the less electronegative atom, they contribute a negative partial charge to that atom. If this contribution is larger than the opposite contribution from the electrons in bonding orbitals, then the net effect will be a small negative partial charge on the *less* electronegative atom.

Molecular symmetry is of the greatest importance in deciding whether a polyatomic molecule is polar or not. Indeed, molecular symmetry is more important than the question of whether or not the atoms in the molecule belong to the same element. Homonuclear polyatomic molecules may be polar if they have low symmetry and the atoms are in inequivalent positions. For instance, the angular molecule ozone, O_3 (**2**), is homonuclear. However, it is polar because the central O atom is different from the outer two (it is bonded to two atoms, they are bonded only to one); moreover, the dipole moments associated with each bond make an angle to each other and do not cancel. Heteronuclear polyatomic molecules may be nonpolar if they have high symmetry, because individual bond dipoles may then cancel. The heteronuclear linear triatomic molecule CO_2, for example, is nonpolar because, although there are partial charges on all three atoms, the dipole moment associated with the OC bond points in the opposite direction to the dipole moment associated with the CO bond, and the two cancel (**3**).

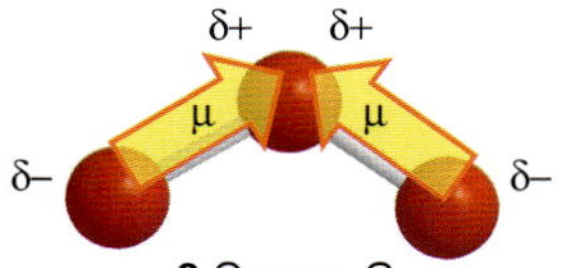

2 Ozone, O_3

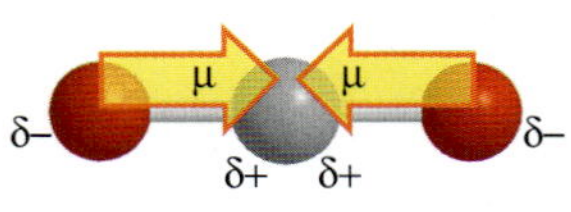

3 Carbon dioxide, CO_2

Self-test 15.1

Use the VSEPR model, which is reviewed in Appendix 4, to judge whether ClF_3 is polar or nonpolar. *Hint.* Predict the structure first.

[*Answer:* polar]

Self-test 15.2

Estimate the ratio of the electric dipole moments of *ortho* (1,2-) and *meta* (1,3-) disubstituted benzenes.

[*Answer:* $\mu(ortho)/\mu(meta) = 1.7$]

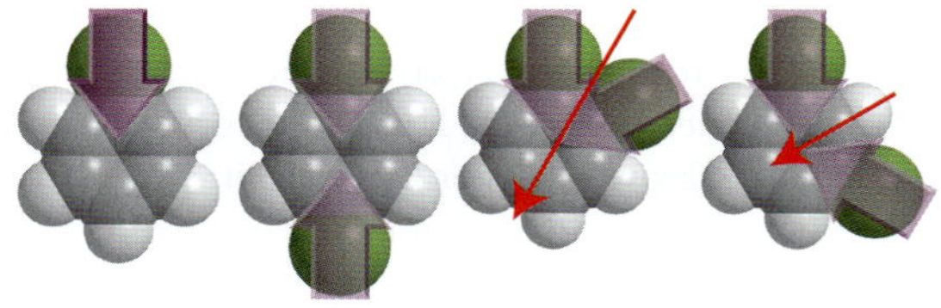

Fig. 15.2 The dipole moments of the dichlorobenzene isomers can be obtained approximately by vectorial addition of two chlorobenzene dipole moments (1.57 D).

To a first approximation, it is possible to resolve the dipole moment of a polyatomic molecule into contributions from various groups of atoms in the molecule and the directions in which these individual contributions lie (Fig. 15.2). Thus, 1,4-dichlorobenzene is nonpolar by symmetry on account of the cancellation of two equal but opposing C—Cl moments (exactly as in carbon dioxide). 1,2-Dichlorobenzene, however, has a dipole moment that is approximately the resultant of two chlorobenzene dipole moments arranged at 60° to each other. This technique of 'vector addition' (see Appendix 2) can be applied with fair success to other series of related molecules, and the resultant μ_{res} of two dipole moments μ_1 and μ_2 that make an angle θ to each other (**4**) is approximately

$$\mu_{res} \approx (\mu_1^2 + \mu_2^2 + 2\mu_1\mu_2 \cos\theta)^{1/2} \qquad (15.3)$$

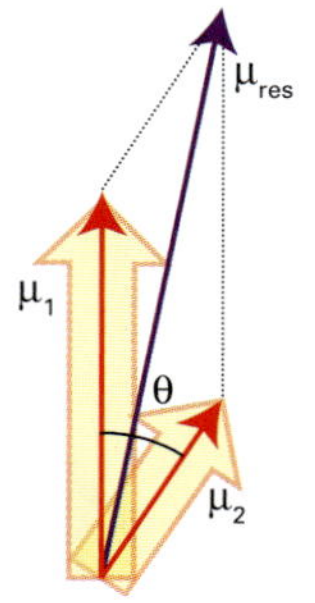

4

A better approach to the calculation of dipole moments is to take into account the locations and magnitudes of the partial charges on all the atoms. These partial charges are included in the output of many molecular structure software packages. The programs calculate the dipole moments of the molecules by noting that an electric dipole moment is actually a vector, $\boldsymbol{\mu}$, with three components, μ_x, μ_y, and μ_z (**5**). The direction of $\boldsymbol{\mu}$ shows the orientation of the dipole in the molecule and the length of the vector is the magnitude, μ, of the dipole moment. In common with all vectors, the magnitude is related to the three components by

$$\mu = (\mu_x^2 + \mu_y^2 + \mu_z^2)^{1/2} \qquad (15.4a)$$

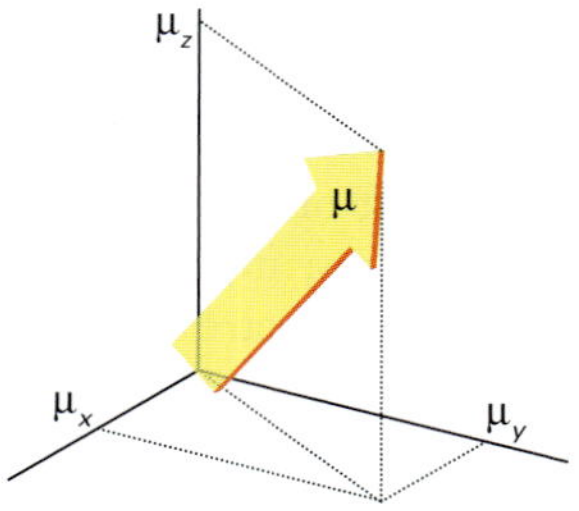

5

To calculate μ we need to calculate the three components and then substitute them into this expression. To calculate the x-component, for instance, we need to know the partial charge on each atom and the atom's x-coordinate relative to a point in the molecule and form the sum

$$\mu_x = \sum_J Q_J x_J \qquad (15.4b)$$

Here, Q_J is the partial charge of atom J, x_J is the x-coordinate of atom J, and the sum is over all the atoms in the molecule. Analogous expressions are used for the y- and z-components. For an electrically neutral molecule, the origin of the coordinates is arbitrary, so it is best chosen to simplify the measurements.

Example 15.1

Calculating a molecular dipole moment

Estimate the electric dipole moment of the peptide group using the partial charges (as multiples of e) in Table 15.1 and the locations of the atoms shown in (**6**).

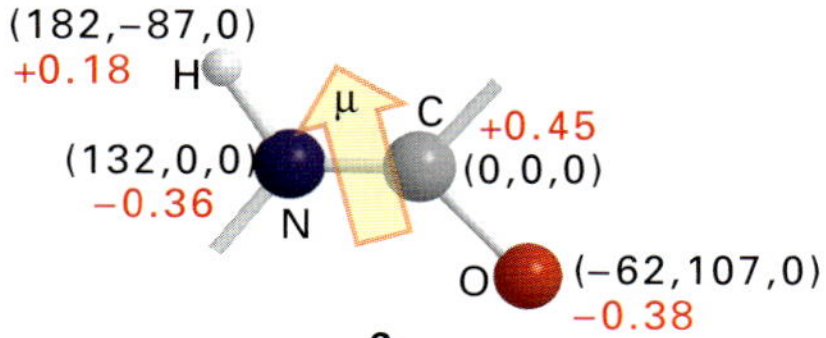

6

Strategy We use eqn 15.4b to calculate each of the components of the dipole moment and then eqn 15.4a to assemble the three components into the magnitude of the dipole moment. Note that the partial charges are multiples of the fundamental charge, $e = 1.609 \times 10^{-19}$ C (see inside front cover).

Solution The expression for μ_x is

$$\begin{aligned}\mu_x &= (-0.36e) \times (132\text{ pm}) + (0.45e) \times (0\text{ pm}) + \\ &\quad (0.18e) \times (182\text{ pm}) + (-0.38e) \times (-62\text{ pm}) = 8.8e\text{ pm} \\ &= 8.8 \times (1.609 \times 10^{-19}\text{ C}) \times (10^{-12}\text{ m}) \\ &= 1.4 \times 10^{-30}\text{ C m}\end{aligned}$$

corresponding to $\mu_x = 0.42$ D. The expression for μ_y is:

$$\begin{aligned}\mu_y &= (-0.36e) \times (0\text{ pm}) + (0.45e) \times (0\text{ pm}) + \\ &\quad (0.18e) \times (-87\text{ pm}) + (-0.38e) \times (107\text{ pm}) \\ &= -56e\text{ pm} = -9.1 \times 10^{-30}\text{ C m}\end{aligned}$$

It follows that $\mu_y = -2.7$ D. Therefore, because $\mu_z = 0$,

$$\mu = \{(0.42\text{ D})^2 + (-2.7\text{ D})^2\}^{1/2} = 2.7\text{ D}$$

We can find the orientation of the dipole moment by arranging an arrow of length 2.7 units of length to have x-, y-, and z-components of 0.42, −2.7, and 0 units; the orientation is superimposed on (**6**).

Self-test 15.3

Calculate the electric dipole moment of formaldehyde, using the information in (**7**).

[*Answer:* −2.3 D]

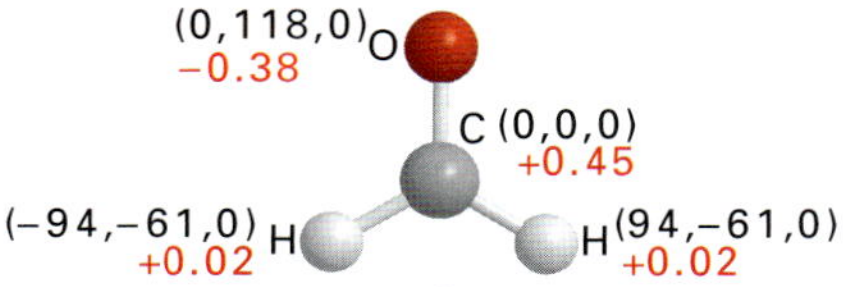

7

15.3 Interactions between dipoles

The potential energy of a dipole μ_1 in the presence of a charge Q_2 is calculated by taking into account the interaction of the charge with the two partial charges of the dipole, one resulting in a repulsion and the other an attraction. The result for the arrangement shown in (**8**) is:

$$V = -\frac{Q_2\mu_1}{4\pi\varepsilon_0 r^2} \qquad (15.5a)$$

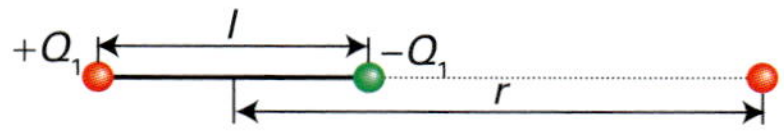

8

Derivation 15.1

The interaction of a charge with a dipole

When the charge and dipole are collinear, as in (**8**), the potential energy is

$$V = \overbrace{\frac{Q_1Q_2}{4\pi\varepsilon_0(r + \frac{1}{2}l)}}^{\text{Repulsion between }+Q_1\text{ and }Q_2} - \overbrace{\frac{Q_1Q_2}{4\pi\varepsilon_0(r - \frac{1}{2}l)}}^{\text{Attraction between }-Q_1\text{ and }Q_2}$$

$$= \frac{Q_1Q_2}{4\pi\varepsilon_0 r\left(1 + \dfrac{l}{2r}\right)} - \frac{Q_1Q_2}{4\pi\varepsilon_0 r\left(1 - \dfrac{l}{2r}\right)}$$

Next, we suppose that the separation of charges in the dipole is much smaller than the distance of the charge Q_2 in the sense that $l/2r \ll 1$. Then we can use (see Appendix 2)

$$\frac{1}{1+x} \approx 1 - x \qquad \frac{1}{1-x} \approx 1 + x$$

to write

$$V = \frac{Q_1Q_2}{4\pi\varepsilon_0 r}\left\{\left(\cancel{1} - \frac{l}{2r}\right) - \left(\cancel{1} + \frac{l}{2r}\right)\right\} = -\frac{Q_1Q_2 l}{4\pi\varepsilon_0 r^2}$$

Now we recognize that $Q_2 l = \mu_2$, the dipole moment of molecule 2, and obtain eqn 15.5a.

A similar calculation for the more general orientation shown in (**9**) gives

$$V = -\frac{\mu_1 Q_2 \cos\theta}{4\pi\varepsilon_0 r^2} \qquad (15.5b)$$

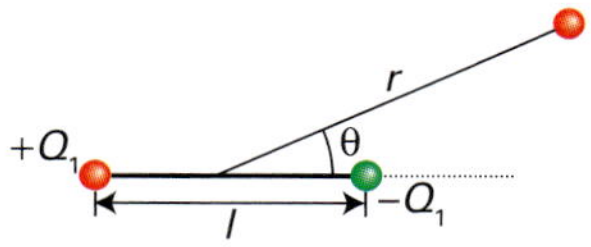

9

where α_2 is the polarizability of molecule 2. The negative sign shows that the interaction is attractive. For a molecule with $\mu = 1$ D (such as HCl) near a molecule of polarizability volume $\alpha' = 1.0 \times 10^{-31}$ m^3 (such as benzene, Table 15.2) the average interaction energy is about −0.8 kJ mol^{-1} when the separation is 0.3 nm. The interaction between a dipole and an induced dipole is another example of a van der Waals interaction that varies as the inverse sixth power of the distance.

15.5 Dispersion interactions

Finally, we consider the interactions between species that have neither a net charge nor a permanent electric dipole moment (such as two Xe atoms in a gas or two nonpolar groups on the peptide residues of a protein). Despite their absence of partial charges, we know that uncharged, nonpolar species can interact because they form condensed phases, such as benzene, liquid hydrogen, and liquid xenon.

The **dispersion interaction**, or **London force**, between nonpolar species arises from the transient dipoles that they possess as a result of fluctuations in the electron density distribution (Fig. 15.5). Suppose, for instance, that the electrons in one molecule flicker into an arrangement that results in partial positive and negative charges and thus gives it an instantaneous dipole moment μ_1. While it exists, this dipole can polarize the other molecule and induce in it an instantaneous dipole moment μ_2. The two dipoles attract each other and the potential energy of the pair is lowered. Although the first molecule will go on to change the size and direction of its dipole (perhaps within 10^{-16} s), the second will follow it, that is, the two dipoles are *correlated* in direction like two meshing gears, with a positive partial charge on one molecule appearing close to a negative partial charge on the other molecule and vice versa. Because of this correlation of the relative positions of the partial charges, and their resulting attractive interaction, the attraction between the two instantaneous dipoles does not average to zero. Instead, it gives rise to a net attractive interaction. Polar molecules interact by a dispersion interaction as well as by dipole–dipole interactions, with the dispersion interaction often dominant.

(a)

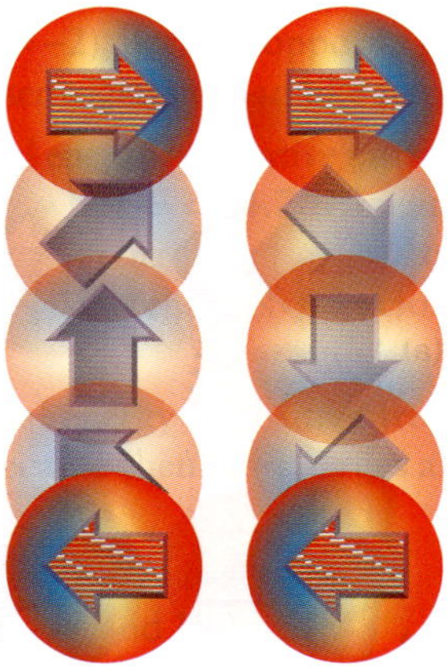

(b)

Fig. 15.5 In the dispersion interaction, an instantaneous dipole on one molecule induces a dipole on another molecule, and the two dipoles then interact to lower the energy. The directions of the two instantaneous dipoles are correlated, and, although they occur in different orientations at different instants, the interaction does not average to zero.

The strength of the dispersion interaction depends on the polarizability of the first molecule because the magnitude of the instantaneous dipole moment μ_1 depends on the looseness of the control that the nuclear charge has over the outer electrons. If that control is loose, the electron distribution can undergo relatively large fluctuations. Moreover, if the control is loose, then the electron distribution can also respond strongly to applied electric fields and hence have a high polarizability. It follows that a high polarizability is a sign of large fluctuations in local charge density. The strength also depends on the polarizability of the second molecule, as that polarizability determines how readily a dipole can be induced in molecule 2 by molecule 1. We therefore expect $V \propto \alpha_1\alpha_2$. The actual calculation of the dispersion interaction is quite involved, but a reasonable approximation to the interaction energy is the **London formula**:

$$V = -\tfrac{2}{3} \times \frac{\alpha_1'\alpha_2'}{r^6} \times \frac{I_1 I_2}{I_1 + I_2} \qquad (15.11)$$

where I_1 and I_2 are the ionization energies of the two molecules. For two CH_4 molecules, $V \approx -5$ kJ mol^{-1} when $r = 0.3$ nm. As usual, we should interpret mathematical expressions and here we see that:

- The potential energy of interaction increases with decreasing ionization energies.

This conclusion may be puzzling at first sight, for the product of the ionization energies appears in the numerator of the right-hand side of eqn 15.11. However, the polarizability is inversely proportional to the ionization energy (Section 15.4), so it follows that $\alpha_1'\alpha_2' \propto (I_1 I_2)^{-1}$ and, for a constant separation r, $V \propto (I_1 + I_2)^{-1}$: the potential energy is inversely proportional to the sum of the ionization energies.

- The potential energy of interaction is proportional to the inverse sixth power of the separation.

We have seen this result for the other interactions considered thus far in this chapter. It is consistent with our previous statement that the potential energy of an attractive van der Waals interaction is commonly proportional to r^{-6}.

The total interaction

So far we have discussed attractive interactions that vary as the inverse sixth power of the separation. However, there are several other types of interaction, both attractive and repulsive, some of which dominate the interactions we have explored when they are present.

15.6 Hydrogen bonding

The strongest intermolecular interaction arises from the formation of a **hydrogen bond**, in which a hydrogen atom lies between two strongly electronegative atoms and binds them together. The bond is normally denoted X—H···Y, with X and Y being nitrogen, oxygen, or fluorine. Unlike the other interactions we have considered, hydrogen bonding is not universal but is restricted to molecules that contain these atoms. A common hydrogen bond is formed between O—H groups and O atoms, as in liquid water and ice. The distance dependence of the hydrogen bond is quite different from the other interactions we have considered, and is best regarded as a 'contact' interaction that turns on when the X—H group is in direct contact with the Y atom.

The most elementary description of the formation of a hydrogen bond is that it is the result of a Coulombic interaction between the partly exposed positive charge of a proton bound to an electron-withdrawing X atom (in the fragment X—H) and the negative charge of a lone pair on the second atom Y, as in $^{\delta-}$X—H$^{\delta+}$···:Y$^{\delta-}$. In Exercise 15.22, you are invited to use the electrostatic model to calculate the dependence of the molar potential energy of interaction on the OOH angle, denoted θ in (**12**), and the results are plotted in Fig. 15.6. We see that, at $\theta = 0$,

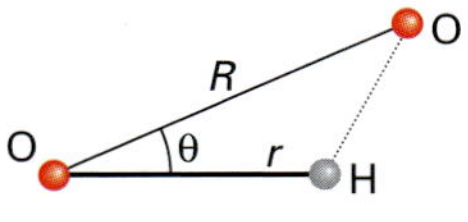

12

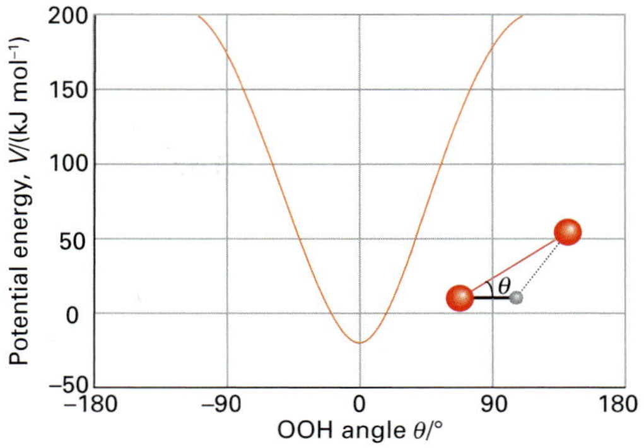

Fig. 15.6 The variation of the energy of interaction (on the electrostatic model) of a hydrogen bond as the angle between the O—H and :O groups is changed.

when the OHO atoms lie in a straight line, the potential energy is -19 kJ mol^{-1}. Note how sharply the energy depends on angle: it is negative only within ±12° of linearity.

Molecular orbital theory provides an alternative description that is more in line with the concept of delocalized bonding and the ability of an electron pair to bind more than one pair of atoms (Section 14.18). Thus, if the X—H bond is regarded as formed from the overlap of an orbital on X, ψ_X, and a hydrogen 1s orbital, ψ_H, and the lone pair on Y occupies an orbital on Y, ψ_Y, then when the two molecules are close together, we can build three molecular orbitals from the three basis orbitals:

$$\psi = c_1\psi_X + c_2\psi_H + c_3\psi_Y$$

One of the molecular orbitals is bonding, one almost nonbonding, and the third antibonding (Fig. 15.7). These three orbitals need to accommodate four

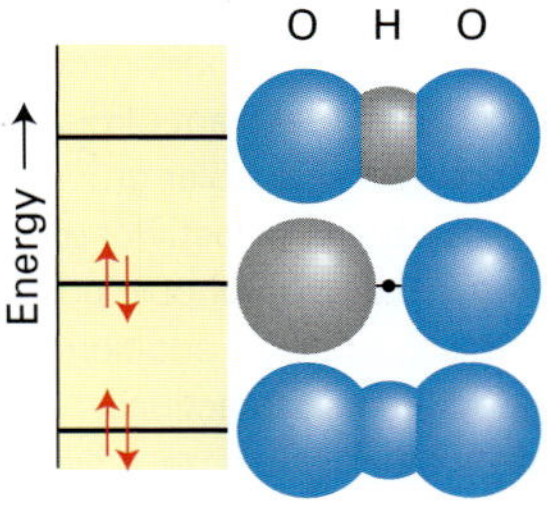

Fig. 15.7 A schematic portrayal of the molecular orbitals that can be formed from an X, H, and Y orbital and that gives rise to an X—H···Y hydrogen bond. The lowest-energy combination is fully bonding, the next nonbonding, and the uppermost is antibonding. The antibonding orbital is not occupied by the electrons provided by the X—H bond and the :Y lone pair, so the configuration shown may result in a net lowering of energy in certain cases (namely when the X and Y atoms are N, O, or F).

electrons (two from the original X—H bond and two from the lone pair of Y), so two enter the bonding orbital and two enter the nonbonding orbital. Because the antibonding orbital remains empty, the net effect —depending on the precise location of the almost nonbonding orbital—may be a lowering of energy.

Experimental evidence and theoretical arguments have been presented in favour of both the electrostatic and molecular orbital models. Recent experiments suggest that the hydrogen bonds in ice have significant covalent character and are best described by a molecular orbital treatment. However, this interpretation of experimental results has been challenged by theoretical studies, which favour the electrostatic model. The matter has not yet been resolved.

Hydrogen-bond formation, which has a typical strength of the order of 20 kJ mol^{-1}, dominates the van der Waals interactions when it can occur. It accounts for the rigidity of molecular solids such as sucrose and ice, the low vapour pressure, high viscosity, and surface tension of liquids such as water, the secondary structure of proteins (the formation of helices and sheets of polypeptide chains), the structure of DNA and hence the transmission of genetic information, and the attachment of drugs to receptors sites in proteins (Box 15.1). Hydrogen bonding

Box 15.1 Molecular recognition

Molecular interactions are responsible for the assembly of many biological structures. Hydrogen bonding and hydrophobic interactions are primarily responsible for the three-dimensional structures of biopolymers, such as proteins, nucleic acids, and cell membranes. The binding of a ligand, or *guest*, to a biopolymer, or *host*, is also governed by molecular interactions. Examples of biological *host–guest complexes* include enzyme–substrate complexes, antigen–antibody complexes, and drug–receptor complexes. In all these cases, a site on the guest contains functional groups that can interact with complementary functional groups of the host. For example, a hydrogen-bond donor group of the guest must be positioned near a hydrogen bond acceptor group of the host for tight binding to occur. It is generally true that many specific intermolecular contacts must be made in a biological host–guest complex and, as a result, a guest binds only chemically similar hosts. The strict rules governing molecular recognition of a guest by a host control every biological process, from metabolism to immunological response, and provide important clues for the design of effective drugs for the treatment of disease.

Interactions between nonpolar groups can be important in the binding of a guest to a host. For example, many active sites of enzymes have hydrophobic pockets that bind nonpolar groups of a substrate. In addition to dispersion, repulsive, and hydrophobic interactions, so-called *π-stacking interactions* are also possible, in which the planar π systems of aromatic macrocycles lie one on top of the other, in a nearly parallel orientation. Such interactions are responsible for the stacking of hydrogen-bonded base pairs in DNA, as shown in the illustration. Some drugs with planar π systems, shown as a rectangle in the illustration, are effective because they intercalate between base pairs through π-stacking interactions, causing the helix to unwind slightly and altering the function of DNA.

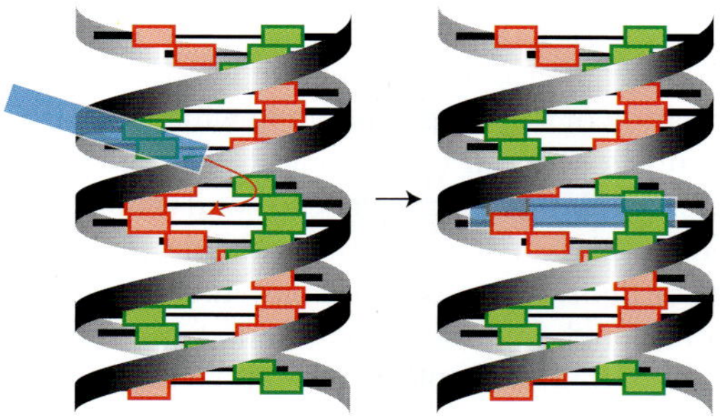

Some drugs with planar π systems, shown by a coloured rectangle, intercalate between the base pairs of DNA.

Coulombic interactions can be important in the interior of a biopolymer host, where the relative permittivity can be much lower than that of the aqueous exterior. For example, at physiological pH, amino acid side chains containing carboxylic acid or amine groups are negatively and positively charged, respectively, and can attract each other. Dipole–dipole interactions are also possible because many of the building blocks of biopolymers are polar, including the peptide link, —CONH— (Example 15.1). However, hydrogen-bonding interactions are by far the most prevalent in biological host–guest complexes. Many effective drugs on the market bind tightly and inhibit the action of enzymes that are associated with the progress of a disease. In many cases, a successful inhibitor will be able to form the same hydrogen bonds with the binding site that the normal substrate of the enzyme can form, except that the drug is chemically inert toward the enzyme. This strategy has been used in the design of drugs for the treatment of acquired immunodeficiency syndrome (AIDS), caused by the human immunodeficiency virus (HIV) (see Exercise 15.26).

Table 15.3
Potential energy of molecular interactions

Interaction type	Distance dependence of potential energy	Typical energy ($kJ\ mol^{-1}$)	Comment
Ion–ion	$1/r$	250	Only between ions
Ion–dipole	$1/r^2$	15	
Dipole–dipole	$1/r^3$	2	Between stationary polar molecules
	$1/r^6$	0.3	Between rotating polar molecules
London (dispersion)	$1/r^6$	2	Between all types of molecules and ions
Hydrogen bonding		20	The interaction is for X—H···Y and occurs on contact for X, Y = N, O, or F

also contributes to the solubility in water of species such as ammonia and compounds containing hydroxyl groups and to the hydration of anions. In this last case, even ions such as Cl^- and HS^- can participate in hydrogen bond formation with water, as their charge enables them to interact with the hydroxylic protons of H_2O.

Table 15.3 summarizes the strengths and distance dependence of the attractive interactions that we have considered so far.

15.7 The hydrophobic effect

There is one further type of interaction that we need to consider: it is an *apparent* force that influences the shape of a macromolecule and that is mediated by the solvent, water. First, we need to understand why hydrocarbon molecules do not dissolve appreciably in water. Experiments indicate that the transfer of a hydrocarbon molecule from a nonpolar solvent into water is often exothermic ($\Delta H < 0$). Therefore, the fact that dissolving is not spontaneous must mean that entropy change is negative ($\Delta S < 0$). For example, the process $CH_4(\text{in } CCl_4) \rightarrow CH_4(aq)$ has $\Delta G = +12\ kJ\ mol^{-1}$, $\Delta H = -10\ kJ\ mol^{-1}$, and $\Delta S = -75\ J\ K^{-1}\ mol^{-1}$ at 298 K. Substances characterized by a positive Gibbs energy of transfer from a nonpolar to a polar solvent are classified as **hydrophobic**.

The origin of the decrease in entropy that prevents hydrocarbons from dissolving in water is the formation of a solvent cage around the hydrophobic molecule (Fig. 15.8). The formation of this cage decreases the entropy of the system because the water molecules must adopt a less disordered arrangement than in the bulk liquid. However, when many solute molecules cluster together, fewer (though larger) cages are required and more solvent molecules are free to move. The net effect of formation of large clusters of hydrophobic molecules is then a decrease in the organization of the solvent and therefore a net *increase* in entropy of the system. This increase in entropy of the solvent is large enough to render spontaneous the association of hydrophobic molecules in a polar solvent.

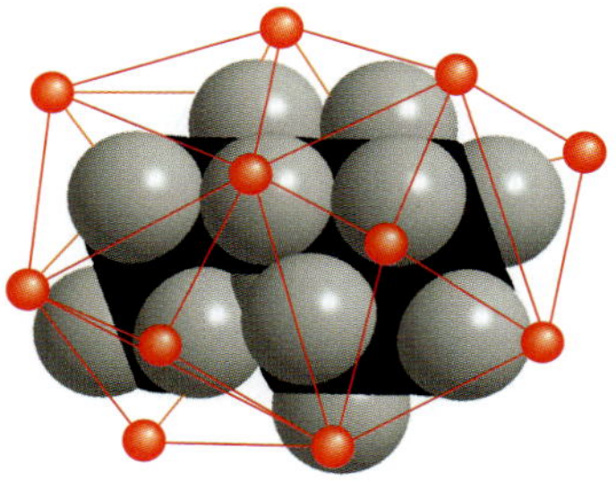

Fig. 15.8 When a hydrocarbon molecule is surrounded by water, the water molecules form a cage called a *clathrate*. As a result of this acquisition of structure, the entropy of the water decreases, so the dispersal of the hydrocarbon into water is entropy-opposed; the coalescence of the hydrocarbon into a single large blob is entropy favoured.

The increase in entropy that results from the decrease in structural demands on the solvent is the origin of the **hydrophobic effect**, which tends to encourage the clustering together of hydrophobic groups in micelles and biopolymers. Thus, the presence of hydrophobic groups in polypeptides results in an increase in structure of the surrounding water and a decrease in entropy. The entropy can increase if the hydrophobic groups are twisted into the interior of the molecule, which liberates the water molecules and results in an increase in their disorder. The hydrophobic interaction is an example of an ordering process, a kind of virtual force, that is mediated by a tendency toward greater disorder of the solvent.

15.8 Modelling the total interaction

The total attractive interaction energy between rotating molecules that cannot participate in hydrogen bonding is the sum of the contributions from the dipole–dipole, dipole–induced-dipole, and dispersion interactions. Only the dispersion interaction contributes if both molecules are nonpolar. All three interactions vary as the inverse sixth power of the separation, so we may write the total attractive contribution to the van der Waals interaction energy as

$$V = -\frac{C}{r^6} \qquad (15.12)$$

where C is a coefficient that depends on the identity of the molecules and the type of interaction between them.

Repulsive terms become important and begin to dominate the attractive forces when molecules are squeezed together (Fig. 15.9), for instance, during the impact of a collision, under the force exerted by a weight pressing on a substance, or simply as a result of the attractive forces drawing the molecules together. These repulsive interactions arise in large measure from the Pauli exclusion principle, which forbids pairs of electrons being in the same region of space. The repulsions increase steeply with decreasing separation in a way that can be deduced only by very extensive, complicated molecular structure calculations. In many cases, however, progress can be made by using a greatly simplified representation of the potential energy, where the details are ignored and the general features expressed by a few adjustable parameters.

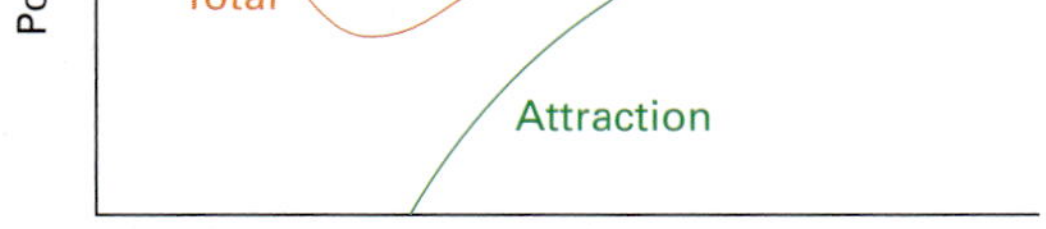

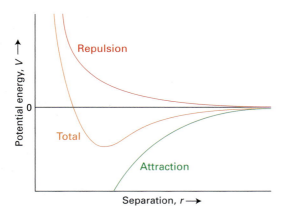

Fig. 15.9 The general form of an intermolecular potential energy curve (the graph of the potential energy of two closed-shell species as the distance between them is changed). The attractive (negative) contribution has a long range, but the repulsive (positive) interaction increases more sharply once the molecules come into contact. The overall potential energy is shown by the yellow line.

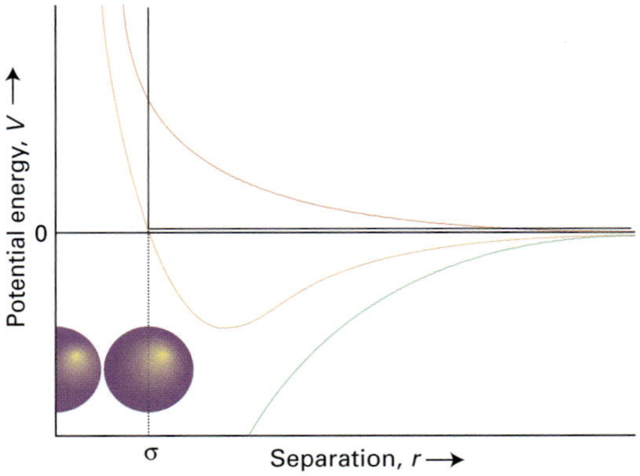

Fig. 15.10 The true intermolecular potential can be modelled in a variety of ways. One of the simplest is this *hard-sphere potential*, in which there is no potential energy of interaction until the two molecules are separated by a distance σ when the potential energy rises abruptly to infinity as the impenetrable hard spheres repel each other.

One such approximation is the **hard-sphere potential**, in which it is assumed that the potential energy rises abruptly to infinity as soon as the particles come within a separation σ (Fig. 15.10):

$$V = \begin{cases} \infty & \text{for } r \le \sigma \\ 0 & \text{for } r > \sigma \end{cases} \qquad (15.13)$$

This very simple potential is surprisingly useful for assessing a number of properties.

Another widely used approximation is to express the short-range repulsive potential energy as inversely proportional to a high power of r:

$$V = +\frac{C^*}{r^n} \qquad (15.14)$$

where C^* is another constant (the star signifies repulsion). Typically, n is set equal to 12, in which case the repulsion dominates the $1/r^6$ attractions strongly at short separations because then $C^*/r^{12} \gg C/r^6$. The sum of the repulsive interaction with $n = 12$ and the attractive interaction given by eqn 15.14 is called the **Lennard-Jones (12,6) potential.** It is normally written in the form

Repulsion | Attraction

$$V = 4\varepsilon\left\{\left(\frac{\sigma}{r}\right)^{12} - \left(\frac{\sigma}{r}\right)^{6}\right\} \qquad (15.15)$$

and is drawn in Fig. 15.11. The two parameters are now ε (epsilon), the depth of the well, and σ, the separation at which $V = 0$. Some typical values are listed in Table 15.4. The well minimum occurs at $r = 2^{1/6}\sigma$. Although the (12,6)-potential has been used in many

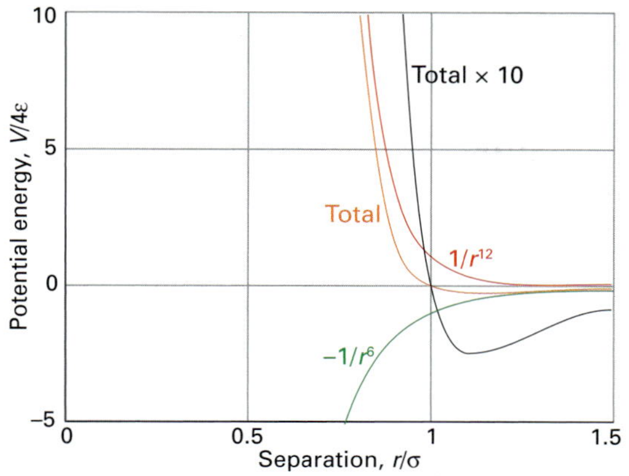

Fig. 15.11 The Lennard-Jones potential is another approximation to the true intermolecular potential energy curves. It models the attractive component by a contribution that is proportional to $1/r^6$, and the repulsive component by a contribution that is proportional to $1/r^{12}$. Specifically, these choices result in the Lennard-Jones (12,6)-potential. Although there are good theoretical reasons for the former, there is plenty of evidence to show that $1/r^{12}$ is only a very poor approximation to the repulsive part of the curve.

Table 15.4

Lennard-Jones parameters for the (12,6) potential

	$\varepsilon/(\text{kJ mol}^{-1})$	σ/pm
Ar	128	342
Br_2	536	427
C_6H_6	454	527
Cl_2	368	412
H_2	34	297
He	11	258
Xe	236	406

calculations, there is plenty of evidence to show that $1/r^{12}$ is a very poor representation of the repulsive potential, and that the exponential form $e^{-r/\sigma}$ is superior. An exponential function is more faithful to the exponential decay of atomic wavefunctions at large distances, and hence to the distance dependence of the overlap that is responsible for repulsion. However, a disadvantage of the exponential form is that it is slower to compute, which is important when considering the interactions between the large numbers of atoms in liquids and macromolecules.

15.9 Molecules in motion

The intermolecular interactions we have described govern the shapes that complicated molecules adopt and the motion of molecules in liquids. We deal with the structural aspects of these interaction in the next chapter. In this section, we consider how to take the interactions into account to describe molecular motion.

In a **molecular dynamics** simulation, the molecule is set in motion by heating it to a specified temperature and the possible trajectories of all atoms under the influence of the intermolecular forces are calculated from Newton's laws of motion. For instance, if the interaction is described by a Lennard-Jones potential, eqn 15.15, the force along the line of centres of two neighbouring molecules is

$$F = -24\varepsilon\left\{\frac{2\sigma^{12}}{r^{13}} - \frac{\sigma^6}{r^7}\right\} \qquad (15.16)$$

The equations of motion are solved numerically, allowing the molecules to adjust their locations and velocities in femtosecond steps (1 fs = 10^{-15} s). The calculation is repeated for tens of thousands of such steps.

The same technique can be used to examine the internal motion of macromolecules, such as the proteins we consider in Chapter 16, and software packages are available that calculate the trajectories of a large number of atoms in three dimensions. The trajectories correspond to the conformations that the molecule can sample at the temperature of the simulation. At very low temperatures, the neighbouring components of the molecule are trapped in wells like that in Fig. 15.11, the atomic motion is restricted, and only a few conformations are possible. At high temperatures, more potential energy barriers can be overcome and more conformations are possible.

In the **Monte Carlo method**, the atoms of a macromolecule or the molecules of a liquid are moved through small but otherwise random distances, and the change in potential energy is calculated. If the potential energy is not greater than before the change, then the new arrangement is accepted. However, if the potential energy is greater than before the change, it is necessary to use a criterion for rejecting or accepting it. To establish this criterion, we use the Boltzmann distribution (Section 22.1), which states that at equilibrium at a temperature T the ratio of populations of two states that differ in energy by ΔE is $e^{-\Delta E/kT}$, where k is Boltzmann's constant. In the Monte Carlo method, the exponential factor is calculated for the new atomic arrangement and compared with a random number between 0 and 1; if the factor is larger than the random number, the new arrangement is accepted; if the factor is not larger, then the new arrangement is rejected and another one is generated instead.

Checklist of key ideas

You should now be familiar with the following concepts.

☐ **1** van de Waals interactions between or within molecules include the attractive and repulsive interactions involving partial electric charges and electron clouds of polar and nonpolar molecules or functional groups, and the repulsive interactions that prevent the complete collapse of matter to densities as high as those characteristic of atomic nuclei.

☐ **2** A polar molecule is a molecule with a permanent electric dipole moment; the magnitude of a dipole moment is the product of the partial charge and the separation.

☐ **3** Dipole moments are approximately additive (as vectors, eqn 15.3).

☐ **4** The polarizability is a measure of the ability of an electric field to induce a dipole moment in a molecule.

☐ **5** A hydrogen bond is an interaction of the form X—H···Y, where X and Y are N, O, or F.

☐ **6** The hydrophobic interaction is an ordering process mediated by a tendency toward greater disorder of the solvent: it causes hydrophobic groups to cluster together.

☐ **7** The Lennard–Jones (6,12) potential is a model of the total intermolecular potential energy.

☐ **8** A molecular dynamics calculation uses Newton's laws of motion to calculate the motion of molecules in a fluid (and the motion of atoms in macromolecules).

☐ **9** A Monte Carlo simulation uses a selection criterion for accepting or rejecting a new arrangement of atoms or molecules.

Table of key equations

The following table summarizes the equations developed in this chapter.

Property	Equation	Comment
Potential energy of interaction between two charges	$V = Q_1Q_2/4\pi\varepsilon r$	Medium of permittivity $\varepsilon = \varepsilon_r\varepsilon_0$
The molecular dipole moment in terms of the: (1) electronegativity difference (2) polarizability	$\mu/\mathrm{D} \approx \Delta\chi$ $\mu = \alpha\mathcal{E}$	Very approximate
Potential energy of a dipole near a charge	$V \propto -1/r^2$	Arrangement as in **9**
Potential energy of interaction between two dipoles	$V \propto -1/r^3$	Arrangement as in **10**
Potential energy of interaction between two polar molecules	$V \propto -1/Tr^6$	Rotating molecules
Potential energy of interaction between a dipole and an induced dipole	$V \propto -\mu_1^2\alpha_2/r^6$	
London formula for the dispersion interaction	$V \propto (\alpha_1'\alpha_2'/r^6) \times \{I_1I_2/(I_1 + I_2)\}$	
Lennard-Jones (6,12) potential	$V = 4\varepsilon\{(\sigma/r)^{12} - (\sigma/r)^6\}$	Definition

Questions and exercises

Discussion questions

15.1 Explain how the permanent dipole moment and the polarizability of a molecule arise and explain how they depend on the structure of the molecule.

15.2 Account for the theoretical conclusion that many attractive interactions between molecules vary with their separation as $1/r^6$.

15.3 Describe how van der Waals interactions depend on the structure of the molecules.

15.4 Describe the formation of a hydrogen bond in terms of (a) electrostatic interactions and (b) molecular orbitals. How would you identify the better model?

15.5 Account for the hydrophobic effect and discuss its manifestations.

15.6 Outline the procedures used to calculate the motion of molecules in molecules in fluids and atoms in molecules.

Exercises

15.1 Estimate the dipole moment of an HF molecule from the electronegativities of the elements and express the answer in debye and coulomb-metres.

15.2 Use the VSEPR model to judge whether PCl_5 is polar.

15.3 The electric dipole moment of toluene (methylbenzene) is 0.40 D. Estimate the dipole moments of the three xylenes (dimethylbenzenes). Which value can you be sure about?

15.4 Calculate the resultant of two dipoles of magnitude 1.20 D and 0.60 D that make an angle 107° to each other.

15.5 From the information in Exercise 15.3, estimate the dipole moments of (a) 1,2,3-trimethylbenzene, (b) 1,2,4-trimethylbenzene, and (c) 1,3,5-trimethylbenzene. Which value can you be sure about?

15.6 At low temperatures a substituted 1,2-dichloroethane molecule can adopt the three conformations (**13**), (**14**), and (**15**) with different probabilities. Suppose that the dipole moment of each bond is 1.50 D. Calculate the mean dipole moment of the molecule when (a) all three conformations are equally likely, (b) only conformation (**14**) occurs, (c) the three conformations occur with probabilities in the ratio 2:1:1 and (d) 1:2:2.

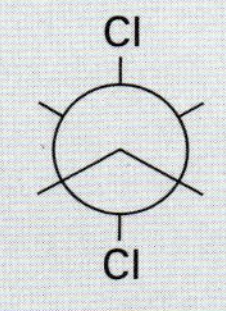

13

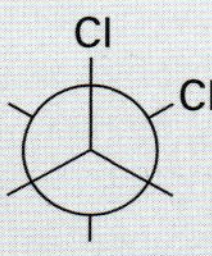

14

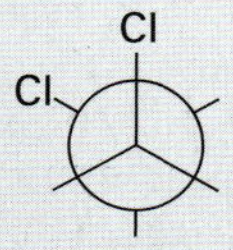

15

15.7 Calculate the magnitude and direction of the dipole moment of the following arrangement of charges in the *xy*-plane: $3e$ at (0,0), $-e$ at (0.32 nm, 0), and $-2e$ at an angle of 20° from the *x*-axis and a distance of 0.23 nm from the origin.

15.8 Calculate the electric dipole moment of a glycine molecule using the partial charges in Table 15.1 and the locations of the atoms shown in (**16**).

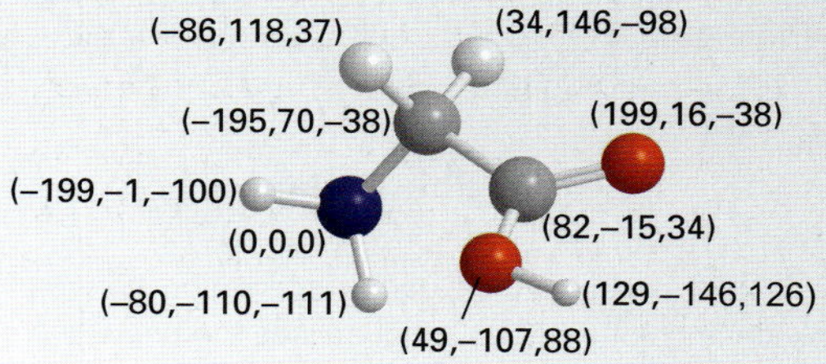

16 Glycine, NH_2CH_2COOH

15.9 (a) Plot the magnitude of the electric dipole moment of hydrogen peroxide as the H—OO—H (azimuthal) angle ϕ changes. Use the dimensions shown in (**17**). (b) Devise a way for depicting how the angle as well as the magnitude changes.

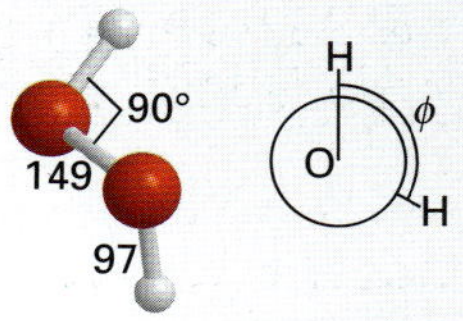

17 Hydrogen peroxide, H_2O_2

15.10 Calculate the molar energy required to reverse the direction of a water molecule located (a) 150 pm, (b) 350 pm from a Li^+ ion. Take the dipole moment of water as 1.85 D.

15.11 Show, by following the procedure in Derivation 15.1, that eqn 15.6 describes the potential energy of two electric dipole moments in the orientation shown in structure (**10**) of the text.

15.12 What is the contribution to the total molar energy of (a) the kinetic energy, (b) the potential energy of interaction between hydrogen chloride molecules in a gas at 298 K when 0.50 mol of molecules is confined to 1.0 dm^3? Is the kinetic theory of gases justifiable in this case?

15.13 (a) What are the units of the polarizability α? (b) Show that the units of polarizability volume are cubic metres (m^3).

15.14 The magnitude of the electric field at a distance r from a point charge Q is equal to $Q/4\pi\varepsilon_0 r^2$. How close to a water molecule (of polarizability volume 1.48×10^{-30} m^3) must a proton approach before the dipole moment it induces is equal to the permanent dipole moment of the molecule (1.85 D)?

15.15 Estimate the energy of the dispersion interaction (use the London formula) for two Ar atoms separated by 1.0 nm.

15.16 Phenylanine (Phe, **18**) is a naturally occurring amino acid with a benzene ring. What is the energy of interaction between its benzene ring and the electric dipole moment of a neighbouring peptide group? Take the distance between the groups as 4.0 nm and treat the benzene ring as benzene itself. Take the dipole moment of the peptide group as 2.7 D.

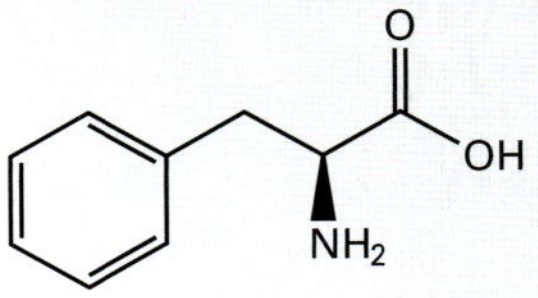

18 Phenylalanine

15.17 Now consider the London interaction between the benzene rings of two Phe residues (see the preceding exercise). Estimate the potential energy of attraction between two such rings (treated as benzene molecules) separated by 4.0 nm. For the ionization energy, use $I = 5.0$ eV.

15.18 In a region of the oxygen-storage protein myoglobin, the OH group of a tyrosine residue is linked by a hydrogen bond to the N atom of a histidine residue in the geometry shown in (**19**). Use the partial charges in Table 15.1 to estimate the potential energy of this interaction.

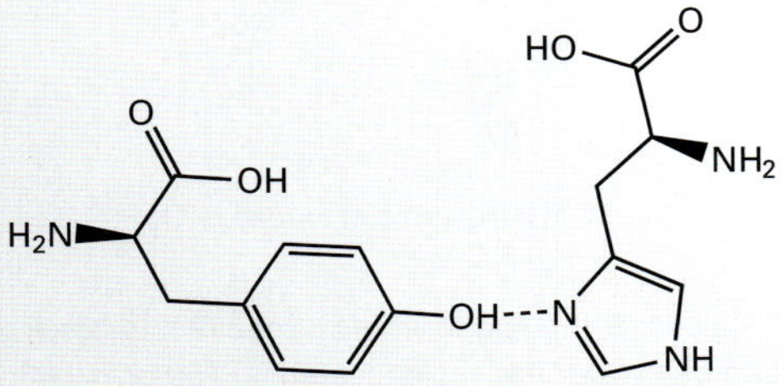

19

15.19 Acetic acid vapour contains a proportion of planar, hydrogen-bonded dimers (**20**). The apparent dipole moment of molecules in pure gaseous acetic acid increases with increasing temperature. Suggest an interpretation of the latter observation.

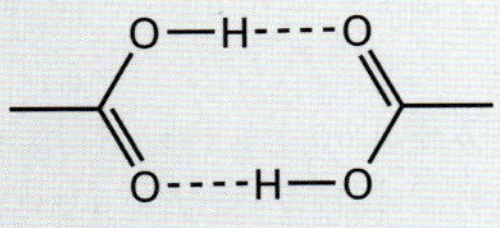

20

15.20 The coordinates of the atoms of an acetic acid dimer are set out in more detail in (**21**). Consider only the Coulombic interactions between the partial charges indicated by the dashed lines and their symmetry-related equivalents. At what distance R does the attraction become attractive?

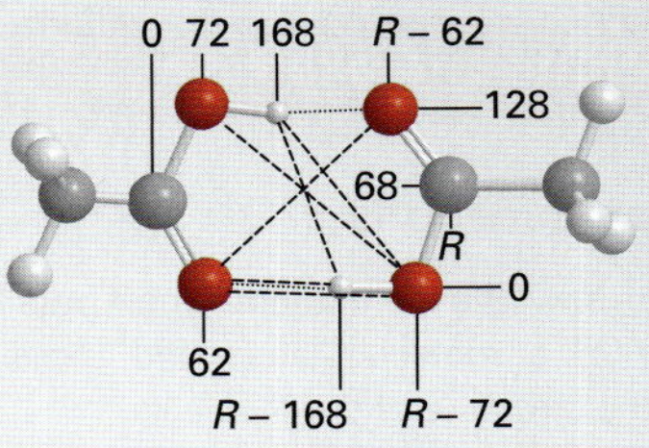

21 $(CH_3COOH)_2$

15.21 The potential energy of a CH_3 group in ethane as it is rotated around the C—C bond can be written $V = \frac{1}{2}V_0(1 + \cos 3\phi)$, where ϕ is the azimuthal angle (**22**) and V_0 = 11.6 kJ mol^{-1}. (a) What is the change in potential energy between the trans and fully eclipsed conformations? (b) Show that for small variations in angle, the torsional (twisting) motion around the C—C bond can be expected to be that of a harmonic oscillator. (d) Estimate the vibrational frequency of this torsional oscillation.

22

15.22 Consider the arrangement shown in (**12**) for a system consisting of an O—H group and an O atom, and then use the electrostatic model of the hydrogen bond to calculate the dependence of the molar potential energy of interaction on the angle θ. Set the partial charges on H and O to 0.45e and −0.83e, respectively, and take R = 200 pm and r = 95.7 pm.

Projects

The symbol ‡ indicates that calculus is required.

15.23‡ Now we explore London interactions in more detail. (a) Given that force is the negative slope of the potential, calculate the distance dependence of the force acting between two non-bonded groups of atoms in a polymeric chain that have a London dispersion interaction with each other. What is the separation at which the force is zero? *Hint.* Calculate the slope by considering the potential energy at R and $R + \delta R$, with $\delta R \ll R$, and evaluating $\{V(R + \delta R) - V(R)\}/\delta R$. You should use the expansion in Derivation 15.1 together with

$$(1 \pm x + \cdots)^6 = 1 \pm 6x + \cdots \qquad (1 \pm x + \cdots)^{12} = 1 \pm 12x + \cdots$$

At the end of the calculation, let δR become vanishingly small. (b) Repeat part (a) now by noting that $F = -dV/dr$ and differentiating the expression for V.

15.24‡ Here we explore alternatives to the Lennard-Jones potential. (a) Suppose you distrusted the Lennard-Jones (12,6) potential for assessing a particular polymer conformation, and replaced the repulsive term by an exponential function of the form $e^{-r/\sigma}$. Sketch the form of the potential energy and locate the distance at which it is a minimum. (b) Use calculus to identify the distance at which the exponential-6 potential described in part (a) is a minimum.

15.25 Molecular orbital calculations may be used to predict structures of intermolecular complexes. Hydrogen bonds between purine and pyrimidine bases are responsible for the

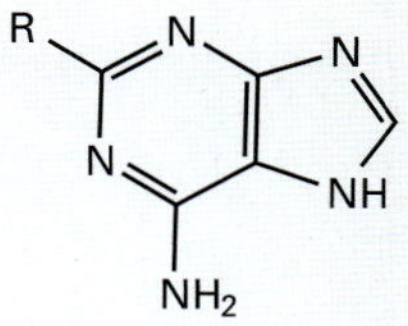

23

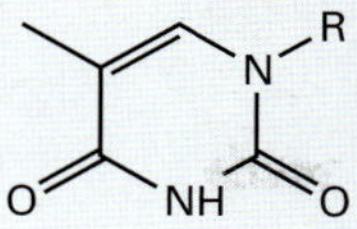

24

double-helix structure of DNA. Consider methyladenine (**23**, with R = CH_3) and methylthymine (**24**, with R = CH_3) as models of two bases that can form hydrogen bonds in DNA. (a) Using molecular modelling software, calculate the partial charges of all atoms in methyladenine and methylthymine. (b) Based on your tabulation of partial charges, identify the atoms in methyladenine and methylthymine that are likely to participate in hydrogen bonds. (c) Draw all possible adenine–thymine pairs that can be linked by hydrogen bonds, keeping in mind that linear arrangements of the A—H···B fragments are preferred. For this step, you may want to use your molecular modelling software to align the molecules properly. (d) Consult a biochemistry textbook and determine which of the pairs that you drew in part (c) occur naturally in DNA molecules.

15.26 For mature HIV particles to form in cells of the host organism, several large proteins coded for by the viral genetic material must be cleaved by a protease enzyme. The drug Crixivan (**25**) is a competitive inhibitor of HIV protease and has several molecular features that optimize binding to the active site of the enzyme. Consult the literature and prepare a brief report summarizing molecular interactions between Crixivan and HIV protease that are thought to be responsible for the drug's efficacy.

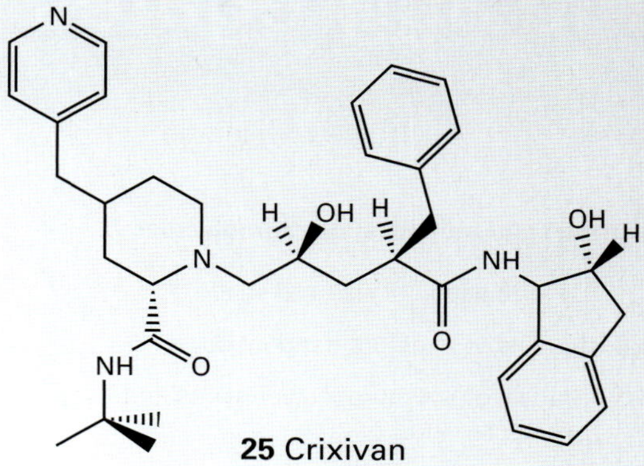

25 Crixivan

Chapter 16

Materials: macromolecules and aggregates

Naturally occurring macromolecules include polysaccharides such as cellulose, polypeptides such as protein enzymes, and polynucleotides such as deoxyribonucleic acid (DNA). Synthetic macromolecules include **polymers** such as nylon and polystyrene that are manufactured by stringing together and in some cases cross-linking smaller units known as **monomers** (Fig. 16.1).

Macromolecules give rise to special problems that include the shapes and the lengths of polymer chains, the determination of their sizes, and the large deviations from ideality of their solutions. Natural macromolecules differ in certain respects from synthetic macromolecules, particularly in their composition and the resulting structure, but the two share a number of common properties. We concentrate on these common properties here. Another level of complexity arises when small molecules group together into large particles in a process called 'self-assembly' and give rise to aggregates. One example is the assembly of haemoglobin from four myoglobin-like polypeptides. A similar type of aggregation gives rise to a variety

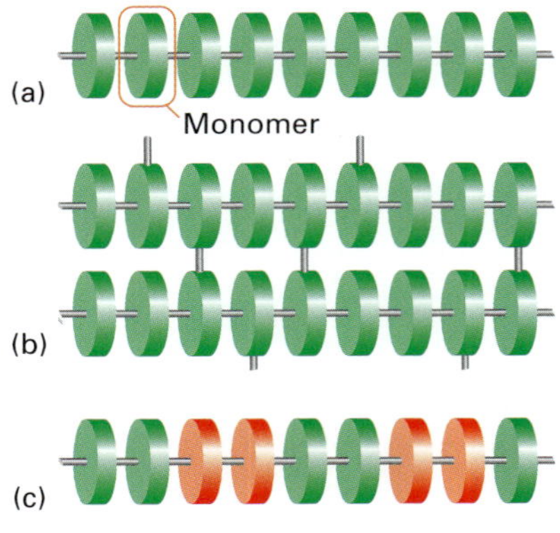

Fig. 16.1 Three varieties of polymer: (a) a simple linear polymer, (b) a cross-linked polymer, and (c) one variety of copolymer (a 'block copolymer').

of **disperse phases**, which include colloids. The properties of these disperse phases resemble to a certain extent the properties of solutions of macromolecules, and we describe their common attributes in the second part of this chapter.

Synthetic and biological macromolecules

Macromolecules provide an interesting and important illustration of how the interactions described in Chapter 15 jointly determine the shape of a molecule and its properties. The overall shape of a polypeptide, for instance, is maintained by a variety of molecular interactions, including van der Waals interactions, hydrogen bonding, and the hydrophobic effect.

16.1 Determination of size and shape

X-ray diffraction, a technique discussed in detail in Chapter 17, can reveal the position of almost every atom, other than hydrogen, even in very large molecules. However, there are several reasons why other techniques must also be used. In the first place, the sample might be a mixture of molecules with different chain lengths and extent of cross-linking, in which case sharp X-ray images are not obtained. Even if all the molecules in the sample are identical, it might prove impossible to obtain a single crystal. Furthermore, although X-ray information about proteins and DNA has shown how immensely interesting and motivating the data can be, it is incomplete. For instance, what can be said about the shape of the molecule in its natural environment, a biological cell? What can be said about the response of its shape to changes in its environment?

Many proteins (and specifically protein enzymes) are **monodisperse**, meaning that they have a single, definite molar mass. There may be small variations, such as one amino acid replacing another, depending on the source of the sample. A synthetic polymer, however, is **polydisperse**, in the sense that a sample is a mixture of molecules with various chain lengths and molar masses. The various techniques that are used to measure molar masses result in different types of mean values of polydisperse systems. The **number-average molar mass**, $\bar{M}_n$, is the value obtained by multiplying each molar mass by the numerical fraction (N_i/N) of molecules of that mass present in the sample:

$$\bar{M}_n = \frac{N_1M_1 + N_2M_2 + \cdots}{N} \tag{16.1a}$$

Here N_i (with $i = 1, 2, \ldots$) is the number of molecules with molar mass M_i and there are N molecules in all. By dividing the terms in the numerator and denominator by Avogadro's constant N_A and writing $N_i/N_A = n_i$ and $N/N_A = n$, we can express this equation in terms of the amounts (in moles) rather than the actual numbers:

$$\bar{M}_n = \frac{n_1M_1 + n_2M_2 + \cdots}{n} \tag{16.1b}$$

The **weight-average molar mass**, $\bar{M}_w$, is the average calculated by multiplying the molar masses of the molecules by the mass fraction (m_i/m) of each one present in the sample:

$$\bar{M}_w = \frac{m_1M_1 + m_2M_2 + \cdots}{m} \tag{16.1c}$$

In this expression, m_i is the total mass of molecules of molar mass M_i and m is the total mass of the sample. In general, these two averages are different and the ratio $\bar{M}_w/\bar{M}_n$ is called the **heterogeneity index** (or 'polydispersity index'). In the determination of protein molar masses we expect the various averages to be the same because unless there has been degradation the sample is monodisperse. A synthetic polymer normally spans a range of molar masses and the different averages yield different values. Typical synthetic materials have $\bar{M}_w/\bar{M}_n \approx 4$. The term 'monodisperse' is conventionally applied to synthetic polymers in which this index is less than 1.1; commercial polyethylene samples might be much more heterogeneous, with an index of close to 30. One consequence of a narrow molar mass distribution for synthetic polymers is often a higher degree of crystallinity in the solid and therefore higher density and melting point. The spread of values is controlled by the choice of catalyst and reaction conditions.

Example 16.1

Determining the heterogeneity index of a polymer sample

Determine the heterogeneity index of a sample of poly(vinyl chloride) from the following data:

Molar mass interval/ (kg mol^{-1})	Average molar mass within interval/(kg mol^{-1})	Mass of sample within interval/g
5–10	7.5	9.6
10–15	12.5	8.7
15–20	17.5	8.9
20–25	22.5	5.6
25–30	27.5	3.1
30–35	32.5	1.7

Strategy Begin by calculating the number-average and weight-average molar masses from eqns 16.1a and 16.1b, respectively. To do so, multiply the molar mass within each interval by the number and mass fractions, respectively, of the molecule in each interval. Obtain the amount (in moles) in each interval by dividing the mass of the sample in each interval by the average molar mass for that interval and then use eqn 16.1b. Finally, use the average molar masses to calculate the heterogeneity index of the sample as the ratio $\bar{M}_w/\bar{M}_n$.

Solution The amounts in each interval are as follows:

Interval	5–10	10–15	15–20
Molar mass/(kg mol^{-1})	7.5	12.5	17.5
Amount/mmol	1.30	0.70	0.51
Interval	20–25	25–30	30–35
Molar mass/(kg mol^{-1})	22.5	27.5	32.5
Amount/mmol	0.25	0.11	0.052
		Total amount/mmol:	2.92

The number-average molar mass is therefore

$$\bar{M}_n/(\text{kg mol}^{-1}) = \frac{1}{2.92}(1.3\times 7.5 + 0.70\times 12.5 + 0.51\times 17.5$$
$$+0.25\times 22.5 + 0.11\times 27.5 + 0.052\times 32.5) = 13$$

The weight-average molar mass is calculated directly from the data by first noting that adding the masses in each interval gives the total mass of the sample, 37.6 g. It follows that:

$$\bar{M}_w/(\text{kg mol}^{-1}) = \frac{1}{37.6}(9.6\times 7.5 + 8.7\times 12.5 +$$
$$8.9\times 17.5 + 5.6\times 22.5 + 3.1\times 27.5 + 1.7\times 32.5) = 16$$

The heterogeneity index is $\bar{M}_w/\bar{M}_n = 1.2$.

Self-test 16.1

The *Z-average molar mass* is defined as

$$\bar{M}_Z = \frac{N_1M_1^3 + N_2M_2^3 + \cdots}{N_1M_1^2 + N_2M_2^2 + \cdots}$$

Evaluate the *Z*-average molar mass of the sample described in Example 16.1.

[*Answer:* 19 kg mol^{-1}]

Number-average molar masses may be determined by measuring the osmotic pressure of polymer solutions (Section 6.7). The upper limit for the reliability of membrane osmometry is about 1000 kDa (1 kDa = 1 kg mol^{-1}). A major problem for macromolecules of relatively low molar mass (less than about 10 kDa), however, is their ability to percolate through the membrane. One consequence of this partial permeability is that membrane osmometry tends to overestimate the average molar mass of a polydisperse mixture. Techniques for the determination of molar mass and polydispersity that are not limited in this way include mass spectrometry, laser light scattering, ultracentrifugation, electrophoresis, and chromatography.

Mass spectrometry is among the most accurate techniques for the determination of molar masses. The procedure consists of ionizing the sample in the gas phase and then measuring the mass-to-charge number ratio (m/z) of all ions. Macromolecules present a challenge because it is difficult to produce gaseous ions of large species without fragmentation. However, **matrix-assisted laser desorption/ionization** (MALDI) has overcome this problem. In this technique, the macromolecule is embedded in a solid matrix composed of an organic material and inorganic salts, such as sodium chloride or silver trifluoroethanoate, $AgCF_3CO_2$. This sample is then irradiated with a pulsed laser. The laser energy, which is absorbed by the matrix, ejects electronically excited matrix ions, cations, and neutral macromolecules, thus creating a dense gas plume above the sample surface. The macromolecule is ionized by collisions and complexation with small cations, such as H^+, Na^+, and Ag^+, and the masses of the resulting ions are determined in a mass spectrometer.

Figure 16.2 shows the MALDI mass spectrum of a polydisperse sample of poly(butyl adipate) (**1**) obtained with NaCl in the matrix. The MALDI technique produces mostly singly charged molecular ions that are not fragmented. Therefore, the multiple peaks in the spectrum arise from polymers of different lengths (different '*N*-mers', where *N* is the number of repeating units), with the intensity of each peak being

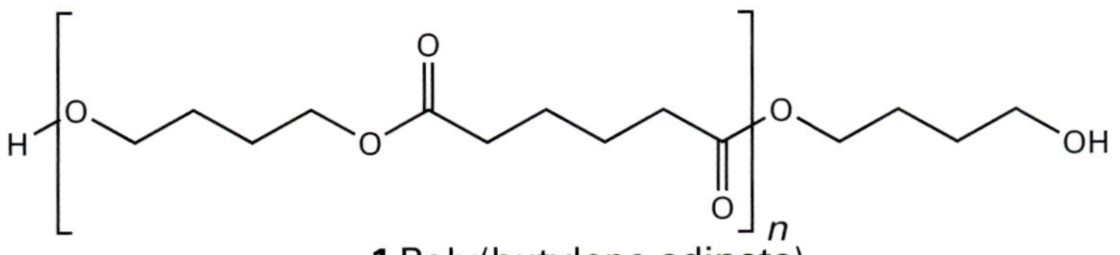

1 Poly(butylene adipate)

proportional to the abundance of each N-mer in the sample. Values of $\bar{M}_n$, $\bar{M}_w$, and the heterogeneity index can be calculated from the data. It is also possible to use the mass spectrum to verify the structure of a polymer, as shown in Example 16.2.

Example 16.2

Interpreting the mass spectrum of a polymer

The mass spectrum in Fig. 16.2 consists of peaks spaced by 200 g mol^{-1}. The peak at 4113 g mol^{-1} corresponds to a polymer with $N = 20$ repeating units. The matrix used contained NaCl. From these data, verify that the sample consists of polymers with the general structure given by (**1**).

Strategy Because each peak corresponds to a different value of N, the molar mass difference, ΔM, between peaks corresponds to the molar mass, M, of the repeating unit (the group inside the brackets in **1**). Furthermore, the molar mass of the terminal groups (the groups outside the brackets in **1**) may be obtained from the molar mass of any peak, by using

$$M(\text{terminal groups}) = M(N\text{-mer}) - N\Delta M - M(\text{cation})$$

where the last term corresponds to the molar mass of the cation that attaches to the macromolecule during ionization.

Solution The value of ΔM is consistent with the molar mass of the repeating unit shown in (**1**), which is 200 g mol^{-1}. The molar mass of the terminal group is calculated by noting that Na^+ is the cation in the matrix:

$$M(\text{terminal group}) = 4113\ \text{g mol}^{-1} - 20(200\ \text{g mol}^{-1}) - 23\ \text{g mol}^{-1} = 90\ \text{g mol}^{-1}$$

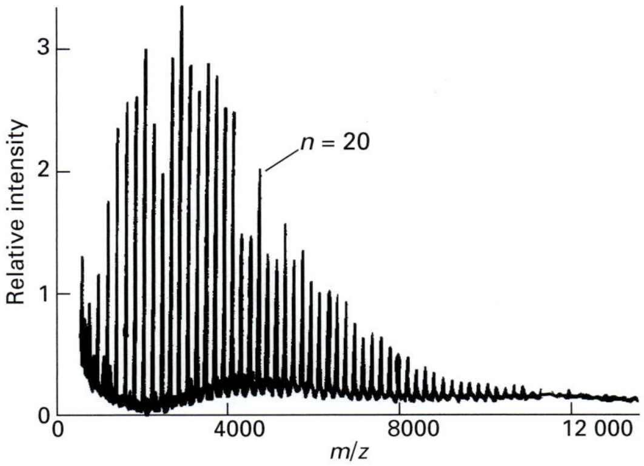

Fig. 16.2 MALDI–TOF spectrum of a sample of poly(butylene adipate) with $\bar{M}_n = 4525$ g mol^{-1}. The illustration and example have been adapted from Mudiman *et al.*, *J. Chem. Educ.*, **74**, 1288 (1997).

The result is consistent with the molar mass of the $-O(CH_2)_4OH$ terminal group (89 g mol^{-1}) plus the molar mass of the —H terminal group (1 g mol^{-1}).

Self-test 16.2

What would be the molar mass of the $N = 20$ polymer if silver trifluoroacetate were used instead of NaCl in the preparation of the matrix?

[*Answer:* 4.2 kg mol^{-1}]

In a gravitational field, heavy particles settle towards the foot of a column of solution by the process called **sedimentation**. The rate of sedimentation depends on the strength of the field and on the masses and shapes of the particles. Spherical molecules (and compact molecules in general) sediment faster than rod-like or extended molecules. For example, DNA helices sediment much faster when they are collapsed into a random coil, so sedimentation rates can be used to study denaturation (the loss of structure). Sedimentation is normally very slow, but it can be accelerated by **ultracentrifugation**, a technique that replaces the gravitational field with a centrifugal field. The effect is achieved in an ultracentrifuge, which is essentially a cylinder that can be rotated at high speed about its axis with a sample in a cell near its periphery (Fig. 16.3). Modern ultracentrifuges can produce accelerations equivalent to about 10^5 that of

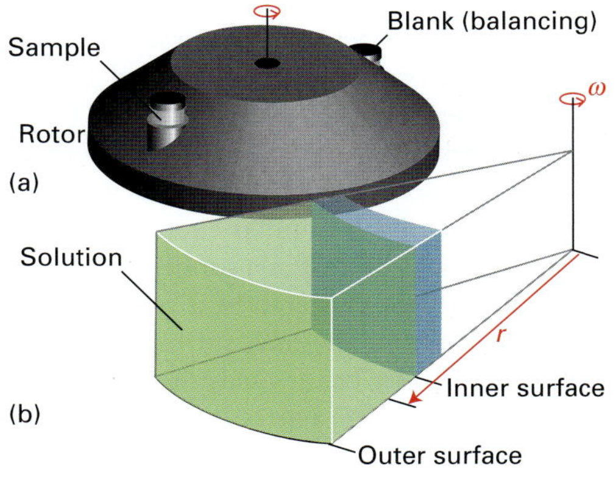

Fig. 16.3 (a) An ultracentrifuge head. The sample on one side is balanced by a blank diametrically opposite. (b) Detail of the sample cavity: the 'top' surface is the inner surface, and the centrifugal force causes sedimentation towards the outer surface; a particle at a radius r experiences a force of magnitude $mr\omega^2$.

gravity ('10^5 g'). Initially the sample is uniform, but the solute molecules move towards the outer edge of the cell at a rate that can be interpreted in terms of the number-average molar mass. In an alternative 'equilibrium' version of the technique, the weight-average molar mass can be obtained from the ratio of concentrations c of the macromolecules at two different radii in a centrifuge operating at angular frequency ω (in radians per second):

$$\bar{M}_w = \frac{2RT}{(r_2^2 - r_1^2)b\omega^2} \ln \frac{c_2}{c_1} \qquad (16.2)$$

Here, b is a factor that takes into account the buoyancy of the medium. The centrifuge is run more slowly in this technique than in the sedimentation rate method to avoid having all the solute pressed in a thin film against the bottom of the cell. At these slower speeds, several days may be needed for equilibrium to be reached.

Many macromolecules, such as DNA, are charged and move in response to an electric field. This motion is called **electrophoresis**. Electrophoretic mobility is a result of a constant drift speed reached by an ion when the electrical driving force is matched by the frictional drag force. Electrophoresis is a very valuable tool in the separation of biopolymers from complex mixtures, such as those resulting from fractionation of biological cells. In **gel electrophoresis**, migration takes place through a gel slab. In **capillary electrophoresis**, the sample is dispersed in a medium (such as methylcellulose) and held in a thin glass or plastic tube with diameters ranging from 20 to 100 μm. The small size of the apparatus makes it easy to dissipate heat when large electric fields are applied. Excellent separations may be effected in minutes rather than hours. Each polymer fraction emerging from the capillary can be characterized further by other techniques, such as MALDI.

Light scattering measurements of polymer size are based on the observation that large particles scatter light very efficiently. A familiar example is the light scattered by specks of dust in a sunbeam. Analysis of the intensity of light scattered by a sample at different angles relative to the incident radiation from a monochromatic laser beam yields the size and molar mass of a polymer, large aggregate (such as a colloid; see Section 16.6), or biological system ranging in size from a protein to a virus.

Dynamic light scattering is used to investigate the diffusion of polymers in solution. Consider two polymer molecules being irradiated by a laser beam. Suppose that at one instant the scattered waves from these particles interfere constructively at the detector, leading to a large signal. However, as the molecules move through the solution, the scattered waves may interfere destructively at a later instant and result in no signal. When this behaviour is extended to a very large number of molecules in solution, it results in fluctuations in light intensity that can be analysed to reveal the molar mass and diffusion coefficient of the polymer.

16.2 Models of structure: random coils

The most likely conformation of a chain of identical units not capable of forming hydrogen bonds or any other type of specific bond is a **random coil**. Polyethylene is a simple example. The random coil model is a helpful starting point for estimating the orders of magnitude of the properties of polymers and denatured proteins in solution.[1]

The simplest model of a random coil is a **freely jointed chain**, in which any bond is free to make any angle with respect to the preceding one (Fig. 16.4). We assume that the residues occupy zero volume, so different parts of the chain can occupy the same region of space. The model is obviously an oversimplification because a bond is actually constrained to a cone of angles around a direction defined by its neighbour and has bulk. In a hypothetical one-dimensional freely jointed chain all the residues lie in a straight line, and the angle between neighbours is either 0° or 180°. The residues in a three-dimensional freely jointed chain are not restricted to lie in a line or a plane.

The **contour length**, R_c, of a polymer is the length of the molecule measured along its backbone from monomer to monomer:

$$R_c = Nl \qquad (16.3a)$$

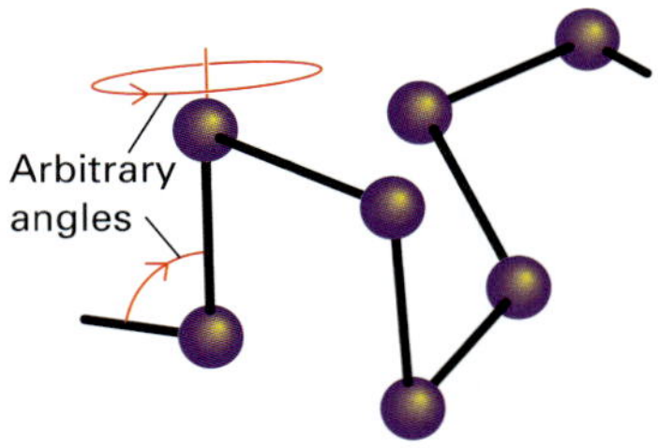

Fig. 16.4 A freely jointed chain is like a three-dimensional random walk, each step being in an arbitrary direction but of the same length.

See an animated version of this figure in the interactive ebook.

[1] For the derivation of the expressions in this section, see our *Physical chemistry* (2006).

The contour length is proportional to the number of monomers, N, in the polymer and the length l occupied by each monomer unit. The radius of the random coil such a molecule forms, however, is proportional only to the square-root of N because the coil consists of steps (neighbouring bonds) that might double back on themselves as the chain grows. Specifically, the **root mean square separation**, $R_{rms} = \langle R^2 \rangle^{1/2}$, is a measure of the average separation of the two ends of a random coil:

$$R_{rms} = N^{1/2} l \qquad (16.3b)$$

Consequently, the volume of the coil increases as $N^{3/2}$. The **radius of gyration**, R_G, of a random coil is the radius of a thin shell (think of a table-tennis ball) that has the same mass as the molecule and the same moment of inertia. The radius of gyration of a table-tennis ball is the same as its actual radius; that of a solid sphere of radius r is $R_G = (\frac{3}{5})^{1/2} r$. For a random coil

$$R_G = \left(\frac{N}{6}\right)^{1/2} l \qquad (16.3c)$$

A brief illustration Consider a polyethylene chain with M = 112 kDa, corresponding to N = 4000. Because l = 154 pm for a C—C bond, we find (by using 10^3 pm = 1 nm)

From eqn 16.3a: $R_c = 4000 \times 154\ \text{pm} = 616\ \text{nm}$

From eqn 16.3b: $R_{rms} = (4000)^{1/2} \times 154\ \text{pm} = 9.74\ \text{nm}$

From eqn 16.3c: $R_G = \left(\frac{4000}{6}\right)^{1/2} \times 154\ \text{pm} = 3.98\ \text{nm}$

The random-coil model ignores the role of the solvent: a poor solvent will tend to cause the coil to tighten so that solute–solvent contacts are minimized; a good solvent does the opposite. Therefore, calculations based on this model are better regarded as lower bounds to the dimensions for a polymer in a good solvent and as an upper bound for a polymer in a poor solvent. The model is most reliable for a polymer in a bulk solid sample, where the coil is likely to have its natural dimensions.

A random coil is the least structured conformation of a polymer chain in the sense that it can be achieved in the greatest possible number of ways (in contrast, for instance, to the straight chain conformation, which can be achieved in only one way) and corresponds to the state of greatest entropy. Any stretching of the coil introduces order and reduces the entropy. Conversely, the formation of a random coil from a more extended form is a spontaneous process (provided enthalpy contributions do not interfere). The change in **conformational entropy**, the entropy arising from the arrangement of bonds, when a one-dimensional coil containing N bonds of length l is stretched or compressed by nl is

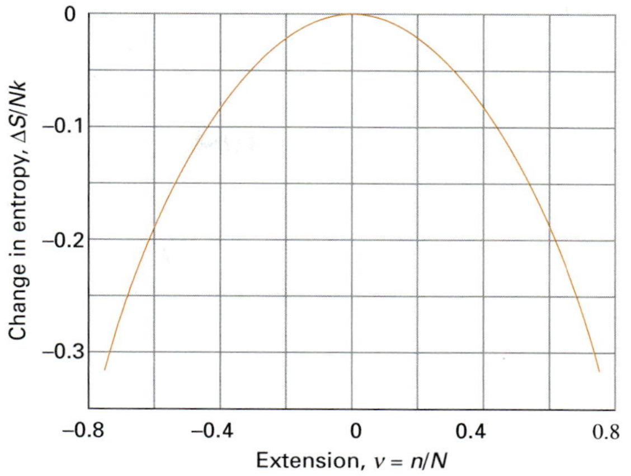

Fig. 16.5 The change in molar entropy of a one-dimensional perfect elastomer as its extension changes; ν = 1 corresponds to complete extension; ν = 0, the conformation of highest entropy, corresponds to the random coil.

$$\Delta S = -\tfrac{1}{2} kN \ln\{(1+\nu)^{1+\nu}(1-\nu)^{1-\nu}\} \qquad \nu = \frac{n}{N} \qquad (16.4)$$

where k is Boltzmann's constant. This function is plotted in Fig. 16.5, and we see that minimum extension—fully coiled ($n = 0$)—corresponds to maximum entropy. This spontaneous tendency to form a coil is responsible for the tendency of rubber (or at least, an ideal rubber with no intermolecular interactions) to spring back into shape after being stretched.

16.3 Models of structure: polypeptides and polynucleotides

Polypeptides are almost at the opposite end of the scale of structure from random coils, for they can become highly ordered: they need to be, for in biology structure is almost synonymous with function. We need to distinguish four levels of structure. The **primary structure** of a biopolymer is the sequence of its monomer units: this sequence is determined by valence forces in the sense that the monomers are linked by covalent bonds. For polypeptides, the primary structure is an ordered list of the amino acid residues. The **secondary structure** of a polypeptide is the spatial arrangement of the polypeptide chain—its twisting into a specific shape —under the influence of interactions between the various peptide residues (the amino acid groups).

We can rationalize the secondary structures of proteins in large part in terms of the hydrogen bonds between the —NH— and —CO— groups of the

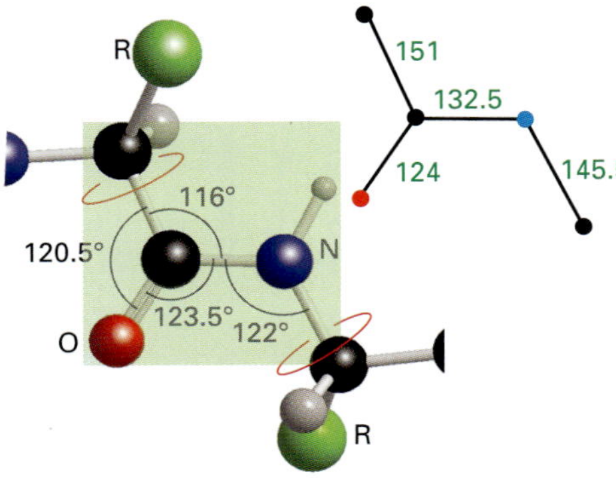

Fig. 16.6 The dimensions that characterize the peptide link. The C—NH—CO—C atoms define a plane (the C—N bond has partial double-bond character), but there is rotational freedom around the C—CO and N—C bonds.

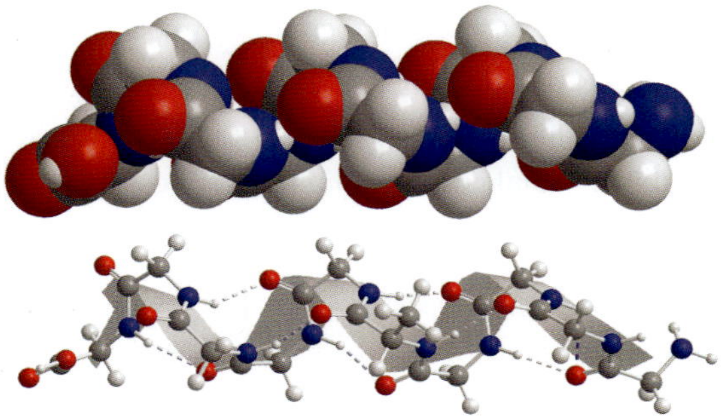

Fig. 16.7 The polypeptide α-helix, with poly-L-glycine as an example. There are 3.6 residues per turn, and a translation along the helix of 150 pm per residue, giving a pitch of 540 pm. The diameter (ignoring side chains) is about 600 pm.

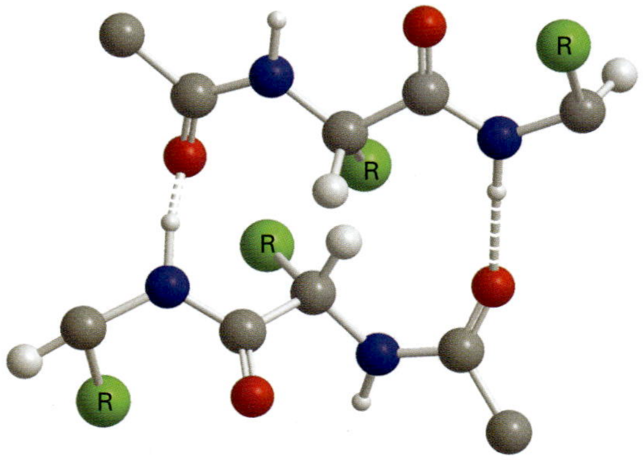

Fig. 16.8 An antiparallel β-sheet in which the N—H—O atoms of the hydrogen bonds form a nearly straight line.

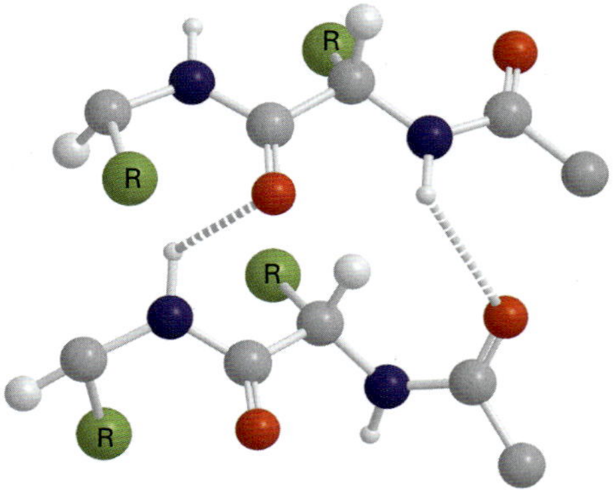

Fig. 16.9 A parallel β-sheet in which the N—H—O atoms of the hydrogen bonds are not as well aligned as in the antiparallel version.

peptide links (Fig. 16.6). These bonds lead to two principal structures. One, which is stabilized by hydrogen bonding between peptide links of the same chain, is the **α-helix**. The other, which is stabilized by hydrogen bonding, links to different chains or more distant parts of the same chain, is the **β-sheet** (or β-*pleated sheet*).

The α-helix is illustrated in Fig. 16.7. Each turn of the helix contains 3.6 amino acid residues, so there are 18 residues in 5 turns of the helix. The pitch of a single turn (the lateral movement corresponding to one complete rotation) is 544 pm. The N—H···O bonds lie parallel to the axis and link every fifth group (so residue i is linked to residues $i - 4$ and $i + 4$). There is freedom for the helix to be arranged as either a right- or a left-handed screw, but the overwhelming majority of natural polypeptides are right-handed on account of the preponderance of the L-configuration of the naturally occurring amino acids. It turns out, in agreement with experience, that a right-handed α-helix of L-amino acids has a marginally lower energy than a left-handed helix of the same acids. A β-sheet is formed by hydrogen bonding between two extended polypeptide chains. Some of the side chains lie above the sheet and some lie below it. Two types of structures can be distinguished from the pattern of hydrogen bonding between the constituent chains: (a) in an **antiparallel β-sheet** (Fig. 16.8), the N—H···O atoms of the hydrogen bonds form a straight line; (b) in a **parallel β-sheet** (Fig. 16.9), the N—H···O atoms of the hydrogen bonds are not perfectly aligned.

Helical and sheet-like polypeptide chains are folded into a **tertiary structure** if there are other bonding influences between the residues of the chain that are strong enough to overcome the interactions responsible for the secondary structure. The folding influences include —S—S— **disulfide links**, van der Waals interactions, hydrophobic interactions, ionic interactions (which depend on the pH), and strong hydrogen bonds (such as O—H···O).

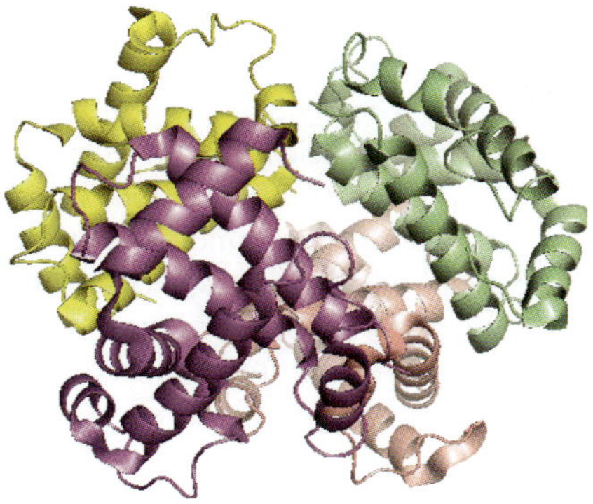

Fig. 16.10 A haemoglobin molecule consists of four myoglobin-like units.

Proteins with $M > 50$ kDa are often found to be aggregates of two or more polypeptide chains. The possibility of such **quaternary structure** often confuses the determination of their molar masses because different techniques might give values differing by factors of 2 or more. Haemoglobin, which consists of four myoglobin-like chains (Fig. 16.10), is an example of a quaternary structure. Myoglobin is an oxygen-storage protein. The subtle differences that arise when four such molecules coalesce to form haemoglobin result in the latter being an oxygen transport protein, able to load O_2 cooperatively and to unload it co-operatively too (see Box 7.2).

Deoxyribonucleic acid (DNA) and ribonucleic acid (RNA), which are key components of the mechanism of storage and transfer of genetic information in biological cells, are *polynucleotides*. The backbones of these molecules consist of alternating sugar and phosphate groups, and one of the bases adenine (A), cytosine (C), guanine (G), and thymine (T, found in DNA only), and uracil (U, found in RNA only) is attached to each sugar. In B-DNA, the most common form of DNA in biological cells, two polynucleotide chains held together by A—T and C—G base pairs (**2** and **3**) wind around each other to form a right-handed double helix (Fig. 16.11). The structure is stabilized further by the π-stacking interactions mentioned in Box 15.1. In contrast, RNA exists primarily

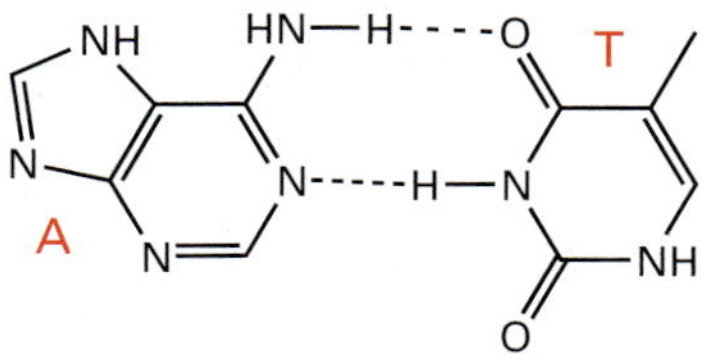

2 A–T base pair

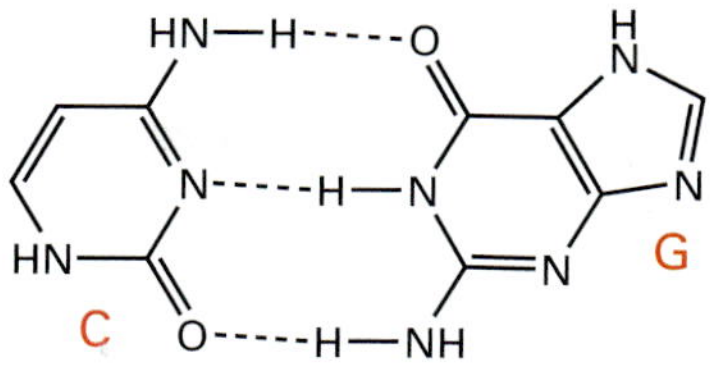

3 C–G base pair

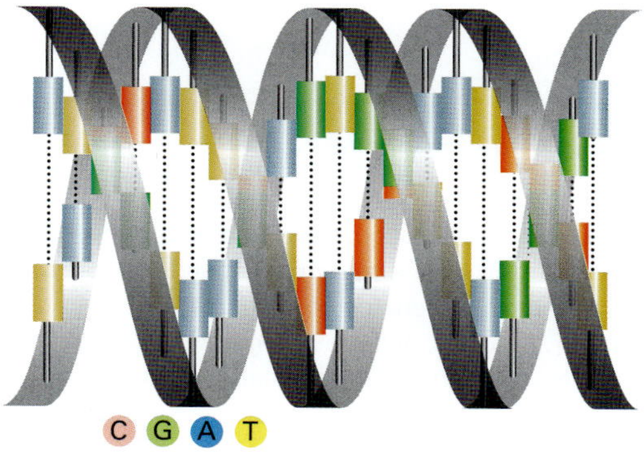

Fig. 16.11 DNA double helix, in which two polynucleotide chains are linked together by hydrogen bonds between adenine (A) and thymine (T) and between cytosine (C) and guanine (G).

as single chains that can fold into complex structures by formation of A—U and G—C base pairs.

Biopolymer **denaturation**, or loss of structure, can be caused by several means, and different aspects of structure may be affected. Denaturation at the secondary level is brought about by agents that destroy hydrogen bonds. Thermal motion may be sufficient, in which case denaturation is a kind of intramolecular melting. When eggs are cooked the albumin is denatured irreversibly, and the protein collapses into a structure resembling a random coil. The **helix–coil transition** of polypeptides is sharp, like ordinary melting, because it is a cooperative process in the sense that when one hydrogen bond has been broken it is easier to break its neighbours, and then even easier to break theirs, and so on. The disruption cascades down the helix, and the transition occurs sharply. Denaturation may also be brought about chemically. For instance, a solvent that forms stronger hydrogen bonds than those within the helix will compete successfully for the NH and CO groups. Acids and bases can cause denaturation by protonation or deprotonation of various groups.

In contemporary physical chemistry and molecular biophysics, a great deal of work is being done on the rationalization and prediction of the structures of biomolecules such as the polypeptides and nucleic acids described here, using the interactions described in Chapter 15 (Box 15.1).

16.4 Mechanical properties of polymers

Synthetic polymers are classified broadly as *elastomers*, *fibres*, and *plastics*, depending on their **crystallinity**, the degree of three-dimensional long-range order attained in the solid state.

An **elastomer** is a flexible polymer that can expand or contract easily upon application of an external force. Elastomers are polymers with numerous cross-links that pull them back into their original shape when a stress is removed. The weak directional constraints on silicon–oxygen bonds is responsible for the high elasticity of silicones. A **perfect elastomer**, a polymer in which the internal energy is independent of the extension of the random coil, can be modelled as a freely jointed chain.

Box 16.1 The prediction of protein structure

A polypeptide chain adopts a conformation corresponding to a minimum Gibbs energy, which depends on the *conformational energy*, the energy of interaction between different parts of the chain, and the energy of interaction between the chain and surrounding solvent molecules. In the aqueous environment of biological cells, the outer surface of a protein molecule is covered by a mobile sheath of water molecules, and its interior contains pockets of water molecules. These water molecules play an important role in determining the conformation that the chain adopts through hydrophobic interactions and hydrogen bonding to amino acids in the chain.

The simplest calculations of the conformational energy of a polypeptide chain ignore entropy and solvent effects and concentrate on the total potential energy of all the interactions between nonbonded atoms. For example, as remarked in the text, these calculations predict that a right-handed α-helix of L-amino acids is marginally more stable than a left-handed helix of the same amino acids.

To calculate the energy of a conformation, we need to make use of many of the molecular interactions described in Chapter 15, and also of some additional interactions:

1. *Bond stretching.* Bonds are not rigid, and it may be advantageous for some bonds to stretch and others to be compressed slightly as parts of the chain press against one another. If we liken the bond to a spring, then the potential energy takes the form (see Section 12.9):

$$V_{\text{stretch}} = \tfrac{1}{2}k_{\text{stretch}}(R - R_e)^2$$

where R_e is the equilibrium bond length and k_{stretch} is the force constant, a measure of the stiffness of the bond in question.

2. *Bond bending.* An O—C—H bond angle (or some other angle) may open out or close in slightly to enable the molecule as a whole to fit together better. If the equilibrium bond angle is θ_e, we write

$$V_{\text{bend}} = \tfrac{1}{2}k_{\text{bend}}(\theta - \theta_e)^2$$

where k_{bend} is the force constant, a measure of how difficult it is to change the bond angle.

3. *Bond torsion.* There is a barrier to internal rotation of one bond relative to another (just like the barrier to internal rotation in ethane). Because the planar peptide link is relatively rigid, the geometry of a polypeptide chain can be specified by the two angles that two neighbouring planar peptide links make to each other. The first illustration shows the two angles ϕ and ψ commonly used to specify this relative orientation. The sign convention is that a positive angle means that the front atom must be rotated clockwise to bring it into an eclipsed position relative to the rear atom. For an all-*trans* form of the chain, all ϕ and ψ are 180°. A helix is obtained when all the ϕ are equal and when all the ψ are equal. For a right-handed α-helix, all $\phi = -57°$ and all $\psi = -47°$. For a left-handed α-helix, both angles are positive. For an antiparallel β-sheet, $\phi = -139°$, $\psi = 113°$. The torsional contribution to the total potential energy is

$$V_{\text{torsion}} = A(1 + \cos 3\phi) + B(1 + \cos 3\psi)$$

in which A and B are constants of the order of 1 kJ mol^{-1}. Because only two angles are needed to specify the conformation of a helix, and they range from −180° to +180°, the torsional potential energy of the entire molecule can be represented on a *Ramachandran plot*, a contour diagram in which one axis represents ϕ and the other represents ψ.

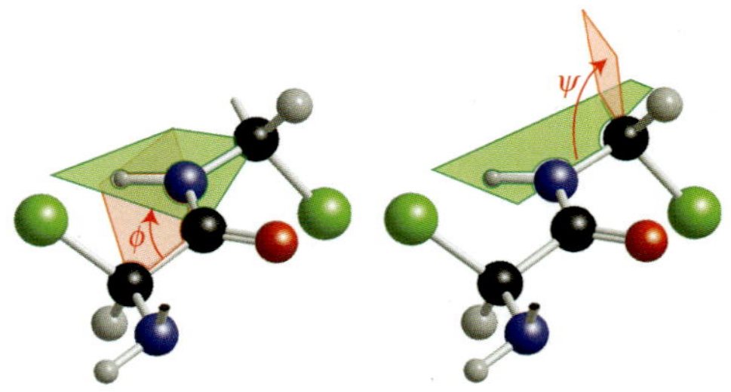

The definition of the torsional angles ψ and ϕ between two peptide units.

4. *Interaction between partial charges*. If the partial charges Q_i and Q_j on the atoms *i* and *j* are known, a Coulombic contribution of the form $1/r$ can be included:

$$V_{\text{Coulomb}} = \frac{Q_iQ_j}{4\pi\varepsilon r}$$

where ε is the permittivity of the medium in which the charges are embedded. Charges of $-0.28e$ and $+0.28e$ are assigned to N and H, respectively, and $-0.39e$ and $+0.39e$ to O and C, respectively. The interaction between partial charges does away with the need to take dipole–dipole interactions into account, for they are taken care of by dealing with each partial charge explicitly.

5. *Dispersive and repulsive interactions*. The interaction energy of two atoms separated by a distance *r* (which we know once ϕ and ψ are specified) can be given by the Lennard-Jones (12,6) form (Section 15.8):

$$V_{\text{LJ}} = \frac{C}{r^{12}} - \frac{D}{r^6}$$

6. *Hydrogen bonding*. In some models of structure, the interaction between partial charges is judged to take into account the effect of hydrogen bonding. In other models, hydrogen bonding is added as another interaction of the form

$$V_{\text{H bonding}} = \frac{E}{r^{12}} - \frac{F}{r^{10}}$$

The total potential energy of a given conformation (ϕ,ψ) can be calculated by summing the contributions given by the preceding equations for all bond angles (including torsional angles) and pairs of atoms in the molecule. The procedure is known as a *molecular mechanics* simulation and is automated in commercially available molecular modelling software. For large molecules, plots of potential energy versus bond distance or bond angle often show several local minima and a global minimum (see the second illustration). The software packages include schemes for modifying the locations of the atoms and searching for these minima systematically.

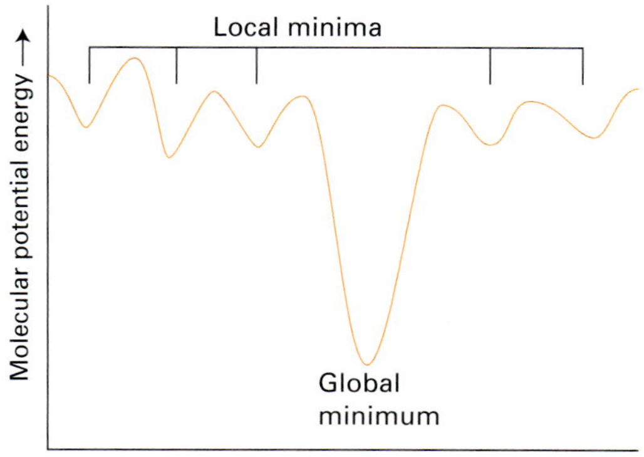

For large molecules, a plot of potential energy against the molecular geometry often shows several local minima and a global minimum.

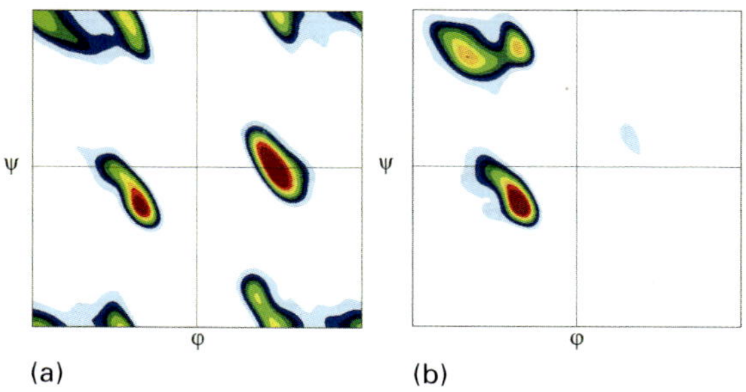

Contour plots of potential energy against the angles ψ and ϕ, also known as a Ramachandran diagram, for (a) a glycyl residue of a polypeptide chain and (b) an alanyl residue. (Hovmoller, et al., *Acta Crystallogr.* **D58**, 768 (2002).)

The third illustration shows the potential energy contours for the helical form of polypeptide chains formed from the nonchiral amino acid glycine (R = H) and the chiral amino acid L-alanine (R = CH_3). The contours were computed by summing all the contributions described above for each choice of angles, and then plotting contours of equal potential energy. The glycine map is symmetrical, with minima of equal depth at $\phi = -80°$, $\psi = +90°$ and at $\phi = +80°$, $\psi = -90°$. In contrast, the map for L-alanine is unsymmetrical, and there are three distinct low-energy conformations (marked I, II, III). The minima of regions I and II lie close to the angles typical of right- and left-handed α-helices, but the former has a lower minimum, which is consistent with the formation of right-handed helices from the naturally occurring L-amino acids.

The structure corresponding to the global minimum of a molecular mechanics simulation is a snapshot of the molecule at $T = 0$ because only the potential energy is included in the calculation; contributions to the total energy from kinetic energy are excluded. In a *molecular dynamics* simulation, the molecule is set in motion by heating it to a specified temperature. The possible trajectories of all atoms under the influence of the intermolecular potentials are then calculated by integration of Newton's equations of motion. These trajectories correspond to the conformations that the molecule can sample at the temperature of the simulation. At very low temperatures, the molecule cannot overcome some of the potential energy barriers described above, atomic motion is restricted, and only a few conformations are possible. At high temperatures, more potential energy barriers can be overcome and more conformations are possible. Therefore, molecular dynamics calculations are useful tools for the visualization of the flexibility of polymers.

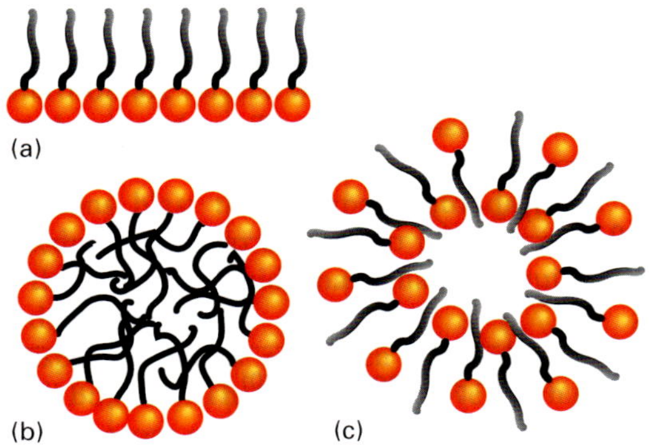

Fig. 16.21 Amphiphilic molecules form a variety of related structures in water: (a) a monolayer, (b) a spherical micelle, (c) a bilayer vesicle.

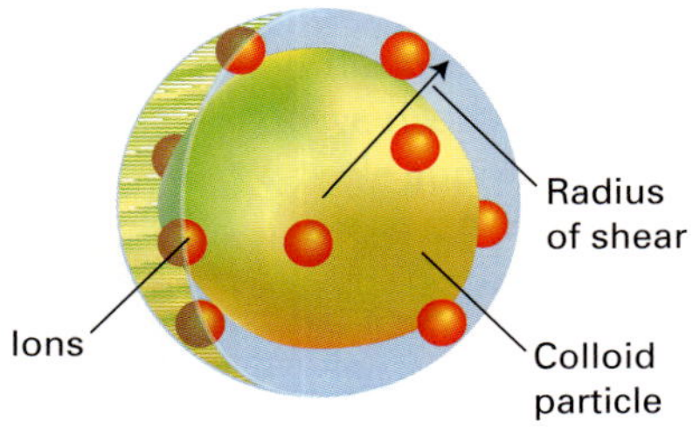

Fig. 16.22 The definition of the radius of shear for a colloidal particle. The spheres are ions attached to the surface of the particle.

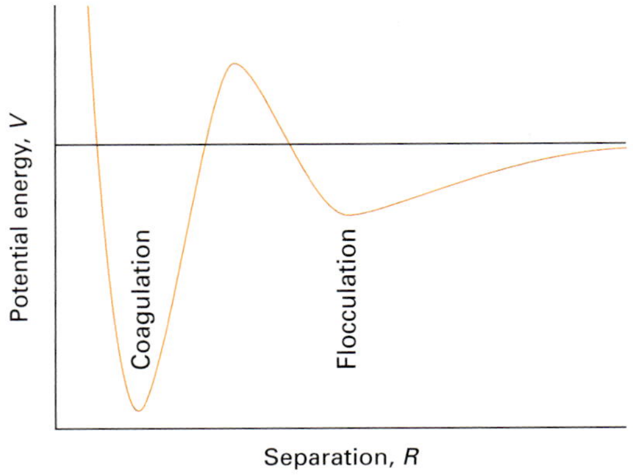

Fig. 16.23 The potential energy of interaction of two colloidal particles varies with distance as shown here. The shallow outer well represents the van der Waals interactions between the particles and accounts for flocculation; the deep inner well represents the merging—the coagulation—of the particles.

their hydrocarbon interiors. For this reason, micellar systems are used as detergents and drug carriers, and for organic synthesis, froth flotation, and petroleum recovery. They can be perceived as a part of a family of similar structures formed when amphiphilic substances are present in water (Fig. 16.21). A **monolayer** forms at the air/water interface, with the hydrophilic head groups facing the water. Micelles are like monolayers that enclose a region. A **bilayer vesicle** is like a double-micelle, with an inward pointing inner surface of molecules surrounded by an outward pointing outer layer. The 'flat' version of a bilayer vesicle is the analogue of a cell membrane.

16.8 The electric double layer

Apart from the physical stabilization of disperse systems, a major source of kinetic nonlability is the existence of an electric charge on the surfaces of the colloidal particles. On account of this charge, ions of opposite charge tend to cluster nearby.

Two regions of charge must be distinguished. First, there is a fairly immobile layer of ions that stick tightly to the surface of the colloidal particle, and that may include water molecules (if that is the support medium). The radius of the sphere that captures this rigid layer is called the **radius of shear**, and is the major factor determining the mobility of the particles (Fig. 16.22). The electric potential at the radius of shear relative to its value in the distant, bulk medium is called the **electrokinetic potential**, ζ (zeta). The charged unit attracts an oppositely charged ionic atmosphere. The inner shell of charge and the outer atmosphere jointly constitute the **electric double layer**.

At high concentrations of ions of high charge number, the atmosphere is dense and the potential falls to its bulk value within a short distance. In this case there is little electrostatic repulsion to hinder the close approach of two colloid particles. As a result, **flocculation**, the aggregation of the colloidal particles, occurs as a consequence of the van der Waals forces (Fig. 16.23). Flocculation is often reversible, and should be distinguished from **coagulation**, which is the irreversible collapse of the colloid into a bulk phase. When river water containing colloidal clay flows into the sea, the brine induces coagulation and is a major cause of silting in estuaries.

Metal oxide and sulfide sols have charges that depend on the pH; sulfur and the noble metals tend to be negatively charged. Naturally occurring macromolecules also acquire a charge when dispersed in water, and an important feature of proteins and other natural macromolecules is that their overall charge depends on the pH of the medium. For instance, in acid environments protons attach to basic groups and the net charge of the macromolecule is positive; in basic media the net charge is negative as a result of proton loss. At the **isoelectric point**, the pH is such that there is no net charge on the macromolecule.

The primary role of the electric double layer is to render the colloid kinetically nonlabile. Colliding colloidal particles break through the double layer and coalesce only if the collision is sufficiently energetic to disrupt the layers of ions and solvating molecules, or if thermal motion has stirred away the surface accumulation of charge. This kind of disruption of the double layer may occur at high temperatures, which is one reason why sols precipitate when they are heated. The protective role of the double layer is the reason why it is important not to remove all the ions (other than those needed to ensure overall electrical neutrality) when a colloid is being purified by dialysis, and why proteins coagulate most readily at their isoelectric point.

The presence of charge on colloidal particles and natural macromolecules also permits us to control their motion, as in dialysis and electrophoresis. Apart from its application to the determination of molar mass, electrophoresis has several analytical and technological applications. One analytical application is to the separation of different macromolecules, as discussed in Section 16.1. Technical applications include silent ink-jet printers, the painting of objects by airborne charged paint droplets, and electrophoretic rubber forming by deposition of charged rubber molecules on anodes formed into the shape of the desired product (for example, surgical gloves).

16.9 Liquid surfaces and surfactants

Liquid surfaces are mobile interfaces where solutes might gather and influence its properties. The smooth surface of stationary liquids is due to the imbalance of forces, for whereas a molecule in the interior of a bulk sample experiences attractions from all directions, those at the surface experience only inward forces. A molecule at an air–liquid surface has a higher potential energy than one in the bulk because it interacts with fewer neighbours, so work must be done to bring a molecule from the bulk into the surface layer. The work required to increase the area of surface by $\Delta\sigma$ is proportional to that increase and we write $w = \gamma\Delta\sigma$, where the constant of proportionality γ is called the **surface tension**. For $\gamma\Delta\sigma$ to be expressed in joules, γ must be in newtons per metre, N m^{-1} (because then N m^{-1} × m^2 = N m = J). Some values of the surface tension are given in Table 16.1. Broadly speaking, surface tensions are high when there are strong forces acting between the molecules or atoms, as in water and mercury. Surface tensions

Table 16.1
Surface tensions of liquids at 293 K

	γ/(mN m^{-1})
Benzene	28.88
Carbon tetrachloride	27.0
Ethanol	22.8
Hexane	18.4
Mercury	472
Methanol	22.6
Water	72.75
	72.0 at 25°C
	58.0 at 100°C

typically decrease as the temperature is raised and vanish at the boiling point.

We saw in Chapter 4 that nonexpansion work (work that does not involve volume expansion against an external pressure) can be identified with a change in Gibbs energy, ΔG, so we can write

$$\Delta G = \gamma\Delta\sigma \qquad (16.7)$$

This is the link between surface properties and thermodynamics.

One consequence of eqn 16.7 is that the pressure is different on either side of a curved liquid surface (as in a droplet or a cavity in a liquid). As we show in Derivation 16.1, the pressure on either side of a spherical surface of radius r is given by the **Laplace equation:**

$$p_{\text{concave}} = p_{\text{convex}} + \frac{2\gamma}{r} \qquad (16.8)$$

This equation tells us that the pressure just inside a curved surface (on the convex side, Fig. 16.24) is lower than that just outside the surface and that the difference is greater the greater the surface tension of the liquid.

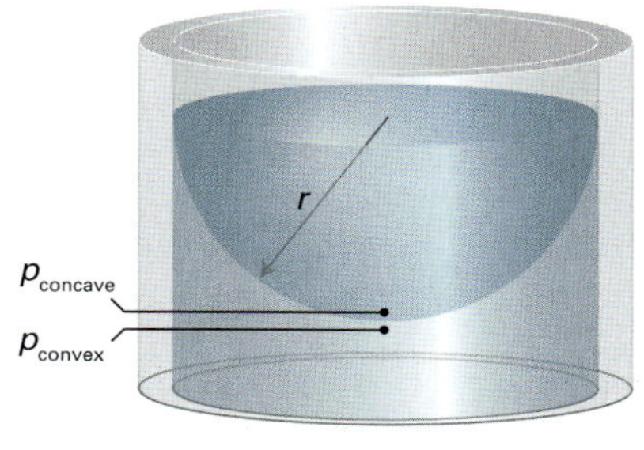

Fig. 16.24 The pressure just inside a curved surface is lower than that just outside the surface; the difference is greater the greater the surface tension of the liquid.

Derivation 16.1

The Laplace equation

If the pressure inside a spherical cavity is $p_{concave}$, the total force (which is pressure × area) acting on the wall of the cavity of area $4\pi r^2$ is $4\pi r^2 p_{concave}$. The force tending to compress the cavity is the sum of effects due to the pressure outside the cavity—on the convex side of the surface—and the surface tension. The former gives rise to a force $4\pi r^2 p_{convex}$. The force due to the surface tension is calculated as follows. The change in surface area when the radius of the cavity increases from r to $r + dr$ is

$$d\sigma = 4\pi(r + dr)^2 - 4\pi r^2$$
$$= 4\pi(r^2 + 2r dr + dr^2) - 4\pi r^2 \approx 8\pi r dr$$

(As usual in calculus, we neglect higher powers of infinitesimal quantities.) From eqn 16.7 the work done when the cavity expands by this amount is $dw = 8\pi\gamma r dr$. However, because work is force × distance, it follows that the force opposing the expansion through dr is $F = 8\pi\gamma r$. The total inward force is therefore $4\pi r^2 p_{convex} + 8\pi\gamma r$. When the inward and outward forces are balanced,

$$4\pi r^2 p_{concave} = 4\pi r^2 p_{convex} + 8\pi\gamma r$$

This relation can now be rearranged into eqn 16.8 by dividing both sides by $4\pi r^2$.

The difference in pressure across a curved interface has a number of consequences. One is that it gives rise to **capillary action**, in which a liquid climbs up the interior of a narrow tube. As can be seen from Fig. 16.25, the pressure just below the meniscus of a liquid in a narrow tube is less, by $2\gamma/r$, than the atmospheric pressure so the liquid is pushed up the tube until the hydrostatic pressure (Section 0.5), which is equal to ρgh, where ρ is the mass density of the liquid, g is the acceleration of free fall, and h is the height of the column, cancels the reduction in pressure due to the curvature. That is, the liquid rises to a height at which $\rho gh = 2\gamma/r$, and therefore

$$h = \frac{2\gamma}{\rho g r} \qquad (16.9)$$

This expression gives a simple method for estimating the surface tension of a liquid (by rearranging it into $\gamma = \frac{1}{2}\rho g r h$).

A brief illustration If water at 25°C rises to a height of 7.36 cm in a capillary tube of internal radius 0.20 mm, the surface tension is

$$\gamma = \tfrac{1}{2}(997.1\ \text{kg m}^{-3}) \times (9.81\ \text{m s}^{-2}) \times (7.36 \times 10^{-2}\ \text{m}) \times (2.0 \times 10^{-4}\ \text{m})$$
$$= 7.2 \times 10^{-2}\ \text{kg s}^{-2} = 7.2 \times 10^{-2}\ \text{N m}^{-1}$$

This value could be reported as 72 mN m^{-1}.

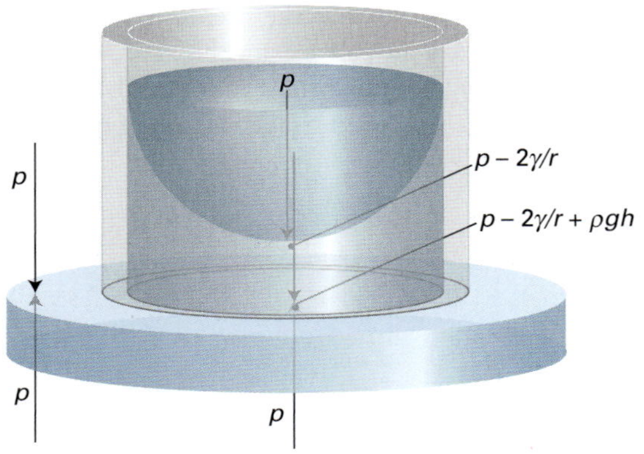

Fig. 16.25 When a capillary tube is first stood in a liquid the liquid climbs up the walls, so curving the surface. It continues to rise until the total pressure at the foot of the column (which arises from the atmosphere, the effect of curvature, and the hydrostatic contribution) is equal to the atmospheric pressure.

The surface tension of a liquid changes markedly if a surfactant is present. Amphiphilic molecules accumulate at the water–air surface with their hydrophobic tails exposed to the air to minimize interaction with the water. Their accumulation at the surface relative to the bulk is reported as the **surface excess,** Γ (uppercase gamma). In a simple case where no surfactant appears in the vapour above the surface, this quantity is measured by noting the total amount of surfactant in a sample of the liquid, n_{total}, and subtracting from that total the amount known to be in the bulk solution, $n_{solution}$, from measurement of its concentration. Then

$$\Gamma = \frac{n_{total} - n_{solution}}{\sigma} \qquad (16.10)$$

where σ is the area of the surface. Below the critical micelle concentration the slope of a plot of surface tension against the logarithm of the concentration is equal to $-RT\Gamma$, so Γ can be determined.[3] Above the CMC the surface tension is independent of the concentration of surfactant, so the CMC can be determined graphically (Fig. 16.26).

Pure liquids do not form foams: the Gibbs energy increases when a surface is formed, so there is always a spontaneous tendency for a cavity in a liquid to collapse. A bubble in boiling water will rise to the

[3] See our *Physical chemistry* (2006).

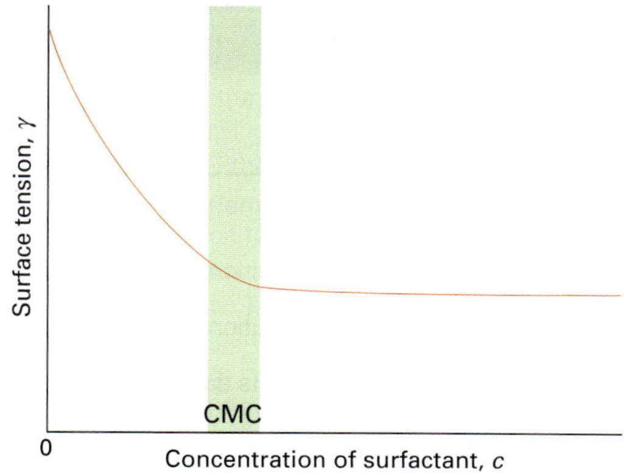

Fig. 16.26 The variation of surface tension with concentration of surfactant.

surface and break when it arrives. If a surfactant is present, however, there is a smaller pressure difference between its interior and the surroundings (because the surface tension is lower) and the surface is stabilized by the surface excess of surfactants. A bubble in a surfactant solution will rise to the surface after it has been formed and will survive at the surface. It will be joined by others, and a foam will develop. The structure of that foam is itself a highly interesting mathematical problem, for the originally spherical bubbles deform into polyhedra that minimize the total surface area. The most common polyhedra are predicted mathematically to have 13.4 sides, and indeed it is observed that most have 14 sides, with the second most abundant having 12.

Checklist of key ideas

You should now be familiar with the following topics.

☐ **1** Many proteins (specifically protein enzymes) are monodisperse, a synthetic polymer is polydisperse.

☐ **2** Techniques for the determination of the mean molar masses of macromolecules include osmometry, mass spectrometry (as MALDI), sedimentation rates and equilibria, gel and capillary electrophoresis, and laser light scattering.

☐ **3** The least structured model of a macromolecule is as a random coil.

☐ **4** The primary primary structure of a biopolymer is the sequence of its monomer units.

☐ **5** The secondary structure of a protein is the spatial arrangement of the polypeptide chain and includes the α-helix and β-sheet.

☐ **6** Helical and sheet-like polypeptide chains are folded into a tertiary structure by bonding influences between the residues of the chain.

☐ **7** Some macromolecules have a quaternary structure as aggregates of two or more polypeptide chains.

☐ **8** Protein denaturation is loss of structure; a helix–coil transition is a cooperative process.

☐ **9** Synthetic polymers are classified as elastomers, fibres, and plastics.

☐ **10** A perfect elastomer is a polymer in which the internal energy is independent of the extension of the random coil; for small extensions a random coil model obeys a Hooke's law restoring force.

☐ **11** Synthetic polymers undergo a transition from a state of high to low chain mobility at the glass transition temperature, T_g.

☐ **12** A mesophase is a bulk phase that is intermediate in character between a solid and a liquid.

☐ **13** A disperse system is a dispersion of small particles of one material in another.

☐ **14** Liquid crystals are classified as smectic, nematic, or cholesteric.

☐ **15** A surfactant is a species that accumulates at the interface of two phases or substances and modifies the properties of the surface.

☐ **16** The radius of shear is the radius of the sphere that captures the rigid layer of charge attached to a colloid particle.

☐ **17** The electrokinetic potential is the electric potential at the radius of shear relative to its value in the distant, bulk medium.

☐ **18** The inner shell of charge and the outer atmosphere jointly constitute the electric double layer.

☐ **19** Many colloid particles are thermodynamically unstable but kinetically nonlabile.

☐ **20** Surface tension is a measure of the work needed to produce a liquid surface.

☐ **21** The pressure on the convex side of a curved surface is lower than that on the concave side; the difference gives rise to capillary action.

☐ **22** The accumulation of a surfactant at a surface lowers the surface tension.

16.20‡ Here we explore elastomers in quantitative detail. (a) Estimate the force required to expand a random coil (a perfect elastomer) consisting of 1000 links by 10 per cent of its fully coiled state at 300 K. (b) The restoring force acting when a random coil is extended by d*x* is related to the conformational entropy by $F = -T\mathrm{d}S/\mathrm{d}x$. Use this expression to deduce eqn 16.5a and eqn 16.5b.

16.21 The following exercises give you a sense of the calculations that must be done when predicting the structure of a biological polymer with the techniques summarized in Box 16.1. (a) Theoretical studies have estimated that the lumiflavin isoalloazine ring system (**4**) has an energy minimum at the bending angle of 15°, but that it requires only 8.5 kJ mol^{-1} to increase the angle to 30°. If there are no other compensating interactions, what is the force constant for lumiflavin bending? (b) The equilibrium bond length of a carbon–carbon single bond is 152 pm. Given a C—C force constant of 400 N m^{-1}, how much energy, in kilojoules per mole, would it take to stretch the bond to 165 pm?

16.22 Here we explore the dynamical properties of biological membranes. (a) Lipid diffusion in a cell plasma membrane occurs with a diffusion constant of 1.0×10^{-8} cm^2 s^{-1} and the same lipid in a lipid bilayer has a diffusion constant of 1.0×10^{-7} cm^2 s^{-1}. How long will it take the lipid to diffuse 10 nm in a plasma membrane and a lipid bilayer? (b) Organisms are capable of biosynthesizing lipids of different composition so that cell membranes have melting temperatures close to the ambient temperature. Keeping in mind that structural elements that prevent alignment of the hydrophobic chains in the gel phase lead to low melting temperatures, explain why bacterial and plant cells grown at low temperatures synthesize more phospholipids with chains containing C=C bonds than do cells grown at higher temperatures.

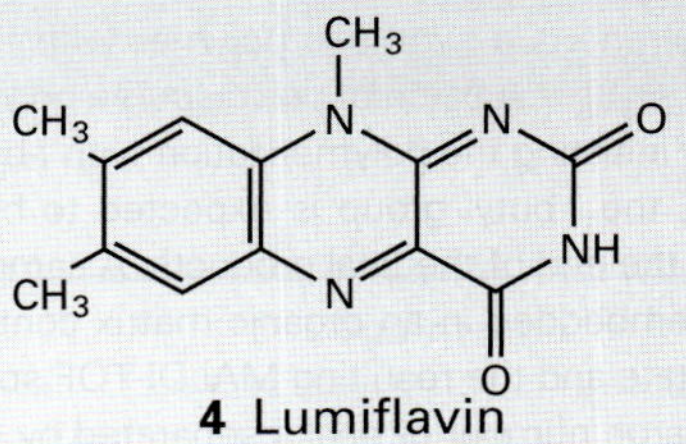

4 Lumiflavin

Chapter 17

Metallic, ionic, and covalent solids

Modern chemistry is closely concerned with the properties of solids. Apart from their intrinsic usefulness for construction, modern solids have made possible the semiconductor revolution and recent advances in ceramics have given rise to the hope that we may now be on the verge of a superconductor revolution. Advances in our understanding of electron mobility in solids are also useful in biology, where electron transport is responsible for many biochemical processes, particularly photosynthesis and respiration.

The principal technique for investigating the arrangements of atoms in condensed phases, primarily crystalline solids, is X-ray diffraction, but nuclear magnetic resonance (NMR, Chapter 21) is now also making significant contributions. Information from X-ray diffraction and NMR is the basis of much of molecular biology, so the material presented here is the foundation for our discussion of biomolecular structures in Chapter 16. In each case, the observed crystal structure is Nature's solution to the problem of condensing objects of various shapes into an aggregate of minimum energy and, for temperatures above zero, of minimum Gibbs energy.

Bonding in solids

The bonding within a solid may be of various kinds. Simplest of all (in principle) is the bonding in an elemental **metallic solid**, in which electrons are delocalized over arrays of identical cations and bind the whole together into a rigid but malleable structure. Because the delocalized electrons can accommodate bonding patterns with very little directional character, the crystal structures of metals are determined largely by the geometrical problem of packing spherical atoms into a dense, orderly array. In an **ionic solid**, the ions (in general, of different radii, and not always spherical)

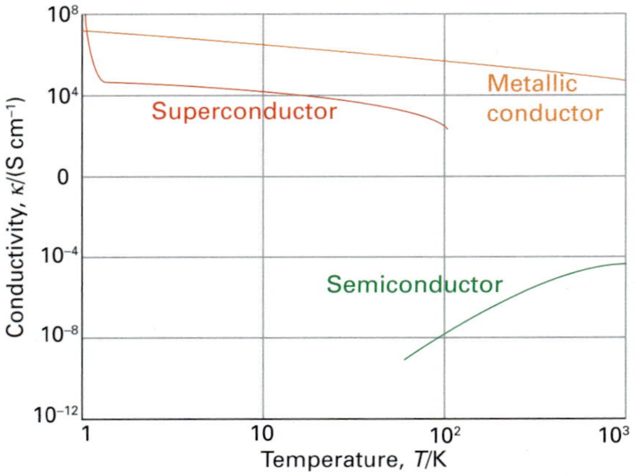

Fig. 17.1 The typical variation with temperature of the electrical conductivities of different classes of electronic conductor.

are held together by their Coulombic interaction, and pack together to give an electrically neutral structure. In a **covalent solid** (or *network solid*), covalent bonds in a definite spatial orientation link the atoms in a network extending through the crystal. The stereochemical demands of valence now override the geometrical problem of packing spheres together, and elaborate and extensive structures may be formed. Important examples of covalent solids are diamond and graphite (Section 17.8). **Molecular solids**, which are the subject of the overwhelming majority of modern structural determinations, consist of discrete molecules attracted to one another by the interactions described in Chapter 15.

Some solids—notably the metals—conduct electricity because they have mobile electrons. These **electronic conductors** are classified on the basis of the variation of their electrical conductivity with temperature (Fig. 17.1):

- A **metallic conductor** is an electronic conductor with a conductivity that *decreases* as the temperature is raised.
- A **semiconductor** is an electronic conductor with a conductivity that *increases* as the temperature is raised.

Metallic conductors include the metallic elements, their alloys, and graphite (parallel to the graphene planes). Some organic solids are metallic conductors. Semiconductors include silicon, diamond, and gallium arsenide. A semiconductor generally has a lower conductivity than that typical of metals, but the magnitude of the conductivity is not relevant to the distinction. It is conventional to classify substances with very low electrical conductivities, such as most ionic solids, as **insulators**. We shall use this term, but it is one of convenience rather than one of fundamental significance. **Superconductors** are substances that conduct electricity with zero resistance. The mechanism of superconductivity in metals at very low (liquid helium) temperatures is well understood: that of the potentially more useful **high-temperature superconductors** (HTSC), which are ceramics mixed oxides such as $YBa_2Cu_3O_7$, is still unresolved.

17.1 The band theory of solids

Metallic and ionic solids can both be treated by molecular orbital theory. The advantage of that approach is that we can then see both types of solid as two extremes of a single kind. In each case, the electrons responsible for the bonding are delocalized throughout the solid (like in a benzene molecule, but on a much bigger scale). In an elemental metal, the electrons can be found on all the atoms with equal probability, which matches the primitive picture of a metal as consisting of cations embedded in a nearly uniform electron 'sea'. In an ionic solid the wavefunctions occupied by the delocalized electrons are almost entirely concentrated on the anions, so the Cl atoms in NaCl, for instance, are present as Cl^- ions and the Na atoms, which have low valence electron density, are present as Na^+ ions.

To set up the molecular orbital theory of solids we shall consider initially a single, infinitely long line of identical atoms, each one having one s orbital available for forming molecular orbitals (as in sodium). One atom of the solid contributes one s orbital with a certain energy (Fig. 17.2). When a second atom is brought up it forms a bonding and antibonding orbital. The orbital of the third atom overlaps its nearest neighbour (and only slightly the next-nearest), and three molecular orbitals are formed from these three atomic orbitals. The fourth atom leads to the formation of a fourth molecular orbital. At this stage we can begin to see that the general effect of bringing up successive atoms is to spread the range of energies covered by the molecular orbitals, and also to fill in the range of energies with more and more orbitals (one more for each additional atom). When N atoms have been added to the line, there are N molecular orbitals covering a band of finite width. The lowest-energy orbital of this band is fully bonding and the highest-energy orbital is fully antibonding between adjacent atoms (Fig. 17.3). In the Hückel

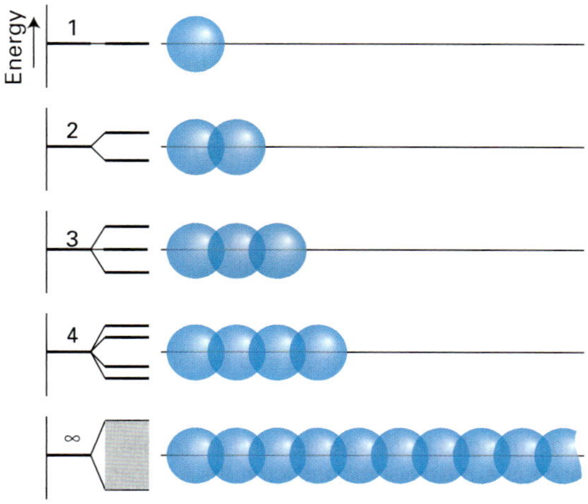

Fig. 17.2 The formation of a band of N molecular orbitals by successive addition of N atoms to a line. Note that the band remains of finite width, and although it looks continuous when N is large, it consists of N different orbitals.

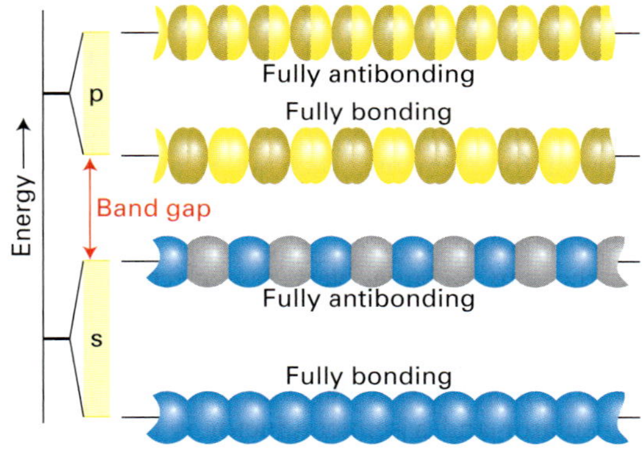

Fig. 17.3 The overlap of s orbitals gives rise to an s band, and the overlap of p-orbitals gives rise to a p band. In this case the s and p orbitals of the atoms are so widely spaced that there is a band gap. In many cases the separation is less, and the bands overlap.

approximation (Section 14.16), the energies of the orbitals are given by

$$E_k = \alpha + 2\beta\cos\left(\frac{k\pi}{N+1}\right) \qquad k = 1, 2, \ldots, N \quad (17.1)$$

where α is approximately equal to (the negative of) the ionization energy of the atom and β is a negative quantity that represents the lowering of energy due to interaction between the atoms. As N becomes infinite, the separation between neighbouring levels, $E_{k+1} - E_k$ goes to zero but, as shown in Derivation 17.1, the width of the band, $E_N - E_1$, becomes 4β, a finite quantity.

Derivation 17.1

The width of a band

The energy of the level with $k = 1$ is

$$E_1 = \alpha + 2\beta\cos\left(\frac{\pi}{N+1}\right)$$

As N becomes infinite, the cosine term becomes cos 0, which is equal to 1. Therefore, in this limit, $E_1 = \alpha + 2\beta$. When k has its maximum value of N,

$$E_N = \alpha + 2\beta\cos\left(\frac{N\pi}{N+1}\right)$$

As N approaches infinity, we can ignore the 1 in the denominator, and the cosine term becomes cos π, which is equal to −1. Therefore, in this limit, $E_N = \alpha - 2\beta$. The difference between the upper and lower energies of the band is therefore 4β.

A band formed from overlap of s orbitals is called an **s band.** If the atoms have p orbitals available, then the same procedure leads to a **p band** (as in the upper half of Fig. 17.3, with different values of α and β in eqn 17.1). If the atomic p orbitals lie higher in energy than the s orbitals, then the p band lies higher than the s band, and there may be a **band gap**, a range of energies for which no molecular orbitals exist. If the separation of the atomic orbitals is not large, the two types of band might overlap.

17.2 The occupation of bands

Now consider the electronic structure of a solid formed from atoms each of which is able to contribute one electron (for example, the alkali metals). There are N atomic orbitals and therefore N molecular orbitals squashed into a band of finite width. There are N electrons to accommodate; they form pairs that occupy the lowest $\frac{1}{2}N$ molecular orbitals (Fig. 17.4). The highest occupied molecular orbital is called the

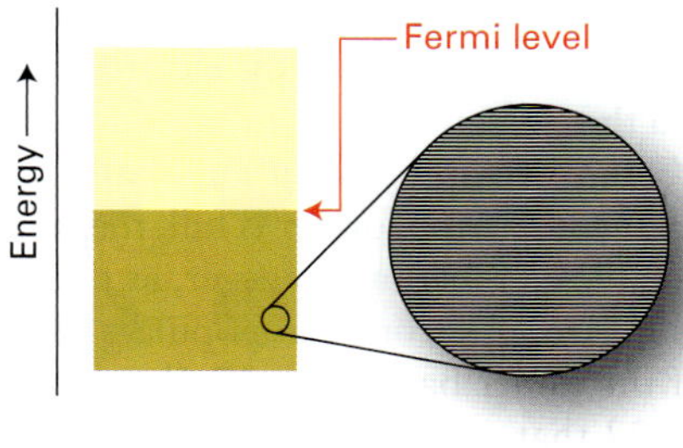

Fig. 17.4 When N electrons occupy a band of N orbitals, it is only half-full and the electrons near the Fermi level (the top of the filled levels) are mobile.

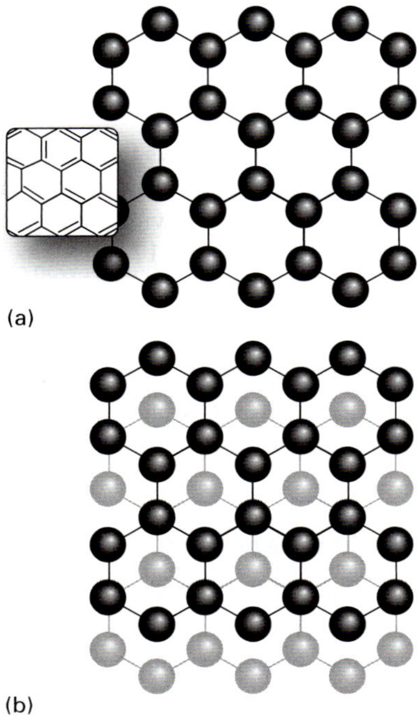

Fig. 17.15 Graphite consists of flat planes of hexagons of carbon atoms lying above one another. (a) The arrangement of carbon atoms in a sheet; (b) the relative arrangement of neighbouring sheets. When impurities are present, the planes can slide over one another easily. Graphite conducts well within the planes but less well perpendicular to the planes.

In graphite, σ bonds between sp^2-hybridized carbon atoms form hexagonal rings that, when repeated throughout a plane, give rise to sheets (Fig. 17.15). Because the sheets can slide against each other (especially when impurities are present), graphite is used widely as a lubricant.

The electrical properties of diamond and graphite are determined by differences in the bonding patterns in these solids. Graphite is an electronic conductor because electrons are free to move through bands formed by the overlap of partially filled, unhybridized p orbitals that are perpendicular to the hexagonal sheets. This band model explains the experimental observation that graphite conducts electricity well within the sheets but less well between them. The electrical conductivity of these 'graphene' (graphite-like) sheets of carbon atoms is now being considered in the design of nanometer-sized electronic devices (Box 17.1). We see from Fig. 17.14 that delocalized π networks are not possible in diamond, which—in contrast to graphite—is an insulator (more precisely, a large-band-gap semiconductor).

17.9 Magnetic properties of solids

The magnetic properties of solids are determined by interactions between the spins of its electrons. Some materials are magnetic and others may become magnetized when placed in an external magnetic field. A bulk sample exposed to a magnetic field of strength $\mathcal{H}$ acquires a **magnetization**, $\mathcal{M}$, which is proportional to $\mathcal{H}$:

$$\mathcal{M} = \chi\mathcal{H} \tag{17.6}$$

where χ is the dimensionless **volume magnetic susceptibility** (Table 17.3). We can think of the magnetization as contributing to the density of lines of force in the material (Fig. 17.16). Materials for which χ is negative are called **diamagnetic** and tend to move out of a magnetic field; the density of lines of force within them is lower than in a vacuum. Those for which χ is positive are called **paramagnetic**; they tend to move into a magnetic field and the density of lines of force within them is greater than in a vacuum.

Box 17.1 Nanowires

A great deal of research effort is now being expended in the fabrication of nanometre-sized assemblies of atoms and molecules that can be used as tiny building blocks in a variety of technological applications. The future economic impact of **nanotechnology**, the aggregate of applications of devices built from nanometre-sized components, could be very significant. For example, increased demand for very small digital electronic devices has driven the design of ever-smaller and more powerful microprocessors. However, there is an upper limit on the density of electronic circuits that can be incorporated into silicon-based chips with current fabrication technologies. As the ability to process data increases with the number of circuits in a chip, it follows that soon chips and the devices that use them will have to become bigger if processing power is to increase indefinitely. One way to circumvent this problem is to fabricate devices from nanometre-sized components. Another advantage of making nanometre-sized electronic devices, or *nanodevices*, is the possibility of using quantum-mechanical effects. For example, electron tunnelling between two conducting regions separated by a thin insulating region can increase the speed of electron conduction and, consequently, the data processing speed in a digital nanoprocessor.

The study of nanodevices can also advance our basic understanding of chemical reactions. Nanometre-sized chemical reactors can serve as laboratories for the study of chemical reactions in constrained environments. Some of these reactions could comprise the foundation for the construction of nanometre-sized chemical sensors, with potential applications in medicine. For example, nanodevices with carefully designed biochemical properties could replace viruses and bacteria as the active species in vaccines.

A number of techniques have already been developed for the fabrication of nanometre-sized structures. The synthesis of *nanowires*, nanometre-sized atomic assemblies that conduct electricity, is a major step in the fabrication of nanodevices. An important type of nanowire is based on *carbon nanotubes*, thin cylinders of carbon atoms that are both mechanically strong and highly conducting. In recent years, methods for selective synthesis of nanotubes have been developed and they consist of different ways to condense a carbon plasma either in the presence or absence of a catalyst. The simplest structural motif is called a *single-walled nanotube* (SWNT) and in shown in the first illustration. In a SWNT, sp^2-hybridized carbon atoms form hexagonal rings reminiscent of the structure of the carbon sheets found in graphite (see the first illustration). The tubes have diameters of between 1 and 2 nm and lengths of several micrometres. The features shown in the illustration have been confirmed by direct visualization with scanning tunneling microscopy. A *multiwalled nanotube* (MWNT) consists of several concentric SWNTs and its diameter varies between 0.4 and 25 nm.

The origin of electrical conductivity in carbon nanotubes is the delocalization of π electrons that occupy unhydridized p orbitals, just as in graphite (Section 17.8). Recent studies have shown a correlation between structure and conductivity in SWNTs. The illustration shows a SWNT that is a semiconductor. If the hexagons are rotated by 60°, the resulting SWNT is a metallic conductor.

Silicon nanowires can be made by focusing a pulsed laser beam onto a solid target composed of silicon and iron. The laser ejects Fe and Si atoms from the surface of the target, forming a vapour phase that can condense into liquid $FeSi_n$ nanoclusters at sufficiently low temperatures. The phase diagram for this complex mixture shows that solid silicon and liquid $FeSi_n$ coexist at temperatures higher than 1473 K. Hence, it is possible to precipitate solid silicon from the mixture if the experimental conditions are controlled to maintain the $FeSi_n$ nanoclusters in a liquid state that is saturated with silicon. It is observed that the silicon precipitate consists of nanowires with diameters of about 10 nm and lengths greater than 1 μm.

Nanowires are also fabricated by *molecular beam epitaxy* (MBE), in which gaseous atoms or molecules are sprayed on to a crystalline surface in an evacuated chamber. Through careful control of the chamber temperature and of the spraying process, it is possible to create nanometre-sized assemblies with specific shapes. For example, the second illustration shows an image of germanium nanowires on a silicon surface. The wires are about 2 nm high, 10–32 nm wide, and 10–600 nm long. It is also possible to deposit *quantum dots*, nanometre-sized boxes or spheres of atoms, on a surface. Semiconducting quantum dots could be important building blocks of nanometre-sized lasers.

Direct manipulation of atoms on a surface also leads to the formation of nanowires. The Coulomb attraction between an atom and the tip of a scanning tunneling microscope (STM, Section 18.2) can be exploited to move atoms along a surface, arranging them into patterns, such as wires.

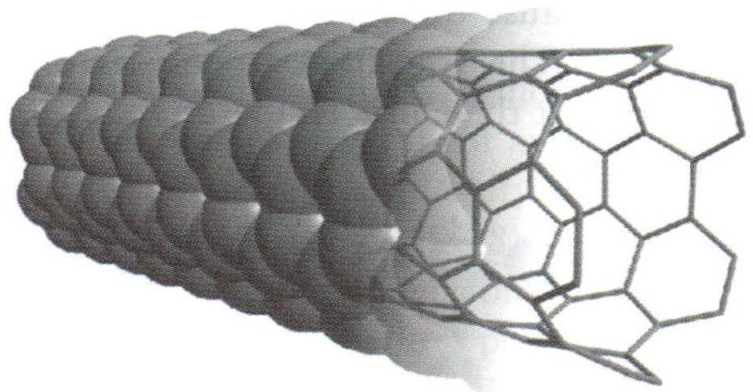

In a single-walled nanotube (SWNT), sp^2-hybridized carbon atoms form hexagonal rings that grow as tubes with diameters between 0.4 and 2 nm and lengths of several micrometres.

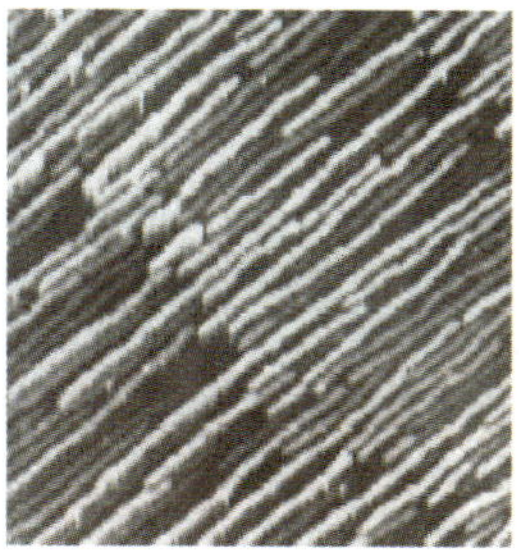

Germanium nanowires fabricated onto a silicon surface by molecular beam epitaxy and imaged by atomic force microscopy. Reproduced with permission from T. Ogino et al. *Acc. Chem. Res.* **32**, 447 (1999).

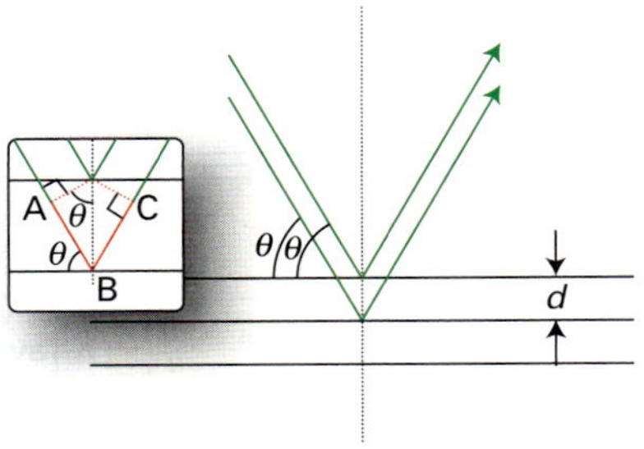

Fig. 17.29 The derivation of Bragg's law treats each lattice plane as reflecting the incident radiation. The path lengths differ by AB + BC, which depends on the glancing angle θ. Constructive interference (a 'reflection') occurs when AB + BC is equal to an integral number of wavelengths.

The path-length difference of the two rays shown in the illustration is

$$AB + BC = 2d \sin \theta$$

where θ is the **glancing angle**. When the path-length difference is equal to one wavelength ($AB + BC = \lambda$), the reflected waves interfere constructively. It follows that a reflection should be observed when the glancing angle satisfies **Bragg's law:**

$$\lambda = 2d \sin \theta \qquad (17.8)$$

The primary use of Bragg's law is to determine the spacing between the layers of atoms, for once the angle θ corresponding to a reflection has been determined, d may readily be calculated.

A note on good practice You will often see Bragg's law in the form $n\lambda = 2d \sin \theta$, where n, an integer, is the 'order' of the diffraction. The modern tendency is to omit n and to ascribe the diffraction to planes of separation d/n instead (recall the discussion in Example 17.2).

Example 17.3

Using Bragg's law

A reflection from the (111) planes of a cubic crystal was observed at a glancing angle of 11.2° when Cu K_α X-rays of wavelength 154 pm were used. What is the length of the side of the unit cell?

Strategy We can find the separation, d, of the lattice planes from eqn 17.9 and the data. Then we find the length of the side of the unit cell by using eqn 17.8. Because the unit cell is cubic, $a = b = c$, so eqn 17.8 simplifies to

$$\frac{1}{d^2} = \frac{h^2}{a^2} + \frac{k^2}{a^2} + \frac{l^2}{a^2} = \frac{h^2 + k^2 + l^2}{a^2}$$

which rearranges to $a^2 = d^2 \times (h^2 + k^2 + l^2)$ and therefore to

$$a = d \times (h^2 + k^2 + l^2)^{1/2}$$

Solution According to Bragg's law, the separation of the (111) planes responsible for the diffraction is

$$d = \frac{\lambda}{2 \sin \theta} = \frac{154 \text{ pm}}{2 \sin 11.2°}$$

It then follows that with $h = k = l = 1$,

$$a = \frac{154 \text{ pm}}{2 \sin 11.2°} \times 3^{1/2} = 687 \text{ pm}$$

Self-test 17.6

Calculate the angle at which the same lattice will give a reflection from the (123) planes.

[*Answer:* 24.8°]

17.14 Experimental techniques

Laue's original method consisted of passing a beam of X-rays of a wide range of wavelengths into a single crystal and recording the diffraction pattern photographically. The idea behind the approach was that a crystal might not be suitably orientated to act as a diffraction grating for a single wavelength, but whatever its orientation Bragg's law would be satisfied for at least one of the wavelengths when a range of wavelengths is present in the beam.

An alternative technique was developed by Peter Debye and Paul Scherrer and independently by Albert Hull. They used monochromatic (single-frequency) X-rays and a powdered sample. When the sample is a powder, we can be sure that some of the randomly distributed crystallites will be orientated so as to satisfy Bragg's law. For example, some of them will be orientated so that their (111) planes, of spacing d, give rise to a reflection at a particular angle and others will be orientated so that their (230) planes give rise to a reflection at a different angle. Each set of (hkl) planes gives rise to reflections at a different angle. In the modern version of the technique, which uses a **powder diffractometer**, the sample is spread on a flat plate and the diffraction pattern is monitored electronically. The major application is for qualitative analysis because the diffraction pattern is a kind of fingerprint and may be recognizable by reference to a library of patterns (Fig. 17.30). The technique is also used for the characterization of substances that cannot be crystallized or for the initial determination of the dimensions and symmetries of unit cells.

Modern X-ray diffraction, which utilizes an **X-ray diffractometer** (Fig. 17.31), is now a highly sophisticated technique. By far the most detailed information

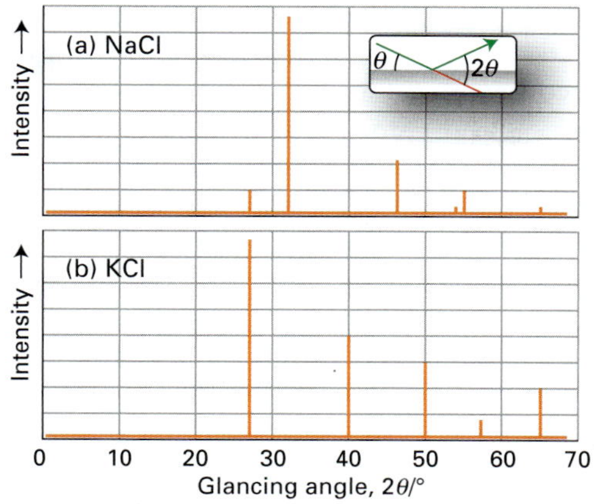

Fig. 17.30 Typical X-ray powder diffraction patterns (for (a) sodium chloride, (b) potassium chloride) that can be used to identify the material and determine the size of its unit cell.

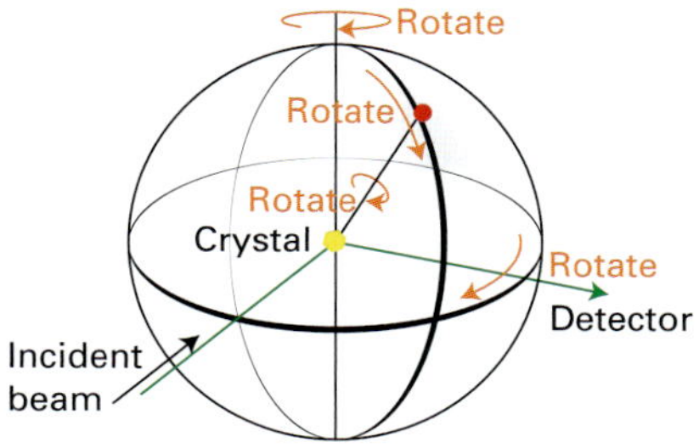

Fig. 17.31 A four-circle diffractometer. The settings of the orientations of the components is controlled by computer; each reflection is monitored in turn, and their intensities are recorded.

comes from developments of the techniques pioneered by the Braggs, in which a single crystal is employed as the diffracting object and a monochromatic beam of X-rays is used to generate the diffraction pattern. The single crystal, which may be only a fraction of a millimetre in length, is rotated relative to the beam and the diffraction pattern is monitored and recorded electronically for each crystal orientation. The primary data is therefore a set of intensities arising from the Miller planes (hkl), with each set of planes giving a reflection of intensity I_{hkl}. For our purposes, we focus on the ($h00$) planes and write the intensities I_h.

To derive the structure of the crystal from the intensities we need to convert them to the *amplitude* of the wave responsible for the signal. Because the intensity of electromagnetic radiation is given by the square of the amplitude, we need to form the **structure factors** $F_h = (I_h)^{1/2}$. Here is the first difficulty: we do not know the sign to take. For instance, if $I_h = 4$, then F_h can be either +2 or −2. This ambiguity is the **phase problem** of X-ray diffraction. However, once we have the structure factors, we can calculate the electron density $\rho(x)$ by forming the following sum:

$$\rho(x) = \frac{1}{V}\left\{F_0 + 2\sum_{h=1}^{\infty} F_h \cos(2h\pi x)\right\} \qquad (17.9)$$

where V is the volume of the unit cell. This expression is called a **Fourier synthesis** of the electron density: we show how it is used in Example 17.4. The point to note is that low values of the index h give the major features of the structure (they correspond to long-wavelength cosine terms) whereas the high values give the fine detail (short-wavelength cosine terms). Clearly, if we do not know the sign of F_h, we do not know whether the corresponding term in the sum is positive or negative and we get different electron densities, and hence crystal structures, for different choices of sign.

Example 17.4

Constructing the electron density

The following intensities were obtained in an experiment on an organic solid:

h	0	1	2	3	4	5	6	7	8	9
I_h	256	100	5	1	50	100	8	10	5	10

h	10	11	12	13	14	15
I_h	40	25	9	4	4	9

Construct the electron density along the x direction.

Strategy Begin by finding the structure factors from the corresponding values of I_h by using $F_h = (I_h)^{1/2}$. Then use eqn 17.19 to plot the electron density as $V\rho(x)$ against x. However, because F_h can be either positive or negative, you will need to make guesses about the signs of F_h, generate different plots for different guesses, and assess the plausibility of each guess.

Solution To find the structure factors, we take square-roots of the intensities:

h	0	1	2	3	4	5	6	7	8	9
F_h	±16.0	±10	±2.2	±1	±7.1	±10	±2.8	±3.2	±2.2	±3.2

h	10	11	12	13	14	15
F_h	±6.3	±5	±3	±2	±2	±3

Suppose we guess that the signs alternate + − + − ⋯ ; then according to eqn 17.19 the electron density is

$$V\rho(x) = 16 - 20\cos(2\pi x) + 4.4\cos(4\pi x) - \cdots - 6\cos(30\pi x)$$

This function is shown in Fig. 17.32 as the green line, and the locations of several types of atom are easy to identify as peaks in the electron density. If we guess + signs up to $h = 5$ and − signs thereafter; the electron density is

$$V\rho(x) = 16 + 20\cos(2\pi x) + 4.4\cos(4\pi x) + \cdots - 6\cos(30\pi x)$$

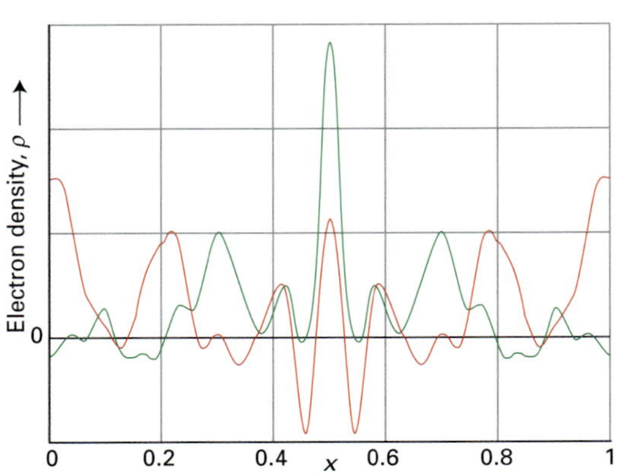

Fig. 17.32 The Fourier synthesis of the electron density of a one-dimensional crystal using the data in the illustration. Green: using alternating signs for the structure factors; red: using positive signs for h up to 5, then negative signs.

This density is shown in Fig. 17.32 as the red line. This structure has more regions of illegal negative electron density, so is less plausible than the structure obtained from the first choice of phases.

Self-test 17.7

In an X-ray investigation, the following structure factors were determined. Construct the electron density along the corresponding direction.

h	0	1	2	3	4	5	6	7	8	9
F_{h00}	10	−10	8	−8	6	−6	4	−4	2	−2

[*Answer:* Fig. 17.33]

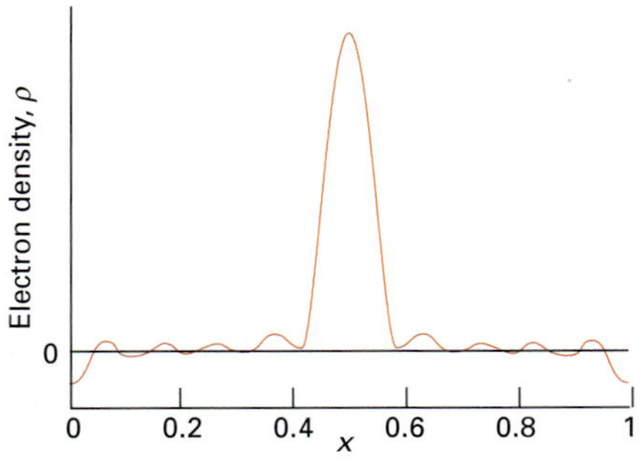

Fig. 17.33 The electron density calculated from data given in Self-test 17.7.

The phase problem can be overcome to some extent by the method of **isomorphous replacement**, in which heavy atoms are introduced into the crystal. The technique relies on the fact that the scattering of X-rays is caused by the oscillations an incoming electromagnetic wave generates in the electrons of atoms, and heavy atoms give rise to stronger scattering than light atoms. So heavy atoms dominate the diffraction pattern and greatly simplify its interpretation. The phase problem can also be resolved by judging whether the calculated structure is chemically plausible, whether the electron density is positive throughout, and by using more refined mathematical techniques. Huge numbers of crystal structures have been determined in this way. In the following sections we review how some of them can be rationalized. For metals and monatomic ions we can model the atoms and ions as hard spheres, and consider how such spheres can be stacked together in a regular, electrically neutral array.

17.15 Metal crystals

Most metallic elements crystallize in one of three simple forms, two of which can be explained in terms of stacking spheres to give the closest possible packing. In such **close-packed structures** the spheres representing the atoms are packed together with least waste of space and each sphere has the greatest possible number of nearest neighbours.

We can form a close-packed layer of identical spheres, one with maximum utilization of space, as shown in Fig. 17.34a. Then we can form a second close-packed layer by placing spheres in the depressions of the first layer (Fig. 17.34b). The third layer may be added in either of two ways, both of which result in the same degree of close packing. In one, the spheres are placed so that they reproduce the first layer (Fig. 17.34c), to give an ABA pattern of layers. Alternatively, the spheres may be placed over the gaps in the first layer (Fig. 17.34d), so giving an ABC pattern.

Two types of structures are formed if the two stacking patterns are repeated. The spheres are **hexagonally close-packed** (hcp) if the ABA pattern is repeated to give the sequence of layers ABABAB.... The name reflects the symmetry of the unit cell (Fig. 17.35). Metals with hcp structures include beryllium, cadmium, cobalt, manganese, titanium, and zinc. Solid helium (which forms only under pressure) also adopts this arrangement of atoms. Alternatively, the spheres are **cubic close-packed** (ccp) if the ABC pattern is repeated to give the sequence of layers ABCABC.... Here too, the name reflects the symmetry of the unit cell (Fig. 17.36). Metals with this structure include silver, aluminium, gold, calcium, copper, nickel, lead, and platinum. The noble gases other than helium also adopt a ccp structure.

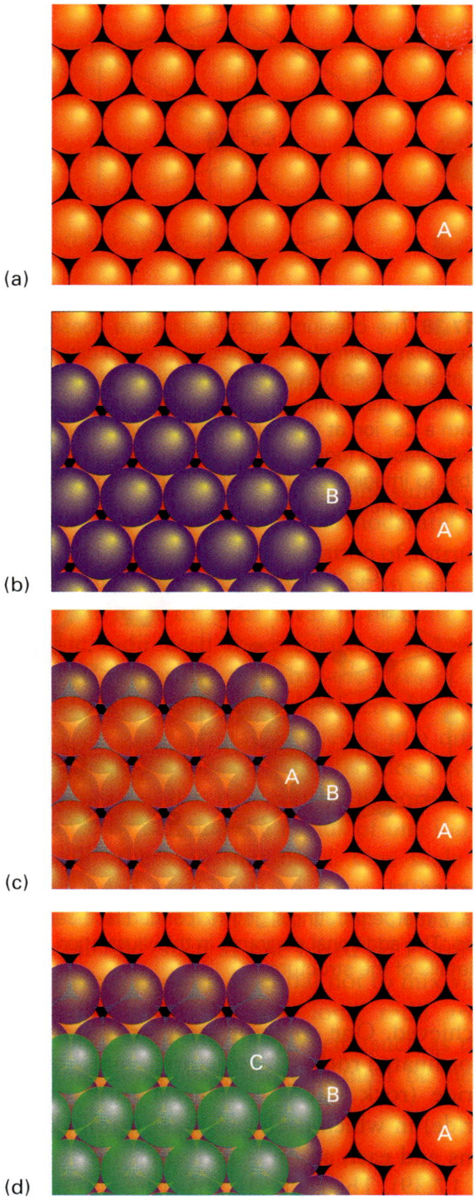

Fig. 17.34 The close-packing of identical spheres. (a) The first layer of close-packed spheres. (b) The second layer of close-packed spheres occupies the dips of the first layer. The two layers are the AB component of the structure. (c) The third layer of close-packed spheres might occupy the dips lying directly above the spheres in the first layer, resulting in an ABA structure. (d) Alternatively, the third layer might lie in the dips that are not above the spheres in the first layer, resulting in an ABC structure.

The compactness of the ccp and hcp structures is indicated by their **coordination number**, the number of atoms immediately surrounding any selected atom, which is 12 in both cases. Another measure of their

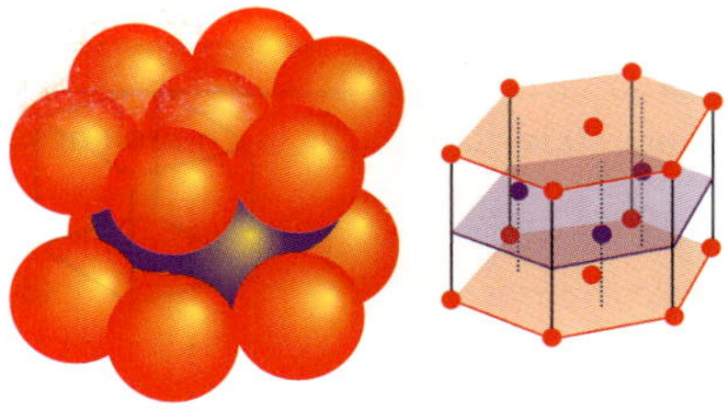

Fig. 17.35 A hexagonal close-packed structure. The tinting of the spheres (denoting the three layers of atoms) is the same as in Fig. 17.33.

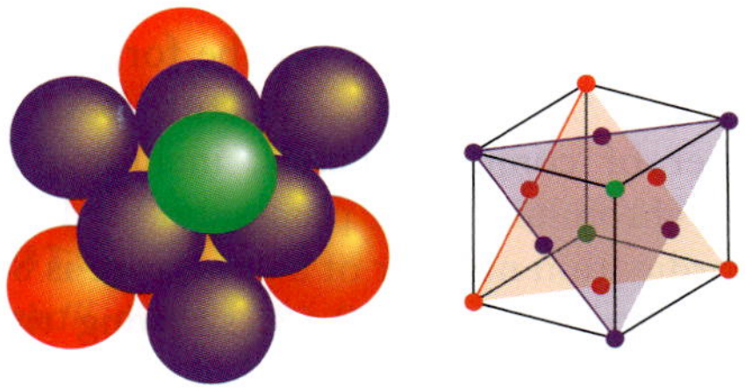

Fig. 17.36 A cubic close-packed structure. The tinting of the spheres is the same as in Fig. 17.33.

compactness is the **packing fraction**, the fraction of space occupied by the spheres, which is 0.740. That is, in a close-packed solid of identical hard spheres, 74.0 per cent of the available space is occupied and only 26.0 per cent of the total volume is empty space.

The fact that many metals are close-packed accounts for one of their common characteristics, their high density. However, there is a difference between ccp and hcp metals. In cubic close packing, the faces of the cubes extend throughout the solid, and give rise to a **slip plane**. Careful analysis of the ccp structure shows that there are eight slip planes in various orientations, whereas an hcp structure has only one set of slip planes. When the metal is under stress, the layers of atoms may slip past one another along a slip plane. Because a ccp metal has more slip planes than an hcp metal, a ccp metal is more malleable than an hcp metal. Thus, copper, which is ccp, is highly malleable, but zinc, which is hcp, is more brittle. It must be borne in mind, however, that metals in real use are not single crystals: they are polycrystalline, with numerous grain-like regions and defects that permeate the structure. Much of metallurgy is associated with the control of the density of grains and grain boundaries.

A number of common metals adopt structures that are not close-packed, which suggests that directional covalent bonding between neighbouring atoms is beginning to influence the structure and impose a specific geometrical arrangement. One such arrangement results in a **body-centred cubic** (bcc) lattice, with

Table of key equations

The following table summarizes the equations developed in this chapter.

Property	Equation	Comment
Born–Mayer equation	$\Delta H_L^{\ominus} = \lvert z_1 z_2 \rvert \times (N_A e^2/4\pi\varepsilon_0 d) \times (1 - d^*/d) \times A$	Coulomb interactions dominant
Magnetization of a bulk sample exposed to a magnetic field	$\mathcal{M} = \chi \mathcal{H}$	
Separation d of neighbouring planes	$1/d^2 = h^2/a^2 + k^2/b^2 + l^2/c^2$	Orthogonal lattice
Bragg's law	$\lambda = 2d \sin\theta$	
Radius ratio	$\gamma = r_{smaller}/r_{larger}$	

Questions and exercises

Discussion questions

17.1 Explain how metallic conductors, semiconductors, and insulators are identified and explain their properties in terms of band theory. Why is graphite an electronic conductor and diamond an insulator?

17.2 Explain how planes of lattice points are labelled.

17.3 Describe the consequences of the phase problem in determining structure factors and how the problem is overcome.

17.4 Describe the structures of elemental metallic solids in terms of the packing of hard spheres.

17.5 Describe the caesium-chloride and rock-salt structures. How does the radius-ratio rule help in the classification of a structure into each type?

17.6 Describe the different types of magnetism that materials can display and account for their origins.

Exercises

17.1 Classify as n-type or p-type a semiconductor formed by doping (a) germanium with phosphorus, (b) germanium with indium.

17.2 The electrical resistance of a sample increased from 100 Ω to 120 Ω when the temperature was changed from 0°C to 100°C. Is the substance a metallic conductor or a semiconductor?

17.3 The energy levels of N atoms in the Hückel approximation are given by (eqn 17.1):

$$E_k = \alpha + 2\beta\cos\frac{k\pi}{N+1} \qquad k = 1, 2, \ldots, N$$

If the atoms are arranged in a ring, the energy levels are given by:

$$E_k = \alpha + 2\beta\cos\frac{2k\pi}{N} \qquad k = 0, \pm1, \pm2, \ldots, \pm\tfrac{1}{2}N$$

(for N even). Discuss the consequences, if any, of joining the ends of an initially straight length of material.

17.4 The tip of a scanning tunnelling microscope can be used to move atoms on a surface. The movement of atoms and ions depends on their ability to leave one position and stick to another, and therefore on the energy changes that occur. As an illustration, consider a two-dimensional square lattice of univalent positive and negative ions separated by 200 pm, and consider a cation on top of this array. Calculate, by direct summation, its Coulombic interaction when it is in an empty lattice point directly above and anion.

17.5 Describe the bonding in magnesium oxide, CaO, in terms of bands composed of Ca and O atomic orbitals. How does this model justify the ionic model of this compound?

17.6 Calculate the lattice enthalpy of CaO from the following data:

	$\Delta H/(\text{kJ mol}^{-1})$
Sublimation of Ca(s)	+178
Ionization of Ca(g) to Ca^{2+}(g)	+1735
Dissociation of O_2(g)	+249
Electron attachment to O(g)	−141
Electron attachment to O^-(g)	+844
Formation of CaO(s) from Ca(s) and O_2(g)	−635

17.7 Calculate the lattice enthalpy of SrI_2 from the following data:

	$\Delta H/(\text{kJ mol}^{-1})$
Sublimation of Sr(s)	+164.4
Ionization of Sr(g) to Sr^{2+}(g)	+1626.1
Sublimation of I_2(s)	+62.4
Dissociation of I_2(g)	+75.6
Electron attachment to I(g)	−303.8
Formation of SrI_2(s) from Sr(s) and I_2(s)	−828.9

17.8 Calculate the potential energy of an ion at the centre of a diffuse 'spherical crystal' in which concentric spheres of ions of opposite charge surround the ion and the numbers of ions on the spherical surfaces fall away rapidly with distance. Let successive spheres lie at radii *d*, 2*d*, ... and the number of ions (all of the same charge) on each successive sphere is inversely proportional to the radius of the sphere. You will need the following sum:

$$1-\frac{1}{2^2}+\frac{1}{3^2}-\frac{1}{4^2}+\cdots=\frac{\pi^2}{12}$$

17.9 Estimate the ratio of the lattice enthalpies of SrO and CaO from the Born–Meyer equation by using the ionic radii in Table 15.4.

17.10 Type I superconductors show abrupt loss of superconductivity when an applied magnetic field exceeds a critical value $\mathcal{H}_c$ that depends on temperature and T_c as

$$\mathcal{H}_c(T)=\mathcal{H}_c(0)\left(1-\frac{T^2}{T_c^2}\right)$$

where $\mathcal{H}_c(0)$ is the value of $\mathcal{H}_c$ as $T\to 0$. Lead has $T_c = 7.19$ K and $\mathcal{H}_c = 63.9$ kA m^{-1}. At what temperature does lead become superconducting in a magnetic field of 20.0 kA m^{-1}?

17.11 Draw a set of points as a rectangular array based on unit cells of side *a* and *b*, and mark the planes with Miller indices (10), (01), (11), (12), (23), (41), (41).

17.12 Repeat Exercise 17.11 for an array of points in which the *a* and *b* axes make 60° to each other.

17.13 In a certain unit cell, planes cut through the crystal axes at (2*a*, 3*b*, *c*), (*a*, *b*, *c*), (6*a*, 3*b*, 3*c*), (2*a*, −3*b*, −3*c*). Identify the Miller indices of the planes.

17.14 Draw an orthorhombic unit cell and mark on it the (100), (010), (001), (011), (101), and (101) planes.

17.15 Draw a triclinic unit cell and mark on it the (100), (010), (001), (011), (101), and (101) planes.

17.16 Calculate the separations of the planes (111), (211), and (100) in a crystal in which the cubic unit cell has sides of length 572 pm.

17.17 Calculate the separations of the planes (123) and (236) in an orthorhombic crystal in which the unit cell has sides of lengths 784, 633, and 454 pm.

17.18 The glancing angle of a Bragg reflection from a set of crystal planes separated by 97.3 pm is 19.85°. Calculate the wavelength of the X-rays.

17.19 The separation of (100) planes of lithium metal is 350 pm and its density is 0.53 g cm^{-3}. Is the structure of lithium fcc or bcc?

17.20 Copper crystallizes in an fcc structure with unit cells of side 361 pm. (a) Predict the appearance of the powder diffraction pattern using 154 pm radiation. (b) Calculate the density of copper on the basis of this information.

17.21 Construct the electron density along the *x*-axis of a crystal given the following structure factors:

h	0	1	2	3	4	5	6	7	8	9
F_h	+30.0	+8.2	+6.5	+4.1	+5.5	−2.4	+5.4	+3.2	−1.0	+1.1

h	10	11	12	13	14	15
F_h	+6.5	+5.2	−4.3	−1.2	+0.1	+2.1

17.22 Calculate the packing fraction of a stack of cylinders.

17.23 Calculate the packing fraction of a cubic close-packed structure.

17.24 Suppose a virus can be regarded as a sphere and that it stacks together in a hexagonal close-packed arrangement. If the density of the virus is the same as that of water (1.00 g cm^{-3}), what is the density of the solid?

17.25 How many (a) nearest neighbours, (b) next-nearest neighbours are there in a body-centred cubic structure? What are their distances if the side of the cube is 600 nm?

17.26 How many (a) nearest neighbours, (b) next-nearest neighbours are there in a cubic close-packed structure? What are their distances if the side of the cube is 600 nm?

17.27 The thermal and mechanical processing of materials is an important step in ensuring that they have the appropriate physical properties for their intended application. Suppose a metallic element underwent a phase transition in which its crystal structure changed from cubic close-packed to body-centred cubic. (a) Would it become more or less dense? (b) By what factor would its density change?

17.28 The compound Rb_3TlF_6 has a tetragonal unit cell with dimensions $a = 651$ pm and $c = 934$ pm. Calculate the volume of the unit cell and the density of the solid.

17.29 The orthorhombic unit cell of $NiSO_4$ has the dimensions $a = 634$ pm, $b = 784$ pm, and $c = 516$ pm, and the density of the solid is estimated as 3.9 g cm^{-3}. Determine the number of formula units per unit cell and calculate a more precise value of the density.

17.30 The unit cells of $SbCl_3$ are orthorhombic with dimensions $a = 812$ pm, $b = 947$ pm, and $c = 637$ pm. Calculate the spacing of (a) the (321) planes, (b) the (642) planes.

17.31 Use the radius-ratio rule to predict the kind of crystal structure expected for magnesium oxide.

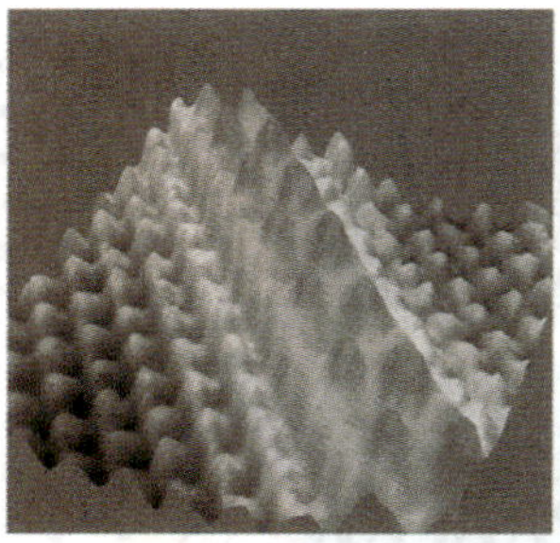

Fig. 18.12 An STM image of caesium atoms on a gallium arsenide surface.

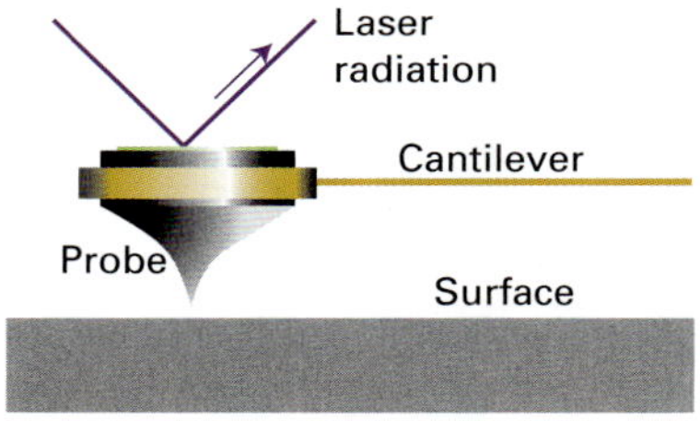

Fig. 18.14 In atomic force microscopy, a laser beam is used to monitor the tiny changes in position of a probe as it is attracted to or repelled from atoms on a surface.

monitored. Because the tunnelling probability is very sensitive to the size of the gap, the microscope can detect tiny, atom-scale variations in the height of the surface. An example of the kind of image obtained with a clean surface is shown in Fig. 18.12, where the cliff is only one atom high. Figure 18.13 shows the dissociation of SiH_3 adsorbed onto a Si(001) surface into adsorbed SiH_2 units and H atoms. The tip of the STM can also be used to manipulate adsorbed atoms on a surface, making possible the fabrication of complex and yet very tiny structures, such as nanometre-sized electronic devices.

In *atomic force microscopy* (AFM) a sharpened stylus attached to a beam is scanned across the surface. The force exerted by the surface and any adsorbate pushes or pulls on the stylus and deflects the beam (Fig. 18.14). The deflection is monitored by using a laser beam. Because no current is needed between the sample and the probe, the technique can be applied to nonconducting surfaces too. A spectacular demonstration of the power of AFM is given in Fig. 18.15, which shows individual DNA molecules on a solid surface.

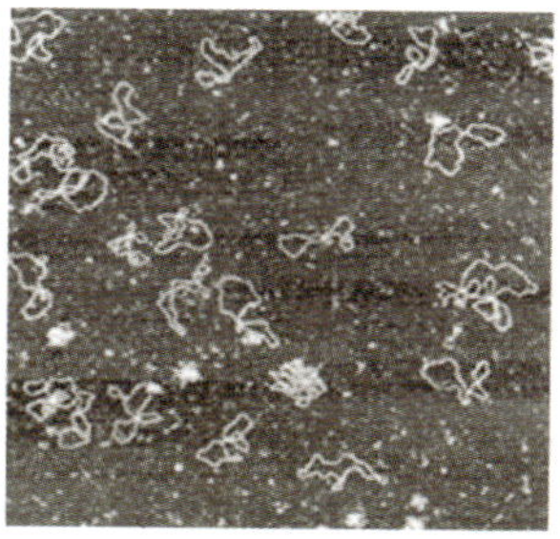

Fig. 18.15 An AFM image of bacterial DNA plasmids on a mica surface. (Courtesy of Veeco Instruments.)

The extent of adsorption

The extent of surface coverage is normally expressed as the **fractional coverage,** θ (theta):

$$\theta = \frac{\text{number of adsorption sites occupied}}{\text{number of adsorption sites available}} \qquad (18.2)$$

The fractional coverage can be inferred from the volume of adsorbate adsorbed by $\theta = V/V_\infty$, where

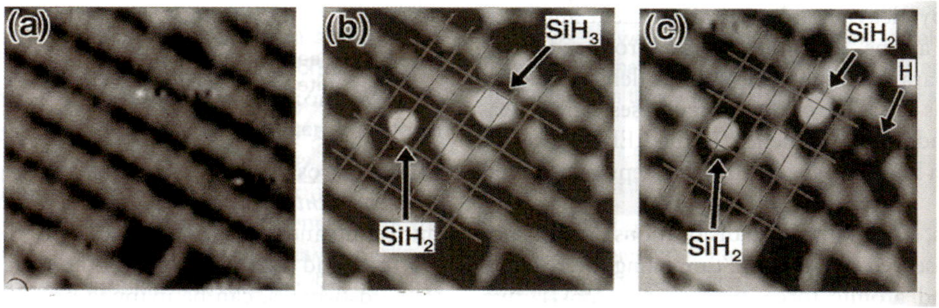

Fig. 18.13 Visualization by STM of the reaction $SiH_3 \rightarrow SiH_2 + H$ on a 4.7 nm × 4.7 nm area of a Si(001) surface. (a) The Si(001) surface before exposure to Si_2H_6(g). (b) Adsorbed Si_2H_6 dissociates into SiH_2(surface), on the left of the image, and SiH_3(surface), on the right. (c) After 8 min, SiH_3(surface) dissociates to SiH_2(surface) and H(surface). Reproduced with permission from Y. Wang, M.J. Bronikowski, and R.J. Hamers, *Surface Science* **64**, 311 (1994).

V_∞ is the volume of adsorbate corresponding to complete monolayer coverage. In each case, the volumes in the definition of θ are those of the free gas measured under the same conditions of temperature and pressure, not the volume the adsorbed gas occupies when attached to the surface. The **rate of adsorption** is the rate of change of surface coverage and is measured by observing the change of fractional coverage with time.

Among the principal techniques for measuring the rate of desorption are flow methods, in which the sample itself acts as a pump because adsorption removes molecules from the gas. One commonly used technique is therefore to monitor the rates of flow of gas into and out of the system: the difference is the rate of gas uptake by the sample. In **flash desorption** the sample is suddenly heated (electrically) and the resulting rise of pressure is interpreted in terms of the amount of adsorbate originally on the sample. The interpretation may be confused by the desorption of a compound (for example, WO_3 from oxygen on tungsten). **Surface plasmon resonance** (SPR) is a technique in which the kinetics and thermodynamics of surface processes, particularly of biological systems, are monitored by detecting the effect of adsorption and desorption on the refractive index of a gold substrate. **Gravimetry**, in which the sample is weighed on a microbalance during the experiment, can also be used. A common instrument for gravimetric measurements is the **quartz crystal microbalance** (QCM), in which the mass of a sample adsorbed to the surface of a quartz crystal is related to changes in the latter's mechanical properties. The key principle behind the operation of a QCM is the ability of a quartz crystal to vibrate at a characteristic frequency when an oscillating electric field is applied. The vibrational frequency decreases when material is spread over the surface of the crystal and the change in frequency is proportional to the mass of material. Masses as small as a few nanograms ($1\ \mathrm{ng} = 10^{-9}\ \mathrm{g}$) can be measured reliably in this way.

18.3 Physisorption and chemisorption

Molecules and atoms can attach to surfaces in two ways, although there is no clear frontier between the two types of adsorption. In **physisorption** (an abbreviation of 'physical adsorption'), there is a van der Waals interaction between the adsorbate and the substrate (for example, a dispersion or a dipolar interaction of the kind responsible for the condensation of vapours to liquids). The energy released when a molecule is physisorbed is of the same order of magnitude as the enthalpy of condensation. Such small energies can be absorbed as vibrations of the lattice and dissipated as thermal motion, and a molecule bouncing across the surface will gradually lose its energy and finally adsorb to it in the process called **accommodation**. The enthalpy of physisorption can be measured by monitoring the rise in temperature of a sample of known heat capacity, and typical values are in the region of $-20\ \mathrm{kJ\ mol^{-1}}$ (Table 18.1). This small enthalpy change is insufficient to lead to bond breaking, so a physisorbed molecule retains its identity but might be distorted. Enthalpies of physisorption may also be measured by observing the temperature dependence of the parameters that occur in the adsorption isotherm (Section 18.4).

In **chemisorption** (an abbreviation of 'chemical adsorption'), the molecules (or atoms) adsorb to the surface by forming a chemical (usually covalent) bond and tend to find sites that maximize their coordination number with the substrate. The enthalpy of chemisorption is much more negative than that for physisorption, and typical values are in the region of $-200\ \mathrm{kJ\ mol^{-1}}$ (Table 18.2). The distance between

Table 18.1

Maximum observed enthalpies of physisorption, $\Delta_{ads}H^{\ominus}_{ads}/(\mathrm{kJ\ mol^{-1}})$

CH_4	−21
CO	−25
H_2	−84
H_2O	−59
N_2	−21
NH_3	−38
O_2	−21

Table 18.2

Enthalpies of chemisorption, $\Delta_{ads}H^{\ominus}_{ads}/(\mathrm{kJ\ mol^{-1}})$

Adsorbate	Adsorbent (substrate)			
	Cr	Fe	Ni	Pt
C_2H_4	−427	−285	−243	
CO		−192		
H_2	−188	−134		
NH_3		−188	−155	
O_2				−293

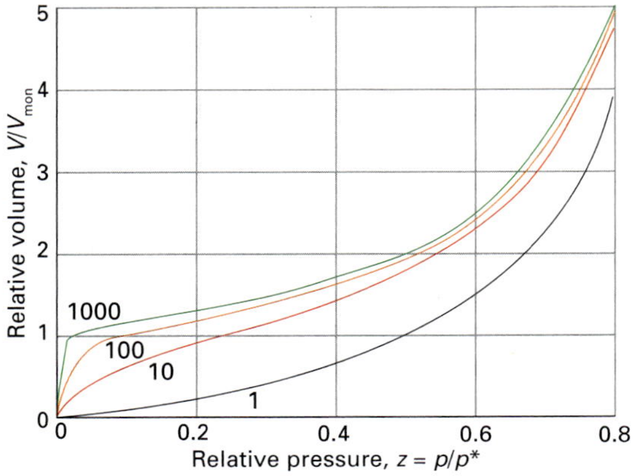

Fig. 18.21 Plots of the BET isotherm for different values of c. The value of V/V_{mon} rises indefinitely because the adsorbate may condense on the covered substrate surface.

becomes important and the extent of coverage rises without limit. A BET isotherm is not accurate at all pressures, but it is widely used in industry to determine the surface areas of solids. When $c \gg 1$ and $cz \gg 1$, which is the case when the enthalpy of desorption from the substrate is very high, the BET isotherm takes the form

$$\frac{V}{V_{mon}} \approx \frac{cz}{(1-z)cz} = \frac{1}{1-z} \qquad (18.10)$$

This expression is applicable to unreactive gases on polar surfaces, for which $c \approx 10^2$.

The BET isotherm is reasonably reliable in the range $0.8 < \theta < 2$, and so provides a reasonably reliable technique for measuring the surface area of solids (corresponding to $\theta = 1$). The adsorbate is typically nitrogen gas.

Example 18.3

Using the BET isotherm to determine the area of a surface

The amount of N_2 adsorbed on 0.30 g of silica at 77 K (the normal boiling point of nitrogen) was determined by measuring the volume adsorbed and then using the perfect gas law to calculate the amount. The following values were obtained:

p/Torr	100	200	300	400
n/mmol	0.90	1.10	1.40	1.90

Determine the value of c and the number of adsorption sites on the sample.

Strategy To use the BET isotherm, we first take the reciprocal of both sides of eqn 18.7a:

$$\frac{V_{mon}}{V} = \frac{(1-z)\{1-(1-c)z\}}{cz} = \frac{(1-z)-(1-z)(1-c)z}{cz} = \frac{1-z}{cz} - \frac{(1-c)(1-z)}{c}$$

The ratio V_{mon}/V can be set equal to n_{mol}/n. Then we multiply both sides by $z/n_{mon}(1-z)$, to obtain

$$\underbrace{\frac{z}{(1-z)n}}_{y} = \underbrace{\frac{1}{cn_{mon}}}_{\text{intercept}} - \underbrace{\frac{(1-c)}{cn_{mon}}}_{\text{slope}} \times z \qquad (18.11)$$

Use $z = p/p^*$, with $p^* = 760$ Torr (because the vapour pressure of a substance at its normal boiling point is 1 atm). This expression is the equation of a straight line when the left-hand side is plotted against $1/z$, with an intercept at $1/cn_{mon}$ and a slope $(c-1)/cn_{mon}$. From the intercept and slope the values of c and n_{mon} can be determined.

Solution Draw up the following table:

p/Torr	100	200	300	400
$z = p/p^*$	0.132	0.263	0.395	0.526
$z/(1-z)(n/\text{mmol})$	0.17	0.32	0.47	0.58

Figure 18.22 shows a plot of the data. The intercept is at 0.039, so $1/c(n_{mon}/\text{mmol}) = 0.039$ and therefore $cn_{mon} = 15$ mmol, The slope is 1.1, so $(c-1)/c(n_{mon}/\text{mmol}) = 1.1$, and therefore

$$c - 1 = 1.1 \times 15 = 16$$

and hence $c = 17$ and $n_{mon} = (15\text{ mmol})/16 = 0.94$ mmol. This amount corresponds to

$$N_{mon} = n_{mon}N_A = (9.4 \times 10^{-4}\text{ mol}) \times (6.022 \times 10^{23}\text{ mol}^{-1}) = 5.7 \times 10^{20}$$

adsorption sites.

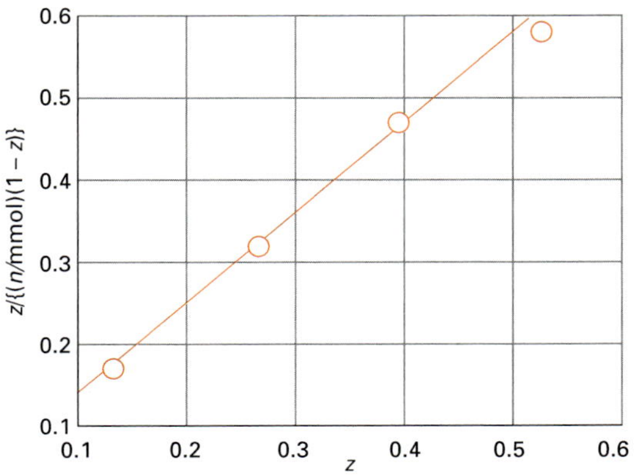

Fig. 18.22 The plot of the data in Example 18.3.

Self-test 18.4

Repeat the analysis using the following data for a different sample of silica:

p/Torr	100	150	200	250	300	350
n/mmol	1.28	1.55	1.79	2.05	2.33	2.67

[*Answer:* n_{mon} = 1.54 mmol, N_{mon} = 9.27 × 10^{20}]

18.5 The rates of surface processes

Figure 18.23 shows how the potential energy of a molecule varies with its distance above the adsorption site. As the molecule approaches the surface its potential energy decreases as it becomes physisorbed into the **precursor state** for chemisorption. Dissociation into fragments often takes place as a molecule moves into its chemisorbed state, and after an initial increase of energy as the bonds stretch there is a sharp decrease as the adsorbate–substrate bonds reach their full strength. Even if the molecule does not fragment, there is likely to be an initial increase of potential energy as the bonds adjust when the molecule approaches the surface.

In most cases, therefore, we can expect there to be a potential energy barrier separating the precursor and chemisorbed states. This barrier, though, might be low and might not rise above the energy of a distant, stationary molecule (as in Fig. 18.23a). In this case, chemisorption is not an activated process and can be expected to be rapid. Many gas adsorptions on clean metals appear to be nonactivated. In some cases the barrier rises above the zero axis (as in Fig 18.23b); such chemisorptions are activated and slower than the nonactivated kind. An example is the adsorption of H_2 on copper, which has an activation energy in the region of 20–40 kJ mol^{-1}.

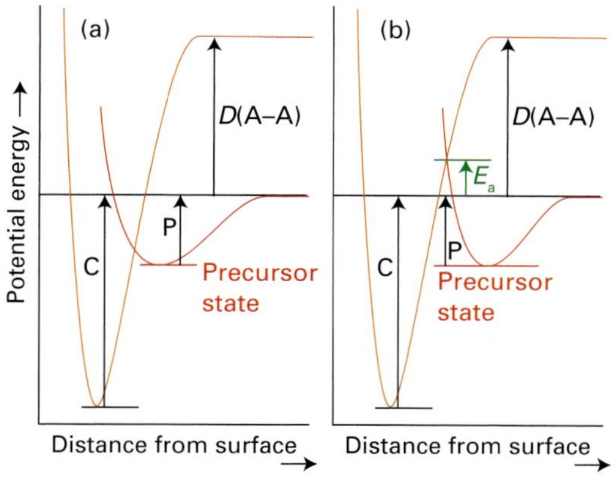

Fig. 18.23 The potential energy profiles for the dissociative chemisorption of an A_2 molecule. In each case, P is the enthalpy of (nondissociative) physisorption and C that for chemisorption (at $T = 0$). The relative locations of the curves determine whether the chemisorption is (a) not activated or (b) activated.

One point that emerges from this discussion is that rates are not good criteria for distinguishing between physisorption and chemisorption. Chemisorption can be fast if the activation energy is small or zero; but it may be slow if the activation energy is large. Physisorption is usually fast, but it can appear to be slow if adsorption is taking place on a porous medium.

The rate at which a surface is covered by adsorbate depends on the ability of the substrate to dissipate the energy of the incoming molecule as thermal motion as it crashes on to the surface. If the energy is not dissipated quickly, the molecule migrates over the surface until a vibration expels it into the overlying gas or it reaches an edge. The proportion of collisions with the surface that successfully lead to adsorption is called the **sticking probability**, s:

$$s = \frac{\text{rate of adsorption of particles by the surface}}{\text{rate of collision of particles with the surface}} \tag{18.12}$$

The denominator can be calculated from kinetic theory (by using eqn 18.1), and the numerator can be measured by observing the rate of change of pressure. Values of s vary widely. For example, at room temperature CO has s in the range 0.1–1.0 for several d-metal surfaces, suggesting that almost every collision sticks, but for N_2 on rhenium $s < 10^{-2}$, indicating that more than a hundred collisions are needed before one molecule sticks successfully.

Desorption is always an activated process because the molecules have to be lifted from the foot of a potential well. A physisorbed molecule vibrates in its shallow potential well, and might shake itself off the surface after a short time. The temperature dependence of the first-order rate of departure can be expected to be Arrhenius-like,

$$k_d = Ae^{-E_d/RT} \tag{18.13}$$

where A is a pre-exponential factor (obtained from the intercept of an Arrhenius plot, Section 10.9, at $1/T = 0$) and the activation energy for desorption, E_d, is likely to be comparable to the enthalpy of physisorption. In the discussion of half-lives of first-order reactions (Section 10.8) we saw that $t_{1/2}$ = (ln 2)/k; so for desorption, the half-life for remaining on the surface has a temperature dependence given by

$$t_{1/2} = \frac{\ln 2}{k_d} = \tau_0 e^{E_d/RT} \qquad \tau_0 = \frac{\ln 2}{A} \tag{18.14}$$

(Note the positive sign in the exponent: the half-life *decreases* as the temperature is raised.) If we suppose that $1/\tau_0$ is approximately the same as the vibrational frequency of the weak molecule–surface bond (about 10^{12} Hz) and $E_d \approx 25$ kJ mol^{-1}, then residence half-lives of around 10 ns are predicted at room temperature. Lifetimes close to 1 s are obtained only by lowering the temperature to about 100 K. For chemisorption, with $E_d = 100$ kJ mol^{-1} and guessing that $\tau_0 = 10^{-14}$ s (because the adsorbate–substrate bond is quite stiff), we expect a residence half-life of about 3×10^3 s (about an hour) at room temperature, decreasing to 1 s at about 350 K.

One way to measure the desorption activation energy is to monitor the rate of increase in pressure when the sample is maintained at a series of temperatures and then to attempt to make an Arrhenius plot. A more sophisticated technique is **temperature programmed desorption** (TPD) or **thermal desorption spectroscopy** (TDS). The basic observation is a surge in desorption rate (as monitored by a mass spectrometer) when the temperature is raised linearly to the temperature at which desorption occurs rapidly; but once the desorption has occurred there is no more adsorbate to escape from the surface, so the desorption flux falls again as the temperature continues to rise. The TPD spectrum, the plot of desorption flux against temperature, therefore shows a peak, the location of which depends on the desorption activation energy. There are three maxima in the example shown in Fig. 18.24, indicating the presence of three adsorption sites with different activation energies.

In many cases only a single desorption activation energy (and a single peak in the TPD spectrum) is observed. When several peaks are observed they might correspond to adsorption on different crystal planes or to multilayer adsorption. For instance, Cd atoms on tungsten show two desorption activation energies, one of 18 kJ mol^{-1} and the other of 90 kJ mol^{-1}. The explanation is that the more tightly bound Cd atoms are attached directly to the substrate, and the less strongly bound are in a layer (or layers) above the primary overlayer. Another example of a system showing two desorption activation energies is CO on tungsten, the values being 120 kJ mol^{-1} and 300 kJ mol^{-1}. The explanation is believed to be the existence of two types of metal–adsorbate binding site, one involving a simple M—CO bond, the other adsorption with dissociation into individually adsorbed C and O atoms.

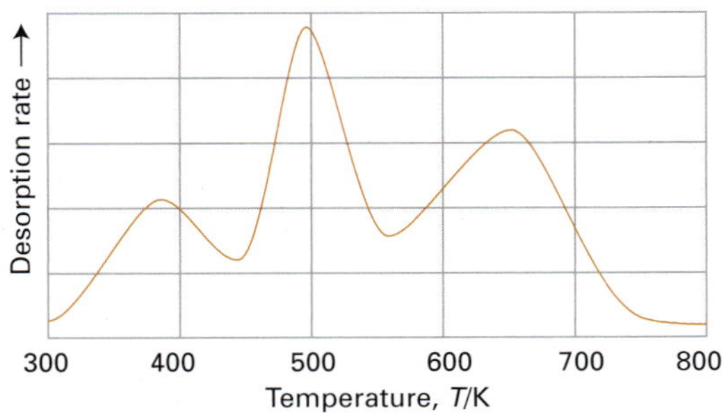

Fig. 18.24 The flash desorption spectrum of H_2 on the (100) face of tungsten. The three peaks indicate the presence of three sites with different adsorption enthalpies and therefore different desorption activation energies. (P. W. Tamm and L. D. Schmidt, *J. Chem. Phys.*, **51**, 5352 (1969).)

Catalytic activity at surfaces

We saw in Chapter 11 that a catalyst acts by providing an alternative reaction path with a lower activation energy. A catalyst does not disturb the final equilibrium composition of the system, only the rate at which that equilibrium is approached. In this section we shall consider **heterogeneous catalysis**, in which the catalyst and the reagents are in different phases. A common example is a solid introduced as a heterogeneous catalyst into a gas-phase reaction. Many industrial processes make use of heterogeneous catalysts, which include platinum, rhodium, zeolites, and various metal oxides, but increasingly attention is turning to homogeneous catalysts, partly because they are easier to cool. However, their use typically requires additional separation steps, and such catalysts are generally immobilized on a support, in which case they become heterogeneous. In general, heterogeneous catalysts are highly selective and to find an appropriate catalyst each reaction must be investigated individually. Computational procedures are beginning to be a fruitful source of prediction of catalytic activity.

A metal acts as a heterogeneous catalyst for certain gas-phase reactions by providing a surface to which a reactant can attach by chemisorption. For example, hydrogen molecules may attach as atoms to a nickel surface and these atoms react much more readily with another species (such as an alkene) than the original molecules. The chemisorption step therefore results in a reaction pathway with a lower activation energy than in the absence of the catalyst. Note that chemisorption is normally required for catalytic activity: physisorption might precede chemisorption but is not itself sufficient.

18.6 Mechanisms of heterogeneous catalysis

Heterogeneous catalysis normally depends on at least one reactant being adsorbed (usually chemisorbed) and modified to a form in which it readily undergoes reaction. Often this modification takes the form of a fragmentation of the reactant molecules. The **catalyst ensemble** is the minimum arrangement of atoms at the surface active site that can be used to model the action of the catalyst. It may be determined, for instance, by diluting the active metal with a chemically inert metal and observing the catalytic activity of the resulting alloy. In this way it has been found, for instance, that as many as 12 neighbouring Ni atoms are needed for the cleavage of the C—C bond in the conversion of ethane to methane.

The decomposition of phosphine (PH_3) on tungsten is first-order at low pressures and zeroth-order at high pressures. To account for these observations, we write down a plausible rate law in terms of an adsorption isotherm and explore its form in the limits of high and low pressure. If the rate is supposed to be proportional to the surface coverage and we suppose that θ is given by the Langmuir isotherm, we would write

$$\text{Rate} = k_r\theta = \frac{k_r Kp}{1+Kp} \qquad (18.15)$$

where p is the pressure of phosphine and k_r is a rate constant. When the pressure is so low that $Kp \ll 1$, we can neglect Kp in the denominator and obtain

$$\text{Rate} = k_r Kp \qquad (18.16a)$$

and the decomposition is first-order. When $Kp \gg 1$, we can neglect the 1 in the denominator, whereupon the Kp terms cancel and we are left with

$$\text{Rate} = k_r \qquad (18.16b)$$

and the decomposition is zeroth-order. Many heterogeneous reactions are first-order, which indicates that the rate-determining stage is the adsorption process.

Self-test 18.5

Suggest the form of the rate law for the deuteration of NH_3 in which D_2 adsorbs dissociatively but not extensively (that is, $Kp \ll 1$, with p the partial pressure of D_2), and NH_3 (with partial pressure p') adsorbs at different sites.

[*Answer:* Rate $= k_r(Kp)^{1/2}K'p'/(1 + K'p')$]

In the **Langmuir–Hinshelwood mechanism** (LH mechanism) of surface-catalysed reactions, the reaction takes place by encounters between molecular fragments and atoms adsorbed on the surface. We therefore expect the rate law to be overall second-order in the extent of surface coverage:

$$\text{A} + \text{B} \rightarrow \text{P} \qquad v = k_r\theta_A\theta_B$$

Insertion of the appropriate isotherms for A and B then gives the reaction rate in terms of the partial pressures of the reactants. For example, if A and B follow the adsorption isotherms given in eqn 18.5, then the rate law can be expected to be

$$v = \frac{k_r K_A K_B p_A p_B}{(1 + K_A p_A + K_B p_B)^2} \qquad (18.17)$$

The parameters K in the isotherms and the rate constant k_r are all temperature dependent, so the overall temperature dependence of the rate may be strongly non-Arrhenius, in the sense that the reaction rate is unlikely to be proportional to $e^{-E_a/RT}$. The LH mechanism is dominant for the catalytic oxidation of CO to CO_2 on the (111) surface of platinum.

In the **Eley–Rideal mechanism** (ER mechanism) of a surface-catalysed reaction, a gas-phase molecule collides with another molecule already adsorbed on the surface. We can therefore expect the rate of formation of product to be proportional to the partial pressure, p_B, of the nonadsorbed gas B and the extent of surface coverage, θ_A, of the adsorbed gas A. It follows that the rate law should be

$$\text{A} + \text{B} \rightarrow \text{P} \qquad v = k_r p_B\theta_A$$

The rate constant, k_r, might be much larger than for the uncatalysed gas-phase reaction because the reaction on the surface has a low activation energy and the adsorption itself is often not activated. If we know the adsorption isotherm for A, we can express the rate law in terms of its partial pressure, p_A. For example, if the adsorption of A follows a Langmuir isotherm in the pressure range of interest, then the rate law would be

$$v = \frac{k_r K p_A p_B}{1 + K p_A} \qquad (18.18)$$

If A were a diatomic molecule that adsorbed as atoms, then we would substitute the isotherm given in eqn 18.5 instead.

According to eqn 18.18, when the partial pressure of A is high (in the sense $Kp_A \gg 1$) there is almost complete surface coverage, and the rate law is

$$v \approx \frac{k_r K p_A p_B}{K p_A} = k_r p_B \qquad (18.19a)$$

Now the rate-determining step is the collision of B with the adsorbed fragments. When the pressure of A is low ($Kp_A \ll 1$), perhaps because of its reaction, the rate law becomes

$$v \approx \frac{k_r K p_A p_B}{1} = k_r K p_A p_B \qquad (18.19b)$$

Now the extent of surface coverage is rate determining.

Almost all thermal surface-catalysed reactions are thought to take place by the LH mechanism, but a number of reactions with an ER mechanism have also been identified from molecular beam investigations. For example, the reaction between H(g) and D(ad) to form HD(g) is thought to be by an ER mechanism involving the direct collision and pick-up of the adsorbed D atom by the incident H atom. However, the two mechanisms should really be thought of as ideal limits, and all reactions lie somewhere between the two and show features of both.

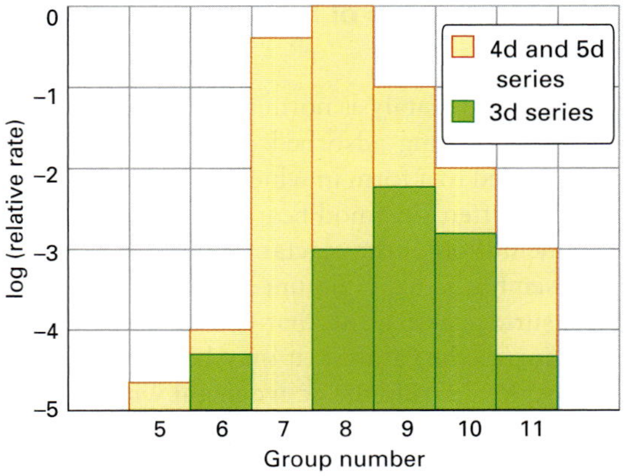

Fig. 18.25 A 'volcano curve' of catalytic activity arises because although the reactants must adsorb reasonably strongly, they must not adsorb so strongly that they are immobilized. The green and yellow rectangles correspond to the 3d and (4d, 5d) series of metals. The group numbers relate to the periodic table (see inside back cover).

18.7 Examples of heterogeneous catalysis

Almost the whole of modern chemical industry depends on the development, selection, and application of catalysts (Table 18.3). All we can hope to do in this section is to give a brief indication of some of the problems involved. Other than the ones we consider, these problems include the danger of the catalyst being poisoned by by-products or impurities, and economic considerations relating to cost, regeneration, and lifetime.

The activity of a catalyst depends on the strength of chemisorption as indicated by the 'volcano' curve in Fig. 18.25 (which is so called on account of its general shape; but note that the vertical axis is logarithmic, so the high activities are very much higher than the low activities). To be active, the catalyst should be extensively covered by adsorbate, which is the case if chemisorption is strong. On the other hand, if the strength of the substrate–adsorbate bond becomes too great, then the activity declines either because the other reactant molecules cannot react with the adsorbate or because the adsorbate molecules are immobilized on the surface. This pattern of behaviour suggests that the activity of a catalyst should initially increase with strength of adsorption (as measured, for instance, by the enthalpy of adsorption) and then decline, and that the most active catalysts should be those lying near the summit of the volcano. Most active metals are those that lie close to the middle of the d block.

Because heterogeneous catalysis is a surface phenomenon, it is essential to achieve high surface areas. Thus, solid catalysts may be finely divided or structures with internal channels and cavities (as in zeolites; see

Table 18.3
Properties of catalysts

Catalyst	Function	Examples
Metals	Hydrogenation Dehydrogenation	Fe, Ni, Pt, Ag
Semiconducting oxides and sulfides	Oxidation Desulfurization	NiO, ZnO, MgO, Bi_2O_3/MoO_3, MoS_2
Insulating oxides	Dehydration	Al_2O_3, SiO_2, MgO
Acids	Polymerization Isomerization Cracking Alkylation	H_3PO_4, H_2SO_4 SiO_2/Al_2O_3, zeolites

Table 18.4
Chemisorption abilities

	O_2	C_2H_2	C_2H_4	CO	H_2	CO_2	N_2
Ti, Cr, Mo, Fe	+	+	+	+	+	+	+
Ni, Co	+	+	+	+	+	+	–
Pd, Pt	+	+	+	+	+	–	–
Mn, Cu	+	+	+	+	±	–	–
Al, Au	+	+	+	+	–	–	–
Li, Na, K	+	+	–	–	–	–	–
Mg, Ag, Zn, Pb	+	–	–	–	–	–	–

+, Strong chemisorption; ±, chemisorption; – no chemisorption

below). Inactive catalyst supports are used to stabilize catalytic nanoparticles dispersed over them.

Many metals are suitable for adsorbing gases, and the general order of adsorption strengths decreases along the series O_2, C_2H_2, C_2H_4, CO, H_2, CO_2, N_2. Some of these molecules adsorb dissociatively (for example, H_2). Elements from the d block, such as iron, vanadium, and chromium, show a strong activity towards all these gases, but manganese and copper are unable to adsorb N_2 and CO_2. Metals towards the left of the periodic table (for example, magnesium and lithium) can adsorb (and, in fact, react with) only the most active gas (O_2). These trends are summarized in Table 18.4.

As an example of catalytic action, consider the hydrogenation of alkenes. The alkene (**1**) adsorbs by forming two bonds with the surface (**2**), and on the same surface there may be adsorbed H atoms. When an encounter occurs, one of the alkene–surface bonds is broken (forming **3** or **4**) and later an encounter with a second H atom releases the fully hydrogenated hydrocarbon, which is the thermodynamically more stable species. The evidence for a two-stage reaction is the appearance of different isomeric alkenes in the mixture. The formation of isomers comes about because while the hydrocarbon chain is waving about over the surface of the metal, an atom in the chain might chemisorb again to form (**5**) and then desorb to (**6**), an isomer of the original molecule. The new alkene would not be formed if the two hydrogen atoms attached simultaneously.

Catalytic oxidation is widely used in industry and in pollution control. Although in some cases it is desirable to achieve complete oxidation (as in the production of nitric acid from ammonia), in others partial oxidation is the aim. For example, the

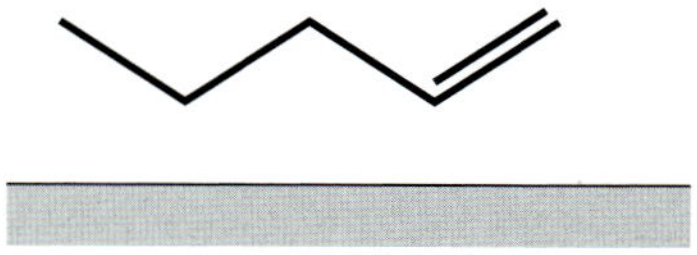

1

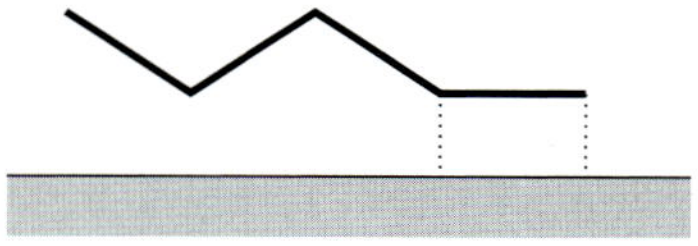

2

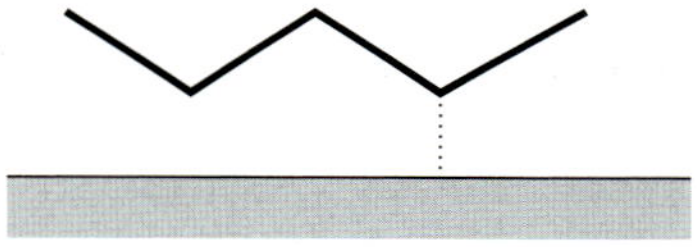

3

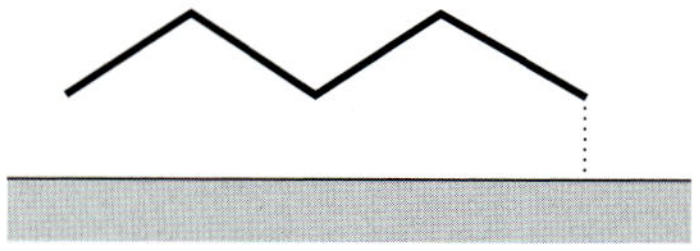

4

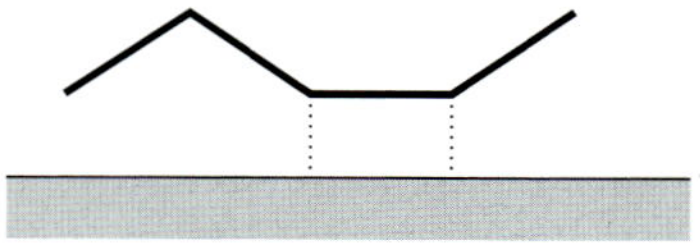

5

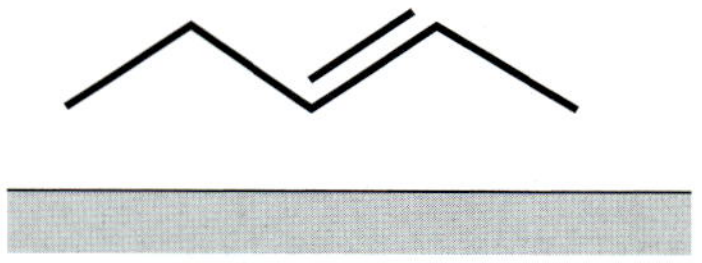

6

complete oxidation of propene to carbon dioxide and water is wasteful, but its partial oxidation to propenal (acrolein, CH_2=CHCHO) is the start of important industrial processes. Likewise, the controlled oxidations of ethene to ethanol, ethanal (acetaldehyde), and (in the presence of chlorine) to chloroethene (vinyl chloride, for the manufacture of PVC), are the initial stages of very important chemical industries.

Some of these reactions are catalysed by d-metal oxides of various kinds. The physical chemistry of oxide surfaces is very complex, as can be appreciated by considering what happens during the oxidation of propene to propenal on bismuth molybdate. The first stage is the adsorption of the propene molecule with loss of a hydrogen to form the propenyl (allyl) radical, CH_2=$CHCH_2$·. An O atom in the surface can now transfer to this radical, leading to the formation of propenal and its desorption from the surface. The H atom also escapes with a surface O atom, and goes on to form H_2O, which leaves the surface. The surface is left with vacancies and metal ions in lower oxidation states. These vacancies are attacked by O_2 molecules in the overlying gas, which then chemisorb as O_2^- ions, so reforming the catalyst. This sequence of events, which is called the **Mars van Krevelen mechanism**, involves great upheavals of the surface, and some materials break up under the stress.

Many of the small organic molecules used in the preparation of all kinds of chemical products come from petroleum. These small building blocks of polymers, and petrochemicals in general, are usually cut from the long-chain hydrocarbons drawn from the Earth as petroleum. The catalytically induced fragmentation of the long-chain hydrocarbons is called **cracking**, and is often brought about on silica–alumina catalysts. These catalysts act by forming unstable carbocations, which dissociate and rearrange to more highly branched isomers. These branched isomers burn more smoothly and efficiently in internal combustion engines, and are used to produce higher octane fuels.

Catalytic **reforming** uses a dual-function catalyst, such as a dispersion of platinum and acidic alumina. The platinum provides the metal function, and brings about dehydrogenation and hydrogenation. The alumina provides the acidic function, being able to form carbocations from alkenes. The sequence of events in catalytic reforming shows up very clearly the complications that must be unravelled if a reaction as important as this is to be understood and improved. The first step is the attachment of the long-chain hydrocarbon by chemisorption to the platinum. In this process first one and then a second H atom is lost, and an alkene is formed. The alkene migrates to a Brønsted acid site, where it accepts a proton and attaches to the surface as a carbocation. This carbocation can undergo several different reactions. It can break into two, isomerize into a more highly branched form, or undergo varieties of ring closure. Then, the adsorbed molecule loses a proton, escapes from the surface, and migrates (possibly through the gas) as an alkene to a metal part of the catalyst where it is hydrogenated. We end up with a rich selection of smaller molecules that can be withdrawn, fractionated, and then used as raw materials for other products.

The concept of a solid surface has been extended in recent years with the availability of **microporous materials**, in which the surface effectively extends deep inside the solid. Zeolites are microporous aluminosilicates with the general formula $\{[M^{n+}]_{x/n}\cdot[H_2O]_m\}\{[AlO_2]_x[SiO_2]_y\}^{x-}$, where M^{n+} cations and H_2O molecules bind inside the cavities, or pores, of the Al—O—Si framework (Fig. 18.26). Small neutral molecules, such as CO_2, NH_3, and hydrocarbons (including aromatic compounds), can also adsorb to the internal surfaces and we shall see that this partially accounts for the utility of zeolites as catalysts.

Some zeolites for which M = H^+ are very strong acids and catalyse a variety of reactions that are of particular importance to the petrochemical industry. Examples include the dehydration of methanol

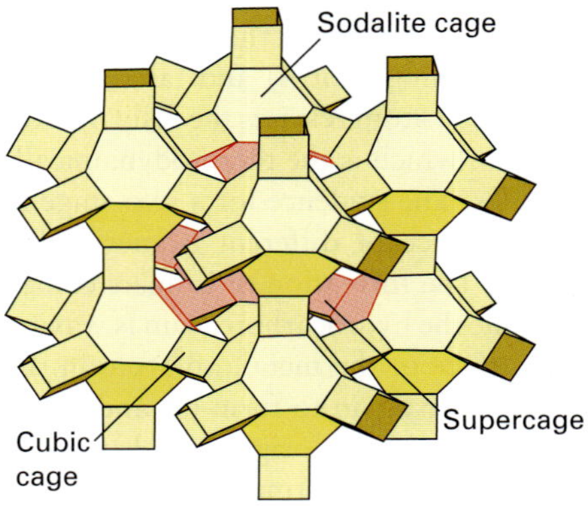

Fig. 18.26 A framework representation of the general layout of the Si, Al, and O atoms in a zeolite material. Each vertex corresponds to a Si or Al atom and each edge corresponds to the approximate location of a O atom. Note the large central pore, which can hold cations, water molecules, or other small molecules.

to form hydrocarbons such as gasoline and other fuels:

$$x\ CH_3OH \xrightarrow{\text{zeolite}} (CH_2)_x + x\ H_2O$$

and the isomerization of 1,3-dimethylbenzene (*m*-xylene) to 1,4-dimethylbenzene (*p*-xylene). The catalytically important form of these acidic zeolites may be either a Brønsted acid (**7**) or a Lewis acid (**8**). Like enzymes, a zeolite catalyst with a specific composition and structure is very selective toward certain reactants and products because only molecules of certain sizes can enter and exit the pores in which catalysis occurs. It is also possible that zeolites derive their selectivity from the ability to bind and to stabilize only transition states that fit properly in the pores. The analysis of the mechanism of zeolite catalysis is greatly facilitated by computer simulation of microporous systems, which shows how molecules fit in the pores, migrate through the connecting tunnels, and react at the appropriate active sites.

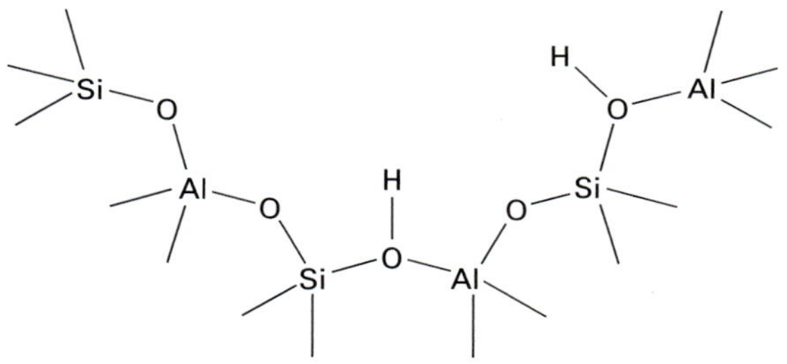

7

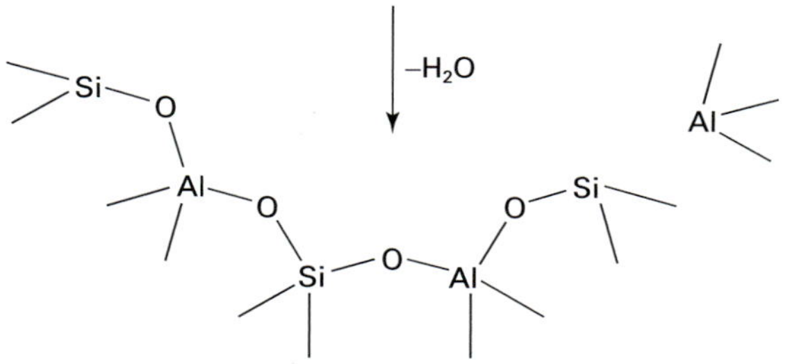

8

Processes at electrodes

A very special kind of surface is that of an electrode in contact with an electrolyte. Studies of processes on electrode surfaces are of enormous importance in electrochemistry where they give information about the rate of electron transfer between the electrode and electroactive species in solution, and are essential to the improvement of the performance of batteries and fuel cells (Box 18.1). Detailed knowledge of the factors that determine the rate of electron transfer leads to a better understanding of power production in batteries and of electron conduction in metals, semiconductors, and nanometre-sized electronic devices. Indeed, the economic consequences of electrode processes are almost incalculable. Most of the modern methods of generating electricity are inefficient, and the development of fuel cells could enhance our production and deployment of energy, not least by the reduction of the generation of polluting nitrogen oxides. Today, we produce energy inefficiently to produce goods that then decay by corrosion. Each step of this wasteful sequence could be improved by discovering more about the kinetics of electrochemical processes. Similarly, the techniques of organic and inorganic electrosynthesis, where an electrode is an active component of an industrial process, depend on intimate understanding of the kinetics of the processes taking place at electrodes.

18.8 The electrode–solution interface

Whereas most of the preceding discussion focused on the gas solid interface, we now have to turn our attention to a metallic conductor immersed in an aqueous solution of ions. The most primitive model of the boundary between the solid and liquid phases is an **electrical double layer**, which consists of a sheet of positive charge at the surface of the electrode and a sheet of negative charge next to it in the solution (or vice versa). This arrangement creates an electrical potential difference, called the **Galvani potential difference**, between the bulk of the electrode and the bulk of the solution. For simplicity in the following, we shall identify the Galvani potential difference with what in Chapter 9 we called the electrode potential.

We can construct a more detailed picture of the interface by speculating about the arrangement of ions and electric dipoles in the solution. In the **Helmholtz layer model** of the interface the solvated ions arrange themselves along the surface of the electrode but are held away from it by their hydration spheres (Fig. 18.27). The location of the sheet of ionic charge, which is called the **outer Helmholtz plane** (OHP), is identified as the plane running through the solvated ions. In this simple model, the electrical potential changes linearly within the layer bounded by the electrode surface on one side and the OHP on the other. In a refinement of this model, ions that have discarded their solvating molecules and have become attached to the electrode surface by chemical bonds are regarded as forming the **inner Helmholtz plane** (IHP). The Helmholtz layer model ignores the

Box 18.1 Fuel cells

A fuel cell operates like a conventional galvanic cell with the exception that the reactants are supplied from outside rather than forming an integral part of its construction.

A fundamental and important example of a fuel cell is the hydrogen/oxygen cell, such as those used in the space shuttle and in some prototype vehicles. One of the electrolytes used is concentrated aqueous potassium hydroxide maintained at 200°C and 20–40 atm; the electrodes may be porous nickel in the form of sheets of compressed powder. The cathode reaction is the reduction

$$O_2(g) + 2\,H_2O(l) + 4\,e^- \rightarrow 4\,OH^-(aq) \qquad E^{\ominus} = +0.40\text{ V}$$

and the anode reaction is the oxidation

$$H_2(g) + 2\,OH^-(aq) \rightarrow 2\,H_2O(l) + 2\,e^-$$

For the corresponding reduction, $E^{\ominus} = -0.83$ V. Because the overall reaction

$$2\,H_2(g) + O_2(g) \rightarrow 2\,H_2O(l) \qquad E^{\ominus}_{cell} = +1.23\text{ V}$$

is exothermic as well as spontaneous, it is less favourable thermodynamically at 200°C than at 25°C, so the cell potential is lower at the higher temperature. However, the increased pressure compensates for the increased temperature, and at 200°C and 40 atm $E_{cell} \approx +1.2$ V.

One advantage of the hydrogen/oxygen system is the large exchange-current density of the hydrogen reaction. Unfortunately, the oxygen reaction has an exchange-current density of only about 0.1 nA cm^{-2}, which limits the current available from the cell. One way round the difficulty is to use a catalytic surface with a large surface area. One type of highly developed fuel cell has phosphoric acid as the electrolyte and operates with hydrogen and air at about 200°C; the hydrogen is obtained from a reforming reaction on natural gas

Anode: $2\,H_2(g) \rightarrow 4\,H^+(aq) + 4\,e^-$

Cathode: $O_2(g) + 4\,H^+(aq) + 4\,e^- \rightarrow 2\,H_2O(l)$

This fuel cell has shown promise for *combined heat and power systems* (CHP systems). In such systems, the waste heat is used to heat buildings or to do work. Efficiency in a CHP plant can reach 80 per cent. The power output of batteries of such cells has reached the order of 10 MW. Although hydrogen gas is an attractive fuel, it has disadvantages for mobile applications: it is difficult to store and dangerous to handle. One possibility for portable fuel cells is to store the hydrogen in carbon nanotubes. It has been shown that carbon nanofibres in herringbone patterns can store huge amounts of hydrogen and result in energy densities twice that of gasoline.

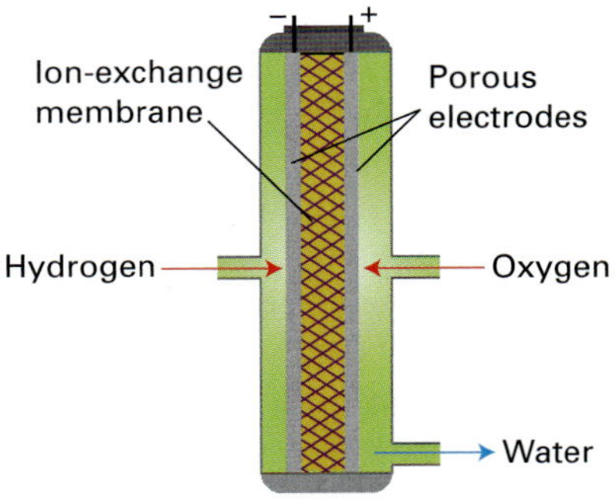

A single cell of a hydrogen/oxygen fuel cell. In practice, a stack of many cells is used.

Cells with molten carbonate electrolytes at about 600°C can make use of natural gas directly. Until these materials have been developed, one attractive fuel is methanol, which is easy to handle and is rich in hydrogen atoms:

Anode: $CH_3OH(l) + 6\,OH^-(aq) \rightarrow 5\,H_2O(l) + CO_2(g) + 6\,e^-$

Cathode: $O_2(g) + 2\,H_2O(l) + 4\,e^- \rightarrow 4\,OH^-(aq)$

One disadvantage of methanol, however, is the phenomenon of 'electro-osmotic drag' in which protons moving through the polymer electrolyte membrane separating the anode and cathode carry water and methanol with them into the cathode compartment where the potential is sufficient to oxidize CH_3OH to CO_2, so reducing the efficiency of the cell.

Solid-state electrolytes are also used. They include one version in which the electrolyte is a solid polymeric ionic conductor at about 100°C (as in the illustration), but in current versions it requires very pure hydrogen to operate successfully. Solid ionic conducting oxide cells operate at about 1000°C and can use hydrocarbons directly as fuel.

A *biofuel cell* is like a conventional fuel cell but in place of a platinum catalyst it uses enzymes or even whole organisms. The electricity will be extracted through organic molecules that can support the transfer of electrons. One application will be as the power source for medical implants, such as pacemakers, perhaps using the glucose present in the bloodstream as the fuel.

disrupting effect of thermal motion, which tends to break up and disperse the rigid outer plane of charge. In the **Gouy–Chapman model** of the **diffuse double layer,** the disordering effect of thermal motion is taken into account in much the same way as the Debye–Hückel model describes the ionic atmosphere of an ion (Section 9.1) with the latter's single central ion replaced by an infinite, plane electrode (Fig. 18.28).

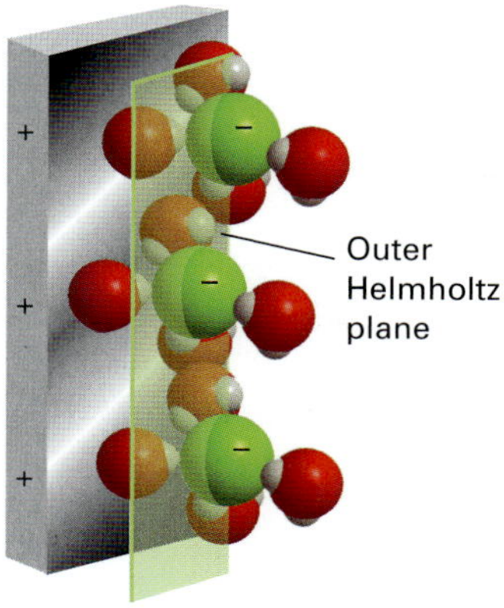

Fig. 18.27 A simple model of the electrode solution interface treats it as two rigid planes of charge. One plane, the outer Helmholtz plane (OHP), is due to the ions with their solvating molecules and the other plane is that of the electrode itself.

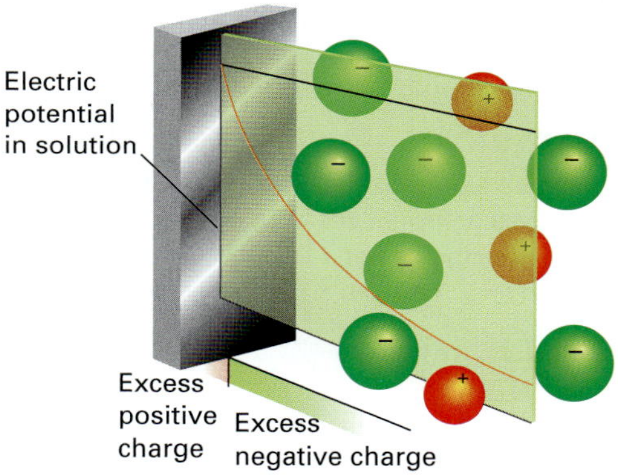

Fig. 18.28 The Gouy–Chapman model of the electrical double layer treats the outer region as an atmosphere of counter-charge, similar to the Debye–Hückel theory of ion atmospheres. The plot of electrical potential against distance from the electrode surface shows the meaning of the diffuse double layer (see text for details).

18.9 The rate of electron transfer

We shall consider a reaction at the electrode in which an ion is reduced by the transfer of a single electron in the rate-determining step. The last phrase is important: in the deposition of cadmium, for instance, only one electron is transferred in the rate-determining step even though overall the deposition involves the transfer of two electrons. The quantity we focus on is the **current density**, j, the electric current flowing through a region of an electrode divided by the area of the region. An analysis of the effect of the Galvani potential difference at the electrode on the current density using a version of transition-state theory (Section 10.11) leads to the **Butler–Volmer equation:**[1]

$$j = j_0\{e^{(1-\alpha)f\eta} - e^{-\alpha f\eta}\} \tag{18.20}$$

We have written $f = F/RT$, where F is Faraday's constant (Section 9.7; at 298 K, $f = 38.9\ V^{-1}$). The quantity η (eta) is the **overpotential:**

$$\eta = E' - E \tag{18.21}$$

where E is the electrode potential at equilibrium, when there is no net flow of current, and E' is the electrode potential when a current is being drawn from the cell. The quantity α is the **transfer coefficient**, and is an indication of where the transition state between the reduced and oxidized forms of the electroactive species in solution is reactant-like ($\alpha = 0$) or product-like ($\alpha = 1$): typical values are close to 0.5. The quantity j_0 is the **exchange-current density**, the magnitude of the equal but opposite current densities when the electrode is at equilibrium. As usual in chemistry, equilibrium is dynamic, so even though there may be no net flow of current at an electrode, there are matching inward and outward flows of electrons. Figure 18.29 shows how eqn 18.20 predicts the current density to depend on the overpotential for different values of the transfer coefficient.

When the overpotential is so small that $f\eta \ll 1$ (in practice, η less than about 0.01 V) the exponentials in eqn 18.20 can be expanded by using $e^x = 1 + x + \cdots$ and $e^{-x} = 1 - x + \cdots$ to give

$$j = j_0\{1 + (1-\alpha)f\eta + \cdots - (1 - \alpha f\eta + \cdots)\} \approx j_0 f\eta \tag{18.22}$$

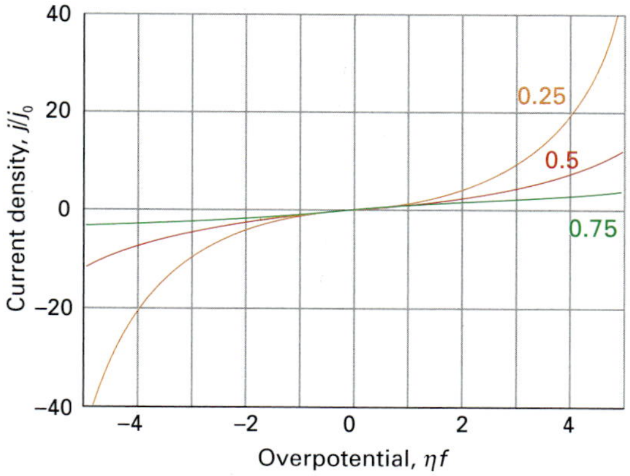

Fig. 18.29 The dependence of the current density on the overpotential for different values of the transfer coefficient.

[1] For a derivation of this equation, see our *Physical chemistry* (2006).

This equation shows that the current density is proportional to the overpotential, so at low overpotentials the interface behaves like a conductor that obeys Ohm's law, which states that the current is proportional to the potential difference.

Self-test 18.6

The exchange current density of a $Pt(s)|H_2(g)|H^+(aq)$ electrode at 298 K is 0.79 mA cm^{-2}. What is the current through an electrode of total area 5.0 cm^2 when the overpotential is +5.0 mV?

[*Answer:* 0.77 mA]

When the overpotential is large and positive (in practice, $\eta \geq 0.12$ V), the second exponential in eqn 18.20 is much smaller than the first, and may be neglected. For instance, if $\eta = 0.2$ V and $\alpha = 0.5$, $e^{-\alpha f\eta} = 0.02$ whereas $e^{(1-\alpha)f\eta} = 49$. Then (ignoring signs, which indicate the direction of the current)

$$j = j_0 e^{(1-\alpha)f\eta}$$

By taking logarithms of both sides we obtain

$$\ln j = \ln j_0 + (1-\alpha)f\eta \qquad (18.23a)$$

If instead the overpotential is large but negative (in practice, $\eta \leq -0.12$ V), the first exponential in eqn 18.20 may be neglected. Then

$$j = j_0 e^{-\alpha f\eta}$$

so

$$\ln j = \ln j_0 - \alpha f\eta \qquad (18.23b)$$

The plot of the logarithm of the current density against the overpotential is called a **Tafel plot** the slope gives the value of α and the intercept at $\eta = 0$ gives the exchange-current density.

Example 18.4

Interpreting a Tafel plot

The data below refer to the anodic current through a platinum electrode of area 2.0 cm^2 in contact with an Fe^{3+},Fe^{2+} aqueous solution at 298 K. Calculate the exchange-current density and the transfer coefficient for the electrode process.

η/mV	50	100	150	200	250
I/mA	8.8	25.0	58.0	131	298

Strategy The anodic process is the oxidation $Fe^{2+}(aq) \rightarrow Fe^{3+}(aq) + e^-$. To analyse the data, we make a Tafel plot (of $\ln j$ against η) using the anodic form (eqn 18.23a). The intercept at $\eta = 0$ is $\ln j_0$ and the slope is $(1-\alpha)f$.

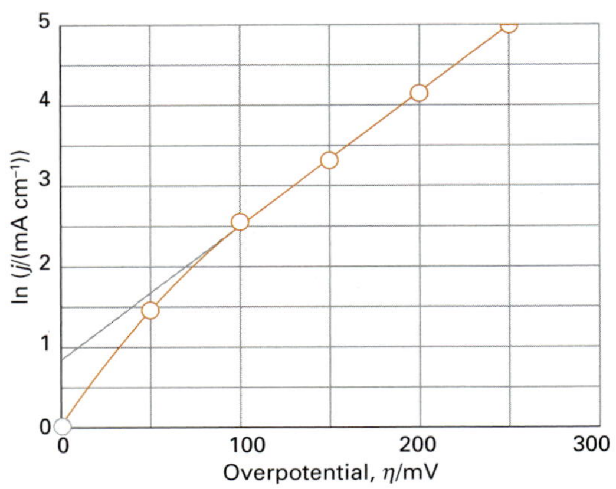

Fig. 18.30 A Tafel plot is used to measure the exchange current density (given by the extrapolated intercept at $\eta = 0$) and the transfer coefficient (from the slope). The data are from Example 18.3.

Solution Draw up the following table:

η/mV	50	100	150	200	250
j/(mA cm^{-2})	4.4	12.5	29.0	65.5	149
$\ln(j/(\text{mA cm}^{-2}))$	1.48	2.53	3.37	4.18	5.00

The points are plotted in Fig. 18.30. The high-overpotential region gives a straight line of intercept 0.88 and slope 0.0165. From the former it follows that $\ln(j_0/(\text{mA cm}^{-2})) = 0.88$, so $j_0 = 2.4$ mA cm^{-2}. From the latter,

$$(1-\alpha)\frac{F}{RT} = 0.0165 \text{ m V}^{-1}$$

so $\alpha = 0.58$. Note that the Tafel plot is nonlinear for $\eta <$ 100 mV; in this region $\alpha f\eta = 2.3$ and the approximation that $\alpha f\eta \gg 1$ fails.

Self-test 18.7

Repeat the analysis using the following cathodic current data:

η/mV	−50	−100	−150	−200	−250	−300
I/mA	0.3	1.5	6.4	27.6	118.6	510

[*Answer:* $\alpha = 0.75$, $j_0 = 0.041$ mA cm^{-2}]

Some experimental values for the Butler–Volmer parameters are given in Table 18.5. From them we can see that exchange-current densities vary over a very wide range. For example, the N_2/N_3^- couple on platinum has $j_0 = 10^{-76}$ A cm^{-2}, whereas the H^+/H_2 couple on platinum has $j_0 = 8 \times 10^{-4}$ A cm^{-2}, a difference of 73 orders of magnitude. Exchange currents are generally large when the redox process involves

Table 18.5

Exchange current densities and transfer coefficients at 298 K

Reaction	Electrode	$j/(\mathrm{A\ cm^{-2}})$	α
$2\ H^+ + 2\ e^- \rightarrow H_2$	Pt	7.9×10^{-4}	
	Ni	6.3×10^{-6}	0.58
	Pb	5.0×10^{-12}	
	Hg	7.9×10^{-13}	0.50
$Fe^{3+} + e^- \rightarrow Fe^{2+}$	Pt	2.5×10^{-3}	0.58
$Ce^{4+} + e^- \rightarrow Ce^{3+}$	Pt	4.0×10^{-5}	0.75

no bond breaking (as in the $[Fe(CN)_6]^{3-}/[Fe(CN)_6]^{4-}$ couple) or if only weak bonds are broken (as in Cl_2/Cl^-). They are generally small when more than one electron needs to be transferred, or when multiple or strong bonds are broken, as in the N_2/N_3^- couple and in redox reactions of organic compounds.

Electrodes with potentials that change only slightly when a current passes through them are classified as **nonpolarizable**. Those with strongly current-dependent potentials are classified as **polarizable**. From the linearized equation (eqn 18.23) it is clear that the criterion for low polarizability is high exchange current density (so η may be small even though j is large). The calomel and $H_2|Pt$ electrodes are both highly nonpolarizable, which is one reason why they are so extensively used as reference electrodes in electrochemistry.

18.10 Voltammetry

One of the assumptions in the derivation of the Butler–Volmer equation is the negligible conversion of the electroactive species at low current densities, resulting in uniformity of concentration near the electrode. This assumption fails at high current densities because the consumption of electroactive species close to the electrode results in a concentration gradient. The diffusion of the species towards the electrode from the bulk is slow and may become rate determining; a larger overpotential is then needed to produce a given current. This effect is called **concentration polarization**. Concentration polarization is important in the interpretation of **voltammetry**, the study of the current through an electrode as a function of the applied potential difference.

The kind of output from **linear-sweep voltammetry** is illustrated in Fig. 18.31. Initially, the absolute value of the potential is low, and the current is due to the migration of ions in the solution. However,

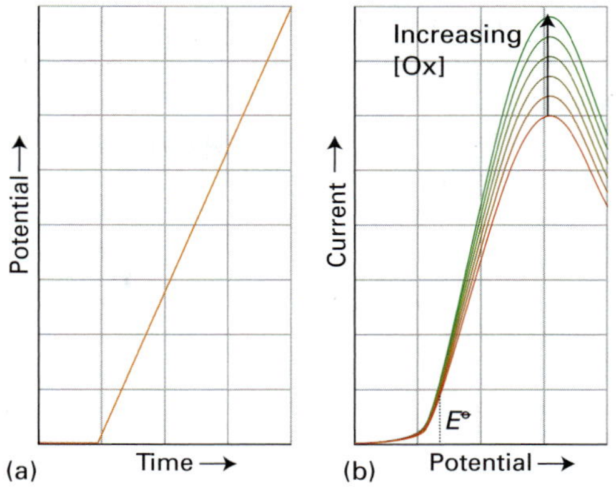

Fig. 18.31 (a) The change of potential with time and (b) the resulting current/potential curve in a voltammetry experiment. The peak value of the current density is proportional to the concentration of electroactive species (for instance, [Ox]) in solution.

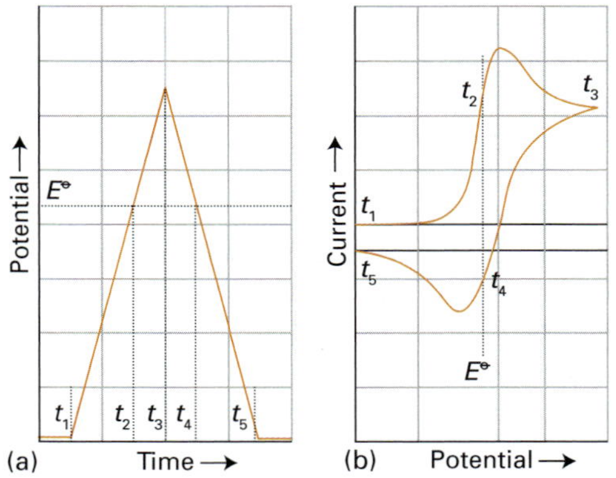

Fig. 18.32 (a) The change of potential with time and (b) the resulting current/potential curve in a cyclic voltammetry experiment.

See an animated version of this figure in the interactive ebook.

as the potential approaches the reduction potential of the reducible solute, the current grows. Soon after the potential exceeds the reduction potential the current rises and reaches a maximum value. This maximum current is proportional to the molar concentration of the species, so that concentration can be determined from the peak height after subtraction of an extrapolated baseline.

In **cyclic voltammetry** the potential is applied with a triangular waveform (linearly up, then linearly down) and the current is monitored. A typical cyclic voltammogram is shown in Fig. 18.32. The shape of the curve is initially like that of a linear sweep experiment, but after reversal of the sweep there is a rapid change in current on account of the high concentration of oxidizable species close to the electrode that was generated on the reductive sweep. When the potential is close to the value required to oxidize the reduced species, there is a substantial current until all the oxidation is complete, and the current returns to zero. Cyclic voltammetry data are obtained at scan rates of about 50 mV s^{-1}, so a scan over a range of 2 V takes about 80 s.

When the reduction reaction at the electrode can be reversed, as in the case of the $[Fe(CN)_6]^{3}/[Fe(CN)_6]^{4-}$ couple, the cyclic voltammogram is broadly symmetric about the standard potential of the couple (as in Fig. 18.32). The scan is initiated with $[Fe(CN)_6]^{3-}$ present in solution, and as the potential approaches $E^{\ominus}$ for the couple, the $[Fe(CN)_6]^{3-}$ near the electrode is reduced and current begins to flow. As the potential continues to change, the current begins to decline again because all the $[Fe(CN)_6]^{3-}$ near the electrode has been reduced and the current reaches its limiting value. The potential is now returned linearly to its initial value, and the reverse series of events occurs with the $[Fe(CN)_6]^{4-}$ produced during the forward scan now undergoing oxidation. The peak of current lies on the other side of $E^{\ominus}$, so the species present and its standard potential can be identified, as indicated in the illustration, by noting the locations of the two peaks.

The overall shape of the curve gives details of the kinetics of the electrode process and the change in shape as the rate of change of potential is altered gives information on the rates of the processes involved. For example, the matching peak on the return phase of the potential sweep may be missing, which indicates that the oxidation (or reduction) is irreversible. The appearance of the curve may also depend on the timescale of the sweep, for if the sweep is too fast some processes might not have time to occur. This style of analysis is illustrated in Example 18.5.

Example 18.5

Analysing a cyclic voltammetry experiment

The electroreduction of *p*-bromonitrobenzene in liquid ammonia is believed to occur by the following mechanism:

$$BrC_6H_4NO_2 + e^- \rightarrow BrC_6H_4NO_2^-$$
$$BrC_6H_4NO_2^- \rightarrow \cdot C_6H_4NO_2 + Br^-$$
$$\cdot C_6H_4NO_2 + e^- \rightarrow C_6H_4NO_2^-$$
$$C_6H_4NO_2^- + H^+ \rightarrow C_6H_5NO_2$$

Suggest the likely form of the cyclic voltammogram expected on the basis of this mechanism.

Strategy Decide which steps are likely to be reversible on the timescale of the potential sweep: such processes will give symmetrical voltammograms. Irreversible processes will give unsymmetrical shapes as reduction (or oxidation) might not occur. However, at fast sweep rates, an intermediate might not have time to react, and a reversible shape will be observed.

Solution At slow sweep rates, the second reaction has time to occur, and a curve typical of a two-electron reduction will be observed, but there will be no oxidation peak on the second half of the cycle because the product, $C_6H_5NO_2$, cannot be oxidized (Fig. 18.33a). At fast sweep rates, the second reaction does not have time to take place before oxidation of the $BrC_6H_4NO_2^-$ intermediate starts to occur during the reverse scan, so the voltammogram will be typical of a reversible one-electron reduction (Fig. 18.33b).

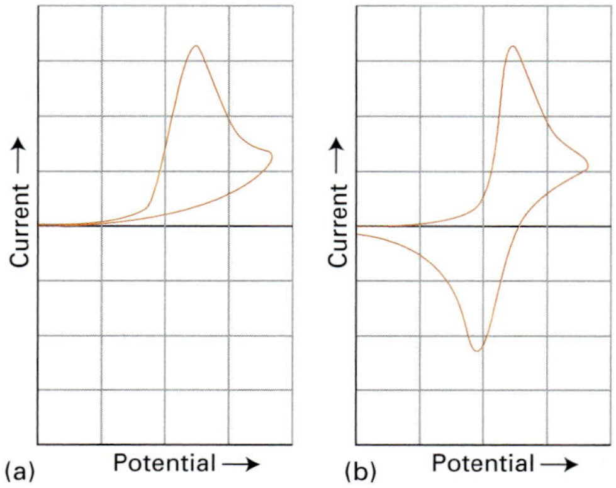

Fig. 18.33 (a) When a nonreversible step in a reaction mechanism has time to occur, the cyclic voltammogram may not show the reverse oxidation or reduction peak. (b) However, if the rate of sweep is increased, the return step may be caused to occur before the irreversible step has had time to intervene, and a typical 'reversible' voltammogram is obtained.

Self-test 18.8

Suggest an interpretation of the cyclic voltammogram shown in Fig. 18.34. The electroactive material is ClC_6H_4CN in acid solution; after reduction to $ClC_6H_4CN^-$, the radical anion may form C_6H_5CN irreversibly.

[*Answer:* $ClC_6H_4CN + e^- \rightleftharpoons ClC_6H_4CN^-$, $ClC_6H_4CN^- + H^+ + e^- \rightarrow C_6H_5CN + Cl^-$, $C_6H_5CN + e^- \rightleftharpoons C_6H_5CN^-$]

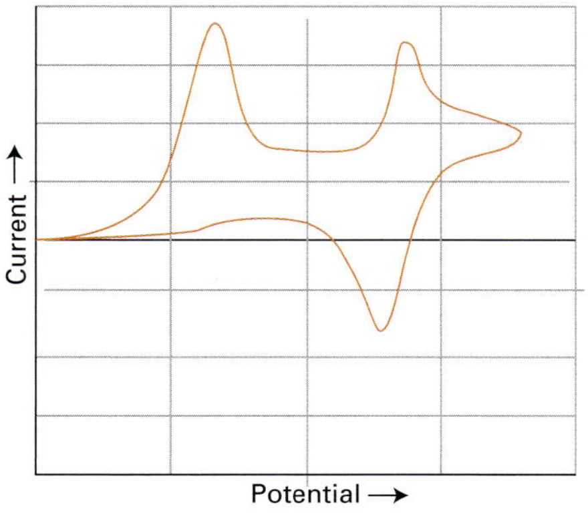

Fig. 18.34 The cyclic voltammogram referred to in Self-test 18.7.

18.11 Electrolysis

To induce current to flow through an electrolytic cell and bring about a nonspontaneous cell reaction, the applied potential difference must exceed the cell potential by at least the **cell overpotential.** The cell overpotential is the sum of the overpotentials at the two electrodes and the ohmic drop (IR_s, where R_s is the internal resistance of the cell) due to the current through the electrolyte. The additional potential needed to achieve a detectable rate of reaction may need to be large when the exchange-current density at the electrodes is small.

The rate of gas evolution or metal deposition during electrolysis can be estimated from the Butler–Volmer equation and tables of exchange-current densities. The exchange-current density depends strongly on the nature of the electrode surface, and changes in the course of the electrodeposition of one metal on another. A very crude criterion is that significant evolution or deposition occurs only if the overpotential exceeds about 0.6 V.

A glance at Table 18.5 shows the wide range of exchange-current densities for a metal/hydrogen electrode. The most sluggish exchange currents occur for lead and mercury, and the value of 1 pA cm^{-2} corresponds to a monolayer of atoms being replaced in about 5 a (a is the SI symbol for annum, year). For such systems, a high overpotential is needed to induce significant hydrogen evolution. In contrast, the value for platinum (1 mA cm^{-2}) corresponds to a monolayer being replaced in 0.1 s, so gas evolution occurs for a much lower overpotential.

Checklist of key ideas

You should now be familiar with the following concepts.

- ☐ **1** Adsorption is the attachment of molecules to a surface; the substance that adsorbs is the adsorbate and the underlying material is the adsorbent or substrate. The reverse of adsorption is desorption.
- ☐ **2** Techniques for studying surface composition and structure include scanning tunnelling microscopy (STM), atomic force microscopy (AFM), photoemission spectroscopy, Auger electron spectroscopy (AES), and low-energy electron diffraction (LEED).
- ☐ **3** The fractional coverage, θ, is the ratio of the number of occupied sites to the number of available sites.
- ☐ **4** Techniques for studying the rates of surface processes include flash desorption, surface plasmon resonance (SPR), and gravimetry by using a quartz crystal microbalance (QCM).
- ☐ **5** Physisorption is adsorption by a van der Waals interaction; chemisorption is adsorption by formation of a chemical (usually covalent) bond.
- ☐ **6** The isosteric enthalpy of adsorption is determined from a plot of ln K against $1/T$.
- ☐ **7** The sticking probability, s, is the proportion of collisions with the surface that successfully lead to adsorption.
- ☐ **8** Desorption is an activated process with half-life $t_{1/2} = \tau_0 e^{E_d/RT}$; the desorption activation energy is measured by temperature programmed desorption (TPD) or thermal desorption spectroscopy (TDS).
- ☐ **9** In the Langmuir–Hinshelwood mechanism (LH mechanism) of surface-catalysed reactions, the reaction takes place by encounters between molecular fragments and atoms adsorbed on the surface.

How would the current density at this electrode depend on the overpotential of the same set of magnitudes but of opposite sign?

18.27 The following current–voltage data are for an indium anode relative to a standard hydrogen electrode at 293 K:

$-E$/V	0.388	0.365	0.350	0.335
j/(A m^{-2})	0	0.590	1.438	3.507

Use the data to calculate the transfer coefficient and the exchange current density. What is the cathodic current density when the potential is 0.365 V?

18.28 The following data are for the overpotential for H_2 evolution with a mercury electrode in dilute aqueous solutions of H_2SO_4 at 25°C. Determine the exchange current density and transfer coefficient, α.

η/V	0.60	0.65	0.73	0.79
j/(mA m^{-2})	2.9	6.3	28	100
η/V	0.84	0.89	0.93	0.96
j/(mA m^{-2})	250	630	1650	3300

Explain any deviations from the result expected from the Tafel equation.

18.29 The illustrations below are four different examples of voltammograms. Identify the processes occurring in each system. In each case the vertical axis is the current and the horizontal axis is the (negative) electrode potential.

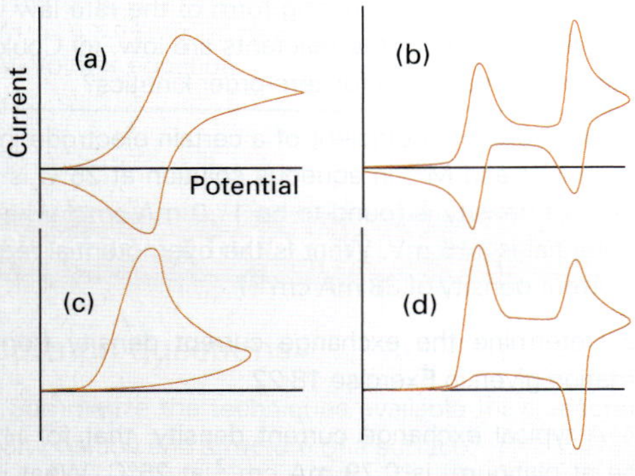

Fig. 18.35 The cyclic voltammogram referred to in Exercise 18.29.

Projects

The symbol ‡ indicates that calculus is required.

18.30‡ Here we explore atomic force microscopy (AFM) quantitatively. (a) We saw in Chapter 9 that the potential energy of interaction between two charges Q_1 and Q_2 separated by a distance r is $V = Q_1Q_2/4\pi\varepsilon_0 r$. To get an idea of the magnitudes of forces measured by AFM, calculate the force acting between two electrons separated by 0.50 nm. By what factor does the force drop if the distance between the electrons increases to 0.60 nm? *Hint*: The relation between force and potential energy is $F = -dV/dr$. (b) Suppose that the interaction probed by an AFM experiment can be expressed as a Lennard-Jones potential (Section 15.8): how does the force vary with distance?

18.31‡ The differential form of the van 't Hoff equation for the temperature dependence of equilibrium constants is $d(\ln K)/dT = \Delta_r H^{\ominus}/RT^2$. Find the corresponding expression for the temperature dependence of the pressure corresponding to a given fractional coverage on the basis of the Langmuir isotherm.

18.32 Here we explore further the design and operation of fuel cells (a) Calculate the thermodynamic limit to the cell potential of fuel cells operating on (i) hydrogen and oxygen, (ii) methane and air. Use the Gibbs energy information in the Data section, and take the species to be in their standard states at 25°C. (b) The reaction $2\,H^+ + 2\,e^- \rightarrow H_2$ is important for the operation of hydrogen/oxygen fuel cells. Use the data in Table 18.5 for the exchange current density and transfer coefficient for the reaction $2\,H^+ + 2\,e^- \rightarrow H_2$ on nickel at 25°C to determine what current density would be needed to obtain an overpotential of 0.20 V as calculated from (i) the Butler–Volmer equation and (ii) the Tafel equation? Is the validity of the Tafel approximation affected at higher overpotentials (of 0.4 V and more)?

Chapter 19

Spectroscopy: molecular rotations and vibrations

Spectroscopy is the analysis of the electromagnetic radiation emitted, absorbed, or scattered by molecules. We saw in Chapter 13 that photons act as messengers from inside atoms and that we can use atomic spectra to obtain detailed information about electronic structure. Photons of radiation ranging from radio waves to the ultraviolet also bring information to us about molecules. The difference between molecular and atomic spectroscopy, however, is that the energy of a molecule can change not only as a result of electronic transitions but also because it can make transitions between its rotational and vibrational states. Molecular spectra are more complicated but they contain more information, including electronic energy levels, bond lengths, bond angles, and bond strength. Molecular spectroscopy is used to analyse materials and to monitor changing concentrations in kinetic studies (Section 10.1).

As in the discussion of atomic spectra, the energy of a photon emitted or absorbed, and therefore the frequency, ν (nu), of the radiation emitted or absorbed, is given by the Bohr frequency condition (Section 13.1):

$$h\nu = |E_1 - E_2| \tag{19.1}$$

Here, E_1 and E_2 are the energies of the two states between which the transition occurs and h is Planck's constant. This relation is often expressed in terms of the wavelength, λ (lambda), of the radiation by using the relation

$$\lambda = \frac{c}{\nu} \tag{19.2a}$$

where c is the speed of light, or in terms of the wavenumber, $\tilde{\nu}$ (nu tilde):

$$\tilde{\nu} = \frac{1}{\lambda} = \frac{\nu}{c} \tag{19.2b}$$

The units of wavenumber are almost always chosen as reciprocal centimetres (cm^{-1}), so we can picture the wavenumber of radiation as the number of complete wavelengths per centimetre. The chart in Fig. 12.2 summarizes the frequencies, wavelengths, and wavenumbers of the various regions of the electromagnetic spectrum.

A note on good practice You will often hear people speak of 'a frequency as so many wavenumbers'. This usage is doubly wrong. First, *frequency* and *wavenumber* are two distinct physical observables with different units, and should be distinguished. Second, 'wavenumber' is not a unit, it is an observable with the dimensions of 1/length and commonly reported in reciprocal centimetres (cm^{-1}).

In this chapter, which explores rotational and vibrational spectroscopy, we first establish the allowed rotational and vibrational energies of molecules and then discuss the transitions between the corresponding states.

Rotational spectroscopy

Very little energy is needed to change the state of rotation of a molecule, and the electromagnetic radiation emitted or absorbed lies in the microwave region, with wavelengths of the order of 0.1–1 cm and frequencies close to 10 GHz. The rotational spectroscopy of gas-phase samples is therefore also known as **microwave spectroscopy**. To achieve sufficient absorption, the path lengths of gaseous samples must be very long, of the order of metres. Long path lengths are achieved by multiple passage of the beam between two parallel mirrors at each end of the sample cavity. A *klystron* (which is also used in radar installations and microwave ovens) or, more commonly now, a semiconductor device known as a *Gunn diode*, is used to generate microwaves. A microwave detector is typically a *crystal diode* consisting of a tungsten tip in contact with a semiconductor, such as germanium, silicon, or gallium arsenide. The intensity of the radiation arriving at the detector is usually modulated, because alternating signals are easier to amplify than a steady signal. In most cases the beam is chopped by a rotating shutter. Gaseous samples are essential for rotational (microwave) spectroscopy, for in that phase molecules rotate freely.

19.1 The rotational energy levels of molecules

To a first approximation, the rotational states of molecules are based on a model system called a **rigid rotor**, a body that is not distorted by the stress of rotation. The simplest type of rigid rotor is called a **linear rotor**, and corresponds to a linear molecule, such as HCl, CO_2, or HC≡CH that is supposed not to be able to bend or stretch under the stress of rotation. When the Schrödinger equation is solved for a linear rotor (see Further information 19.1), the energies are found to be

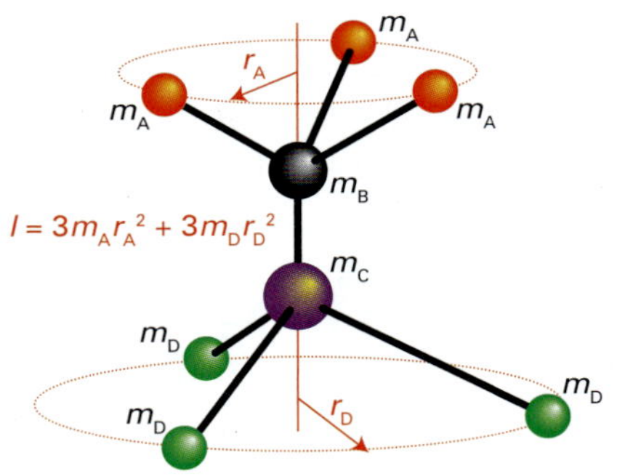

Fig. 19.1 The definition of moment of inertia. In this molecule there are three identical atoms attached to the B atom and three different but mutually identical atoms attached to the C atom. In this example, the centre of mass lies on an axis passing through the B and C atom, and the perpendicular distances are measured from this axis.

$$E_J = hBJ(J+1) \qquad J = 0, 1, 2, \ldots \tag{19.3}$$

where J is the **rotational quantum number**. The constant B (a frequency, with the units hertz, Hz) is called the **rotational constant** of the molecule, and is defined as

$$B = \frac{\hbar}{4\pi I} \tag{19.4}$$

where I is the **moment of inertia** of the molecule. The moment of inertia of a molecule is the mass of each atom multiplied by the square of its distance from the axis of rotation (Fig. 19.1):

$$I = \sum_i m_i r_i^2 \tag{19.5}$$

The moment of inertia plays a role in rotation analogous to the role played by mass in translation. A body with a high moment of inertia (like that of a flywheel or a heavy molecule) undergoes only a small rotational acceleration when a twisting force (a torque) is applied, but a body with a small moment of inertia undergoes a large acceleration when subjected to the same torque. Table 19.1 gives the expressions for the moments of inertia of various types of molecules in terms of the masses of their atoms and their bond lengths and bond angles.

Table 19.1

*Moments of inertia**

1. *Diatomic molecules*

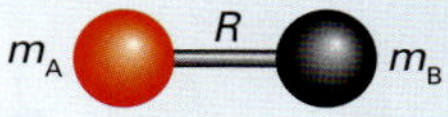

$$I = \mu R^2 \qquad \mu = \frac{m_A m_B}{m}$$

2. *Triatomic linear rotors*

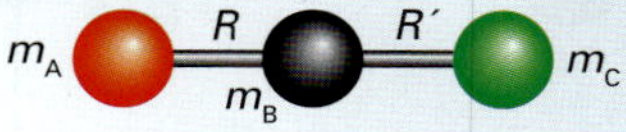

$$I = m_A R^2 + m_C R'^2 - \frac{(m_A R - m_C R')^2}{m}$$

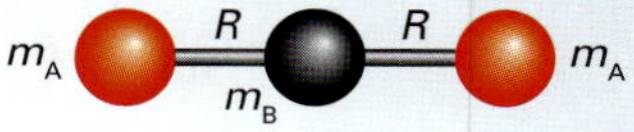

$$I = 2m_A R^2$$

3. *Symmetric rotors*

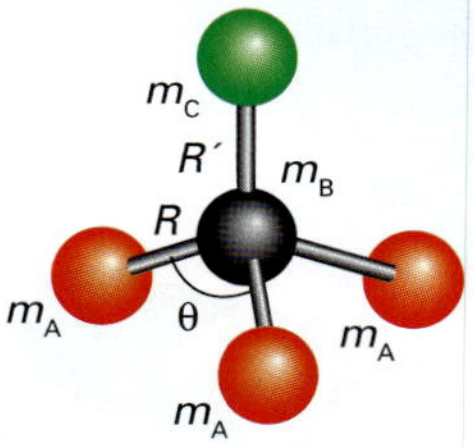

$$I_{\parallel} = 2m_A(1 - \cos\theta)R^2$$

$$I_{\perp} = m_A(1 - \cos\theta)R^2 + \frac{m_A}{m}(m_B + m_C)(1 + 2\cos\theta)R^2$$

$$+ \frac{m_C}{m}\{(3m_A + m_B)R' + 6m_A R[\tfrac{1}{3}(1 + 2\cos\theta)]^{1/2}\}R'$$

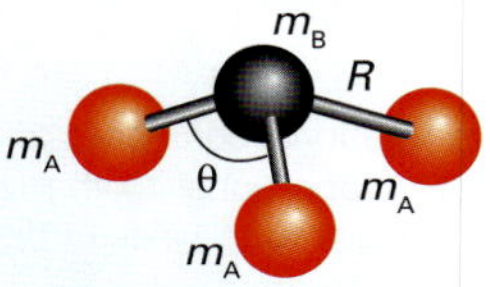

$$I_{\parallel} = 2m_A(1 - \cos\theta)R^2$$

$$I_{\perp} = m_A(1 - \cos\theta)R^2 + \frac{m_A}{m}(m_B + m_C)(1 + 2\cos\theta)R^2$$

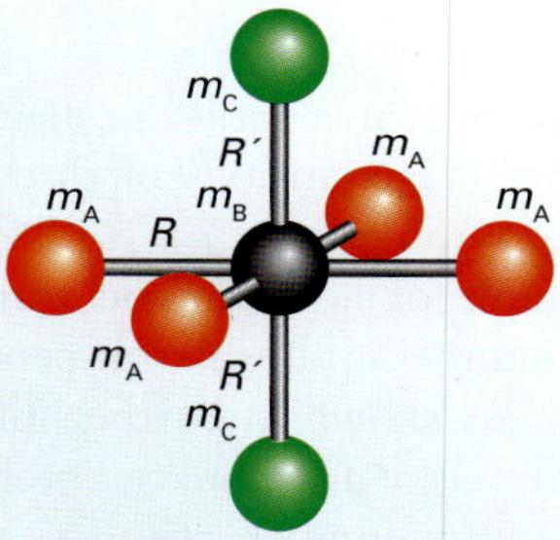

$$I_{\parallel} = 4m_A R^2$$

$$I_{\perp} = 2m_A R^2 + 2m_C R'^2$$

4. *Spherical rotors*

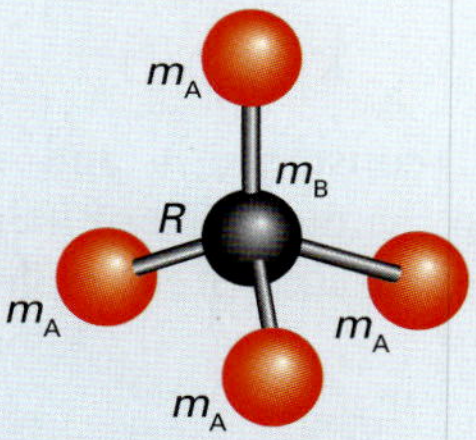

$$I = \tfrac{8}{3} m_A R^2$$

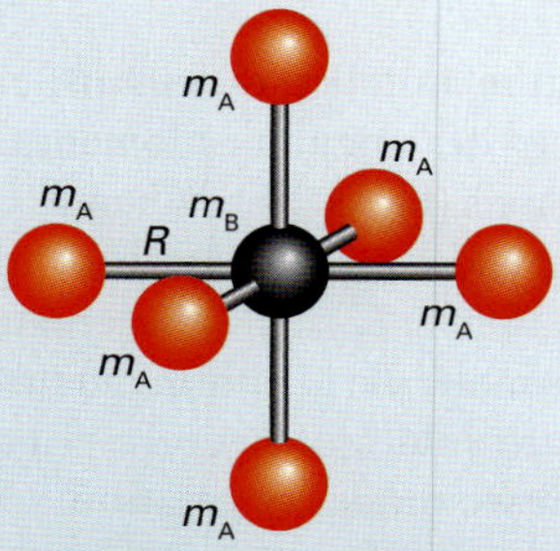

$$I = 4m_A R^2$$

* In each case, m is the total mass of the molecule.

A brief illustration To calculate the moment of inertia of a $^{12}C^{16}O_2$ molecule, the axis of rotation is perpendicular to the axis of the molecule and passes through the C atom (the centre of mass of the molecule). It follows that

$$I = m_O R^2 + 0 + m_O R^2 = 2m_O R^2$$

where R is the CO bond length. We now use $m_O = 16.00m_u$ (where the atomic mass constant unit is $m_u = 1.660\ 54 \times 10^{-27}$ kg) and $R = 116$ pm, and find

$$\begin{aligned} I &= 2 \times (16.00m_u) \times (1.16 \times 10^{-10}\ \text{m})^2 \\ &= 2 \times (16.00 \times 1.660\ 54 \times 10^{-27}\ \text{kg}) \times (1.16 \times 10^{-10}\ \text{m})^2 \\ &= 7.15 \times 10^{-46}\ \text{kg m}^2 \end{aligned}$$

A note on good practice To calculate the moment of inertia precisely, we need to specify the nuclide. Also, the mass to use is the actual atomic mass, not the element's molar mass. Nuclide masses are reported as multiples of the atomic mass constant (a constant, not a unit), so we write, for instance, $16.00m_u$ not $16.00\ m_u$. The atomic mass constant has replaced the atomic mass unit, u, which was treated as a unit (with a mass written, for instance, as 16.00 u).

Figure 19.2 shows the energy levels predicted by eqn 19.3: note that the separation of neighbouring levels increases with J. Note also that, because J may be 0 (Section 12.8), the lowest possible energy is 0: there is no zero-point rotational energy for molecules. The rotational quantum number also specifies the angular momentum of the molecule (classically, a measure of its rate of rotation): a molecule with $J = 0$ has zero angular momentum, and as J increases, so does the molecule's angular momentum. In general, the rotational angular momentum is $\{J(J+1)\}^{1/2}\hbar$, the same relation as that between the orbital angular momentum of an electron and the quantum number l.

A number of nonlinear molecules can be modelled as a **symmetric rotor**, a rigid rotor in which the moments of inertia about two axes are the same but different from a third (and all three are nonzero). The

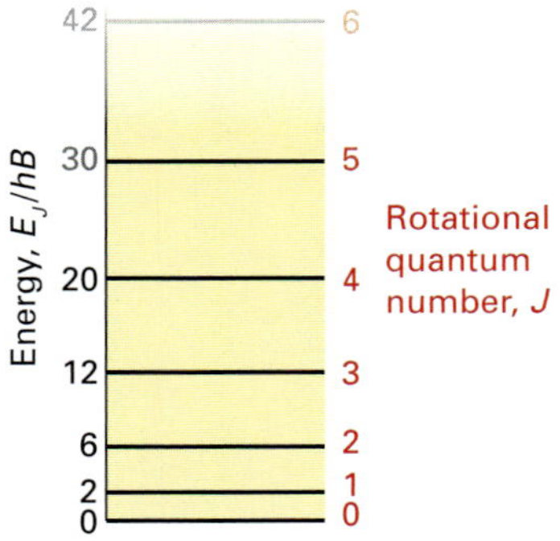

Fig. 19.2 The energy levels of a linear rigid rotor as multiples of hB.

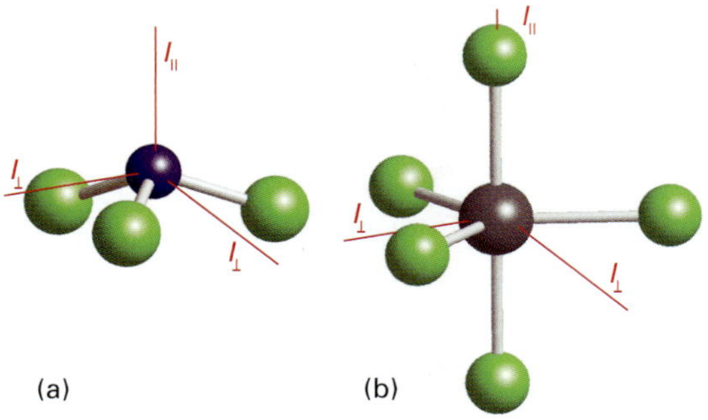

Fig. 19.3 The two different moments of inertia of (a) a trigonal pyramidal molecule and (b) a trigonal bipyramidal molecule.

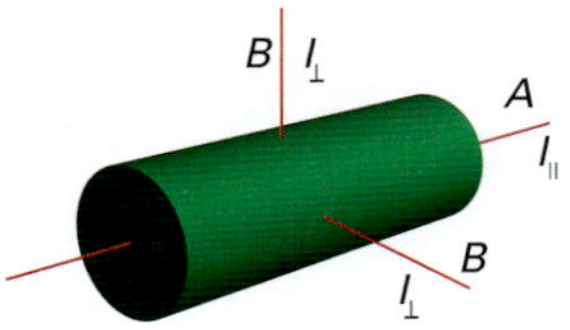

Fig. 19.4 The two rotational constants of a symmetric rotor, which are inversely proportional to the moments of inertia parallel and perpendicular to the axis of the molecule.

formal criterion of a molecule being a symmetric rotor is that it has an axis of three-fold or higher symmetry.

An example of a symmetric rotor is ammonia, NH_3, and another is phosphorus pentachloride, PCl_5 (Fig. 19.3). As shown in Further information 19.1, the energy levels of a symmetric rotor are determined by two quantum numbers, J and K, and are

$$E_{J,K} = hBJ(J+1) + h(A-B)K^2$$
$$J = 0, 1, 2, \ldots \qquad K = J, J-1, \ldots, -J \qquad (19.6)$$

The rotational constants A and B are inversely proportional to the moments of inertia parallel and perpendicular to the axis of the molecule (Fig. 19.4):

$$A = \frac{\hbar}{4\pi I_{\parallel}} \qquad B = \frac{\hbar}{4\pi I_{\perp}} \qquad (19.7)$$

The quantum number K tells us, through $K\hbar$, the component of angular momentum around the molecular axis (Fig. 19.5). When $K = 0$, the molecule is rotating end-over-end and not at all around its own axis. When $K = \pm J$ (the greatest values in its range), the molecule is rotating mainly about its axis. Intermediate values of K correspond to a combination of the two modes of rotation.

A special case of a symmetric rotor is a **spherical rotor**, a rigid body with three equal moments of inertia

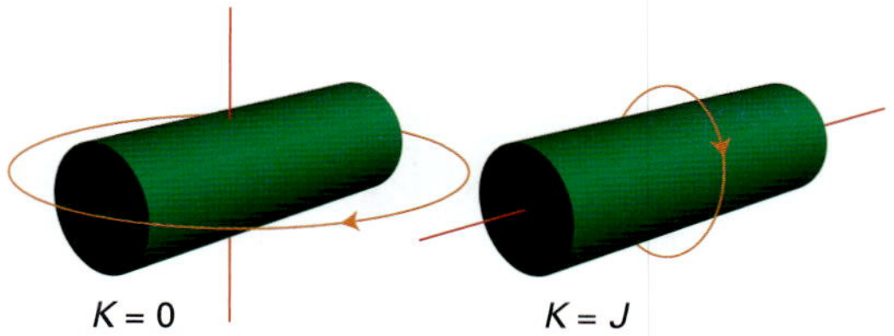

Fig. 19.5 When $K = 0$ for a symmetric rotor, the entire motion of the molecule is around an axis perpendicular to the symmetry axis of the rotor. When the value of $|K|$ is close to J, almost all the motion is around the symmetry axis.

(like a sphere). Tetrahedral, octahedral, and icosahedral molecules (CH_4, SF_6, and C_{60}, for instance) are spherical rotors. Their energy levels are very simple: when $I_{\parallel} = I_{\perp}$, the rotational constants A and B are equal and eqn 19.6 simplifies to eqn 19.3.

There is one final remark. Molecules are not really *rigid* rotors: they distort under the stress of rotation. As their bond lengths increase, their energy levels become slightly closer together. This effect is taken into account by supposing that eqn 19.3 can be modified to

$$E_J = hBJ(J+1) - hDJ^2(J+1)^2 \qquad (19.8)$$

The parameter D is the **centrifugal distortion constant.** It is large when the bond is easily stretched, and so its magnitude is related to the force constants of bonds, a measure of their rigidity (Section 19.6).

19.2 The populations of rotational states

The question we now address is the relative numbers of molecules in each rotational state, for that will affect the appearance of the rotational spectrum. There are two considerations: first, whether there are restrictions on the rotational state in which a molecule can exist and second, how the molecules are distributed over the permitted rotational states at a given temperature.

Not all the rotational states of symmetrical molecules, like H_2 and CO_2, are permitted. The elimination of certain states is a consequence of the Pauli exclusion principle that, as we saw in Chapter 13, also forbids the occurrence of certain atomic states (such as those with three electrons in one orbital, or two electrons with the same spin in the same orbital). The restrictions on the permitted rotational states due to the Pauli principle can be traced to the effect of nuclear spin called **nuclear statistics.**

To understand how the Pauli exclusion principle excludes certain rotational states, we need to express the principle in a more general way than in Further information 13.1, which referred only to electrons. The most general form of the Pauli principle states

> When any two indistinguishable fermions are interchanged, the wavefunction must change sign; when any two indistinguishable bosons are interchanged, the wavefunction must remain the same.

(Bosons are particles with integral spin; fermions are particles with half-integral spin; Section 13.6.) In short, if A and B are indistinguishable particles, then

For fermions: $\psi(B,A) = -\psi(A,B)$

For bosons: $\psi(B,A) = \psi(A,B)$

The 'fermion' part of this principle implies the Pauli *exclusion* principle, as we saw in Chapter 13. However, in this form it is more general and has wider implications.

Consider a CO_2 molecule (more precisely, a CO_2 molecule in which both O atoms are identical, as in $^{16}OC^{16}O$), which we denote O_ACO_B. When the rotates through 180°, it becomes O_BCO_A, with the two O atoms interchanged. The nuclear spin of oxygen-16 is zero, so it is a boson, and therefore the wavefunction must remain unchanged by this interchange. However, when *any* molecule is rotated through 180°, its wavefunction changes by a factor of $(-1)^J$. To see why that is so, we have drawn the first few wavefunctions for a particle travelling on a ring in Fig. 19.6, and we see that a rotation of 180° leaves wavefunctions with $J = 0, 2, \ldots$ unchanged but changes the sign of those with $J = 1, 3, \ldots$. The only way for the two requirements (that the wavefunction does not change sign and the fact that it changes by a factor of $(-1)^J$) to be consistent is for J to be restricted

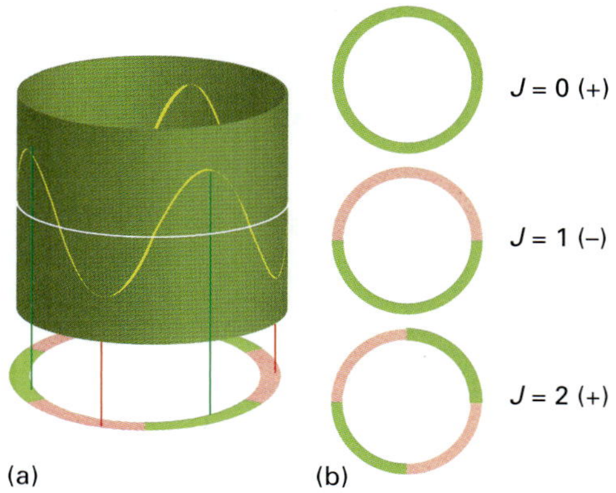

Fig. 19.6 The phases of the wavefunctions of a particle on a ring for the first few states: note that the parity of the wavefunction (its behaviour under inversion through the centre of the ring) is even, odd, even, ...

to even values. That is, a CO_2 molecule can exist only in the rotational states with $J = 0, 2, 4, \ldots$.

The analysis of the implications of nuclear statistics is more complex for molecules in which the nuclei have nonzero spin (which includes H_2, with its spin-$\frac{1}{2}$ nuclei) because the permitted rotational states depend on the relative orientation of the nuclear spins. However, the results can be expressed quite simply:

$$\frac{\text{Number of ways of achieving odd } J}{\text{Number of ways of achieving even } J} = \begin{cases} (I+1)/I \text{ for half-integral spin nuclei} \\ I/(I+1) \text{ for integral spin nuclei} \end{cases}$$

where I is the nuclear spin quantum number. For H_2, with its spin-$\frac{1}{2}$ nuclei ($I = \frac{1}{2}$), the ratio is 3:1. For D_2 and N_2, with their spin-1 nuclei ($I = 1$, where D is deuterium, ^2H), the ratio is 1:2.

The even-J rotational states of H_2 are allowed when the two nuclear spins are antiparallel (↑↓) and the odd-J states are allowed when the nuclear spins are parallel (↑↑). Different relative nuclear spin orientations change into one another only very slowly, so an H_2 molecule with parallel nuclear spins remains distinct from one with paired nuclear spins for long periods. The two forms of hydrogen can be separated by physical techniques, and stored. The form with parallel nuclear spins is called *ortho-hydrogen* and the form with paired nuclear spins is called *para-hydrogen* (remember: *para* for *paired*). Because *ortho*-hydrogen cannot exist in a state with $J = 0$, it continues to rotate at very low temperatures and has an effective rotational zero-point energy (Fig. 19.7). This energy is of some concern to manufacturers of liquid hydrogen, for the slow conversion of *ortho*-hydrogen into *para*-hydrogen (which can exist with $J = 0$) as nuclear spins slowly realign releases rotational energy, which vaporizes the liquid. Techniques are used to accelerate the conversion of *ortho*-hydrogen to *para*-hydrogen to avoid this problem. One such technique is to pass

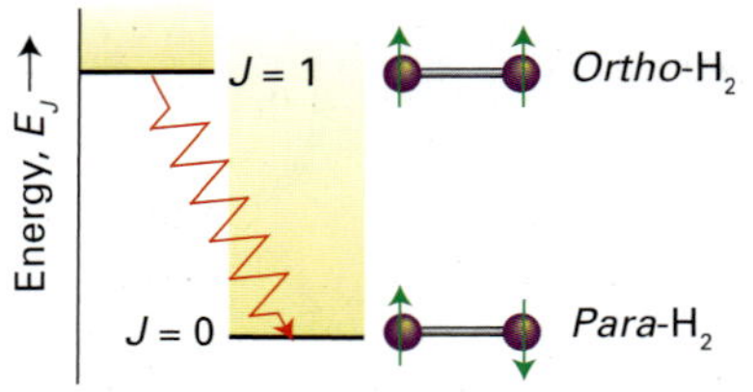

Fig. 19.7 When hydrogen is cooled, the molecules with parallel nuclear spins accumulate in their lowest available rotational state, the one with $J = 0$. They can enter the lowest rotational state ($J = 0$) only if the spins change their relative orientation and become antiparallel. This is a slow process under normal circumstances, so energy is slowly released.

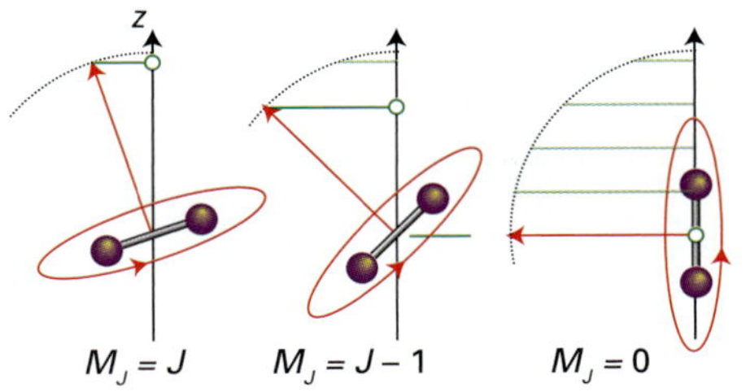

Fig. 19.8 The significance of the quantum number M_J (in this case, for $J = 4$): it indicates the orientation of the molecular rotational angular momentum with respect to an external axis.

See an animated version of this figure in the interactive ebook.

hydrogen over a metal surface: the molecules adsorb on the surface as atoms, which then recombine in the lower-energy *para*-hydrogen form.

Next, we need to consider how the permitted rotational states are populated. Because the rotational states of molecules are close together in energy, we can expect many states to be occupied at ordinary temperatures. However, we have to take into account the degeneracy of the rotational levels, because although a given *state* may have a low population, there may be many states of the same energy, and the total population of an *energy level* may be quite large.

For a linear molecule, which is the only type we consider, the angular momentum of the molecule may have $2J + 1$ different orientations with respect to an external axis, each designated by the value of the quantum number $M_J = J, J-1, \ldots, -J$ (Fig. 19.8, just as in atoms, there are $2l + 1$ orientations of the orbital angular momentum, one corresponding to each permitted value of m_l). The energy of the molecule is independent of its plane of rotation, so all $2J + 1$ states have the same energy, and therefore a level with a given value of J is $(2J + 1)$-fold degenerate. We shall see in Section 22.3 that each of these individual states has a population that is proportional to the Boltzmann factor, $e^{-E_J/kT}$ with $E_J = hBJ(J+1)$, so the total population of a given *level* is

$$P_J \propto (2J+1)e^{-hBJ(J+1)/kT} \qquad (19.9)$$

Figure 19.9 shows how this population varies with J. As shown in Derivation 19.1, it passes through a maximum at

$$J_{\max} = \left(\frac{kT}{2hB}\right)^{1/2} - \tfrac{1}{2} \qquad (19.10)$$

For a typical linear molecule (for example, OCS, with $B = 6$ GHz) at room temperature, $J_{\max} = 22$.

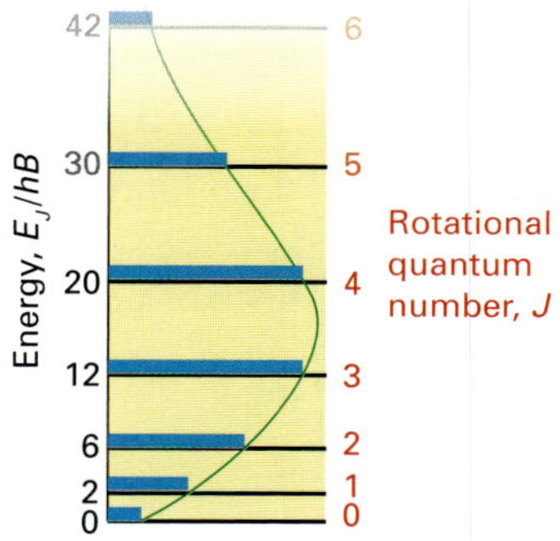

Fig. 19.9 The thermal equilibrium relative populations of the rotational energy levels of a linear rotor.

Broadly speaking, then, the absorption spectrum of the molecule should show a similar distribution of intensities.

Derivation 19.1

The most populated level

Here we need to find the value of J for which P_J is a maximum. To proceed, we recall that to find the value of x corresponding to the extremum (maximum or minimum) of any function $f(x)$, we differentiate the function, set the result equal to zero, and solve the equation for x. Applying this procedure to eqn 19.9, we obtain:

$$\frac{d}{dJ}(2J+1)e^{-hBJ(J+1)/kT}$$

Use $d uv/dx = (du/dx)v + u(dv/dx)$

$$= \frac{d(2J+1)}{dJ}e^{-hBJ(J+1)/kT} + (2J+1)\frac{de^{-hBJ(J+1)/kT}}{dJ}$$

Use $de^{ax}/dx = ae^{ax}$

$$= 2e^{-hBJ(J+1)/kT} - \frac{hB(2J+1)^2}{kT}e^{-hBJ(J+1)/kT}$$

$$= \left\{2 - \frac{hB(2J+1)^2}{kT}\right\}e^{-hBJ(J+1)/kT} = 0$$

Therefore, after setting J in the above expression to J_{max}, we need to solve

$$2 - \frac{hB(2J_{max}+1)^2}{kT} = 0$$

which gives eqn 19.10.

19.3 Rotational transitions: microwave spectroscopy

Whether or not a transition can be driven by or drive the oscillations of the surrounding electromagnetic field depends on a quantity called the **transition dipole moment**. This quantity is a measure of the dipole moment associated with the shift of electric charge that accompanies a transition (Fig. 19.10). The intensity of the transition is proportional to the square of the associated transition dipole moment. A large transition dipole moment indicates that the transition gives a strong 'thump' to the electromagnetic field, and conversely that the electromagnetic field interacts strongly with the molecule. A **selection rule** is a statement about when a transition dipole may be nonzero. There are two parts to a selection rule. A **gross selection rule** specifies the general features a molecule must have if it is to have a spectrum of a given kind. A transition that is permitted by a specific selection rule is classified as **allowed**. Transitions that are disallowed by a specific selection rule are called **forbidden**. Forbidden transitions sometimes occur weakly because the selection rule is based on an approximation that turns out to be slightly invalid.

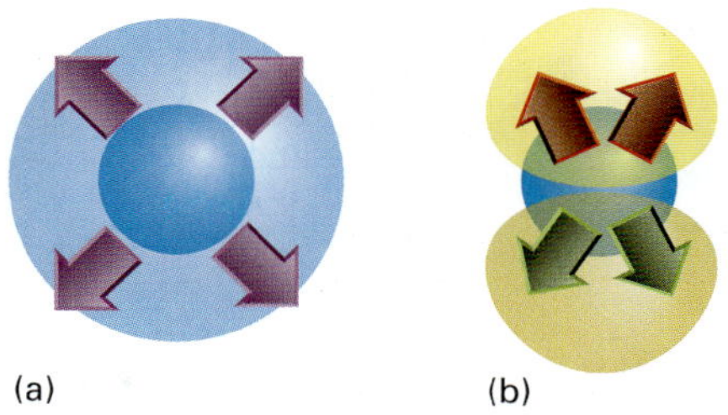

Fig. 19.10 The transition moment is a measure of the magnitude of the shift in charge during a transition. (a) A spherical redistribution of charge as in this transition has no associated dipole moment, and does not give rise to electromagnetic radiation. (b) This redistribution of charge has an associated dipole moment.

The gross selection rule for rotational transitions is that *the molecule must be polar*. The classical basis of this rule is that a stationary observer watching a rotating polar molecule sees its partial charges moving backwards and forwards and their motion shakes the electromagnetic field into oscillation (Fig. 19.11). Because the molecule must be polar, it follows that tetrahedral (CH_4, for instance), octahedral (SF_6), symmetric linear (CO_2), and homonuclear diatomic (H_2) molecules do not have rotational spectra. On the other hand, heteronuclear diatomic (HCl) and less symmetrical polar polyatomic molecules (NH_3) are polar and do have rotational spectra. We say that polar molecules are **rotationally active** whereas nonpolar molecules are **rotationally inactive.**

The specific selection rules for rotational transitions are

$$\Delta J = \pm 1 \qquad \Delta K = 0$$

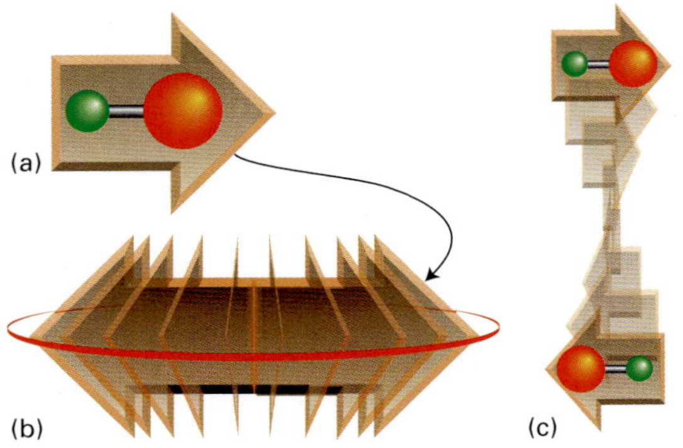

Fig. 19.11 To an external observer, (a) a rotating polar molecule has (b) an electric dipole (the arrow) that (c) appears to oscillate. This oscillating dipole can interact with the electromagnetic field.

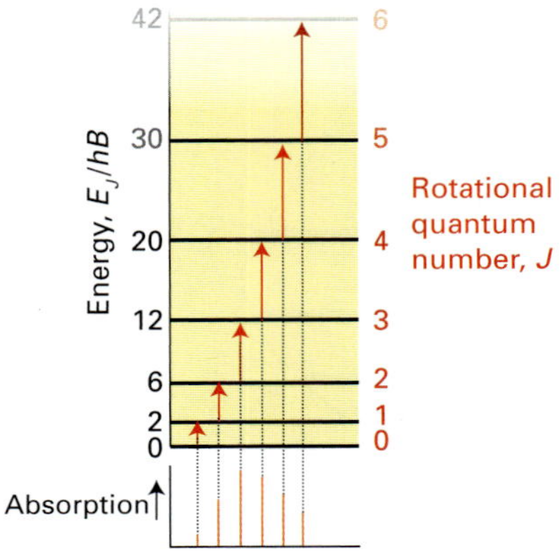

Fig. 19.12 The allowed rotational transitions (shown as absorptions) for a linear molecule.

The first of these selection rules can be traced, like the rule $\Delta l = \pm 1$ for atoms (Section 13.7), to the conservation of angular momentum when a photon is absorbed or created. A photon is a spin-1 particle, and when one is absorbed or created the angular momentum of the molecule must change by a compensating amount. Because J is a measure of the angular momentum of the molecule, J can change only by ± 1 (for pure rotational transitions, $\Delta J = +1$ corresponds to absorption, $\Delta J = -1$ to emission). The second selection rule ($\Delta K = 0$; that is, the quantum number K may not change) can be traced to the fact that the dipole moment of a polar molecule does not move when a molecule rotates around its symmetry axis (think of NH_3 rotating around its three-fold axis). As a result, there can be no acceleration or deceleration of the rotation of the molecule about that axis by the absorption or emission of electromagnetic radiation.

When a rigid molecule changes its rotational quantum number from J to $J + 1$ in an absorption, the change in rotational energy of the molecule is

$$\Delta E = E_{J+1} - E_J = hB(J+1)(J+2) - hBJ(J+1) = 2hB(J+1)$$

The energies of these transitions are $2hB$, $4hB$, $6hB$, ... The frequency of the radiation absorbed in a transition starting from the level J is therefore

$$\nu_J = 2B(J+1) \qquad (19.11a)$$

and the lines occur at $2B$, $4B$, $6B$, ... The intensity distribution will be like that Fig. 19.12 with a maximum intensity at $\nu_{J_{max}}$, with J_{max} given by eqn 19.10. A rotational spectrum of a polar linear molecule (HCl) and of a polar symmetric rotor (NH_3), therefore consists of a series of lines at frequencies separated by $2B$ (Fig. 19.12). If centrifugal distortion is significant, then we use eqn 19.8 in the same way, and find

$$\nu_J = 2B(J+1) - 4D(J+1)^3 \qquad (19.11b)$$

Now the lines converge as J increases. To determine B and D, we divide both sides by $J + 1$, to obtain

$y = b + mx$

$$\frac{\nu_J}{J+1} = 2B - 4D(J+1)^2 \qquad (19.12)$$

Therefore, by plotting the $\nu_J/(J + 1)$ against $(J + 1)^2$, we should get a straight line with intercept $2B$ and slope $-4D$ (see Exercise 19.19).

A note on good practice It is often sensible to formulate an expression that, when plotted, results in a straight line (a line with the equation $y = b + mx$). Then deviations—and deficiencies in the model—are easiest to identify. If the straight line looks plausible, the data should be analysed statistically, by doing a linear regression (least squares) analysis of the data.

Example 19.1

Estimating the frequency of a rotational transition

Estimate the frequency of the $J = 0 \rightarrow 1$ transition of the $^1H^{35}Cl$ molecule. The masses of the two atoms are $1.008m_u = 1.673 \times 10^{-27}$ kg and $34.969m_u = 5.807 \times 10^{-26}$ kg, respectively, and the equilibrium bond length is 127.4 pm.

Strategy The calculation depends on the value of B, which we obtain by substituting the data into eqn 19.4. The frequency of the transition is $2B$.

Solution The moment of inertia of the molecule is

$$I = \mu R^2$$

$$= \frac{(1.673\times10^{-27}\ \text{kg})\times(5.807\times10^{-26}\ \text{kg})}{(1.673\times10^{-27}\ \text{kg})+(5.807\times10^{-26}\ \text{kg})} \times (1.274\times10^{-10}\ \text{m})^2$$

$$= 2.639\times10^{-47}\ \text{kg m}^2$$

Therefore, the rotational constant is

$$B = \frac{\hbar}{4\pi I}$$

$$= \frac{1.054\,57\times10^{-34}\ \text{J s}}{4\pi\times(2.639\times10^{-47}\ \text{kg m}^2)} = 3.180\times10^{11}\ \text{s}^{-1}$$

or 318.0 GHz (1 GHz = 10^9 Hz). It follows that the frequency of the transition is

$$\nu = 2B = 636.0\ \text{GHz}$$

This frequency corresponds to the wavelength 0.4712 mm.

Self-test 19.1

What is the frequency and wavelength of the same transition in the $^2H^{35}Cl$ molecule? The mass of ^2H is $2.014m_u = 3.344 \times 10^{-27}$ kg. Before commencing the calculation, decide whether the frequency should be higher or lower than for $^1H^{35}Cl$.

[*Answer:* 327.0 GHz, 0.9167 mm]

Once we have measured the separation between adjacent lines in a rotational spectrum of a molecule and converted it to B, we can use the value of B to obtain a value for the moment of inertia $I_\perp$. For a diatomic molecule, we can convert that value to a value of the bond length, R, by using eqn 19.5. Highly accurate bond lengths can be obtained in this way. In some cases, isotopic substitution can help. A classic case is the determination of the two bond lengths in the molecule OCS. Analysis of the microwave spectrum of this linear molecule gives a single quantity, the rotational constant, and from this single quantity we cannot deduce the two different bond lengths. However, by recording the absorption of the two isotopomers (molecules of different isotopic composition) $^{16}O^{12}C^{33}S$ and $^{16}O^{12}C^{34}S$ and assuming that isotopic substitution leaves the bond lengths unchanged, we get two pieces of information, the moment of inertia of each isotopomer, and it is now possible to determine the two bond lengths (see Exercise 19.34).

19.4 Linewidths

Spectral lines are not infinitely narrow. An important broadening process in gaseous samples is the **Doppler effect,** in which radiation is shifted in frequency when the source is moving towards or away from the observer. Molecules reach high speeds in all directions in a gas, and a stationary observer detects the corresponding Doppler-shifted range of frequencies. Some molecules approach the observer, some move away; some move quickly, others slowly. The detected spectroscopic 'line' is the absorption or emission profile arising from all the resulting Doppler shifts. The profile reflects the Maxwell distribution of molecular speeds (Section 1.6) towards or away from the observer, and the outcome is that we observe a bell-shaped Gaussian curve (a curve of the form e^{-x^2}, Fig. 19.13). When the temperature is T and the molar mass of the molecule is M, the width of the line at half its maximum height (the 'width at half-height') is

$$\delta\lambda = \frac{2\lambda}{c}\left(\frac{2RT\ln 2}{M}\right)^{1/2} \qquad (19.13)$$

which is best remembered as $\delta\lambda \propto (T/M)^{1/2}$. The Doppler width increases with temperature because the molecules acquire a wider range of speeds. Therefore, to obtain spectra of maximum sharpness, it is best to work with cold gaseous samples.

Another source of line broadening is the finite lifetime of the states involved in the transition. When the Schrödinger equation is solved for a system that is changing with time, it is found that the states of the system do not have precisely defined energies. If the time constant for the decay of a state is τ (tau),

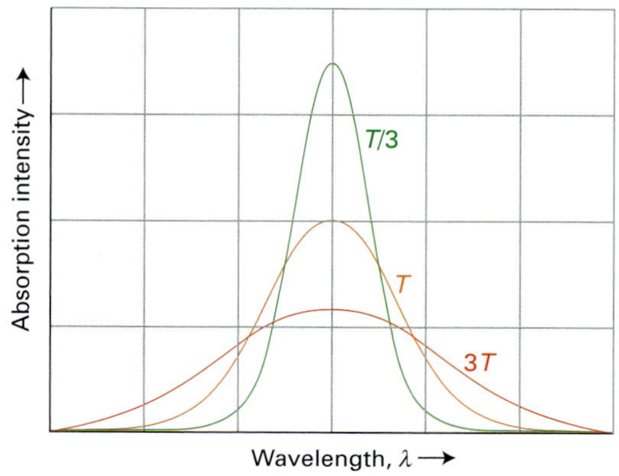

Fig. 19.13 The shape of a Doppler-broadened spectral line reflects the Maxwell distribution of speeds in the sample at the temperature of the experiment. Notice that the line broadens as the temperature is increased. The width at half-height is given by eqn 11.7.

which is called the **lifetime** of the state, then its energy levels are blurred by δE, with

$$E \approx \frac{\hbar}{\tau} \qquad (19.14a)$$

and where the decay of the state is assumed to be exponential and proportional to $e^{-t/\tau}$. We see that the shorter the lifetime of a state, the less well defined is its energy. The energy spread inherent to the states of systems that have finite lifetimes is called **lifetime broadening.** When we express the energy spread as a wavenumber by writing $\delta E = hc\delta\tilde{\nu}$ and use the values of the fundamental constants, the practical form of this relation becomes

$$\tilde{\nu} \approx \frac{5.3 \text{ cm}^{-1}}{\tau/\text{ps}} \qquad (19.14b)$$

Only if τ is infinite can the energy of a state be specified exactly (with $\delta E = 0$). However, no excited state has an infinite lifetime; therefore, all states are subject to some lifetime broadening, and the shorter the lifetimes of the states involved in a transition, the broader the spectral lines.

A note on good practice Lifetime broadening is sometimes referred to as 'uncertainty broadening' because eqn 19.14 can be written as $\tau\delta E \approx \hbar$, which resembles the form of a Heisenberg uncertainty principle for energy and time. However, there are technical reasons for not regarding this expression as a true uncertainty principle and the term 'lifetime broadening' is to be preferred.

Self-test 19.2

What is the width (expressed as a wavenumber) of a transition from a state with a lifetime of 5.0 ps?

[*Answer:* 1.1 cm^{-1}]

Two processes are principally responsible for the finite lifetimes of excited states, and hence for the widths of transitions to or from them. The dominant one is **collisional deactivation**, which arises from collisions between molecules or with the walls of the container. If the collisional lifetime is τ_{col}, then the resulting collisional linewidth is $\delta E_{\text{col}} \approx \hbar/\tau_{\text{col}}$. In gases, the collisional lifetime can be lengthened, and the broadening—which in this case is also called *pressure broadening*—minimized, by working at low pressures. The second contribution is **spontaneous emission** (see Further information 20.1), the emission of radiation when an excited state collapses into a lower state. The rate of spontaneous emission depends on details of the wavefunctions of the excited and lower states. Because the rate of spontaneous emission cannot be changed (without changing the molecule), it is a natural limit to the lifetime of an excited state. The resulting lifetime broadening is the **natural linewidth** of the transition.

The natural linewidth of a transition cannot be changed by modifying the temperature or pressure. Natural linewidths depend strongly on the transition frequency ν (they increase as ν^3), so low-frequency transitions (such as the microwave transitions of rotational spectroscopy) have very small natural linewidths; for such transitions, collisional and Doppler line-broadening processes are dominant.

19.5 Rotational Raman spectra

In **Raman spectroscopy**, molecular energy levels are explored by examining the frequencies present in the radiation scattered by molecules. In a typical experiment, a monochromatic incident laser beam is passed through the sample and the radiation scattered from the front face of the sample is monitored (Fig. 19.14). About 1 in 10^7 of the incident photons collide with the molecules, give up some of their energy, and emerge with a lower energy. These scattered photons constitute the lower-frequency **Stokes radiation** from the sample. Other incident photons may collect energy from the molecules (if they are already excited), and emerge as higher-frequency **anti-Stokes radiation.** The component of radiation scattered into the forward direction without change of frequency is called **Rayleigh radiation.**

Lasers are used as the radiation sources in Raman spectrometers for two reasons. First, the shifts in frequency of the scattered radiation from the incident radiation are quite small, so highly monochromatic radiation from a laser is required if the shifts are to be

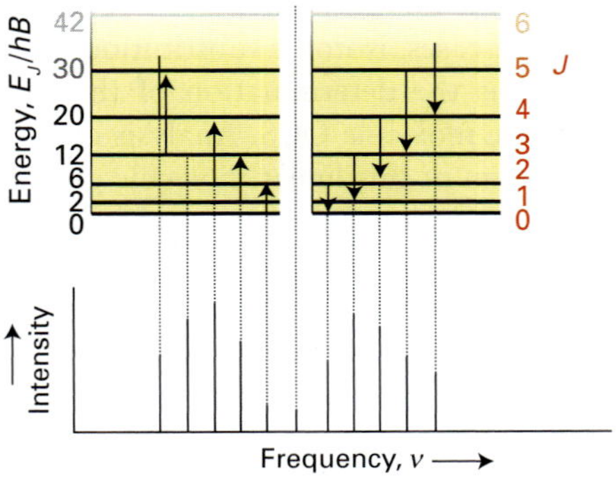

Fig. 19.14 The transitions responsible for the Stokes and anti-Stokes lines of a rotational Raman spectrum of a linear molecule.

observed. Second, the intensity of scattered radiation is low, so intense incident beams, such as those from a laser, are needed. Raman spectra may be examined by using visible and ultraviolet lasers, in which case a *diffraction grating*, a device that separates a beam of electromagnetic radiation into its component wavenumbers, is used to distinguish between Rayleigh, Stokes, and anti-Stokes radiation.

The gross selection rule for rotational Raman spectra is that the polarizability of *the molecule must be anisotropic*. We saw in Section 15.4 that the polarizability of a molecule is a measure of the extent to which an applied electric field can induce an electric dipole moment ($\mu = \alpha\mathcal{E}$). The *anisotropy* of this polarizability is its variation with the orientation of the molecule. Tetrahedral (CH_4), octahedral (SF_6), and icosahedral (C_{60}) molecules, like all spherical rotors, have the same polarizability regardless of their orientations, so these molecules are **rotationally Raman inactive**: they do not have rotational Raman spectra. All other molecules, including homonuclear diatomic molecules such as H_2, are **rotationally Raman active**.

The specific selection rules for the rotational Raman transitions of linear molecules (the only ones we consider) are

$$\Delta J = +2 \text{ (Stokes lines)} \qquad \Delta J = -2 \text{ (anti-Stokes lines)}$$

It follows that the change in energy when a rigid rotor makes the transition $J \rightarrow J + 2$ is

$$\begin{aligned}\Delta E = E_{J+2} - E_J &= hB(J+2)(J+3) - hBJ(J+1) \\ &= 2hB(2J+3) \qquad (19.15)\end{aligned}$$

Therefore, when a photon scatters from molecules in the rotational states $J = 0, 1, 2, \ldots$, and transfers some of its energy to the molecule, the energy of the photon is decreased by $6hB, 10hB, 14hB, \ldots$ and its frequency is reduced by $6B, 10B, 14B, \ldots$ from the frequency of the incident radiation. If the photon acquires energy during the collision, then a similar argument shows that the anti-Stokes lines occur with frequencies $6B, 10B, 14B, \ldots$ higher than the incident radiation (Fig. 19.14). It follows that from a measurement of the separation of the Raman lines, we can determine the value of B and hence calculate the bond length. Because homonuclear diatomic species are rotationally Raman active, this technique can be applied to them as well as to heteronuclear species.

There is an important qualification of these remarks for symmetrical molecules, such as H_2 and $C^{16}O_2$. We saw in Section 19.2 that nuclear statistics either rules out certain states or leads to an alternation of populations. We saw, for instance, that $C^{16}O_2$ can exist only in states with even values of J. As a result, its rotational Raman spectrum consists of lines at $6B, 14B, 22B, \ldots$ and separated by $8B$ because the lines starting from odd values of J are missing. For molecules with nonzero nuclear spin, all the Raman lines are present but they show an alternation of intensities: for H_2, the odd-J lines are three times more intense than the even-J lines, whereas for D_2 and N_2, even-J lines are twice as intense as the odd-J lines.

Vibrational spectroscopy

All molecules are capable of vibrating, and complicated molecules may do so in a large number of different modes. Even a benzene molecule, with 12 atoms, can vibrate in 30 different modes, some of which involve the periodic swelling and shrinking of the ring and others its buckling into various distorted shapes. A molecule as big as a protein can vibrate in tens of thousands of different ways, twisting, stretching, and buckling in different regions and in different manners. Vibrations can be excited by the absorption of electromagnetic radiation. Observing the frequencies at which this absorption occurs gives very valuable information about the identity of the molecule and provides quantitative information about the flexibility of its bonds.

19.6 The vibrations of molecules

We base our discussion on Fig. 19.15, which shows a typical potential energy curve (it is a reproduction of Fig. 14.1) of a diatomic molecule as its bond is

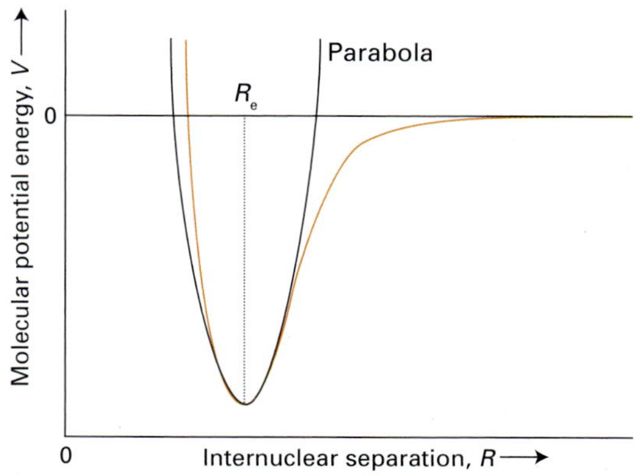

Fig. 19.15 A molecular potential energy curve can be approximated by a parabola near the bottom of the well. A parabolic potential results in harmonic oscillation. At high vibrational excitation energies the parabolic approximation is poor.

lengthened by pulling one atom away from the other or pressing it into the other. In regions close to the equilibrium bond length R_e (at the minimum of the curve) we can approximate the potential energy by a parabola (a curve of the form $y = x^2$), and write

$$V = \tfrac{1}{2}k(R - R_e)^2 \qquad (19.16)$$

where k is the force constant of the bond (units: newton per metre, N m^{-1}), as in the discussion of vibrations in Section 12.11. The steeper the walls of the potential (the stiffer the bond), the greater is the force constant.

The potential energy in eqn 19.16 has the same form as that for the harmonic oscillator (Section 12.9), so we can use the solutions of the Schrödinger equation given there. The only complication is that both atoms joined by the bond move, so the 'mass' of the oscillator has to be interpreted carefully. Detailed calculation shows that for two atoms of masses m_A and m_B joined by a bond of force constant k, the energy levels are

$$E_v = (v + \tfrac{1}{2})h\nu \qquad v = 0, 1, 2, \ldots \qquad (19.17a)$$

where

$$\nu = \frac{1}{2\pi}\left(\frac{k}{\mu}\right)^{1/2} \qquad \mu = \frac{m_A m_B}{m_A + m_B} \qquad (19.17b)$$

and μ is called the **effective mass** of the molecule, a measure of the quantity of matter moved during the vibration. The effective masses of polyatomic molecules are complicated combinations of the atomic masses. The reduced mass, which only for diatomic molecules is coincidentally the same as the effective mass, is a quantity that occurs in the separation of the internal motion of a molecule from its overall translation. Vibrational transitions are commonly expressed as a wavenumber (in reciprocal centimetres), so it is often convenient to write eqn 19.17a as

$$E_v = (v + \tfrac{1}{2})hc\tilde{\nu} \qquad v = 0, 1, 2, \ldots \qquad (19.17c)$$

with $\tilde{\nu} = \nu/c$. Figure 19.16 (a repeat of Fig. 12.31) illustrates these energy levels: we see that they form a uniform ladder of separation $hc\tilde{\nu}$ between neighbours.

A note on good practice We have previously warned about the importance of distinguishing between the quantum number v (vee) and the frequency ν (nu).

At first sight it might be puzzling that the effective mass appears rather than the total mass of the two atoms. However, the presence of μ is physically plausible. If atom A were as heavy as a brick wall, it would not move at all during the vibration and the vibrational frequency would be determined by the lighter, mobile atom. Indeed, if A were a brick wall, we could neglect m_B compared with m_A in the denominator of μ and find $\mu \approx m_B$, the mass of the lighter atom. This is approximately the case in HI, for example, where the I atom barely moves and $\mu \approx m_H$. In the case of a homonuclear diatomic molecule, for which $m_A = m_B = m$, the effective mass is half the mass of one atom: $\mu = \tfrac{1}{2}m$.

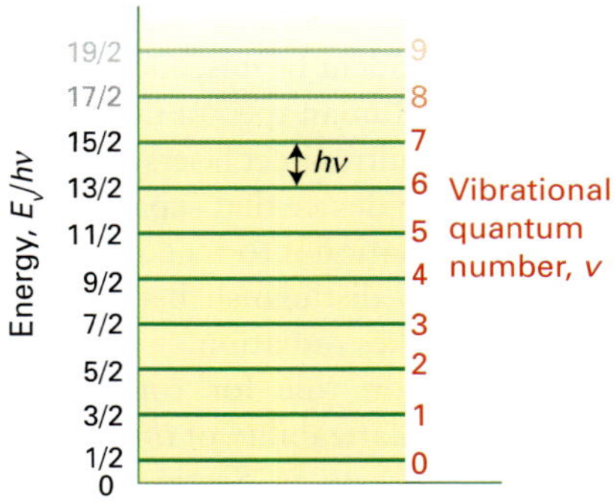

Fig. 19.16 The energy levels of an harmonic oscillator. The quantum number v ranges from 0 to infinity, and the permitted energy levels form a uniform ladder with spacing $h\nu$.

Self-test 19.3

An $^{1}H^{35}Cl$ molecule has a force constant of 516 N m^{-1}, a reasonably typical value. Calculate (a) the vibrational frequency, ν, and (b) the wavenumber, $\tilde{\nu}$, of the molecule and (c) the energy separation between any two neighbouring vibrational energy levels.

[*Answer:* (a) 89.7 THz; (b) 2992 cm^{-1}, (c) 59.4 zJ]

19.7 Vibrational transitions

Because a typical vibrational excitation energy is of the order of 0.01–0.1 aJ, the frequency of the radiation should be of the order of 10^{13}–10^{14} Hz (from $\Delta E = h\nu$). This frequency corresponds to infrared radiation, so vibrational transitions are observed by **infrared spectroscopy**. As we have remarked, in infrared spectroscopy, transitions are normally expressed in terms of their wavenumbers and lie typically in the range 300–3000 cm^{-1}.

The gross selection rule for vibrational spectra is that *the electric dipole moment of the molecule must change during the vibration*. The basis of this rule is that the molecule can shake the electromagnetic field into oscillation only if it has an electric dipole

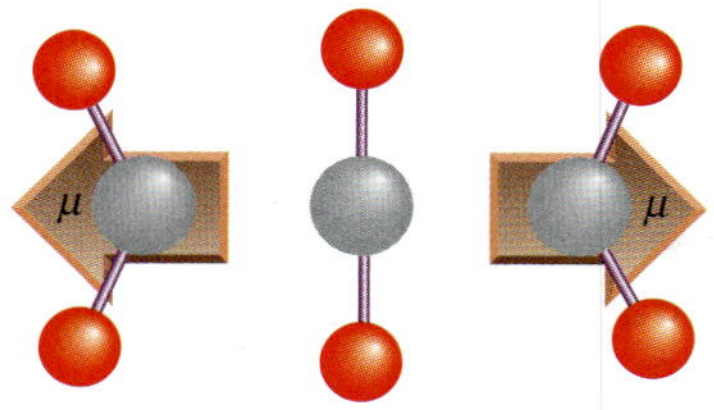

Fig. 19.17 The oscillation of a molecule, even if it is nonpolar, may result in an oscillating dipole that can interact with the electromagnetic field. Here we see a representation of a bending mode of CO_2.

moment that oscillates as the molecule vibrates (Fig. 19.17). The molecule need not have a permanent dipole: the rule requires only a *change* in dipole moment, possibly from zero. The stretching motion of a homonuclear diatomic molecule does not change its electric dipole moment from zero, so the vibrations of such molecules neither absorb nor generate radiation. We say that homonuclear diatomic molecules are **infrared inactive**, because their dipole moments remain zero however long the bond. Heteronuclear diatomic molecules, which have a dipole moment that changes as the bond lengthens and contracts, are **infrared active**.

Example 19.2

Using the gross selection rule

State which of the following molecules are infrared active: N_2, CO_2, OCS, H_2O, $CH_2{=}CH_2$, C_6H_6.

Strategy Molecules that are infrared active (that is, have vibrational spectra) have dipole moments that change during the course of a vibration. Therefore, judge whether a distortion of the molecule can change its dipole moment (including changing it from zero).

Solution All the molecules except N_2 possess at least one vibrational mode that results in a change of dipole moment, so all except N_2 are infrared active. It should be noted that not all the modes of complicated molecules are infrared active. For example, a vibration of CO_2 in which the O—C—O bonds stretch and contract symmetrically is inactive because it leaves the dipole moment unchanged (at zero). A bending motion of the molecule, however, is active and can absorb radiation.

Self-test 19.4

Repeat the question for H_2, NO, and N_2O.

[*Answer:* NO, and N_2O]

The specific selection rule for vibrational transitions is

$$\Delta v = \pm 1$$

The change in energy for the transition from a state with quantum number v to one with quantum number $v+1$ is

$$\Delta E = (v+\tfrac{3}{2})hc\tilde{\nu} - (v+\tfrac{1}{2})hc\tilde{\nu} = hc\tilde{\nu} \qquad (19.18)$$

It follows that absorption occurs when the incident radiation provides photons with this energy, and therefore when the incident radiation has a wavenumber given by eqn 19.17c. Molecules with stiff bonds (large k) joining atoms with low masses (small μ) have high vibrational wavenumbers. Bending modes are usually less stiff than stretching modes, so bends typically occur at lower wavenumbers than stretches in a spectrum.

At room temperature, almost all the molecules are in their vibrational ground states initially (the state with $v = 0$). Therefore, the most important spectral transition is from $v = 0$ to $v = 1$.

A brief illustration It follows from the calculation of $\tilde{\nu}$ for HCl (in Self-test 19.3), that $\tilde{\nu} = 2992\ \text{cm}^{-1}$, so the infrared spectrum of the molecule will be an absorption at that frequency. The corresponding frequency and wavelength are 89.7 THz (1 THz = 10^{12} Hz) and 3.34 μm, respectively.

Self-test 19.5

The force constant of the bond in the CO group of a peptide link is approximately 1.2 kN m^{-1}. At what wavenumber would you expect it to absorb? [*Hint:* For the effective mass, treat the group as a $^{12}C^{16}O$ molecule.]

[*Answer:* at approximately 1720 cm^{-1}]

19.8 Anharmonicity

The vibrational terms in eqn 19.18 are only approximate because they are based on a parabolic approximation to the actual potential energy curve. A parabola cannot be correct at all extensions because it does not allow a molecule to dissociate. At high vibrational excitations the swing of the atoms (more precisely, the spread of vibrational wavefunction) allows the molecule to explore regions of the potential energy curve where the parabolic approximation is poor. The motion then becomes **anharmonic**, in the sense that the restoring force is no longer proportional to the displacement. Because the actual curve is less confining than a parabola, we can anticipate that the energy levels become less widely spaced at high excitation, just as the energy levels of a particle in a box get closer together as the length of the box is increased.

The convergence of levels at high vibrational quantum numbers is expressed by replacing eqn 19.17c by

$$E_v = (v + \tfrac{1}{2})hc\tilde{\nu} - (v + \tfrac{1}{2})^2 hc\tilde{\nu}x_e + \cdots \qquad (19.19)$$

Where x_e is the **anharmonicity constant.** Anharmonicity also accounts for the appearance of additional weak absorption lines called **overtones** corresponding to the transitions with $\Delta v = +2, +3, \ldots$. These overtones appear because the usual selection rule is derived from the properties of harmonic oscillator wavefunctions, which are only approximately valid when anharmonicity is present. Overtones in a vibrational spectrum can appear in the near-infrared region and *overtone spectroscopy* is a technique used by analytical chemists in the characterization of food.

19.9 The technique

The source in an infrared spectrometer typically produces radiation spanning a range of frequencies. For the far infrared ($35\ \text{cm}^{-1} < \tilde{\nu} < 200\ \text{cm}^{-1}$), the source is commonly a mercury arc inside a quartz envelope, most of the radiation being generated by the hot quartz. A *Nernst filament* or *globar* is used to generate radiation in the mid-infrared ($200\ \text{cm}^{-1} < \tilde{\nu} < 4000\ \text{cm}^{-1}$): it consists of a heated ceramic filament containing rare-earth (lanthanoid) oxides, and emits radiation as it is heated.

Modern spectrometers, particularly those operating in the infrared and near-infrared, now almost always use Fourier transform techniques of spectral detection and analysis. The heart of a **Fourier transform spectrometer** is a *Michelson interferometer*, a device for analysing the frequencies present in a composite signal. The total signal from a sample is like a chord played on a piano, and the Fourier transform of the signal is equivalent to the separation of the chord into its individual notes, its spectrum. A major advantage of the Fourier transform procedure is that all the radiation emitted by the source is monitored continuously. That is in contrast to a spectrometer in which a monochromator discards most of the generated radiation. As a result, Fourier transform spectrometers have a higher sensitivity than conventional spectrometers.

The most common detectors found in commercial infrared spectrometers are sensitive in the mid-infrared region. An example is the mercury cadmium telluride (MCT) detector, a *photovoltaic device* for which the potential difference changes upon exposure to infrared radiation.

19.10 Vibrational Raman spectra of diatomic molecules

In **vibrational Raman spectroscopy** the incident photon leaves some of its energy in the vibrational modes of the molecule it strikes, or collects additional energy from a vibration that has already been excited.

The gross selection rule for vibrational Raman transitions is that *the molecular polarizability must change as the molecule vibrates*. The polarizability plays a role in vibrational Raman spectroscopy because the molecule must be squeezed and stretched by the incident radiation in order that a vibrational excitation may occur during the photon–molecule collision. Both homonuclear and heteronuclear diatomic molecules swell and contract during a vibration, and the control of the nuclei over the electrons, and hence the molecular polarizability, changes too. Both types of diatomic molecule are therefore vibrationally Raman active.

The specific selection rule for vibrational Raman transitions is the same as for infrared transitions ($\Delta v = \pm 1$). The photons that are scattered with a lower wavenumber than that of the incident light, the Stokes lines, are those for which $\Delta v = +1$. The Stokes lines are more intense than the anti-Stokes lines (for which $\Delta v = -1$), because very few molecules are in an excited vibrational state initially.

The information available from vibrational Raman spectra adds to that from infrared spectroscopy because homonuclear diatomic molecules can also be studied. The spectra can be interpreted in terms of the force constants, dissociation energies, and bond lengths, and some of the information obtained is included in Table 19.2.

19.11 The vibrations of polyatomic molecules

How many modes of vibration are there in a polyatomic molecule? We can answer this question by thinking about how each atom may change its location, and we show in Derivation 19.2 that

Nonlinear molecules: *Number of vibrational modes* $= 3N - 6$

Linear molecules: *Number of vibrational modes* $= 3N - 5$

A brief illustration A water molecule, H_2O, is triatomic and nonlinear, and has three modes of vibration. Naphthalene, $C_{10}H_8$, has 48 distinct modes of vibration. Any diatomic molecule ($N = 2$) has one vibrational mode; carbon dioxide ($N = 3$) has four vibrational modes.

Derivation 19.2

The number of normal modes

Each atom may move along any of three perpendicular axes. Therefore, the total number of such displacements in a molecule consisting of N atoms is $3N$. Three of these displacements correspond to movement of the centre of mass of the molecule, so these three displacements correspond to the translational motion of the molecule as a whole. The remaining $3N - 3$ displacements are 'internal' modes of the molecule that leave its centre of mass unchanged. Three angles are needed to specify the orientation of a nonlinear molecule in space (Fig. 19.18).

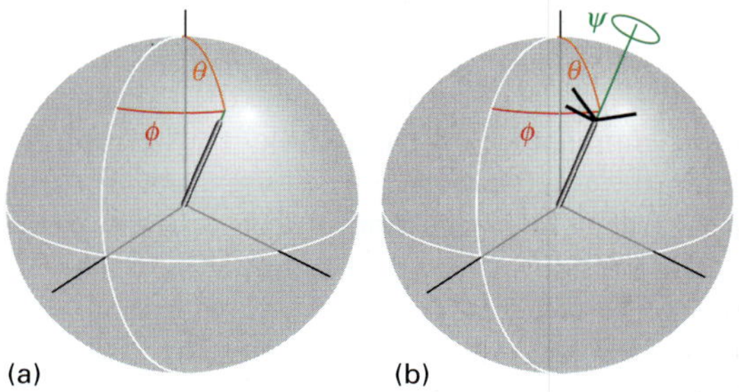

Fig. 19.18 (a) The orientation of a linear molecule requires the specification of two angles (the latitude and longitude of its axis). (b) The orientation of a nonlinear molecule requires the specification of three angles (the latitude and longitude of its axis and the angle of twist—the azimuthal angle—around that axis).

Therefore three of the $3N - 3$ internal displacements leave all bond angles and bond lengths unchanged but change the orientation of the molecule as a whole. These three displacements are therefore rotations. That leaves $3N - 6$ displacements that change neither the centre of mass of the molecule nor the orientation of the molecule in space. These $3N - 6$ displacements are the vibrational modes. A similar calculation for a linear molecule, which requires only two angles to specify its orientation in space, gives $3N - 5$ as the number of vibrational modes.

Table 19.2

Properties of diatomic molecules

	$\tilde{\nu}$/cm^{-1}	R_e/pm	k/(N m^{-1})	D/(kJ mol^{-1})
$^1H_2^+$	2322	106	160	256
$^1H_2$	4401	74	575	432
$^2H_2$	3118	74	577	440
$^1H^{19}F$	4138	92	955	564
$^1H^{35}Cl$	2991	127	516	428
$^1H^{81}Br$	2649	141	412	363
$^1H^{127}I$	2308	161	314	295
$^{14}N_2$	2358	110	2294	942
$^{16}O_2$	1580	121	1177	494
$^{19}F_2$	892	142	445	154
$^{35}Cl_2$	560	199	323	239

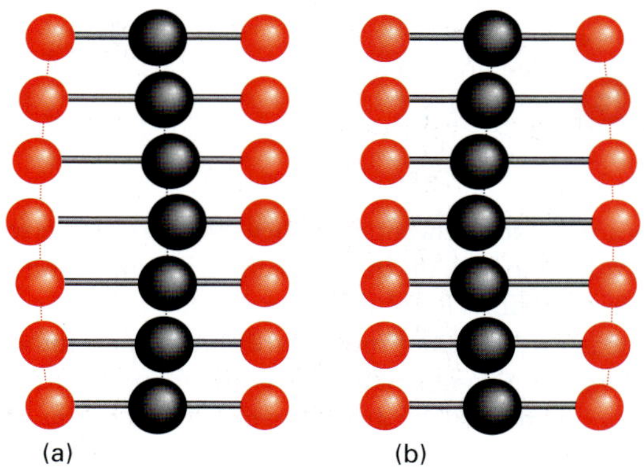

Fig. 19.19 The stretching vibrations of a CO_2 molecule can be represented in a number of ways. In this representation, (a) one O=C bond vibrates and the remaining O atom is stationary, and (b) the C=O bond vibrates while the other O atom is stationary. Because the stationary atom is linked to the C atom, it does not remain stationary for long. That is, if one vibration begins, it rapidly stimulates the other to occur.

See an animated version of this figure in the interactive ebook.

The description of the vibrational motion of a polyatomic molecule is much simpler if we consider combinations of the stretching and bending motions of individual bonds. For example, although we could describe two of the four vibrations of a CO_2 molecule as individual carbon–oxygen bond stretches, ν_L and ν_R in Fig. 19.19, the description of the motion is much simpler if we use two combinations of these vibrations. One combination is ν_1 in Fig. 19.20: this combination is the **symmetric stretch.** The other combination is ν_3, the **antisymmetric stretch,** in which the two O atoms always move in the same directions and opposite to the C atom. The two modes are independent in the sense that if one is excited, then its motion

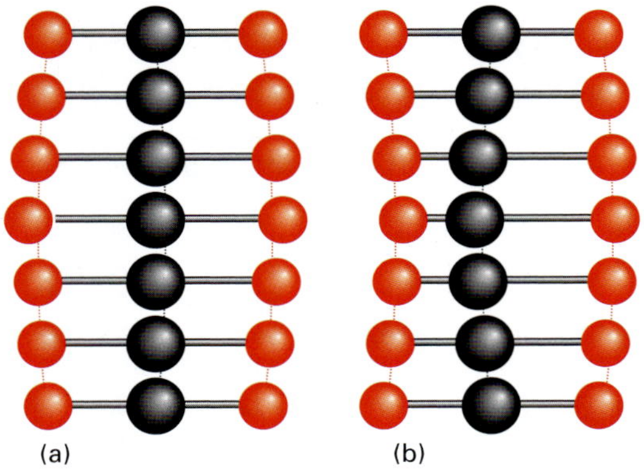

Fig. 19.20 Alternatively, linear combinations of the two modes can be taken to give these two normal modes of the molecule. The mode in (a) is the symmetric stretch and that in (b) is the antisymmetric stretch. The two modes are independent, and if either of them is stimulated, the other remains unexcited. Normal modes greatly simplify the description of the vibrations of the molecule.

See an animated version of this figure in the interactive ebook.

does not excite the other. They are two of the four 'normal modes' of the molecule, its independent, collective vibrational displacements. The two other (degenerate) normal modes are the **bending modes**, ν_2. In general, a **normal mode** is an independent, synchronous motion of atoms or groups of atoms that may be excited without leading to the excitation of any other normal mode.

Self-test 19.6

How many normal modes of vibration are there in (a) ethyne (HC≡CH) and (b) a protein molecule of 4000 atoms?

[*Answer:* (a) 7, (b) 11 994]

The four normal modes of CO_2, and the $3N - 6$ (or $3N - 5$) normal modes of polyatomic molecules in general, are the key to the description of molecular vibrations. Each normal mode behaves like an independent harmonic oscillator and the energies of the vibrational levels are given by the same expression as in eqn 19.17, but with an effective mass that depends on the extent to which each of the atoms contributes to the vibration. Atoms that do not move, such as the C atom in the symmetric stretch of CO_2, do not contribute to the effective mass. The force constant also depends in a complicated way on the extent to which bonds bend and stretch during a vibration. Typically, a normal mode that is largely a bending motion has a lower force constant (and hence a lower frequency) than a normal mode that is largely a stretching motion.

The gross selection rule for the infrared activity of a normal mode is that *the motion corresponding to a normal mode must give rise to a changing dipole moment*. Deciding whether this is so can sometimes be done by inspection. For example, the symmetric stretch of CO_2 leaves the dipole moment unchanged (at zero), so this mode is infrared inactive and makes no contribution to the molecule's infrared spectrum. The antisymmetric stretch, however, changes the dipole moment because the molecule becomes unsymmetrical as it vibrates, so this mode is infrared active. The fact that the mode does absorb infrared radiation enables carbon dioxide to act as a 'greenhouse gas' by absorbing infrared radiation emitted from the surface of the Earth (Box 19.1). Because the dipole moment change is parallel to the molecular axis in the antisymmetric stretching mode, the transitions arising from this mode are classified as **parallel bands** in the spectrum. Both bending modes are also infrared active: they are accompanied by a changing dipole perpendicular to the molecular axis (as in Fig. 19.20), so transitions involving them lead to a **perpendicular band** in the spectrum.

Self-test 19.7

State the ways in which the infrared spectrum of dinitrogen oxide (nitrous oxide, N_2O) will differ from that of carbon dioxide.

[*Answer:* different frequencies on account of different atomic masses and force constants; all four modes infrared active]

Some of the normal modes of organic molecules can be regarded as motions of individual functional groups. Others cannot be regarded as localized in this way and are better regarded as collective motions of the molecule as a whole. The latter are generally of relatively low frequency, and occur at wavenumbers below about 1500 cm^{-1} in the spectrum. The resulting whole-molecule region of the absorption spectrum is called the **fingerprint region** of the spectrum, as it is characteristic of the molecule. The matching of the fingerprint region with a spectrum

Box 19.1 Climate change*

Solar energy strikes the top of the Earth's atmosphere at a rate of 343 W m^{-2}. About 30 per cent of this energy is reflected back into space by the Earth or the atmosphere. The Earth–atmosphere system absorbs the remaining energy and re-emits it into space as black-body radiation, with most of the intensity being carried by infrared radiation in the range 200–2500 cm^{-1} (4–50 μm). The Earth's average temperature is maintained by an energy balance between solar radiation absorbed by the Earth and black-body radiation emitted by the Earth.

The trapping of infrared radiation by certain gases in the atmosphere is known as the *greenhouse effect*, so called because it warms the Earth as if the planet were enclosed in a huge greenhouse. The result is that the natural greenhouse effect raises the average surface temperature well above the freezing point of water and creates an environment in which life is possible. The major constituents to the Earth's atmosphere, O_2 and N_2, do not contribute to the greenhouse effect because homonuclear diatomic molecules cannot absorb infrared radiation. However, the minor atmospheric gases, water vapour and CO_2, do absorb infrared radiation and hence are responsible for the greenhouse effect (see the first illustration). Water vapour absorbs strongly in the ranges 1300–1900 cm^{-1} (5.3–7.7 μm) and 3550–3900 cm^{-1} (2.6–2.8 μm), whereas CO_2 shows strong absorption in the ranges 500–725 cm^{-1} (14–20 μm) and 2250–2400 cm^{-1} (4.2–4.4 μm).

Increases in the levels of greenhouse gases, which also include methane, dinitrogen oxide, ozone, and certain chlorofluorocarbons, as a result of human activity have the potential to enhance the natural greenhouse effect, leading to significant warming of the planet. This problem is referred to as *global warming*, which we now explore in some detail.

The concentration of water vapour in the atmosphere has remained steady over time, but concentrations of some other greenhouse gases are rising. From about the year 1000 until about 1750, the CO_2 concentration remained fairly stable, but, since then, it has increased by 28 per cent. The concentration of methane, CH_4, has more than doubled during this time and is now at its highest level for 160 000 years (160 ka; a is the SI unit denoting 1 year). Studies of air pockets in ice cores taken from Antarctica show that increases in the concentration of both atmospheric CO_2 and CH_4 over the past 160 ka correlate well with increases in the global surface temperature.

Human activities are primarily responsible for the rising concentrations of atmospheric CO_2 and CH_4. Most of the atmospheric CO_2 comes from the burning of hydrocarbon fuels, which began on a large scale with the Industrial Revolution in the middle of the nineteenth century. The additional methane comes mainly from the petroleum industry and from agriculture.

The temperature of the surface of the Earth has increased by about 0.8°C since the middle of the nineteenth century (see the second illustration). In 2007, the Intergovernmental Panel on Climate Change (IPCC) estimated that our continued reliance on hydrocarbon fuels, coupled to current trends in population growth, could result in an additional increase of 1–3°C in the temperature of the Earth by 2100, relative to

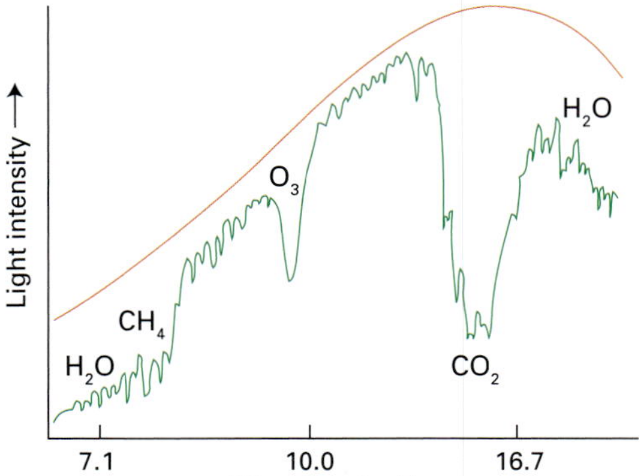

The intensity of infrared radiation that would be lost from Earth in the absence of greenhouse gases is shown by the smooth line. The jagged line is the intensity of the radiation actually emitted. The maximum wavelength of radiation absorbed by each greenhouse gas is indicated.

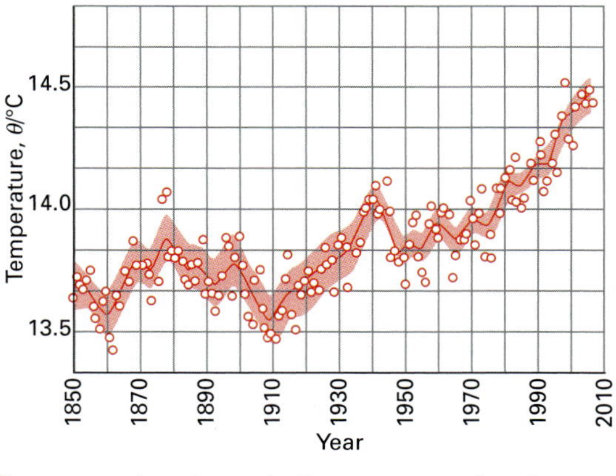

The progressive change in the average surface temperature of the Earth, based on the IPCC report in 2007.

* This box is based on, but updated from, a similar contribution initially prepared by Loretta Jones and appearing in *Chemical Principles*, Peter Atkins and Loretta Jones, W.H. Freeman and Co., New York (2008).

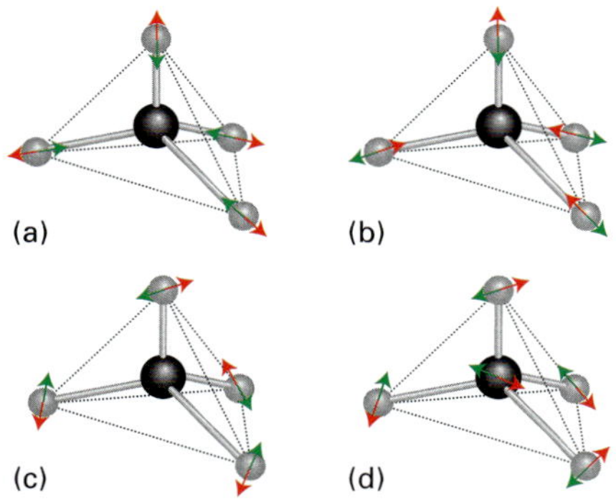

Some of the normal modes of vibration of CH_4. An arrow indicates the direction of motion of an atom during the vibration.

See an animated version of this figure in the interactive ebook.

the surface temperature in 2000. Furthermore, the rate of temperature change is likely to be greater than at any time in the last 10 ka. To place a temperature rise of 3°C in perspective, note that the average temperature of the Earth during the last ice age was only 6°C colder than at present. Just as cooling the planet (for example, during an ice age) can lead to detrimental effects on ecosystems, so too can a dramatic warming. One example of a significant change in the environment caused by a temperature increase of 3°C is a rise in sea level by about 0.5 m, which is sufficient to alter weather patterns and submerge currently coastal ecosystems.

Computer projections for the next 200 years predict further increases in atmospheric CO_2 levels and suggest that, to maintain CO_2 at its current concentration, we would have to reduce hydrocarbon fuel consumption immediately. Clearly, in order to reverse global warming trends, we need to develop alternatives to fossil fuels, such as hydrogen (which can be used in fuel cells, Box 9.2) and solar energy technologies.

of a known compound in a library of infrared spectra is a very powerful way of confirming the presence of a particular substance.

The characteristic vibrations of functional groups that occur outside the fingerprint region are very useful for the identification of an unknown compound. Most of these vibrations can be regarded as stretching modes, for the lower frequency bending modes usually occur in the fingerprint region and so are less readily identified. The characteristic wavenumbers of some functional groups are listed in Table 19.3.

Table 19.3
Typical vibrational wavenumbers

Vibration type	$\tilde{\nu}$/cm^{-1}
C—H	2850–2960
C—H	1340–1465
C—C stretch, bend	700–1250
C=C stretch	1620–1680
C≡C stretch	2100–2260
O—H stretch	3590–3650
C=O stretch	1640–1780
C≡N stretch	2215–2275
N—H stretch	3200–3500
Hydrogen bonds	3200–3570

Example 19.3

Interpreting an infrared spectrum

The infrared spectrum of an organic compound is shown in Fig. 19.21. Suggest an identification.

Strategy Some of the features at wavenumbers above 1500 cm^{-1} can be identified by comparison with the data in Table 19.3.

Solution (a) C—H stretch of a benzene ring, indicating a substituted benzene; (b) carboxylic acid O—H stretch, indicating a carboxylic acid; (c) the strong absorption of a conjugated C≡C group, indicating a substituted alkyne; (d) this strong absorption is also characteristic of a carboxylic acid that is conjugated to a carbon–carbon multiple bond; (e) a characteristic vibration of a benzene ring,

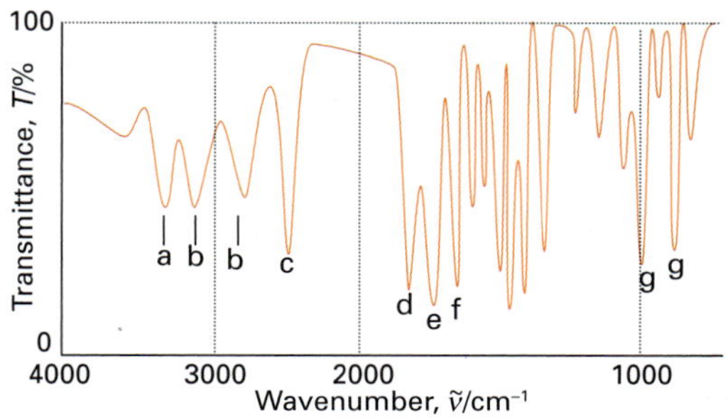

Fig. 19.21 A typical infrared absorption spectrum taken by forming a sample into a disk with potassium bromide. As explained in the example, the substance can be identified as $O_2NC_6H_4$—C≡C—COOH.

confirming the deduction drawn from (a); (f) a characteristic absorption of a nitro group ($—NO_2$) connected to a multiply bonded carbon–carbon system, suggesting a nitro-substituted benzene. The molecule contains as components a benzene ring, an aromatic carbon–carbon bond, a —COOH group, and a $—NO_2$ group. The molecule is in fact $O_2N—C_6H_4—C≡C—COOH$. A more detailed analysis and comparison of the fingerprint region shows it to be the 1,4-isomer.

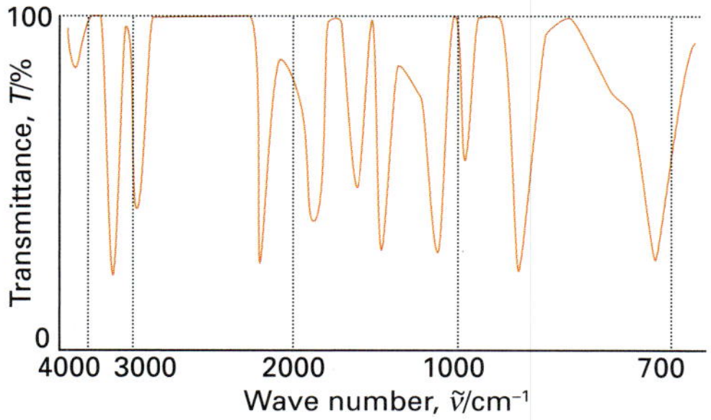

Fig. 19.22 The spectrum considered in Self-test 19.8.

Self-test 19.8

Suggest an identification of the organic compound responsible for the spectrum shown in Fig. 19.22. [*Hint:* The molecular formula of the compound is C_3H_5ClO.]

[*Answer:* $CH_2{=}CClCH_2OH$]

19.12 Vibration–rotation spectra

The vibrational spectra of gas-phase molecules are more complicated than this discussion implies, because the excitation of a vibration also results in the excitation of rotation. The effect is rather like what happens when ice skaters throw out or draw in their arms: they rotate more slowly or more rapidly. The effect on the spectrum is to break the single line resulting from a vibrational transition into a multitude of lines with separations between neighbours that depend on the rotational constant of the molecule.

To establish the so-called 'band structure' of a vibrational transition, we begin by writing the expressions for the vibrational and rotational levels. For a linear molecule (the only type we consider), we combine eqns 19.3 and 19.17 and write

$$E_{v,J} = (v + \tfrac{1}{2})h\nu + hBJ(J+1) \qquad (19.20)$$

(For this part of the discussion it is simpler to express vibrational transitions as frequencies rather than wavenumbers, but the conversion between them is

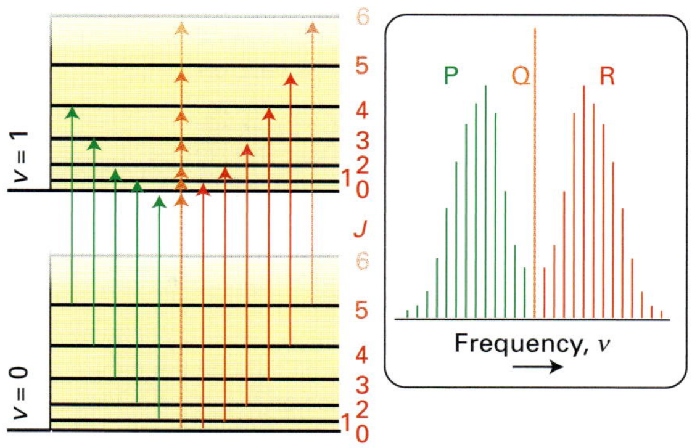

Fig. 19.23 The formation of P, Q, and R branches in a vibration–rotation spectrum. The intensities reflect the populations of the initial rotational levels.

straightforward.) Next, we apply the selection rules. Provided the molecule is polar, or at least acquires a dipole moment in a vibrational transition (as when CO_2 bends or undergoes an asymmetric stretch), the rotational quantum number may change by ±1 or (in some cases, see below) 0. The absorptions then fall into three groups called **branches** of the spectrum.

P branch, transitions with $\Delta J = -1$: $\nu_J = \nu - 2BJ$
Q branch, transitions with $\Delta J = 0$: $\nu_J = \nu$
R branch, transitions with $\Delta J = +1$: $\nu_J = \nu + 2B(J+1)$

Figure 19.23 shows the resulting appearance of the branches of a typical spectrum. The separation between the lines in the P and R branches of a vibrational transition is $2B$. Therefore, the bond length can be deduced without needing to take a pure rotational microwave spectrum. However, the latter is more precise.

A brief comment The Q branch is not always allowed. For example, it is observed in the spectrum of NO, but not in the spectrum of HCl: the difference can be traced to the fact that NO, with an electron in a π orbital, has electronic angular momentum around its internuclear axis but HCl does not.[1]

19.13 Vibrational Raman spectra of polyatomic molecules

The gross selection rule for the vibrational Raman spectrum of a polyatomic molecule is that *the normal mode of vibration is accompanied by a changing polarizability*. However, it is often quite difficult to judge by inspection when this is so. The symmetric stretch of CO_2, for example, alternately swells and contracts the molecule: this motion changes its polarizability, so the mode is Raman active. The other

[1] For more information, see our *Physical Chemistry* (2006).

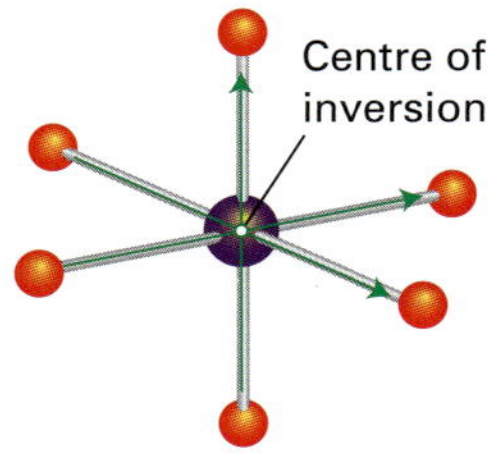

Fig. 19.24 In an inversion operation, we consider every point in a molecule, and project them all through the centre of the molecule out to an equal distance on the other side.

modes of CO_2 are Raman inactive because the polarizability does not change as the atoms move collectively. A very simple explanation (which is not reliable in all cases) is that the polarizability of a molecule depends on its size, and whereas the symmetric stretch changes the size of the molecule neither the antisymmetric stretch nor the bending modes do—at least, to a first approximation.

In some cases it is possible to make use of a very general rule about the infrared and Raman activity of vibrational modes:

> The **exclusion rule** states that if the molecule has a centre of inversion, then no modes can be both infrared and Raman active.

(A mode may be inactive in both.) A molecule has a centre of inversion if it looks unchanged when each atom is projected through a single point and out an equal distance on the other side (Fig. 19.24). Because we can often judge intuitively when a mode changes the molecular dipole moment, we can use this rule to identify modes that are not Raman active. The rule applies to CO_2 but to neither H_2O nor CH_4 because they have no centre of symmetry. Thus, both the antisymmetric stretch and the bending modes of CO_2 are infrared active, so we know at once that they are Raman inactive, as we asserted above.

Self-test 19.9

One vibrational mode of benzene is a 'breathing mode' in which the ring alternately expands and contracts. May it be vibrationally Raman active?

[*Answer:* yes]

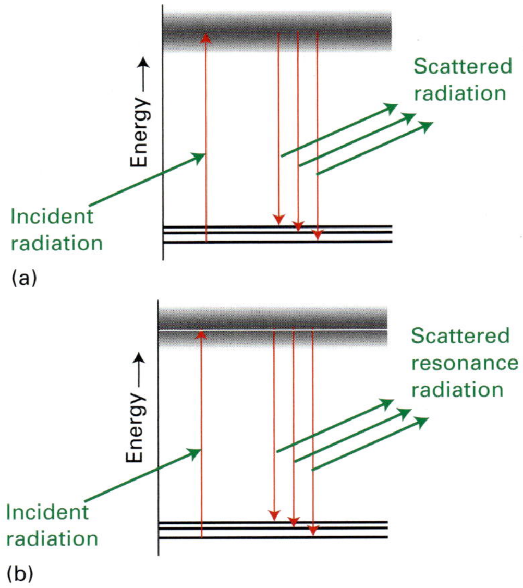

Fig. 19.25 (a) In Raman spectroscopy, an incident photon is scattered from a molecule with either an increase in frequency (if the radiation collects energy from the molecule) or—as shown here—with a lower frequency if it loses energy to the molecule. The process can be regarded as taking place by an excitation of the molecule to a wide range of states (represented by the shaded band), and the subsequent return of the molecule to a lower state; the net energy change is then carried away by the photon. (b) In the *resonance Raman effect*, the incident radiation has a frequency corresponding to an actual electronic excitation of the molecule. A photon is emitted when the excited state returns to a state close to the ground state.

A modification of the basic Raman effect involves using incident radiation that nearly coincides with the frequency of an electronic transition of the sample (Fig. 19.25). The technique is then called **resonance Raman spectroscopy.** It is characterized by a much greater intensity in the scattered radiation. Furthermore, because it is often the case that only a few vibrational modes contribute to the more intense scattering, the spectrum is greatly simplified. Resonance Raman spectroscopy is used to study biological molecules that absorb strongly in the ultraviolet and visible regions of the spectrum. Examples include the haem co-factors in haemoglobin and the cytochromes and the pigments β-carotene and chlorophyll, which capture solar energy during plant photosynthesis.

Checklist of key ideas

You should now be familiar with the following concepts.

☐ **1** The populations of rotational energy levels are given by the Boltzmann distribution in connection with noting the degeneracy of each level.

☐ **2** The intensity of a transition is proportional to the square of the transition dipole moment.

☐ **3** A selection rule is a statement about when the transition dipole may be nonzero.

☐ **4** A gross selection rule specifies the general features a molecule must have if it is to have a spectrum of a given kind.

☐ **5** A specific selection rule is a statement about which changes in quantum number may occur in a transition.

☐ **6** The gross selection rule for rotational transitions is that the molecule must be polar. The specific selection rules are in the following table.

☐ **7** The Pauli principle states for fermions $\psi(\mathrm{B,A}) = -\psi(\mathrm{A,B})$ and for bosons $\psi(\mathrm{B,A}) = \psi(\mathrm{A,B})$. The consequences of the Pauli principle for rotational states are called nuclear statistics.

☐ **8** The rotational spectrum of a polar linear molecule and of a polar symmetric rotor consists of a series of lines at frequencies separated by $2B$.

☐ **9** One contribution to the linewidth is the Doppler effect; another contribution is lifetime broadening.

☐ **10** In a Raman spectrum lines shifted to lower frequency than the incident radiation are called Stokes lines and lines shifted to higher frequency are called anti-Stokes lines.

☐ **11** The gross selection rule for rotational Raman spectra is that the polarizability of the molecule must be anisotropic. The specific selection rules are in the table of key equations below.

☐ **12** The gross selection rule for vibrational spectra is that the electric dipole moment of the molecule must change during the vibration. The specific selection rules are in the table of key equations below.

☐ **13** The number of vibrational modes of nonlinear molecules is $3N-6$; for linear molecules the number is $3N-5$.

☐ **14** Rotational transitions accompany vibrational transitions and split the spectrum into a P branch ($\Delta J = -1$), a Q branch ($\Delta J = 0$), and an R branch ($\Delta J = +1$). A Q branch is observed only when the molecule possesses angular momentum around its axis.

☐ **15** The gross selection rule for the vibrational Raman spectrum of a polyatomic molecule is that the normal mode of vibration is accompanied by a changing polarizability.

☐ **16** The exclusion rule states that if the molecule has a centre of inversion, then no modes can be both infrared and Raman active.

☐ **17** In resonance Raman spectroscopy, radiation that nearly coincides with the frequency of an electronic transition is used to excite the sample and the result is a much greater intensity in the scattered radiation.

19.15 Suppose that hydrogen is replaced by deuterium in $^1H^{35}Cl$. Would you expect the $J = 1 \leftarrow 0$ transition to move to higher or lower wavenumber?

19.16 The wavenumber of the incident radiation in a Raman spectrometer is 20 623 cm^{-1}. What is the wavenumber of the scattered Stokes radiation for the $J = 4 \leftarrow 2$ transition of $^{16}O_2$?

19.17 The rotational constant of $^{12}C^{16}O_2$ (from Raman spectroscopy) is 11.70 GHz. What is the CO bond length in the molecule?

19.18 The microwave spectrum of $^1H^{127}I$ consists of a series of lines separated by 384 GHz. Compute its bond length. What would be the separation of the lines in $^2H^{127}I$?

19.19 The following wavenumbers are observed in the rotational spectrum of OCS: 1.217 105 4 cm^{-1}, 1.1.622 800 5 cm^{-1}, 2.028 488 3 cm^{-1}, and 2.434 170 8 cm^{-1}. Use the graphical procedure implied by eqn 19.12 to infer the values of B and D for this molecule.

19.20 The microwave spectrum of $^{16}O^{12}CS$ gave absorption lines (in GHz) as follows:

J	1	2	3	4
^{32}S	24.325 92	36.488 82	48.651 64	60.814 08
^{34}S	23.732 33		47.462 40	

Assume that the bond lengths are unchanged by substitution and calculate the CO and CS bond lengths in OCS. *Hint:* The moment of inertia of a linear molecule of the form ABC is

$$I = m_A R_{AB}^2 + m_C R_{BC}^2 - \frac{(m_A R_{AB} - m_C R_{BC})^2}{m_A + m_B + m_C}$$

where r_{AB} and r_{BC} are the A–B and B–C bond lengths, respectively.

19.21 What is the Doppler-shifted wavelength of a red (660 nm) traffic light approached at 65 mph? At what speed would it appear green (520 nm)?

19.22 A spectral line of $^{48}Ti^{8+}$ in a distant star was found to be shifted from 654.2 nm to 706.5 nm and to be broadened to 61.8 pm. What is the speed of recession and the surface temperature of the star?

19.23 Estimate the lifetime of a state that gives rise to a line of width (a) 0.10 cm^{-1}, (b) 1.0 cm^{-1}, (c) 1.0 GHz.

19.24 A molecule in a liquid undergoes about 1.0×10^{13} collisions in each second. Suppose that (a) every collision is effective in deactivating the molecule vibrationally and (b) that one collision in 200 is effective. Calculate the width (in cm^{-1}) of vibrational transitions in the molecule.

19.25 Suppose the C=O group in a peptide bond can be regarded as isolated from the rest of the molecule. Given the force constant of the bond in a carbonyl group is 908 N m^{-1}, calculate the vibrational frequency of (a) $^{12}C{=}^{16}O$, (b) $^{13}C{=}^{16}O$.

19.26 The wavenumber of the fundamental vibrational transition of Cl_2 is 565 cm^{-1}. Calculate the force constant of the bond.

19.27 The hydrogen halides have the following fundamental vibrational wavenumbers:

	HF	HCl	HBr	HI
$\tilde{\nu}$/cm^{-1}	4141.3	2988.9	2649.7	2309.5

Calculate the force constants of the hydrogen–halogen bonds.

19.28 From the data in Exercise 19.26, predict the fundamental vibrational wavenumbers of the deuterium halides.

19.29 Infrared absorption by $^1H^{81}Br$ gives rise to an R branch from $v = 0$. What is the wavenumber of the line originating from the rotational state with $J = 2$?

19.30 Which of the following molecules may show infrared absorption spectra: (a) H_2; (b) HCl; (c) CO_2; (d) H_2O; (e) CH_3CH_3; (f) CH_4; (g) CH_3Cl; (h) N_2?

19.31 How many normal modes of vibration are there for (a) NO_2, (b) N_2O, (c) cyclohexane, (d) hexane?

19.32 Consider the vibrational mode that corresponds to the uniform expansion of the benzene ring. Is it (a) Raman, (b) infrared active?

19.33 Suppose that three conformations are proposed for the nonlinear molecule H_2O_2 (**1**, **2**, and **3**). The infrared absorption spectrum of gaseous H_2O_2 has bands at 870, 1370, 2869, and 3417 cm^{-1}. The Raman spectrum of the same sample has bands at 877, 1408, 1435, and 3407 cm^{-1}. All bands correspond to fundamental vibrational wavenumbers and you may assume that: (i) the 870 and 877 cm^{-1} bands arise from the same normal mode, and (ii) the 3417 and 3407 cm^{-1} bands arise from the same normal mode. (a) If H_2O_2 were linear, how many normal modes of vibration would it have? (b) Determine which of the proposed conformations is inconsistent with the spectroscopic data. Explain your reasoning.

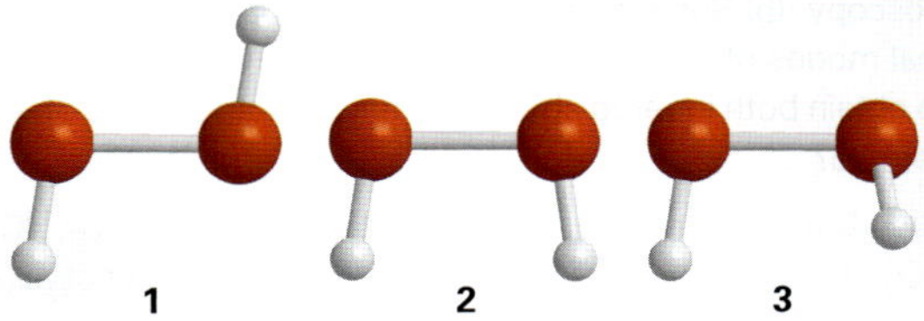

1 **2** **3**

Projects

The symbol ‡ indicates that calculus is required.

19.34‡ The most populated rotational energy level of a linear rotor is given in eqn 19.10. What is the most populated rotational level of a spherical rotor, given that its degeneracy is $(2J + 1)^2$?

19.35 The protein haemerythrin (hemerythrin, Her) is responsible for binding and carrying O_2 in some invertebrates. Each protein molecule has two Fe^{2+} ions that are in very close proximity and work together to bind one molecule of O_2. The Fe_2O_2 group of oxygenated haemerythrin is coloured and has an electronic absorption band at 500 nm. (a) The resonance

Raman spectrum of oxygenated haemerythrin obtained with laser excitation at 500 nm has a band at 844 cm^{-1} that has been attributed to the O—O stretching mode of bound $^{16}O_2$. Why is resonance Raman spectroscopy and not infrared spectroscopy the method of choice for the study of the binding of O_2 to haemerythrin? (b) Proof that the 844 cm^{-1} band in the resonance Raman spectrum of oxygenated haemerythrin arises from a bound O_2 species may be obtained by conducting experiments on samples of haemerythrin that have been mixed with $^{18}O_2$, instead of $^{16}O_2$. Predict the fundamental vibrational wavenumber of the $^{18}O—^{18}O$ stretching mode in a sample of haemerythrin that has been treated with $^{18}O_2$. (c) The fundamental vibrational wavenumbers for the O—O stretching modes of O_2, O_2^- (superoxide anion), and O_2^{2-} (peroxide anion) are 1555, 1107, and 878 cm^{-1}, respectively. (i) Explain this trend in terms of the electronic structures of O_2, O_2^-, and O_2^{2-}. (ii) What are the bond orders of O_2, O_2^-, and O_2^{2-}? (d) Based on the data given in part (c), which of the following species best describes the Fe_2O_2 group of haemerythrin: $Fe^{2+}{}_2O_2$, $Fe^{2+}Fe^{3+}O_2^-$, or $Fe^{3+}{}_2O_2^{2-}$? Explain your reasoning. (e) The resonance Raman spectrum of haemerythrin mixed with $^{16}O^{18}O$ has two bands that can be attributed to the O—O stretching mode of bound oxygen. Discuss how this observation may be used to exclude one or more of the four proposed schemes (**4–7**) for binding of O_2 to the Fe_2 site of haemerythrin.

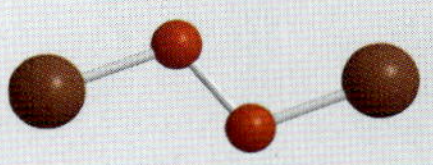

4

5

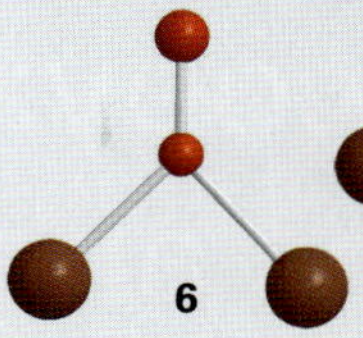

6

7

19.36 We saw in Box 19.1 that water, carbon dioxide, and methane are able to absorb some of the Earth's infrared emissions whereas nitrogen and oxygen cannot. The computational methods discussed in Section 14.17 can also be used to simulate vibrational spectra and from the results of the calculation it is possible to determine the correspondence between a vibrational frequency and the atomic displacements that give rise to a normal mode. (a) Using molecular modelling software and the computational method of your instructor's choice, investigate and depict pictorially the vibrational normal modes of CH_4, CO_2, and H_2O in the gas phase. (b) Which vibrational modes of CH_4, CO_2, and H_2O are responsible for absorption of infrared radiation?

Chapter 20

Spectroscopy: electronic transitions and photochemistry

The energy needed to change the occupation of orbitals in a molecule is of the order of several electronvolts. Consequently, the photons emitted or absorbed when such changes occur lie in the visible and ultraviolet regions of the spectrum, which spread from about 14 000 cm^{-1} for red light to 21 000 cm^{-1} for blue, and on to 50 000 cm^{-1} for ultraviolet radiation (Table 20.1).

A brief comment The electronvolt (eV) is a convenient unit for expressing changes in electronic energy: 1 eV = $e \times 1$ V = 1.60×10^{-19} J (corresponding to 96.5 kJ mol^{-1}). Think of it as the energy needed to move an electron through a potential difference of 1 V.

Many of the colours of the objects in the world around us, including the green of vegetation, the colours of flowers and of synthetic dyes, and the colours of pigments and minerals, stem from transitions in which an electron makes a transition from one orbital of a molecule or ion into another. The change in the distribution of probability density of an electron that takes place when chlorophyll absorbs red and blue light (leaving green to be reflected) is the primary energy harvesting step by which our planet captures energy from the Sun and uses it to drive the nonspontaneous reactions of photosynthesis. In some cases the relocation of an electron may be so extensive that it results in the breaking of a bond and the dissociation of the molecule: such processes give rise to the numerous reactions of photochemistry, including the reactions that sustain or damage the atmosphere.

Ultraviolet and visible spectra

White light is a mixture of light of all different colours. The removal, by absorption, of any one of these colours from white light results in the

Table 20.1
Colour, frequency, and energy of light

Colour	λ/nm	$\nu/(10^{14}$ Hz)	$\tilde{\nu}$/(104 cm^{-1})	E/eV	E/(kJ mol^{-1})
Infrared	>1000	<3.00	<1.00	<1.24	<120
Red	700	4.28	1.43	1.77	171
Orange	620	4.84	1.61	2.00	193
Yellow	580	5.17	1.72	2.14	206
Green	530	5.66	1.89	2.34	226
Blue	470	6.38	2.13	2.64	254
Violet	420	7.14	2.38	2.95	285
Near ultraviolet	300	10.0	3.3	4.15	400
Far ultraviolet	200	15.0	5.00	6.20	598

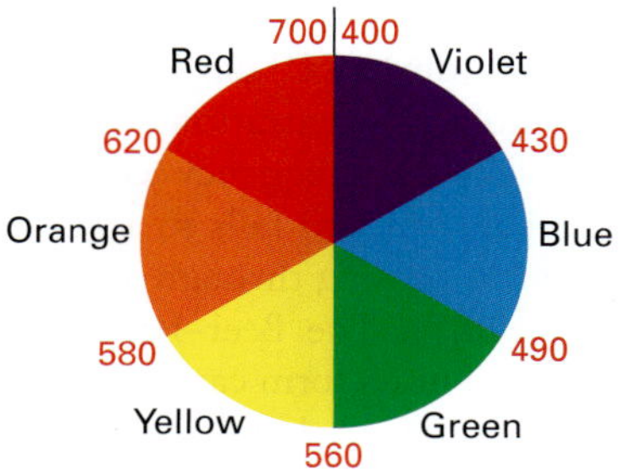

Fig. 20.1 An artist's colour wheel: complementary colours are opposite one another on a diameter. The numbers correspond to wavelengths of light in nm.

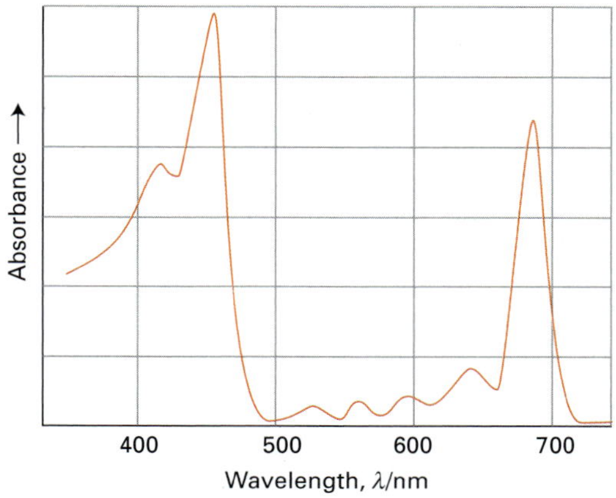

Fig. 20.2 The absorption spectrum of chlorophyll in the visible region. Note that it absorbs in the red and blue regions, and that green light is not absorbed.

complementary colour being observed. For instance, the absorption of red light from white light by an object results in that object appearing green, the complementary colour of red. Conversely, the absorption of green results in the object appearing red. The pairs of complementary colours are neatly summarized by the artist's colour wheel shown in Fig. 20.1, where complementary colours lie opposite one another along a diameter.

It should be stressed, however, that the perception of colour is a very subtle phenomenon. Although an object may appear green because it absorbs red light, it may also appear green because it absorbs all colours from the incident light *except* green. This is the origin of the colour of vegetation, because chlorophyll absorbs in two regions of the spectrum, leaving green to be reflected (Fig. 20.2). Moreover, an absorption band may be very broad, and although it may be a maximum at one particular wavelength, it may have a long tail that spreads into other regions (Fig. 20.3). In such cases, it is very difficult to predict the perceived colour from the location of the absorption maximum.

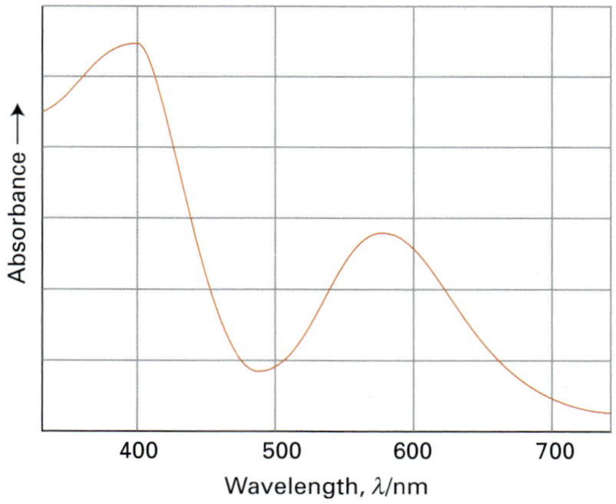

Fig. 20.3 An electronic absorption of a species in solution is typically very broad and consists of several broad bands.

20.1 Practical considerations

For the visible region of the spectrum, a *tungsten–iodine lamp* is used as the source in an absorption spectrometer: it gives out intense white light. A discharge through deuterium gas or xenon in quartz is still widely used for the near ultraviolet.

The simplest dispersing element is a glass or quartz prism, but modern instruments use a diffraction grating. For work in the visible region of the spectrum, the device consists of a glass or ceramic plate into which fine grooves have been cut about 1000 nm apart (a spacing comparable to the wavelength of visible light) and covered with a reflective aluminium coating. The grating causes interference between waves reflected from its surface, and constructive interference occurs at specific angles that depend on the frequency of the radiation being used. Thus, each wavelength of light is directed into a specific direction (Fig. 20.4). In a *monochromator*, a narrow exit slit allows only a narrow range of wavelengths to reach the detector. Turning the grating around an axis perpendicular to the incident and diffracted beams allows different wavelengths to be analysed; in this way, the absorption spectrum is built up one narrow wavelength range at a time.

Detectors may consist of a single radiation-sensing element or of several small elements arranged in one or two-dimensional arrays. A common detector is a *photodiode*, a solid-state device that conducts electricity when struck by photons because light-induced electron transfer reactions in the detector material create mobile charge carriers (negatively charged electrons and positively charged 'holes'). Silicon is sensitive in the visible region. A *charge-coupled device* (CCD) is a two-dimensional array of several million photodiode detectors. With a CCD, a wide range of wavelengths that emerge from a polychromator are detected simultaneously, thus eliminating the need to measure light intensity one narrow wavelength range at a time. CCD detectors are used widely to monitor absorption, emission, and Raman scattering.

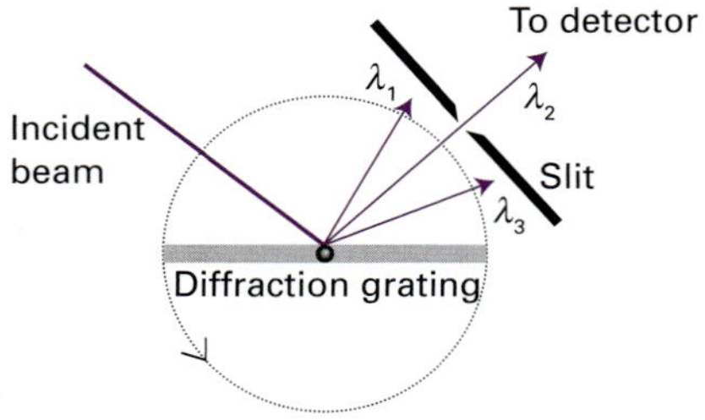

Fig. 20.4 A beam of light is dispersed by a diffraction grating into three component wavelengths λ_1, λ_2, and λ_3. In the configuration shown, only radiation with λ_2 passes through a narrow slit and reaches the detector. Rotating the diffraction grating in the direction shown by the double arrows allows λ_1 and λ_3 to reach the detector.

20.2 Absorption intensities

The intensity of absorption of radiation at a particular wavelength is related to the concentration [J] of the absorbing species by the Beer–Lambert law:

$$I = I_0 10^{-\varepsilon[\mathrm{J}]L} \tag{20.1}$$

I_0 and I are the incident and transmitted intensities, respectively, L is the length of the sample, and ε (epsilon) is the **molar absorption coefficient** (formerly and still widely the 'extinction coefficient'), with dimensions of 1/(molar concentration × length). Typical values of ε for strong transitions are of the order of 10^4–10^5 dm^3 mol^{-1} cm^{-1}, indicating that in a solution of molar concentration 0.01 mol dm^{-3} the intensity of light (of frequency corresponding to the maximum absorption) falls to 10 per cent of its initial value after passing through about 0.1 mm of solution (Fig. 20.5). The Beer–Lambert law is an empirical result, but its form can be justified by considering the passage of light through a uniform, absorbing medium (Further information 20.1).

The **absorbance** $A = \varepsilon[\mathrm{J}]L$ of a sample is measured by using the incident and final intensities of a light beam and using eqn 20.1 in the form

$$A = \log \frac{I_0}{I} \tag{20.2}$$

(The logarithm is a common logarithm, to the base 10.) It is common to report the absorption of radiation in terms of the **transmittance**, T, of a sample at a given frequency, where

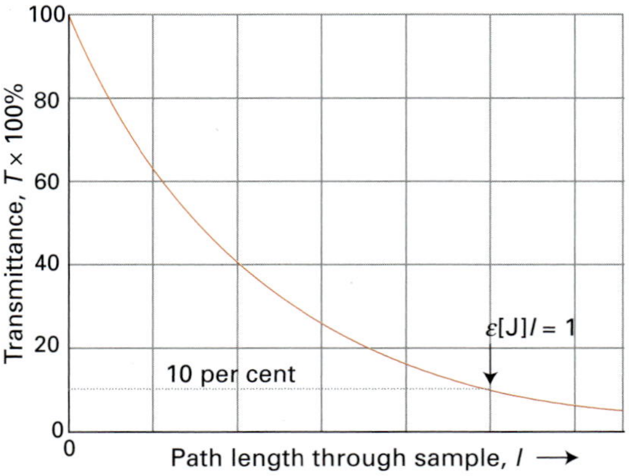

Fig. 20.5 The intensity of light transmitted by an absorbing sample decreases exponentially with the path length through the sample.

$$T = \frac{I}{I_0} \qquad (20.3)$$

Thus, $A = -\log T$. From the value of A (or T) we can obtain the concentration of the absorbing species by using

$$[\mathrm{J}] = \frac{A}{\varepsilon L} \qquad (20.4)$$

which is obtained by taking logarithms of both sides of eqn 20.1 and then using eqn 20.2.

Example 20.1

Using the molar absorption coefficient

Radiation of wavelength 256 nm passed through 1.0 mm of a solution that contained benzene at a concentration of 0.050 mol dm^{-3} in a transparent solvent. The light intensity is reduced to 16 per cent of its initial value (so $T = 0.16$). Calculate the absorbance and the molar absorption coefficient of the benzene. What would be the transmittance through a cell of thickness 2.0 mm?

Strategy With $A = -\log T$, eqn 20.4 can be rearranged into

$$\varepsilon = -\frac{\log T}{[\mathrm{J}]L}$$

For the transmittance through the thicker cell, we use the value of ε calculated here and $T = 10^{-A}$.

Solution The molar absorption coefficient is

$$\varepsilon = -\frac{\log 0.16}{(0.050\ \mathrm{mol\ dm^{-3}}) \times (1.0\ \mathrm{mm})} = 16\ \mathrm{dm^3\ mol^{-1}\ mm^{-1}}$$

These units are convenient for the rest of the calculation (but the outcome could be reported as 1.6×10^2 dm^3 mol^{-1} cm^{-1} if desired). The absorbance is

$$A = -\log 0.16 = 0.80$$

The absorbance of a sample of length 2.0 mm is

$$A = \varepsilon[\mathrm{J}]L = (16\ \mathrm{dm^3\ mol^{-1}\ mm^{-1}}) \times (0.050\ \mathrm{mol\ dm^{-3}}) \times (2.0\ \mathrm{mm}) = 1.6$$

It follows that the transmittance is then

$$T = 10^{-A} = 10^{-1.6} = 0.025$$

That is, the emergent light is reduced to 2.5 per cent of its incident intensity.

Self-test 20.1

The transmittance of an aqueous solution that contained Cu^{2+} ions at a molar concentration of 0.10 mol dm^{-3} was measured as 0.30 at 600 nm in a cell of length 5.0 mm. Calculate the molar absorption coefficient of Cu^{2+}(aq) at that wavelength, and the absorbance of the solution. What would be the transmittance through a cell of length 1.0 mm?

[*Answer:* 10 dm^3 mol^{-1} cm^{-1}, $A = 0.52$, $T = 0.79$]

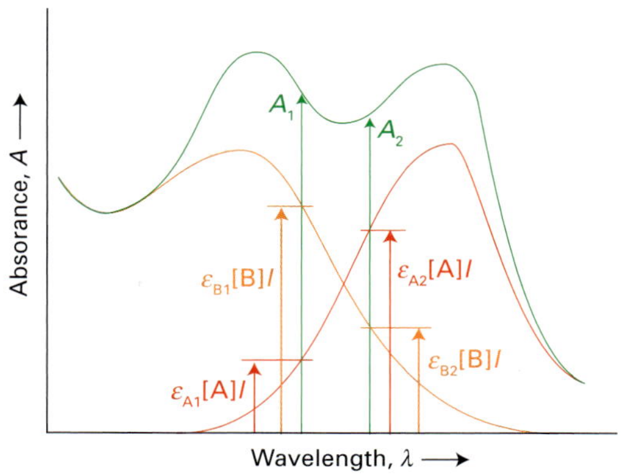

Fig. 20.6 The concentrations of two absorbing species in a mixture can be determined from their molar absorption coefficients and the measurement of their absorbances at two different wavelengths lying within their joint absorption region.

Measurements at two wavelengths can be used to find the individual concentrations of two components A and B in a mixture. For this analysis, we write the total absorbance at a given wavelength as

$$A = A_A + A_B = \varepsilon_A[\mathrm{A}]L + \varepsilon_B[\mathrm{B}]L = (\varepsilon_A[\mathrm{A}] + \varepsilon_B[\mathrm{B}])L$$

Then, for two measurements of the total absorbance at wavelengths λ_1 and λ_2 at which the molar absorption coefficients are ε_1 and ε_2 (Fig. 20.6), we have

$$A_1 = (\varepsilon_{A1}[\mathrm{A}] + \varepsilon_{B1}[\mathrm{B}])L \qquad A_2 = (\varepsilon_{A2}[\mathrm{A}] + \varepsilon_{B2}[\mathrm{B}])L$$

As shown in Derivation 20.1, these two simultaneous equations can be solved for the two unknowns, the molar concentrations of A and B:

$$[\mathrm{A}] = \frac{\varepsilon_{B2}A_1 - \varepsilon_{B1}A_2}{(\varepsilon_{A1}\varepsilon_{B2} - \varepsilon_{A2}\varepsilon_{B1})L} \qquad (20.5a)$$

$$[\mathrm{B}] = \frac{\varepsilon_{A1}A_2 - \varepsilon_{A2}A_1}{(\varepsilon_{A1}\varepsilon_{B2} - \varepsilon_{A2}\varepsilon_{B1})L} \qquad (20.5b)$$

Derivation 20.1

Determining concentrations in a mixture

The two equations to solve for [A] and [B] are

$$\varepsilon_{A1}[\mathrm{A}]L + \varepsilon_{B1}[\mathrm{B}]L = A_1 \qquad \varepsilon_{A2}[\mathrm{A}]L + \varepsilon_{B2}[\mathrm{B}]L = A_2$$

To match the two second terms, multiply the first by ε_{B2} and the second by ε_{B1}, to obtain

$$\varepsilon_{B2}\varepsilon_{A1}[\mathrm{A}]L + \varepsilon_{B2}\varepsilon_{B1}[\mathrm{B}]L = \varepsilon_{B2}A_1$$
$$\varepsilon_{B1}\varepsilon_{A2}[\mathrm{A}]L + \varepsilon_{B1}\varepsilon_{B2}[\mathrm{B}]L = \varepsilon_{B1}A_2$$

When the second is subtracted from the first, we obtain

$$\varepsilon_{B2}\varepsilon_{A1}[\mathrm{A}]L - \varepsilon_{B1}\varepsilon_{A2}[\mathrm{A}]L = \varepsilon_{B2}A_1 - \varepsilon_{B1}A_2$$

which rearranges into eqn 20.5a. To obtain eqn 20.5b, repeat the process by multiplying the first equation by ε_{A2} and the second by ε_{A1} so that the [A] terms cancel when the two equations are subtracted.

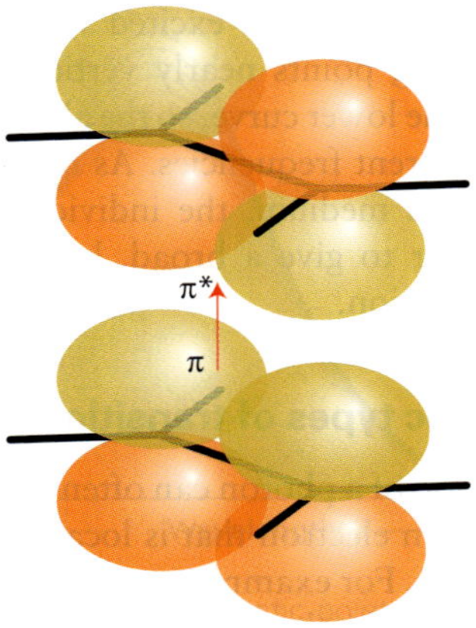

Fig. 20.12 A carbon–carbon double bond acts as a chromophore. One of its important transitions is the π-to-π* transition illustrated here, in which an electron is promoted from a π orbital to the corresponding antibonding orbital.

visible region of the spectrum. Many of the reds and yellows of vegetation are due to transitions of this kind. For example, the carotenes that are present in green leaves (but are concealed by the intense absorption of the chlorophyll until the latter decays) collect some of the solar radiation incident on the leaf by a π-to-π* transition in their long conjugated hydrocarbon chains. A similar type of absorption is responsible for the primary process of vision (Box 20.1).

A d-metal complex may absorb light as a result of the transfer of an electron from the ligands into the d orbitals of the central atom, or vice versa. In such **charge-transfer transitions** the electron moves through a considerable distance, which means that the redistribution of charge as measured by the transition dipole moment may be large and the absorption correspondingly intense. This mode of chromophore activity is shown by the permanganate ion, MnO_4^-:

Box 20.1 Vision

The eye is an exquisite photochemical organ that acts as a transducer, converting radiant energy into electrical signals that travel along neurons. Here we concentrate on the events taking place in the human eye, but similar processes occur in all animals. Indeed, a single type of protein, rhodopsin, is the primary receptor for light throughout the animal kingdom, which indicates that vision emerged very early in evolutionary history, no doubt because of its enormous value for survival.

Photons enter the eye through the cornea, pass through the ocular fluid that fills the eye, and fall on the retina. The ocular fluid is principally water, and passage of light through this medium is largely responsible for the *chromatic aberration* of the eye, the blurring of the image as a result of different frequencies being brought to slightly different focuses. The chromatic aberration is reduced to some extent by the tinted region called the *macular pigment* that covers part of the retina. The pigments in this region are the carotene-like xanthophylls (**B1**), which remove some of the blue light and hence help to sharpen the image. They also protect the photoreceptor molecules from too great a flux of potentially dangerous high-energy photons. The xanthophylls have delocalized electrons that spread along the chain of conjugated double bonds, and the π-to-π* transition lies in the visible.

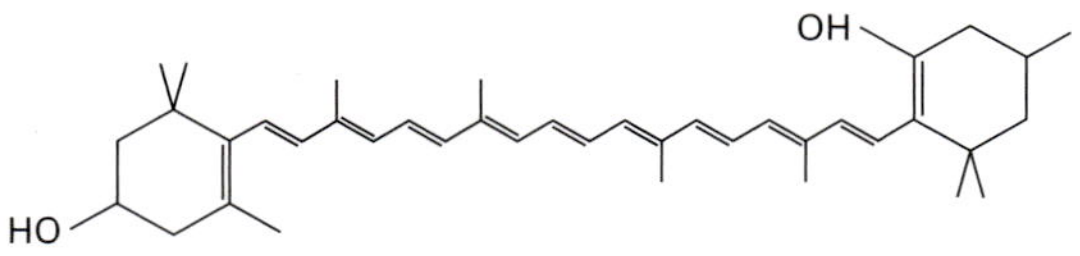

B1 A xanthophyll

About 57 per cent of the photons that enter the eye reach the retina; the rest are scattered or absorbed by the ocular fluid. Here the primary act of vision takes place, in which the chromophore of a rhodopsin molecule absorbs a photon in another π-to-π* transition. A rhodopsin molecule consists of an opsin protein molecule to which is attached a 11-*cis*-retinal molecule (**B2**). The latter resembles half a carotene molecule, showing Nature's economy in its use of available materials. The attachment is by the formation of a Schiff's base, utilizing the —CHO group of the chromophore. The free 11-*cis*-retinal molecule absorbs in the ultraviolet, but attachment to the opsin protein molecule shifts the absorption into the visible region. The rhodopsin molecules are situated in the membranes of special cells (the 'rods' and the 'cones') that cover the retina. The opsin molecule is anchored into the cell membrane by two hydrophobic groups and largely surrounds the chromophore (see the illustration).

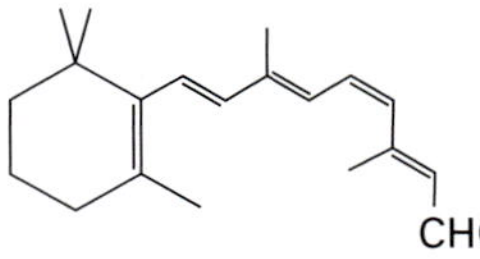

B2 11-*cis*-Retinal

Immediately after the absorption of a photon, the 11-*cis*-retinal molecule undergoes photoisomerization into all-*trans*-retinal (**B3**). Photoisomerization takes about 200 fs and about 67 pigment molecules isomerize for every

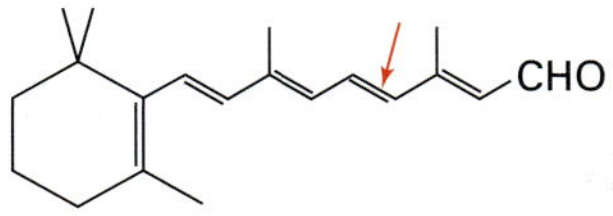

B3 All-*cis*-retinal

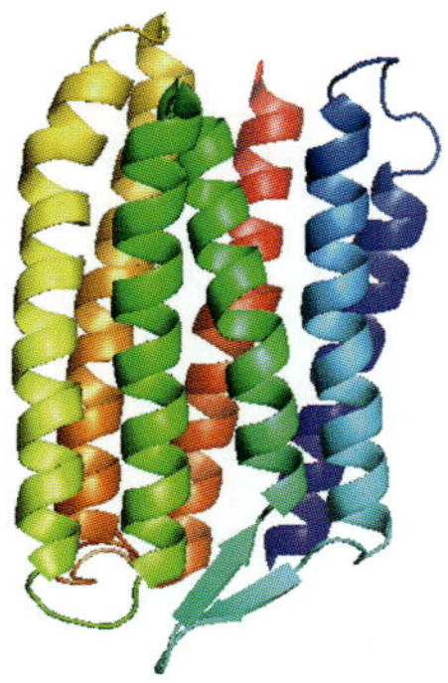

The structure of the rhodopsin molecule.

100 photons that are absorbed. The process is able to occur because the π-to-π* excitation of an electron loosens one of the π bonds (the one indicated by the arrow in the diagram), its torsional rigidity is lost, and one part of the molecule swings round into its new position. At that point, the molecule returns to its ground state, but is now trapped in its new conformation. The straightened tail of the all-*trans*-retinal results in the molecule taking up more space than 11-*cis*-retinal did, so the molecule presses against the coils of the opsin molecule that surrounds it. Thus, in about 0.25–0.50 ms from the initial absorption event, the rhodopsin molecule is activated.

Now a sequence of biochemical events—the *biochemical cascade*—converts the altered configuration of the rhodopsin molecule into a pulse of electric potential that travels through the optical nerve into the optical cortex, where it is interpreted as a signal and incorporated into the web of events we call 'vision'. At the same time, the resting state of the rhodopsin molecule is restored by a series of nonradiative chemical events powered by ATP. The process involves the escape of all-*trans*-retinal as all-*trans*-retinol (in which —CHO has been reduced to $—CH_2OH$) from the opsin molecule by a process catalysed by the enzyme rhodopsin kinase and the attachment of another protein molecule, arrestin. The free all-*trans*-retinol molecule now undergoes enzyme-catalysed isomerization into 11-*cis*-retinol followed by dehydrogenation to form 11-*cis*-retinal, which is then delivered back into an opsin molecule. At this point, the cycle of excitation, photoisomerization, and regeneration is ready to begin again.

the charge redistribution that accompanies the migration of an electron from the O atoms to the central Mn atom accounts for its intense purple colour (resulting from absorption in the range 420–700 nm).

Self-test 20.2

Estimate the wavelength of maximum absorption for a transition of energy 4.3 eV.

[*Answer:* 288 nm]

Radiative and nonradiative decay

In most cases, the excitation energy of a molecule that has absorbed a photon is degraded into the disordered thermal motion of its surroundings in a process known as **internal conversion** (IC). However, one process by which an electronically excited molecule can discard its excess energy is by **radiative decay**, in which an electron relaxes back into a lower energy orbital and in the process generates a photon. As a result, and if the emitted radiation is in the visible region of the spectrum, an observer sees the sample glowing.

There are two principal modes of radiative decay, fluorescence and phosphorescence (Fig. 20.13). In **fluorescence**, the spontaneously emitted radiation ceases very soon (within nanoseconds) after the exciting radiation is extinguished. In **phosphorescence**, the spontaneous emission may persist for long periods—even hours, but characteristically seconds or fractions of seconds. The difference suggests that fluorescence is an immediate conversion of absorbed light into re-emitted radiant energy and that phosphorescence involves the storage of energy in a reservoir from which it slowly leaks.

Other than thermal degradation, a nonradiative fate for an electronically excited molecule is **dissociation**, or fragmentation (Fig. 20.14). The onset of dissociation can be detected in an absorption spectrum by seeing that the vibrational structure of a band terminates at a certain energy. Absorption occurs in a continuous band above this **dissociation limit**, the highest frequency before the onset of continuous absorption, because the final state is unquantized

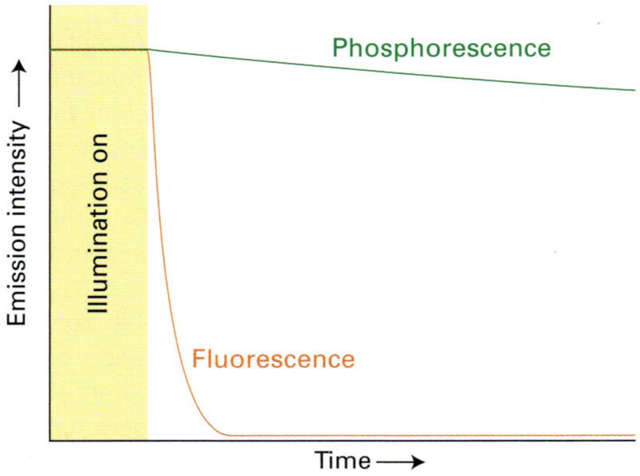

Fig. 20.13 The empirical (observation-based) distinction between fluorescence and phosphorescence is that the former is extinguished very quickly after the exciting source is removed, whereas the latter continues with relatively slowly diminishing intensity.

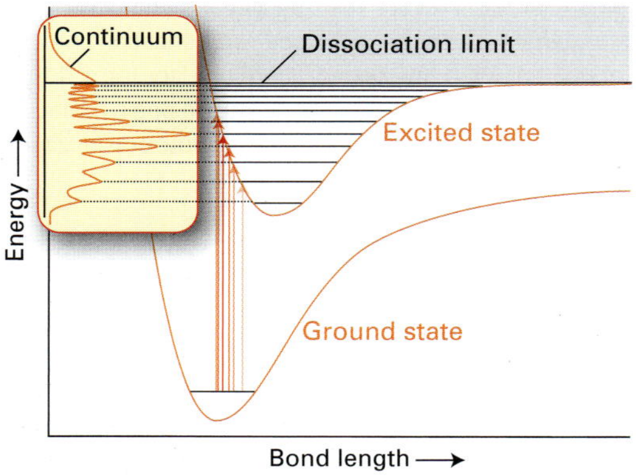

Fig. 20.14 When absorption occurs to unbound states of the upper electronic state, the molecule dissociates and the absorption is a continuum. Below the dissociation limit the electronic spectrum has a normal vibrational structure.

translational motion of the fragments. Locating the dissociation limit is a valuable way of determining the bond dissociation energy. Dissociation may also occur if the electronic transition takes place directly to a purely repulsive state, one that shows no minimum corresponding to bonding.

20.5 Fluorescence

Figure 20.15 is a simple example of a **Jablonski diagram**, a schematic portrayal of molecular electronic and vibrational energy levels, which shows the sequence of steps involved in fluorescence. The initial absorption takes the molecule to an excited electronic state, and if the absorption spectrum were monitored

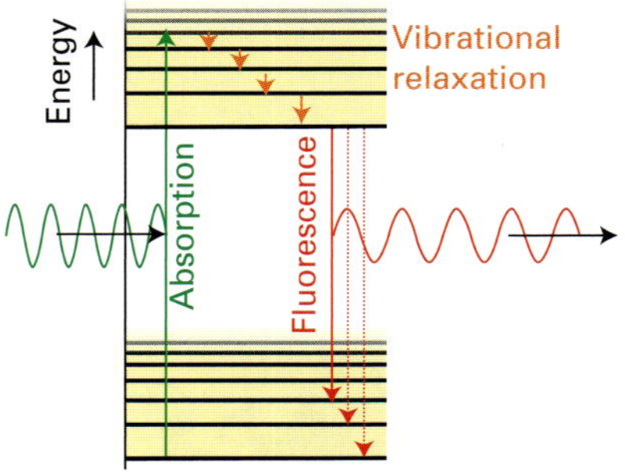

Fig. 20.15 A Jablonski diagram showing the sequence of steps leading to fluorescence. After the initial absorption the upper vibrational states undergo radiationless decay—the process of vibrational relaxation—by giving up energy to the surroundings. A radiative transition then occurs from the ground state of the upper electronic state. In practice, the separation of the ground states of the electronic states (the heavy horizontal lines) is 10 to 100 times greater than the separation of the vibrational levels.

See an animated version of this figure in the interactive ebook.

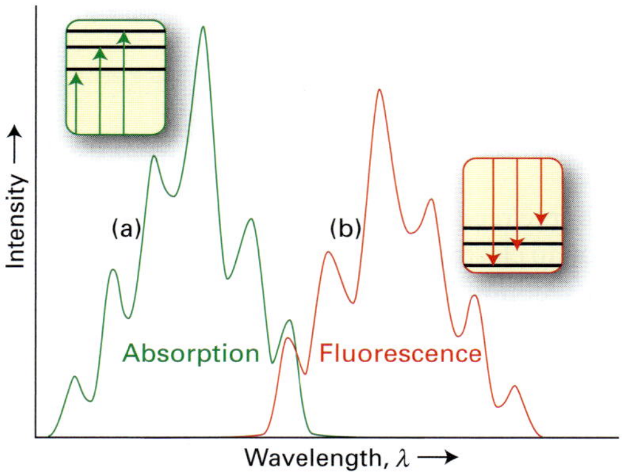

Fig. 20.16 The absorption spectrum (a) shows a vibrational structure characteristic of the upper state. The fluorescence spectrum (b) shows a structure characteristic of the lower state; it is also displaced to lower frequencies and resembles a mirror image of the absorption.

it would look like the one shown in Fig. 20.16a. The excited molecule is subjected to collisions with the surrounding molecules, and as it gives up energy it steps down the ladder of vibrational levels. The surrounding molecules, however, might be unable to accept the larger energy needed to lower the molecule to the ground electronic state. The excited state might therefore survive long enough to generate a photon and emit the remaining excess energy as radiation. The downward electronic transition is **vertical**, which means in accord with the Franck–Condon principle, and the fluorescence spectrum has a vibrational structure characteristic of the lower electronic state (Fig. 20.16b).

Fluorescence occurs at a lower frequency than that of the incident radiation because the fluorescence radiation is emitted after some vibrational energy has been lost to the surroundings. The vivid oranges and greens of fluorescent dyes are an everyday manifestation of this effect: they absorb in the ultraviolet and fluoresce in the visible. The mechanism also suggests that the intensity of the fluorescence ought to depend on the ability of the surrounding molecules, such as those of a solvent, to accept the electronic and vibrational quanta. It is indeed found that a solvent composed of molecules with widely spaced vibrational levels (such as water) may be able to accept the large quantum of electronic energy and so decrease the intensity of the solute's fluorescence.

Fluorescence is a very important for studying molecular biology and photochemical processes in general. We deal with it at length in Sections 20.11 and 20.12 in the context of photochemically induced rate processes.

20.6 Phosphorescence

Figure 20.17 is a Jablonski diagram showing the events leading to phosphorescence. The first steps are the same as in fluorescence, but the presence of a triplet state plays a decisive role. In a **triplet state** two electrons in different orbitals have parallel spins: the

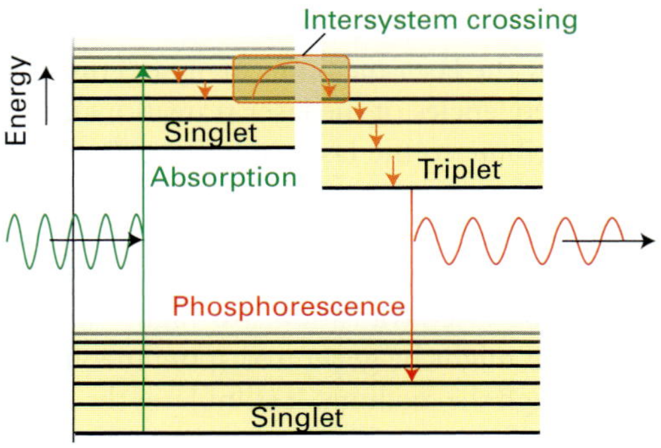

Fig. 20.17 The sequence of steps leading to phosphorescence. The important step is the intersystem crossing from an excited singlet to an excited triplet state. The triplet state acts as a slowly radiating reservoir because the return to the ground state is very slow.

See an animated version of this figure in the interactive ebook.

ground state of O_2, which was discussed in Section 14.13 is an example. The name 'triplet' reflects the (quantum-mechanical) fact that the total spin of two parallel electron spins (↑↑) can adopt only three orientations with respect to an axis. An ordinary spin-paired state (↑↓) is called a **singlet state** because there is only one orientation in space for such a pair of spins. In the language introduced in Section 13.17, a triplet state has $S = 1$ and M_S has one of the three values +1, 0, and −1; a singlet state has $S = 0$ and M_S has the single value 0.

The ground state of a typical phosphorescent molecule is a singlet because its electrons are all paired; the excited state to which the absorption excites the molecule is also a singlet. The peculiar feature of a phosphorescent molecule, however, is that it possesses an excited triplet state of an energy similar to that of the excited singlet state and into which the excited singlet state may convert. Hence, if there is a mechanism for unpairing two electron spins (and so converting ↑↓ into ↑↑), then the molecule may undergo **intersystem crossing** (ISC) and become a triplet state. The unpairing of electron spins is possible if the molecule contains a heavy atom, such as an atom of sulfur, with strong spin–orbit coupling (Section 13.18). Then, the angular momentum needed to convert a singlet state into a triplet state may be acquired from the orbital motion of the electrons.

After an excited singlet molecule crosses into a triplet state, it continues to discard energy into the surroundings and to step down the ladder of vibrational states. However, it is now stepping down the triplet's ladder and at the lowest vibrational energy level it is trapped. The surroundings cannot extract the final, large quantum of electronic excitation energy. Moreover, the molecule cannot radiate its energy because return to the ground state is forbidden: a triplet state cannot convert into a singlet state because the spin of one electron cannot reverse in direction relative to the other electron during a transition ($\Delta S = 0$ for electronic transition). The radiative transition, however, is not totally forbidden because the spin–orbit coupling responsible for the intersystem crossing also breaks this rule. The molecules are therefore able to emit weakly and the emission may continue long after the original excited state was formed.

The mechanism of phosphorescence summarized in Fig. 20.17 accounts for the observation that the excitation energy seems to become trapped in a slowly leaking reservoir. It also suggests, as is confirmed experimentally, that phosphorescence should be most intense from solid samples: energy transfer is then less efficient and the intersystem crossing has time to occur as the singlet excited state loses vibrational energy. The mechanism also suggests that the phosphorescence efficiency should depend on the presence of a moderately heavy atom—with its ability to flip electron spins—which is in fact the case.

20.7 Lasers

The word *laser* is an acronym formed from *l*ight *a*mplification by *s*timulated *e*mission of *r*adiation. As this name suggests, it is a process that depends on *stimulated* emission as distinct from the spontaneous emission processes characteristic of fluorescence and phosphorescence. In **stimulated emission**, an excited state is stimulated to emit a photon by the presence of radiation of the same frequency, and the more photons there are present, the greater the probability of the emission (for details, see Further information 20.2). To picture the process, we can think of the oscillations of the electromagnetic field as periodically distorting the excited molecule at the frequency of the transition and hence encouraging the molecule to generate a photon of the same frequency. The essential feature of laser action is the strong **gain**, or growth of intensity, that results: the more photons present of the appropriate frequency, the more photons of that frequency the excited molecules will be stimulated to form, and so the laser medium fills with photons. These photons then escape either continuously or in pulses.

One requirement for laser action is the existence of an excited state that has a long enough lifetime for it to participate in stimulated emission. Another requirement is the existence of a greater population in the upper state than in the lower state where the transition terminates. Because at thermal equilibrium the population is greater in the lower energy state, it is necessary to achieve a **population inversion** in which there are more molecules in the upper state than in the lower.

Figure 20.18 illustrates one way to achieve population inversion indirectly through an intermediate state I. Thus, the molecule is excited to I, which then gives up some of its energy nonradiatively (by passing energy on to vibrations of the surroundings) and changes into a lower state B; the laser transition is the return of B to a lower state A. Because four levels are involved overall, this arrangement leads to a **four-level laser**. One advantage of this arrangement is that the population inversion of the A and B levels is easier to achieve than when the lower state is the heavily populated ground state. The transition from X to I is caused by an intense flash of light in the process called **pumping**. In some cases the pumping flash is achieved with an electric discharge through xenon or with the radiation from another laser.

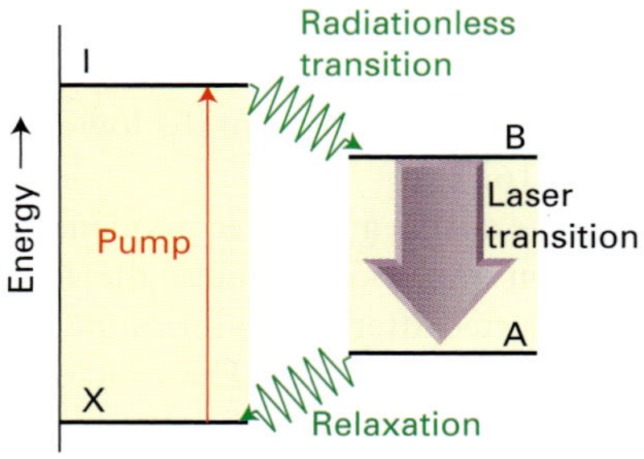

Fig. 20.18 The transitions involved in a four-level laser. Because the laser transition terminates in an excited state (A), the population inversion between A and B is much easier to achieve than when the lower state of the laser transition is the ground state.

In practice, the laser medium is confined to a cavity that ensures that only certain photons of a particular frequency, direction of travel, and state of polarization are generated abundantly. The cavity is essentially a region between two mirrors, which reflect the light back and forth. This arrangement can be regarded as a version of the particle in a box, with the particle now being a photon. As in the treatment of a particle in a box (Section 12.7), the only wavelengths that can be sustained satisfy $N \times \frac{1}{2}\lambda = L$, where N is an integer and L is the length of the cavity. That is, only an integral number of half-wavelengths fit into the cavity; all other waves undergo destructive interference with themselves. In addition, not all wavelengths that can be sustained by the cavity are amplified by the laser medium (many fall outside the range of frequencies of the laser transitions), so only a few contribute to the laser radiation. These wavelengths are the **resonant modes** of the laser.

Photons with the correct wavelength for the resonant modes of the cavity and the correct frequency to stimulate the laser transition are highly amplified. One photon might be generated spontaneously, and travel through the medium. It stimulates the emission of another photon, which in turn stimulates more (Fig. 20.19). The cascade of energy builds up rapidly, and soon the cavity is an intense reservoir of radiation at all the resonant modes it can sustain. Some of this radiation can be withdrawn if one of the mirrors is partially transmitting.

The resonant modes of the cavity have various natural characteristics, and to some extent may be selected. Only photons that are travelling strictly

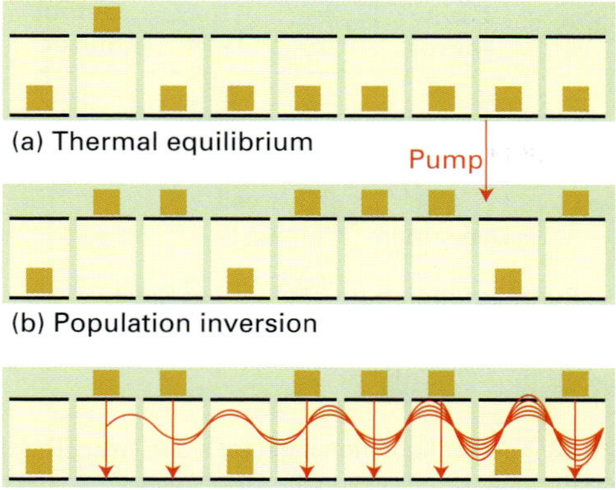

Fig. 20.19 A schematic illustration of the steps leading to laser action. (a) At thermal equilibrium, more atoms are in the ground state. (b) When the initial state absorbs, the populations are inverted (the atoms are pumped to the excited state). (c) A cascade of radiation then occurs, as one emitted photon stimulates another atom to emit, and so on. The radiation is coherent (phases in step).

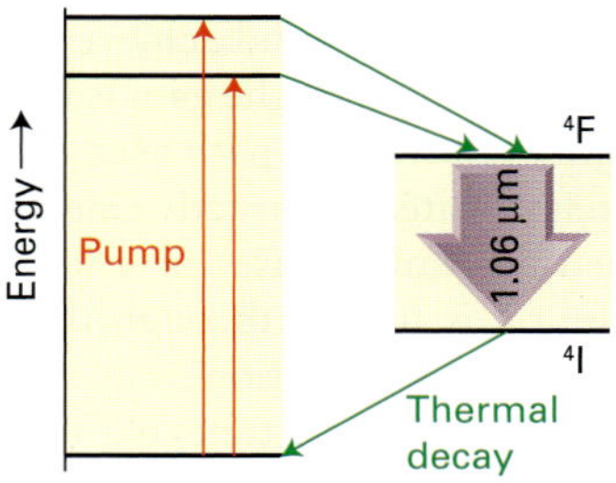

Fig. 20.20 The transitions involved in a neodymium laser. The laser action takes place between two excited states, and the population inversion is easier to achieve than in the ruby laser.

parallel to the axis of the cavity undergo more than a couple of reflections, so only they are amplified, all others simply vanishing into the surroundings. Hence, laser light generally forms a beam with very low divergence. It may also be polarized, with its electric vector in a particular plane (or in some other state of polarization), by including a polarizing filter into the cavity or by making use of polarized transitions in a solid medium.

A *neodymium laser* is an example of a four-level laser (Fig. 20.20). In one form it consists of Nd^{3+} ions at low concentration in yttrium aluminium garnet (YAG, specifically $Y_3Al_5O_{12}$), and is then known as a *Nd-YAG laser*. This laser operates at a number of wavelengths in the infrared, the band at 1064 nm being most common. The transition at 1064 nm is very efficient and the laser is capable of substantial power output. The power is great enough that focusing the beam on to a material may lead to the observation of **nonlinear optical phenomena**, which arise from changes in the optical properties of the substance in the presence of an intense electric field from electromagnetic radiation. A useful nonlinear optical phenomenon is **frequency doubling**, or *second-harmonic generation*, in which an intense laser beam is converted to radiation with twice (and in general a multiple) of its initial frequency as it passes though a suitable material. Frequency doubling and tripling of a Nd-YAG laser produce green light at 532 nm and ultraviolet radiation at 355 nm, respectively. In *diode lasers*, of the type used in CD players and bar-code readers, the light emission at a p–n junction (Section 17.3) is sustained by sweeping away the electrons that fall into the holes of the p-type semiconductor. This process is arranged to occur in a cavity formed by making use of the abrupt difference in refractive index between the different components of the junction, and the radiation trapped in the cavity enhances the production of more radiation. One widely used material is GaAs doped with aluminium, which produces 780 nm red laser radiation and is widely used in CD players. The new generation of DVD players that use blue rather than red laser radiation, so allowing a greater superficial density of information, use GaN as the active material.

Because *gas lasers* can be cooled by a rapid flow of the gas through the cavity, they can be used to generate high powers. The *carbon dioxide laser*, which produces radiation between 9.2 μm and 10.8 μm, with the strongest emission at 10.6 μm, in the infrared makes use of vibrational transitions (Fig. 20.21). Most of the working gas is nitrogen, which becomes

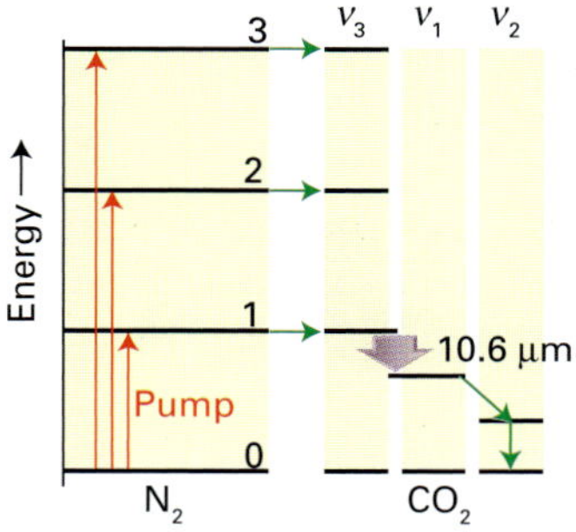

Fig. 20.21 The transitions involved in a carbon dioxide laser. The pumping also depends on the coincidental matching of energy separations; in this case the vibrationally excited N_2 molecules have excess energies that correspond to a vibrational excitation of the antisymmetric stretch of CO_2. The laser transition is from $v_3 = 1$ to $v_1 = 1$.

vibrationally excited by electronic and ionic collisions in an electric discharge. The vibrational levels happen to coincide with the ladder of antisymmetric stretch (ν_3, see Fig. 19.20) levels of CO_2, which pick up the energy during a collision. Laser action then occurs from the lowest excited level of ν_3 to the lowest excited level of the symmetric stretch (ν_1), which has remained unpopulated during the collisions. Such high powers of radiation can be achieved that carbon dioxide lasers can be used to cut steel for ship-building.

The population inversion needed for laser action is achieved in a more underhand way in *exciplex lasers*, for in these (as we shall see) the lower state does not effectively exist. This odd situation is achieved by forming an **exciplex**, a combination of two atoms (or molecules) that survives only in an excited state and that dissociates as soon as the excitation energy has been discarded. The term 'excimer laser' is also widely encountered and used loosely when 'exciplex laser' is more appropriate. An exciplex has the form AB*, whereas an excimer, an excited dimer, is AA*. An example of an exciplex laser is a mixture of xenon, chlorine, and neon. An electric discharge through the mixture produces excited Cl atoms, which attach to the Xe atoms to give the exciplex XeCl*. The exciplex survives for about 10 ns, which is time for it to participate in laser action at 308 nm (in the ultraviolet). As soon as XeCl* has discarded a photon, the atoms separate because the molecular potential energy curve of the ground state is dissociative, and the ground state of the exciplex cannot become populated (Fig. 20.22).

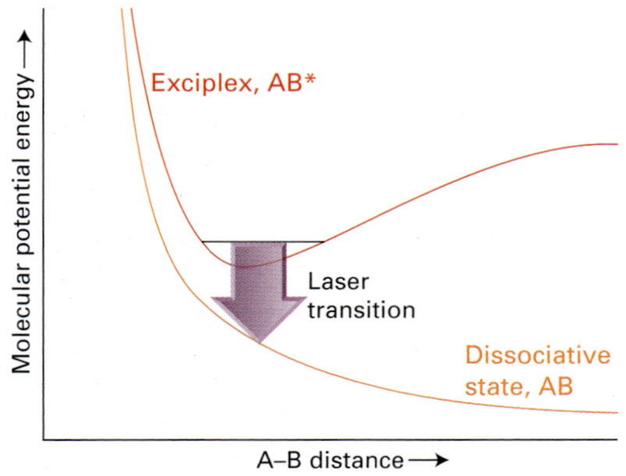

Fig. 20.22 The molecular potential energy curves for an exciplex. The species can survive only as an excited state, because on discarding its energy it enters the lower, dissociative state. Because only the upper state can exist, there is never any population in the lower state.

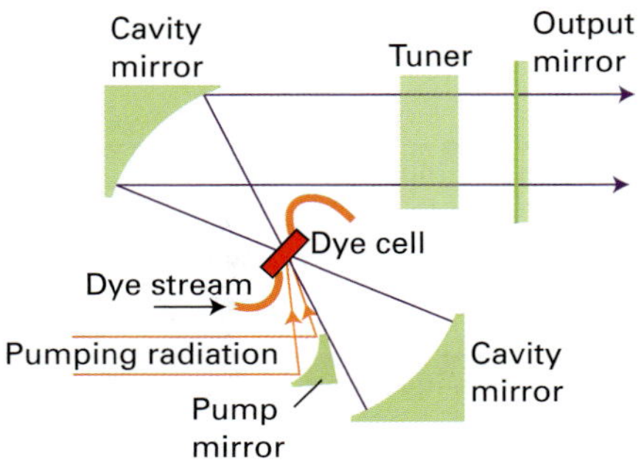

Fig. 20.23 The configuration used for a dye laser. The dye is flowed through the cell inside the laser cavity. The flow helps to keep it cool and prevents degradation.

The wavelengths of lasers can be selected in a variety of ways. An early solution was to use a *dye laser*, which has broad spectral characteristics because the solvent broadens the vibrational structure of the transitions into bands. Hence, it is possible to scan the wavelength continuously (by rotating the diffraction grating in the cavity) and achieve laser action at any chosen wavelength. As the gain is very high, only a short length of the optical path need be through the dye. The excited states of the active medium, the dye, are sustained by another laser or a flash lamp, and the dye solution is flowed through the laser cavity to avoid thermal degradation (Fig. 20.23). More modern solutions are to use tuneable Ti:sapphire (for wavelengths in the range 650–1100 nm) and white light lasers.

20.8 Applications of lasers in chemistry

Laser radiation has a number of advantages for applications in chemistry. One advantage is its highly monochromatic character, which enables very precise spectroscopic observations to be made. Another advantage is the ability of laser radiation to be produced in very short pulses (currently, as brief as about 1 fs): as a result, very fast chemical events, such as the individual transfers of atoms during a chemical reaction, can be followed (see Box 10.1). Laser radiation is also very intense, which reduces the time needed for spectroscopic observations. Raman spectroscopy (Chapter 19) was revitalized by the introduction of lasers because the intense beam increases the intensity of scattered radiation, so the use of laser sources increases the sensitivity of Raman spectroscopy. A well-defined beam also implies that the detector can be designed to collect only the radiation that has passed through a sample, and can be screened much more effectively against the stray scattered light that can obscure the Raman signal. Laser light can be delivered through fibre optics for use in transportable systems and focused to such small diameters that *Raman microscopy* can be used to study submicrometre particles. The monochromaticity of laser radiation is also a great advantage, for it makes possible the observation of scattered light that differs by only fractions of reciprocal centimetres from the incident radiation. Such high resolution is particularly useful for observing the rotational structure of Raman lines because rotational transitions are of the order of a few reciprocal centimetres.

The large number of photons in an incident beam generated by a laser gives rise to a qualitatively different branch of spectroscopy, for the photon density is so high that more than one photon may be absorbed by a single molecule and give rise to **multiphoton processes.** Because the selection rules for multiphoton processes are different, states inaccessible by conventional one-photon spectroscopy become observable.

The monochromatic character of laser radiation allows us to excite specific states with very high precision. One consequence of state specificity is that the illumination of a sample may be efficient in stimulating a photochemical reaction, because its frequency can be tuned exactly to an absorption. The specific excitation of a particular excited state of a molecule may greatly enhance the rate of a reaction even at low temperatures. As we saw in Chapter 10, the rate of a reaction is increased by raising the temperature because the energies of the various modes of motion of the molecule are enhanced. However, this enhancement increases the energy of all the modes, even those that do not contribute appreciably to the reaction rate. With a laser we can excite the kinetically significant mode, so rate enhancement is achieved most efficiently. An example is the reaction

$$BCl_3 + C_6H_6 \rightarrow C_6H_5\text{—}BCl_2 + HCl$$

which normally proceeds only above 600°C in the presence of a catalyst; exposure to 10.6 μm CO_2 laser radiation results in the formation of products at room temperature without a catalyst. The commercial potential of this procedure is considerable, provided laser photons can be produced sufficiently cheaply, because heat-sensitive compounds, such as pharmaceuticals, may perhaps be made at lower temperatures than in conventional reactions.

Laser isotope separation is possible because two isotopomers (species that differ only in their isotopic composition), have slightly different energy levels and hence slightly different absorption frequencies. At least two absorption processes are required. In the first step, a photon excites an atom to a higher state; in the second step, a photon achieves photoionization from that state (Fig. 20.24). The energy separation between the two states involved in the first step depends on the nuclear mass. Therefore, if the laser radiation is tuned to the appropriate frequency, only one of the isotopomers will undergo excitation and hence be available for photoionization in the second step. An example of this procedure is the photoionization of uranium vapour, in which the incident laser is tuned to excite ^{235}U but not ^{238}U. The ^{235}U atoms in the atomic beam are ionized in the two-step process; they are then attracted to a negatively charged electrode, and may be collected (Fig. 20.25).

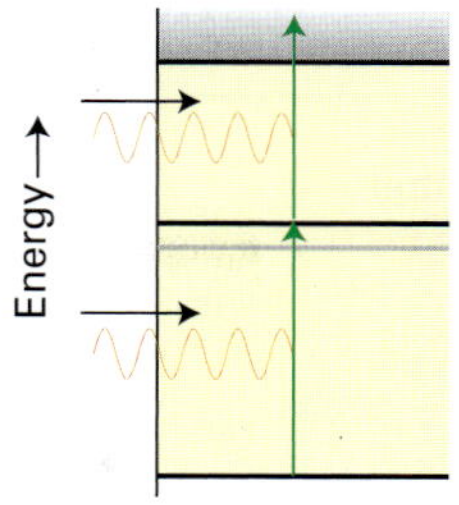

Fig. 20.24 In one method of isotope separation, one photon excites an isotopomer to an excited state, and then a second photon achieves photoionization. The success of the first step depends on the nuclear mass.

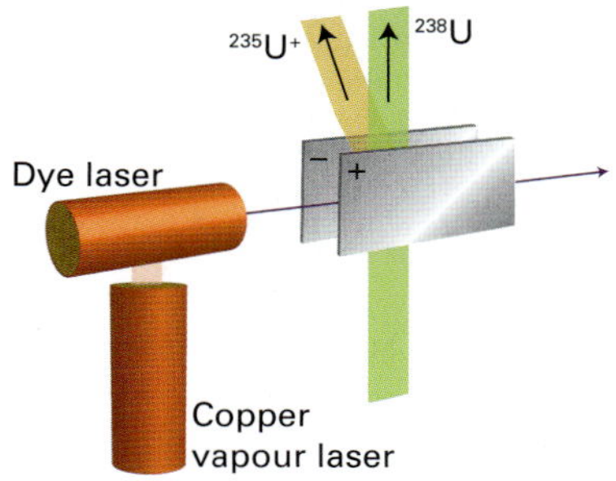

Fig. 20.25 An experimental arrangement for isotope separation. The dye laser, which is pumped by a copper-vapour laser, photoionizes the U atoms selectively according to their mass, and the ions are deflected by the electric field applied between the plates.

The ability of lasers to produce pulses of very short duration is particularly useful in chemistry when we want to monitor processes in time. In **time-resolved spectroscopy**, laser pulses are used to obtain the absorption, emission, or Raman spectrum of reactants, intermediates, products, and even transition

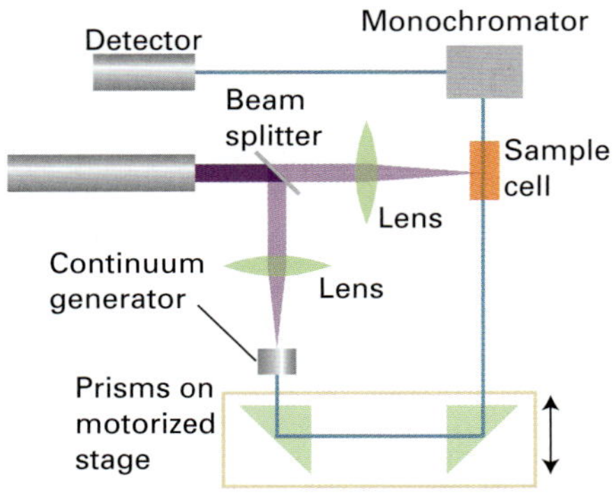

Fig. 20.26 A configuration used for time-resolved absorption spectroscopy, in which the same pulsed laser is used to generate a monochromatic pump pulse and, after continuum generation in a suitable liquid, a 'white' light probe pulse. The time delay between the pump and probe pulses may be varied.

states of reactions. Lasers that produce nanosecond pulses are generally suitable for the observation of reactions with rates controlled by the speed with which reactants can move through a fluid medium. However, femtosecond to picosecond laser pulses are needed to study energy transfer, molecular rotations, vibrations, and conversion from one mode of motion to another. The arrangement shown in Fig. 20.26 is often used to study ultrafast chemical reactions that can be initiated by light. An intense but brief laser pulse, the *pump*, promotes a molecule A to an excited electronic state A* that can either emit a photon (as fluorescence or phosphorescence) or react with another species B to yield a product C. The rates of appearance and disappearance of the various species are determined by observing time-dependent changes in the absorption spectrum of the sample during the course of the reaction. This observation is made by passing a weak pulse of white light, the *probe*, through the sample at different times after the laser pulse. Pulsed 'white' light can be generated directly from the laser pulse by the nonlinear optical phenomenon of *continuum generation*, in which focusing an ultrashort laser pulse on a vessel containing a liquid such as water or carbon tetrachloride results in an outgoing beam with a wide distribution of frequencies. A time delay between the strong laser pulse and the 'white' light pulse can be introduced by allowing one of the beams to travel a longer distance before reaching the sample. For example, a difference in travel distance of $\Delta d = 3$ mm corresponds to a time delay $\Delta t = \Delta d/c \approx 10$ ps between two beams, where c is the speed of light.

Photoelectron spectroscopy

The exposure of a molecule to high-frequency radiation can result in the ejection of an electron. This **photoejection** is the basis of another type of spectroscopy in which we monitor the energies of the ejected photoelectrons. If the incident radiation has frequency ν, the photon that causes photoejection has energy $h\nu$. If the ionization energy of the molecule is I, the difference in energy, $h\nu - I$, is carried away as kinetic energy. Because the kinetic energy of an electron of speed v is $\frac{1}{2}m_e v^2$, we can write

$$h\nu = I + \tfrac{1}{2}m_e v^2 \qquad (20.7)$$

Therefore, by monitoring the velocity of the photoelectron, and knowing the frequency of the incident radiation, we can determine the ionization energy of the molecule and hence the strength with which the electron was bound (Fig. 20.27). In this context the 'ionization energy' of the molecule has different values depending on the orbital that the photoelectron occupied, and the slower the ejected electron, the lower in energy the orbital from which it was ejected. The apparatus is a modification of a mass spectrometer (Fig. 20.28), in which the velocity of the photoelectrons is measured by determining the strength of the electric field required to bend their paths on to the detector.

Self-test 20.3

What is the velocity of photoelectrons that are ejected from a molecule with radiation of energy 21 eV (from a helium discharge lamp) and are known to come from an orbital of ionization energy 12 eV?

[*Answer:* 1.8×10^3 km s^{-1}]

Figure 20.29 shows a typical photoelectron spectrum (of HBr). If we disregard the fine structure, we see that the HBr lines fall into two main groups. The least tightly bound electrons (with the lowest ionization energies and hence highest kinetic energies when ejected) are those in the lone pairs of the Br atom. The next ionization energy lies at 15.2 eV, and corresponds to the removal of an electron from the H—Br σ bond.

The HBr spectrum shows that ejection of a σ electron is accompanied by a considerable amount of vibrational excitation. The Franck–Condon principle would account for this observation if ejection were accompanied by an appreciable change of equilibrium

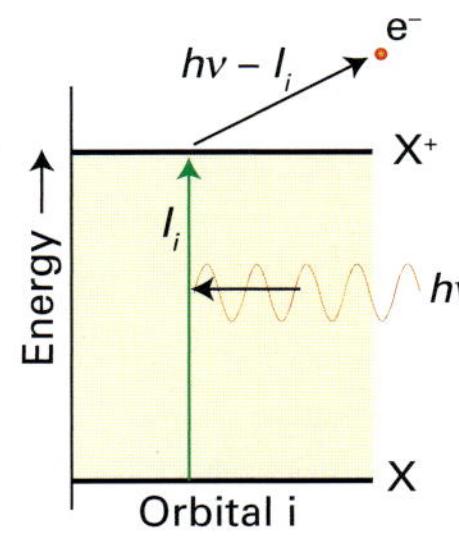

Fig. 20.27 The basic principle of photoelectron spectroscopy. An incoming photon of known energy collides with an electron in one of the orbitals and expels it with a kinetic energy that is equal to the difference between the energy supplied by the photon and the ionization energy from the occupied orbital. An electron from an orbital with a low ionization energy will emerge with a high kinetic energy (and high speed) whereas an electron from an orbital with a high ionization energy will be ejected with a low kinetic energy (and low speed).

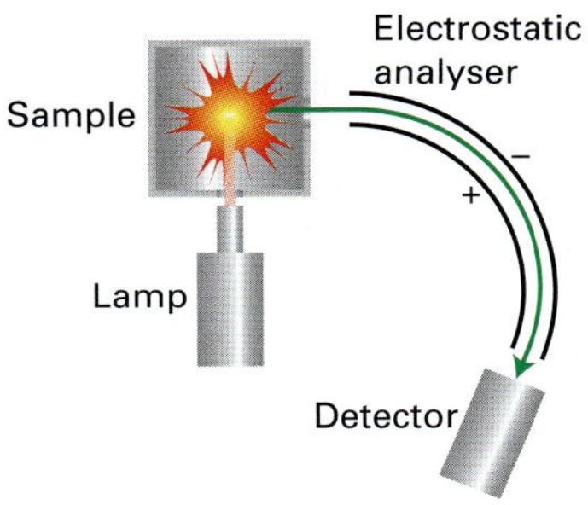

Fig. 20.28 A photoelectron spectrometer consists of a source of ionizing radiation (such as a helium discharge lamp for UPS and an X-ray source for XPS), an electrostatic analyser, and an electron detector. The deflection of the electron paths caused by the analyser depends on their speed.

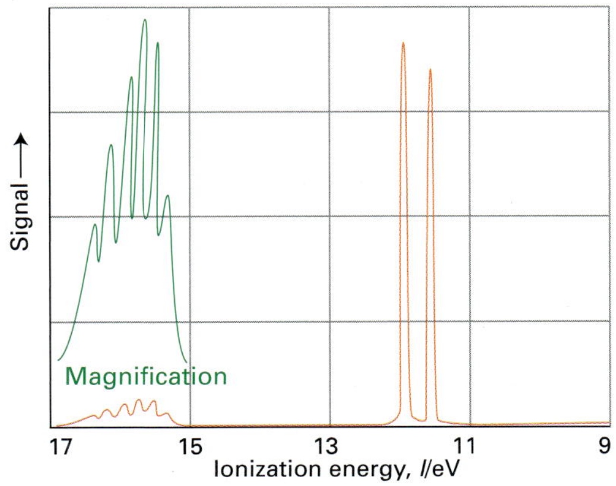

Fig. 20.29 The photoelectron spectrum of HBr. The lowest ionization energy band corresponds to the ionization of a Br lone-pair electron. The higher ionization energy band corresponds to the ionization of a bonding electron. The structure on the latter is due to the vibrational excitation of HBr^+ that results from the ionization.

bond length between HBr and HBr^+: if that is so, the ion is formed in a bond-compressed state, which is consistent with the important bonding effect of the σ electrons. The lack of much vibrational structure in the other band is consistent with the nonbonding role of the $Br4p_x$ and $Br4p_y$ lone-pair electrons, for the equilibrium bond length is little changed when one is removed.

A brief illustration The highest kinetic energy electrons in the spectrum of H_2O using 21.22 eV He radiation are at about 9 eV and show a large vibrational spacing of 0.41 eV (1 eV = 8065.5 cm^{-1}). Because 0.41 eV corresponds to 3.3×10^3 cm^{-1}, which is similar to the wavenumber of the symmetric stretching mode of the neutral H_2O molecule (3652 cm^{-1}), we can suspect that the electron is ejected from an orbital that has little influence on the bonding in the molecule. That is, photoejection is from a largely nonbonding orbital.

Self-test 20.4

In the same spectrum of H_2O, the band near 7.0 eV shows a long vibrational series with spacing 0.125 eV. The bending mode of H_2O lies at 1596 cm^{-1}. What conclusions can you draw about the characteristics of the orbital occupied by the photoelectron?

[*Answer:* The electron contributes to long-distance HH bonding across the molecule]

Photochemistry

Photochemical reactions are reactions that are initiated by the absorption of light. The most important of all are the photochemical processes that capture the Sun's radiant energy. Some of these reactions lead to the heating of the atmosphere during the daytime by absorption in the ultraviolet region as a result of reactions like those depicted in Fig. 20.30. Others include the absorption of red and blue light by chlorophyll and the subsequent use of the energy to bring about the synthesis of carbohydrates from carbon dioxide and water (Box 20.2). Indeed, without photochemical processes the world would be simply a warm, sterile, rock.

20.9 Quantum yield

A molecule may acquire enough energy to react by absorbing a photon. However, not every excited molecule may form a specific primary product (atoms, radicals, or ions, for instance) because there are many ways in which the excitation may be lost other

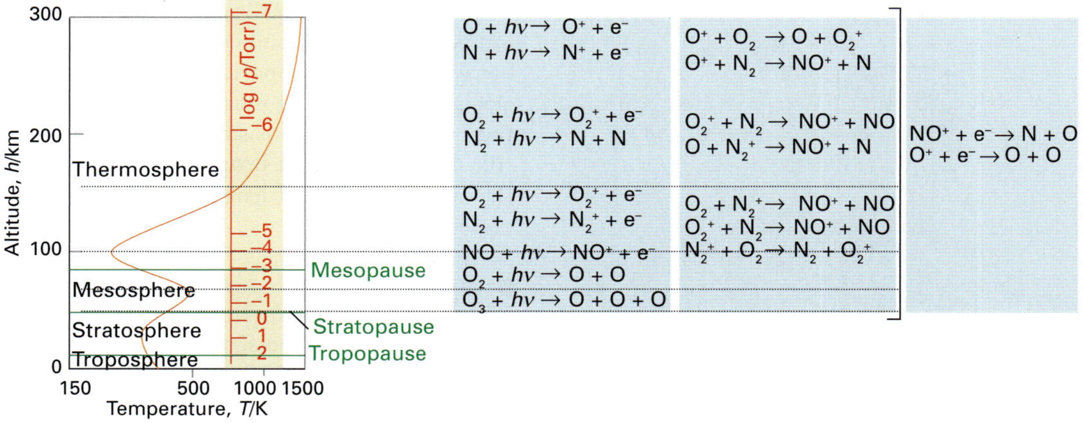

Fig. 20.30 The temperature profile through the atmosphere and some of the reactions that take place in each region.

Box 20.2 Photosynthesis

Up to 1 kW m^{-2} of radiation from the Sun reaches the Earth's surface, with the exact intensity depending on latitude, time of day, and weather. A large proportion of solar radiation with wavelengths below 400 nm and above 1000 nm is absorbed by atmospheric gases such as ozone and O_2, which absorb ultraviolet radiation, and CO_2 and H_2O, which absorb infrared radiation. As a result, plants, algae, and some species of bacteria evolved photosynthetic apparatus that captures visible and near-infrared radiation. Plants use radiation in the wavelength range 400–700 nm to drive the endergonic reduction of CO_2 to glucose, with concomitant oxidation of water to O_2 ($\Delta_r G^{\ominus} = +2880$ kJ mol^{-1}).

Plant photosynthesis takes place in the *chloroplast*, a special organelle of the plant cell. Electrons flow from reductant to oxidant via a series of electrochemical reactions that are coupled to the synthesis of ATP. In the chloroplast, chlorophylls *a* and *b* and carotenoids (of which *β*-carotene is an example) bind to proteins called *light-harvesting complexes*, which absorb solar energy and transfer it to protein complexes known as *reaction centres*, where light-induced electron transfer reactions occur. The combination of a light-harvesting complex and a reaction centre complex is called a *photosystem*, and plants have two: photosystem I and photosystem II.

In photosystems I and II, absorption of a photon raises a chlorophyll or carotenoid molecule to an excited singlet state. The initial energy and electron transfer events of photosynthesis are under tight kinetic control and the efficient capture of solar energy stems from rapid quenching of the excited singlet state of chlorophyll by processes that occur with relaxation times that are much shorter than the fluorescence lifetime, which is about 5 ns in diethyl ether at room temperature. Time-resolved spectroscopic data show that within 0.1–5 ps of absorption of light by a chlorophyll molecule in a light-harvesting complex, the energy hops to a nearby pigment by the Förster mechanism. About 100–200 ps later, which corresponds to thousands of hops within the complex, more than 90 per cent of the absorbed energy reaches the reaction centre. The absorption of energy from light decreases the reduction potential of special dimers of chlorophyll *a* molecules known as P700 (in photosystem I) and P680 (in photosystem II). In their excited states, P680 and P700 initiate electron-transfer reactions that culminate in the oxidation of water to O_2 and the reduction of $NADP^+$ to NADPH. The initial electron transfer steps are fast and compete effectively with chlorophyll's fluorescence. For example, the transfer of an electron from the excited singlet state of P680 occurs within 3 ps. Experiments show that for each molecule of NADPH formed in the chloroplast of green plants, one molecule of ATP is synthesized. Finally, the ATP and NADPH molecules participate in the *Calvin–Benson cycle*, a sequence of enzyme-controlled reactions that leads to the reduction of CO_2 to glucose in the chloroplast.

In summary, plant photosynthesis uses solar energy to transfer electrons from a poor reductant (water) to carbon dioxide. In the process, high-energy molecules (carbohydrates, such as glucose) are synthesized in the cell. Animals feed on the carbohydrates derived from photosynthesis. The O_2 released by photosynthesis as a waste product is used to oxidize carbohydrates to CO_2. This reaction drives biological processes, such as biosynthesis, muscle contraction, cell division, and nerve conduction. Hence, the sustenance of life on Earth depends on a tightly regulated carbon–oxygen cycle that is driven by solar energy.

than by dissociation or ionization. We therefore speak of the **primary quantum yield**, ϕ (phi), which is the number of events (physical changes or chemical reactions) that lead to primary products (photons, atoms or ions, for instance) divided by the number of photons absorbed by the molecule in the same time interval:

$$\phi = \frac{\text{number of events}}{\text{number of photons absorbed}} \qquad (20.8)$$

If each molecule that absorbs a photon undergoes dissociation (for instance), then $\phi = 1$. If none does, because the excitation energy is lost before the molecule has time to dissociate, then $\phi = 0$.

If we divide the numerator and denominator of eqn 20.8 by the time interval during which the photochemical event occurs, we see that primary quantum yield is also the rate of radiation-induced primary events, v, divided by the rate of photon absorption, I_{abs}:

$$\phi = \frac{v}{I_{abs}} \qquad (20.9)$$

A molecule in an excited state must either decay to the ground state or form a photochemical product. Therefore, the total number of molecules deactivated by radiative processes, nonradiative processes, and photochemical reactions must be equal to the number of excited species produced by absorption of light. We conclude that the sum of primary quantum yields ϕ_i for *all* physical changes and photochemical reactions i, which are referred to as **channels**, must be equal to 1, regardless of the number of reactions involving the excited state. It follows that

$$\sum_i \phi_i = \sum_i \frac{v_i}{I_{abs}} = 1 \qquad (20.10)$$

where v_i is the rate of product formation in channel i.

One successfully excited molecule might initiate the consumption of more than one reactant molecule. We therefore need to introduce the **overall quantum yield, Φ** (upper case phi), which is the number of reactant molecules that react for each photon absorbed. In the photolysis of HI, for example, the processes are

$$HI + h\nu \rightarrow H + I$$
$$H + HI \rightarrow H_2 + I$$
$$I + I + M \rightarrow I_2 + M$$

where M is an inert 'third body' that removes the energy released when the I—I bond forms. The overall quantum yield is 2 because the absorption of one photon leads to the destruction of two HI molecules. In a photochemically initiated chain reaction, Φ may be very large, and values of about 10^4 are common. In such cases the chain reaction acts as a chemical amplifier of the initial absorption step.

Example 20.2

Using the quantum yield

The overall quantum yield for the formation of ethene from 4-heptanone with 313 nm light is 0.21. How many molecules of 4-heptanone per second, and what chemical amount per second, are destroyed when the sample is irradiated with a 50 W, 313 nm source under conditions of total absorption?

Strategy We need to determine the rate of emission of photons of the lamp as the number of photons emitted by the lamp per second. Because all photons are absorbed (by assertion), this quantity is also I_{abs}, which we calculate by dividing the power (joules per second) by the energy of a single photon ($E = h\nu$, with $\nu = c/\lambda$). From eqn 20.9, with Φ in place of ϕ, the rate of the photochemical reaction (the number of molecules destroyed per second) is I_{abs} multiplied by the overall quantum yield, Φ.

Solution A source of power P (the rate at which energy is supplied) generates photons at a rate given by P divided by the energy of each photon, E. The energy of a photon of wavelength λ is $E = hc/\lambda$. Therefore,

$$\text{Rate of photon production} = I_{abs} = \frac{P}{(hc/\lambda)} = \frac{P\lambda}{hc}$$

We now substitute the data:

$$\text{Rate of photon production} = \frac{(50\ \text{J s}^{-1}) \times (313 \times 10^{-9}\ \text{m})}{(6.6261 \times 10^{-34}\ \text{J s}) \times (2.9979 \times 10^{8}\ \text{m s}^{-1})} = 7.9 \times 10^{19}\ \text{s}^{-1}$$

The number of 4-heptanone molecules destroyed per second is therefore 0.21 times this quantity, or $1.7 \times 10^{19}\ \text{s}^{-1}$, corresponding (after multiplication by N_A) to 28 μmol s^{-1}.

Self-test 20.5

The overall quantum yield for another reaction at 290 nm is 0.30. For what length of time must irradiation with a 100 W source continue in order to destroy 1.0 mol of molecules?

[*Answer:* 3.8 h]

20.10 Mechanisms of photochemical reactions

As an example of how to incorporate the photochemical activation step into a mechanism, consider the photochemical activation of the reaction

$$H_2(g) + Br_2(g) \rightarrow 2\,HBr(g)$$

which we described in Section 11.15. In place of the first step in the thermal reaction we have

$$Br_2 + h\nu \rightarrow Br + Br$$

$$\text{Rate of consumption of } Br_2 = \text{Rate of photon absorption} = I_{abs}$$

Because the thermal reaction mechanism had $k_a[Br_2]$ in place of I_{abs} for the equivalent of this step, it follows that I_{abs} should take the place of $k_a[Br_2]$ in the rate law we derived for the thermal reaction scheme. Therefore, from eqn 11.35 we can write

$$\text{Rate of formation of HBr} = \frac{2k_b(1/k_d)^{1/2}[H_2][Br_2]I_{abs}^{1/2}}{[Br_2]+(k_c/k_b')[HBr]} \quad (20.11)$$

Although the details of this expression are complicated, the essential prediction is clear: the reaction rate should depend on the square root of the absorbed light intensity (which is proportional to the rate of photon absorption). This prediction is confirmed experimentally.

20.11 The kinetics of decay of excited states

In many cases, proper description of the rates and mechanisms of photochemical reactions also requires knowledge of such processes as fluorescence and phosphorescence that can deactivate an excited state before the reaction has a chance to occur. Electronic transitions caused by absorption of ultraviolet and visible radiation occur within 10^{-16}–10^{-15} s. We expect, then, that the upper limit for the rate constant of a first-order photochemical reaction is about $10^{16}\,s^{-1}$. Fluorescence is slower than absorption, with typical time constants of 10^{-12}–10^{-6} s. Therefore, the excited singlet state can initiate very fast photochemical reactions in the femtosecond (10^{-15} s) to picosecond (10^{-12} s) timescale. Examples of such ultrafast reactions are the initial events of vision (Box 20.1) and photosynthesis (Box 20.2). Typical intersystem crossing and phosphorescence time constants for large organic molecules are 10^{-12}–10^{-4} s and 10^{-6}–10^{-1} s, respectively. As a consequence, excited triplet states are photochemically important. Indeed, because phosphorescence decay is several orders of magnitude slower than most typical reactions, species in excited triplet states can undergo a very large number of collisions with other reactants before deactivation.

We begin our exploration of the interplay between reaction rates and excited state decay rates by considering the mechanism of deactivation of an excited singlet state in the absence of a chemical reaction. The following steps are involved:

Process	Equation	Rate
Absorption	$S + h\nu_i \rightarrow S^*$	I_{abs}
Fluorescence	$S^* \rightarrow S + h\nu_f$	$k_f[S^*]$
Intersystem crossing	$S^* \rightarrow T^*$	$k_{ISC}[S^*]$
Internal conversion	$S^* \rightarrow S$	$k_{IC}[S^*]$

in which S is a singlet state absorbing species, S* an excited singlet state, T* an excited triplet state, and $h\nu_i$ and $h\nu_f$ denote the incident and fluorescent photons, respectively. The **branching ratio** is used to express the ratio of concentrations of products when several outcomes are possible. It follows that after the exciting light has been turned off and S* is no longer formed,

$$\text{Rate of decay of } S^* = k_f[S^*] + k_{ISC}[S^*] + k_{IC}[S^*] = (k_f + k_{ISC} + k_{IC})[S^*] \quad (20.12)$$

It follows that the excited state decays by a first-order process, so when the light is turned off, [S*] varies with time t as

$$[S^*]_t = [S^*]_0 e^{-(k_f+k_{ISC}+k_{IC})t/\tau_{obs}} = [S^*]_0 e^{-t/\tau_{obs}} \quad (20.13a)$$

where the **observed fluorescence lifetime**, τ_{obs}, is

$$\tau_{obs} = \frac{1}{k_f + k_{ISC} + k_{IC}} \quad (20.13b)$$

(Note that the observed lifetime is not simply the sum of the individual lifetimes, $\tau_f = 1/k_f$, etc.) It then follows, as we show in Derivation 20.2, that the quantum yield of fluorescence is

$$\phi_f = \frac{k_f}{k_f + k_{ISC} + k_{IC}} = k_f\tau_{obs} \quad (20.14)$$

To measure τ_{obs} a sample is excited with a short light pulse from a laser at a wavelength where S absorbs strongly. Then, the exponential decay of the fluorescence intensity after the pulse is monitored with a fast detector system. The rate constant k_f can be determined through $k_f = \phi_f/\tau_{obs}$, with ϕ_f measured by using a steady-state technique that compares the fluorescence properties of the sample of interest with that of a sample with known fluorescence quantum yield.

Derivation 20.2

The quantum yield of fluorescence

The rate of formation of S* when the illumination is present is I_{abs} and its rate of loss is given by eqn 20.12. When a steady state has been reached the net rate of formation of S* is zero, so we can write

$$\text{Rate of change of } [S^*] = I_{abs} - (k_f + k_{ISC} + k_{IC})[S^*] = 0$$

Consequently,

$$I_{abs} = (k_f + k_{ISC} + k_{IC})[S^*]$$

Because (from eqn 20.9)

$$\phi_f = \frac{v_f}{I_{abs}}$$

where v_f is the rate at which fluorescence photons are generated and $v_f = k_f[S^*]$, it follows that

$$\phi_f = \frac{k_f[S^*]}{(k_f + k_{ISC} + k_{IC})[S^*]} = \frac{k_f}{k_f + k_{ISC} + k_{IC}}$$

Equation 20.14 then follows by using eqn 20.13b.

20.12 Fluorescence quenching

Now we consider the kinetic information about photochemical processes that can be obtained by studying the effect of a molecule that can remove the excitation energy from the fluorescent molecule and that therefore acts to **quench** the fluorescence.

Quenching may be either a desired process, such as energy or electron transfer, or an undesired side reaction that can decrease the quantum yield of a desired photochemical process. As shown in Derivation 20.3, the relation between the fluorescence quantum yields ϕ_f and ϕ in the absence and presence, respectively, of a quencher Q at a molar concentration [Q] is given by the **Stern–Volmer equation:**

$$\frac{\phi_f}{\phi} = 1 + \tau_{obs} k_Q[Q] \qquad (20.15a)$$

This equation tells us that a plot of ϕ_f/ϕ against [Q] should be a straight line with slope $\tau_{obs}k_Q$. Such a plot is called a **Stern–Volmer plot** (Fig. 20.31). The method is quite general and may also be applied to the quenching of phosphorescence emission. An alternative version is obtained by noting from eqn 20.14 ($\phi_f = k_f\tau_{obs}$) that $\phi_f/\phi = \tau_{obs}/\tau_{obs}(Q)$, where $\tau_{obs}(Q)$ is the observed fluorescence lifetime in the presence of the quencher:

$$\frac{\tau_{obs}}{\tau_{obs}(Q)} = 1 + \tau_{obs} k_Q[Q]$$

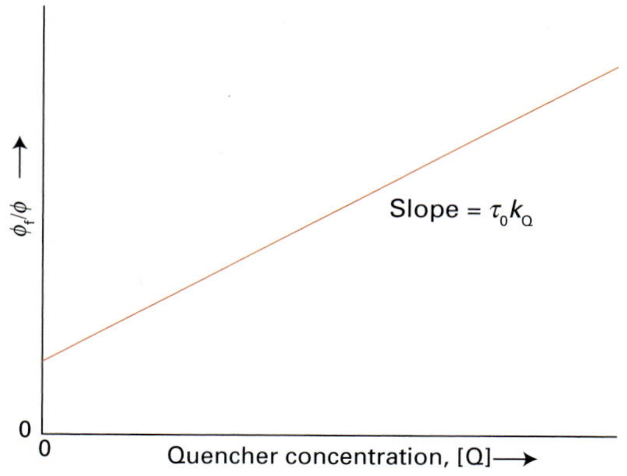

Fig. 20.31 The format of a Stern–Volmer plot and the interpretation of the slope in terms of the rate constant for quenching and the observed fluorescence lifetime in the absence of quenching.

After division by τ_{obs}, this expression becomes

$$\frac{1}{\tau_{obs}(Q)} = \frac{1}{\tau_{obs}} + k_Q[Q] \qquad (20.15b)$$

Thus, a plot of $1/\tau_{obs}(Q)$ against [Q] has an intercept of $1/\tau_{obs}$ and a slope of k_Q.

A brief illustration The quenching of the complex $Ru(bpy)_3^{2+}$ fluorescence by $Fe(H_2O)_6^{3+}$ in acidic solution was monitored by measuring emission lifetimes at 600 nm:

$[Fe(H_2O)_6^{3+}]/(10^{-4}$ mol dm$^{-3})$	0	1.6	4.7	7	9.4
$\tau_{obs}(Q)/(10^{-7}$ s)	6	4.1	3.4	3	2.17

Figure 20.32 shows a plot of $1/\tau$ against $[Fe^{3+}]$ and the results of a least-squares fit to eqn 20.15b. The slope of the line is $k_Q = 2.8 \times 10^9$ dm^3 mol^{-1} s^{-1}. This result is typical of a diffusion-controlled process. In Section 11.12 we saw that a diffusion-controlled rate constant is approximately equal to $8RT/3\eta$, where η is the viscosity of the medium; for water at 25°C, this expression evaluates to 7.6×10^9 dm^3 mol^{-1} s^{-1}.

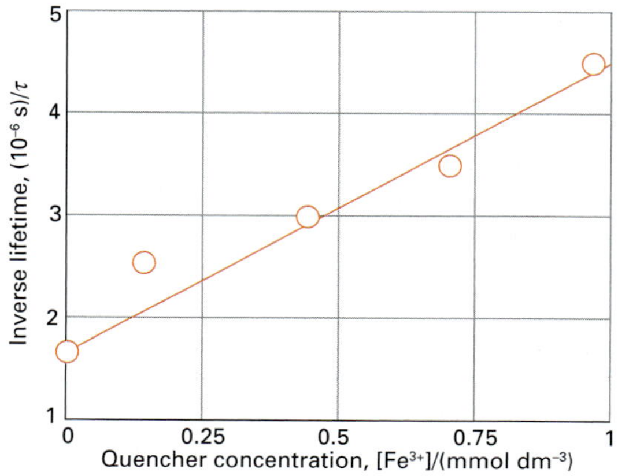

Fig. 20.32 The Stern–Volmer plot of the data.

Self-test 20.6

From the data above, predict the value of $[Fe^{3+}]$ required to decrease the intensity of $Ru(bpy)_3^{2+}$ emission to 50 per cent of the unquenched value.

[*Answer:* 6.0×10^{-4} mol dm^{-3}]

Derivation 20.3

The Stern–Volmer equation

The addition of a quencher, Q, opens an additional channel for deactivation of S*:

Quenching: $S^* + Q \rightarrow S + Q$

Rate of quenching $= k_Q[Q][S^*]$

The steady-state approximation for [S*] now gives:

$$\text{Rate of change of } [S^*] = I_{abs} - (k_f + k_{IC} + k_{ISC} + k_{IC} + k_Q[Q])[S^*] = 0$$

and the fluorescence quantum yield in the presence of the quencher is

$$\phi = \frac{k_f}{k_f + k_{ISC} + k_{IC} + k_Q[Q]}$$

It follows from this expression and eqn 20.14 that the ratio ϕ_f/ϕ is

$$\frac{\phi_f}{\phi} = \left(\frac{k_f}{k_f + k_{ISC} + k_{IC}}\right) \times \left(\frac{k_f + k_{ISC} + k_{IC} + k_Q[Q]}{k_f}\right)$$

$$= \frac{k_f + k_{ISC} + k_{IC} + k_Q[Q]}{k_f + k_{ISC} + k_{IC}}$$

$$= 1 + \frac{k_Q}{k_f + k_{ISC} + k_{IC}}[Q]$$

After using eqn 20.13b, this expression simplifies to eqn 20.15.

Three common mechanisms for quenching of an excited singlet (or triplet) state are:

Collisional deactivation: $S^* + Q \rightarrow S + Q$

Electron transfer: $S^* + Q \rightarrow S^+ + Q^-$ or $S^- + Q^+$

Resonance energy transfer: $S^* + Q \rightarrow S + Q^*$

Collisional quenching is particularly efficient when Q is a heavy species, such as an iodide ion, which receives energy from S* and then decays nonradiatively to the ground state. This fact may be used to determine the accessibility of amino acid residues of a folded protein to solvent. For example, fluorescence from a tryptophan residue ($\lambda_{abs} \approx 290$ nm, $\lambda_{fluor} \approx 350$ nm) is quenched by an iodide ion when the residue is on the surface of the protein and hence accessible to the solvent. Conversely, residues in the hydrophobic interior of the protein are not quenched effectively by I^-.

The quenching rate constant itself does not give much insight into the mechanism of quenching apart from suggesting that it is diffusion controlled. However, according to the **Marcus theory** of electron transfer, which was proposed by R. A. Marcus in 1965, the rates of electron transfer (from ground or excited states) depend on:

1. The distance between the donor and acceptor, with electron transfer becoming more effcient as the distance between donor and acceptor decreases.
2. The reaction Gibbs energy, $\Delta_r G$, with electron transfer becoming more efficient as the reaction becomes more exergonic. For example, efficient photooxidation of S requires that the reduction potential of S* be lower than the reduction potential of Q.
3. The 'reorganization energy', the energy cost incurred by molecular rearrangements of donor, acceptor, and medium during electron transfer. The electron transfer rate is predicted to increase as this reorganization energy is matched more closely by the reaction Gibbs energy.

Electron transfer can be studied by time-resolved spectroscopy. The oxidized and reduced products often have electronic absorption spectra distinct from those of their neutral parent compounds. Therefore, the rapid appearance of such known features in the absorption spectrum after excitation by a laser pulse may be taken as an indication of quenching by electron transfer.

To understand resonance energy transfer we note that in an absorption process the incident electromagnetic radiation induces a transition electric dipole moment in S. When that excited state collapses back to the ground state the resulting transition dipole can induce a corresponding transition dipole moment in a neighbouring Q molecule. It does so with an efficiency, η_T, that can be expressed in terms of the fluorescence lifetimes in the absence and the presence of the quencher:

$$\eta_T = \frac{\tau_{obs} - \tau_{obs}(Q)}{\tau_{obs}} \qquad (20.16)$$

According to the **Förster theory** of resonance energy transfer, which was proposed by T. Förster in 1959, for donor–acceptor (S—Q) systems that are held rigidly either by covalent bonds or by a protein 'scaffold', η_T, increases with decreasing distance, R, according to

Table 20.2
*Values of R_0 for some donor–acceptor pairs**

Donor	Acceptor	R_0/nm
Naphthalene	Dansyl	2.2
Dansyl	ODR	4.3
Pyrene	Coumarin	3.9
IAEDANS	FITC	4.9
Tryptophan	IAEDANS	2.2
Tryptophan	Haem	2.9

*Abbreviations:

Dansyl,	5-dimethylamino-l-naphthalenesulfonic
FITC	fluorescein-5-isothiocyanate
IEADANS	5-(((2-iodoacetyl)amino)ethyl)amino)naphthalene-1-sulfonic acid
ODR	octadecyl-rhodamine

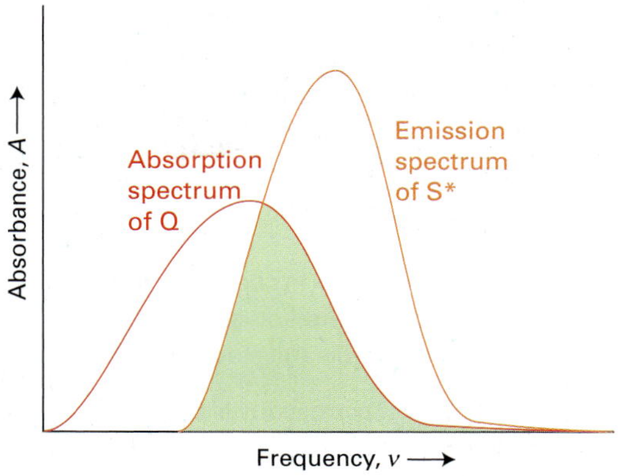

Fig. 20.33 According to the Förster theory, the rate of energy transfer from a molecule S* in an excited state to a quencher molecule Q is optimized at radiation frequencies in which the emission spectrum of S* overlaps with the absorption spectrum of Q, as shown in the shaded region.

$$\eta_T = \frac{R_0^6}{R_0^6 + R^6} \qquad (20.17)$$

where R_0 is a parameter (with dimensions of distance) that is characteristic of each donor–acceptor pair. Equation 20.17 has been verified experimentally and values of R_0 are available for a number of donor–acceptor pairs (Table 20.2). According to the Förster theory, for a given separation, the efficiency is greatest when the emission spectrum of the donor molecule overlaps significantly with the absorption spectrum of the acceptor. In the overlap region, photons emitted by the donor can be absorbed resonantly by the acceptor, as the energy gaps of the two molecules then match (Fig. 20.33).

The dependence of η_T on R forms the basis of **fluorescence resonance energy transfer** (FRET), a technique that can be used to measure distances in biological systems. In a typical FRET experiment, a site on a biopolymer or membrane is labelled covalently with an energy donor and another site is labelled covalently with an energy acceptor. The distance between the labels is then calculated from the known value of R_0 and eqn 20.15. Several tests have shown that the FRET technique is useful for measuring distances ranging from 1 to 9 nm.

A brief illustration The energy donor 1.5-I AEDANS (A, **2**) has a fluorescence quantum yield of 0.75 in aqueous solution. The visual pigment 11-*cis*-retinal (Box 20.1) is a quencher of fluorescence and R_0 = 5.4 nm for the A-*cis*-retinal pair. When an amino acid on the surface of the protein rhodopsin, which binds 11-*cis*-retinal in its interior, was labelled covalently with A, the fluorescence quantum yield of the label decreased to 0.68. From eqn 20.16, noting that $\tau_{obs}(Q)/\tau_{obs} = \phi/\phi_f$, we calculate $\varepsilon_T = 1 - (0.68/0.75) = 0.093$ and from eqn 20.17 we then calculate $R = 7.9$ nm, which is taken as the distance between the surface of the protein and 11-*cis*-retinal. ●

Checklist of key ideas

You should now be familiar with the following concepts.

☐ **1** An isosbestic point corresponds to a wavelength at which the total absorbance of a binary mixture is the same for all compositions.

☐ **2** The Franck–Condon principle states that because nuclei are so much more massive than electrons, an electronic transition takes place faster than the nuclei can respond.

☐ **3** A chromophore is a group with characteristic optical absorption: they include d-metal complexes, the carbonyl group, and the carbon–carbon double bond.

☐ **4** In fluorescence, the spontaneously emitted radiation ceases almost immediately (within nanoseconds) after the exciting radiation is extinguished.

☐ **5** In phosphorescence, the spontaneous emission may persist for long periods; the process involves intersystem crossing into a triplet state.

☐ **6** Laser action depends on the achievement of population inversion and the stimulated emission of radiation.

☐ **7** Applications of lasers in chemistry include Raman spectroscopy, time-resolved spectroscopy, the study of multiphoton and state-specific processes.

☐ **8** Photoelectron spectroscopy is based on the photoejection of an electron by ultraviolet radiation or X-rays.

☐ **9** The primary quantum yield of a photochemical reaction is the number of reactant molecules producing specified primary products for each photon absorbed; the overall quantum yield is the number of reactant molecules that react for each photon absorbed.

☐ **10** A Stern–Volmer plot is used to analyse the kinetics of fluorescence quenching in solution.

☐ **11** Collisional deactivation, electron transfer, and resonance energy transfer are common fluorescence quenching processes. The rate constants of electron and resonance energy transfer decrease with increasing separation between donor and acceptor molecules.

Table of key equations

The following table summarizes the equations developed in the text.

Description	Equation	Comment
Beer–Lambert law	$I = I_n e^{-\varepsilon[J]L}$	Uniform solution
Absorbance	$A = \log(I_0/I)$	
Photoelectron spectroscopy	$h\nu = I + \frac{1}{2}m_e\upsilon^2$	
Observed fluorescence lifetime	$\tau_{obs} = 1/(k_f + k_{ISC} + k_{IS})$	
Fluorescence quantum yield	$\phi_f = k_f\tau_{obs}$	
Stern—Volmer equation	$\phi_f/\phi = 1 + \tau_{obs}k_Q[Q]$	
	$1/\tau_{obs}(Q) = 1/\tau_{obs} + k_Q[Q]$	Rigid S—Q separation, R
Quenching efficiency	$\eta_T = (\tau_{obs} - \tau_{obs}(Q))/\tau_{obs}$	
Förster theory	$\eta_T = R_0^6/(R_0^6 - R^6)$	

Further information 20.1

The Beer–Lambert law

The Beer–Lambert law is an empirical result. However, it is simple to account for its form. We think of the sample as consisting of a stack of infinitesimal slices, like sliced bread (Fig. 20.34). The thickness of each layer is dx. The change in intensity, dI, that occurs when electromagnetic radiation passes through one particular slice is proportional to the

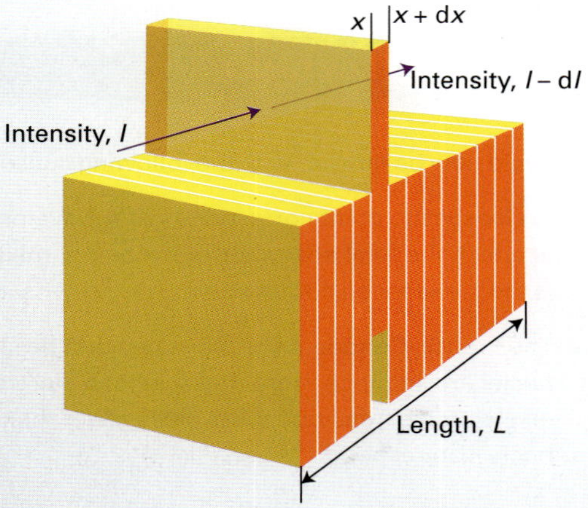

Fig 20.34. (*right*) To establish the Beer–Lambert law, the sample is supposed to be sliced into a large number of planes. The reduction in intensity caused by one plane is proportional to the intensity incident on it (after passing through the preceding planes), the thickness of the plane, and the concentration of absorbing species.

thickness of the slice, the concentration of the absorber J, and the intensity of the incident radiation at that slice of the sample, so $dI \propto [J]I dx$. Because dI is negative (the intensity is reduced by absorption), we can write

$$dI = -\kappa[J]I dx$$

where κ (kappa) is the proportionality coefficient. Division by I gives

$$\frac{dI}{I} = -\kappa[J]dx$$

This expression applies to each successive slice. To obtain the intensity that emerges from a sample of thickness L when the intensity incident on one face of the sample is I_0, we sum all the successive changes. Because a sum over infinitesimally small increments is an integral, we write:

$$\int_{I_0}^{I} \frac{dI}{I} = -\kappa \int_0^L [J]dx$$

If the concentration is uniform, [J] is independent of location and may be taken outside the integral on the right; the expression then integrates to

$$\ln \frac{I}{I_0} = -\kappa[J]L$$

Because the relation between natural and common logarithms is $\ln x = (\ln 10)\log x$, we can write $\varepsilon = \kappa/\ln 10$ and obtain

$$\log \frac{I}{I_0} = -\varepsilon[J]L$$

which, on substituting $A = \log(I_0/I) = -\log(I/I_0)$, is the Beer–Lambert law (eqn 20.1).

Further information 20.2

The Einstein transition probabilities

The intensity of an absorption line is related to the rate at which energy from electromagnetic radiation at a specified frequency is absorbed by a molecule. Einstein identified three contributions to the rates of transitions between states. *Stimulated absorption* is the transition from a low-energy state to one of higher energy that is driven by the electromagnetic field oscillating at the transition frequency. He reasoned that the more intense the electromagnetic field (the more intense the incident radiation), the greater the rate at which transitions are induced and hence the stronger the absorption by the sample, so he wrote the rate of stimulated absorption as

$$\text{Rate of stimulated absorption} = NB\rho$$

where N is the number of molecules in the lower state, the constant B is the *Einstein coefficient of stimulated absorption*, and $\rho\Delta\nu$ is the energy density of radiation in the frequency range ν to $\nu + \Delta\nu$, with ν as the frequency of the transition. For the time being, we can treat B as an empirical parameter that characterizes the transition: if B is large, then a given intensity of incident radiation will induce transitions strongly and the sample will be strongly absorbing.

Einstein considered that the radiation was also able to induce the molecule in the upper state to undergo a transition to the lower state, and hence to generate a photon of frequency ν. Thus, he wrote the rate of this stimulated emission as

$$\text{Rate of stimulated emission} = N'B'\rho$$

where N' is the number of molecules in the excited state and B' is the *Einstein coefficient of stimulated emission*. Note that only radiation of the same frequency as the transition can stimulate an excited state to fall to a lower state. However, he realized that stimulated emission was not the only means by which the excited state could generate radiation and return to the lower state, and suggested that an excited state could undergo *spontaneous emission* at a rate that is independent of the intensity of the radiation (of any frequency) that is already present. Einstein therefore wrote the total rate of transition from the upper to the lower state as

$$\text{Overall rate of emission} = N'(A + B'\rho)$$

The constant A is the *Einstein coefficient of spontaneous emission*. It can be shown that the coefficients of stimulated absorption and emission are equal, and that the coefficient of spontaneous emission is related to them by[1]

$$A = \left(\frac{8\pi h\nu^3}{c^3}\right)B$$

The equality of the coefficients of stimulated emission and absorption implies that if two states happen to have equal populations, then the rate of stimulated emission is equal to the rate of stimulated absorption, and there is then no net absorption. The drop in the value of A with decreasing frequency implies that spontaneous emission can be largely ignored at the relatively low frequencies of rotational and vibrational transitions, and the intensities of these transitions can be discussed in terms of stimulated emission and absorption. Then the net rate of absorption is given by

$$\text{Net rate of absorption} = NB\rho - N'B'\rho = (N - N')B\rho$$

and is proportional to the population difference of the two states involved in the transition. In Chapter 22, we shall see

[1] See our *Physical chemistry* (2006) for the derivation.

solvent, the phosphorescence intensity varied with amine concentration as shown below. A time-resolved laser spectroscopy experiment had also shown that the half-life of the fluorescence in the absence of quencher is 29 μs. What is the value of k_q?

$[Q]/(\text{mol dm}^{-3})$	0.0010	0.0050	0.0100
I_f/(arbitrary units)	0.41	0.25	0.16

20.21 The quenching of tryptophan fluorescence by dissolved O_2 gas was monitored by measuring emission lifetimes at 348 nm in aqueous solutions. Determine the quenching rate constant for this process from the following data:

$[O_2]/(10^{-2}\text{ mol dm}^{-3})$	0	2.3	5.5	8	10.8
τ/ns	2.6	1.5	0.92	0.71	0.57

20.22 The fluorescence of a solution of a plant pigment illuminated by 330 nm radiation was studied in the presence of a quenching agent, with the following results:

$[Q]/(\text{mmol dm}^{-3})$	1.0	2.0	3.0	4.0	5.0
I_f/I_{abs}	0.31	0.18	0.13	0.10	0.081

In a second series of experiments, the incident radiation was extinguished and the lifetime of the decay of the fluorescence was observed:

$[Q]/(\text{mmol dm}^{-3})$	1.0	2.0	3.0	4.0	5.0
τ/ns	76	45	32	25	20

Determine the quenching rate constant and the half-life of the fluorescence.

20.23 The fluorescence lifetime in the absence of a quencher is 1.4 ns and in the presence of a quencher it is 0.8 ns. What is the quenching efficiency?

20.24 The following data refer to a family of compounds with the general composition A—B_n—C in which the distance R between A and C was varied by increasing the number of B units in the linker:

R/nm	1.2	1.5	1.8	2.8	3.1
η_T	0.99	0.94	0.97	0.82	0.74
R/nm	3.4	3.7	4.0	4.3	4.6
η_T	0.65	0.40	0.28	0.24	0.16

Are the data described adequately by the Förster theory (eqn 20.17)? If so, what is the value of R_0 for the A—C pair?

20.25 Light-induced degradation of molecules, also called *photobleaching*, is a serious problem in *fluorescence microscopy*, in which a specimen (such as a biological cell) labelled with a fluorescent dye is observed under an optical microscope. A molecule of a dye commonly used to label biopolymers can withstand about 10^6 excitations by photons before light-induced reactions destroy its π system and the molecule no longer fluoresces. For how long will a single dye molecule fluoresce while being excited by 1.0 mW of 488 nm radiation from an argon ion laser? You may assume that the dye has an absorption spectrum that peaks at 488 nm and that every photon delivered by the laser is absorbed by the molecule.

Projects

20.26 Here we explore vision in more detail. (a) The flux of visible photons reaching Earth from the North Star is about $4\times10^3\text{ mm}^{-2}\text{ s}^{-1}$. Of these photons, 30 per cent are absorbed or scattered by the atmosphere and 25 per cent of the surviving photons are scattered by the surface of the cornea of the eye. A further 9 per cent are absorbed inside the cornea. The area of the pupil at night is about 40 mm^2 and the response time of the eye is about 0.1 s. Of the photons passing through the pupil, about 43 per cent are absorbed in the ocular medium. How many photons from the North Star are focused on to the retina in 0.1 s? For a continuation of this story, see R. W. Rodieck, *The first steps in seeing*, Sinauer (1998). (b) In the free-electron molecular orbital theory of electronic structure, the π electrons in a conjugated molecule are treated as noninteracting particles in a box of length equal to the length of the conjugated system. On the basis of this model, at what wavelength would you expect all-*trans*-retinal to absorb? Take the mean carbon–carbon bond length to be 140 pm.

20.27 Now we explore the energy and electron-transfer events of photosynthesis. (a) In light-harvesting complexes, the fluorescence of a chlorophyll molecule is quenched by nearby chlorophyll molecules. Given that for a pair of chlorophyll *a* molecules R_0 = 5.6 nm, by what distance should two chlorophyll *a* molecules be separated to shorten the fluorescence lifetime from 1 ns (a typical value for monomeric chlorophyll *a* in organic solvents) to 10 ps? (b) The light-induced electron-transfer reactions in photosynthesis occur because chlorophyll molecules (whether in monomeric or dimeric forms) are better reducing agents in their electronic excited states. Justify this observation with the help of molecular orbital theory.

Chapter 21

Spectroscopy: magnetic resonance

One of the most widely used and helpful forms of spectroscopy, and a technique that has transformed the practice of chemistry and its dependent disciplines, makes use of an effect that is familiar from classical physics. When two pendulums are joined by the same slightly flexible support and one is set in motion, the other is forced into oscillation by the motion of the common axle, and energy flows between the two. The energy transfer occurs most efficiently when the frequencies of the two oscillators are identical. The condition of strong effective coupling when the frequencies are identical is called **resonance**, and the excitation energy is said to 'resonate' between the coupled oscillators.

Resonance is the basis of a number of everyday phenomena, including the response of radios to the weak oscillations of the electromagnetic field generated by a distant transmitter. In this chapter we explore a spectroscopic application that when originally developed (and in some cases still) depends on matching a set of energy levels to a source of monochromatic radiation in the radio-frequency and microwave ranges and observing the strong absorption by nuclei and electrons, respectively, that occurs at resonance. All spectroscopy is a form of resonant coupling between the electromagnetic field and the molecules; what distinguishes magnetic resonance is that the energy levels themselves are modified by the application of a magnetic field.

Principles of magnetic resonance

The application of resonance that we describe here depends on the fact that electrons and many nuclei possess spin angular momentum (Table 21.1). An

Table 21.1

Nuclear constitution and the nuclear spin quantum number

Number of protons	Number of neutrons	I
even	even	0
odd	odd	integer (1, 2, 3, ...)
even	odd	half-integer ($\frac{1}{2}, \frac{3}{2}, \frac{5}{2}, \ldots$)
odd	even	half-integer ($\frac{1}{2}, \frac{3}{2}, \frac{5}{2}, \ldots$)

electron (with spin quantum number $s = \frac{1}{2}$) in a magnetic field can take two orientations, corresponding to $m_s = +\frac{1}{2}$ (denoted α or ↑) and $m_s = -\frac{1}{2}$ (denoted β or ↓). A nucleus with **nuclear spin quantum number** I (the analogue of s for electrons, and that may be an integer or a half-integer) may take $2I + 1$ different orientations relative to an arbitrary axis. These orientations are distinguished by the quantum number m_I, which may take on the values $m_I = I, I - 1, \ldots, -I$. A proton has $I = \frac{1}{2}$ (the same spin as an electron) and may adopt either of two orientations ($m_I = +\frac{1}{2}$ and $-\frac{1}{2}$). A ^{14}N nucleus has $I = 1$ and may adopt any of three orientations ($m_I = +1, 0, -1$). Spin-$\frac{1}{2}$ nuclei include protons (^1H), ^{13}C, ^{19}F, and ^{31}P nuclei. As for electrons, the state with $m_I = +\frac{1}{2}$ (↑) is denoted α, and that with $m_I = -\frac{1}{2}$ (↓) is denoted β.

21.1 Electrons and nuclei in magnetic fields

An electron possesses a magnetic moment due to its spin and this moment interacts with an external magnetic field. That is, an electron behaves like a tiny magnet. The orientation of this magnet is determined by the value of m_s and in a magnetic field $\mathcal{B}$ the two orientations have different energies. These energies are given by

$$E_{m_s} = -g_e \gamma \hbar \mathcal{B} m_s \qquad (21.1)$$

where γ is the **magnetogyric ratio** of the electron

$$\gamma = -\frac{e}{2m_e} \qquad (21.2)$$

and g_e is a factor, the **g-value of the electron**, which is close to 2.0023 for a free electron. The 2 comes from Dirac's relativistic theory of the electron; the 0.0023 comes from additional correction terms.

The energies are sometimes expressed in terms of the **Bohr magneton**

$$\mu_B = \frac{e\hbar}{2m_e} \qquad \mu_B = 9.274 \times 10^{-24}\ \mathrm{J\ T^{-1}} \qquad (21.3)$$

a fundamental unit of magnetism, where the symbol T denotes the unit tesla, which is used to express the intensity of a magnetic field (more precisely, the magnetic induction, with $1\ \mathrm{T} = 1\ \mathrm{kg\ s^{-2}\ A^{-1}}$). Now we can write

$$E_{m_s} = g_e \mu_B \mathcal{B} m_s \qquad (21.4)$$

For an electron, the β state lies below the α state.

A brief comment Classically, the energy of a magnetic moment $\boldsymbol{m}$ in a magnetic field $\boldsymbol{\mathcal{B}}$ is $E = -\boldsymbol{m}.\boldsymbol{\mathcal{B}}$. Equation 21.1 is the quantum-mechanical version of the classical expression, with the field along the z direction and the magnetic moment equal to $g_e \gamma s_z$ and $s_z = m_s \hbar$.

A nucleus with nonzero spin also has a magnetic moment and behaves like a tiny magnet. The orientation of this magnet is determined by the value of m_I, and in a magnetic field $\mathcal{B}$ the $2I + 1$ orientations of the nucleus have different energies. These energies are given by

$$E_{mI} = -\gamma_N \hbar \mathcal{B} m_I \qquad (21.5)$$

where γ_N is the **nuclear magnetogyric ratio.** For spin-$\frac{1}{2}$ nuclei with positive magnetogyric ratios (such as ^1H), the α state lies below the β state. The energy is sometimes written in terms of the **nuclear magneton,** μ_N,

$$\mu_N = \frac{e\hbar}{2m_p} \qquad \mu_N = 5.051 \times 10^{-27}\ \mathrm{J\ T^{-1}} \qquad (21.6)$$

and an empirical constant called the **nuclear g-factor,** g_I, when it becomes

$$E_{mI} = -g_I \mu_N \mathcal{B} m_I \qquad (21.7)$$

Nuclear g-factors are experimentally determined dimensionless quantities with values typically between −6 and +6. Positive values of γ_N (and g_I) indicate that the North pole of the nuclear magnet lies in the same direction as the nuclear spin (this is the case for protons). Negative values indicate that the magnet points in the opposite direction. A nuclear magnet is about 2000 times weaker than the magnet associated with electron spin. Two very common nuclei, ^{12}C and ^{16}O, have zero spin and hence are not affected by external magnetic fields.

The energy separation of the two spin states of an electron (Fig. 21.1) is

$$\begin{aligned}\Delta E &= E_\alpha - E_\beta = (\tfrac{1}{2})g_e \mu_B \mathcal{B} - (-\tfrac{1}{2}g_e \mu_B \mathcal{B}) \\ &= g_e \mu_B \mathcal{B}\end{aligned} \qquad (21.8)$$

We shall see in Section 22.1 in the discussion of the Boltzmann distribution that the populations of the α

Table 21.2

Nuclear spin properties

Nucleus	Natural abundance per cent	Spin, I	$\gamma_N/(10^7\ T^{-1}\ s^{-1})$
^1H	99.98	$\frac{1}{2}$	26.752
^2H (D)	0.0156	1	4.1067
^{12}C	98.99	0	–
^{13}C	1.11	$\frac{1}{2}$	6.7272
^{14}N	99.64	1	1.9328
^{16}O	99.96	0	–
^{17}O	0.037	$\frac{5}{2}$	−3.627
^{19}F	100	$\frac{1}{2}$	25.177
^{31}P	100	$\frac{1}{2}$	10.840
^{35}Cl	75.4	$\frac{3}{2}$	2.624
^{37}Cl	24.6	$\frac{3}{2}$	2.184

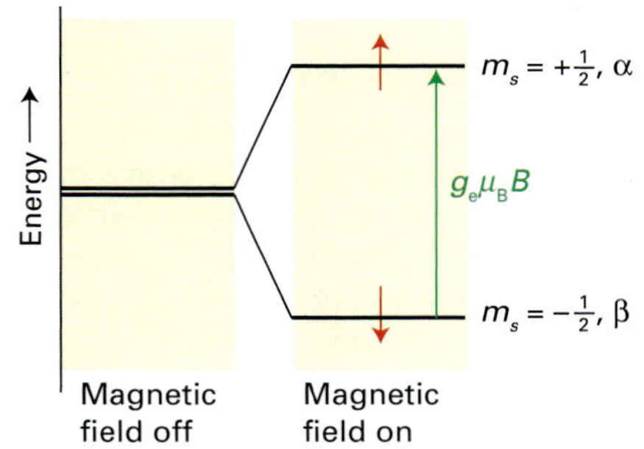

Fig. 21.1 The energy levels of an electron in a magnetic field. Resonance occurs when the energy separation of the levels matches the energy of the photons in the electromagnetic field.

and β states, N_α and N_β, are proportional to $e^{-E_\alpha/kT}$ and $e^{-E_\beta/kT}$, respectively, so the ratio of populations at equilibrium is

$$\frac{N_\alpha}{N_\beta} = e^{-(E_\alpha - E_\beta)/kT} \qquad (21.9a)$$

Because $E_\alpha - E_\beta > 0$ (for an electron the β state lies below the α state), $N_\alpha/N_\beta < 1$ and there are slightly more β spins than α spins. We show in Derivation 21.1 that

$$N_\beta - N_\alpha \approx \frac{N g_e \mu_B \mathcal{B}}{2kT} \qquad (21.9b)$$

In a field of 1.0 T at 300 K, $(N_\beta - N_\alpha)N \approx 0.0022$, so there is an imbalance of populations of only about 2 electrons in a thousand (see Exercise 21.5).

Derivation 21.1

The population difference

To write an expression for the population difference, we begin with eqn 21.9a, written as

$$\frac{N_\alpha}{N_\beta} = e^{-(E_\alpha - E_\beta)/kT} \approx 1 - \frac{E_\alpha - E_\beta}{kT} = 1 - \frac{g_e \mu_B \mathcal{B}_{cal}}{kT}$$

where we have used the fact that an exponential function used here is $e^{-x} = 1 - x + \frac{1}{2}x^2 - \cdots$. If $x \ll 1$, then $e^{-x} \approx 1 - x$. The expansion of the exponential term is appropriate for $\Delta E \ll kT$, a condition usually met for electron and nuclear spins. It follows that

$$\frac{N_\beta - N_\alpha}{N_\beta + N_\alpha} = \frac{N_\beta(1 - N_\alpha/N_\beta)}{N_\beta(1 + N_\alpha/N_\beta)} = \frac{1 - N_\alpha/N_\beta}{1 + N_\alpha/N_\beta}$$

$$\approx \frac{1 - (1 - g_e\mu_B\mathcal{B}_{cal}/kT)}{1 + (1 - g_e\mu_B\mathcal{B}_{cal}/kT)} \approx \frac{g_e\mu_B\mathcal{B}_{cal}/kT}{2}$$

Then, with $N_\alpha + N_\beta = N$, the total number of spins, we have eqn 21.9b.

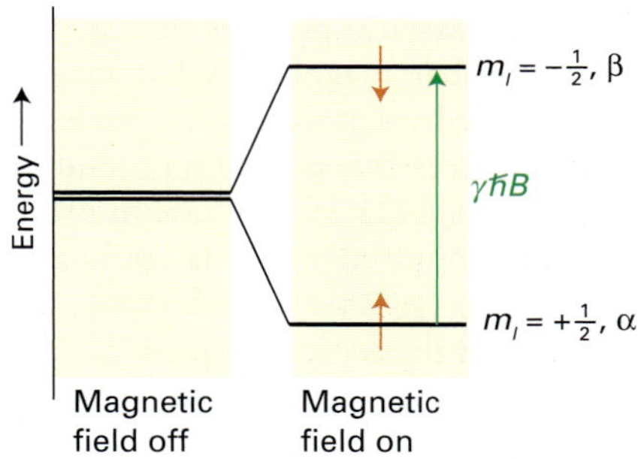

Fig. 21.2 The energy levels of a spin-$\frac{1}{2}$ nucleus (e.g. ^1H or ^{13}C) in a magnetic field. Resonance occurs when the energy separation of the levels matches the energy of the photons in the electromagnetic field.

If the sample is exposed to radiation of frequency ν, the energy separations come into resonance with the radiation when the frequency satisfies the **resonance condition:**

$$h\nu = g_e\mu_B\mathcal{B} \qquad \text{or} \qquad \nu = \frac{g_e\mu_B\mathcal{B}}{h} \qquad (21.10)$$

At resonance there is strong coupling between the electron spin and the radiation, and strong absorption occurs as the spins flip from β (low energy) to α (high energy).

The behaviour of nuclei is very similar. The energy separation of the two states of a spin-$\frac{1}{2}$ nucleus (Fig. 21.2) is

$$\Delta E = E_\beta - E_\alpha = \tfrac{1}{2}\gamma_N\hbar\mathcal{B} - (-\tfrac{1}{2}\gamma_N\hbar\mathcal{B}) = \gamma_N\hbar\mathcal{B} \qquad (21.11)$$

Because for nuclei with positive γ_N the α state lies below the β state, $E_\beta - E_\alpha > 0$ and it follows from eqn 21.9 that $N_\beta/N_\alpha < 1$: there are slightly more α spins than β spins (the opposite of an electron) and in the same way as in Derivation 21.1 we can write

$$N_\alpha - N_\beta \approx \frac{N\gamma_N\hbar\mathcal{B}}{2kT} \qquad (21.12)$$

For protons in a field of 10 T at 300 K, $(N_\alpha - N_\beta)/N \approx 3 \times 10^{-5}$, so even in such a strong field there is only a tiny imbalance of population of about 30 in a million.

As for electrons, if the sample is exposed to radiation of frequency ν, the energy separations come into resonance with the radiation when the frequency satisfies the resonance condition:

$$h\nu = \gamma_N\hbar\mathcal{B} \qquad \text{or} \qquad \nu = \frac{\gamma_N\mathcal{B}}{2\pi} \qquad (21.13)$$

At resonance there is strong coupling between the nuclear spins and the radiation, and absorption occurs as the spins flip from α (low energy) to β (high energy).

It is sometimes useful to compare the quantum-mechanical and classical pictures of magnetic nuclei pictured as tiny bar magnets. A bar magnet in an externally applied magnetic field undergoes the motion called **precession** as it twists round the direction of the field (Fig. 21.3). The rate of precession is proportional to the strength of the applied field, and is in fact equal to $(\gamma_N/2\pi)\mathcal{B}$, which in this context is called the **Larmor precession frequency**. That is, resonance absorption occurs when the Larmor precession frequency is the same as the frequency of the applied electromagnetic field.

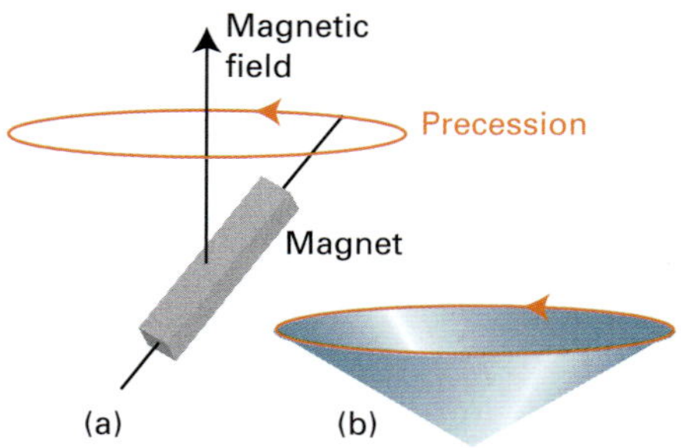

Fig. 21.3 A bar magnet in a magnetic field undergoes the motion called *precession*. A nuclear spin (and an electron spin) has an associated magnetic moment, and behaves in the same way. The frequency of precession is called the Larmor precession frequency, and is proportional to the applied field and the magnitude of the magnetic moment.

See an animated version of this figure in the interactive ebook.

Self-test 21.1

Calculate the frequency at which radiation comes into resonance with proton spins in a 12 T magnetic field.

[*Answer:* 510 MHz]

The intensity of an NMR transition depends on a number of factors. We show in Derivation 21.2 that

$$\text{Intensity} \propto (N_\alpha - N_\beta)\mathcal{B} \propto \mathcal{B}^2 \qquad (21.14)$$

It follows that decreasing the temperature increases the intensity by increasing the population difference. The intensity can also be enhanced significantly by increasing the strength of the applied magnetic field, making spectrometers operating at high fields highly desirable. Similar arguments apply to EPR transitions.

Derivation 21.2

Intensities in NMR spectra

From the general considerations of transition intensities in Further information 20.2, we know that the rate of absorption of electromagnetic radiation is proportional to the population of the lower energy state (N_α in the case of a proton NMR transition) and the rate of stimulated emission is proportional to the population of the upper state (N_β). At the low frequencies typical of magnetic resonance, we can neglect spontaneous emission as it is very slow. Therefore, the net rate of absorption is proportional to the difference in populations, and we can write

$$\text{Rate of transition} \propto N_\alpha - N_\beta$$

The intensity of absorption, the rate at which energy is absorbed, is proportional to the product of the rate of transition (the rate at which photons are absorbed) and the energy of each photon, and the latter is proportional to the frequency ν of the incident radiation. At resonance, this frequency is proportional to the applied magnetic field, so we can write

$$\text{Intensity of absorption} \propto (N_\alpha - N_\beta)\mathcal{B}$$

with the population difference proportional to the field (eqn 21.12).

21.2 The technique

In its simplest form, **nuclear magnetic resonance** (NMR) is observation of the frequency at which magnetic nuclei in molecules come into resonance with an electromagnetic field when the molecule is

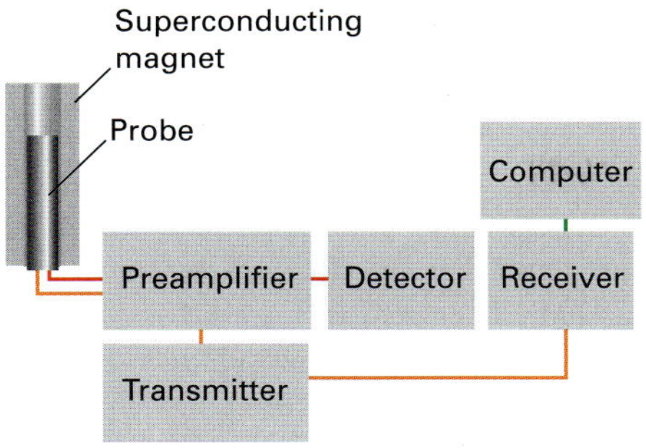

Fig. 21.4 The layout of a typical NMR spectrometer. The link from the transmitter to the detector indicates that the high frequency of the transmitter is subtracted from the high-frequency received signal to give a low-frequency signal for processing.

exposed to a strong magnetic field. When applied to proton spins, the technique is occasionally called **proton magnetic resonance** (^{1}H-NMR). In the early days of the technique the only nuclei that could be studied were protons (which behave like relatively strong magnets because γ_N is large), but now a wide variety of nuclei (especially ^{13}C, ^{31}P, and ^{15}N) are investigated routinely.

An NMR spectrometer consists of a magnet that can produce a uniform, intense field and the appropriate sources of radiofrequency radiation (Fig. 21.4). In simple instruments the magnetic field is provided by an electromagnet; for serious work, a superconducting magnet capable of producing fields of the order of 10 T and more is used. (A magnetic field of 10 T is very strong: a small magnet, for example, gives a magnetic field of only a few millitesla.) The use of high magnetic fields has two advantages. One is that, as we have seen, the field increases the intensities of transitions. Secondly, a high field simplifies the appearance of certain spectra. Proton resonance occurs at about 400 MHz in fields of 9.4 T, so NMR is a radio-frequency technique (400 MHz corresponds to a wavelength of 75 cm).

Fourier transform NMR (FT-NMR) is the most common technique used in modern magnetic resonance. The sample is held in a strong magnetic field generated by a superconducting magnet and exposed to one or more carefully controlled brief bursts of radiofrequency radiation. This radiation changes the orientations of the nuclear spins in a controlled way, and the radiofrequency radiation they emit as they return to equilibrium is monitored and analysed mathematically (the latter is the 'Fourier transform' part of the technique). The detected radiation contains all the information in the spectrum obtained by the earlier technique, but it is a much more efficient way of obtaining the spectrum and hence is much more sensitive. Moreover, by choosing different sequences of exciting pulses, the data can be analysed much more closely.

The resonance technique for electrons in a magnetic field is called **electron paramagnetic resonance** (EPR) or **electron spin resonance** (ESR). Because electron magnetic moments are much bigger than nuclear magnetic moments, even quite modest fields can require high frequencies to achieve resonance. Much work is done using fields of about 0.3 T, when resonance occurs at about 9 GHz, corresponding to 3 cm ('X-band') microwave radiation or at about 1 T, when resonance occurs at about 35 GHz, corresponding to about 9 mm ('Q-band') microwave radiation. Electron paramagnetic resonance is much more limited than NMR because it is applicable only to species with unpaired electrons, which include radicals (perhaps prepared by radiation damage or photolysis) and d-metal complexes, including such biologically active species as haemoglobin. However, it gives valuable information about electron distributions and can be used to monitor, for instance, the uptake of oxygen by haemoglobin and biological electron transfer processes.

Both Fourier-transform (FT) and continuous-wave (CW) EPR spectrometers are available. The FT-EPR instrument is like an FT-NMR spectrometer except that pulses of microwaves are used to excite electron spins in the sample. The layout of the more common CW-EPR spectrometer is shown in Fig. 21.5. It consists of a microwave source (a klystron or a Gunn oscillator), a cavity in which the sample is inserted in a glass or quartz container, a microwave detector, and an electromagnet with a field that can be varied in the region of 0.3 T (X-band) or 1 T (Q-band).

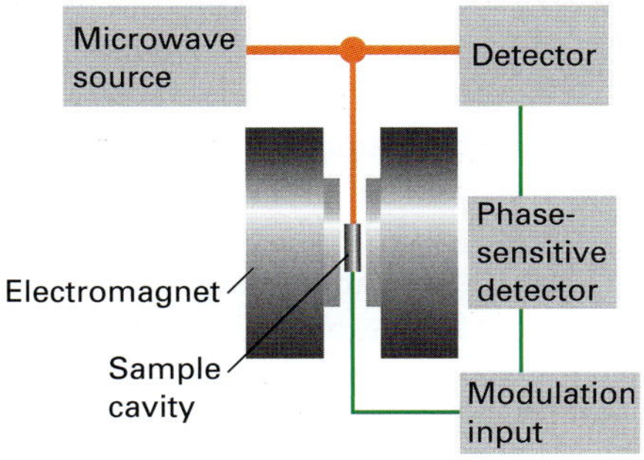

Fig. 21.5 The layout of a continuous-wave EPR spectrometer. A typical magnetic field is 0.3 T, which requires microwaves of frequency 9 GHz (wavelength 3 cm) for resonance.

The information in NMR spectra

Nuclear spins interact with the *local* magnetic field, the field in their immediate vicinity. The local field may differ from the applied field either on account of the local electronic structure of the molecule or because there is another magnetic nucleus nearby.

21.3 The chemical shift

The applied magnetic field can induce a circulating motion of the electrons in the molecule, and that motion gives rise to a small additional magnetic field, $\mathcal{B}_{add}$. This additional field is proportional to the applied field, and it is conventional to express it as

$$\mathcal{B}_{add} = -\sigma\mathcal{B} \tag{21.15}$$

where the dimensionless quantity σ is the **shielding constant.** The shielding constant may be positive or negative according to whether the induced field adds to or subtracts from the applied field. The ability of the applied field to induce the circulation of electrons through the nuclear framework of the molecule depends on the details of the electronic structure near the magnetic nucleus of interest, so nuclei in different chemical groups have different shielding constants.

Because the total local field is

$$\mathcal{B}_{loc} = \mathcal{B} + \mathcal{B}_{add} = (1-\sigma)\mathcal{B}$$

the resonance condition is

$$\nu = \frac{\gamma_N \mathcal{B}_{loc\,cal}}{2\pi} = \frac{\gamma_N}{2\pi}(1-\sigma)\mathcal{B} \tag{21.16}$$

Because σ varies with the environment, different nuclei (even of the same element in different parts of a molecule) come into resonance at different frequencies.

The **chemical shift** of a nucleus is the difference between its resonance frequency and that of a reference standard. The standard for protons is the proton resonance in tetramethylsilane, $Si(CH_3)_4$, commonly referred to as TMS, which bristles with protons and dissolves without reaction in many solutions. Other references are used for other nuclei. For ^{13}C, the reference frequency is the ^{13}C resonance in TMS, and for ^{31}P it is the ^{31}P resonance in 85 per cent $H_3PO_4(aq)$. The separation of the resonance of a particular group of nuclei from the standard increases with the strength of the applied magnetic field because the induced field is proportional to the applied field, and the stronger the latter, the greater the shift.

Chemical shifts are reported on the **δ scale**, which is defined as

$$\delta = \frac{\nu - \nu^\circ}{\nu^\circ} \times 10^6 \tag{21.17}$$

where ν° is the resonance frequency of the standard. The advantage of the δ scale is that shifts reported on it are independent of the applied field (because both numerator and denominator are proportional to the applied field). The resonance frequencies themselves, however, do depend on the applied field through

$$\nu = \nu^\circ + (\nu^\circ/10^6)\delta \tag{21.18}$$

A note on good practice In much of the literature, chemical shifts are reported in parts per million, ppm, in recognition of the factor of 10^6 in the definition; this is unnecessary. If you see '$\delta = 10$ ppm', interpret it, and use it in eqn 21.18, as $\delta = 10$.

A brief illustration A nucleus with $\delta = 1.00$ in a spectrometer operating at 500 MHz, a '500 MHz NMR spectrometer', will have a shift relative to the reference equal to

$$\nu - \nu^\circ = (500\text{ MHz}/10^6) \times 1.00 = (500\text{ Hz}) \times 1.00 = 500\text{ Hz}$$

because 1 MHz = 10^6 Hz. In a spectrometer operating at 100 MHz, the shift relative to the reference would be only 100 Hz.

Self-test 21.2

What is the shift of the resonance from TMS of a group of nuclei with $\delta = 3.50$ and an operating frequency of 350 MHz?

[*Answer:* 1.23 kHz]

If $\delta > 0$, we say that the nucleus is **deshielded;** if $\delta < 0$, then it is **shielded.** A positive δ indicates that the resonance frequency of the group of nuclei in question is higher than that of the standard. Hence $\delta > 0$ indicates that the local magnetic field is stronger than that experienced by the nuclei in the standard under the same conditions. Figure 21.6 shows some typical chemical shifts.

A brief illustration The existence of a chemical shift explains the general features of the spectrum of ethanol shown in Fig. 21.7. The CH_3 protons form one group of nuclei with $\delta = 1$. The two CH_2 are in a different part of the molecule, experience a different local magnetic field, and hence resonate at $\delta = 3$. Finally, the OH proton is in another environment, and has a chemical shift of $\delta = 4$.

A note on good practice Traditionally, NMR spectra are plotted with δ increasing from right to left. Consequently, in a given applied magnetic field the resonance frequency also increases from right to left.

We can use the relative intensities of the signal (the areas under the absorption lines) to help distinguish

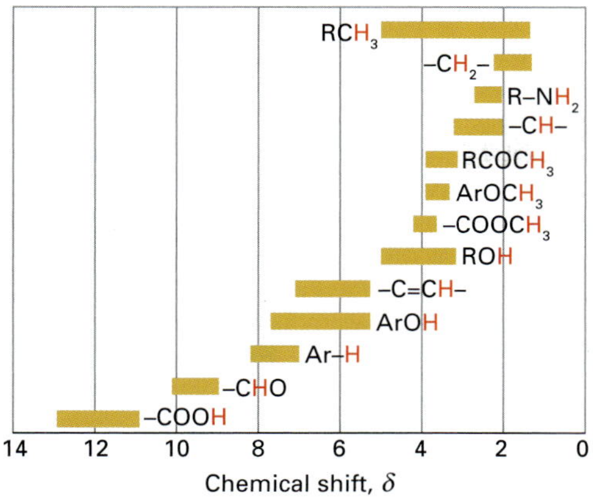

Fig. 21.6 The range of typical chemical shifts for ^{1}H resonances.

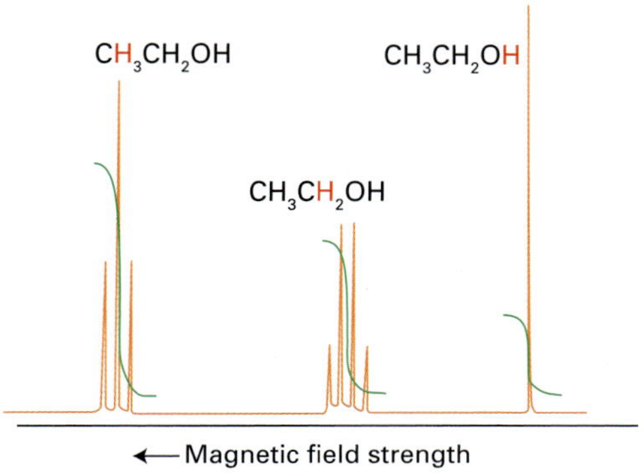

Fig. 21.7 The NMR spectrum of ethanol. The bold letters denote the protons giving rise to the resonance peak, and the step-like curves are the integrated signals for each group of lines.

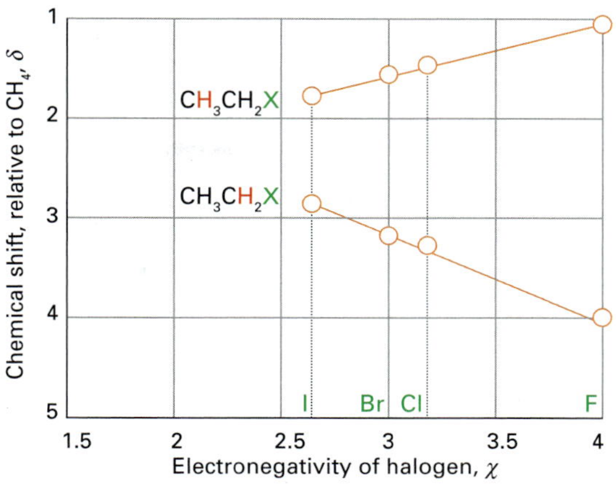

Fig. 21.8 The variation of chemical shift with the electronegativity of the halogen in the haloalkanes. Note that although the chemical shift of the immediately adjacent protons becomes more positive (the protons are deshielded) as the electronegativity increases, that of the next-nearest protons decreases.

which group of lines corresponds to which chemical group, and spectrometers can **integrate** the absorption—that is, determine the areas under the absorption signal—automatically (as is shown in Fig. 21.7). In ethanol the group intensities are in the ratio 3:2:1 because there are three CH_3 protons, two CH_2 protons, and one OH proton in each molecule. Counting the number of magnetic nuclei as well as noting their chemical shifts is valuable analytically because it helps us identify the compound present in a sample and to identify substances in different environments (Box 21.1).

The observed shielding constant is the sum of three contributions:

$$\sigma = \sigma(\text{local}) + \sigma(\text{neighbour}) + \sigma(\text{solvent}) \quad (21.19)$$

The **local contribution**, σ(local), is essentially the contribution of the electrons of the atom that contains the nucleus in question. The **neighbouring group contribution**, σ(neighbour), is the contribution from the groups of atoms that form the rest of the molecule. The **solvent contribution**, σ(solvent), is the contribution from the solvent molecules.

The local contribution is broadly proportional to the electron density of the atom containing the nucleus of interest. It follows that the shielding is decreased if the electron density on the atom is reduced by the influence of an electronegative atom nearby. That reduction in shielding translates into an increase in deshielding, and hence to an increase in the chemical shift δ as the electronegativity of a neighbouring atom increases (Fig. 21.8). That is, as the electronegativity increases, δ decreases. Another contribution to σ(local) arises from the ability of the applied field to force the electrons to circulate through the molecule by making use of orbitals that are unoccupied in the ground state. This contribution is large in molecules with low-lying excited states and is dominant for atoms other than hydrogen. It is zero in free atoms and around the axes of linear molecules (such as ethyne, HC≡CH) where the electrons can circulate freely, because a field applied along the internuclear axis is unable to force them into other orbitals.

The neighbouring group contribution arises from the currents induced in nearby groups of atoms. The strength of the additional magnetic field the proton experiences is inversely proportional to the cube of the distance r between H and X. A special case of a neighbouring group effect is found in aromatic compounds. The strong anisotropy of the magnetic susceptibility of the benzene ring is ascribed to the ability of the field to induce a **ring current**, a circulation

Box 21.1 Magnetic resonance imaging

One of the most striking applications of nuclear magnetic resonance is in medicine. *Magnetic resonance imaging* (MRI) is a portrayal of the concentrations of protons in a solid object. The technique relies on the application of specific pulse sequences to an object in an inhomogeneous magnetic field (a field with values that vary inside the sample).

If an object containing hydrogen nuclei (a tube of water or a human body) is placed in an NMR spectrometer and exposed to a *homogeneous* magnetic field (a field that has the same value throughout the sample), then a single resonance signal will be detected. Now consider a flask of water in a magnetic field that varies linearly in the z direction according to $\mathcal{B}_0 + \mathcal{G}_z z$, where $\mathcal{G}_z$ if the field gradient along the z direction (see the first illustration). Then the water protons will be resonant at the frequencies

$$\nu(z) = \frac{\gamma_N}{2}(\mathcal{B}_0 + \mathcal{G}_z z)$$

(Similar equations may be written for gradients along the x and y directions.) Exposing the sample to radiation with frequency $\nu(z)$ will result in a signal with an intensity that is proportional to the numbers of protons at the position z. This is an example of *slice selection*, the use of radio-frequency radiation that excites nuclei in a specific region, or slice, of the sample. It follows that the intensity of the NMR signal will be a projection of the numbers of protons on a line parallel to the field gradient. The image of a three-dimensional object such as a flask of water can be obtained if the slice selection technique is applied at different orientations (see the first illustration). In *projection reconstruction*, the projections can be analysed on a computer to reconstruct the three-dimensional distribution of protons in the object.

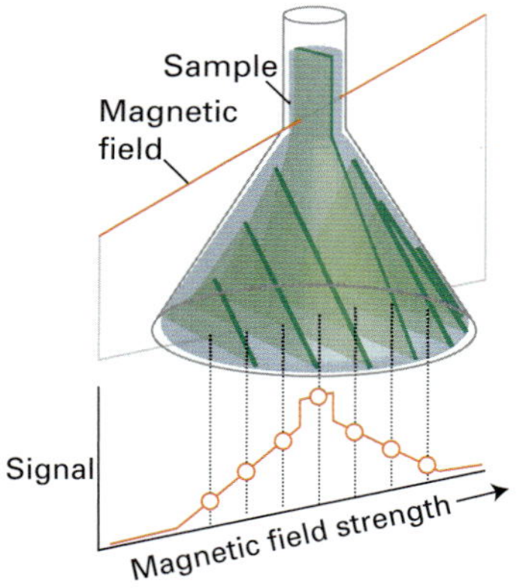

In a magnetic field that varies linearly over a sample, all the protons within a given slice (that is, at a given field value) come into resonance and give a signal of the corresponding intensity. The resulting intensity pattern is a map of the numbers in all the slices, and portrays the shape of the sample. Changing the orientation of the field shows the shape along the corresponding direction, and computer manipulation can be used to build up the three-dimensional shape of the sample.

A common problem with the techniques described above is image contrast, which must be optimized in order to show spatial variations in water content in the sample. One strategy for solving this problem takes advantage of the fact that the relaxation times of water protons are shorter for water in biological tissues than for the pure liquid. Furthermore, relaxation times from water protons are also different in healthy and diseased tissues. A *T_1-weighted image* is obtained by obtaining data before spin–lattice relaxation can return the spins in the sample to equilibrium. Under these conditions, differences in signal intensities are directly related to differences in T_1. A *T_2-weighted image* is obtained by collecting data after the system has relaxed extensively, though not completely. In this way, signal intensities are strongly dependent on variations in T_2. However, allowing so much of the decay to occur leads to weak signals even for those protons with long spin–spin relaxation times. Another strategy involves the use of *contrast agents*, paramagnetic compounds that shorten the relaxation times of nearby protons. The technique is particularly useful in enhancing image contrast and in diagnosing disease if the contrast agent is distributed differently in healthy and diseased tissues.

The MRI technique is used widely to detect physiological abnormalities and to observe metabolic processes. With *functional MRI*, blood flow in different regions of the brain can be studied and related to the mental activities of the subject. The special advantage of MRI is that it can image *soft* tissues (see the second illustration), whereas X-rays are largely used for imaging hard, bony structures and abnormally dense regions, such as tumours. In fact, the invisibility of hard structures in MRI is an advantage, as it allows the imaging of structures encased by bone, such as the brain and the spinal cord. X-rays are known to be dangerous on account of the ionization they cause; the high magnetic fields used in MRI may also be dangerous, but apart from anecdotes about the extraction of loose fillings from teeth, there is no convincing evidence of their harmfulness, and the technique is considered safe.

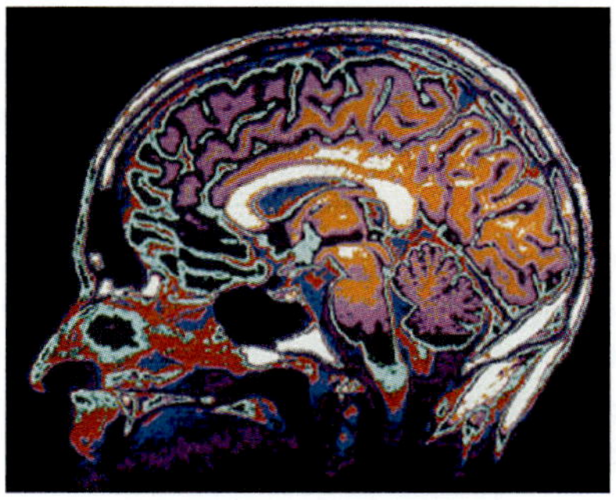

The great advantage of MRI is that it can display soft tissue, such as in this cross-section through a patient's head. [Courtesy of the University of Manitoba.]

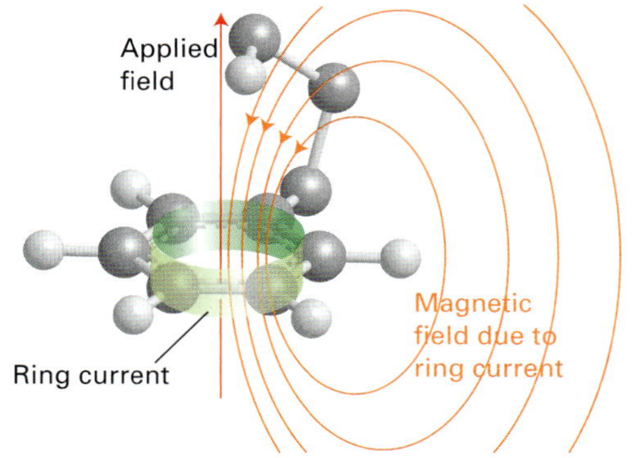

Fig. 21.9 The shielding and deshielding effects of the ring current induced in the benzene ring by the applied field. Protons attached to the ring are deshielded but a proton attached to a substituent that projects above the ring is shielded.

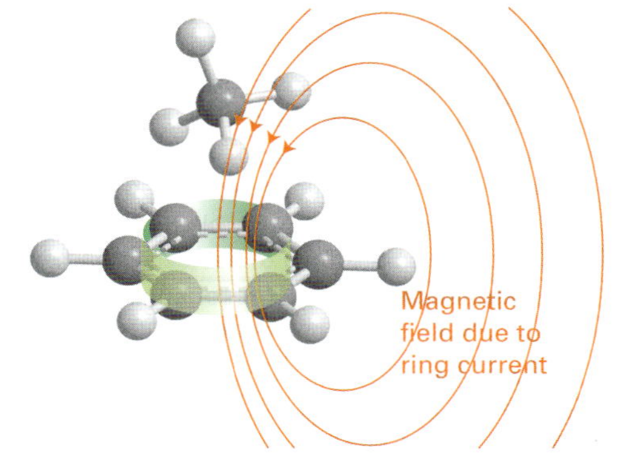

Fig. 21.10 An aromatic solvent (benzene here) can give rise to local currents that shield or deshield a proton in a solvent molecule. In this relative orientation of the solvent and solute, the proton on the solute molecule is shielded.

of electrons around the ring, when it is applied perpendicular to the molecular plane. Protons in the plane are deshielded (Fig. 21.9), but any that happen to lie above or below the plane (as members of substituents of the ring) are shielded.

A solvent can influence the local magnetic field experienced by a nucleus in a variety of ways. Some of these effects arise from specific interactions between the solute and the solvent (such as hydrogen-bond formation and other forms of Lewis acid–base complex formation). The magnetic susceptibility of the solvent molecules, especially if they are aromatic, can also be the source of a local magnetic field. Moreover, if there are steric interactions that result in a loose but specific interaction between a solute molecule and a solvent molecule, then protons in the solute molecule may experience shielding or deshielding effects according to their location relative to the solvent molecule (Fig. 21.10). We shall see that the NMR spectra of species that contain protons with widely different chemical shifts are easier to interpret than those in which the shifts are similar, so the appropriate choice of solvent may help to simplify the appearance and interpretation of a spectrum.

21.4 The fine structure

The splitting of the groups of resonances into individual lines in Fig. 21.7 is called the **fine structure** of the spectrum. It arises because each magnetic nucleus contributes to the local field experienced by the other nuclei and modifies their resonance frequencies. The strength of the interaction is expressed in terms of the **spin–spin coupling constant**, J, and reported in hertz (Hz). Spin coupling constants are an intrinsic

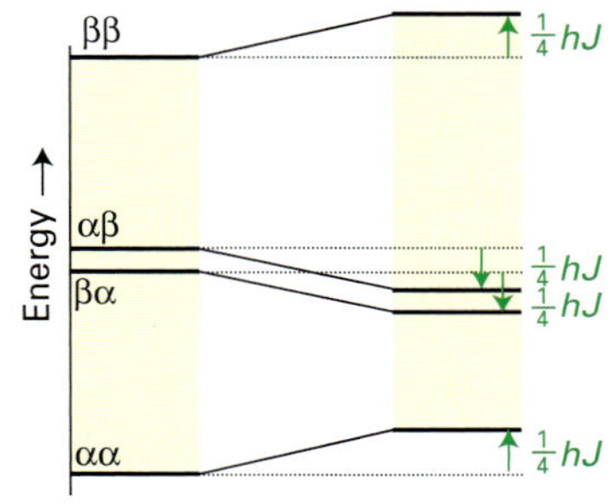

Fig. 21.11 The energy levels of a two-proton system in the presence of a magnetic field. The levels on the left apply in the absence of spin–spin coupling. Those on the right are the result of allowing for spin–spin coupling. The only allowed transitions differ in frequency by J.

property of the molecule and independent of the strength of the applied field.

In NMR, letters far apart in the alphabet (typically A and X) are used to indicate nuclei with very different chemical shifts; letters close together (such as A and B) are used for nuclei with similar chemical shifts. Let's consider first a molecule that contains two spin-$\frac{1}{2}$ nuclei A and X. First, neglect spin–spin coupling. The total energy of two protons in a magnetic field $\mathcal{B}$ is the sum of two terms like eqn 21.11 but with $\mathcal{B}$ modified to $(1-\sigma)\mathcal{B}$:

$$E = -\gamma_N\hbar(1-\sigma_A)\mathcal{B}m_A - \gamma_N\hbar(1-\sigma_X)\mathcal{B}m_X$$

Here σ_A and σ_X are the shielding constants of A and X. The four energy levels predicted by this formula are shown on the left of Fig. 21.11. The spin–spin coupling energy is normally written

$$E_{\text{spin–spin}} = hJm_Am_X \qquad (21.20)$$

A brief illustration The investigation of H—N—C—H couplings in polypeptides can help reveal their conformation. For $^3J_{HH}$ coupling in such a group, $A = +5.1$ Hz, $B = -1.4$ Hz, and $C = +3.2$ Hz. For an α-helix, ϕ is close to 120°, which would give $^3J_{HH} \approx 4$ Hz. For a β-sheet, ϕ is close to 180°, which would give $^3J_{HH} \approx 10$ Hz. Consequently, small coupling constants indicate an α-helix, whereas large couplings indicate a β-sheet.

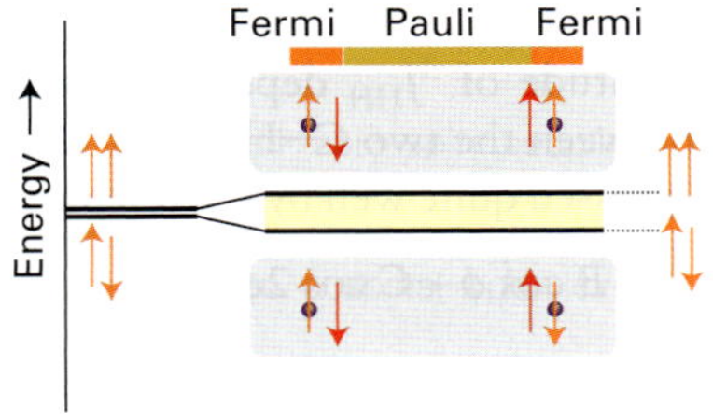

Fig. 21.17 The polarization mechanism for $^1J_{HH}$ spin–spin coupling. The two arrangements have slightly different energies. In this case, J is positive, corresponding to a lower energy when the nuclear spins are antiparallel.

Spin–spin coupling in molecules in solution can be explained in terms of the **polarization mechanism,** in which the interaction is transmitted through the bonds. The simplest case to consider is that of $^1J_{XY}$, where X and Y are spin-$\frac{1}{2}$ nuclei joined by an electron-pair bond (Fig. 21.17). The coupling mechanism depends on the fact that in some atoms it is favourable for the nucleus and a nearby electron spin to be parallel (both α or both β), but in others it is favourable for them to be antiparallel (one α and the other β). The electron–nucleus coupling is magnetic in origin, and may be either a dipolar interaction (Section 15.3) between the magnetic moments of the electron and nuclear spins or a **Fermi contact interaction,** an interaction that depends on the very close approach of an electron to the nucleus and hence can occur only if the electron occupies an s orbital. We shall suppose that it is energetically favourable for an electron spin and a nuclear spin to be antiparallel (as is the case for a proton and an electron in a hydrogen atom), either $\alpha_e\beta_N$ or $\beta_e\alpha_N$, where we are using the labels e and N to distinguish the electron and nucleus spins.

If the X nucleus is α_X, a β electron of the bonding pair will tend to be found nearby (because that is energetically favourable for it). The second electron in the bond, which must have α spin if the other is β, will be found mainly at the far end of the bond (because electrons tend to stay apart to reduce their mutual repulsion). Because it is energetically favourable for the spin of Y to be antiparallel to an electron spin, a Y nucleus with β spin has a lower energy than a Y nucleus with α spin:

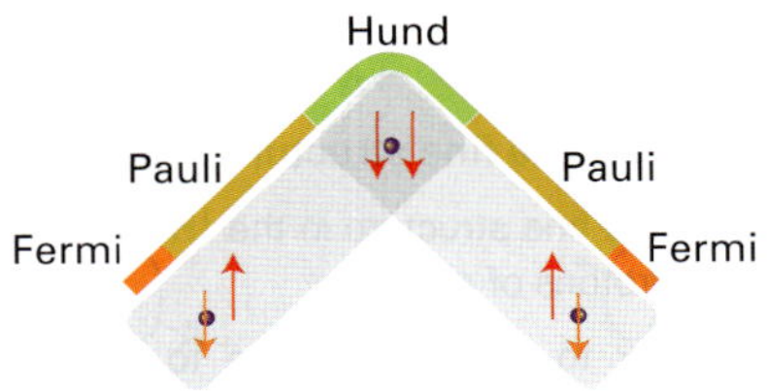

Fig. 21.18 The polarization mechanism for $^2J_{HH}$ spin–spin coupling. The spin information is transmitted from one bond to the next by a version of the mechanism that accounts for the lower energy of electrons with parallel spins in different atomic orbitals (Hund's rule of maximum multiplicity). In this case, $J < 0$, corresponding to a lower energy when the nuclear spins are parallel.

Low energy: $\alpha_X\beta_e \ldots \alpha_e\beta_Y$
High energy: $\alpha_X\beta_e \ldots \alpha_e\alpha_Y$

The opposite is true when X is β, for now the α spin of Y has the lower energy:

Low energy: $\beta_X\alpha_e \ldots \beta_e\alpha_Y$
High energy: $\beta_X\alpha_e \ldots \beta_e\beta_Y$

In other words, antiparallel arrangements of nuclear spins ($\alpha_X\beta_Y$ and $\beta_X\alpha_Y$) lie lower in energy than parallel arrangements ($\alpha_X\alpha_Y$ and $\beta_X\beta_Y$) as a result of their magnetic coupling with the bond electrons. That is, $^1J_{HH}$ is positive, for then hJm_Xm_Y is negative when m_X and m_Y have opposite signs.

To account for the value of $^2J_{XY}$, as in H—C—H, we need a mechanism that can transmit the spin alignments through the central C atom (which may be ^{12}C, with no nuclear spin of its own). In this case (Fig. 21.18), an X nucleus with α spin polarizes the electrons in its bond, and the α electron is likely to be found closer to the C nucleus. The more favourable arrangement of two electrons on the same atom is with their spins parallel (Hund's rule, Section 13.4d), so the more favourable arrangement is for the α electron of the neighbouring bond to be close to the C nucleus. Consequently, the β electron of that bond is more likely to be found close to the Y nucleus, and therefore that nucleus will have a lower energy if it is α:

Low energy: $\alpha_X\beta_e \ldots \alpha_e[C]\alpha_e \ldots \beta_e\alpha_Y$
High energy: $\alpha_X\beta_e \ldots \alpha_e[C]\alpha_e \ldots \beta_e\beta_Y$
Low energy: $\beta_X\alpha_e \ldots \beta_e[C]\beta_e \ldots \alpha_e\beta_Y$
High energy: $\beta_X\alpha_e \ldots \beta_e[C]\beta_e \ldots \alpha_e\alpha_Y$

Hence, according to this mechanism, the energy of Y will be obtained if its spin is parallel ($\alpha_X\alpha_Y$ and $\beta_X\beta_Y$) to that of X. That is, $^2J_{HH}$ is negative for then hJm_Xm_Y is negative when m_X and m_Y have the same sign.

The coupling of nuclear spin to electron spin by the Fermi contact interaction is most important for proton spins, but it is not necessarily the most important mechanism for other nuclei. These nuclei may also interact by a dipolar mechanism with the electron magnetic moments and with their orbital motion, and there is no simple way of specifying whether J will be positive or negative.

21.5 Spin relaxation

As resonant absorption continues, the population of the upper state rises to match that of the lower state. From eqn 21.14, we can expect the intensity of the absorption signal to decrease with time as the populations of the spin states equalize. This decrease due to the progressive equalization of populations is called **saturation.**

The fact that saturation is often not observed, especially when the radio-frequency power is kept low, must mean that there are nonradiative processes by which β nuclear spins can release energy to become α spins again, and hence help to maintain the population difference between the two sites. The nonradiative return to an equilibrium distribution of populations in a system (eqn 21.9a) is an aspect of the process called **relaxation.** If we were to imagine forming a system of spins in which all the nuclei were in their β state, then the system returns exponentially to the equilibrium distribution (a small excess of α spins over β spins) with a time constant called the **spin–lattice relaxation time,** T_1 (Fig. 21.19).

However, there is another, more subtle aspect of relaxation. Consider the classical picture of magnetic nuclei (Section 21.1) and imagine that somehow

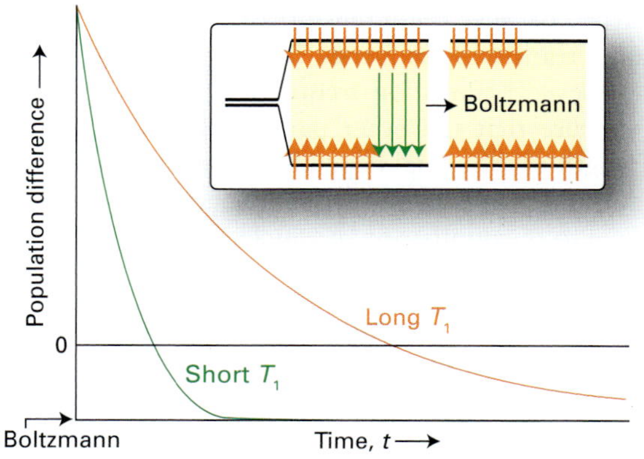

Fig. 21.19 The spin–lattice relaxation time is the time constant for the exponential return of the population of the spin states to their equilibrium (Boltzmann) distribution.

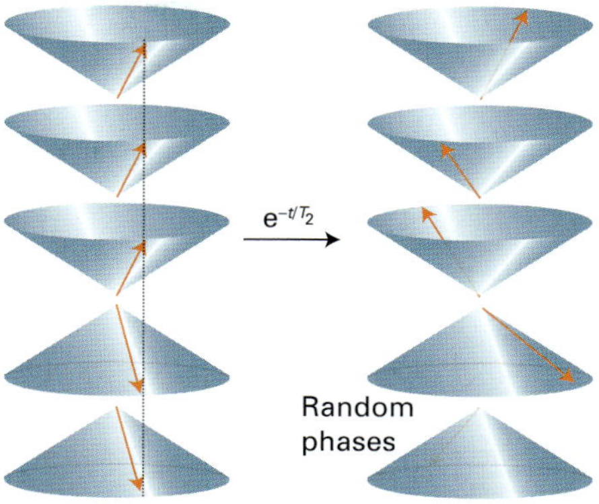

Fig. 21.20 The spin–spin relaxation time is the time constant for the exponential return of the spins to a random distribution around the direction of the magnetic field. No change in populations of the two spin states is involved in this type of relaxation, so no energy is transferred from the spins to the surroundings.

See an animated version of this figure in the interactive ebook.

we have arranged all the spins in a sample to have exactly the same angle around the field direction at an instant. If each spin has a slightly different Larmor frequency (because they experience slightly different local magnetic fields), they will gradually fan out. At thermal equilibrium, all the bar magnets lie at *random* angles round the direction of the applied field, and the time constant for the exponential return of the system into this random arrangement is called the **spin–spin relaxation time,** T_2 (Fig. 21.20). For spins to be truly at thermal equilibrium, not only is the ratio of populations of the spin states given by eqn 21.9a, but the spin orientations must be random around the field direction.

What causes each type of relaxation? In each case the spins are responding to local magnetic fields that act to twist them into different orientations. However, there is a crucial difference between the two processes.

The best kind of local magnetic field for inducing a transition from β to α (as in spin–lattice relaxation) is one that fluctuates at a frequency close to the resonance frequency. Such a field can arise from the tumbling motion of the molecule in the fluid sample. If the tumbling motion of the molecule is slow compared to the resonance frequency, it will give rise to a fluctuating magnetic field that oscillates too slowly to induce transitions, so T_1 will be long. If the molecule tumbles much faster than the resonance frequency, then it will give rise to a fluctuating magnetic field that oscillates too rapidly to induce transitions,

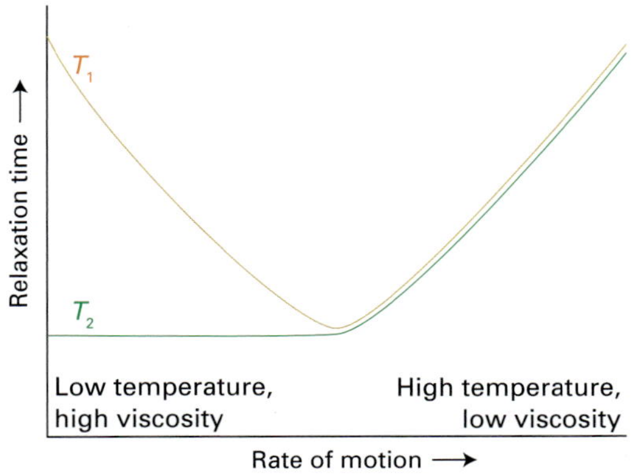

Fig. 21.21 The variation of the two relaxation times with the rate at which the molecules move (either by tumbling or migrating through the solution). The horizontal axis can be interpreted as representing temperature or viscosity. Note that at rapid rates of motion, the two relaxation times coincide.

so T_1 will again be long. Only if the molecule tumbles at about the resonance frequency will the fluctuating magnetic field be able to induce transitions effectively, and only then will T_1 be short. The rate of molecular tumbling increases with temperature and with reducing viscosity of the solvent, so we can expect a dependence like that shown in Fig. 21.21.

The best kind of local magnetic field for causing spin–spin relaxation is one that does not change very rapidly. Then each molecule in the sample lingers in its particular local magnetic environment for a long time, and the orientations of the spins have time to become randomized around the applied field direction. If the molecules move rapidly from one magnetic environment to another, the effects of different magnetic fields average out, and the randomization does not take place as quickly. In other words, slow molecular motion corresponds to short T_2 and fast motion corresponds to long T_2 (as shown in Fig. 21.21). Detailed calculation shows that when the motion is fast, the two relaxation times are equal, as has been drawn in the illustration.

Spin relaxation studies—using advanced techniques that use complicated sequences of pulses of radio-frequency energy to stimulate the spins into special orientations, and then monitoring their return to equilibrium—have two main applications. First, they reveal information about the mobility of molecules or parts of molecules. For example, by studying spin-relaxation times of protons in the hydrocarbon chains of micelles and bilayers it is possible to build up a detailed picture of the motion of these chains, and hence come to an understanding of the dynamics of cell membranes. Second, relaxation times depend on the separation of the nucleus from the source of the magnetic field that is causing its relaxation: that source may be another magnetic nucleus in the same molecule. By studying the relaxation times, the internuclear distances within the molecule can be determined and used to build up a model of its shape.

21.6 Proton decoupling

Nuclear magnetic resonance spectroscopy is used widely to characterize newly synthesized organic compounds. Consequently, in addition to proton NMR spectra, ^{13}C-NMR spectra are obtained routinely. Carbon-13 is a *dilute spin species* in the sense that it is unlikely that more than one ^{13}C nucleus will be found in any given small molecule (provided the sample has not been enriched with that isotope; the natural abundance of ^{13}C is only 1.1 per cent). Even in large molecules, although more than one ^{13}C nucleus may be present, it is unlikely that they will be close enough to give an observable splitting. Hence, it is not normally necessary to take into account ^{13}C—^{13}C spin–spin coupling within a molecule.

Protons are *abundant-spin species* in the sense that a molecule is likely to contain many of them. If we were observing a ^{13}C-NMR spectrum, we would obtain a very complex spectrum on account of the coupling of the one ^{13}C nucleus with many of the protons that are present. To avoid this difficulty, ^{13}C-NMR spectra are normally observed using the technique of **proton decoupling**. Thus, if the CH_3 protons of ethanol are irradiated with a second, strong, resonant radio-frequency pulse, they undergo rapid spin reorientations and the ^{13}C nucleus senses an average orientation. As a result, its resonance is a single line and not a 1:3:3:1 quartet. Proton decoupling has the additional advantage of enhancing sensitivity, because the intensity is concentrated into a single transition frequency instead of being spread over several transition frequencies. If care is taken to ensure that the other parameters on which the strength of the signal depends are kept constant, the intensities of proton-decoupled spectra are proportional to the number of ^{13}C nuclei present.

21.7 Conformational conversion and chemical exchange

The appearance of an NMR spectrum is changed if magnetic nuclei can jump rapidly between different environments. Consider a molecule, such as *N*,*N*-dimethylformamide, that can jump between conformations; in its case, the methyl shifts depend on

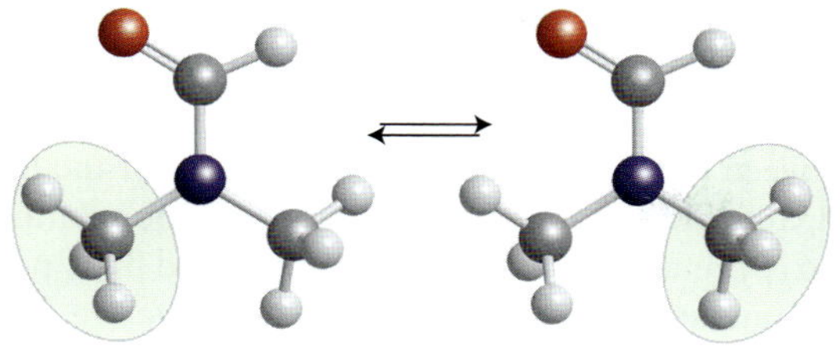

Fig. 21.22 When a molecule changes from one conformation to another, the positions of its protons are interchanged and jump between magnetically distinct environments.

whether they are *cis* or *trans* to the carbonyl group (Fig. 21.22). When the jumping rate is low, the spectrum shows two sets of lines, one each from molecules in each conformation. When the interconversion is fast, the spectrum shows a single line at the mean of the two chemical shifts. At intermediate inversion rates, the line is very broad. This maximum broadening occurs when the lifetime, τ (tau), of a conformation gives rise to a linewidth that is comparable to the difference of resonance frequencies, $\Delta\nu$, and both broadened lines blend together into a very broad line. Coalescence of the two lines occurs when

$$\tau = \frac{2^{1/2}}{\pi\Delta\nu} \qquad (21.22)$$

Example 21.2
Interpreting line broadening

The NO group in *N,N*-dimethylnitrosamine, $(CH_3)_2N{-}NO$, rotates about the N—N bond and, as a result, the magnetic environments of the two CH_3 groups are interchanged. In a 600 MHz spectrometer the two CH_3 resonances are separated by 390 Hz. At what rate of interconversion will the resonance collapse to a single line?

Strategy Use eqn 21.22 for the average lifetimes of the conformations. The rate of interconversion is the inverse of their lifetime.

Solution With $\Delta\nu = 390$ Hz,

$$\tau = \frac{2^{1/2}}{\pi \times (390\ \mathrm{s^{-1}})} = 1.2\ \mathrm{ms}$$

It follows that the signal will collapse to a single line when the interconversion rate exceeds about 830 $\mathrm{s^{-1}}$.

Self-test 21.5

What would you deduce from the observation of a single line from the same molecule in a 300 MHz spectrometer?

[*Answer:* Conformation lifetime less than 2.3 ms]

A similar explanation accounts for the loss of fine structure in solvents able to exchange protons with the sample. For example, hydroxyl protons are able to exchange with water protons. When this **chemical exchange** occurs, a molecule ROH with an α-spin proton (we write this ROH_α) rapidly converts to ROH_β and then perhaps to ROH_α again because the protons provided by the solvent molecules in successive exchanges have random spin orientations. Therefore, instead of seeing a spectrum composed of contributions from both ROH_α and ROH_β molecules (that is, a spectrum showing a doublet structure due to the OH proton) we see a spectrum that shows no splitting caused by coupling of the OH proton (as in Fig. 21.7). The effect is observed when the lifetime of a molecule due to this chemical exchange is so short that the lifetime broadening is greater than the doublet splitting. Because this splitting is often very small (a few hertz), a proton must remain attached to the same molecule for longer than about 0.1 s for the splitting to be observable. In water, the exchange rate is much faster than that, so alcohols show no splitting from the OH protons. In dry dimethylsulfoxide (DMSO, $(CH_3)_2SO$), the exchange rate may be slow enough for the splitting to be detected.

21.8 The nuclear Overhauser effect

An effect that makes use of spin relaxation is of considerable usefulness for the determination of the conformations of proteins and other biological macromolecules in their natural aqueous environment. To introduce the effect, we consider a very simple AX system in which the two spins interact by a magnetic dipole–dipole interaction. We expect two lines in the spectrum, one from A and the other from X. However, when we irradiate the system with radio-frequency radiation at the resonance frequency of X using such a high intensity that we saturate the transition, we find that the A resonance is modified. It may be enhanced, diminished, or even converted into an emission rather than an absorption. That modification of one resonance by saturation of another is called the **nuclear Overhauser effect** (NOE).

To understand the effect, we need to think about the populations of the four levels of an AX system (Fig. 21.23). At thermal equilibrium, the population of the $\alpha_A\alpha_X$ level is the greatest, and that of the $\beta_A\beta_X$ level is the least; the other two levels have the same energy and an intermediate population. The thermal equilibrium absorption intensities reflect these populations, as the figure shows. Now consider the combined effect of saturating the X transition and spin

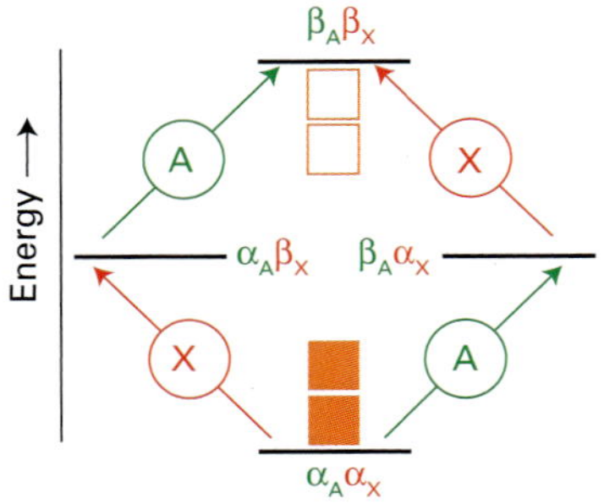

Fig. 21.23 The energy levels of an AX system and an indication of their relative populations. Each orange square above the line represents an excess population and each white square below the line represents a population deficit. The transitions of A and X are marked.

relaxation. When we saturate the X transition, the populations of the X levels are equalized, but at this stage there is no change in the populations of the A levels. If that were all there were to happen, all we would see would be the loss of the X resonance and no effect on the A resonance.

Now consider the effect of spin relaxation. Relaxation can occur in a variety of ways if there is a dipolar interaction between the A and X spins. One possibility is for the magnetic field acting between the two spins to cause them both to flop from β to α, so the $\alpha_A\alpha_X$ and $\beta_A\beta_X$ states regain their thermal equilibrium populations. However, the populations of the $\alpha_A\beta_X$ and $\beta_A\alpha_X$ levels remain unchanged at the values characteristic of saturation. As we see from Fig. 21.24, the population difference between the states joined by transitions of A is now greater than at equilibrium, so the resonance absorption is enhanced. Another possibility is for the dipolar interaction between the two spins to cause α to flip to β and β to flop to α. This transition equilibrates the populations of $\alpha_A\beta_X$ and $\beta_A\alpha_X$ but leaves the $\alpha_A\alpha_X$ and $\beta_A\beta_X$ populations unchanged (Fig. 21.25). Now we see from the illustration that the population differences in the states involved in the A transitions are decreased, so the resonance absorption is diminished.

Which effect wins? Does NOE enhance the A absorption or does it diminish it? As in the discussion of relaxation times in Section 21.5, the efficiency of the intensity-enhancing $\beta_A\beta_X \leftrightarrow \alpha_A\alpha_X$ relaxation is high if the dipole field is modulated at the transition frequency, which in this case is close to 2ν; likewise, the efficiency of the intensity-diminishing $\alpha_A\beta_X \leftrightarrow \beta_A\alpha_X$ relaxation is high if the dipole field is stationary (because there is no frequency difference between the initial and final states). A large molecule rotates so slowly that there is very little motion at 2ν, so we expect intensity decrease (Fig. 21.26). A small molecule rotating rapidly can be expected to have substantial motion at 2ν, and a consequent enhancement of the signal. In practice, the enhancement lies somewhere between the two extremes and is reported in terms of the parameter η (eta), where

$$\eta = \frac{I - I_0}{I_0} \qquad (21.23)$$

Here I_0 is the normal intensity and I is NOE intensity of a particular transition; theoretically, η lies between -1 (diminution) and $+\frac{1}{2}$ (enhancement).

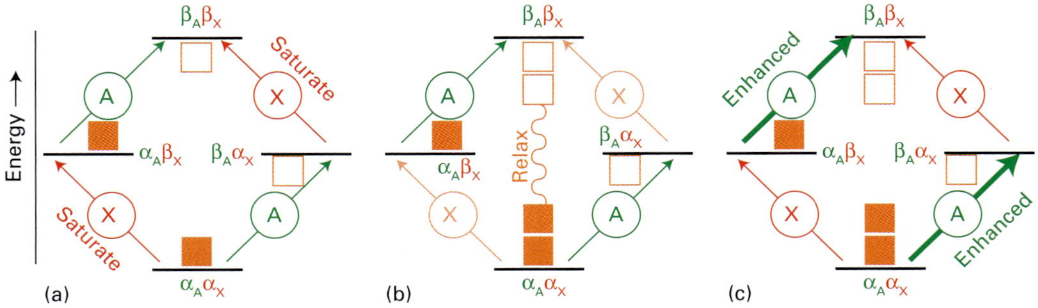

Fig. 21.24 (a) When the X transition is saturated, the populations of its two states are equalized and the population excess and deficit become as shown (using the same symbols as in Fig. 21.23). (b) Dipole–dipole relaxation relaxes the populations of the highest and lowest states, and they regain their original populations. (c) The A transitions reflect the difference in populations resulting from the preceding changes, and are enhanced compared with those shown in Fig. 21.23.

See an animated version of this figure in the interactive ebook.

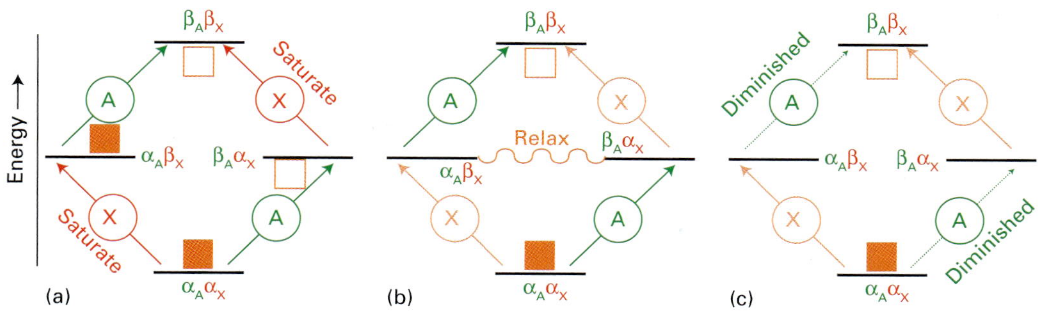

Fig. 21.25 (a) When the X transition is saturated, just as in Fig. 21.24 the populations of its two states are equalized and the population excess and deficit become as shown. (b) Dipole–dipole relaxation relaxes the populations of the two intermediate states, and they regain their original populations. (c) The A transitions reflect the difference in populations resulting from the preceding changes, and are diminished compared with those shown in Fig. 21.23.

See an animated version of this figure in the interactive ebook.

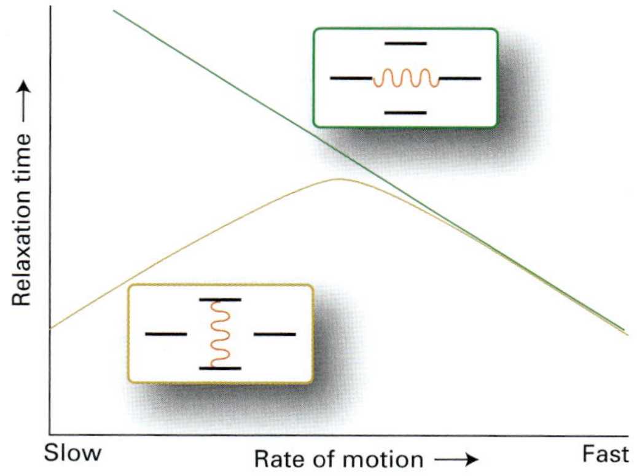

Fig. 21.26 The relaxation rates of the two types of relaxation (as indicated by the small diagrams) as a function of the tumbling rate of the molecule.

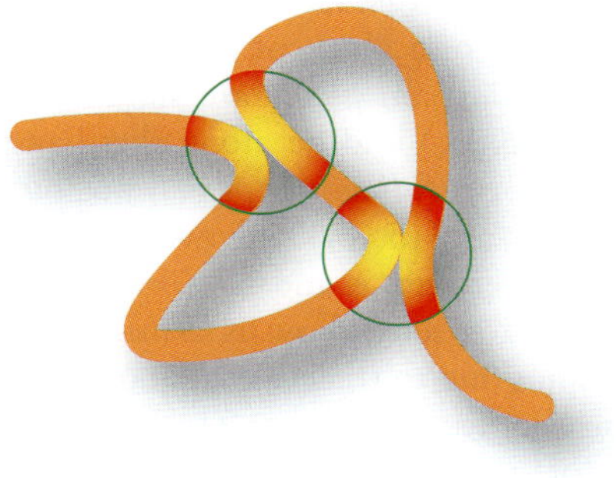

Fig. 21.27 If a NOE experiment shows that the protons within each of the two circles are coupled by a dipolar interaction, we can be confident that those protons are close together, and therefore infer the conformation of the polypeptide chain.

The value of η depends strongly on the separation of the two spins involved in the NOE, for the strength of the dipolar interaction between two spins separated by a distance r is proportional to $1/r^3$ and its effect depends on the square of that strength, and therefore on $1/r^6$. This sharp dependence on separation is used to build up a picture of the conformation of a protein by using NOE to identify which nuclei can be regarded as neighbours (Fig. 21.27). The enormous importance of this procedure is that we can determine the conformation of polypeptides in an aqueous environment and do not need to try to make the single crystals that are essential for an X-ray diffraction investigation.

21.9 Two-dimensional NMR

An NMR spectrum contains a great deal of information and, if many spins are present, is very complex, for the fine structure of different groups of lines can overlap. The complexity would be reduced if we could use two axes to display the data, with resonances belonging to different groups lying at different locations on the second axis. This separation is essentially what is achieved in **two-dimensional NMR.**

Much modern NMR work makes use of techniques such as **correlation spectroscopy** (COSY) in which a clever choice of pulses and Fourier transformation techniques makes it possible to determine all spin–spin couplings in a molecule. The COSY spectrum of an

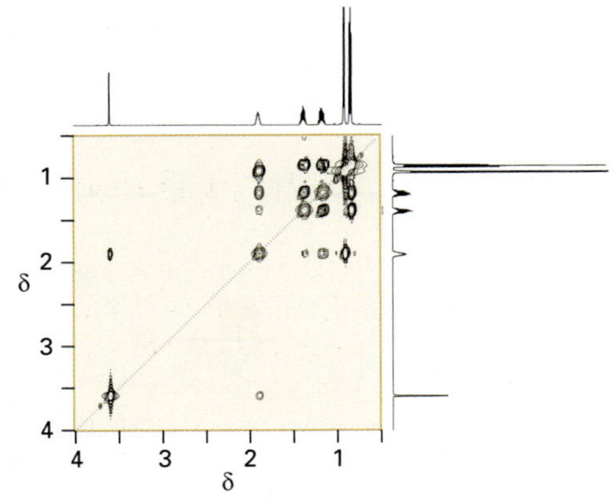

Fig. 21.28 Proton COSY spectrum of 1-nitropropane. The off-diagonal peaks show that the CH_3 protons are coupled to the central CH_2 protons, which in turn are coupled to the terminal CH_2NO_2 protons. (Spectrum provided by Prof. G. Morris.)

AX system contains four groups of signals centred on the two chemical shifts. Each group shows fine structure, consisting of a block of four signals separated by J_{AX}. The *diagonal peaks* are signals centred on (δ_A,δ_A) and (δ_X,δ_X) and lie along the diagonal The *cross peaks* (or *off-diagonal peaks*) are signals centred on (δ_A,δ_X) and (δ_X,δ_A) and owe their existence to the coupling between A and X. Consequently, cross peaks in COSY spectra allow us to map the couplings between spins and to trace out the bonding network in complex molecules. Figure 21.28 shows a simple example of a proton COSY spectrum of 1-nitropropane, $CH_3CH_2CH_2NO_2$.

Although information from two-dimensional NMR spectroscopy is trivial in an AX system, it can be of enormous help in the interpretation of more complex spectra. For example, a complex spectrum from a synthetic polymer or a protein would be impossible to interpret in one-dimensional NMR but can be interpreted reasonably rapidly by two-dimensional NMR.

21.10 Solid-state NMR

The principal difficulty with the application of NMR to solids is the low-resolution characteristic of solid samples, leading to broadening and coalescence of spectral lines into features that do not reveal useful information. Nevertheless, there are good reasons for seeking to overcome these difficulties. They include the possibility that a compound of interest is unstable in solution or that it is insoluble, so conventional solution NMR cannot be employed. Moreover, many species are intrinsically interesting as solids and it is important to determine their structural and dynamical features. Synthetic polymers are particularly interesting in this regard, and information can be obtained about the arrangement of molecules, their conformations, and the motion of different parts of the chain. This kind of information is crucial to an interpretation of the bulk properties of the polymer in terms of its molecular characteristics. Similarly, inorganic substances, such as the zeolites that are used as molecular sieves and shape-selective catalysts, can be studied using solid-state NMR, and structural problems can be resolved that cannot be tackled by X-ray diffraction.

Problems of resolution and linewidth are not the only features that plague NMR studies of solids, but the rewards are so great that considerable efforts have been made to overcome them and have achieved notable success. Because molecular rotation has almost ceased (except in special cases, including 'plastic crystals' in which the molecules continue to tumble), spin–lattice relaxation times are very long but spin–spin relaxation times are very short. Hence, in a pulse experiment, there is a need for lengthy delays—of several seconds—between successive pulses so that the spin system has time to revert to equilibrium. Even gathering the murky information may therefore be a lengthy process. Moreover, because lines are so broad, very high powers of radio-frequency radiation may be required to achieve saturation. Whereas solution pulse NMR uses transmitters of a few tens of watts, solid-state NMR may require transmitters rated at several hundreds of watts.

There are two principal contributions to the linewidths of solids. One is the direct dipolar interaction between nuclear spins. As we saw in the discussion of spin–spin coupling, a nuclear magnetic moment will give rise to a local magnetic field, which points in different directions at different locations around the nucleus. If we are interested only in the component parallel to the direction of the applied magnetic field (because only this component has a significant effect), then we can use a classical expression to write the magnitude of the local magnetic field as

$$\mathcal{B}_{\text{loc}} = -\frac{\gamma\hbar\mu_0 m_I}{4\pi R^3}(1 - 3\cos^2\theta) \qquad (21.24)$$

Unlike in solution, this field is not motionally averaged to zero. Many nuclei may contribute to the total local field experienced by a nucleus of interest, and different nuclei in a sample may experience a wide range of fields. Typical dipole fields are of the order of 10^{-3} T, which corresponds to splittings and linewidths of the order of 10^4 Hz.

A second source of linewidth is the *anisotropy*, or dependence on orientation, of the chemical shift. We have seen that chemical shifts arise from the ability of the applied field to generate electron currents in molecules. In general, this ability depends on the orientation of the molecule relative to the applied field. In solution, when the molecule is tumbling rapidly, only the average value of the chemical shift is relevant. However, the anisotropy is not averaged to zero for stationary molecules in a solid, and molecules in different orientations have resonances at different frequencies. The chemical shift anisotropy also varies with the angle between the applied field and the principal axis of the molecule as $1 - 3\cos^2\theta$.

Fortunately, there are techniques available for reducing the linewidths of solid samples. One technique, **magic-angle spinning** (MAS), takes note of the $1 - 3\cos^2\theta$ dependence of both the dipole–dipole interaction and the chemical shift anisotropy. The 'magic angle' is the angle at which $1 - 3\cos^2\theta = 0$, and corresponds to 54.74°. In the technique, the sample is spun at high speed at the magic angle to the applied field (Fig. 21.29). All the dipolar interactions and the anisotropies average to the value they would have at the magic angle; but at that angle they are zero. The difficulty with MAS is that the spinning frequency must not be less than the width of the spectrum, which is of the order of kilohertz. However, gas-driven sample spinners that can be rotated at up to 25 kHz are now routinely available, and a considerable body of work has been done.

Pulsed techniques similar to those described in the previous section may also be used to reduce linewidths. The dipolar field of protons, for instance, may be reduced by a decoupling procedure. However, because the range of coupling strengths is so large, radio-frequency power of the order of 1 kW is required. Elaborate pulse sequences have also been devised that reduce linewidths by averaging procedures that make use of twisting the magnetization vector through an elaborate series of angles.

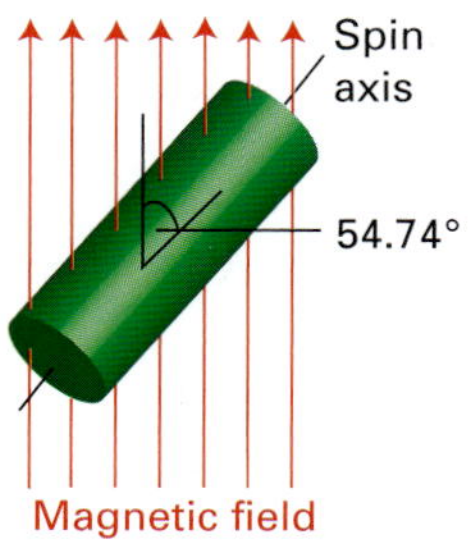

Fig. 21.29 In magic angle spinning, the sample spins at 54.74° (that is, arccos $\frac{1}{3}^{1/2}$) to the applied magnetic field. Rapid motion at this angle averages dipole–dipole interactions and chemical shift anisotropies to zero.

The information in EPR spectra

An EPR spectrum is obtained by monitoring the microwave absorption as the field is changed, and a typical spectrum (of the benzene radical anion, $C_6H_6^-$) is shown in Fig. 21.30. The peculiar appearance of the spectrum, which is in fact the first derivative of the absorption, arises from the detection technique, which is sensitive to the slope of the absorption curve (Fig. 21.31).

21.11 The *g*-value

Equation 21.10 gives the resonance frequency for a transition between the $m_s = -\frac{1}{2}$ and the $m_s = +\frac{1}{2}$ levels

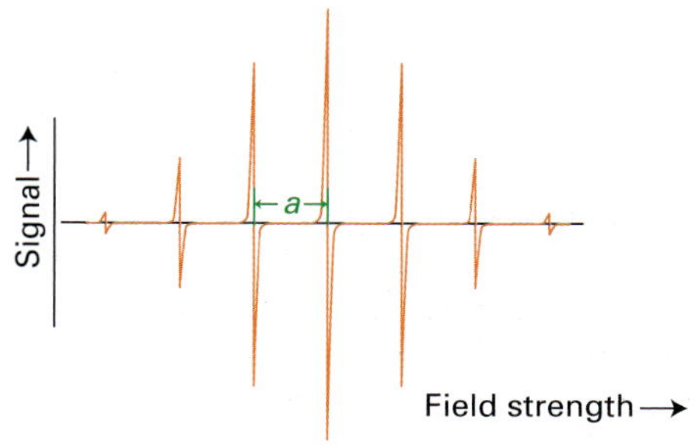

Fig. 21.30 The ESR spectrum of the benzene radical anion, $C_6H_6^-$, in fluid solution. *a* is the hyperfine splitting of the spectrum; the centre of the spectrum is determined by the *g*-value of the radical.

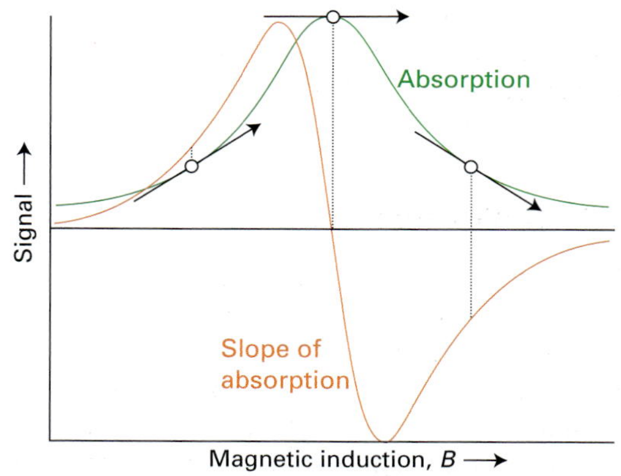

Fig. 21.31 When phase-sensitive detection is used, the signal is the first derivative of the absorption intensity. Note that the peak of the absorption corresponds to the point where the derivative passes through zero.

of a 'free' electron in terms of the g-value $g_e \approx 2.0023$. The magnetic moment of an unpaired electron in a radical also interacts with an external field, but the g-value is different from that of a free electron on account of local magnetic fields induced in the molecular framework of the radical. Consequently, the resonance condition is normally written as

$$h\nu = g\mu_B\mathcal{B} \tag{21.25}$$

where g is the **g-value** of the radical. Many organic radicals have g-values close to 2.0027; inorganic radicals have g-values typically in the range 1.9–2.1; paramagnetic d-metal complexes have g-values in a wider range (for example, 0 to 6).

The deviation of g from $g_e = 2.0023$ depends on the ability of the applied field to induce local electron currents in the radical, and therefore its value gives some information about electronic structure. In that sense, the g-value in EPR plays a similar role to the shielding constants in NMR. However, because g-values differ very little from g_e in many radicals (for instance, 2.003 for H, 1.999 for NO_2, and 2.01 for ClO_2), its main use in chemical applications is to aid the identification of the species present in a sample.

A brief illustration The centre of the EPR spectrum of the methyl radical occurred at 329.40 mT in a spectrometer operating at 9.2330 GHz (the so-called 'X-band' of the microwave spectrum). Its g-value is therefore

$$g = \frac{h\nu}{\mu_B\mathcal{B}_{cal}} = \frac{(6.626\,08\times10^{-34}\ \text{J s})\times(\overbrace{9.2330\times10^{9}\ \text{s}^{-1}}^{9.2330\ \text{GHz}})}{(9.2740\times10^{-24}\ \text{J T}^{-1})\times(0.329\,40\ \text{T})}$$
$$= 2.0027$$

Self-test 21.6

At what magnetic field would the methyl radical come into resonance in a spectrometer operating at 34.000 GHz (the so-called 'Q-band' of the microwave spectrum)?

[*Answer:* 1.213 T]

21.12 Hyperfine structure

The most important features of EPR spectra are their **hyperfine structure**, the splitting of individual resonance lines into components. In general in spectroscopy, the term 'hyperfine structure' means the structure of a spectrum that can be traced to interactions of the electrons with nuclei other than as a result of the latter's point electric charge. The source of the hyperfine structure in EPR is the magnetic interaction between the electron spin and the magnetic dipole moments of the nuclei present in the radical.

Consider the effect on the EPR spectrum of a single H nucleus located somewhere in a radical. The proton spin is a source of magnetic field, and depending on the orientation of the nuclear spin, the field it generates adds to or subtracts from the applied field. The total local field is therefore

$$\mathcal{B}_{loc} = \mathcal{B} + am_I \qquad m_I = \pm\tfrac{1}{2} \tag{21.26}$$

where a is the **hyperfine coupling constant.** Half the radicals in a sample have $m_I = +\frac{1}{2}$, so half resonate when the applied field satisfies the condition

$$h\nu = g\mu_B(\mathcal{B}+\tfrac{1}{2}a), \quad \text{or } \mathcal{B} = \frac{h\nu}{g\mu_B} - \tfrac{1}{2}a \tag{21.27a}$$

The other half (which have $m_I = -\frac{1}{2}$) resonate when

$$h\nu = g\mu_B(\mathcal{B}-\tfrac{1}{2}a), \quad \text{or } \mathcal{B} = \frac{h\nu}{g\mu_B} + \tfrac{1}{2}a \tag{21.27b}$$

Therefore, instead of a single line, the spectrum shows two lines of half the original intensity separated by a and centred on the field determined by g (Fig. 21.32).

If the radical contains an ^{14}N atom ($I = 1$), its EPR spectrum consists of three lines of equal intensity, because the ^{14}N nucleus has three possible spin orientations, and each spin orientation is possessed by one third of all the radicals in the sample. In general, a

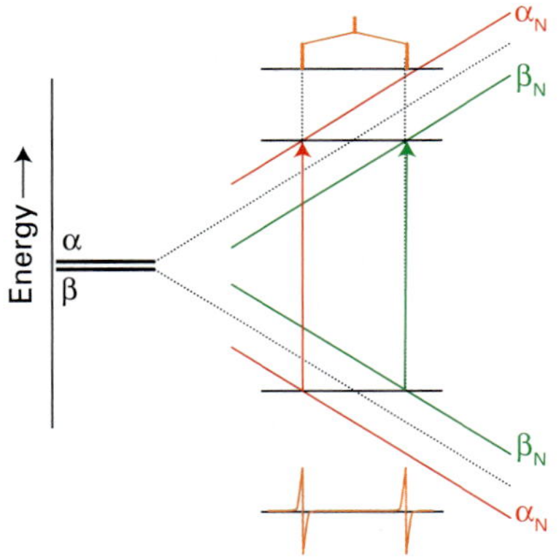

Fig. 21.32 The hyperfine interaction between an electron and a spin-$\frac{1}{2}$ nucleus results in four energy levels in place of the original two. As a result, the spectrum consists of two lines (of equal intensity) instead of one. The intensity distribution can be summarized by a simple stick diagram. The diagonal lines show the energies of the states as the applied field is increased, and resonance occurs when the separation of states matches the fixed energy of the microwave photon.

spin-I nucleus splits the spectrum into $2I + 1$ hyperfine lines of equal intensity.

When there are several magnetic nuclei present in the radical, each one contributes to the hyperfine structure. In the case of equivalent protons (for example, the two CH_2 protons in the radical CH_3CH_2) some of the hyperfine lines are coincident. It is not hard to show that if the radical contains N equivalent protons, then there are $N + 1$ hyperfine lines with an intensity distribution given by Pascal's triangle (Section 21.4). The spectrum of the benzene radical anion in Fig. 21.30, which has seven lines with intensity ratio 1:6:15:20:15:6:1, is consistent with a radical containing six equivalent protons. More generally, if the radical contains N equivalent nuclei with spin quantum number I, then there are $2NI + 1$ hyperfine lines with an intensity distribution given by modified versions of Pascal's triangle (see the exercises).

Example 21.3

Predicting the hyperfine structure of an EPR spectrum

A radical contains one ^{14}N nucleus ($I = 1$) with hyperfine constant 1.61 mT and two equivalent protons ($I = \frac{1}{2}$) with hyperfine constant 0.35 mT. Predict the form of the EPR spectrum.

Strategy We should consider the hyperfine structure that arises from each type of nucleus or group of equivalent nuclei in succession. So, split a line with one nucleus, then each of those lines is split by a second nucleus (or group of nuclei), and so on. It is best to start with the nucleus with the largest hyperfine splitting; however, any choice could be made, and the order in which nuclei are considered does not affect the conclusion.

Solution The ^{14}N nucleus gives three hyperfine lines of equal intensity separated by 1.61 mT. Each line is split into doublets of spacing 0.35 mT by the first proton, and each line of these doublets is split into doublets with the same 0.35 mT splitting (Fig. 21.33). The central lines of each split doublet coincide, so the proton splitting gives 1:2:1 triplets of internal splitting 0.35 mT. Therefore, the spectrum consists of three equivalent 1:2:1 triplets.

Self-test 21.7

Predict the form of the EPR spectrum of a radical containing three equivalent ^{14}N nuclei.

[*Answer:* Fig. 21.34]

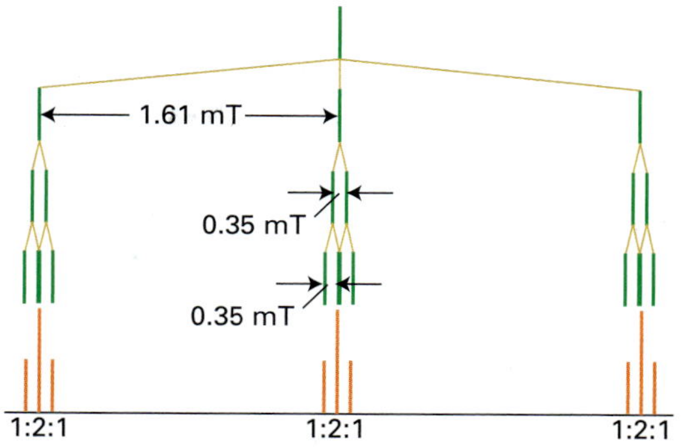

Fig. 21.33 The analysis of the hyperfine structure of radicals containing one ^{14}N nucleus ($I = 1$) and two equivalent protons.

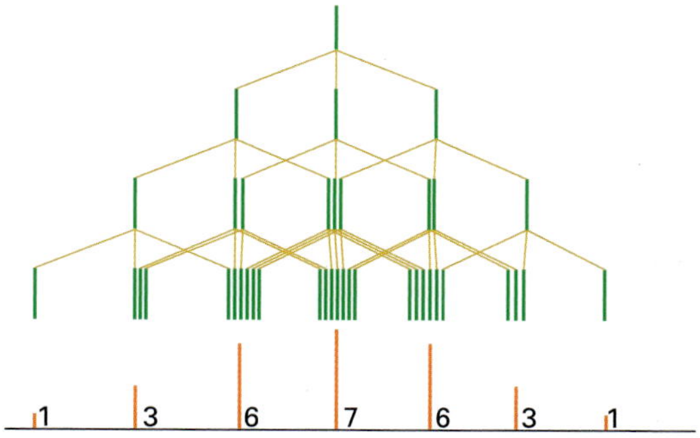

Fig. 21.34 The analysis of the hyperfine structure of radicals containing three equivalent ^{14}N nuclei.

The hyperfine structure of an EPR spectrum is a kind of fingerprint that helps to identify the radicals present in a sample. The interaction between the unpaired electron and the hydrogen nucleus responsible for hyperfine structure is either a dipolar interaction or the Fermi contact interaction described in Section 21.4. In the case of the contact interaction, the magnitude of the splitting depends on the distribution of the unpaired electron near the magnetic nuclei present, so the spectrum can be used to map the molecular orbital occupied by the unpaired electron. For example, because the hyperfine splitting in $C_6H_6^-$ is 0.375 mT, and one proton is close to a C atom with one-sixth the unpaired electron density (because the electron is spread uniformly around the ring), the hyperfine splitting caused by a proton in the electron spin entirely confined to a single adjacent C atom should be $6 \times 0.375\ \text{mT} = 2.25\ \text{mT}$. If in another aromatic radical we find a hyperfine splitting constant a, then the **spin density**, ρ (rho), the probability that an unpaired electron is on the atom, can be calculated from the **McConnell equation:**

$$a = Q\rho \tag{21.28}$$

with $Q = 2.25$ mT. In this equation, ρ is the spin density on a C atom and a is the hyperfine splitting observed for the H atom to which it is attached.

Checklist of key ideas

You should now be familiar with the following concepts.

☐ **1** Resonance is the condition of strong effective coupling when the frequencies of two oscillators are identical.

☐ **2** Nuclear magnetic resonance (NMR) is the observation of the frequency at which magnetic nuclei in molecules come into resonance with an electromagnetic field when the molecule is exposed to a strong magnetic field; NMR is a radiofrequency technique.

☐ **3** Electron paramagnetic resonance (EPR) or electron spin resonance (ESR) is the observation of the frequency at which an electron spin comes into resonance with an electromagnetic field when the molecule is exposed to a strong magnetic field; EPR is a microwave technique.

☐ **4** The intensity of an NMR or EPR transition is increases with the difference in population of α and β states and the strength of the applied magnetic field (as $\mathcal{B}^2$).

☐ **5** The chemical shift of a nucleus is the difference between its resonance frequency and that of a reference standard.

☐ **6** The observed shielding constant is the sum of a local contribution, a neighbouring group contribution, and a solvent contribution.

☐ **7** The fine structure of an NMR spectrum is the splitting of the groups of resonances into individual lines; the strength of the interaction is expressed in terms of the spin–spin coupling constant, J.

☐ **8** N equivalent spin-$\frac{1}{2}$ nuclei split the resonance of a nearby spin or group of equivalent spins into $N+1$ lines with an intensity distribution given by Pascal's triangle.

☐ **9** Spin–spin coupling in molecules in solution can be explained in terms of the polarization mechanism, in which the interaction is transmitted through the bonds.

☐ **10** The Fermi contact interaction is a magnetic interaction that depends on the very close approach of an electron to the nucleus and can occur only if the electron occupies an s orbital.

☐ **11** Relaxation is the nonradiative return to an equilibrium distribution of populations in a system with random relative spin orientations; the system returns exponentially to the equilibrium distribution with a time constant called the spin–lattice relaxation time, T_1.

☐ **12** The spin–spin relaxation time, T_2, is the time constant for the exponential return of the system into random relative orientations.

☐ **13** In proton decoupling of ^{13}C-NMR spectra, protons are made to undergo rapid spin reorientations and the ^{13}C nucleus senses an average orientation. As a result, its resonance is a single line and not a group of lines.

☐ **14** Coalescence of the two lines occurs in conformational interchange or chemical exchange when the lifetime, τ, of the states is related to their resonance frequency difference, $\Delta\boldsymbol{\nu}$.

☐ **15** The nuclear Overhauser effect (NOE) is the modification of one resonance by the saturation of another.

☐ **16** In two-dimensional NMR, spectra are displayed in two axes, with resonances belonging to different groups lying at different locations on the second axis. An example of a two-dimensional NMR technique is correlation spectroscopy (COSY), in which all spin–spin couplings in a molecule are determined.

☐ **17** In solid-state NMR, the spectra of solids are simplified by such techniques as magic angle spinning.

☐ **18** The EPR resonance condition is written $h\nu = g\mu_B\mathcal{B}$, where g is the g-value of the radical; the deviation of g from $g_e = 2.0023$ depends on the ability of the applied field to induce local electron currents in the radical.

☐ **19** The hyperfine structure of an EPR spectrum is its splitting of individual resonance lines into components by the magnetic interaction of the electron and nuclei with spin.

Table of key equations

The following table summarizes the equations developed in this chapter.

Property	Equation	Comment
Energy of an electron in a magnetic field	$E_{m_s} = -g\gamma_e\hbar\mathcal{B}m_s$	$m_s = \pm\frac{1}{2}$, $\gamma_e = -e/2m_e$
Energy of a nucleus in a magnetic field	$E_{m_I} = -\gamma_N\hbar\mathcal{B}m_I$	$m_I = I, I-1, \ldots, -I$
Resonance condition for an electron in a magnetic field	$h\nu = g\mu_B\mathcal{B}$	No hyperfine interaction
Resonance condition for a nucleus in a magnetic field	$h\nu = \gamma_N\hbar\mathcal{B}$	No local interactions
Spin–spin coupling	hJm_Am_X	AX system
δ scale	$\delta = (\nu - \nu^\circ) \times 10^6/\nu^\circ$	
Coalescence lifetime	$\tau = 2^{1/2}/\pi\Delta\nu$	
Karplus equation	$^3J_{HH} = A + B\cos\phi + C\cos 2\phi$	
McConnell equation	$a = Q\rho$ with $Q = 2.25$ mT	H atom attached to aromatic ring

Questions and exercises

Discussion questions

21.1 Discuss the origins of the local, neighbouring group, and solvent contributions to the shielding constant.

21.2 Suggest a reason why the relaxation times of ^{13}C nuclei are typically much longer than those of ^1H nuclei.

21.3 Suggest a reason why the spin–lattice relaxation time of benzene (a small molecule) in a mobile, deuterated hydrocarbon solvent increases, whereas that of a polymer decreases.

21.4 Discuss how the Fermi contact interaction and the polarization mechanism contribute to spin–spin couplings in NMR.

21.5 Discuss the origin of the nuclear Overhauser effect and how it can be used to measure distances between protons in a polymer.

21.6 Explain how the EPR spectrum of an organic radical can be used to identify the molecular orbital occupied by the unpaired electron.

21.7 The hyperfine interaction of a π electron of an aromatic ring with a methyl group attached to the ring varies as the methyl group rotates. Suggest a mechanism for the interaction.

Exercises

21.1 Calculate the energy separation between the spin states of an electron in a magnetic field of 0.250 T.

21.2 The nucleus ^{32}S has $I = \frac{3}{2}$ and $g_I = 0.4289$. Calculate the energies of the nuclear spin states in a magnetic field of 6.000 T.

21.3 Equations 21.5 and 21.7 define the g-value and the magnetogyric ratio of a nucleus. Given that g is a dimensionless number, what are the units of γ_N expressed in (a) tesla and hertz, (b) SI base units?

21.4 The magnetogyric ratio of ^{31}P is $1.0840 \times 10^8\ T^{-1}\ s^{-1}$. What is the g-value of the nucleus?

21.5 Calculate the value of $(N_\beta - N_\alpha)/N$ for electrons in a field of (a) 0.40 T, (b) 1.2 T.

21.6 Calculate the resonance frequency and the corresponding wavelength for an electron in a magnetic field of 0.330 T, the magnetic field commonly used in EPR.

21.7 Calculate the value of $(N_\alpha - N_\beta)/N$ for (a) protons, (b) carbon-13 nuclei in a field of 8.5 T.

21.8 The magnetogyric ratio of ^{19}F is $2.5177 \times 10^8\ T^{-1}\ s^{-1}$. Calculate the frequency of the nuclear transition in a field of 7.500 T.

21.9 Calculate the resonance frequency of a ^{14}N nucleus ($I = 1$, $g_I = 0.4036$) in a 14.20 T magnetic field.

21.10 Calculate the magnetic field needed to satisfy the resonance condition for unshielded protons in a 800.0 MHz radio-frequency field.

21.11 What is the shift of the resonance from TMS of a group of protons with $\delta = 6.33$ in a polypeptide in a spectrometer operating at 500.0 MHz?

21.12 The chemical shift of the CH_3 protons in acetaldehyde (ethanal) is $\delta = 2.20$ and that of the CHO proton is 9.80. What is the difference in local magnetic field between the two regions of the molecule when the applied field is (a) 1.2 T, (b) 5.0 T?

21.13 Use the information in Fig. 21.6 to state the splitting (in hertz, Hz) between the methyl and aldehydic proton resonances in a spectrometer operating at (a) 300 MHz, (b) 750 MHz.

21.14 What would be the nuclear magnetic resonance spectrum for a proton resonance line that was split by interaction with seven identical protons?

21.15 What would be the nuclear magnetic resonance spectrum for a proton resonance line that was split by interaction with (a) two, (b) three equivalent nitrogen nuclei (the spin of a nitrogen nucleus is 1)?

21.16 Repeat the analysis in Section 21.4 for an AX_2 spin-$\frac{1}{2}$ system and deduce the pattern of lines expected in the spectrum.

21.17 Sketch the appearance of the ^{1}H-NMR spectrum of acetaldehyde (ethanal) using $J = 2.90$ Hz and the data in Fig. 21.6 in a spectrometer operating at (a) 300 MHz, (b) 550 MHz.

21.18 Sketch the form of the ^{19}F-NMR spectra of a natural sample of $^{10}BF_4^-$ and $^{11}BF_4^-$.

21.19 Sketch the form of an $A_3M_2X_4$ spectrum, where A, M, and X are protons with distinctly different chemical shifts and $J_{AM} > J_{AX} > J_{MX}$.

21.20 Formulate the version of Pascal's triangle that you would expect to represent the fine structure in an NMR spectrum for a collection of N spin-1 nuclei, with N up to 5.

21.21 Formulate the version of Pascal's triangle that you would expect to represent the fine structure in an NMR spectrum for a collection of N spin-$\frac{3}{2}$ nuclei, with N up to 5.

21.22 A proton jumps between two sites with $\delta = 2.7$ and $\delta = 4.8$. At what rate of interconversion will the two signals collapse to a single line in a spectrometer operating at 550 MHz?

21.23 NMR spectroscopy may be used to determine the equilibrium constant for dissociation of a complex between a small molecule, such as an enzyme inhibitor I, and a protein, such as an enzyme E:

$$EI \rightleftharpoons E + I \qquad K_I = [E][I]/[EI]$$

In the limit of slow chemical exchange, the NMR spectrum of a proton in I would consist of two resonances: one at ν_I for free I and another at ν_{EI} for bound I. When chemical exchange is fast, the NMR spectrum of the same proton in I consists of a single peak with a resonance frequency ν given by:

$$\nu = f_I\nu_I + f_{EI}\nu_{EI}$$

where $f_I = [I]/([I] + [EI])$ and $f_{EI} = [EI]/([I] + [EI])$ are, respectively, the fractions of free I and bound I. For the purposes of analysing the data, it is also useful to define the frequency differences $\delta\nu = \nu - \nu_I$ and $\Delta\nu = \nu_{EI} - \nu_I$. Show that when the initial concentration of I, $[I]_0$, is much greater than the initial concentration of E, $[E]_0$, a plot of $[I]_0$ versus $\delta\nu^{-1}$ is a straight line with slope $[E]_0\Delta\nu$ and y-intercept $-K_I$.

21.24 The centre of the EPR spectrum of atomic hydrogen lies at 329.12 mT in a spectrometer operating at 9.2231 GHz. What is the g-value of the atom?

21.25 A radical containing two equivalent protons shows a three-line spectrum with an intensity distribution 1:2:1. The lines occur at 330.2 mT, 332.5 mT, and 334.8 mT. What is the hyperfine coupling constant for each proton? What is the g-value of the radical given that the spectrometer is operating at 9.319 GHz?

21.26 Predict the intensity distribution in the hyperfine lines of the EPR spectra of (a) $\cdot CH_3$, (b) $\cdot CD_3$.

21.27 The benzene radical anion has $g = 2.0025$. At what field should you search for resonance in a spectrometer operating at (a) 9.302 GHz, (b) 33.67 GHz?

21.28 The EPR spectrum of a radical with two equivalent nuclei of a particular kind is split into five lines of intensity ratio 1:2:3:2:1. What is the spin of the nuclei?

21.29 Formulate the version of Pascal's triangle that you would expect to represent the hyperfine structure in an EPR spectrum for a collection of N spin-$\frac{3}{2}$ nuclei, with N up to 5.

21.30 The hyperfine coupling constants observed in the radical anions (**3**), (**4**), and (**5**) are shown (in millitesla, mT). Use the McConnell equation to map the probability of finding the unpaired electron in the π orbital on each C atom.

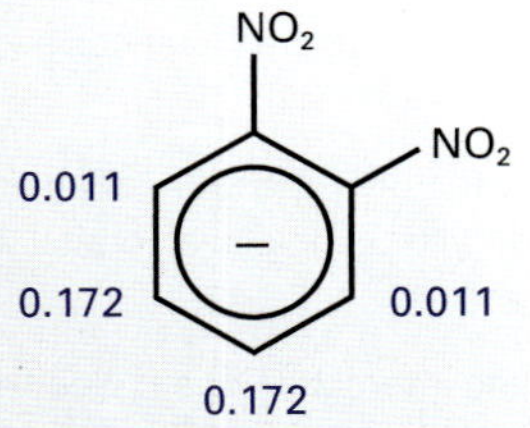

3

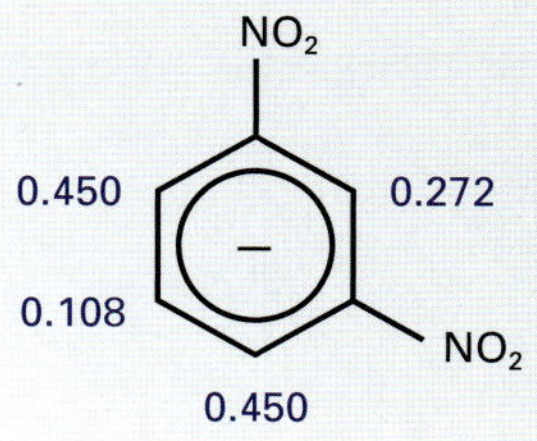

4

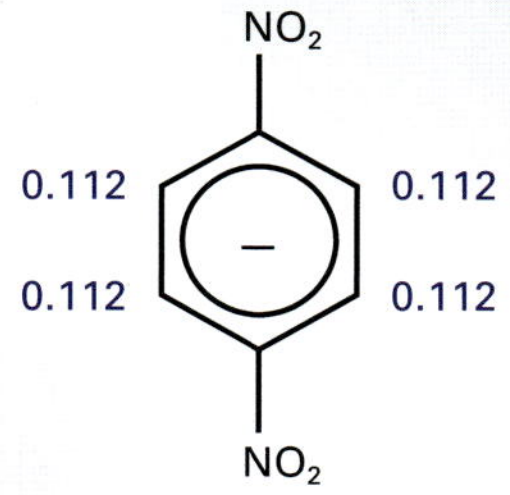

5

Projects

The symbol ‡ indicates that calculus is required.

21.31‡ Show that the coupling constant as expressed by the Karplus equation passes through a minimum when cos $\phi = B/4C$. *Hint:* Use calculus: evaluate the first derivative with respect to ϕ and set the result equal to 0. To confirm that the extremum is a minimum, go on to evaluate the second derivative and show that it is positive.

21.32 Here we explore magnetic resonance imaging in more detail. (a) You are designing an MRI spectrometer. What field gradient (in microtesla per metre, $\mu T\ m^{-1}$) is required to produce a separation of 100 Hz between two protons separated by the long diameter of a human kidney (taken as 8 cm) given that they are in environments with $\delta = 3.4$? The radiofrequency field of the spectrometer is at 400 MHz and the applied field is 9.4 T. (b) Suppose a uniform disk-shaped organ is in a linear field gradient, and that the MRI signal is proportional to the number of protons in a slice of width δx at each horizontal distance x from the centre of the disk. Sketch the shape of the absorption intensity for the MRI image of the disk before any computer manipulation has been carried out.

Chapter 22

Statistical thermodynamics

There are two great rivers in physical chemistry. One is the river of thermodynamics, which deals with the relations between bulk properties of matter, particularly properties related to the transfer of energy. The other is the river of molecular structure, including spectroscopy, which deals with the structures and properties of individual atoms and molecules. These two great rivers flow together in the part of physical chemistry called **statistical thermodynamics**, which shows how thermodynamic properties emerge from the properties of atoms and molecules. The first half of this book dealt with thermodynamic properties; the second half has dealt with atomic and molecular structure and its investigation. This is the chapter where the two great rivers merge.

A great problem with statistical thermodynamics is that it is highly mathematical. Many of the derivations—even the most fundamental—are beyond the scope of this text.[1] All we can hope to see is some of the key concepts and the key results. Where possible the treatment will be qualitative.

A brief comment In this chapter, some *Examples, Self-tests,* and the *brief illustrations* require calculus: they are marked with the symbol ‡.

The partition function

The key concept of quantum mechanics is the existence of a wavefunction that contains in principle all the dynamical information about a system, such as its energy, the electron density, the dipole moment, and so on. Once we know the wavefunction of an atom or molecule, we can extract from it all the dynamical information possible about the system—provided we know how to manipulate it. There is a

[1] See our *Physical chemistry* (2006), for details.

similar concept in statistical thermodynamics. The **partition function**, q, contains all the *thermo*dynamic information about the system, such as its internal energy, entropy, heat capacity, and so on. Our task here is to see how to calculate the partition function and how to extract the information it contains.

22.1 The Boltzmann distribution

The single most important result in the whole of statistical thermodynamics is the **Boltzmann distribution**, the formula that tells us how to calculate the numbers of molecules in each state of a system at any temperature:

$$N_i = \frac{Ne^{-E_i/kT}}{q} \tag{22.1}$$

Here N_i is the number of molecules in a state with energy E_i, N is the total number of molecules, k is Boltzmann's constant, a fundamental constant with the value 1.381×10^{-23} J K^{-1}, and T is the absolute temperature. Boltzmann's constant k and the gas constant R are related by $R = N_A k$. The term in the denominator, q, is the **partition function**:

$$q = \sum_i e^{-E_i/kT} = e^{-E_0/kT} + e^{-E_1/kT} + \cdots \tag{22.2}$$

where the sum is over all the states of the system. We shall have much more to say about q later, and see how it can be calculated and given physical meaning. At this stage it is just a kind of normalizing factor, for it ensures that $\sum_i N_i = N$.

The conceptual basis of the derivation of eqn 22.1 is very simple. We imagine a stack of energy levels arranged like bookshelves, one above the other. Then we imagine being blindfolded and throwing balls (the molecules) at the shelves (the energy levels) and letting them land on the available shelves entirely at random, apart from one condition.[1] That condition is that the total energy, E, of the final arrangement must have the actual energy of the sample of matter we are seeking to describe. So, provided the temperature is above absolute zero, not all the balls are allowed to land on the bottom shelf, for that would give a total energy of zero. Some of the balls may land on the bottom shelf, but there must be others ending up on higher shelves to ensure that the total energy is E. If we imagine throwing 100 balls at a set of shelves, then we will end up with one particular valid distribution. If we repeated the experiment with the same number of balls, we would end up with a different but still valid distribution. If we went on repeating the experiment, we would get many different distributions, but some of them would occur more often than others (Fig. 22.1).

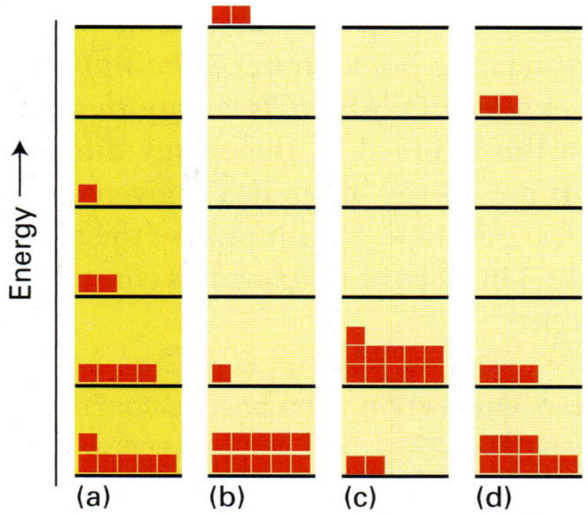

Fig. 22.1 The derivation of the Boltzmann distribution involves imagining that the molecules of a system (the squares) are distributed at random over the available energy levels subject to the requirements that the number of molecules and the total energy is constant, and then looking for the most probable arrangement. Of the four shown here, the numbers of ways of achieving each arrangement are (a) 181 180, (b) 858, (c) 78, (d) 12 870.[2] The number of ways of achieving (a) is by far the greatest, so this distribution is the most probable; it corresponds to the Boltzmann distribution.

When this game is analysed mathematically, it turns out that the *most probable* distribution—the arrangement that turns up most often—is that given by eqn 22.1. In other words, *the Boltzmann distribution is the outcome of blind chance occupation of energy levels, subject to the requirement that the total energy has a particular value.* When we deal with about 10^{23} molecules and repeat the experiment millions of times, the Boltzmann distribution turns out to be very accurate, and we can use it with confidence for all typical samples of matter.

The simplest application of the Boltzmann distribution is to calculate the relative numbers of molecules in two states separated in energy by ΔE. Suppose the energies of the two states are E_1 and E_2, then from eqn 22.1 we can write

Boltzmann distribution

$$\frac{N_2}{N_1} = \frac{\cancel{N} e^{-E_2/kT}/\cancel{q}}{\cancel{N} e^{-E_1/kT}/\cancel{q}} = \frac{e^{-E_2/kT}}{e^{-E_1/kT}} = e^{-(E_2-E_1)/kT} = e^{-\Delta E/kT} \tag{22.3}$$

Boltzmann distribution; ΔE

[1] Don't press this analogy too far; it is intended just to set a visualizable image and actually has little relation to the way that actual molecules become distributed!

[2] To calculate the number of ways, W, of arranging N molecules with N_1 in state 1, N_2 in state 2, etc., use $W = N!/N_1!N_2!\ldots$, with $n! = n(n-1)(n-2)\ldots1$, and $0! = 1$.

where $\Delta E = E_2 - E_1$. This important result tells us that *the relative population of the upper state decreases exponentially with its energy above the lower state*. In this expression, the energy difference is in joules. If the energy difference is given in joules (or kilojoules) per mole, we simply use the gas constant in eqn 22.3 in place of Boltzmann's constant (because $R = kN_A$).

A brief illustration The boat conformation of cyclohexane (**1**) lies 22 kJ mol^{-1} higher in energy than the chair conformation (**2**). To find the relative populations of the two conformations in a sample of cyclohexane at 20°C, we set $\Delta E = 22$ kJ mol^{-1} and $T = 293$ K and use eqn 22.3 with R in the exponent. We obtain:

$$\frac{N_{\text{boat}}}{N_{\text{chair}}} = e^{-\frac{2.2\times10^4\ \text{J mol}^{-1}}{(8.3145\ \text{J K}^{-1}\ \text{mol}^{-1})\times(293\ \text{K})}} = e^{-\frac{2.2\times10^4}{8.3145\times293}}$$

$$= 1.2\times10^{-4}$$

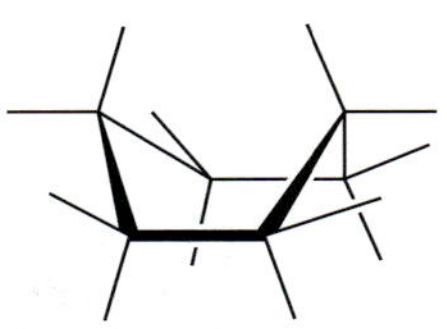

1

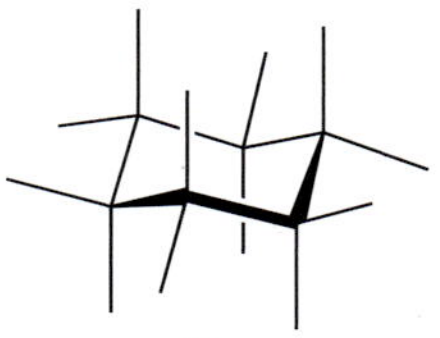

2

A note on good practice Note how the units cancel in the exponent: as always, you will avoid serious error (for instance, using k instead of R) by writing the units and ensuring that they cancel. Because exponentials are very sensitive to the numerical value of the exponent, do not round the intermediate steps but store them in your calculator until the last step of the calculation. Finally, note how the number of significant figures in the answer (two) does not exceed the number in the data.

One very important feature of the Boltzmann distribution is that it applies to the populations of *states*. We have seen that in some cases (the hydrogen atom and rotating molecules are examples) several different states have the same energy. That is, some energy levels are *degenerate* (Section 12.7). The Boltzmann distribution can be used to calculate, for instance, the number of hydrogen atoms at a temperature T that have their electron in a $2p_x$ orbital. Because a $2p_y$ orbital has exactly the same energy, the number of atoms with an electron in a $2p_y$ orbital is the same as the number with an electron in a $2p_x$ orbital. The same is true of atoms with an electron in a $2p_z$ orbital. Therefore, if we want the *total* number of atoms with electrons in 2p orbitals, we have to multiply the number in *one* of them by a factor of 3.

In general, if the degeneracy of an energy level (that is, the number of states of that energy) is g, we use a factor of g to get the population of the *level* (as distinct from an individual *state*). It is obviously very important to decide whether we wish to express the population of an individual state or the population of an entire degenerate energy level. We shall denote levels by L, so in terms of levels the Boltzmann distribution and the partition function are

$$N_L = \frac{Ng_L e^{-E_L/kT}}{q} \qquad q = \sum_L g_L e^{-E_L/kT} \qquad (22.4)$$

where N_L is the total number of molecules in the level L (the sum of populations of all the states of that level), g_L is its degeneracy, and E_L is its energy.

A brief illustration We saw in Section 19.1 that the rotational energy of a linear rotor is $hBJ(J+1)$ and that the degeneracy of each level is $2J+1$. Because the degeneracy of the level with $J=2$ (and energy $6hB$) is 5 and that of the level with $J=1$ (and energy $2hB$) is 3, the relative numbers of molecules with $J=2$ and 1 is

Boltzmann distribution | Degeneracy, $g = 2J+1$

$$\frac{N_2}{N_1} = \frac{Ng_2 e^{-E_2/kT}/q}{Ng_1 e^{-E_1/kT}/q} = \frac{g_2}{g_1} e^{-(E_2-E_1)/kT} = \frac{5}{3} e^{-4hB/kT}$$

Boltzmann distribution | Energy, $E = hBJ(J+1)$

For HCl, $B = 318.0$ GHz, so at 25°C (corresponding to 298 K), this ratio works out as 1.36: there are *more* molecules in the level with $J=2$ than in the level with $J=1$, even though $J=2$ corresponds to a higher energy. Each individual *state* with $J=2$ has a lower population than each state with $J=1$, but there are more states in the level with $J=2$.

One important convention that we adopt (largely for convenience) is that *all energies are measured relative to the ground state*. That is, we set the ground-state energy equal to zero, even if there is a zero-point energy. For instance, the energies of the states of a harmonic oscillator are measured from zero for the ground state:

Actual energies: $E = \frac{1}{2}h\nu, \frac{3}{2}h\nu, \frac{5}{2}h\nu, \ldots$

Our convention: $E = 0, h\nu, 2h\nu, \ldots$

Likewise, the energies of the hydrogen atom are measured from zero for the 1s orbital:

Actual energies: $E = -hcR_H, -\frac{1}{4}hcR_H, -\frac{1}{9}hcR_H, \ldots$

Our convention: $E = 0, \frac{3}{4}hcR_H, \frac{8}{9}hcR_H, \ldots$

This convention greatly simplifies our interpretation of the significance of q.

22.2 The interpretation of the partition function

When we are interested only in the relative populations of levels and states, we do not need to know the partition function because it cancels in eqn 22.4. However, if we want to know the actual population of a state, then we use eqn 22.1, which requires us to know q. We also need to know q when we derive thermodynamic functions, as we shall see.

The definition of q is the sum over states (not levels; remember that there may be several states of the same energy), as given in eqn 22.2. We can write out the first few terms as follows:

$$q = 1 + e^{-E_1/kT} + e^{-E_2/kT} + e^{-E_3/kT} + \cdots$$

The first term is 1 because the energy of the ground state (E_0) is 0, according to our convention, and $e^0 = 1$. In principle, we just substitute the values of the energies, evaluate each term for the temperature of interest, and add them together to get q. However, that procedure does not give much insight.

To see the physical significance of q, let's suppose first that $T = 0$. Then, because $e^{-\infty} = 0$, all terms other than the first are equal to 0, and $q = 1$. At $T = 0$ only the ground state is occupied and (provided that state is nondegenerate) $q = 1$. Now consider the other extreme: a temperature so high that all the $E_i/kT = 0$. Then, because $e^0 = 1$, the partition function is $q \approx 1 + 1 + 1 + 1 + \cdots = N_{states}$, where N_{states} is the total number of states of the molecule. That is, at very high temperatures, all the states of the system are thermally accessible. It follows that if the molecule has an infinite number of states, then q rises to infinity as T approaches infinity. We should begin to suspect that the partition function is telling us the number of states that are occupied at a given temperature.

Now consider an intermediate temperature, at which only some of the states are occupied significantly. Suppose that the temperature is such that kT is large compared to E_1 and E_2 but small compared to E_3 and all subsequent terms (Fig. 22.2). Because E_1/kT and E_2/kT are both small compared to 1, and $e^{-x} \approx 1$ when x is very small, the first three terms are all close to 1. However, because E_3/kT is large compared to 1, and $e^{-x} \approx 0$ when x is large, all the remaining terms are close to 0. Therefore,

$$q = 1 + \underbrace{\boxed{e^{-E_1/kT}}}_{1} + \underbrace{\boxed{e^{-E_2/kT}}}_{1} + \underbrace{\boxed{e^{-E_3/kT}}}_{0} + \underbrace{\boxed{e^{-E_4/kT}}}_{0} + \underbrace{\boxed{e^{-E_5/kT}}}_{0} + \cdots$$

and $q \approx 1 + 1 + 1 + 0 + \cdots = 3$. Once again, we see that the partition function is telling us the number of significantly occupied states at the temperature of interest. That is the principal meaning of the partition function: *q tells us the number of thermally accessible states at the temperature of interest.*

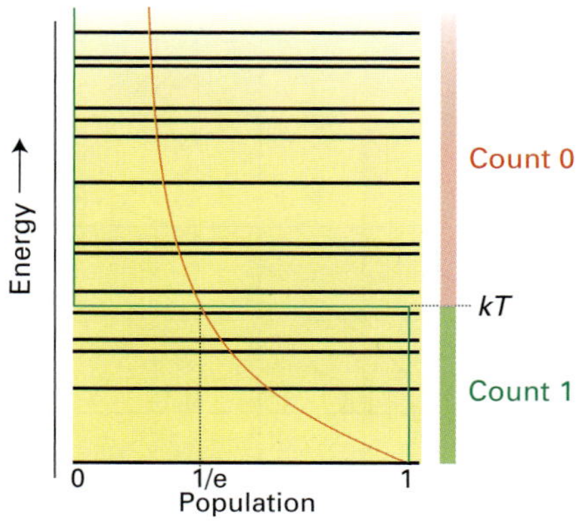

Fig. 22.2 The partition function is a measure of the number of thermally accessible states. Thus, for all states with $E < kT$ the exponential term is reasonably close to 1, whereas for all states with $E > kT$ the exponential term is close to 0. The states with $E < kT$ are significantly thermally accessible.

Once we grasp the significance of q, statistical thermodynamics becomes much easier to understand. We can anticipate, even before we do any calculations, that q increases with temperature, because more states become accessible as the temperature is raised. At low temperatures q is small, and falls to 1 as the temperature approaches absolute zero (when only one state, the ground state, is accessible and we are supposing that that state is nondegenerate). Molecules with numerous, closely spaced energy levels (like the rotational states of bulky molecules) can be expected to have very large partition functions. Molecules with widely spaced energy levels can be expected to have small partition functions, because only the few lowest states will be occupied at low temperatures.

Example 22.1

Calculating a partition function

Calculate the partition function for the cyclohexane molecule, confining attention to the chair and boat conformations mentioned in the preceding brief illustration. Show how the partition function varies with temperature.

Strategy Whenever calculating a partition function, start at the definition in eqn 22.2 and write out the individual terms. Remember to set the ground-state energy equal to 0. When the energies of states are given in joules (or kilojoules) per mole, replace the k in the definition of q by $R = N_A k$.

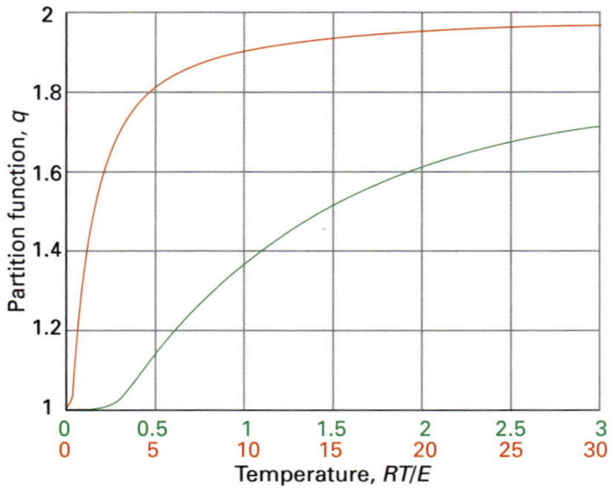

Fig. 22.3 The partition function for a two-level system with states at the energies 0 and E. At 20°C (293 K) and for $E = 22$ kJ mol^{-1}, $RT/E = 0.11$, where $q = 1.0001$. Note how the partition function rises from 1 and approaches 2 at high temperatures.

Solution There are only two states, so the partition function has only two terms. The energy of the chair form is set at 0 and that of the boat form is $E = 22$ kJ mol^{-1}. Therefore:

$$q = 1 + e^{-E/RT}$$

$$= 1 + e^{-\frac{2.2\times10^4\ \mathrm{J\,mol^{-1}}}{(8.3145\ \mathrm{J\,K^{-1}\,mol^{-1}})\times T}} = 1 + e^{-(2646\ \mathrm{K})/T}$$

This function is plotted in Fig. 22.3. We see that it rises from $q = 1$ (only the chair form is accessible at $T = 0$, when $(2646\ \mathrm{K})/T = \infty$ and $e^{-\infty} = 0$) to $q = 2$ at $T = \infty$ (when $(2646\ \mathrm{K})/T = 0$ and $e^0 = 1$; both states are thermally accessible at high temperatures). At 20°C, $q = 1.0001$. As we saw earlier, the boat form is only slightly populated and so q differs very little from 1.

A note on good practice Note how the units are treated in the exponent: the units of E and R cancel apart from K^{-1} in the denominator, which becomes K in the numerator (in the form 2646 K), which will cancel the units K of T when values of the latter are introduced. You will sometimes see an expression like '$q = 1 + e^{-2646/T}$, with T in kelvins' (or, worse, 'with T the absolute temperature'); retention of the units, as we show, is completely unambiguous and therefore better practice.

Self-Test 22.1

The ground configuration of a fluorine atom gives rise to a ^{2}P term with two levels, the $J = \frac{3}{2}$ level (of degeneracy 4) and the $J = \frac{1}{2}$ level (of degeneracy 2) at an energy corresponding to 404.0 cm^{-1} above the ground state. Write down an expression for the partition function and plot it as a function of temperature. *Hint*: The notation used here was introduced in Section 13.17. Take $E = hc\tilde{\nu}$ for the energy of the upper level. In this instance, the ground state is degenerate.

[*Answer*: $q = 4 + 2e^{-hc\tilde{\nu}/kT}$; Fig. 22.4]

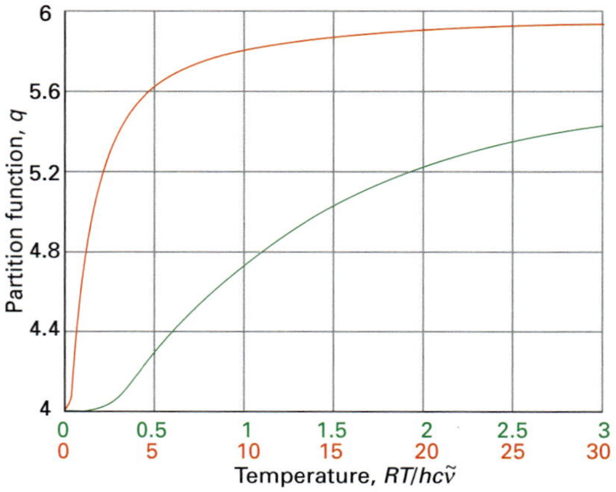

Fig. 22.4 The partition function for the six-level system treated in Self-test 22.1. Note how q rises from 4 (when only the four states of the $^3P_{3/2}$ level are occupied) and approaches 6 (when the two states of the $^3P_{1/2}$ level are also accessible). At 20°C, $kT/hc\tilde{\nu} = 0.504$, corresponding to $q = 5.21$.

22.3 Examples of partition functions

In a number of cases it is possible to derive simple closed expressions for partition functions. For example, the energy levels of a harmonic oscillator form a simple ladder-like array (Fig. 22.5). If we set the energy of the lowest state equal to zero, the energies of the states are

$$E_0 = 0, \quad E_1 = h\nu, \quad E_2 = 2h\nu, \quad E_3 = 3h\nu, \text{ etc.}$$

Therefore, the **vibrational partition function** is

$$q = 1 + e^{-h\nu/kT} + e^{-2h\nu/kT} + e^{-3h\nu/kT} + \cdots$$

$$= 1 + e^{-h\nu/kT} + (e^{-h\nu/kT})^2 + (e^{-h\nu/kT})^3 + \cdots$$

Use $e^{nx} = (e^x)^n$

The sum of the infinite series $1 + x + x^2 + \cdots$ is $1/(1 - x)$, so with $x = e^{-h\nu/kT}$,

$$q = \frac{1}{1 - e^{-h\nu/kT}} \tag{22.5}$$

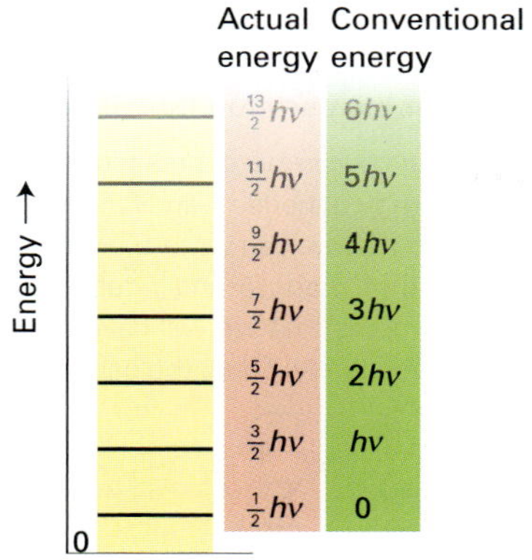

Fig. 22.5 The energy levels of an harmonic oscillator. When calculating a partition function, set the zero of energy at the lowest level, as shown on the right.

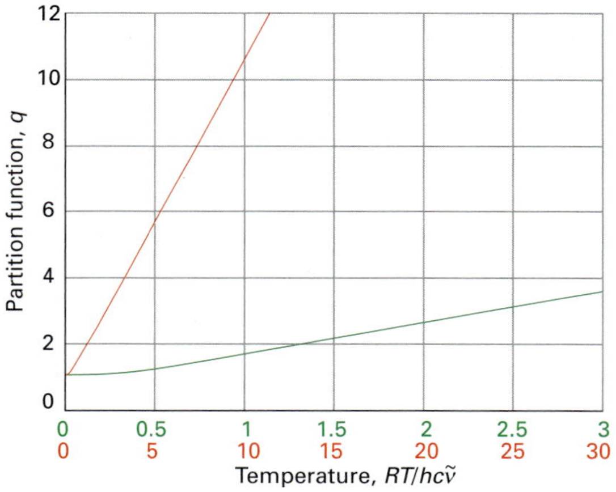

Fig. 22.6 The partition function for an harmonic oscillator. For an oscillator with $\tilde{\nu} = 1000\ \text{cm}^{-1}$, at 20°C, $kT/hc\tilde{\nu} = 0.204$, corresponding to $q = 1.01$.

Equation 22.5 is the partition function for a harmonic oscillator or any vibrating diatomic molecule. Figure 22.6 shows how q varies with temperature. Note that $q = 1$ at $T = 0$, when only the lowest state is occupied, and that as T becomes high, so q becomes infinite because all the states of the infinite ladder are thermally accessible. At room temperature, and for typical molecular vibrational frequencies, q is very close to 1 because only the vibrational ground state is occupied (see Exercise 22.10). When the temperature is so high that $h\nu/kT \ll 1$, eqn 22.4 can be simplified considerably by writing $e^{-x} \approx 1 - x$:

$$q \approx \frac{1}{1-(1-h\nu/kT)} = \frac{1}{h\nu/kT} = \frac{kT}{h\nu} \qquad (22.6)$$

This expression is consistent with the interpretation of q as the number of thermally accessible states: the average energy of a harmonic oscillator is kT, according to classical physics, and the separation of energy levels is $h\nu$, so about $kT/h\nu$ states must be occupied. (Think of a ladder with rungs separated by $h\nu$: to reach kT we need a ladder with $kT/h\nu$ rungs.)

We can carry out similar calculations for certain other types of motion. For example, suppose a molecule of mass m is confined in a flask of volume V at a temperature T, then (as shown in Further information 22.1) to a good approximation for typical containers and $T > 0$, the **translational partition function** is

$$q = \frac{(2\pi mkT)^{3/2}V}{h^3} \qquad (22.7)$$

We see that the partition function increases with temperature, as we have come to expect. However, notice that q also increases with the volume of the flask. That we should expect too: the energy levels of a particle in a box become closer together as the size of the box increases (Section 12.9), so at a given temperature, more states are thermally accessible.

A brief illustration Suppose we have an O_2 molecule (of mass $32m_u$) in a flask of volume 100 cm^3 at 20°C. Its translational partition function is

$$q = \frac{\{2\pi \times \overbrace{32 \times (1.661 \times 10^{-27}\ \text{kg})}^{m} \times \overbrace{(1.381 \times 10^{-23}\ \text{J K}^{-1}) \times (298\ \text{K})}^{kT}\}^{3/2} \times \overbrace{(1.00 \times 10^{-4}\ \text{m}^3)}^{V}}{(6.626 \times 10^{-34}\ \text{J s})^3}$$
$$= 9.67 \times 10^{25}$$

Note that a huge number of translational states are accessible at room temperature. This result is consistent with the derivation of eqn 22.7, which assumed that the translational energy levels form a near continuum in containers of macroscopic size.

A note on good practice All the units must cancel because all partition functions are dimensionless numbers. Here, because 1 J = 1 kg m^2 s^{-2}, the units cancel as follows:

$$\frac{(\text{kg J K}^{-1}\ \text{K})^{3/2}\ \text{m}^3}{(\text{J s})^3} = \frac{(\text{kg kg m}^2\ \text{s}^{-2})^{3/2}\ \text{m}^3}{(\text{kg m}^2\ \text{s}^{-2}\ \text{s})^3}$$
$$= \frac{(\text{kg m s}^{-1})^3\ \text{m}^3}{(\text{kg m}^2\ \text{s}^{-1})^3} = \frac{\text{kg}^3\ \text{m}^6\ \text{s}^{-3}}{\text{kg}^3\ \text{m}^6\ \text{s}^{-3}}$$
$$= 1$$

It might seem tedious to do this cancellation explicitly, but it is a very good way of making sure that you have set up the numerical calculation correctly.

The **rotational partition function** can also be approximated when the temperature is high enough for many rotational states to be occupied. For a linear rotor it turns out (see Further information 22.1) that for heavy molecules and $T > 0$,

$$q = \frac{kT}{\sigma hB} \tag{22.8}$$

In this expression, B is the rotational constant (Section 19.1) and σ is the **symmetry number:** $\sigma = 1$ for an unsymmetrical linear rotor (such as HCl or HCN) and $\sigma = 2$ for a symmetrical linear rotor (such as H_2 or CO_2). The symmetry number reflects the fact that an unsymmetrical molecule is distinguishable after rotation by 180° but a symmetrical molecule is not. When evaluating q we have to count only distinguishable states, and a symmetrical molecule has fewer distinguishable states than a less symmetrical molecule. The rotational partition function of HCl at 25°C works out to 19.6 (see Exercise 22.15), so about 20 rotational states (not levels: remember the $(2J + 1)$-fold degeneracy of each rotational level; 20 states corresponds to about the first 4 levels) are significantly occupied at that temperature.

No closed form can be given for the **electronic partition function**, the partition function for the distribution of electrons over their available states. However, for closed-shell molecules the excited states are so high in energy that only the ground state is occupied, and for them $q = 1$. Special care has to be taken for atoms and molecules that do not have closed shells (as we saw in Self-test 22.1 for a fluorine atom).

22.4 The molecular partition function

The energy of a molecule can be approximated as the sum of contributions from its different modes of motion (translation, rotation, and vibration), the distribution of electrons, and the electronic and nuclear spin. Given that the energy is a sum of independent contributions, we show in Derivation 22.1 that the partition function is a product of contributions:

$$q = q^{\mathrm{T}} q^{\mathrm{R}} q^{\mathrm{V}} q^{\mathrm{E}} q^{\mathrm{S}} \tag{22.9}$$

where T denotes translation, R rotation, V vibration, E the electronic contribution, and S the spin contribution. The contribution from electronic spin is important in atoms or molecules containing unpaired electrons. For example, consider the Cs atom, which has one unpaired electron. We saw in Chapters 13 and 20 that the two spin states of this unpaired electron are equally occupied in the absence of any magnetic field, so it contributes a factor of 2 to the molecular partition function.

Derivation 22.1

Factorization of the partition function

Suppose the energy can be expressed as the sum of contributions from two modes A and B (such as vibration and rotation), and that we can write $E_{i,j} = E_i^{\mathrm{A}} + E_j^{\mathrm{B}}$, where i denotes a state of mode A and j denotes a state of mode B and the sums that we will have to do are over both i and j independently. Then, the partition function is

$$q = \sum_{i,j} e^{-E_{i,j}/kT} \overset{E_{i,j} = E_i + E_j}{=} \sum_{i,j} e^{-E_i^A/kT - E_j^B/kT}$$

$$\overset{\text{use } e^{x+y} = e^x e^y}{=} \sum_{i,j} e^{-E_i^A/kT} e^{-E_j^B/kT} = \underbrace{\sum_i e^{-E_i^A/kT}}_{q^A} \underbrace{\sum_j e^{-E_j^B/kT}}_{q^B}$$

$$= q^{\mathrm{A}} q^{\mathrm{B}}$$

This argument is readily extended to three and more modes, as in eqn 22.9.

Thermodynamic properties

The principal reason for calculating the partition function is to use it to calculate thermodynamic properties of systems as small as atoms and as large as biopolymers. There are two fundamental relations we need. We can deal with First Law quantities (such as heat capacity and enthalpy) once we know how to calculate the internal energy. We can deal with Second Law quantities (such as the Gibbs energy and equilibrium constants) once we know how to calculate the entropy.

22.5 The internal energy and the heat capacity

To calculate the total energy, E, of the system, we note the energy of each state (E_i), multiply that energy by the number of molecules in the state (N_i), and then add together all these products:

$$E = N_0 E_0 + N_1 E_1 + N_2 E_2 + \cdots = \sum_i N_i E_i$$

However, the Boltzmann distribution tells us the number of molecules in each state of a system, so we can replace the N_i in this expression by the expression in eqn 22.1:

$$E = \sum_i \underbrace{\frac{N e^{-E_i/kT}}{q}}_{\text{Population of state } i} \times \underbrace{E_i}_{\text{Energy of state } i} = \frac{N}{q} \sum_i E_i e^{-E_i/kT} \tag{22.10}$$

If we know the individual energies of the states (from spectroscopy, for instance), then we just substitute their values into this expression. However, there is a much simpler method—or at least a much more succinct formula—available when we have an expression for the partition function, such as those given in Section 22.3. In Derivation 22.2 we show that the energy is related to the slope of q plotted against T:

$$E = \frac{NkT^2}{q} \times \text{ slope of } q \text{ plotted against } T \quad (22.11)$$

Derivation 22.2

The internal energy from the partition function

The sum on the right of eqn 22.10 resembles the definition of the partition function, but differs from it by having the E_i factor multiplying each term. However, we can recognize (by using the rules of differentiation set out in Appendix A2.4) that

(Use $\mathrm{d}e^{f(x)}/\mathrm{d}x = e^{f(x)} \times \mathrm{d}f/\mathrm{d}x$) (Use $\mathrm{d}(1/x)/\mathrm{d}x = -1/x^2$)

$$\frac{\mathrm{d}}{\mathrm{d}T}e^{-E_i/kT} = e^{-E_i/kT} \times \frac{\mathrm{d}}{\mathrm{d}T}\left(-\frac{E_i}{kT}\right) = \frac{E_i}{kT^2}e^{-E_i/kT}$$

In other words,

$$E_i e^{-E_i/kT} = kT^2 \frac{\mathrm{d}}{\mathrm{d}T}e^{-E_i/kT}$$

With this substitution, the expression for the total energy becomes

$$E = \frac{N}{q}\sum_i kT^2 \frac{\mathrm{d}}{\mathrm{d}T}e^{-E_i/kT} = \frac{NkT^2}{q}\frac{\mathrm{d}}{\mathrm{d}T}\underbrace{\sum_i e^{-E_i/kT}}_{q}$$

because the sum of derivatives is the derivative of the sum. Magically (or, more precisely, mathematically), the expression for the partition function has appeared, so we can write

$$E = \frac{NkT^2}{q}\frac{\mathrm{d}q}{\mathrm{d}T}$$

which, because $\mathrm{d}q/\mathrm{d}T$ is the slope of a graph of q plotted against T, is eqn 22.11.

The remarkable feature of eqn 22.11 is that it is an expression for the total energy in terms of the partition function alone. The partition function is starting to fulfil its promise to deliver all thermodynamic information about the system.

There is one more detail to take into account before we use eqn 22.11. Recall that we have set the zero of energy at the energy of the lowest state of the molecule. However, the internal energy of the system might be nonzero on account of zero-point energy, and the E in eqn 22.11 is the energy *above* the zero-point energy. That is, the internal energy at a temperature T is

$$U = U(0) + E \quad (22.12)$$

with E given by eqn 22.11.

‡Example 22.2

Calculating the internal energy

Calculate the molar internal energy of a monatomic gas.

Strategy The only mode of motion of a monatomic gas is translation (we ignore electronic excitation). Therefore, substitute the translational partition function in eqn 22.7 into eqn 22.11 (using the precise mathematical form given in Derivation 22.2, and then insert the result into eqn 22.12. The partition function has the form $q = aT^{3/2}$, where a is a collection of constants.

Solution First, we need the first derivative of q with respect to T:

($\mathrm{d}x^n/\mathrm{d}x = nx^{n-1}$ with $n = \frac{3}{2}$)

$$\frac{\mathrm{d}q}{\mathrm{d}T} = \frac{\mathrm{d}}{\mathrm{d}T}(aT^{3/2}) = \tfrac{3}{2}aT^{1/2}$$

When we substitute this result into eqn 22.11 we get

$$E = \frac{NkT^2}{q} \times \frac{\mathrm{d}q}{\mathrm{d}T} = \frac{NkT^2}{aT^{3/2}} \times \tfrac{3}{2}aT^{1/2} = \tfrac{3}{2}NkT$$

The molar internal energy is obtained by replacing N by Avogadro's constant and using eqn 22.12:

$$U_m = U_m(0) + \tfrac{3}{2}N_A kT = U_m(0) + \tfrac{3}{2}RT$$

The term $U_m(0)$ contains all the contributions from the binding energy of the electrons and of the nucleons in the nucleus. The term $\frac{3}{2}RT$ is the contribution to the internal energy from the translational motion of the atoms in their container.

Self-test 22.2

Calculate the molar internal energy of a gas of diatomic molecules.

[*Answer:* $U_m = U_m(0) + \frac{5}{2}RT$]

Once we have calculated the internal energy of a sample of molecules, it is a simple matter to calculate the heat capacity. It should be recalled that the heat capacity at constant volume, C_V, is defined as the slope of the plot of internal energy against temperature:

$$C_V = \frac{\Delta U}{\Delta T} \quad \text{at constant volume}$$

Therefore, all we need do is to evaluate the slope of the expression for U obtained from the partition function.

‡A brief illustration The slope of U with respect to T is actually the first derivative:

$$C_V = \frac{dU}{dT} \quad \text{at constant volume}$$

(Remember from Section 2.7 that a more sophisticated notation for this expression is $C_V = (\partial U/\partial T)_V$.) The constant-volume molar heat capacity of a monatomic gas is therefore obtained by substituting the molar internal energy, $U_m = U_m(0) + \frac{3}{2}RT$ into this expression:

$$C_{V,m} = \frac{d}{dT}(U_m(0) + \tfrac{3}{2}RT) = \tfrac{3}{2}R$$

To calculate $C_{p,m}$, we use eqn 2.19 ($C_{p,m} - C_{V,m} = R$) and obtain $C_{p,m} = \frac{5}{2}R$.

‡Self-test 22.3

Calculate the contribution to the molar constant-volume heat capacity of a two-state system, like the chair–boat interconversion of cyclohexane (Section 22.1) and show how the heat capacity varies with temperature.

[*Answer:* $C_{V,m} = R(E/RT)^2 e^{E/RT}/(1 + e^{E/RT})^2$, Fig. 22.7]

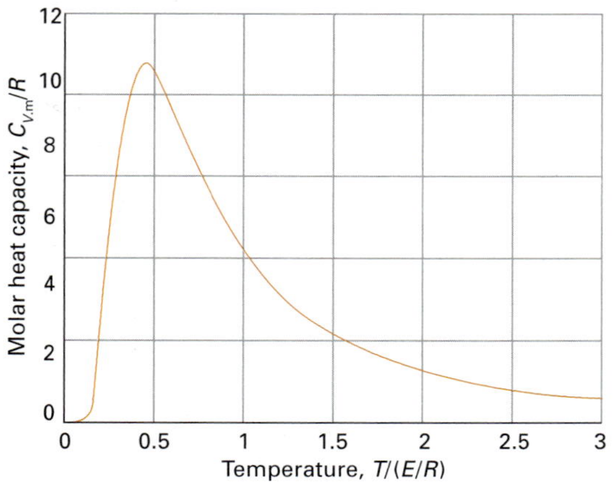

Fig. 22.7 The variation of the heat capacity of a two-level system with states at energies 0 and E. Note how the heat capacity is zero at $T = 0$, passes through a maximum at $T = 0.417E/R$, and approaches 0 at high temperatures.

22.6 The entropy and the Gibbs energy

The entry point into the calculation of properties arising from the Second Law of thermodynamics is the proposal made by Boltzmann that the entropy of a system can be calculated from the expression

$$S = k \ln W \tag{22.13}$$

Here W is the number of different ways in which the molecules of a system can be arranged yet result in the same total energy. Its formal name is the **weight** of a configuration. This expression is the **Boltzmann formula** for the entropy. The entropy is zero if there is only one way of achieving a given total energy (because $\ln 1 = 0$). The entropy is high if there are many ways of achieving the same energy.

In most cases, $W = 1$ at $T = 0$ because there is only one way of achieving zero energy: put all the molecules into the same, lowest state. Therefore, $S = 0$ at $T = 0$, in accord with the Third Law of thermodynamics (Section 4.7). In certain cases, though, W may differ from 1 at $T = 0$. This is the case if positional or orientational disorder survives down to absolute zero because there is no energy advantage in adopting a particular orientation. For instance, there may be no energy difference between the arrangements ... AB AB AB ... and ... BA AB BA ..., so $W > 1$ even at $T = 0$. If $S > 0$ at $T = 0$ we say that the substance has a **residual entropy**. Ice has a residual molar entropy of 3.4 J K^{-1} mol^{-1}. It stems from the disorder in the hydrogen bonds between neighbouring water molecules: a given O atom has two short O—H bonds and two long O···H bonds to its neighbours, but there is a degree of randomness in which two bonds are short and two are long (see Fig. 5.17).

A brief illustration Consider a sample of solid carbon monoxide containing N CO molecules. We saw in Section 4.9 that CO has a very small dipole moment. In fact, the dipolar interactions between CO molecules are so weak in a solid that even at $T = 0$ they lie either head-to-tail or head-to-head with approximately equal energies to give randomly orientated arrangements such as ... CO CO OC CO OC OC .... with the same energy. Because each CO molecule can lie in either of two orientations (CO or OC) with equal energy, there are $2 \times 2 \times 2 \times \ldots = 2^N$ ways of achieving the same energy. The residual entropy of the sample, its entropy at $T = 0$, where there is this orientation disorder but no motional disorder, is therefore $S = k \ln 2^N = Nk \ln 2$. The molar entropy is therefore $S_m = N_A k \ln 2 = R \ln 2$, or 5.8 J K^{-1} mol^{-1}, which is close to the experimental value of 5 J K^{-1} mol^{-1}.

Boltzmann went on to show that there is a close relation between the entropy and the partition function: both are measures of the number of arrangements available to the molecules. The precise connection for *distinguishable* molecules (those locked in place in a solid) is[3]

$$S = \frac{U - U(0)}{T} + Nk \ln q \qquad (22.14a)$$

The analogous term for *indistinguishable* molecules (identical molecules free to move, as in a gas) is

$$S = \frac{U - U(0)}{T} + Nk \ln q - Nk(\ln N - 1) \qquad (22.14b)$$

Because we can calculate the first term on the right from q, we now have a method for calculating the entropy of any system of noninteracting molecules once we know its partition function.

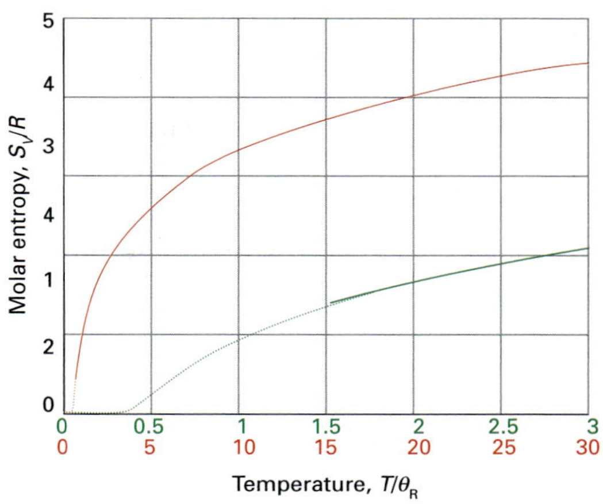

Fig. 22.8 The variation of the rotational contribution to the molar entropy with temperature. Note that eqn 22.6 is valid only for high temperatures, so the formula derived in Example 22.3 cannot be used at low temperatures (so we have terminated the curves before the equation becomes invalid). The dotted lines show the correct behaviour.

Example 22.3

Calculating the entropy

Calculate the contribution that rotational motion makes to the molar entropy of a gas of HCl molecules at 25°C.

Strategy We have already calculated the contribution to the internal energy (Self-test 22.2), and we have the rotational partition function in eqn 22.6 (with $\sigma = 1$). We need to combine the two parts. We use eqn 22.14a because we are concentrating on the internal motion (the rotation) of the molecules, not their translational motion.

Solution We substitute $U - U(0) = RT$ and $q = kT/hB$ into eqn 22.14a, and obtain (for $T > 0$)

$$S_m = \frac{RT}{T} + R \ln \frac{kT}{hB} = R\left(1 + \ln \frac{kT}{hB}\right)$$

Notice that the entropy increases with temperature (Fig. 22.8). At a given temperature, the entropy is larger the smaller the value of B. That is, bulky molecules (which have large moments of inertia and therefore small rotational constants) have a higher rotational entropy than small molecules. Substitution of the numerical values gives $S_m = 3.98R$, or 33.1 J K^{-1} mol^{-1}.

Self-test 22.4

The rotational partition function of an ethene molecule is 661 at 25°C. What is the contribution of rotation to its molar entropy?

[*Answer:* 7.49*R*]

The Gibbs energy, G, was central to most of the thermodynamic discussions in the early chapters of this book, so to show that statistical thermodynamics is really useful we have to see how to calculate G from the partition function, q. We shall confine our attention to a perfect gas, because it is difficult to take molecular interactions into account, and in Derivation 22.3 we show that for a gas of N molecules

$$G - G(0) = -NkT \ln \frac{q}{N} \qquad (22.15)$$

Derivation 22.3

Calculating the Gibbs energy from the partition function

To set up the calculation, we go back to first principles. The Gibbs energy is defined as $G = H - TS$, and the enthalpy, H, is defined as $H = U + pV$. Therefore

$$G = U - TS + pV$$

For a perfect gas we can replace pV by $nRT = NkT$ (because $N = nN_A$ and $R = N_A k$), and note that at $T = 0$, $G(0) = U(0)$ (because the terms TS and NkT vanish at $T = 0$). Therefore,

$$G - G(0) = U - U(0) - TS + NkT$$

Now we substitute eqn 22.14b for S, and obtain

$$G - G(0) = -NkT \ln q + kT(N \ln N - N) + NkT$$
$$= -NkT(\ln q - \ln N)$$

Then, because $\ln q - \ln N = \ln(q/N)$, we obtain eqn 22.15.

[3] For a derivation, see our *Physical chemistry* (2006).

We can convert eqn 22.15 into an expression for the molar Gibbs energy. First, we write $N = nN_A$, and it becomes

$$G - G(0) = -nN_A kT \ln \frac{q}{nN_A}$$

Then we introduce the **molar partition function**, $q_m = q/n$, with units 1/mole (mol^{-1}). On dividing both sides of the preceding equation by n, we get

$$G_m - G_m(0) = -RT \ln \frac{q_m}{N_A} \qquad (22.16)$$

Example 22.4

Calculating the Gibbs energy

Calculate the molar Gibbs energy of a monatomic perfect gas and express it in terms of the pressure of the gas.

Strategy The calculation is based on eqn 22.16 with $q_m = q/n$. All we need to know is the translational partition function, which is given in eqn 22.5. Convert from V to p by using the perfect gas law.

Solution When we substitute $q_m = (2\pi mkT)^{3/2}V/nh^3$ into eqn 22.16 we get

q_m^T

$$G_m - G_m(0) = -RT \ln \frac{(2\pi mkT)^{3/2}V}{nh^3 N_A}$$

Next, we replace V by nRT/p (notice that the ns cancel), and obtain (after a little tidying up, including writing $R = kN_A$)

$V = nRT/p,\ R = N_Ak$

$$G_m - G_m(0) = -RT \ln \frac{(2\pi m)^{3/2}(kT)^{3/2}}{nh^3 N_A} \frac{nN_A kT}{p}$$

$$= -RT \ln \frac{(2\pi m)^{3/2}(kT)^{5/2}}{ph^3} \qquad (22.17)$$

$-\ln x = \ln(1/x)$

$$= RT \ln ap \qquad a = \frac{h^3}{(2\pi m)^{3/2}(kT)^{5/2}}$$

The Gibbs energy increases logarithmically (as $\ln p$) as p increases, just as we saw in Section 5.2 (eqn 5.3b).

Self-test 22.5

Ignore vibration and write the molar partition function of a diatomic molecule as $q_m^T q^R$ (see eqn 22.8). What is the molar Gibbs energy of such a gas?

[*Answer:* As in eqn 22.17, but with a replaced by $a\sigma hB/kT$]

The only further piece of information we require is the expression for the *standard* molar Gibbs energy, for that played such an important role in the discussion of equilibrium properties. All we need to do is to use the partition function calculated at $p^\ominus$. For instance, for a monatomic gas, we use p = 1 bar in eqn 22.17 and obtain the standard value of the molar Gibbs energy. In general, we write

$$G_m^\ominus - G_m^\ominus(0) = -RT \ln \frac{q_m^\ominus}{N_A} \qquad (22.18)$$

where the standard state sign on q simply reminds us to calculate its value at $p^\ominus$; to do so, we use $V_m^\ominus = RT/p^\ominus$ wherever it appears in $q_m^\ominus$. We shall see an example of that in the following section.

22.7 The statistical basis of chemical equilibrium

We can obtain a deeper insight into the origin and significance of that most chemical of quantities, the equilibrium constant K, by considering the Boltzmann distribution of molecules over the available states of a system composed of reactants and products. When atoms can exchange partners, as in a reaction, the available states of the system include arrangements in which the atoms are present in the form of reactants and in the form of products: these arrangements have their characteristic sets of energy levels, but the Boltzmann distribution does not distinguish between their identities, only their energies. The atoms distribute themselves over both sets of energy levels in accord with the Boltzmann distribution (Fig. 22.9). At a given temperature, there will be a specific

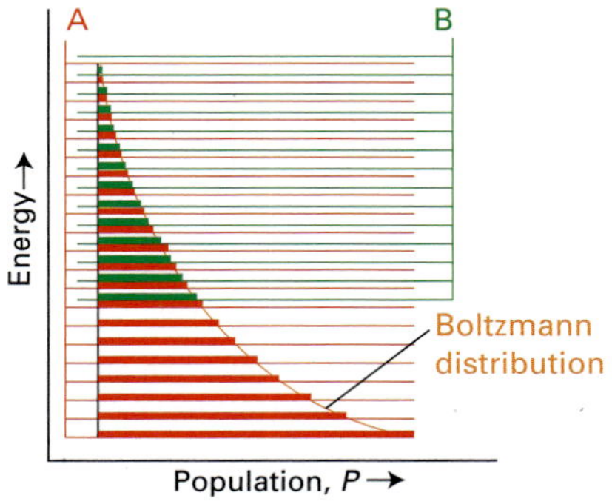

Fig. 22.9 The Boltzmann distribution of populations over the energy levels of two species A and B with similar densities of energy levels; the reaction A → B is endothermic in this example. The bulk of the population is associated with the species A, so that species is dominant at equilibrium.

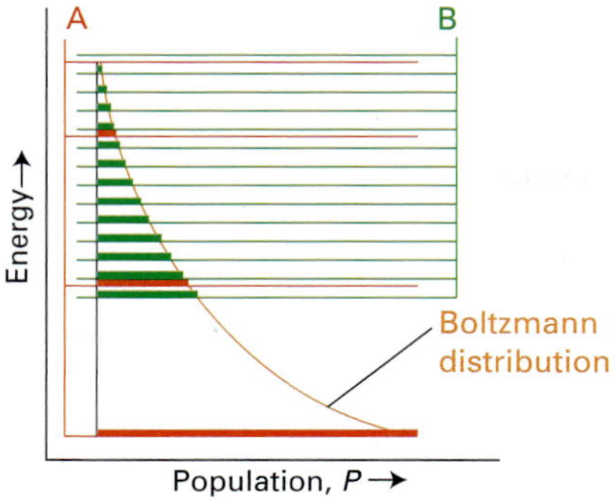

Fig. 22.10 Even though the reaction A → B is endothermic, the density of energy levels in B is so much greater than that in A, that the population associated with B is greater than that associated with A, so B is dominant at equilibrium.

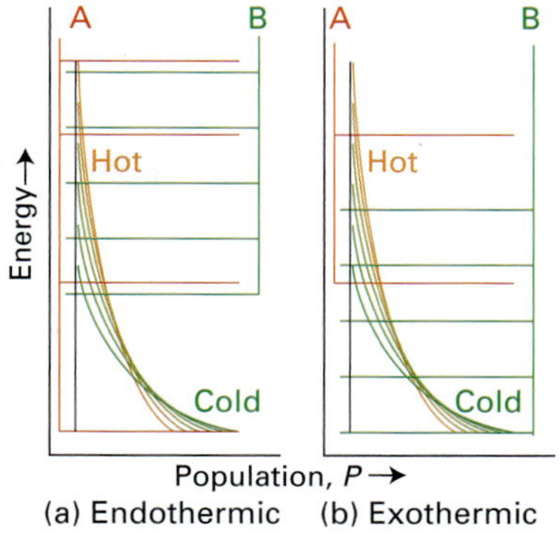

Fig. 22.11 The effect of temperature on a chemical equilibrium can be interpreted in terms of the change in the Boltzmann distribution with temperature and the effect of that change in the population of the species. (a) In an endothermic reaction, the population of B increases at the expense of A as the temperature is raised. (b) In an exothermic reaction, the opposite happens.

distribution of populations, and hence a specific composition of the reaction mixture.

It can be appreciated from Fig. 22.9 that if the reactants and products both have similar arrays of molecular energy levels, then the dominant species in a reaction mixture at equilibrium will be the species with the lower set of energy levels. However, the fact that the equilibrium constant is related to the Gibbs energy ($\ln K = -\Delta_r G^{\circ}/RT$) is a signal that entropy plays a role as well as energy. Its role can be appreciated by referring to Fig. 22.10. We see that although the B energy levels lie higher than the A energy levels, in this instance they are much more closely spaced. As a result, their total population may be considerable and B could even dominate in the reaction mixture at equilibrium. Closely spaced energy levels correlate with a high entropy (see eqn 22.14), so in this case we see that entropy effects dominate adverse energy effects. That is, a positive reaction enthalpy results in a lowering of the equilibrium constant (that is, an endothermic reaction can be expected to have an equilibrium composition that favours the reactants). However, if there is positive reaction entropy, then the equilibrium composition may favour products, despite the endothermic character of the reaction.

Statistical principles also give us insight into the temperature dependence of the equilibrium constant. In Section 7.8, we saw that for a reaction that is exothermic under standard conditions ($\Delta_r H^{\circ} < 0$), K decreases as the temperature rises. The opposite occurs in the case of endothermic reactions. The typical arrangement of energy levels for an endothermic reaction is shown in Fig. 22.11a. When the temperature is increased, the Boltzmann distribution adjusts and the populations change as shown. The change corresponds to an increased population of the higher energy states at the expense of the population of the lower-energy states. We see that the states that arise from the B molecules become more populated at the expense of the A molecules. Therefore, the total population of B states increases, and B becomes more abundant in the equilibrium mixture. Conversely, if the reaction is exothermic (Fig. 22.11b), then an increase in temperature increases the population of the A states (which start at higher energy) at the expense of the B states, so the reactants become more abundant.

22.8 The calculation of the equilibrium constant

We can go beyond the qualitative picture developed above by writing a statistical thermodynamic expression for the equilibrium constant. We show in Further Information 22.2 that, for the equilibrium $A(g) + B(g) \rightleftharpoons C(g)$,

$$K = \frac{q_m^{\circ}(C)N_A}{q_m^{\circ}(A)q_m^{\circ}(B)} e^{-\Delta E/RT} \tag{22.19}$$

where ΔE is the difference in energy between the ground state of the product and that of the reactants. This expression is easy to remember: it has the same form as the equilibrium constant written in terms of the activities (Section 7.6), but with q_m°/N_A

replacing each activity (and an additional exponential factor):

$$K = \frac{\boxed{p_C/p^{\ominus}}\quad\boxed{q_m^{\ominus}(C)/N_A}}{\boxed{(p_A/p^{\ominus})}\boxed{(p_B/p^{\ominus})}\boxed{(q_m^{\ominus}(A)/N_A)}\boxed{(q_m^{\ominus}(B)/N_A)}}e^{-\Delta E/RT}$$

When we cancel the N_A, we get eqn 22.19.

Equation 22.19 is quite extraordinary, for it provides a key link between partition functions, which can be derived from spectroscopy, and the equilibrium constant, which is central to the analysis of chemical reactions at equilibrium. It represents the merging of the two rivers that have flowed through this text.

Example 22.5

Calculating an equilibrium constant

Calculate the equilibrium constant for the gas-phase ionization $Cs(g) \rightleftharpoons Cs^+(g) + e^-(g)$ at 500 K.

Strategy This is a reaction of the form $A(g) \rightleftharpoons B(g) + C(g)$ rather than $A(g) + B(g) \rightleftharpoons C(g)$, so we need to modify eqn 22.19 slightly, but the form to use should be clear. Analyse each species individually, and write its partition function as the product of partition functions for each mode of motion. Evaluate these partition functions at the standard pressure (1 bar), and combine them as specified in eqn 22.19. For the difference in energy ΔE, use the ionization energy of $Cs(g)$.

Solution The equilibrium constant is

$$K = \frac{q_m^{\ominus}(Cs^+,g)q_m^{\ominus}(e^-,g)}{q_m^{\ominus}(Cs,g)N_A}e^{-\Delta E/RT}$$

Note how, in this instance, Avogadro's constant appears in the denominator: its units, mol^{-1}, ensure that K is dimensionless. The electron has translational motion, so we need its translational partition function. We saw in Section 22.4 that the spin states contribute a factor of 2 to the molecular partition function. Therefore

$$q_m^{\ominus}(e^-,g) = \overbrace{2}^{q^{spin}} \times \overbrace{\frac{(2\pi m_e kT)^{3/2}V^{\ominus}}{nh^3}}^{q_m^T}$$

$$\underbrace{}_{V^{\ominus} = nRT/p^{\ominus}}$$

$$= \frac{2(2\pi m_e kT)^{3/2}RT}{p^{\ominus}h^3}$$

The Cs^+ ion, a closed shell species, has only translational freedom:

$$q_m^{\ominus}(Cs^+,g) = \frac{(2\pi m_{Cs}kT)^{3/2}RT}{p^{\ominus}h^3}$$

The partition function of the Cs atom has a translational and a spin contribution, as we saw in Section 22.4:

$$q_m^{\ominus}(Cs,g) = \overbrace{2}^{q^{spin}} \times \overbrace{\frac{(2\pi m_{Cs}kT)^{3/2}RT}{p^{\ominus}h^3}}^{q_m^T}$$

(We are not distinguishing the masses of the atoms Cs atom and the Cs^+ ion.) Then, with $\Delta E = I$, the ionization energy of the atom, we find

$$K = \frac{\overbrace{\{(2\pi m_{Cs}kT)^{3/2}RT/p^{\ominus}h^3\}}^{q_m^{\ominus}(Cs^+,g)} \times \overbrace{\{2(2\pi m_e kT)^{3/2}RT/p^{\ominus}h^3\}}^{q_m^{\ominus}(e^-,g)}}{\underbrace{\{2(2\pi m_{Cs}kT)^{3/2}RT/p^{\ominus}h^3\}}_{q_m^{\ominus}(Cs,g)} \times N_A} \times \overbrace{e^{-I/RT}}^{\Delta E}$$

$$= \frac{(2\pi m_e kT)^{3/2}RT}{p^{\ominus}h^3 N_A} \times e^{-I/RT} = \frac{(2\pi m_e kT)^{3/2}kT}{p^{\ominus}h^3} \times e^{-I/RT}$$

$$\underbrace{}_{R = N_A k}$$

$$= \frac{(2\pi m_e)^{3/2}(kT)^{5/2}}{p^{\ominus}h^3} \times e^{-I/RT}$$

When we substitute the data (only the ionization energy is specific to the element), we find:

$$K = \frac{(2\pi \times \overbrace{9.109 \times 10^{-31} \text{ kg}}^{m_e})^{3/2} \times (\overbrace{1.381 \times 10^{-23} \text{ J K}^{-1} \times 1000 \text{ K}}^{kT})^{5/2}}{(\underbrace{10^5 \text{ Pa}}_{p^{\ominus}}) \times (6.626 \times 10^{-34} \text{ J s})^3}$$

$$\times e^{-\frac{3.76 \times 10^5 \text{ J mol}^{-1}}{(8.314 \text{ J K}^{-1} \text{ mol}^{-1}) \times (1000 \text{ K})}}$$

$$= 2.42 \times 10^{-19}$$

A note on good practice Verify that the units do in fact all cancel (use $1 \text{ J} = 1 \text{ kg m}^2 \text{ s}^{-2}$ and $1 \text{ Pa} = 1 \text{ kg m}^{-1} \text{ s}^{-2}$). The K calculated by the procedure described here is the thermodynamic equilibrium constant, which for gases is expressed in terms of the partial pressures of the reactants and products (relative to the standard pressure), $K = (p_{Cs^+}/p^{\ominus})(p_{e^-}/p^{\ominus})/(p_{Cs}/p^{\ominus}) = p_{Cs^+}p_{e^-}/p_{Cs}p^{\ominus}$

Self-test 22.6

Calculate the equilibrium constant for the dissociation $Na_2(g) \rightleftharpoons 2 Na(g)$ at 1000 K. You will need the following information about $Na_2(g)$: $B = 46.38$ MHz, $\tilde{v} = 159.2 \text{ cm}^{-1}$, and the dissociation energy is 70.4 kJ mol^{-1}. The ground state of a sodium atom is ^2S.

[*Answer*: 2.42]

Checklist of key ideas

You should now be familiar with the following concepts.

☐ **1** The Boltzmann distribution gives the numbers of molecules in each state of a system at any temperature.

☐ **2** The partition function is an indication of the number of thermally accessible states at the temperature of interest.

☐ **3** The molecular partition function is the product of contributions from translation, rotation, vibration, electronic and spin distributions: $q = q^T q^R q^V q^E q^S$.

☐ **4** The electronic partition function is $q^E = 1$ for closed-shell molecules with high-energy excited states.

Table of key equations

The following table summarizes the equations that have been developed in the text.

Description	Equation	Comment
Boltzmann distribution	$N_i = Ne^{-E_i/kT}/q$	
Partition function	$q = \sum_i e^{-E_i/kT}$	Independent molecules
Vibrational partition function	$q^V = 1/(1 - e^{-h\nu/kT})$	Harmonic oscillator
Translational partition function	$q^T = (2\pi mkT)^{3/2}V/h^3$	Perfect gas, $T > 0$
Rotational partition function	$q^R = kT/\sigma hB$	Linear rotor, $kT > \sigma hB$
Internal energy	$U = U(0) + E$ $E = (NkT^2/q)(dq/dT)$	Independent molecules
Boltzmann formula	$S = k \ln W$	
Entropy	$S = \{U - U(0)\}/T + Nk \ln q$	Independent, distinguishable molecules
Standard molar	$S = \{U - U(0)\}/T + Nk \ln q - Nk(\ln N - 1)$	Independent, indistinguishable molecules
Gibbs energy	$G_m^\ominus - G_m^\ominus(0) = -RT \ln(q_m^\ominus/N_A)$	Independent, indistinguishable molecules
Equilibrium constant	$K = \{q_m^\ominus(C)N_A/q_m^\ominus(A)q_m^\ominus(B)\}e^{-\Delta E/kT}$	$A(g) + B(g) \rightleftharpoons C(g)$, perfect gases

Further information 22.1

The calculation of partition functions

1. The translational partition function

Consider a particle of mass m in a rectangular box of sides X, Y, Z. Each direction can be treated independently and then the total partition function obtained by multiplying together the partition functions for each direction. The same strategy was used to write an expression for the molecular partition function by multiplying the contributions from (independent) modes of molecular motion.

The energy levels of a molecule of mass m in a container of length X are given by eqn 12.8 with $L = X$:

$$E_n = \frac{n^2h^2}{8mX^2} \qquad n = 1, 2, \ldots$$

The lowest level ($n = 1$) has energy $h^2/8mX^2$, so the energies relative to that level are

$$\varepsilon_n = (n^2 - 1)\varepsilon \qquad \varepsilon = h^2/8mX^2$$

The sum to evaluate is therefore

$$q_X = \sum_{n=1}^{\infty} e^{-(n^2-1)\varepsilon/kT}$$

The translational energy levels are very close together in a container the size of a typical laboratory vessel; therefore, the sum can be approximated by an integral:

$$q_X = \int_1^\infty e^{-(n^2-1)\varepsilon/kT}dn$$

The extension of the lower limit to $n = 0$ and the replacement of $n^2 - 1$ by n^2 introduces negligible error but turns the integral into standard form. We make the substitution $x^2 = n^2\varepsilon/kT$, implying $dn = dx/(\varepsilon/kT)^{1/2}$, and therefore that

$$q_X = \left(\frac{kT}{\varepsilon}\right)^{1/2}\int_0^\infty e^{-x^2}dx = \left(\frac{kT}{\varepsilon}\right)^{1/2}\left(\frac{\pi^{1/2}}{2}\right) = \left(\frac{2\pi mkT}{h^2}\right)^{1/2}X$$

The same expression applies to the other dimensions of a rectangular box of sides Y and Z, so

$$q^T = q_X q_Y q_Z = \left(\frac{2\pi mkT}{h^2}\right)^{3/2}XYZ = \left(\frac{2\pi mkT}{h^2}\right)^{3/2}V$$

where $V = XYZ$ is the volume of the box.

2. The rotational partition function

The rotational partition function of a nonsymmetrical (AB) linear rigid rotor is

$$q^R = \sum_J (2J+1)e^{-hBJ(J+1)/kT}$$

where the sum is over the rotational energy levels and the factor $2J + 1$ takes into account the degeneracy of the levels. When many rotational states are occupied and kT is much larger than the separation between neighbouring states, we can approximate the sum by an integral:

$$q^R = \int_0^\infty (2J+1)e^{-hBJ(J+1)/kT}dJ$$

Although this integral looks complicated, it can be evaluated without much effort by noticing that it can also be written as

$$q^{rot} = \frac{kT}{hB}\int_0^\infty \left(\frac{d}{dJ}e^{-hBJ(J+1)/kT}\right)dJ$$

Then, because the integral of a derivative of a function is the function itself,

$$q^R = -\frac{kT}{hB}e^{-hBJ(J+1)/kT}\Big|_0^\infty = \frac{kT}{hB}$$

For a homonuclear diatomic molecule, which looks the same after rotation by 180°, we have to divide this result by 2 to avoid double-counting of states, so in general

$$q^R = \frac{kT}{\sigma hB}$$

where $\sigma = 1$ for heteronuclear diatomic molecules and 2 for homonuclear diatomic molecules.

Further information 22.2

The equilibrium constant from the partition function

We know from thermodynamics (Section 7.6) that the equilibrium constant for a reaction is related to the standard reaction Gibbs energy by

$$\Delta_r G^\ominus = -RT\ln K$$

For the reaction $A(g) + B(g) \rightleftharpoons C(g)$,

$$\Delta_r G^\ominus = G_m^\ominus(C) - \{G_m^\ominus(A) + G_m^\ominus(B)\}$$

Equation 22.16 is an expression for each of these standard molar Gibbs energies in terms of the partition function of each species, so we can write

$$\Delta_r G^\ominus = \left\{G_m^\ominus(C,0) - RT\ln\frac{q_m^\ominus(C)}{N_A}\right\} - \left\{G_m^\ominus(A,0) - RT\ln\frac{q_m^\ominus(A)}{N_A}\right\} - \left\{G_m^\ominus(B,0) - RT\ln\frac{q_m^\ominus(B)}{N_A}\right\}$$

The first term in each of the braces is just the difference in ground-state energies because $G = U$ at $T = 0$, so

$$G_m^\ominus(C,0) - \{G_m^\ominus(A,0) + G_m^\ominus(B,0)\} = U_m^\ominus(C,0) - \{U_m^\ominus(A,0) + U_m^\ominus(B,0)\} = \Delta E$$

The three logarithms can be combined by using $\ln x - \ln y - \ln z = \ln(x/yz)$, to obtain

$$\ln\frac{q_m^\ominus(C)}{N_A} - \left\{\ln\frac{q_m^\ominus(A)}{N_A} + \ln\frac{q_m^\ominus(B)}{N_A}\right\} = \ln\frac{q_m^\ominus(C)N_A}{q_m^\ominus(A)q_m^\ominus(B)}$$

At this stage we have reached

$$\Delta_r G^\ominus = \Delta E - RT\ln\frac{q_m^\ominus(C)N_A}{q_m^\ominus(A)q_m^\ominus(B)}$$

The ΔE can be brought inside the logarithm by writing

$$\Delta E = -RT\ln e^{-\Delta E/RT}$$

(Because $\ln e^x = x$). Therefore

$$\Delta_r G^\ominus = -RT\ln e^{-\Delta E/RT} - RT\ln\frac{q_m^\ominus(C)N_A}{q_m^\ominus(A)q_m^\ominus(B)} = -RT\ln\left\{\frac{q_m^\ominus(C)N_A}{q_m^\ominus(A)q_m^\ominus(B)}e^{-\Delta E/RT}\right\}$$

All we have to do now is to compare this expression with the thermodynamic expression, $\Delta_r G^\ominus = -RT\ln K$, and see that the term in parentheses is the expression for K (eqn 22.19).

Questions and exercises

Discussion questions

22.1 Outline the derivation of the Boltzmann distribution.

22.2 What is temperature?

22.3 Describe the physical significance of the molecular partition function.

22.4 When are particles of the same composition identical and when are they not?

22.5 Explain how the internal energy and entropy of a system composed of two levels vary with temperature.

22.6 Justify the identification of the statistical entropy with the thermodynamic entropy.

22.7 Explain the origin of the residual entropy.

22.8 Use concepts of statistical thermodynamics to describe the molecular features that determine the magnitudes of equilibrium constants and their variation with temperature.

Exercises

22.1 Suppose polyethylene molecules in solution can exist either as a single version of a random coil (that is, ignore the fact that a random coil can be achieved in many different ways) or fully stretched out, with the latter conformation 2.4 kJ mol^{-1} higher in energy. What is the ratio of the two conformations at 20°C? See Exercise 22.30 for an elaboration of this exercise.

22.2 What is the ratio of populations of proton spin orientations in a magnetic field of (a) 1.5 T, (b) 15 T in a sample at 20°C? *Hint:* For the energy difference, refer to Chapter 21.

22.3 What is the ratio of populations of electron spin orientations in a magnetic field of 0.33 T in a sample at 20°C? *Hint:* For the energy difference, refer to Chapter 21.

22.4 Calculate the ratio of populations of CO_2 molecules with $J=4$ and $J=2$ at 25°C. The rotational constant of CO_2 is 11.70 GHz. *Hint:* Molecular rotations are discussed in Chapter 19.

22.5 Calculate the ratio of populations of CH_4 molecules with $J=4$ and $J=2$ at 25°C. The rotational constant of CH_4 is 157 GHz. *Hint:* The degeneracy of a spherical rotor in a state with quantum number J is $(2J+1)^2$.

22.6 (a) Write down the expression for the partition function of a molecule that has three energy levels at 0, 2ε, and 5ε with degeneracies 1, 6, and 3, respectively. What are the values of q at (b) $T=0$, (c) $T=\infty$?

22.7 The ground configuration of carbon gives rise to a triplet with the three levels $^3P_0$, $^3P_1$, and $^3P_2$ at wavenumbers 0, 16.4, and 43.5 cm^{-1}, respectively. Evaluate the partition function of carbon at (a) 10 K, (b) 298 K. *Hint:* Remember that a level with quantum number J has $2J+1$ states.

22.8 The ground configuration of oxygen gives rise to the three levels $^3P_2$, $^3P_1$, and $^3P_0$ at wavenumbers 0, 158.5, and 226.5 cm^{-1}, respectively. (a) Before doing any calculation, state the value of the partition function at $T=0$. (b) Evaluate the partition function at 298 K and confirm that its value at $T=0$ is what you anticipated in (a).

22.9 Evaluate the vibrational partition function for HBr at 298 K. For data, see Table 19.2. Above what temperature is the high-temperature approximation (eqn 22.6) in error by 10 per cent or less?

22.10 A CO_2 molecule has four vibrational modes with wavenumbers 1388 cm^{-1}, 2349 cm^{-1}, and 667 cm^{-1} (the last being a doubly degenerate bending motion). Calculate the total vibrational partition function at (a) 500 K, (b) 1000 K.

22.11 Evaluate the translational partition function of (a) N_2, (b) gaseous CS_2 in a flask of volume 10.0 cm^3. Why is one so much larger than the other?

22.12 Evaluate the translational partition function at 298 K of (a) a methane molecule trapped in the pore of a zeolite catalyst: take the pore to be spherical with a radius that allows the molecule to move through 1 nm in any direction (that is, the *effective* diameter is 1 nm), (b) a methane molecule in a flask of volume 100 cm^3.

22.13 Evaluate the rotational partition function of HBr ($\tilde{B}$ = 8.465 cm^{-1}) at 298 K (a) by direct summation of the energy levels, (b) by using the high-temperature approximation.

22.14 Repeat the previous exercise at different temperatures (use mathematical software) and determine the temperature at which the approximate formula is 10 per cent in error.

22.15 Evaluate the rotational partition function at 298 K of (a) $^1H^{35}Cl$, for which the rotational constant is 318 GHz, (b) $^{12}C^{16}O_2$, for which the rotational constant is 11.70 GHz.

22.16 N_2O and CO_2 have similar rotational constants (12.6 and 11.7 GHz, respectively) but strikingly different rotational partition functions. Why?

22.17 Derive an expression for the energy of a molecule that has three energy levels at 0, ε, and 3ε with degeneracies 1, 5, and 3, respectively.

22.18 The states arising from the ground configuration of a carbon atom are described in Exercise 22.7. (a) Derive an expression for the electronic contribution to the molar internal energy and plot it as a function of temperature. (b) Evaluate the expression at 25°C.

22.19 (a) Derive an expression for the electronic contribution to the molar heat capacity of an oxygen atom and plot it as a function of temperature. (b) Evaluate the expression at 25°C. The structure of the atom is described in Exercise 22.8.

22.20 Suppose that the $FClO_3$ molecule can take up any of four orientations in the solid at $T = 0$. What is its residual molar entropy?

22.21 An average human DNA molecule has 5×10^8 base pairs (rungs on the DNA ladder) of four different kinds. If each rung were a random choice of one of these four possibilities, what would be the residual entropy associated with this typical DNA molecule?

22.22 Calculate the molar entropy of nitrogen (N_2) at 298 K. *Hint:* Ignore the vibration of the molecule. Write the overall partition function as the product of the translational and rotational partition functions. For data, see Table 19.1.

22.23 Without carrying out an explicit calculation, explain the relative values of the standard molar entropies (at 298 K) of the following substances: (a) Ne(g) (146 J K^{-1} mol^{-1}) compared with Xe(g) (170 J K^{-1} mol^{-1}); (b) H_2O(g) (189 J K^{-1} mol^{-1}) compared with D_2O(g) (198 J K^{-1} mol^{-1}); (c) C(diamond) (2.4 J K^{-1} mol^{-1}) compared with C(graphite) (5.7 J K^{-1} mol^{-1}).

22.24 Estimate the change in molar entropy when a micelle consisting of 100 molecules disperses. *Hint:* Treat the transition as the expansion of a gas-like substance that initially occupies a volume $V_{micelle}$ and spreads into a volume $V_{solution}$. What does this model neglect?

22.25 Calculate the standard molar Gibbs energy of carbon dioxide at 298 K relative to its value at $T = 0$.

22.26 Write down the expression for the equilibrium constant of the reaction $N_2(g) + 3\,H_2(g) \rightleftharpoons 2\,NH_3(g)$ in terms of the molecular partition functions of the species.

22.27 Calculate the equilibrium constant for the ionization equilibrium of sodium atoms at 1000 K.

22.28 Calculate the equilibrium constant for the dissociation of I_2(g) at 500 K.

Projects

The symbol ‡ indicates that calculus is required.

22.29‡ Here we use statistical thermodynamics to calculate the internal energy and heat capacity of a system (such as the surface of an atomic solid) modelled as a collection of harmonic oscillators. (a) Derive an expression for the internal energy of a collection of harmonic oscillators. Deduce from your expression the high-temperature approximation and identify the temperature above which it is reliable. *Hint:* Substitute eqn 22.5 for the partition function into eqn 22.11 for the energy. (b) Now find an expression for the heat capacity of the oscillators and its high-temperature limit.

22.30 The very first exercise invited you to neglect the fact that a random coil can be achieved in many different ways. Repeat that exercise, allowing for this feature. Explore how the ratio of populations varies with the number of units, N, in the polymer.

Appendix 1 Quantities and units

The result of a measurement is a **physical quantity** (such as mass or density) that is reported as a numerical multiple of an agreed **unit:**

physical quantity = numerical value × unit

For example, the mass of an object may be reported as $m = 2.5$ kg and its density as $d = 1.01$ kg dm^{-3} where the units are, respectively, 1 kilogram (1 kg) and 1 kilogram per decimetre cubed (1 kg dm^{-3}). Units are treated like algebraic quantities, and may be multiplied, divided, and cancelled. Thus, the expression (physical quantity)/unit is simply the numerical value of the measurement in the specified units, and hence is a dimensionless quantity. For instance, the mass reported above could be denoted $m/\text{kg} = 2.5$ and the density as $d/(\text{kg dm}^{-3}) = 1.01$.

Physical quantities are denoted by italic or Greek letters (as in m for mass and Π for osmotic pressure). Units are denoted by Roman letters (as in m for metre). In the **International System** of units (SI, from the French Système International d'Unités), the units are formed from seven **base units** listed in Table A1.1. All other physical quantities may be expressed as combinations of these physical quantities and reported in terms of **derived units.** Thus, volume is (length)^3 and may be reported as a multiple of 1 metre cubed (1 m^3), and density, which is mass/volume, may be reported as a multiple of 1 kilogram per metre cubed (1 kg m^{-3}).

Table A1.1
The SI base units

Physical quantity	Symbol for quantity	Base unit
Length	l	metre, m
Mass	m	kilogram, kg
Time	t	second, s
Electric current	I	ampere, A
Thermodynamic temperature	T	kelvin, K
Amount of substance	n	mole, mol
Luminous intensity	I_v	candela, cd

Table A1.2
A selection of derived units

Physical quantity	Derived unit*	Name of derived unt
Force	1 kg m s^{-2}	newton, N
Pressure	1 kg m^{-1} s^{-2}	pascal, Pa
	1 N m^{-2}	
Energy	1 kg m^2 s^{-2}	joule, J
	1 N m	
	1 Pa m^3	
Power	kg m^2 s^{-3}	watt, W
	1 J s^{-1}	

* Equivalent definitions in terms of derived units are given following the definition in terms of base units.

A number of derived units have special names and symbols. The names of units derived from names of people are lower case (as in torr, joule, pascal, and kelvin), but their symbols are upper case (as in Torr, J, Pa, and K). The most important of this kind for our purposes are listed in Table A1.2. In all cases (both for base and derived quantities), the units may be modified by a prefix that denotes a factor of a power of 10. In a perfect world, Greek prefixes of units are upright (as in μm) and sloping for physical properties (as in μ for chemical potential), but available typefaces are not always so obliging. Among the most common prefixes are those listed in Table A1.3. Examples of the use of these prefixes are

$$1\ \text{nm} = 10^{-9}\ \text{m} \quad 1\ \text{ps} = 10^{-12}\ \text{s} \quad 1\ \mu\text{mol} = 10^{-6}\ \text{mol}$$

The kilogram (kg) is anomalous: although it is a base unit, it is interpreted as 10^3 g, and prefixes are attached to the gram (as in 1 mg = 10^{-3} g). Powers of units apply to the prefix as well as the unit they modify:

Table A1.3
Common SI prefixes

Prefix	z	a	f	p	n	μ	m	c	d
Name	zepto	atto	femto	pico	nano	micro	milli	centi	deci
Factor	10^{-21}	10^{-18}	10^{-15}	10^{-12}	10^{-9}	10^{-6}	10^{-3}	10^{-2}	10^{-1}
Prefix	k	M	G	T	P				
Name	kilo	mega	giga	tera	peta				
Factor	10^3	10^6	10^9	10^{12}	10^{15}				

$$1\ \text{cm}^3 = 1\ (\text{cm})^3 = 1\ (10^{-2}\ \text{m})^3 = 10^{-6}\ \text{m}^3$$

Note that 1 cm^3 does not mean 1 $c(m^3)$. When carrying out numerical calculations, it is usually safest to write out the numerical value of an observable as powers of 10.

There are a number of units that are in wide use but are not a part of the International System. Some are exactly equal to multiples of SI units. These include the *litre* (L), which is exactly 10^3 cm^3 (or 1 dm^3) and the *atmosphere* (atm), which is exactly 101.325 kPa. Others rely on the values of fundamental constants, and hence are liable to change when the values of the fundamental constants are modified by more accurate or more precise measurements. Thus, the size of the energy unit *electronvolt* (eV), the energy acquired by an electron that is accelerated through a potential difference of exactly 1 V, depends on the value of the charge of the electron, and the present (2009) conversion factor is 1 eV = $1.602\ 176 \times 10^{-19}$ J. Table A1.4 gives the conversion factors for a number of these convenient units.

Table A1.4
Some common units

Physical quantity	Name of unit	Symbol for unit	Value
Time	minute	min	60 s
	hour	h	3600 s
	day	d	86 400 s
Length	ångström	Å	10^{-10} m
Volume	litre	L, l	1 dm^3
Mass	tonne	t	10^3 kg
Pressure	bar	bar	10^5 Pa
	atmosphere	atm	101.325 kPa
Energy	electronvolt	eV	$1.602\ 176\ 5 \times 10^{-19}$ J
			96.485 31 kJ mol^{-1}

All values in the final column are exact, except for the definition of 1 eV.

Appendix 2 Mathematical techniques

The art of doing mathematics correctly is to do nothing at each step of a calculation. That is, it is permissible to develop an equation by ensuring that the left-hand side of an expression remains equal to the right-hand side. There are several ways of modifying the *appearance* of an expression without upsetting its balance.

Basic procedures

We set the stage for the mathematical arguments in the text by reviewing a few basic procedures, such as manipulation of equations, graphs, logarithms, exponentials, and vectors.

A2.1 Algebraic equations and graphs

The simplest types of equation we have to deal with have the form

$$y = ax + b$$

This expression may be modified by subtracting b from both sides, to give

$$y - b = ax$$

It may be modified further by dividing both sides by a, to give

$$\frac{y-b}{a} = x$$

This series of manipulations is called **rearranging** the expression for y in terms of x to give an expression for x in terms of y. A short cut, as can be seen by inspecting these two steps, is that an added term can be moved through the equals sign provided that as it passes = it changes sign (that happened to b in the example). Similarly, a multiplying factor becomes a divisor (and vice versa) when it passes through the = sign (as happened to a).

There are several more complicated manipulations that are required in certain cases. For example, we can find the values of x that satisfy an equation of the form

$$ax^2 + bx + c = 0$$

or any equation that can be rearranged into this form by the steps we have already illustrated. An equation in which x occurs as its square is called a **quadratic equation**. Its solutions are found by inserting the values of the constants a, b, and c into the expression

$$x = \frac{-b \pm (b^2 - 4ac)^{1/2}}{2a} \qquad \text{(A2.1)}$$

where the two values of x given by this expression (one by using the + sign and the other by using the – sign) are called the two **roots** of the original quadratic equation.

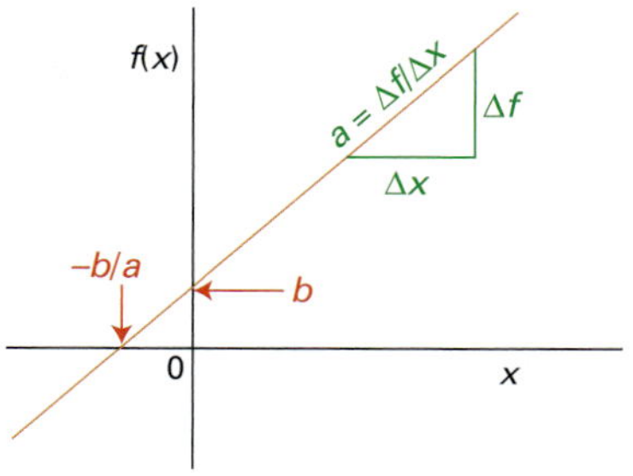

Fig. A2.1 A straight line is described the equation $f(x) = ax + b$, where a is the slope and b is the intercept.

A **function**, f, tells us how something changes as a variable is changed. For example, we might write

$$f(x) = ax + b$$

to show how a property f changes as x is changed. The variation of f with x is best shown by drawing a graph in which f is plotted on the vertical axis and x is plotted horizontally. The graph of the function we have just written is shown in Fig. A2.1. The important point about this graph is that it is **linear** (that is, it is a straight line); its **intercept** with the vertical axis (the value of f when $x = 0$) is b, and its **slope** is a. That is, a straight line has the form

$$f = \text{slope} \times x + \text{intercept}$$

A positive value of a indicates an upward slope from left to right (increasing x); a change of sign of a results in a negative slope, down from left to right. We say that *y varies linearly* with x if the relation between them is $y = ax + b$; we say that y is *proportional* to x if the relation is $y = bx$.

The solutions of the equation $f(x) = 0$ can be visualized graphically: they are the values of x for which f cuts through the horizontal axis (the axis corresponding to $f = 0$). For example, the solution of the quadratic equation given earlier is depicted in Fig. A2.2 (see p. 544). In general, a quadratic equation has a graph that cuts through the horizontal axis at two points (the equation has two roots), a cubic equation (an equation in which x^3 is the highest power of x) cuts through it three times (the equation has three roots), and so on.

A2.2 Logarithms, exponentials, and powers

Some equations are most readily solved by using logarithms and related functions. The **natural logarithm** of a number x is denoted $\ln x$, and is defined as the power to which a certain number designated e must be raised for the result to be equal to x. The number e, which is equal to 2.718 ... may seem to be decidedly unnatural; however, it

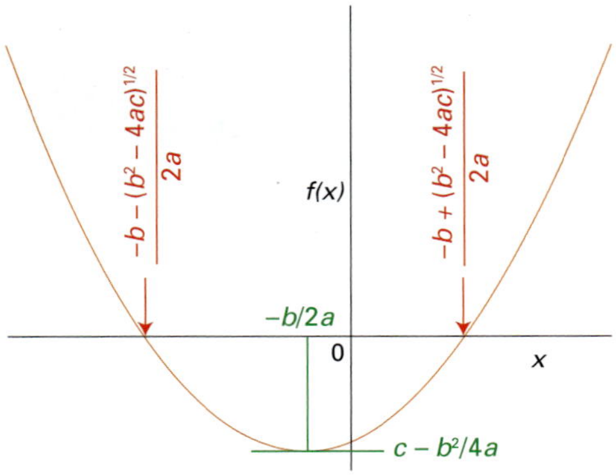

Fig. A2.2 The roots of a quadratic equation are given by the values of x where the parabola intersects the x-axis.

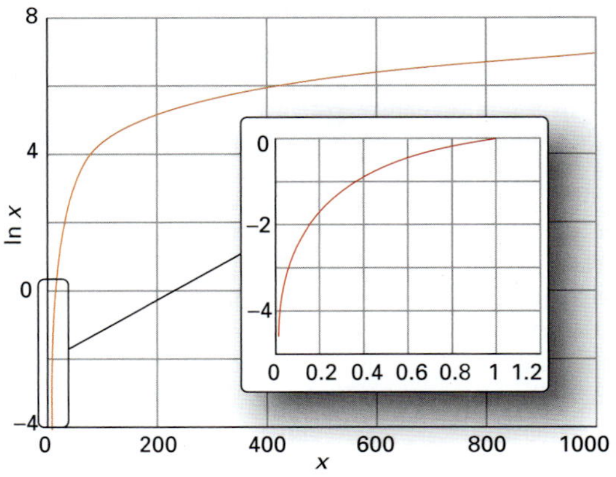

Fig. A2.3 The graph of ln x. Note that ln x approaches $-\infty$ as x approaches 0.

falls out naturally from various manipulations in mathematics and its use greatly simplifies calculations. On a calculator, ln x is obtained simply by entering the number x and pressing the 'ln' key or its equivalent. It follows from the definition of logarithms that

$$\ln x + \ln y = \ln xy \qquad \text{(A2.2a)}$$

$$\ln x - \ln y = \frac{x}{y} \qquad \text{(A2.2b)}$$

$$a \ln x = \ln x^a \qquad \text{(A2.2c)}$$

Thus, ln 5 + ln 3 is the same as ln 15 and ln 6 − ln 2 is the same as ln 3, as may readily be checked with a calculator. The last of these three relations is very useful for finding an awkward root of a number. The **antilogarithm** of a number is the value of x for which the natural logarithm is the number quoted.

A brief illustration Suppose we wanted the fifth root of 28. We write the required root as x, with $x^5 = 28$. We take logarithms of both sides, which gives $\ln x^5 = \ln 28$, and then rewrite the left-hand side of this equation as $5 \ln x$. At this stage we see that we have to solve

$$5 \ln x = \ln 28$$

To do so, we divide both sides by 5, which gives

$$\ln x = \frac{\ln 28}{5} = 0.6664\ldots$$

The natural antilogarithm of a number is obtained by pressing the 'exp' key on a calculator (where 'exp' is an abbreviation for exponential), and in this case the answer is 1.947....

There are a number of useful points to remember about logarithms, and they are summarized in Fig. A2.3. We see how logarithms increase only very slowly as x increases. For instance, when x increases from 1 to 1000, ln x increases from 0 to only 6.9. Another point is that the logarithm of 1 is 0: ln 1 = 0. The logarithms of numbers less than 1 are negative, and in elementary mathematics the logarithms of negative numbers are not defined.

A brief comment The logarithm of a negative number is complex (that is, involves i, the square-root of −1): $\ln(-x) = i\pi + \ln x$.

We also encounter the **common logarithm** of a number, the logarithm compiled with 10 in place of e; they are denoted log x. For example, log 5 is the power to which 10 must be raised to obtain 5, and is 0.698 97.... Common logarithms follow the same rules of addition and subtraction as natural logarithms. They are largely of historical interest now that calculators are so readily available, but they survive in the context of acid–base chemistry and pH. Common and natural logarithms (log and ln, respectively) are related by

$$\ln x = \ln 10 \times \log x = (2.303\ldots) \times \log x \qquad \text{(A2.3)}$$

The **exponential function**, e^x, plays a very special role in the mathematics of chemistry. It is evaluated by entering x and pressing the 'exp' key on a calculator. The following properties are important:

$$e^x \times e^y = e^{x+y} \qquad \text{(A2.4a)}$$

$$\frac{e^x}{e^y} = e^{x-y} \qquad \text{(A2.4b)}$$

$$(e^x)^a = e^{ax} \qquad \text{(A2.4c)}$$

(These relations are the analogues of the relations for logarithms.) A graph of e^x is shown in Fig. A2.4. As we see, it is positive for all values of x. It is less than 1 for all negative values of x, is equal to 1 when $x = 0$, and rises ever more rapidly towards infinity as x increases. This sharply rising character of e^x is the origin of the colloquial expression 'exponentially increasing' widely but loosely used in the media. (Strictly, a function increases exponentially if its rate of change is proportional to its current value.)

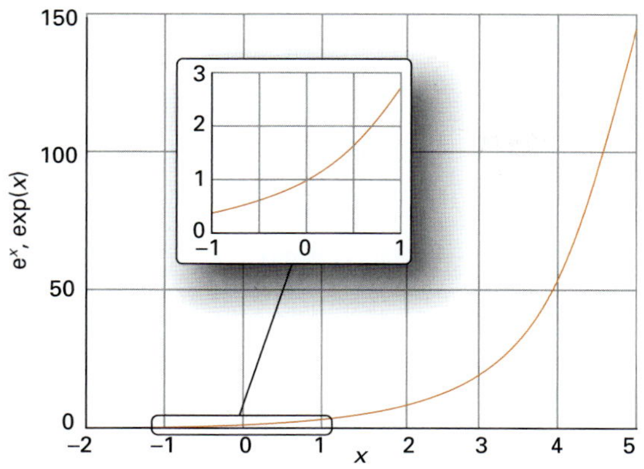

Fig. A2.4 The graph of e^x. Note that e^x approaches 0 as x approaches $-\infty$.

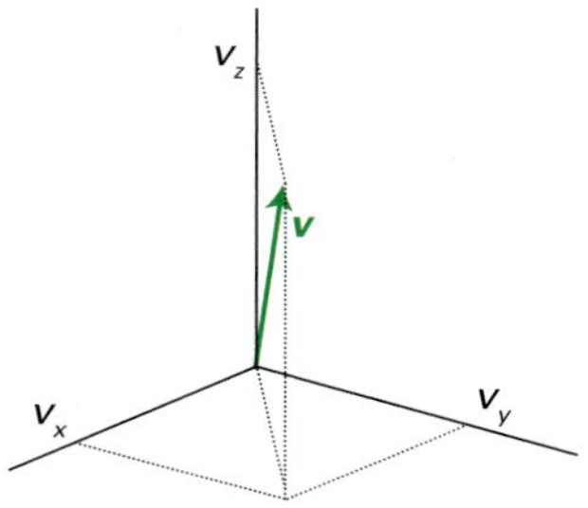

Fig. A2.5 The vector $\boldsymbol{v}$ has components v_x, v_y, and v_z on the *x*-, *y*-, and *z*-axes, respectively.

A2.3 Vectors

A vector quantity has both magnitude and direction. The vector $\boldsymbol{v}$ shown in Fig. A2.5 has components on the x-, y-, and z-axes with magnitudes v_x, v_y, and v_z, respectively. The direction of each of the components is denoted with a plus sign or minus sign. For example, if $v_x = -1.0$, the x-component of the vector $\boldsymbol{v}$ has a magnitude of 1.0 and points in the $-x$ direction. The magnitude of the vector is denoted v or $|\boldsymbol{v}|$ and is given by

$$v = (v_x^2 + v_y^2 + v_z^2)^{1/2} \qquad \text{(A2.5)}$$

Operations involving vectors are not as straightforward as those involving numbers. Here we describe a procedure for adding and subtracting two vectors because such vector operations are important for the discussion of atomic structure and molecular dipole moments.

Consider two vectors $\boldsymbol{v}_1$ and $\boldsymbol{v}_2$ making an angle θ (Fig. A2.6a). The first step in the addition of $\boldsymbol{v}_2$ to $\boldsymbol{v}_1$ consists of joining the tail of $\boldsymbol{v}_2$ to the had of $\boldsymbol{v}_1$, as shown in Fig. A2.6b. In the second step, we draw a vector $\boldsymbol{v}_{\text{res}}$, the **resultant vector,** originating from the tail of $\boldsymbol{v}_1$ to the head of $\boldsymbol{v}_2$, as shown in Fig. A2.6c.

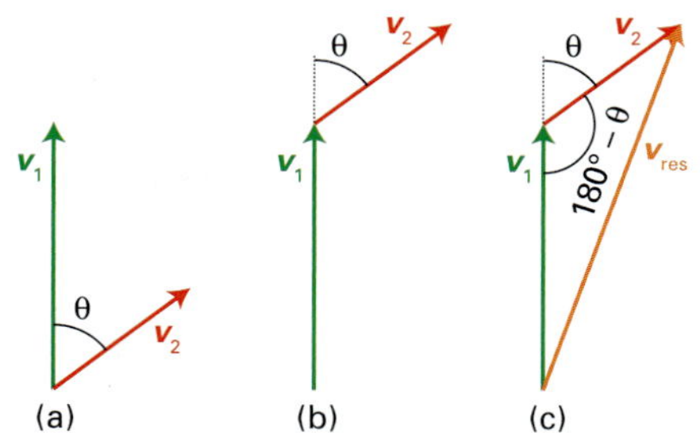

Fig. A2.6 (a) The vectors $\boldsymbol{v}_1$ and $\boldsymbol{v}_2$ make an angle θ. (b) To add $\boldsymbol{v}_2$ to $\boldsymbol{v}_1$, we first join the tail of $\boldsymbol{v}_2$ to the head of $\boldsymbol{v}_1$, making sure that the angle θ between the vectors remains unchanged. (c) To finish the process, we draw the resultant vector $\boldsymbol{v}_{\text{res}}$ by joining the tail of $\boldsymbol{v}_2$ to the head of $\boldsymbol{v}_1$.

Self-test A2.1

Using the same vectors shown in Fig. A2.6a, show that reversing the order of addition leads to the same result. That is, we obtain the same $\boldsymbol{v}_{\text{res}}$ whether we add $\boldsymbol{v}_2$ to $\boldsymbol{v}_1$ or $\boldsymbol{v}_1$ to $\boldsymbol{v}_2$.

[*Answer:* See Fig. A2.6c for the result of adding $\boldsymbol{v}_2$ to $\boldsymbol{v}_1$ and Fig. A2.7 for the result of adding $\boldsymbol{v}_1$ to $\boldsymbol{v}_2$.]

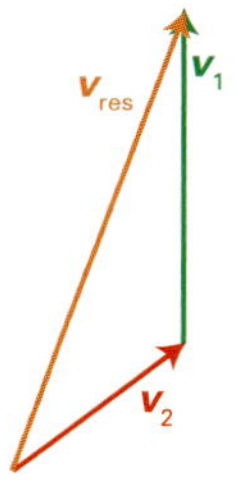

Fig. A2.7 The result of adding the vector $\boldsymbol{v}_1$ to the vector $\boldsymbol{v}_2$, with both vectors defined in Fig. A2.7a. Comparison with the result shown in Fig. A2.7c for the addition of $\boldsymbol{v}_2$ to $\boldsymbol{v}_1$ shows that reversing the order of vector addition does not affect the result.

To calculate the magnitude of $\boldsymbol{v}_{\text{res}}$, we note that $\boldsymbol{v}_1$, $\boldsymbol{v}_2$, and $\boldsymbol{v}_{\text{res}}$ form a triangle and that we know the magnitudes of two of its sides (v_1 and v_2) and of the angle between them ($180° - \theta$; see Fig. A2.6c). To calculate the magnitude of the third side, v_{res}, we make use of the *law of cosines*, which states that:

For a triangle with sides a, b, and c, and angle C facing side c:

$$c^2 = a^2 + b^2 - 2ab \cos C$$

This law is summarized graphically in Fig. A2.8 and its application to the case shown in Fig. A2.6c leads to the expression

$$v_{\text{res}}^2 = v_1^2 + v_2^2 - 2v_1v_2 \cos(180° - \theta)$$

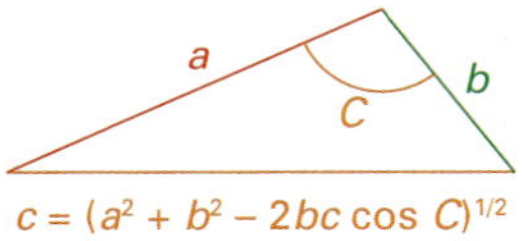

Fig. A2.8 The graphical representation of the law of cosines.

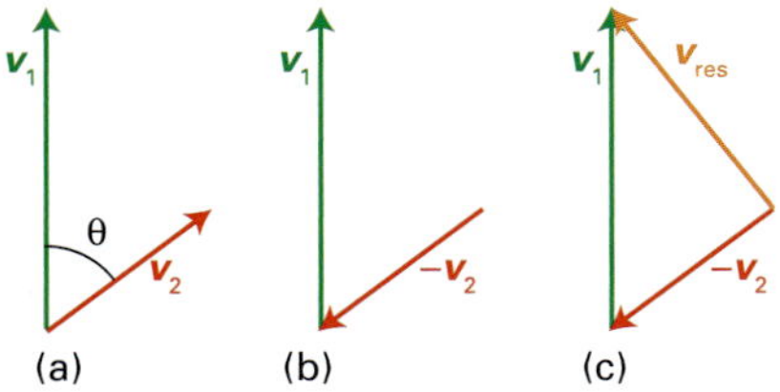

Fig. A2.9 The graphical method for subtraction of (a) the vector $\boldsymbol{v}_2$ from the vector $\boldsymbol{v}_1$ (as shown in Fig. A2.6a) consists of two steps: (b) reversing the direction of $\boldsymbol{v}_2$ to form $-\boldsymbol{v}_2$, and (c) adding $-\boldsymbol{v}_2$ to $\boldsymbol{v}_1$.

Because cos $(180° - \theta) = -\cos\theta$, it follows after taking the square root of both sides of the preceding expression that

$$v_{res} = (v_1^2 + v_2^2 + 2v_1v_2\cos\theta)^{1/2} \quad \text{(A2.6)}$$

which is the result used in Section 15.2 for the addition of two dipole moment vectors.

The subtraction of vectors follows the same principles outlined above for addition. Consider again the vectors shown in Fig. A2.6a. We note that subtraction of $\boldsymbol{v}_2$ from $\boldsymbol{v}_1$ amounts to addition of $-\boldsymbol{v}_2$ to $\boldsymbol{v}_1$. It follows that in the first step of subtraction we draw $-\boldsymbol{v}_2$ by reversing the direction of $\boldsymbol{v}_2$ (Fig. A2.9a). Then, the second step consists of adding the $-\boldsymbol{v}_2$ to $\boldsymbol{v}_1$ by using the strategy shown in Fig. A2.6c: we draw a resultant vector $\boldsymbol{v}_{res}$ by joining the tail of $-\boldsymbol{v}_2$ to the head of $\boldsymbol{v}_1$.

One procedure for multiplying vectors—and the only one we shall discuss here—consists of calculating the **scalar product** (or **dot product**) of two vectors $\boldsymbol{v}_1$ and $\boldsymbol{v}_2$ making an angle θ:

$$\boldsymbol{v}_1 \cdot \boldsymbol{v}_2 = v_1v_2\cos\theta \quad \text{(A2.7)}$$

As its name suggests, the scalar product of two vectors is a scalar (a number) and not a vector. Another procedure, which we do not use in this text, involves calculation of the cross-product of two vectors. Vector division is not defined.

Calculus

Now we turn to techniques of calculus, a branch of mathematics that is used to model a host of physical, chemical, and biological phenomena.

A2.4 Differentiation

Rates of change of functions—slopes—are best discussed in terms of the infinitesimal calculus. The slope of a function, like the slope of a hill, is obtained by dividing the rise of the hill by the horizontal distance (Fig. A2.10). However,

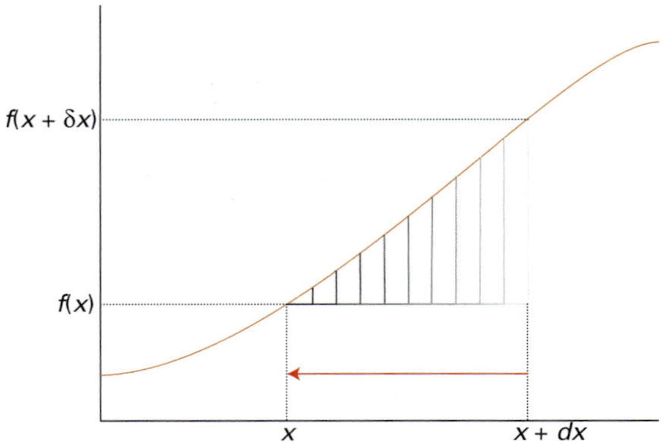

Fig. A2.10 The slope of $f(x)$ at x, df/dx, is obtained by making a series of approximations to the value of $f(x + \delta x) - f(x)$ divided by change in x, denoted δx, and allowing δx to approach 0 (as denoted by the vertical lines getting closer to x).

because the slope may vary from point to point, we should take the horizontal distance between the points as small as possible. In fact, we let it become infinitesimally small—hence the name ***infinitesimal* calculus**. The values of a function f at two locations x and $x + \delta x$ are $f(x)$ and $f(x + \delta x)$, respectively. Therefore, the slope of the function f at x is the vertical distance, which we write δf divided by the horizontal distance, which we write δx:

$$\text{Slope} = \frac{\text{rise in value}}{\text{horizontal distance}} = \frac{\delta f}{\delta x} = \frac{f(x+\delta x) - f(x)}{\delta x}$$

The slope exactly *at x* itself is obtained by letting the horizontal distance become zero, which we write lim $\delta x \to 0$. In this limit, the δ is replaced by a d, and we write

$$\text{Slope at } x = \frac{df}{dx} = \lim_{\delta x \to 0} \frac{f(x+\delta x) - f(x)}{\delta x}$$

To work out the slope of any function, we work out the expression on the right: this process is called **differentiation**. It leads to the following important expressions:

$$\frac{dx^n}{dx} = nx^{n-1} \qquad \frac{de^{ax}}{dx} = ae^{ax} \qquad \frac{d\ln ax}{dx} = \frac{1}{x}$$

$$\frac{d\sin ax}{dx} = a\cos ax \qquad \frac{d\cos ax}{dx} = -a\sin ax$$

Most of the functions encountered in chemistry can be differentiated by using these relations in conjunction with the following rules:

Rule 1. For two functions f and g:

$$d(f+g) = df + dg \quad \text{(A2.8)}$$

Rule 2 (the product rule). For two functions f and g:

$$d(fg) = f\,dg + g\,df \quad \text{(A2.9)}$$

Rule 3 (the quotient rule). For two functions f and g:

$$d\frac{f}{g} = \frac{1}{g}df - \frac{f}{g^2}dg \quad \text{(A2.10)}$$

Rule no. 4 (the chain rule). For a function $f = f(g)$, where $g = g(t)$,

$$\frac{df}{dt} = \frac{df}{dg}\frac{dg}{dt} \quad \text{(A2.11)}$$

In the last rule, $f(g)$ is a 'function of a function', as in $\ln(1 + x^2)$ or $\ln(\sin x)$.

The second derivative of a function, denoted d^2f/dx^2, is calculated by taking the first derivative, df/dx, and then taking the derivative of df/dx. For example, to calculate the second derivative of the function sin ax (where a is a constant), we write

$$\frac{d^2 \sin ax}{dx^2} = \frac{d}{dx}\left(\frac{d \sin ax}{dx}\right) = \frac{d}{dx}(a \cos ax) = -a^2 \sin ax$$

A very useful mathematical procedure involving differentiation consists of finding the value of x corresponding to the extremum (maximum or minimum) of any function $f(x)$. At an extremum the slope of the graph of the function is exactly zero (Fig. A2.11), so to find the value of x at which a maximum or minimum occurs we differentiate the function, set the result equal to zero, and solve the equation for x. To decide whether the function has a maximum or a minimum at this point, we note that the second derivative is an indication of the curvature of a function. Where d^2f/dx^2 is positive, the graph of the function has a ∪ shape; where it is negative, the graph has a ∩ shape.

● **A brief illustration** Consider the function $4x^2 + 3x - 6$. The first derivative is zero when

$$\frac{d}{dx}(4x^2 + 3x - 6) = 8x + 3 = 0 \quad \text{or} \quad x = -\frac{3}{8}$$

Then

$$\frac{d^2}{dx^2}(4x^2 + 3x - 6) = \frac{d}{dx}(8x + 3) = 8 > 0$$

It follows that the function $f(x) = 4x^2 + 3x - 6$ has a minimum at $x = -\frac{3}{8}$. ●

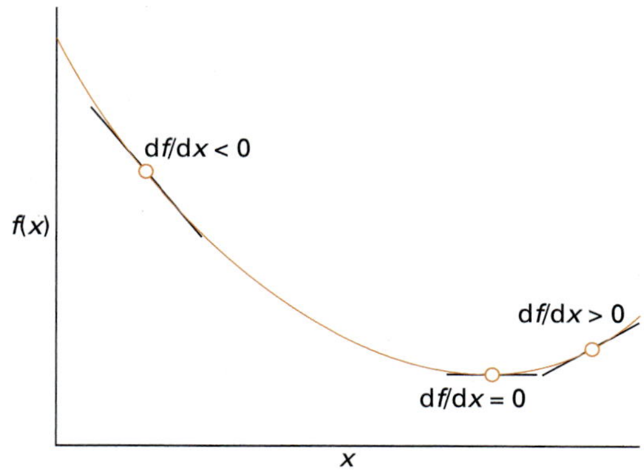

Fig. A2.11 At an extremum, the first derivative of a function is zero. The figure shows the case of a minimum.

A2.5 Power series and Taylor expansions

A **power series** has the form

$$c_0 + c_1(x - a) + c_2(x - a)^2 + \cdots + c_n(x - a)^n + \cdots \quad \text{(A2.12)}$$

where c_n and a are constants. It is often useful to express a function $f(x)$ in the vicinity of $x = a$ as a special power series called the **Taylor series**, or **Taylor expansion**, which has the form:

$$f(x) = f(a) + \left(\frac{df}{dx}\right)_a (x - a) + \frac{1}{2!}\left(\frac{d^2f}{dx^2}\right)_a (x - a)^2 + \cdots + \frac{1}{n!}\left(\frac{d^nf}{dx^n}\right)_a (x - a)^n + \cdots \quad \text{(A2.13)}$$

where n! denotes a **factorial** given by

$$n! = n(n - 1)(n - 2) \ldots 1$$

(By definition, $0! = 1$.) The following Taylor expansions are often useful:

$$\frac{1}{1 + x} = 1 - x + x^2 \ldots$$

$$e^x = 1 + x + \tfrac{1}{2}x^2 + \cdots$$

$$\ln x = (x - 1) - \tfrac{1}{2}(x - 1)^2 + \tfrac{1}{3}(x - 1)^3 - \tfrac{1}{4}(x - 1)^4 + \cdots.$$

$$\ln(1 + x) = x - \tfrac{1}{2}x^2 + \tfrac{1}{3}x^3 \cdots$$

If $x \ll 1$, then $(1 + x)^{-1} \approx 1 - x$, $e^x \approx 1 + x$, and $\ln(1 + x) \approx x$.

A2.6 Integration

The area under a graph of any function f is found by the techniques of **integration**. For instance, the area under the graph of the function f drawn in Fig. A2.12 can be written as the value of f evaluated at a point multiplied by the width of the region, δx, and then all those products $f(x)\delta x$ summed over all the regions:

$$\text{Area between } a \text{ and } b = \sum f(x)\delta x$$

When we allow δx to become infinitesimally small, written dx, and sum an infinite number of strips, we write

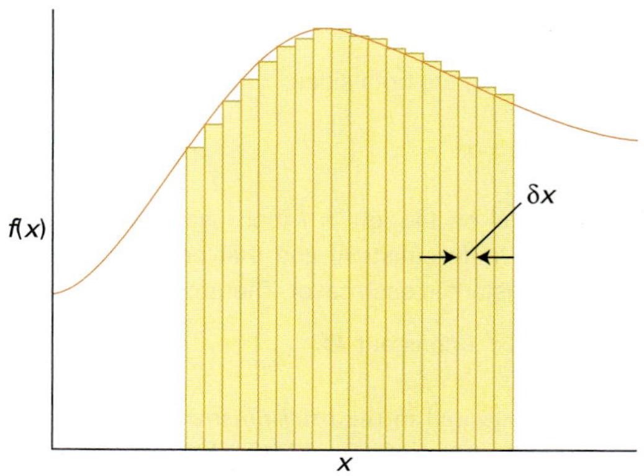

Fig A2.12 The shaded area is equal to the definite integral of $f(x)$ between the limits a and b.

Area between a and $b = \int_a^b f(x)\,dx$

The elongated S symbol on the right is called the **integral** of the function f. When written as $\int$ alone, it is the **indefinite integral** of the function. When written with limits (as in the expression above), it is the **definite integral** of the function. The definite integral is the indefinite integral evaluated at the upper limit (b) minus the indefinite integral evaluated at the lower limit (a).

Some important integrals are

$$\int x^n\,dx = \frac{x^{n+1}}{n+1} \qquad \int e^{ax}\,dx = \frac{e^{ax}}{a} \qquad \int \ln ax\,dx = x\ln ax - x$$

$$\int \sin ax\,dx = -\frac{\cos ax}{a} \qquad \int \cos ax\,dx = \frac{\sin ax}{a}$$

A brief comment Strictly, an indefinite integral should be written with an arbitrary constant on the right, so $\int x\,dx = \frac{1}{2}x^2$ + constant. However, tables of integrals commonly omit the constant. It cancels when the definite integral is evaluated.

It may be verified from these examples—and this is a very deep result of infinitesimal calculus—that *integration is the inverse of differentiation*. That is, if we integrate a function and then differentiate the result, we get back the original function.

A2.6 Differential equations

An **ordinary differential equation** is a relation between derivatives of a function f of one variable x and various functions of x including, perhaps, f itself. For example, if the slope of a function increases in proportion to x, we write

$$\frac{df}{dx} = ax$$

where a is a constant. To solve a differential equation, we have to look for the function f that satisfies it: the process is called **integrating** the equation.

● **A brief illustration** In this case we would multiply each side by dx, to obtain

$$df = ax\,dx$$

and then integrate both sides:

$$\int df = \int ax\,dx$$

The integral on the left is f (because integration is the inverse of differentiation) and that on the right is $\frac{1}{2}ax^2$ (plus a constant in each case). Therefore:

$$f(x) = \tfrac{1}{2}ax^2 + \text{constant} \ ●$$

The result quoted in the *brief illustration* is the **general solution** of the equation (Fig. A2.13). To fix the value of the constant and to find the **particular solution**, we take note of the **boundary conditions** that the function must satisfy, the value that we know the function has at a particular point.

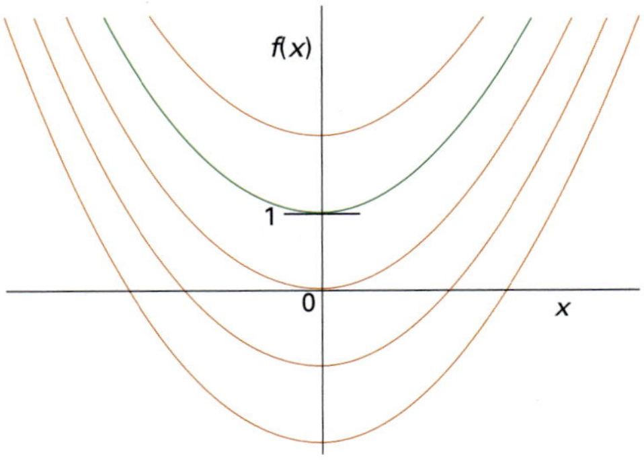

Fig. A2.13 The *general* solution of the differential equation $df/dx = ax$ is any one of the parabolas shown here (and others like them); the *particular* solution, which is identified by the boundary condition that f must satisfy, is shown by the green line.

● **A brief illustration** Thus, if we know that $f(0) = 1$, then we can write

$$1 = \tfrac{1}{2}a + \text{constant}, \qquad \text{so constant} = 1 - \tfrac{1}{2}a$$

The particular solution that satisfies the boundary condition is therefore

$$f(x) = \tfrac{1}{2}ax^2 + 1 - \tfrac{1}{2}a \ ●$$

In chemical kinetics, for instance, we may know that the reaction rate is proportional to the concentration of a reactant, and look for a general solution of the rate equation (a differential equation) which tells us how the concentration varies with time as the reaction proceeds. The particular solution is then obtained by making sure that the concentration has the correct value initially. A boundary condition is called an *initial condition* if the variable is time, as in a rate law.

A differential equation that is expressed in terms of first derivatives is a **first-order differential equation**. Rate laws are first-order differential equations.

A note on good practice Do not confuse this use of the term 'order' with the order of the rate law: even a second-order rate law is a first-order differential equation!

A differential equation that is expressed in terms of second derivatives is a **second-order differential equation**. The Schrödinger equation is a second-order differential equation. The solution of differential equations is a very powerful technique in the physical sciences, but is often very difficult. All the second-order differential equations that occur in this text can be found tabulated in compilations of solutions or can be solved with mathematical software, and the specialized techniques that are needed to establish the form of the solutions may be found in mathematical texts.

Appendix 3 Concepts of physics

Throughout the text we use ideas of classical physics as the basis for discussion of energy exchanges during chemical reactions, atomic and molecular structure, molecular interactions, and spectroscopic techniques. Here we review of concepts of classical mechanics, electromagnetism, and electrostatics.

Classical mechanics

Classical mechanics describes the behaviour of particles in terms of two equations. One expresses the fact that the total energy is constant in the absence of external forces and the other expresses the response of particles to the forces acting on them.

A3.1 Energy

Kinetic energy, E_k, is the energy that a body (a block of matter, an atom, or an electron) possesses by virtue of its motion. The formula for calculating the kinetic energy of a body of mass m that is travelling at a speed v is

$$E_k = \tfrac{1}{2}mv^2 \qquad \text{(A3.1)}$$

This expression shows that a body may have a high kinetic energy if it is heavy (m large) and is travelling rapidly (v large). A stationary body ($v = 0$) has zero kinetic energy, whatever its mass. The energy of a sample of perfect gas is entirely due to the kinetic energy of its molecules: they travel more rapidly (on average) at high temperatures than at low, so raising the temperature of a gas increases the kinetic energy of its molecules.

Potential energy, E_P or V, is the energy that a body has by virtue of its position. A body on the surface of the Earth has a potential energy on account of the gravitational force it experiences: if the body is raised, then its potential energy is increased. There is no general formula for calculating the potential energy of a body because there are several kinds of force. For a body of mass m at a height h above (but close to) the surface of the Earth, the gravitational potential energy is

$$E_p = mgh$$

where g is the acceleration of free fall ($g = 9.81$ m s^{-2}). A heavy object at a certain height has a greater potential energy than a light object at the same height. One very important contribution to the potential energy is encountered when a charged particle is brought up to another charge. In this case the potential energy is inversely proportional to the distance between the charges (see Section A3.3):

$$E_p \propto \frac{1}{r} \quad \text{specifically, } E_p = \frac{Q_1Q_2}{4\pi\varepsilon_0 r}$$

This **Coulomb potential energy** decreases with distance, and two infinitely widely separated charged particles have zero potential energy of interaction. The Coulomb potential energy plays a central role in the structures of atoms, molecules, and solids.

The **total energy**, E, of a body is the sum of its kinetic and potential energies. It is a central feature of physics that *the total energy of a body that is free from external influences is constant*. Thus, a stationary ball at a height h above the surface of the Earth has a potential energy of magnitude mgh; if it is released and begins to fall to the ground, it loses potential energy (as it loses height), but gains the same amount of kinetic energy (and therefore accelerates). Just before it hits the surface, it has lost all its potential energy, and all its energy is kinetic.

The SI unit of energy is the *joule* (J), which is defined as

$$1\text{ J} = 1\text{ kg m}^2\text{ s}^{-2} \qquad \text{(A3.2)}$$

Calories (cal) and kilocalories (kcal) are still encountered in the chemical literature: by definition, 1 cal = 4.184 J. An energy of 1 cal is enough to raise the temperature of 1 g of water by 1°C.

The rate of change of energy is called the **power**, P, expressed as joules per second, or *watt*, W:

$$1\text{ W} = 1\text{ J s}^{-1} \qquad \text{(A3.3)}$$

A3.2 Force

Classical mechanics describes the motion of a particle in terms of its **velocity**, $\boldsymbol{v}$, the rate of change of its position:

$$\boldsymbol{v} = \frac{d\boldsymbol{r}}{dt} \qquad \text{(A3.4)}$$

The velocity is a vector, with both direction and magnitude (see Appendix 2). The magnitude of the velocity is the **speed**, v. The **linear momentum**, $\boldsymbol{p}$, of a particle of mass m is related to its velocity, $\boldsymbol{v}$, by

$$\boldsymbol{p} = m\boldsymbol{v} \qquad \text{(A3.5)}$$

Like the velocity vector, the linear momentum vector points in the direction of travel of the particle (Fig. A3.1). In terms of the linear momentum, the kinetic energy of a particle is

$$E_k = \frac{p^2}{2m} \qquad \text{(A3.6)}$$

The state of motion of a particle is changed by a **force**, $\boldsymbol{F}$. According to Newton's second law of motion, a force changes the momentum of a particle such that the acceleration, $\boldsymbol{a}$, of the particle (its rate of change of velocity, or $d\boldsymbol{v}/dt$) is proportional to the strength of the force:

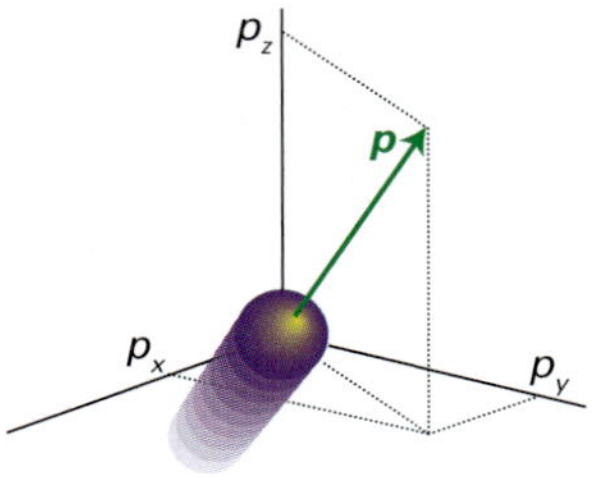

Fig. A3.1 The linear momentum of a particle is a vector property and points in the direction of motion.

$$\text{force} = \text{mass} \times \text{acceleration, or } F = ma = m\frac{\mathrm{d}v}{\mathrm{d}t} \quad \text{(A3.7)}$$

We note that the force and acceleration, like the velocity and momentum, are vectors. The SI unit for expressing the magnitude of a force is the *newton* (N), which is defined as

$$1 \text{ N} = 1 \text{ kg m s}^{-2} \quad \text{(A3.8)}$$

Equation A3.7 shows that a stronger force is required to accelerate a heavy particle by a given amount than to accelerate a light particle by the same amount. A force can be used to change the kinetic energy of a body, by accelerating the body to a higher speed. It may also be used to change the potential energy of a body by moving it to another position (for example, by raising it near the surface of the Earth). The force experienced by a particle free to move in one dimension is related to its potential energy, V, by

$$F = -\frac{\mathrm{d}V}{\mathrm{d}x} \quad \text{(A3.9)}$$

This relation implies that the direction of the force is towards decreasing potential energy (Fig. A3.2).

The **work**, w, done on an object is the product of the distance, s, moved and the force opposing the motion:

$$w = -Fs \quad \text{(A3.10a)}$$

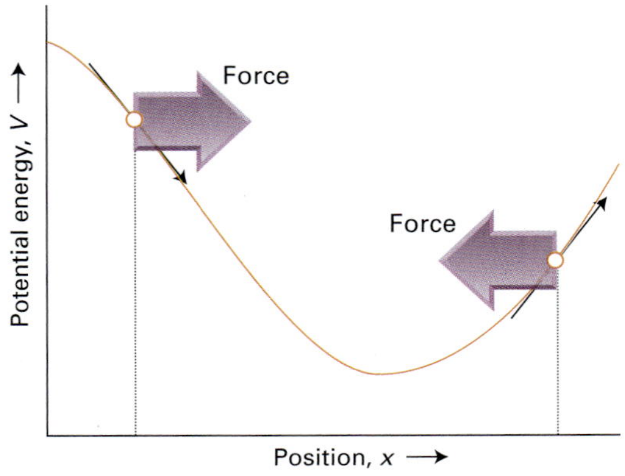

Fig. A3.2 The force acting on a particle is determined by the slope of the potential energy at each point. The force points in the direction of lower potential energy.

It requires a lot of work to move a long distance against a strong opposing force (think of cycling into a strong wind). If the opposing force changes at different points on the path, then we consider the force as a function of position, $F(s)$, and write

$$w = -\int F(s)\,\mathrm{d}s \quad \text{(A3.10b)}$$

The integral is evaluated along the path traversed by the particle.

Electrostatics

Electrostatics is the study of the interactions of stationary electric charges. The elementary charge, the magnitude of charge carried by a single electron or proton, is $e \approx 1.60 \times 10^{-19}$ C. The magnitude of the charge per mole is the Faraday constant: $F = N_A e = 9.65 \times 10^4 \text{ C mol}^{-1}$.

A3.3 The Coulomb interaction

The fundamental expression in electrostatics is the Coulomb potential energy of one charge of magnitude Q at a distance r from another charge Q':

$$V = \frac{QQ'}{4\pi\varepsilon_0 r} \quad \text{(A3.11)}$$

That is, the potential energy is inversely proportional to the separation of the charges. The fundamental constant ε_0 is the **vacuum permittivity**; its value is $\varepsilon_0 = 8.854 \times 10^{-12} \text{ J}^{-1} \text{ C}^2 \text{ m}^{-1}$. With r in metres and the charges in coulombs, the potential energy is in joules. The potential energy is equal to the work that must be done to bring up a charge Q from infinity to a distance r from a charge Q'. The implication is then that the magnitude of the *force* exerted by a charge Q on a charge Q' is inversely proportional to the *square* of their separation:

$$F = \frac{QQ'}{4\pi\varepsilon_0 r^2} \quad \text{(A3.12)}$$

This expression is **Coulomb's inverse-square law of force.**

A3.4 The Coulomb potential

We can express the potential energy of a charge Q in the presence of another charge Q' in terms of the **Coulomb potential,** ϕ, due to Q':

$$V = Q\phi, \qquad \phi = \frac{Q'}{4\pi\varepsilon_0 r} \quad \text{(A3.13)}$$

(Note the distinction between the potential ϕ and the potential energy V.) The units of potential are joules per coulomb (J C^{-1}), so when ϕ is multiplied by a charge in coulombs, the result is in joules. The combination joules per coulomb occurs widely in electrostatics, and is called a *volt*, V:

$$1 \text{ V} = 1 \text{ J C}^{-1}$$

(which implies that 1 V C = 1 J). If there are several charges $Q_1, Q_2, \ldots$ present in the system, then the total potential experienced by the charge Q is the sum of the potential generated by each charge:

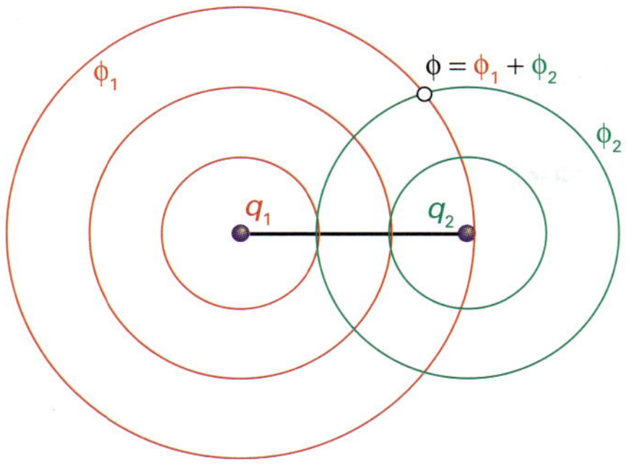

Fig. A3.3 The electric potential at a point is equal to the sum of the potentials due to each charge.

$$\phi = \phi_1 + \phi_2 + \cdots$$

For example, the potential generated by a dipole is the sum of the potentials of the two equal and opposite charges: these potentials do not in general cancel because the point of interest is at different distances from the two charges (Fig. A3.3).

A3.5 Current, resistance, and Ohm's law

The motion of charge gives rise to an **electric current**, I. Electric current is measured in amperes, A, where

$$1\ \text{A} = 1\ \text{C s}^{-1}$$

If the electric charge is that of electrons (as it is through metals and semiconductors), then a current of 1 A represents the flow of 6×10^{18} electrons per second. If the current flows from a region of potential ϕ_i to ϕ_f, through a potential difference $\Delta\phi = \phi_f - \phi_i$, then the rate of doing work is the current (the rate of transfer of charge) multiplied by the potential difference, $I \times \Delta\phi$. The rate of doing work is called **power**, P, so

$$P = I\Delta\phi \qquad \text{(A3.14)}$$

With current in amperes and the potential difference in volts, the power works out in joules per second, or watts, W (Section A3.1).

The total energy supplied in a time t is the power (the energy per second) multiplied by the time:

$$E = Pt = It\Delta\phi$$

The energy is obtained in joules with the current in amperes, the potential difference in volts, and the time in seconds.

The current flowing through a conductor is proportional to the potential difference between the ends of the conductor and inversely proportional to the **resistance**, R, of the conductor:

$$I = \frac{\Delta\phi}{R} \qquad \text{(A3.15)}$$

This empirical relation is called **Ohm's law.** With the current in amperes and the potential difference in volts, the resistance is measured in *ohms*, Ω, with $1\ \Omega = 1\ \text{V A}^{-1}$.

Electromagnetic radiation

Waves are disturbances that travel through space with a finite velocity. Examples of disturbances include the collective motion of water molecules in ocean waves and of gas particles in sound waves. Waves can be characterized by a **wave equation,** a differential equation that describes the motion of the wave in space and time. **Harmonic waves** are waves with displacements that can be expressed as sine or cosine functions. These concepts are used in classical physics to describe the wave character of electromagnetic radiation, which is the focus of the following discussion.

A3.6 The electromagnetic field

In classical physics, electromagnetic radiation is understood in terms of the **electromagnetic field,** an oscillating electric and magnetic disturbance that spreads as a harmonic wave through empty space, the vacuum. The wave travels at a constant speed called the *speed of light*, c, which is about $3 \times 10^8\ \text{m s}^{-1}$. As its name suggests, an electromagnetic field has two components, an **electric field** that acts on charged particles (whether stationary or moving) and a **magnetic field** that acts only on moving charged particles. The electromagnetic field is characterized by a **wavelength,** λ (lambda), the distance between the neighbouring peaks of the wave, and its **frequency,** ν (nu), the number of times per second at which its displacement at a fixed point returns to its original value (Fig. A3.4). The frequency is measured in *hertz*, where $1\ \text{Hz} = 1\ \text{s}^{-1}$. The wavelength and frequency of an electromagnetic wave are related by

$$\lambda\nu = c \qquad \text{(A3.16)}$$

Therefore, the shorter the wavelength, the higher the frequency. The characteristics of a wave are also reported by giving the **wavenumber,** $\tilde{\nu}$ (nu tilde), of the radiation, where

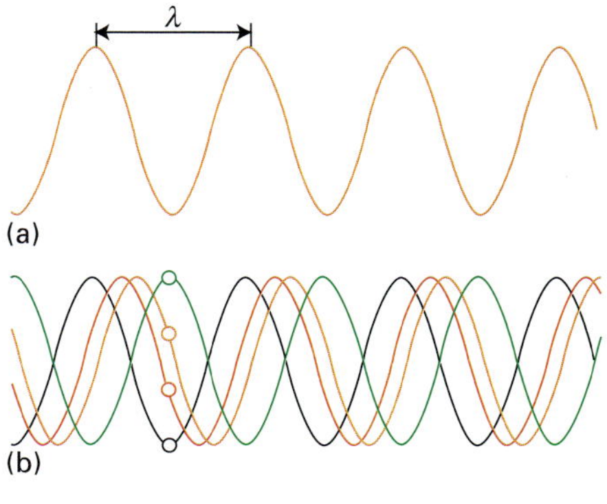

Fig. A3.4 (a) The wavelength, λ (lambda), of a wave is the peak-to-peak distance. (b) The wave is shown travelling to the right at a speed c. At a given location, the instantaneous amplitude of the wave changes through a complete cycle (the four dots show half a cycle) as it passes a given point. The frequency, ν (nu) is the number of cycles per second that occur at a given point. Wavelength and frequency are related by $\lambda\nu = c$.

Table A3.1
*The regions of the electromagnetic spectrum**

Region	Wavelength	Frequency/Hz
Radiofrequency	>30 cm	$<10^9$
Microwave	3 mm to 30 cm	10^9 to 10^{11}
Infrared	1000 nm to 3 mm	10^{11} to 3×10^{14}
Visible	400 nm to 800 nm	4×10^{14} to 8×10^{14}
Ultraviolet	3 nm to 300 nm	10^{15} to 10^{17}
X-rays, γ-rays	<3 nm	$>10^{17}$

* The boundaries of the regions are only approximate.

$$\tilde{\nu} = \frac{\nu}{c} = \frac{1}{\lambda} \qquad \text{(A3.17)}$$

A wavenumber can be interpreted as the number of complete wavelengths in a given length. Wavenumbers are normally reported in reciprocal centimetres (cm^{-1}), so a wavenumber of 5 cm^{-1} indicates that there are 5 complete wavelengths in 1 cm. The classification of the electromagnetic field according to its frequency and wavelength is summarized in Table A3.1.

A3.7 Features of electromagnetic radiation

Consider an electromagnetic disturbance travelling along the x direction with wavelength λ and frequency ν. The functions that describe the oscillating electric field, $\mathcal{E}(x,t)$, and magnetic field, $\mathcal{B}(x,t)$, may be written as

$$\mathcal{E}(x,t) = \mathcal{E}_0\cos\{2\pi\nu t - (2\pi/\lambda)x + \phi\} \qquad \text{(A3.18a)}$$

$$\mathcal{B}(x,t) = \mathcal{B}_0\cos\{2\pi\nu t - (2\pi/\lambda)x + \phi\} \qquad \text{(A3.18b)}$$

where $\mathcal{E}_0$ and $\mathcal{B}_0$ are the amplitudes of the electric and magnetic fields, respectively, and the parameter ϕ is the **phase** of the wave, which varies from $-\pi$ to π and gives the relative location of the peaks of two waves. If two waves, in the same region of space, with the same wavelength are shifted by $\phi = \pi$ or $-\pi$ (so the peaks of one wave coincide with the troughs of the other), then the resultant wave will have diminished amplitudes. The waves are said to interfere destructively. A value of $\phi = 0$ (coincident peaks) corresponds to constructive interference, or the enhancement of the amplitudes. According to classical electromagnetic theory, the intensity of electromagnetic radiation is proportional to the square of the amplitude of the wave. For example, the light detectors used in spectroscopy are commonly based on the interaction between the electric field of the incident radiation and the detecting element, so light intensities are proportional to $\mathcal{E}_0^2$.

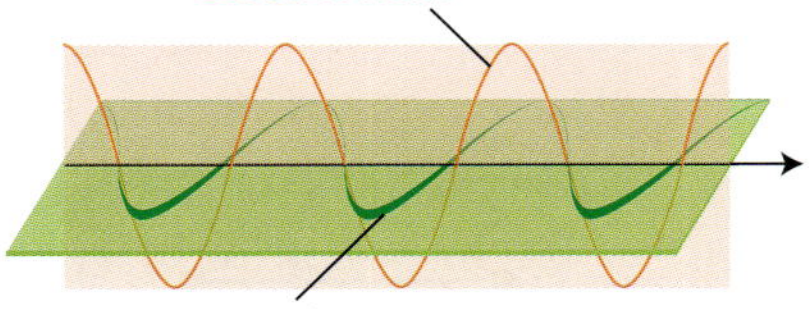

Fig. A3.5 Electromagnetic radiation consists of a wave of electric and magnetic fields perpendicular to the direction of propagation (in this case the x direction), and mutually perpendicular to each other. This illustration shows a plane polarized wave, with the electric and magnetic fields oscillating in the *xy*- and *xz*-planes, respectively.

Equations A3.18a and A3.18b represent electromagnetic radiation that is **plane polarized**; it is so called because the electric and magnetic fields each oscillate in a single plane (in this case the *xy*-plane, Fig. A3.5). The plane of polarization may be orientated in any direction around the direction of propagation (the x direction in Fig. A3.5), with the electric and magnetic fields perpendicular to that direction (and perpendicular to each other). An alternative mode of polarization is **circular polarization**, in which the electric and magnetic fields rotate around the direction of propagation in either a clockwise or a counterclockwise sense but remain perpendicular to it and each other.

According to quantum theory, a ray of frequency ν consists of a stream of **photons**, each one of which has energy

$$E = h\nu \qquad \text{(A3.19)}$$

where h is Planck's constant (Section 12.1). Thus, a photon of high-frequency radiation has more energy than a photon of low-frequency radiation. The greater the intensity of the ray, the greater the number of photons in it. In a vacuum, each photon travels with the speed of light. The frequency of the radiation determines the colour of visible light because different visual receptors in the eye respond to photons of different energy. The relation between colour and frequency is shown in Table A3.2, which also gives the energy carried by each type of photon.

Photons may also be polarized. A plane polarized ray of light consists of plane polarized photons and a circularly polarized ray consists of circularly polarized photons. The latter can be regarded as spinning either clockwise (for left-circularly polarized radiation) or counterclockwise (for right-circularly polarized radiation) about their direction of propagation.

Table A3.2
*Colour, frequency, and wavelength of light**

	Frequency (10^{14} Hz)	Wavelength/nm	Energy of photon/10^{-19} aJ
X-rays and γ-rays	10^3 and above	3 and below	660 and above
Ultraviolet	10	300	6.6
Visible light			
Violet	7.1	420	4.7
Blue	6.4	470	4.2
Green	5.7	530	3.7
Yellow	5.2	580	3.4
Orange	4.8	620	3.2
Red	4.3	700	2.8
Infrared	3.0	1000	1.9
Microwaves and radiowaves	3×10^{-3} Hz and below	3×10^6 and above	2.0×10^{-22} J and below

*The values given are approximate but typical.

Appendix 4 Review of chemical principles

The concepts reviewed below are used throughout the text. They are usually covered in introductory chemistry texts, which should be consulted for further information.

A4.1 Oxidation numbers

A simple way of judging whether a monatomic species has undergone oxidation or reduction is to note if the charge number of the species has changed. For example, an increase in the charge number of a monatomic ion (which corresponds to electron loss), as in the conversion of Fe^{2+} to Fe^{3+}, is an oxidation. A decrease in charge number (to a less positive or more negative value, as a result of electron gain), as in the conversion of Br to Br^-, is a reduction.

It is possible to assign to an atom in a polyatomic species an effective charge number, called the **oxidation number**, N_{ox}. (There is no standard symbol for this quantity.) The oxidation number is defined so that an increase in its value ($\Delta N_{ox} > 0$) corresponds to oxidation, and a decrease ($\Delta N_{ox} < 0$) corresponds to reduction.

An oxidation number is assigned to an element in a compound by supposing that it is present as an ion with a characteristic charge; for instance, oxygen is present as O^{2-} in most of its compounds, and fluorine is present as F^- (Fig. A4.1). The more electronegative element is supposed to be present as the anion. This procedure implies that:

1. The oxidation number of an elemental substance is zero, $N_{ox}(\text{element}) = 0$.
2. The oxidation number of a monatomic ion is equal to the charge number of that ion: $N_{ox}(E^{z\pm}) = \pm z$.
3. The sum of the oxidation numbers of all the atoms in a species is equal to the overall charge number of the species.

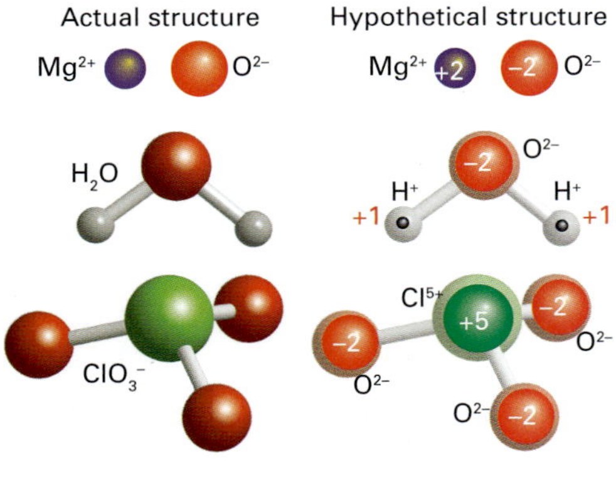

Fig. A4.1 To calculate the oxidation number of an element in an oxide or oxoacid, we suppose that each O atom is present as an O^{2-} ion, and then identify the charge of the element required to give the actual overall charge of the species. The more electronegative element plays a similar role in other compounds.

Thus, hydrogen, oxygen, iron, and all the elements have $N_{ox} = 0$ in their elemental forms; $N_{ox}(Fe^{3+}) = +3$ and $N_{ox}(Br^-) = -1$. It follows that the conversion of Fe to Fe^{3+} is an oxidation (because $\Delta N_{ox} > 0$) and the conversion of Br to Br^- is a reduction (because $\Delta N_{ox} < 0$). The definition of oxidation number and its relation to oxidation and reduction are consistent with the definitions in terms of electron loss and gain.

As an illustration, consider the oxidation numbers of the elements in SO_2 and SO_4^{2-}. The sum of oxidation numbers of the atoms in SO_2 must be 0, so we can write

$$N_{ox}(S) + 2N_{ox}(O) = 0$$

Each O atom has $N_{ox} = -2$. Hence,

$$N_{ox}(S) + 2 \times (-2) = 0$$

which solves to $N_{ox}(S) = +4$. Now consider SO_4^{2-}. The sum of oxidation numbers of the atoms in the ion is −2, so we can write

$$N_{ox}(S) + 4N_{ox}(O) = -2$$

Because $N_{ox}(O) = -2$,

$$N_{ox}(S) + 4 \times (-2) = -2$$

which solves to $N_{ox}(S) = +6$. The sulfur is more highly oxidized in the sulfate ion than in sulfur dioxide.

Self-test A4.1

Calculate the oxidation numbers of the elements in (a) H_2S, (b) PO_4^{3-}, (c) NO_3^-.

[*Answer:* (a) $N_{ox}(H) = +1$, $N_{ox}(S) = -2$; (b) $N_{ox}(P) = +5$, $N_{ox}(O) = -2$; (c) $N_{ox}(N) = +5$, $N_{ox}(O) = -2$]

A4.2 The Lewis theory of covalent bonding

In his original formulation of a theory of the covalent bond, G. N. Lewis proposed that each bond consisted of one electron pair. Each atom in a molecule shared electrons until it had acquired an octet characteristic of a noble gas atom near it in the periodic table. (Hydrogen is an exception: it acquires a duplet of electrons.) Thus, to write down a Lewis structure:

1. Arrange the atoms as they are found in the molecule.
2. Add one electron pair (represented by dots, :) between each bonded atom.
3. Use the remaining electron pairs to complete the octets of all the atoms present either by forming lone pairs or by forming multiple bonds.
4. Replace bonding electron pairs by bond lines (–) but leave lone pairs as dots (:).

A Lewis structure does not (except in very simple cases), portray the actual geometrical structure of the molecule; it is a topological map of the arrangement of bonds.

As an example, consider the Lewis structure of methanol, CH_3OH, in which there are $4 \times 1 + 4 + 6 = 14$ electrons (and hence seven electron pairs) to accommodate. The first step is to write the atoms in the arrangement (**1**); the pale rectangles have been included to indicate which atoms are linked. The next step is to add electron pairs to denote bonds (**2**). The C atom now has a complete octet and all four H atoms have complete duplets. There are two unused electron pairs, which are used as lone pairs to complete the octet of the O atom (**3**). Finally, replace the bonding pairs by lines to indicate bonds (**4**). An example of a species with a multiple bond is acetic acid (**5**).

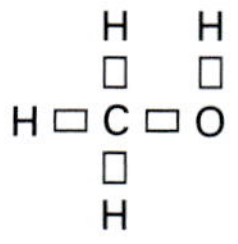

1

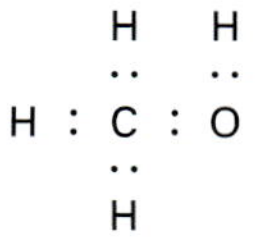

2

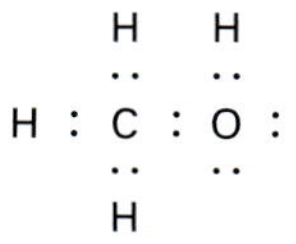

3

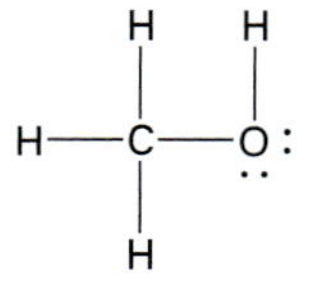

4

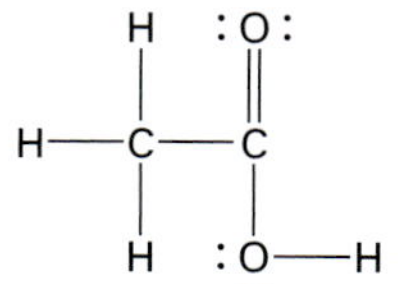

5

In some cases, more than one structure can be written in which the only difference is the location of multiple bonds or lone pairs. In such cases, the molecule's structure is interpreted as a **resonance hybrid**, a quantum-mechanical blend, of the individual structures. Resonance is depicted by a double-headed arrow. For example, the ozone molecule, O_3, is a resonance hybrid of two structures (**6**). Resonance distributes multiple-bond character over the participating atoms.

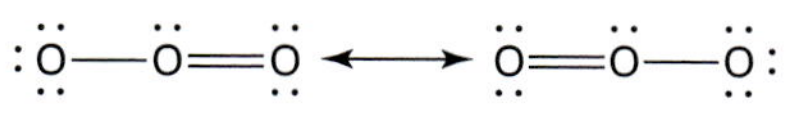

6

Many molecules cannot be written in a way that conforms to the octet rule. Those classified as **hypervalent molecules** require an expansion of the octet. Although it is often stated that octet expansion requires the involvement of d-orbitals, and is therefore confined to Period 3 and subsequent elements, there is good evidence to suggest that octet expansion is a consequence of an atom's size, not its intrinsic orbital structure. Whatever the reason, octet expansion is need to account for the structures of PCl_5 with expansion to ten electrons (**7**), SF_6, expansion to 12 electrons (**8**), and XeO_4, expansion to 16 electrons (**9**). Octet expansion is also encountered in species that do not necessarily require it, but that, if it is permitted, may acquire a lower energy. Thus, of the structures (**10a**) and

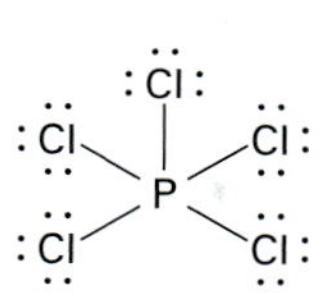

7

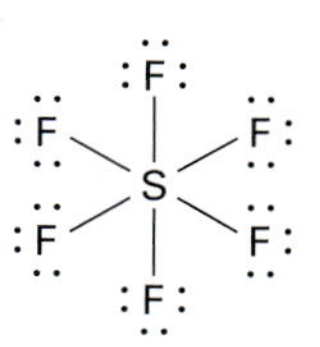

8

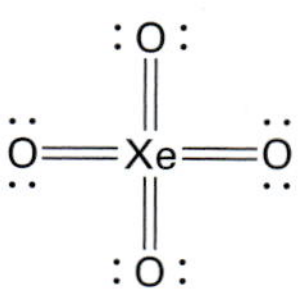

9

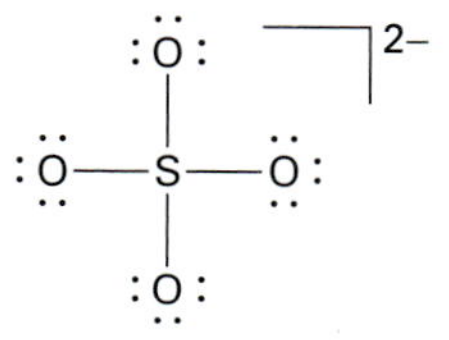

10a

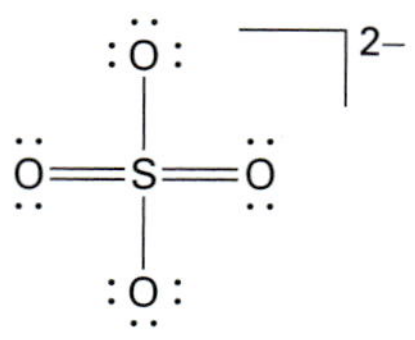

10b

(**10b**) of the SO_4^{2-} ion, the second has a lower energy than the first. The actual structure of the ion is a resonance hybrid of both structures (together with analogous structures with double bonds in different locations), but the latter structure makes the dominant contribution.

Octet completion is not always energetically appropriate. Such is the case with boron trifluoride, BF_3. Two of the possible Lewis structures for this molecule are (**11a**) and (**11b**). In the former, the B atom has an **incomplete octet**. Nevertheless, it has a lower energy than the other structure, as to form the latter structure one F atom has had partially to relinquish an electron pair, which is energetically demanding for such an electronegative element. The actual molecule is a resonance hybrid of the two structures (and of others with the double bond in different locations), but the overwhelming contribution is from the former structure. Consequently, we regard BF_3 as a molecule with an incomplete octet. This feature is responsible for its ability to act as a Lewis acid (an electron pair acceptor).

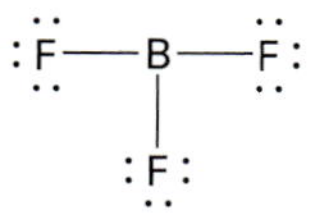

11a

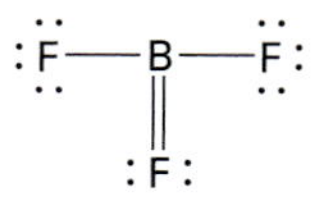

11b

The Lewis approach fails for a class of **electron-deficient compounds**, which are molecules that have too few electrons for a Lewis structure to be written. The most famous example is diborane, B_2H_6, which requires at least seven pairs of electrons to bind the eight atoms together, but it has only twelve valence electrons in all. The structures of such molecules can be explained in terms of molecular orbital theory and the concept of delocalized electron pairs, in which the influence of an electron pair is distributed over several atoms.

A4.3 The VSEPR model

In the **valence-shell electron pair repulsion model** (VSEPR) we focus on a single, central atom and consider the local arrangement of atoms that are linked to it. For example, in considering the H_2O molecule, we concentrate on the electron pairs in the valence shell of the central O atom. This procedure can be extended to molecules in which there is no obvious central atom, such as in benzene, C_6H_6, or hydrogen peroxide, H_2O_2, by focusing attention on a group of atoms, such as a C—CH—C fragment of benzene or an H—O—O fragment of hydrogen peroxide, and considering the arrangement of electron pairs around the central atom of the fragment.

The basic assumption of the VSEPR model is that *the valence-shell electron pairs of the central atom adopt positions that maximize their separations*. Thus, if the atom has four electron pairs in its valence shell, then the pairs adopt a tetrahedral arrangement around the atom; if the atom has five pairs, then the arrangement is trigonal bipyramidal. The arrangements adopted by electron pairs are summarized in Table A4.1.

Table A4.1
Arrangements of electron-dense regions

Number of regions	Arrangement
2	Linear
3	Trigonal planar
4	Tetrahedral
5	Trigonal bipyramidal
6	Octahedral
7	Pentagonal bipyramidal

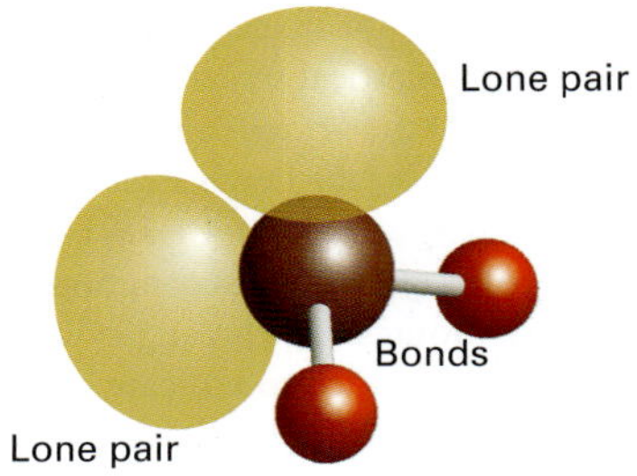

Fig. A4.2 The shape of a molecule is identified by noting the arrangement of its atoms, not its lone pairs. This molecule is angular even though the electron-pair distribution is tetrahedral.

Once the basic shape of the arrangement of electron pairs has been identified, the pairs are identified as bonding or nonbonding. For instance, in the H_2O molecule, two of the tetrahedrally arranged pairs are bonding pairs and two are nonbonding pairs. Then, the shape of the molecule is classified by noting the arrangement of the atoms around the central atom. The H_2O molecule, for instance, has an underlying tetrahedral arrangement of lone pairs, but as only two of the pairs are bonding pairs, the molecule is classified as angular (Fig. A4.2). It is important to keep in mind the distinction between the arrangement of electron pairs and the shape of the resulting molecule: the latter is identified by noting the relative locations of the atoms, not the lone pairs (Fig. A4.3).

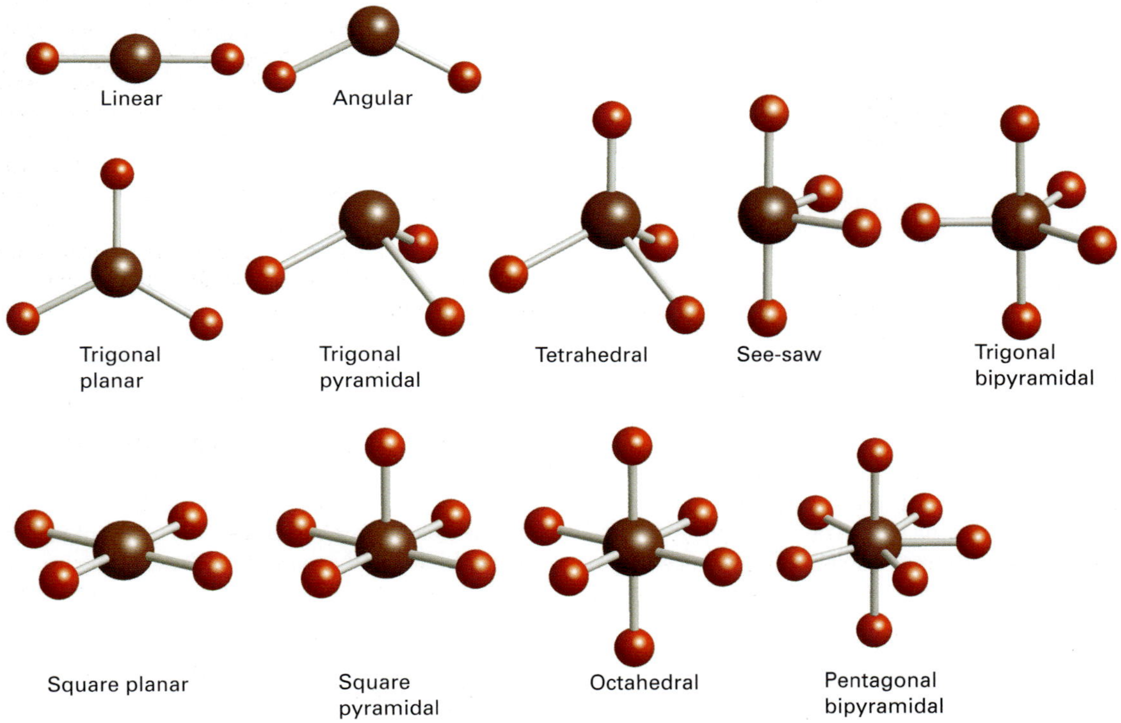

Fig. A4.3 The classification of molecular shapes according to the relative locations of atoms.

For example, to predict the shape of an ethane molecule we concentrate on one of the C atoms initially. That atom has four electron pairs in its valence shell (in the molecule), and they adopt a tetrahedral arrangement. All four electron pairs are bonding: three bond H atoms and the fourth bonds the second C atom. Therefore, the arrangement of atoms is tetrahedral around the C atom. The second C atom has the same environment, so we conclude that the ethane molecule consists of two tetrahedral $-CH_3$ groups (**12**).

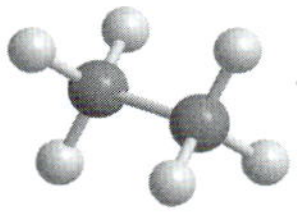

12 Ethane, CH_3CH_3

The next stage in the application of the VSEPR model is to accommodate the greater repelling effect of lone pairs compared with that of bonding pairs. That is, *bonding pairs tend to move away from lone pairs even though that might reduce their separation from other bonding pairs.* The NH_3 molecule provides a simple example. The N atom has four electron pairs in its valence shell and they adopt a tetrahedral arrangement. Three of the pairs are bonding pairs, and the fourth is a lone pair. The basic shape of the molecule is therefore trigonal pyramidal. However, a lower energy is achieved if the three bonding pairs move away from the lone pair, even though they are brought slightly closer together (**13**). We therefore predict an HNH bond angle of slightly less than the tetrahedral angle of 109.5°, which is consistent with the observed angle of 107°.

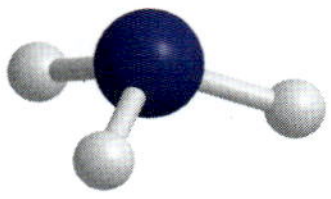

13 Ammonia, NH_3

As an example, consider the shape of an SF_4 molecule. The first step is to write a Lewis (electron dot) structure for the molecule to identify the number of lone pairs in the valence shell of the S atom (**14**). This structure shows that there are five electron pairs on the S atom. Reference to Table A4.1 shows that the five pairs are arranged as a trigonal bipyramid. Four of the pairs are bonding pairs and one is a lone pair. The repulsions stemming from the lone pair are minimized if the lone pair is placed in an equatorial position: then it is close to the axial pairs (**15**), whereas if it had adopted an axial position it would have been close to three equatorial pairs (**16**). Finally, the four bonding pairs

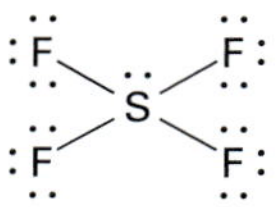

14

15 Sulfur tetrafluoride, SF_4

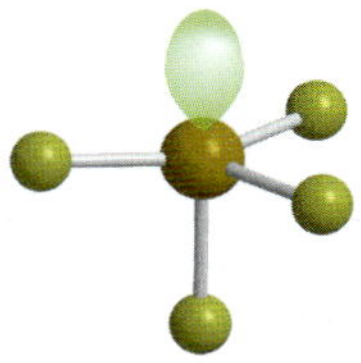

16

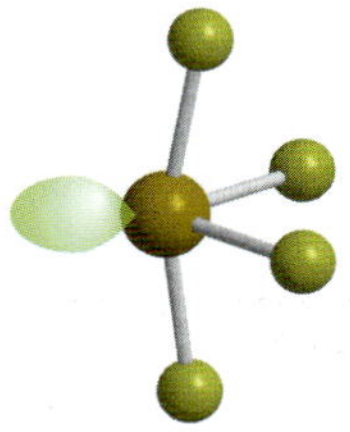

17

are allowed to relax away from the single lone pair, to give a distorted seesaw arrangement (**17**).

To take into account multiple bonds, each set of two or three electron pairs is treated as a single region of high electron density, a kind of 'superpair'. For example, each C atom in an ethene molecule, $CH_2{=}CH_2$, is regarded as having three pairs (one of them the superpair of two electrons pairs of the double bond); these regions of high electron density adopt a trigonal planar arrangement around each atom, so the shape of the molecule is trigonal planar at each C atom (**18**). Another example is the SO_3^{2-} ion: if we adopt the Lewis structure in (**19**), then we see that there are four regions of high electron density around the S atom, indicating a tetrahedral arrangement. One region is a lone pair, so overall the ion is trigonal pyramidal (**20**). We would reach the same conclusion if we adopted the alternative Lewis structure (**21**) in which there are four electron pairs (none of them a 'superpair').

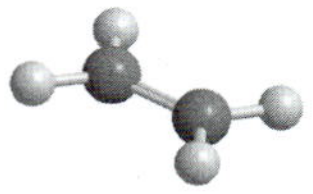

18 Ethene, $CH_2{=}CH_2$

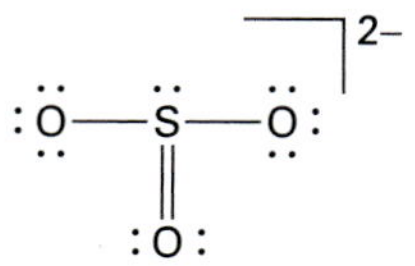

19

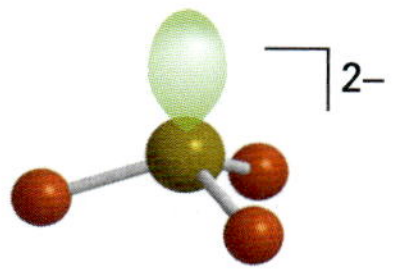

20 Sulfite ion, SO_3^{2-}

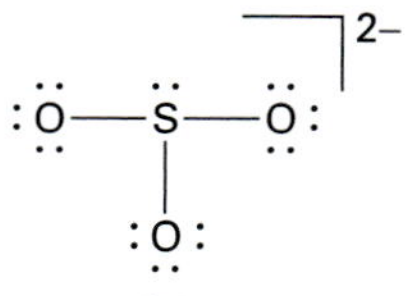

21

Data section

1 Thermodynamic data

Table D1.1 *Thermodynamic data for organic compounds (all values relate to 298.15 K)*

	M/ (g mol^{-1})	$\Delta_f H^{\ominus}$/ (kJ mol^{-1})	$\Delta_f G^{\ominus}$/ (kJ mol^{-1})	$S_m^{\ominus}$/ (J K^{-1} mol^{-1})	$C_{p,m}$/ (J K^{-1} mol^{-1})	$\Delta_c H^{\ominus}$/ (kJ mol^{-1})
C(s) (graphite)	12.011	0	0	5.740	8.527	−393.51
C(s) (diamond)	12.011	+1.895	+2.900	2.377	6.113	−395.40
CO_2(g)	44.010	−393.51	−394.36	213.74	37.11	
Hydrocarbons						
CH_4(g), methane	16.04	−74.81	−50.72	186.26	35.31	−890
CH_3(g), methyl	15.04	+145.69	+147.92	194.2	38.70	
C_2H_2(g), ethyne	26.04	+226.73	+209.20	200.94	43.93	−1300
C_2H_4(g), ethene	28.05	+52.26	+68.15	219.56	43.56	−1411
C_2H_6(g), ethane	30.07	−84.68	−32.82	229.60	52.63	−1560
C_3H_6(g), propene	42.08	+20.42	+62.78	267.05	63.89	−2058
C_3H_6(g), cyclopropane	42.08	+53.30	+104.45	237.55	55.94	−2091
C_3H_8(g), propane	42.10	−103.85	−23.49	269.91	73.5	−2220
C_4H_8(g), 1-butene	56.11	−0.13	+71.39	305.71	85.65	−2717
C_4H_8(g), *cis*-2-butene	56.11	−6.99	+65.95	300.94	78.91	−2710
C_4H_8(g), *trans*-2-butene	56.11	−11.17	+63.06	296.59	87.82	−2707
C_4H_{10}(g), butane	58.13	−126.15	−17.03	310.23	97.45	−2878
C_5H_{12}(g), pentane	72.15	−146.44	−8.20	348.40	120.2	−3537
C_5H_{12}(l)	72.15	−173.1				
C_6H_6(l), benzene	78.12	+49.0	+124.3	173.3	136.1	−3268
C_6H_6(g)	78.12	+82.93	+129.72	269.31	81.67	−3320
C_6H_{12}(l), cyclohexane	84.16	−156	+26.8		156.5	−3902
C_6H_{14}(l), hexane	86.18	−198.7		204.3		−4163
$C_6H_5CH_3$(g), methylbenzene (toluene)	92.14	+50.0	+122.0	320.7	103.6	−3953
C_7H_{16}(l), heptane	100.21	−224.4	+1.0	328.6	224.3	
C_8H_{18}(l), octane	114.23	−249.9	+6.4	361.1		−5471
C_8H_{18}(l), iso-octane	114.23	−255.1				−5461
$C_{10}H_8$(s), naphthalene	128.18	+78.53				−5157
Alcohols and phenols						
CH_3OH(l), methanol	32.04	−238.66	−166.27	126.8	81.6	−726
CH_3OH(g)	32.04	−200.66	−161.96	239.81	43.89	−764
C_2H_5OH(l), ethanol	46.07	−277.69	−174.78	160.7	111.46	−1368
C_2H_5OH(g)	46.07	−235.10	−168.49	282.70	65.44	−1409
C_6H_5OH(s), phenol	94.12	−165.0	−50.9	146.0		−3054
Carboxylic acids, hydroxy acids, and esters						
HCOOH(l), formic	46.03	−424.72	−361.35	128.95	99.04	−255
CH_3COOH(l), acetic	60.05	−484.5	−389.9	159.8	124.3	−875
CH_3COOH(aq)	60.05	−485.76	−396.46	178.7		
$CH_3CO_2^-$(aq)	59.05	−486.01	−369.31	86.6	−6.3	
$(COOH)_2$(s), oxalic	90.04	−827.2			117	−254
C_6H_5COOH(s), benzoic	122.13	−385.1	−245.3	167.6	146.8	−3227
$CH_3CH(OH)COOH$(s), lactic	90.08	−694.0				−1344
$CH_3COOC_2H_5$(l), ethyl acetate	88.11	−479.0	−332.7	259.4	170.1	−2231

Table D1.1 *(continued)*

	M/(g mol^{-1})	$\Delta_f H^{\ominus}$/(kJ mol^{-1})	$\Delta_f G^{\ominus}$/(kJ mol^{-1})	$S_m^{\ominus}$/(J K^{-1} mol^{-1})	$C_{p,m}$/(J K^{-1} mol^{-1})	$\Delta_c H^{\ominus}$/(kJ mol^{-1})
Alkanals and alkanones						
HCHO(g), methanal	30.03	−108.57	−102.53	218.77	35.40	−571
CH_3CHO(l), ethanal	44.05	−192.30	−128.12	160.2		−1166
CH_3CHO(g)	44.05	−166.19	−128.86	250.3	57.3	−1192
CH_3COCH_3(l), propanone	58.08	−248.1	−155.4	200.4	124.7	−1790
Sugars						
$C_6H_{12}O_6$(s), α-D-glucose	180.16	−1274				−2808
$C_6H_{12}O_6$(s), β-D-glucose	180.16	−1268	−910	212		
$C_6H_{12}O_6$(s), β-D-fructose	180.16	−1266				−2810
$C_{12}H_{22}O_{11}$(s), sucrose	342.30	−2222	−1543	360.2		−5645
Nitrogen compounds						
$CO(NH_2)_2$(s), urea	60.06	−333.51	−197.33	104.60	93.14	−632
CH_3NH_2(g), methylamine	31.06	−22.97	+32.16	243.41	53.1	−1085
$C_6H_5NH_2$(l), aniline	93.13	+31.1				−3393
$CH_2(NH_2)COOH$(s), glycine	75.07	−532.9	−373.4	103.5	99.2	−969

Table D1.2 *Thermodynamic data (all values relate to 298.15 K)**

	M/(g mol^{-1})	$\Delta_f H^{\ominus}$/(kJ mol^{-1})	$\Delta_f G^{\ominus}$/(kJ mol^{-1})	$S_m^{\ominus}$/(J K^{-1} mol^{-1})	$C_{p,m}$/(J K^{-1} mol^{-1})
Aluminium (aluminum)					
Al(s)	26.98	0	0	28.33	24.35
Al(l)	26.98	+10.56	+7.20	39.55	24.21
Al(g)	26.98	+326.4	+285.7	164.54	21.38
Al^{3+}(g)	26.98	+5483.17			
Al^{3+}(aq)	26.98	−531	−485	−321.7	
Al_2O_3(s, α)	101.96	−1675.7	−1582.3	50.92	79.04
$AlCl_3$(s)	133.24	−704.2	−628.8	110.67	91.84
Argon					
Ar(g)	39.95	0	0	154.84	20.786
Antimony					
Sb(s)	121.75	0	0	45.69	25.23
SbH_3(g)	153.24	+145.11	+147.75	232.78	41.05
Arsenic					
As(s, α)	74.92	0	0	35.1	24.64
As(g)	74.92	+302.5	+261.0	174.21	20.79
As_4(g)	299.69	+143.9	+92.4	314	
AsH_3(g)	77.95	+66.44	+68.93	222.78	38.07
Barium					
Ba(s)	137.34	0	0	62.8	28.07
Ba(g)	137.34	+180	+146	170.24	20.79
Ba^{2+}(aq)	137.34	−537.64	−560.77	+9.6	
BaO(s)	153.34	−553.5	−525.1	70.43	47.78
$BaCl_2$(s)	208.25	−858.6	−810.4	123.68	75.14

Table D1.2 *(continued)*

	M/(g mol^{-1})	$\Delta_f H^{\ominus}$/(kJ mol^{-1})	$\Delta_f G^{\ominus}$/(kJ mol^{-1})	$S_m^{\ominus}$/(J K^{-1} mol^{-1})	$C_{p,m}$/(J K^{-1} mol^{-1})
Beryllium					
Be(s)	9.01	0	0	9.50	16.44
Be(g)	9.01	+324.3	+286.6	136.27	20.79
Bismuth					
Bi(s)	208.98	0	0	56.74	25.52
Bi(g)	208.98	+207.1	+168.2	187.00	20.79
Bromine					
Br_2(l)	159.82	0	0	152.23	75.689
Br_2(g)	159.82	+30.907	+3.110	245.46	36.02
Br(g)	79.91	+111.88	+82.396	175.02	20.786
Br^-(g)	79.91	−219.07			
Br^-(aq)	79.91	−121.55	−103.96	+82.4	−141.8
HBr(g)	90.92	−36.40	−53.45	198.70	29.142
Cadium					
Cd(s, γ)	112.40	0	0	51.76	25.98
Cd(g)	112.40	+112.01	+77.41	167.75	20.79
Cd^{2+}(aq)	112.40	−75.90	−77.612	−73.2	
CdO(s)	128.40	−258.2	−228.4	54.8	43.43
$CdCO_3$(s)	172.41	−750.6	−669.4	92.5	
Caesium (cesium)					
Cs(s)	132.91	0	0	85.23	32.17
Cs(g)	132.91	+76.06	+49.12	175.60	20.79
Cs^+(aq)	132.91	−258.28	−292.02	+133.05	−10.5
Calcium					
Ca(s)	40.08	0	0	41.42	25.31
Ca(g)	40.08	+178.2	+144.3	154.88	20.786
Ca^{2+}(aq)	40.08	−542.83	−553.58	−53.1	
CaO(s)	56.08	−635.09	−604.03	39.75	42.80
$CaCO_3$(s) (calcite)	100.09	−1206.9	−1128.8	92.9	81.88
$CaCO_3$(s) (aragonite)	100.09	−1207.1	−1127.8	88.7	81.25
CaF_2(s)	78.08	1219.6	−1167.3	68.87	67.03
$CaCl_2$(s)	110.99	−795.8	−748.1	104.6	72.59
$CaBr_2$(s)	199.90	−682.8	−663.6	130	
Carbon (for 'organic' compounds of carbon, see Table D1.1)					
C(s) (graphite)	12.011	0	0	5.740	8.527
C(s) (diamond)	12.011	+1.895	+2.900	2.377	6.133
C(g)	12.011	+716.68	+671.26	158.10	20.838
C_2(g)	24.022	+831.90	+775.89	199.42	43.21
CO(g)	28.011	−110.53	−137.17	197.67	29.14
CO_2(g)	44.010	−393.51	−394.36	213.74	37.11
CO_2(aq)	44.010	−413.80	−385.98	117.6	
H_2CO_3(aq)	62.03	−699.65	−623.08	187.4	
HCO_3^-(aq)	61.02	−691.99	−586.77	+91.2	
CO_3^{2-}(aq)	60.01	−677.14	−527.81	−56.9	
CCl_4(l)	153.82	−135.44	−65.21	216.40	131.75
CS_2(l)	76.14	+89.70	+65.27	151.34	75.7
HCN(g)	27.03	+135.1	+124.7	201.78	35.86
HCN(l)	27.03	+108.87	+124.97	112.84	70.63
CN^-(aq)	26.02	+150.6	+172.4	+94.1	

Table D1.2 *(continued)*

	M/(g mol^{-1})	$\Delta_f H^\ominus$/(kJ mol^{-1})	$\Delta_f G^\ominus$/(kJ mol^{-1})	$S_m^\ominus$/(J K^{-1} mol^{-1})	$C_{p,m}$/(J K^{-1} mol^{-1})
Chlorine					
Cl_2(g)	70.91	0	0	223.07	33.91
Cl(g)	35.45	+121.68	+105.68	165.20	21.840
Cl^-(g)	35.45	−233.13			
Cl^-(aq)	35.45	−167.16	−131.23	+56.5	−136.4
HCl(g)	36.46	−92.31	−95.30	186.91	29.12
HCl(aq)	36.46	−167.16	−131.23	56.5	−136.4
Chromium					
Cr(s)	52.00	0	0	23.77	23.35
Cr(g)	52.00	+396.6	+351.8	174.50	20.79
CrO_4^{2-}(aq)	115.99	−881.15	−727.75	+50.21	
$Cr_2O_7^{2-}$(aq)	215.99	−1490.3	−1301.1	+261.9	
Copper					
Cu(s)	63.54	0	0	33.150	24.44
Cu(g)	63.54	+338.32	+298.58	166.38	20.79
Cu^+(aq)	63.54	+71.67	+49.98	+40.6	
Cu^{2+}(aq)	63.54	+64.77	+65.49	−99.6	
Cu_2O(s)	143.08	−168.6	−146.0	93.14	63.64
CuO(s)	79.54	−157.3	−129.7	42.63	42.30
$CuSO_4$(s)	159.60	−771.36	−661.8	109	100.0
$CuSO_4 \cdot H_2O$(s)	177.62	−1085.8	−918.11	146.0	134
$CuSO_4 \cdot 5H_2O$(s)	249.68	−2279.7	−1879.7	300.4	280
Deuterium					
D_2(g)	4.028	0	0	144.96	29.20
HD(g)	3.022	+0.318	−1.464	143.80	29.196
D_2O(g)	20.028	−249.20	−234.54	198.34	34.27
D_2O(l)	20.028	−294.60	−243.44	75.94	84.35
HDO(g)	19.022	−245.30	−233.11	199.51	33.81
HDO(l)	19.022	−289.89	−241.86	79.29	
Fluorine					
F_2(g)	38.00	0	0	202.78	31.30
F(g)	19.00	+78.99	+61.91	158.75	22.74
F^-(aq)	19.00	−332.63	−278.79	−13.8	−106.7
HF(g)	20.01	−271.1	−273.2	173.78	29.13
Gold					
Au(s)	196.97	0	0	47.40	25.42
Au(g)	196.97	+366.1	+326.3	180.50	20.79
Helium					
He(g)	4.003	0	0	126.15	20.786
Hydrogen (see also deuterium)					
H_2(g)	2.016	0	0	130.684	28.824
H(g)	1.008	+217.97	+203.25	114.71	20.784
H^+(aq)	1.008	0	0	0	0
H_2O(l)	18.015	−285.83	−237.13	69.91	75.291
H_2O(g)	18.015	−241.82	−228.57	188.83	33.58
H_2O_2(l)	34.015	−187.78	−120.35	109.6	89.1

Table D1.2 *(continued)*

	M/(g mol^{-1})	$\Delta_f H^{\ominus}$/(kJ mol^{-1})	$\Delta_f G^{\ominus}$/(kJ mol^{-1})	$S_m^{\ominus}$/(J K^{-1} mol^{-1})	$C_{p,m}$/(J K^{-1} mol^{-1})
Iodine					
I_2(s)	253.81	0	0	116.135	54.44
I_2(g)	253.81	+62.44	+19.33	260.69	36.90
I(g)	126.90	+106.84	+70.25	180.79	20.786
I^-(aq)	126.90	−55.19	−51.57	+111.3	−142.3
HI(g)	127.91	+26.48	+1.70	206.59	29.158
Iron					
Fe(s)	55.85	0	0	27.28	25.10
Fe(g)	55.85	+416.3	+370.7	180.49	25.68
Fe^{2+}(aq)	55.85	−89.1	−78.90	−137.7	
Fe^{3+}(aq)	55.85	−48.5	−4.7	−315.9	
Fe_3O_4(s) (magnetite)	231.54	−1184.4	−1015.4	146.4	143.43
Fe_2O_3(s) (haematite)	159.69	−824.2	−742.2	87.40	103.85
FeS(s, α)	87.91	−100.0	−100.4	60.29	50.54
FeS_2(s)	119.98	−178.2	−166.9	52.93	62.17
Krypton					
Kr(g)	83.80	0	0	164.08	20.786
Lead					
Pb(s)	207.19	0	0	64.81	26.44
Pb(g)	207.19	+195.0	+161.9	175.37	20.79
Pb^{2+}(aq)	207.19	−1.7	−24.43	+10.5	
PbO(s, yellow)	223.19	−217.32	−187.89	68.70	45.77
PbO(s, red)	223.19	−218.99	−188.93	66.5	45.81
PbO_2(s)	239.19	−277.4	−217.33	68.6	64.64
Lithium					
Li(s)	6.94	0	0	29.12	24.77
Li(g)	6.94	+159.37	+126.66	138.77	20.79
Li^+(aq)	6.94	−278.49	−293.31	+13.4	+68.6
Magnesium					
Mg(s)	24.31	0	0	32.68	24.89
Mg(g)	24.31	+147.70	+113.10	148.65	20.786
Mg^{2+}(aq)	24.31	−466.85	−454.8	−138.1	
MgO(s)	40.31	−601.70	−569.43	26.94	37.15
$MgCO_3$(s)	84.32	−1095.8	−1012.1	65.7	75.52
$MgCl_2$(s)	95.22	−641.32	−591.79	89.62	71.38
$MgBr_2$(s)	184.13	−524.3	−503.8	117.2	
Mercury					
Hg(l)	200.59	0	0	76.02	27.983
Hg(g)	200.59	+61.32	+31.82	174.96	20.786
Hg^{2+}(aq)	200.59	+171.1	+164.40	−32.2	
Hg_2^{2+}(aq)	401.18	+172.4	+153.52	+84.5	
HgO(s)	216.59	−90.83	−58.54	70.29	44.06
Hg_2Cl_2(s)	472.09	−265.22	−210.75	192.5	102
$HgCl_2$(s)	271.50	−224.3	−178.6	146.0	
HgS(s, black)	232.65	−53.6	−47.7	88.3	
Neon					
Ne(g)	20.18	0	0	146.33	20.786

Table D1.2 *(continued)*

	$M/(\text{g mol}^{-1})$	$\Delta_f H^\ominus/(\text{kJ mol}^{-1})$	$\Delta_f G^\ominus/(\text{kJ mol}^{-1})$	$S_m^\ominus/(\text{J K}^{-1}\text{ mol}^{-1})$	$C_{p,m}/(\text{J K}^{-1}\text{ mol}^{-1})$
Nitrogen					
N_2(g)	28.013	0	0	191.61	29.125
N(g)	14.007	+472.70	+455.56	153.30	20.786
NO(g)	30.01	+90.25	+86.55	210.76	29.844
N_2O(g)	44.01	+82.05	+104.20	219.85	38.45
NO_2(g)	46.01	+33.18	+51.31	240.06	37.20
N_2O_4(g)	92.01	+9.16	+97.89	304.29	77.28
N_2O_5(s)	108.01	−43.1	+113.9	178.2	143.1
N_2O_5(g)	108.01	+11.3	+115.1	355.7	84.5
HNO_3(l)	63.01	−174.10	−80.71	155.60	109.87
HNO_3(aq)	63.01	−207.36	−111.25	146.4	−86.6
NO_3^-(aq)	62.01	−205.0	−108.74	+146.4	−86.6
NH_3(g)	17.03	−46.11	−16.45	192.45	35.06
NH_3(aq)	17.03	−80.29	−26.50	113.3	
NH_4^+(aq)	18.04	−132.51	−79.31	+113.4	+79.9
NH_2OH(s)	33.03	−114.2			
HN_3(l)	43.03	+264.0	+327.3	140.6	43.68
HN_3(g)	43.03	+294.1	+328.1	238.97	98.87
N_2H_4(l)	32.05	+50.63	+149.43	121.21	139.3
NH_4NO_3(s)	80.04	−365.56	−183.87	151.08	84.1
NH_4Cl(s)	53.49	−314.43	−202.87	94.6	
Oxygen					
O_2(g)	31.999	0	0	205.138	29.355
O(g)	15.999	+249.17	+231.73	161.06	21.912
O_3(g)	47.998	+142.7	+163.2	238.93	39.20
OH^-(aq)	17.007	−229.99	−157.24	−10.75	−148.5
Phosphorus					
P(s, wh)	30.97	0	0	41.09	23.840
P(g)	30.97	+314.64	+278.25	163.19	20.786
P_2(g)	61.95	+144.3	+103.7	218.13	32.05
P_4(g)	123.90	+58.91	+24.44	279.98	67.15
PH_3(g)	34.00	+5.4	+13.4	210.23	37.11
PCl_3(g)	137.33	−287.0	−267.8	311.78	71.84
PCl_3(l)	137.33	−319.7	−272.3	217.1	
PCl_5(g)	208.24	−374.9	−305.0	364.6	112.8
PCl_5(s)	208.24	−443.5			
H_3PO_3(s)	82.00	−964.4			
H_3PO_3(aq)	82.00	−964.8			
H_3PO_4(s)	94.97	−1279.0	−1119.1	110.50	106.06
H_3PO_4(l)	94.97	−1266.9			
H_3PO_4(aq)	94.97	−1277.4	−1018.7	−222	
PO_4^{3-}(aq)	94.97	−1277.4	−1018.7	−222	
P_4O_{10}(s)	283.89	−2984.0	−2697.0	228.86	211.71
P_4O_6(s)	219.89	−1640.1			
Potassium					
K(s)	39.10	0	0	64.18	29.58
K(g)	39.10	+89.24	+60.59	160.336	20.786
K^+(g)	39.10	+514.26			
K^+(aq)	39.10	−252.38	−283.27	+102.5	+21.8
KOH(s)	56.11	−424.76	−379.08	78.9	64.9
KF(s)	58.10	−576.27	−537.75	66.57	49.04
KCl(s)	74.56	−436.75	−409.14	82.59	51.30
KBr(s)	119.01	−393.80	−380.66	95.90	52.30
KI(s)	166.01	−327.90	−324.89	106.32	52.93

Table D1.2 *(continued)*

	M/(g mol^{-1})	$\Delta_f H^\ominus$/(kJ mol^{-1})	$\Delta_f G^\ominus$/(kJ mol^{-1})	$S_m^\ominus$/(J K^{-1} mol^{-1})	$C_{p,m}$/(J K^{-1} mol^{-1})
Silicon					
Si(s)	28.09	0	0	18.83	20.00
Si(g)	28.09	+455.6	+411.3	167.97	22.25
SiO_2(s,α)	60.09	−910.93	−856.64	41.84	44.43
Silver					
Ag(s)	107.87	0	0	42.55	25.351
Ag(g)	107.87	+284.55	+245.65	173.00	20.79
Ag^+(aq)	107.87	+105.58	+77.11	+72.68	+21.8
AgBr(s)	187.78	−100.37	−96.90	107.1	52.38
AgCl(s)	143.32	−127.07	−109.79	96.2	50.79
Ag_2O(s)	231.74	−31.05	−11.20	121.3	65.86
$AgNO_3$(s)	169.88	−124.39	−33.41	140.92	93.05
Sodium					
Na(s)	22.99	0	0	51.21	28.24
Na(g)	22.99	+107.32	+76.76	153.71	20.79
Na^+(aq)	22.99	−240.12	−261.91	+59.0	+46.4
NaOH(s)	40.00	−425.61	−379.49	64.46	59.54
NaCl(s)	58.44	−411.15	−384.14	72.13	50.50
NaBr(s)	102.90	−361.06	−348.98	86.82	51.38
NaI(s)	149.89	−287.78	−286.06	98.53	52.09
Sulfur					
S(s, α) (rhombic)	32.06	0	0	31.80	22.64
S(s, β) (monoclinic)	32.06	+0.33	+0.1	32.6	23.6
S(g)	32.06	+278.81	+238.25	167.82	23.673
S_2(g)	64.13	+128.37	+79.30	228.18	32.47
S^{2-}(aq)	32.06	+33.1	+85.8	−14.6	
SO_2(g)	64.06	−296.83	−300.19	248.22	39.87
SO_3(g)	80.06	−395.72	−371.06	256.76	50.67
H_2SO_4(l)	98.08	−813.99	−690.00	156.90	138.9
H_2SO_4(aq)	98.08	−909.27	−744.53	20.1	−293
SO_4^{2-}(aq)	96.06	−909.27	−744.53	+20.1	−293
HSO_4^-(aq)	97.07	−887.34	−755.91	+131.8	−84
H_2S(g)	34.08	−20.63	−33.56	205.79	34.23
H_2S(aq)	34.08	−39.7	−27.83	121	
HS^-(aq)	33.072	−17.6	+12.08	+62.08	
SF_6(g)	146.05	−1209	−1105.3	291.82	97.28
Tin					
Sn(s,β)	118.69	0	0	51.55	26.99
Sn(g)	118.69	+302.1	+267.3	168.49	20.26
Sn^{2+}(aq)	118.69	−8.8	−27.2	−17	
SnO(s)	134.69	−285.8	−256.8	56.5	44.31
SnO_2(s)	150.69	−580.7	+519.6	52.3	52.59
Xenon					
Xe(g)	131.30	0	0	169.68	20.786
Zinc					
Zn(s)	65.37	0	0	41.63	25.40
Zn(g)	65.37	+130.73	+95.14	160.98	20.79
Zn^{2+}(aq)	65.37	−153.89	−147.06	−112.1	+46
ZnO(s)	81.37	−348.28	−318.30	43.64	40.25

* Entropies and heat capacities of ions are relative to H^+(aq) and are given with a sign.

2 Standard potentials

Table D2.1a *Standard potentials at 298.15 K in electrochemical order*

Reduction half-reaction	$E^\ominus/V$	*Reduction half-reaction*	$E^\ominus/V$
Strongly oxidizing			
$H_4XeO_6 + 2H^+ + 2e^- \rightarrow XeO_3 + 3H_2O$	+3.0	$BrO^- + H_2O + 2e^- \rightarrow Br^- + 2OH^-$	+0.76
$F_2 + 2e^- \rightarrow 2F^-$	+2.87	$Hg_2SO_4 + 2e^- \rightarrow 2Hg + SO_4^{2-}$	+0.62
$O_3 + 2H^+ + 2e^- \rightarrow O_2 + H_2O$	+2.07	$MnO_4^{2-} + 2H_2O + 2e^- \rightarrow MnO_2 + 4OH^-$	+0.60
$S_2O_8^{2-} + 2e^- \rightarrow 2SO_4^{2-}$	+2.05	$MnO_4^- + e^- \rightarrow MnO_4^{2-}$	+0.56
$Ag^{2+} + e^- \rightarrow Ag^+$	+1.98	$I_2 + 2e^- \rightarrow 2I^-$	+0.54
$Co^{3+} + e^- \rightarrow Co^{2+}$	+1.81	$Cu^+ + e^- \rightarrow Cu$	+0.52
$HO_2 + 2H^+ + 2e^- \rightarrow 2H_2O$	+1.78	$I_3^- + 2e^- \rightarrow 3I^-$	+0.53
$Au^+ + e^- \rightarrow Au$	+1.69	$NiOOH + H_2O + e^- \rightarrow Ni(OH)_2 + OH^-$	+0.49
$Pb^{4+} + 2e^- \rightarrow Pb^{2+}$	+1.67	$Ag_2CrO_4 + 2e^- \rightarrow 2Ag + CrO_4^{2-}$	+0.45
$2HClO + 2H^+ + 2e^- \rightarrow Cl_2 + 2H_2O$	+1.63	$O_2 + 2H_2O + 4e^- \rightarrow 4OH^-$	+0.40
$Ce^{4+} + e^- \rightarrow Ce^{3+}$	+1.61	$ClO_4^- + H_2O + 2e^- \rightarrow ClO_3^- + 2OH^-$	+0.36
$2HBrO + 2H^+ + 2e^- \rightarrow Br_2 + 2H$	+1.60	$[Fe(CN)_6]^{3-} + e^- \rightarrow [Fe(CN)_6]^{4-}$	+0.36
$MnO_4^- + 8H^+ + 5e^- \rightarrow Mn^{2+} + 4H_2O$	+1.51	$Cu^{2+} + 2e^- \rightarrow Cu$	+0.34
$Mn^{3+} + e^- \rightarrow Mn^{2+}$	+1.51	$Hg_2Cl_2 + 2e^- \rightarrow 2Hg + 2Cl^-$	+0.27
$Au^{3+} + 3e^- \rightarrow Au$	+1.40	$AgCl + e^- \rightarrow Ag + Cl^-$	+0.22
$Cl_2 + 2e^- \rightarrow 2Cl^-$	+1.36	$Bi^{3+} + 3e^- \rightarrow Bi$	+0.20
$Cr_2O_7^{2-} + 14H^+ + 6e^- \rightarrow 2Cr^{3+} + 7H_2O$	+1.33	$Cu^{2+} + e^- \rightarrow Cu^+$	+0.16
$O_3 + H_2O + 2e^- \rightarrow O_2 + 2OH^-$	+1.24	$Sn^{4+} + 2e^- \rightarrow Sn^{2+}$	+0.15
$O_2 + 4H^+ + 4e^- \rightarrow 2H_2O$	+1.23	$AgBr + e^- \rightarrow Ag + Br^-$	+0.07
$ClO_4^- + 2H^+ + 2e^- \rightarrow ClO_3^- + H_2O$	+1.23	$Ti^{4+} + e^- \rightarrow Ti^{3+}$	0.00
$MnO_2 + 4H^+ + 2e^- \rightarrow Mn^{2+} + 2H_2O$	+1.23	$2H^+ + 2e^- \rightarrow H$	0, by definition
$Br_2 + 2e^- \rightarrow 2Br^-$	+1.09	$Fe^{3+} + 3e^- \rightarrow Fe$	−0.04
$Pu^{4+} + e^- \rightarrow Pu^{3+}$	+0.97	$O_2 + H_2O + 2e^- \rightarrow HO_2^- + OH^-$	−0.08
$NO_3^- + 4H^+ + 3e^- \rightarrow NO + 2H_2O$	+0.96	$Pb^{2+} + 2e^- \rightarrow Pb$	−0.13
$2Hg^{2+} + 2e^- \rightarrow Hg_2^{2+}$	+0.92	$In^+ + e^- \rightarrow In$	−0.14
$ClO^- + H_2O + 2e^- \rightarrow Cl^- + 2OH^-$	+0.89	$Sn^{2+} + 2e^- \rightarrow Sn$	−0.14
$Hg^{2+} + 2e^- \rightarrow Hg$	+0.86	$AgI + e^- \rightarrow Ag + I^-$	−0.15
$NO_3^- + 2H^+ + e^- \rightarrow NO_2 + H_2O$	+0.80	$Ni^{2+} + 2e^- \rightarrow Ni$	−0.23
$Ag^+ + e^- \rightarrow Ag$	+0.80	$Co^{2+} + 2e^- \rightarrow Co$	−0.28
$Hg_2^{2+} + 2e^- \rightarrow 2Hg$	+0.79	$In^{3+} + 3e^- \rightarrow In$	−0.34
$Fe^{3+} + e^- \rightarrow Fe^{2+}$	+0.77	$Tl^+ + e^- \rightarrow Tl$	−0.34
$Ti^{3+} + e^- \rightarrow Ti^{2+}$	−0.37	$PbSO_4 + 2e^- \rightarrow Pb + SO_4^{2-}$	−0.36
$Cd^{2+} + 2e^- \rightarrow Cd$	−0.40	$Ti^{2+} + 2e^- \rightarrow Ti$	−1.63
$In^{2+} + e^- \rightarrow In^+$	−0.40	$Al^{3+} + 3e^- \rightarrow Al$	−1.66
$Cr^{3+} + e^- \rightarrow Cr^{2+}$	−0.41	$U^{3+} + 3e^- \rightarrow U$	−1.79
$Fe^{2+} + 2e^- \rightarrow Fe$	−0.44	$Mg^{2+} + 2e^- \rightarrow Mg$	−2.36
$In^{3+} + 2e^- \rightarrow In^+$	−0.44	$Ce^{3+} + 3e^- \rightarrow Ce$	−2.48
$S + 2e^- \rightarrow S^{2-}$	−0.48	$La^{3+} + 3e^- \rightarrow La$	−2.52
$In^{3+} + e^- \rightarrow In^{2+}$	−0.49	$Na^+ + e^- \rightarrow Na$	−2.71
$U^{4+} + e^- \rightarrow U^{3+}$	−0.61	$Ca^{2+} + 2e^- \rightarrow Ca$	−2.87
$Cr^{3+} + 3e^- \rightarrow Cr$	−0.74	$Sr^{2+} + 2e^- \rightarrow Sr$	−2.89
$Zn^{2+} + 2e^- \rightarrow Zn$	−0.76	$Ba^{2+} + 2e^- \rightarrow Ba$	−2.91
$Cd(OH)_2 + 2e^- \rightarrow Cd + 2OH^-$	−0.81	$Ra^{2+} + 2e^- \rightarrow Ra$	−2.92
$2H_2O + 2e^- \rightarrow H_2 + 2OH^-$	−0.83	$Cs^+ + e^- \rightarrow Cs$	−2.92
$Cr^{2+} + 2e^- \rightarrow Cr$	−0.91	$Rb^+ + e^- \rightarrow Rb$	−2.93
$Mn^{2+} + 2e^- \rightarrow Mn$	−1.18	$K^+ + e^- \rightarrow K$	−2.93
$V^{2+} + 2e^- \rightarrow V$	−1.19	$Li^+ + e^- \rightarrow Li$	−3.05

Table D2.1b *Standard potentials at 298.15 K in alphabetical order*

Reduction half-reaction	$E^\ominus/V$	*Reduction half-reaction*	$E^\ominus/V$
Strongly reducing			
$Ag^+ + e^- \rightarrow Ag$	+0.80	$Co^{2+} + 2e^- \rightarrow Co$	−0.28
$Ag^{2+} + e^- \rightarrow Ag^+$	+1.98	$Co^{3+} + e^- \rightarrow Co^{2+}$	+1.81
$AgBr + e^- \rightarrow Ag + Br^-$	+0.0713	$Cr^{2+} + 2e^- \rightarrow Cr$	−0.91
$AgCl + e^- \rightarrow Ag + Cl^-$	+0.22	$Cr_2O_7^{2-} + 14H^+ + 6e^- \rightarrow 2Cr^{3+} + 7H_2O$	+1.33
$Ag_2CrO_4 + 2e^- \rightarrow 2Ag + CrO_4^{2-}$	+0.45	$Cr^{3+} + 3e^- \rightarrow Cr$	−0.74
$AgF + e^- \rightarrow Ag + F^-$	+0.78	$Cr^{3+} + e^- \rightarrow Cr^{2+}$	−0.41
$AgI + e^- \rightarrow Ag + I^-$	−0.15	$Cs^+ + e^- \rightarrow Cs$	−2.92
$Al^{3+} + 3e^- \rightarrow Al$	−1.66	$Cu^+ + e^- \rightarrow Cu$	+0.52
$Au^+ + e^- \rightarrow Au$	+1.69	$Cu^{2+} + 2e^- \rightarrow Cu$	+0.34
$Au^{3+} + 3e^- \rightarrow Au$	+1.40	$Cu^{2+} + e^- \rightarrow Cu^+$	+0.16
$Ba^{2+} + 2e^- \rightarrow Ba$	−2.91	$F_2 + 2e^- \rightarrow 2F^-$	+2.87
$Be^{2+} + 2e^- \rightarrow Be$	−1.85	$Fe^{2+} + 2e^- \rightarrow Fe$	−0.44
$Bi^{3+} + 3e^- \rightarrow Bi$	+0.20	$Fe^{3+} + 3e^- \rightarrow Fe$	−0.04
$Br_2 + 2e^- \rightarrow 2Br^-$	+1.09	$Fe^{3+} + e^- \rightarrow Fe^{2+}$	+0.77
$BrO^- + H_2O + 2e^- \rightarrow Br^- + 2OH^-$	+0.76	$[Fe(CN)_6]^{3-} + e^- \rightarrow [Fe(CN)_6]^{4-}$	+0.36
$Ca^{2+} + 2e^- \rightarrow Ca$	−2.87	$2H^+ + 2e^- \rightarrow H_2$	0, by definition
$Cd(OH)_2 + 2e^- \rightarrow Cd + 2OH^-$	−0.81	$2H_2O + 2e^- \rightarrow H_2 + 2OH^-$	−0.83
$Cd^{2+} + 2e^- \rightarrow Cd$	−0.40	$2HBrO + 2H^+ + 2e^- \rightarrow Br_2 + 2H_2O$	+1.60
$Ce^{3+} + 3e^- \rightarrow Ce$	−2.48	$2HClO + 2H^+ + 2e^- \rightarrow Cl_2 + 2H_2O$	+1.63
$Ce^{4+} + e^- \rightarrow Ce^{3+}$	+1.61	$H_2O_2 + 2H^+ + 2e^- \rightarrow 2H_2O$	+1.78
$Cl_2 + 2e^- \rightarrow 2Cl^-$	+1.36	$H_4XeO_6 + 2H^+ + 2e^- \rightarrow XeO_3 + 3H_2O$	+3.0
$ClO^- + H_2O + 2e^- \rightarrow Cl^- + 2OH^-$	+0.89	$Hg_2^{2+} + 2e^- \rightarrow 2Hg$	+0.79
$ClO_4^- + 2H^+ + 2e^- \rightarrow ClO_3^- + H_2O$	+1.23	$Hg_2Cl_2 + 2e^- \rightarrow 2Hg + 2Cl^-$	+0.27
$ClO_4^- + H_2O + 2e^- \rightarrow ClO_3^- + 2OH^-$	+0.36	$Hg^{2+} + 2e^- \rightarrow Hg$	+0.86
$2Hg^{2+} + 2e^- \rightarrow Hg_2^{2+}$	+0.92	$O_2 + 4H^+ + 4e^- \rightarrow 2H_2O$	+1.23
$Hg_2SO_4 + 2e^- \rightarrow 2Hg + SO_4^{2-}$	+0.62	$O_2 + e^- \rightarrow O_2^-$	−0.56
$I_2 + 2e^- \rightarrow 2I^-$	+0.54	$O_2 + H_2O + 2e^- \rightarrow HO_2^- + OH^-$	−0.08
$I_3^- + 2e^- \rightarrow 3I^-$	+0.53	$O_3 + 2H^+ + 2e^- \rightarrow O_2 + H_2O$	+2.07
$In^+ + e^- \rightarrow In$	−0.14	$O_3 + H_2O + 2e^- \rightarrow O_2 + 2OH^-$	+1.24
$In^{2+} + e^- \rightarrow In^+$	−0.40	$Pb^{2+} + 2e^- \rightarrow Pb$	−0.13
$In^{3+} + 2e^- \rightarrow In^+$	−0.44	$Pb^{4+} + 2e^- \rightarrow Pb^{2+}$	+1.67
$In^{3+} + 3e^- \rightarrow In$	−0.34	$PbSO_4 + 2e^- \rightarrow Pb + SO_4^{2-}$	−0.36
$In^{3+} + e^- \rightarrow In^{2+}$	−0.49	$Pt^{2+} + 2e^- \rightarrow Pt$	+1.20
$K^+ + e^- \rightarrow K$	−2.93	$Pu^{4+} + e^- \rightarrow Pu^{3+}$	+0.97
$La^{3+} + 3e^- \rightarrow La$	−2.52	$Ra^{2+} + 2e^- \rightarrow Ra$	−2.92
$Li^+ + e^- \rightarrow Li$	−3.05	$Rb^+ + e^- \rightarrow Rb$	−2.93
$Mg^{2+} + 2e^- \rightarrow Mg$	−2.36	$S + 2e^- \rightarrow S^{2-}$	−0.48
$Mn^{2+} + 2e^- \rightarrow Mn$	−1.18	$S_2O_8^{2-} + 2e^- \rightarrow 2SO_4^{2-}$	+2.05
$Mn^{3+} + e^- \rightarrow Mn^{2+}$	+1.51	$Sn^{2+} + 2e^- \rightarrow Sn$	−0.14
$MnO_2 + 4H^+ + 2e^- \rightarrow Mn^{2+} + 2H_2O$	+1.23	$Sn^{4+} + 2e^- \rightarrow Sn^{2+}$	+0.15
$MnO_4^- + 8H^+ + 5e^- \rightarrow Mn^{2+} + 4H_2O$	+1.51	$Sr^{2+} + 2e^- \rightarrow Sr$	−2.89
$MnO_4^- + e^- \rightarrow MnO_4^{2-}$	+0.56	$Ti^{2+} + 2e^- \rightarrow Ti$	−1.63
$MnO_4^{2-} + 2H_2O + 2e^- \rightarrow MnO_2 + 4OH^-$	+0.60	$Ti^{3+} + e^- \rightarrow Ti^{2+}$	−0.37
$Na^+ + e^- \rightarrow Na$	−2.71	$Ti^{4+} + e^- \rightarrow Ti^{3+}$	0.00
$Ni^{2+} + 2e^- \rightarrow Ni$	−0.23	$Tl^+ + e^- \rightarrow Tl$	−0.34
$NiOOH + H_2O + e^- \rightarrow Ni(OH)_2 + OH^-$	+0.49	$U^{3+} + 3e^- \rightarrow U$	−1.79
$NO_3^- + 2H^+ + e^- \rightarrow NO_2 + H_2O$	+0.80	$U^{4+} + e^- \rightarrow U^{3+}$	−0.61
$NO_3^- + 3H^+ + 3e^- \rightarrow NO + 2H_2O$	+0.96	$V^{2+} + 2e^- \rightarrow V$	−1.19
$NO_3^- + H_2O + 2e^- \rightarrow NO_2^- + 2OH^-$	+0.10	$V^{3+} + e^- \rightarrow V^{2+}$	−0.26
$O_2 + 2H_2O + 4e^- \rightarrow 4OH^-$	+0.40	$Zn^{2+} + 2e^- \rightarrow Zn$	−0.76

Index